AQUATIC POLLUTION

✧ ◻ ✧ ◻ ✧

AQUATIC POLLUTION

✧ ¤ ✧ ¤ ✧

AN INTRODUCTORY TEXT
Third Edition

EDWARD A. LAWS
University of Hawaii

John Wiley & Sons, Inc.
New York • Chichester • Weinheim • Brisbane • Singapore • Toronto

Copyright © 2000 by John Wiley & Sons, Inc. All rights reserved.

Published simultaneously in Canada.

This publication is designed to provide accurate and authoritative information in regard to the subject matter covered. It is sold with the understanding that the publisher is not engaged in rendering professional services. If professional advice or other expert assistance is required, the services of a competent professional person should be sought.

Library of Congress Cataloging-in-Publication Data:

Laws, Edward A., 1945–
    Aquatic pollution : an introductory text / Edward A. Laws.—3rd ed.
       p. cm.
    Includes bibliographical references and index.
    ISBN 0-471-34875-9 (paper : acid-free paper)
      1. Water—Pollution. I. Title.
    TD420.L38 2000
    628.1′68—dc21   00-026061

Printed in the United States of America.

10 9 8 7 6 5 4 3 2 1

✦ ⊡ ✦ ⊡ ✦

Contents

✧ ◻ ✧ ◻ ✧

Preface to the Third Edition

The last three decades of the 20th century have been a time of increasing environmental pressure from a human population that now numbers 6 billion persons and is expected to reach 8 billion by 2025. During this time, there has been a growing awareness that the resources of planet Earth are finite and that those resources must ultimately be managed in a sustainable manner. Water is one of the most important of those resources, and its availability for drinking, irrigation, recreation, and as a habitat for fish and other aquatic organisms has become an issue of great concern in many parts of the world. In central Asia, for example, large-scale irrigation of cotton and rice in the lowlands of Turan during the last quarter of the 20th century virtually dried up the Amu-Dar'ja and Syr-Dar'ja rivers, the principal tributaries of the Aral Sea. As a result, the Aral Sea, once the fourth largest lake in the world, shrank by 50% over a period of 30 years. As the Aral Sea dried up, it became more saline, and the once-thriving freshwater commercial fishery collapsed. The productivity of agricultural fields in the surrounding area dramatically declined as a result of the deposition of wind-blown salt from the exposed former seabed.

While the Aral Sea has been characterized as one of the Earth's greatest environmental disasters, it is by no means an isolated example. In a 1999 study supported by the World Bank and the United Nations, the World Commission on Water for the 21st Century concluded that over half of the world's major rivers were going dry or polluted. The report noted that the contamination of these rivers and surrounding watersheds had contributed to the creation of 25 million environmental refugees, a figure for the first time exceeding the world's 21 million war-related refugees.

Globally, about 80% of all diseases in developing countries are currently being spread by consumption of contaminated water. In those regions, waterborne pathogens and pollution kill an estimated 25 million persons each year, about one-third of all deaths in developing countries. The principal culprits are malaria, cholera, and typhoid fever. Even in an advanced nation such as the United States, an average of more than 40,000 persons each year contract cryptosporidiosis, an illness associated with prolonged diarrhea, abdominal pain, weight loss, and fever caused by the waterborne protozoan *Cryptosporidium*. The organism is particularly troublesome because it can infect persons at a very low dose and produces encysted eggs that are not readily killed by chlorination. The eggs can be effectively removed from public water supplies only by filtration.

There have been some success stories. One of the most dramatic has been the recovery of Lake Washington in Seattle following diversion of sewage discharges during the 1960s. The response of the lake was described in the first edition of this book. Follow-up studies have continued to the present day, and an updated report on the status of Lake Washington is included in Chapter 4 of this edition of *Aquatic Pollution*. In 1987 the U.S. Environmental Protection Agency (EPA) and various states surrounding the Chesapeake

Bay embarked on a multifaceted effort to reduce pollution of the bay. Since that time, industrial chemical discharges to the bay have declined by a factor of 3, and the striped bass population is recovering nicely (Chapter 3). Concern over mercury pollution has resulted in more than a threefold reduction in world production of mercury since 1970 (Chapter 12), and restrictions on the use of the pesticide DDT in the United States have resulted in dramatic increases in the numbers of predatory birds such as bald eagles (Chapter 10). Globally, the percentage of paper products accounted for by recycled paper increased from 18% to 36% from 1970 to 2000, and in the Unites States 1,300 companies participated in an EPA-sponsored program that reduced industrial discharges of toxic substances by more than 50% over an eight-year period (Chapter 9). The incidence of large oil spills to the marine environment declined by roughly a factor of 3 from the 1970s to the 1990s (Chapter 13), and participation of volunteers in beach cleanups in the United States almost doubled during the 1990s (Chapter 17).

Not all the news has been good. Although there have been some positive signs from Chesapeake Bay, both the Chesapeake Bay oyster and blue crab populations are far below historical numbers. Despite literally billions of dollars spent to upgrade sewage treatment plants in the Lake Erie watershed, the incidence of Lake Erie beach closures has not declined (Chapter 4). Malaria continues to be a serious public health problem in sub-Saharan Africa and other endemic regions of the world. The dramatic successes achieved by spraying DDT to control mosquitoes have never been repeated. The mosquitoes have developed resistance to DDT and many other pesticides, and no alternative means of pest control has produced satisfactory control of the malaria vector (Chapter 10). Despite significant reductions in nutrient loading of Onondaga Lake (Syracuse, New York), the lake continues to experience serious eutrophication problems and has been characterized as the most polluted lake in the United States (Chapter 6).

Events of recent years have obviously provided many opportunities for persons concerned with the management of aquatic resources to better understand the factors that limit human use and exploitation of those resources. The first edition of *Aquatic Pollution*, published in 1981, took advantage of those opportunities to provide an introduction to the subject of water pollution, with college undergraduates as the primary target audience. The book used issues and examples from the previous 20 years to illustrate the practical application of principles and theories presented in the text to the solution of water pollution problems and to the intelligent management of aquatic resources. The second edition, published in 1991, was in part an updated version of the first edition but expanded the range of topics. The present edition of *Aquatic Pollution* differs in several respects from the first two editions. In general, all information has been updated to reflect recent developments and current understanding. The case studies have, of course, been updated, and several new case studies have been added to illustrate points made in the general discussion. Each chapter now includes a set of questions intended to reinforce important concepts or expand on issues discussed in the chapter. A chapter outline now appears at the beginning of each chapter to serve as a study guide and to help the reader anticipate the flow of the discussion. A glossary has been added to help students learn unfamiliar terms, and there is now a short section on units of measurement.

As in the past, the target audience for *Aquatic Pollution* continues to be undergraduate students, although the text can undoubtedly serve as a useful reference source for professionals. More than ever, educated people need to have some understanding and appreciation of the issues that arise in a discussion of water pollution. These include concepts and principles that are specific to water pollution, but they also include general ecological understanding and awareness of issues such as sustainability and resource management. The material in the book can be covered in two college quarters. Some judicious selection of material will be necessary if the text is to be used for a one-semester course, particularly if the material in the text is augmented with examples of local interest.

I am indebted to Dr. Philip Manor of John Wiley & Sons, who initially encouraged me to undertake the third edition of *Aquatic Pollution*. Without his urging, the third edition would still be on the drawing boards. As in the past, I am indebted to numerous persons who provided information that was not otherwise readily available. Those persons include Ms. Sally Abella, Dr. Daniel Anderson, Dr. Jeff Busch, Dr. Murray Charlton, Ms. Jennifer Day, Dr. David Dolan, Dr. Fred Dobbs, Dr. W. T. Edmondson, Mr. Roger Knight, Mr. Stuart Ludsin, Dr. Charles Madenjian, Mr. Albert Murray, Mr. Jeff Reutter, and Ms. Sharon Thelan. In addition, I am indebted to the staff of the science and technology and government documents sections of Hamilton Library at the University of Hawaii (UH) for their assistance in locating documents. I would like to thank particularly Mr. Brooks Bays and Ms. Nancy Hulbirt of the UH School of Ocean and Earth Science and Technology graphics/illustrations department, who produced many of the figures that appear in this edition. Finally, I am very much indebted to my wife, Stephanie, and my children, Ryan and Jennifer, who provided support and encouragement throughout the preparation of this edition.

EDWARD A. LAWS
Honolulu, Hawaii

✧   ¤   ✧   ¤   ✧

Preface to the Second Edition

The decade of the 1970s was a period of environmental awareness and increasing appreciation of the fact that the quantities and types of waste materials discharged to the environment by human society were creating serious pollution problems that could be solved only by reducing the quantities of waste and/or identifying more satisfactory means of disposal. Concern over water pollution was a major issue at that time, a concern reflected in the United States by the passage of federal legislation such as the Safe Drinking Water Act of 1974 and the Clean Water Act of 1977. There have been some success stories, cases in point being the reversal of cultural eutrophication in Lake Washington following diversion of sewage during the 1960s and the recovery of the southern California brown pelican after the 1972 ban on DDT use in the United States. In other cases, however, efforts to reduce or eliminate water pollution have met with mixed results. Eutrophication in Lake Erie, for example, continues to be a problem, despite the fact that literally billions of dollars have been spent to reduce nutrient inputs to the lake. Over $7 billion have been spent to clean up contaminated aquifers in the United States since the Comprehensive Environment Response, Compensation, and Liability Act was passed in 1980, but according to some authorities the cleanup program has little success to show for the money that has been spent.

The record leaves little doubt that an effective approach to solving water pollution problems will require policy makers and environmentalists to be knowledgeable about scientific issues and cognizant of the efficacy and limitations of corrective measures. It is unrealistic, however, to expect that many of these persons will have PhDs in environmental science, toxicology, oceanography, ecology, or some related field of natural science. PhDs can always be consulted for their opinions on specific environmental issues. It would be desirable, however, for environmentalists and persons involved in establishing water quality policy to be aware of fundamental ecological and toxicological principles relevant to water pollution and of the issues and considerations that influence decision making in specific cases. Indeed such knowledge is desirable for the general public, whose votes and opinions weigh heavily in environmental decision making.

Aquatic Pollution is intended to provide that knowledge. It was written to be a comprehensive introduction to the subject of water pollution, the intended audience being college undergraduates. This second edition, written a little more than 10 years after publication of the first edition, is in part simply an update on the status of particular water pollution problems. As in the first edition, numerous case studies have been used to illustrate concepts and principles discussed in the text. Chapters 1–3 introduce the reader to basic ecological principles relevant to water pollution. Chapter 8 likewise introduces basic principles of toxicology. The remaining chapters are concerned with particular types of water pollution, in most cases involving an introduction to concepts and issues associated with the particular type of pollution followed by one or more case studies. This second edition includes three topics not covered in the first edition, namely acid rain (Chapter 15), groundwater pollution (Chapter 16), and plastics in the sea (Chapter 17). These topics have been the cause of increasing concern in the past decade and certainly deserve a place in a textbook of this kind. Covering all the material in the second edition requires roughly two quarters of classroom time. Some judicious selection of material will probably be necessary if the book is to be used in a one-semester course.

As in the case of the first edition, I am indebted to numerous persons who made available information to me that was not readily available in the literature or helped to answer questions. I would particularly like to thank the staff of the science and technology and government documents sections of Hamilton Library at the University of Hawaii. Their assistance was invaluable. I am also indebted to the Hawaii Institute of Geophysics publications staff, who helped with the preparation of many of the illustrations in the book. Others whose help deserves mention include Dr. Peter Betzer, Dr. W. T. Edmondson, Dr. Paul LaRock, Mr. Jack Huizingh, Dr. Roger Fujioka, Mr. Clifford Henry, Dr. Daniel Anderson, Dr. Jota Kanda, Dr. Nancy Yamaguchi, Dr. John Bardach, Dr. Fred Mackenzie, Dr. James Galloway, Dr. Roland Wollast, and Ms. Christine Curick. I would especially like to thank my wife, Stephanie, and son, Ryan, who provided support and encouragement throughout the preparation of this second edition.

EDWARD A. LAWS
Honolulu, Hawaii
April, 1992

AQUATIC POLLUTION

✧ ¤ ✧ ¤ ✧

Chapter One

✧ ¤ ✧ ¤ ✧

Fundamental Concepts

1. Rationale for the Study of Food Chains/Webs
2. Food Chains
   a. Pathways of Organic Carbon
   b. Predator/Prey Relationships
   c. Use of Food
      (1) biomass
      (2) energy (respiration)
3. Primary Production and Food Chains
   a. Low-energy Carbon → High-energy Carbon

$$\text{energy}$$
$$6CO_2 + 6H_2O \rightarrow C_6H_{12}O_6 + 6O_2$$

   low energy   high energy

   b. Photosynthesis and Chemosynthesis
4. Autotrophs and Heterotrophs
   a. Primary and Secondary Production
5. Grazing Food Chain
   a. Trophic Levels
      (1) plants → herbivores → carnivores
   b. Ecological Efficiencies ≈ 20%
6. Ecological Pyramids
   a. Declining Biomass with Increasing Trophic Level
   b. Length of Food Chains
   c. Nonsteady State Inverted Pyramids
   d. Metabolic Rate and Size Effects
7. Detritus Food Chain
   a. Excretion
   b. Detritivores
   c. Microbial Loop
   d. Balance Between Production and Consumption of Organic Matter
8. Eutrophication
   a. $P > R$
   b. Cultural Eutrophication
9. Food Chain Magnification (Biological Magnification)
10. Food Webs
    a. Simple versus Complex Models
11. Stability and Complexity

THE INTRODUCTION of pollutants into aquatic systems is a perturbation that can set off a complicated series of biological and chemical reactions. Some knowledge and appreciation of basic ecological concepts is necessary to understand and anticipate the nature of those reactions. Let's consider a simple example. Assume that an industry is discharging wastewater into an estuary. The wastewater contains mercury, which is a toxic metal. The mercury in the water reduces the photosynthetic rates of algae in the vicinity of the discharge.

Would the stress on the algae be the extent of the impact? Unfortunately, the answer is no. The reduction of photosynthetic rates would be only the first step. To the extent that photosynthetic rates were lowered, the food supply of herbivores would be reduced, and their biomass and production rates would also be lowered. Furthermore, the herbivores would assimilate some of the mercury absorbed by the algae and become stressed by the presence of the mercury in their tissues. Thus, the herbivores would be affected adversely both by a reduction in their food supply and by the presence of the mercury in their bodies. Using the same logic, it is easy to imagine how animals that preyed on the herbivores could be affected through similar mechanisms and how predator/prey interactions could ultimately spread the mercury to every organism in the water. Obviously, some understanding of the feeding relationships in a natural aquatic system is necessary to appreciate and anticipate the effects of such pollutants.

Now suppose that the mercury discharges ceased. Would the system recover and return to its original condition? Perhaps, but not necessarily. The stability of natural systems to perturbations such as pollutant discharges is a fundamental area of study in systems analysis and a critical consideration in the understanding of pollutant effects. The fact that a natural system is in equilibrium by no means guarantees that the system will return to its original state following a perturbation. To cite a popular example, had a very small meteor struck the Earth 65 million years ago, it is possible that a few dinosaurs might have been killed or injured. However, the dinosaur population probably would have returned to normal within a short time through natural processes. It is now generally agreed, however, that the extinction of all the dinosaurs was probably caused by a very large meteor that struck the Earth about 65 million years ago. Conditions on the Earth for a period of time following that event are believed to have been incompatible with the survival of dinosaurs, the result being that the system did not return to its pre-event status.

SIMPLE FOOD CHAIN THEORY

With this introduction, let's consider some basic ecological principles that relate to the movement and transformation of pollutants within aquatic systems. All animals require food. Food may be burned (respired) to provide energy or incorporated into the animal's body in the form of proteins, fats, carbohydrates, and other compounds to provide essential structural or metabolic components. Plants are by far the most important producers of food in most aquatic systems, although certain bacteria may be significant producers in some parts of the deep sea (Jannasch and Wirsen, 1977). Plants utilize sunlight as an energy source to manufacture organic compounds from carbon dioxide, water, and various inorganic nutrients in a process called *photosynthesis*. For example, a simplified equation describing the manufacture of glucose may be written

$$\text{energy} + \underset{\text{carbon dioxide}}{6CO_2} + \underset{\text{water}}{6H_2O} \underset{\underset{\text{respiration}}{\longleftarrow}}{\overset{\overset{\text{photosynthesis}}{\longrightarrow}}{}} \underset{\text{glucose}}{C_6H_{12}O_6} + \underset{\text{oxygen}}{6O_2}$$

In this case glucose is the organic compound, the adjective *organic* meaning that the compound is found in organisms. If the reaction proceeds from left to right, the energy source is sunlight. Part of this energy is stored chemically in the glucose molecule. If the glucose is then oxidized by burning it with oxygen, the reaction proceeds from right to left and the energy stored in the glucose is released. Some of that energy is made available to the organism mediating the respiratory process and is used to perform various metabolic functions. It is common practice to use either organic carbon (C) or its associated chemical energy content as a metric for food supply, 1 gram of organic carbon being associated with an energy content of 8–11 kilocalories (kcal). All animals have the ability to transform organic compounds from one form to another and hence to convert their food into the compounds they require. However, only plants and certain bacteria have the ability to manufacture organic, high-energy compounds from inorganic, low-energy constituents, and it is this transformation that is referred to as *primary production*. If the energy needed to drive the transformation comes from light, the process is called *photosynthesis*. If the energy is obtained from chemical reactions involving inorganic compounds, the process is called *chemosynthesis*. Only certain types of bacteria and fungi are capable of mediating the latter process. All living organisms depend either directly or indirectly on primary producers as a source of food. Organisms that can produce most or all of the substances they need from inorganic compounds are called *photoautotrophs* or *chemoautotrophs*, depending on whether the energy needed to effect the conversion comes from light or the reactions of inorganic chemicals, respectively. Organisms that lack autotrophic capabilities are called *heterotrophs*. The production of biomass by heterotrophs involves the conversion of some form of organic matter (food) into living biomass and is called *secondary production*.[1] Plants are autotrophs, and animals are heterotrophs. Most bacteria are heterotrophs, although some bacteria do have well-developed photoautotrophic or chemoautotrophic capabilities.

A plant-eating heterotroph, or *herbivore*, may consume food initially produced by a plant. The hervivore may in turn be eaten by another heterotroph, or *primary carnivore*, which converts part of the herbivore biomass into primary carnivore biomass. The primary carnivore may in turn be eaten by another heterotroph, or *secondary carnivore*, which in turn may be eaten by a *tertiary carnivore*, and so forth. Ecologists refer to such a system of successive food transfers as a *food chain*. Each component of the food chain is called a *trophic level*. In the example given, plants would make up the first trophic level, herbivores the second trophic level, primary carnivores the third trophic level, and so forth. Such a food chain is depicted schematically in Figure 1.1.

In most aquatic systems the transfer of food from one trophic level to the next is believed to occur with an efficiency of only about 20%. In other words, the rate at which food is ingested by a trophic level is about five times greater than the rate at which food is passed on to the next trophic level. This efficiency is referred to as an *ecological efficiency*, or more specifically as a *trophic level intake efficiency* (Odum, 1971, p. 76). Ecological efficiencies are generally low because much of the food ingested by a trophic level is either respired to provide energy or excreted because it cannot be incorporated into new trophic level biomass. However, ecological efficiencies are also reduced when, for example, an organism dies from disease or a female fish releases her eggs into the water. Eggs occupy a trophic level that is always lower than that of the organism that produced them.

---

[1]The term *secondary production* has sometimes been taken to mean the production of organisms that consume primary producers (Levinton, 1982) or the production of biomass by animals (Lalli and Parsons, 1993). The definition given here implies that secondary production includes the production of both animal and bacterial biomass by heterotrophic processes and is consistent with Strayer (1988) and Scavia (1988).

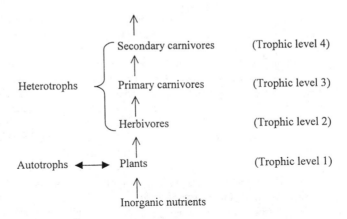

FIGURE 1.1   Diagram of a food chain through trophic level four.

Ecological Pyramids

Because ecological efficiencies are only about 20% in aquatic systems, the flux of food from one trophic level to the next steadily decreases as one moves up the food chain. The result is that the primary production rate is likely to greatly exceed the production of top-level carnivores, the magnitude of the discrepancy depending on the number of trophic levels in the food chain. Ryther (1969) has estimated that there are roughly six trophic levels in typical open ocean marine food chains. On the other hand, some coastal and upwelling areas may have food chains with as few as three trophic levels. This difference stems in part from the fact that the primary producers in open ocean systems are dominated by very small microscopic plants called *phytoplankton*, whereas in coastal and upwelling areas the individual phytoplankton cells tend to be larger, and the cells tend to form chains and gelatinous masses. In the coastal and upwelling areas the primary producers can therefore be efficiently grazed by rather large herbivorous crustaceans such as copepods or even small fish. However, in the open ocean, most of the phytoplankton are much too small to be consumed by crustaceans and small fish, and several intermediate trophic levels therefore separate these two categories of organisms. Regardless of the length of the food chain, the steady decrease in the flux of food to higher and higher trophic levels usually results in a decrease in the biomass of organisms on successively higher trophic levels. Thus, if one were to represent the biomass of each trophic level by a bar whose length was proportional to the biomass of organisms in the trophic level and if one were to lay these bars on top of each other, the resulting figure would look qualitatively like Figure 1.2. Arranged in this way, the bars of trophic level biomass form a pyramid, often referred to as an *ecological pyramid*.

There are two caveats relative to the issue of ecological pyramids. First, although Figure 1.2 correctly depicts the average distribution of biomasses in a food chain, it is quite possible in non-steady-state systems for the biomass distribution in two or more trophic levels to become temporarily inverted. In other words, trophic level biomass increases rather than decreases with increasing trophic level number. For example, in temperate oceans and lakes, a so-called bloom of plant biomass may occur in the spring as the water temperature and average daily solar insolation increase. This plant bloom generally does not occur at a time when the herbivore biomass is large, but the herbivore biomass begins to increase rapidly shortly thereafter in direct response to the increase in herbivore food. Typically, herbivore grazing reduces the plant biomass to a low level. Herbivore biomass peaks

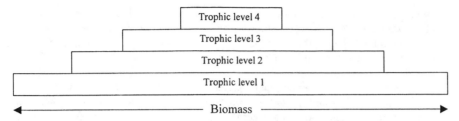

FIGURE 1.2   Trophic level biomass through trophic level four in a hypothetical food chain.

and then declines. The fall in herbivore biomass is caused both by the decrease in herbivore food and by grazing pressure from primary carnivores. Figure 1.3 shows qualitatively how plant and herbivore biomass may vary with time during this period.

A system in which the herbivore biomass is greater than the plant biomass for a short period following the plant bloom is shown in Figure 1.3. Such a condition may exist for a short time in many aquatic systems that are subject to large-scale seasonal cycles. During this period, the first two trophic level biomasses form a so-called inverted pyramid because the second trophic level biomass is greater than that of the first. This situation lasts for only a short time, and the average distribution of biomass is similar to that shown in Figure 1.2. The logical arguments that lead us to expect a normal pyramid of biomass do

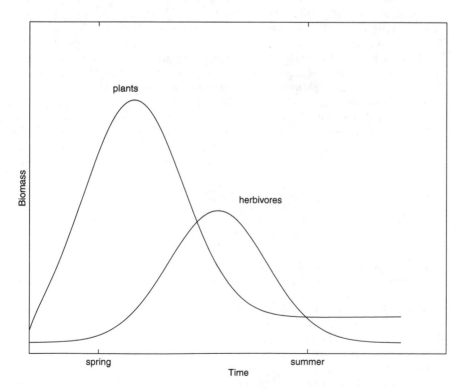

FIGURE 1.3   Biomass of plants and herbivores during spring and early summer in a hypothetical temperate aquatic ecosystem.

not necessarily apply in a non-steady-state system because, over short time intervals, predators may consume more food than prey are producing and hence reduce the prey biomass to a low level. Obviously, this situation cannot persist for long; otherwise, the predators would destroy their food supply. Hence, on the average, one does expect to see a normal pyramid of biomass.

A second point about ecological pyramids concerns the size of organisms making up each trophic level. In general, one expects predators to be larger than prey, and hence higher trophic level organisms should be larger than lower trophic level organisms. This expectation is generally fulfilled, although there are certainly exceptions to the rule (Longhurst, 1991). For example, animals that hunt in groups or packs, such as wolves or killer whales, may kill organisms larger than themselves. However, predators are usually larger than their prey, and as a result, the number of organisms on successively higher trophic levels decreases even more rapidly than the total biomass. Although it is generally true that large organisms consume more food than small organisms, it is also generally true that large organisms consume less food per unit biomass than do small organisms. The relationship between organism size and metabolic rate is such that, if two organisms differ in weight by a factor of 10,000, the larger organism can be expected to consume only 10% as much food per unit body weight as the smaller organism. In other words, the larger organism would consume about 1000 times as much food as the smaller organism, or 1,000/10,000 = 1/10 as much food per unit body weight.

Now consider a case in which the size of individual organisms on two successive trophic levels differs by a factor of 10,000, and the ecological transfer efficiency between the trophic levels is 20%. In this case, a steady-state situation might exist in which the total biomass of the second trophic level is twice that of the first. Although the second trophic level receives only 20% as much food as the first trophic level, the second trophic level needs only 10% as much food to support a given amount of biomass as the first trophic level. Thus, the logical arguments that lead us to expect an ecological pyramid of biomass need not apply to food chains in which the size of organisms on successive trophic levels differs greatly because these arguments implicitly assume the food requirements per unit biomass of all trophic levels to be identical. The fact that normal ecological pyramids of biomass are found in most natural aquatic food chains (e.g., Odum, 1971, p. 80; Sheldon et al., 1972) indicates that differences in organism size on successive trophic levels are not sufficiently great to invert the pyramids. Nevertheless, the difference in successive trophic level biomasses is often less than the factor of 5 that would be expected to result from transfer efficiencies of 20% if all organisms required the same amount of food per unit biomass (see question 1.8). Thus, organism size differences tend to reduce, but not eliminate, the effect of low ecological transfer efficiencies on trophic level biomass structure.

Recycling and the Microbial Loop

The food chain we have discussed up to this point is called the *grazing food chain* because the second and higher trophic levels consist of predators that graze upon prey. Primary producers occupy the first trophic level of the grazing food chain. A very important companion of the grazing food chain in any healthy aquatic system is the *detritus food chain*. The first trophic level in the detritus food chain is the nonliving organic matter produced by living organisms. This nonliving organic matter may exist either as particles or as dissolved organic substances and is referred to as *detritus*. The detritus provides food for a category of organisms called *detritivores*, a designation that includes both bacteria and certain metazoans. Bacteria have no mouthparts, and hence, strictly speaking, must feed entirely on dissolved organics. However, by exuding enzymes, they are able to solubilize and hence make use of particulate material as well. Metazoan detritivores such as benthic worms feed primarily on particulate detritus. Because detritivores are living organisms, they respire and excrete organics, just as do the members of the grazing food chain. The organic compounds excreted by detritivores may be utilized as food by other detritivores, and as a result only the most refractory organic compounds accumulate in the system. Most of the organic matter initially synthesized by the primary producers is ultimately respired, either by organisms in the grazing food chain or by detritivores. Animals or protozoans consume the detritivores, and in this way some of the organic C excreted by the grazing food chain is recycled back into the grazing food chain. The process is illustrated schematically in Figure 1.4. The portion of the detritus food chain involving dissolved organics, bacteria, and protozoans is often referred to as the *microbial loop* and is believed to account for much of the degradation of detritus in aquatic systems.

It is apparent from Figure 1.4 that the grazing food chain and the detritus food chain are interconnected and do not function independently of each other. The interaction between the two food chains is a mutualistic one, i.e., favorable to both and obligatory. The grazing food chain benefits the detritus food chain by excreting much of the organic matter needed by the detritivores for food; the detritus food chain benefits the grazing food chain by removing potentially toxic waste products excreted by both food chains. An approximate balance between the anabolism and catabolism of organic matter is essential to the maintenance of a stable aquatic ecosystem. In a system in which primary production on the average exceeds respiration, organic matter in the form of either plant or animal biomass or detritus will accumulate in the system. Eventually, the whole system may fill up with organic sediments. In fact exactly this process does occur, although often at a very slow rate, in most freshwater habitats and in some marine basins. This gradual accumulation of organic debris results in part from the fact that some detritus is rather refractory and is not broken down efficiently by detritivores. On the other hand, if respiration exceeds primary production, then a net consumption of biomass is occurring within the system. Such a system cannot persist unless it is subsidized by an external input of organic compounds, as, for example, from stream runoff.

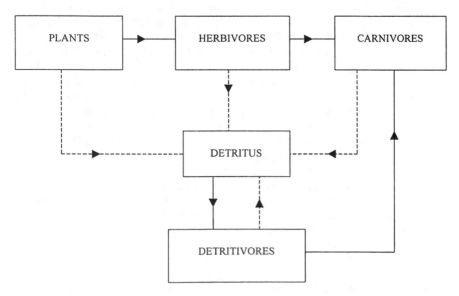

FIGURE 1.4　Box model of the grazing and detritus food chains and the interactions between the two food chains. Solid lines represent feeding relationships. Dashed lines represent excretion.

It is important to realize that primary producers and detritivores utilize the waste products resulting from respiration and excretion, respectively, to create living biomass. For example, CO_2, which is a direct product of respiration, is the source of carbon for primary production. Ammonia (as ammonium ions), which is excreted by many aquatic organisms, can be directly assimilated by primary producers as a source of nitrogen (N) for the production of proteins and nucleic acids. Waste products can be, and often are, toxic to the organisms that produce them. However, in a well-balanced ecosystem, waste products never reach high concentrations because they are constantly being utilized as a source of food by other organisms in the system. Detritivores play a crucial recycling role in aquatic systems by consuming organic wastes and converting them to inorganic forms that are utilized by primary producers. The grazing food chain utilizes the organic matter synthesized by the primary producers and releases part of it in the form of detritus, which in turn provides the food for the detritus food chain.

Due to this internal recycling, there is a tendency for both organic and inorganic compounds to accumulate in aquatic systems. Inorganic C can, of course, escape to the atmosphere as CO_2, and inorganic nitrogen may similarly escape as ammonia or N_2. However, under normal circumstances, the latter escape routes are not very efficient for N, and removal of organic compounds and essential nutrients via washout rarely occurs with 100% efficiency. The accumulation of refractory organic debris in the sediments and buildup of organic matter and nutrient concentrations in the water column are natural processes in most aquatic systems. Associated with these phenomena are increases in the rates of primary production and respiration and a decrease in the depth of the system caused by sediment accumulation. The whole process is referred to as *eutrophication*. Eutrophication eventually causes most lakes to fill up with sediments after a period of perhaps hundreds, thousands, or even tens of thousands of years. Sediments do accumulate at the bottom of the ocean, but they are removed by tectonic processes at subduction zones at rates that approximately balance their rate of formation. Obviously, there is no danger that the oceans will fill up with sediments. However, some regions of the ocean are much more productive than others, and this fact directly reflects the relative efficiency with which essential nutrients are recycled by the grazing and detritus food chain in different parts of the ocean.

Eutrophication is sometimes considered to be an unnatural phenomenon, but the imbalance between photosynthesis (P) and respiration (R) associated with eutrophication is nothing new. It was a fact of life on Earth literally billions of years ago.[2] The atmosphere of the Earth was initially devoid of oxygen (O_2), and the O_2 in the atmosphere and ocean today is the product of photosynthesis. The first primitive plants evolved in the ocean, where the water shielded them from ultraviolet radiation. The O_2 produced by those plants eventually accumulated in the ocean and atmosphere, and photochemical reactions in the atmosphere converted some of the O_2 to ozone. The ozone in the atmosphere then became a shield against ultraviolet radiation. It was only after the establishment of this ozone shield that organisms were able to leave the ocean and evolve on land. Thus, the very habitability of the terrestrial environment today depends on the fact that there was an excess of photosynthesis over respiration on a grand scale during the early evolution of life on Earth. However, the imbalance between P and R has had other profound implications. O_2 is one product of photosynthesis. The other product is organic matter. The imbalance between P and R during the geologic history of the Earth has resulted in the accumulation of both O_2 and organic matter. Oil and coal deposits are obvious manifestations of the imbalance between P and R over geologic time.

Any unnatural acceleration of the eutrophication process due to the activities of humans is called *cultural eutrophication*. Cultural eutrophication can be caused, for example,

---

[2]The Earth is approximately 4.5 billion years old. Primitive forms of life began to appear about 3.5–4.0 billion years ago.

by the discharging of sewage containing a high concentration of nutrients and organic matter. Instances of cultural eutrophication constitute one of the most common and widespread examples of water pollution problems. We will explore a few of these examples in detail in Chapter 4.

Food Chain Magnification

Respiration and excretion obviously play a critical role in controlling the flux of organic and inorganic materials between the grazing and detritus food chains. However, from the standpoint of water pollution, respiration and excretion are also important in determining the movement of pollutants both between and within these same food chains. If the pollutant is biodegradable, it may, of course, be catabolized and rendered harmless. However, if the pollutant is nonbiodegradable, it may be passed from prey to predator and in this way be spread throughout the grazing food chain. If some of the pollutant is excreted, then it may spread to the detritus food chain as well. One of the most important applications of food chain theory to water pollution problems has been the effort to explain how these transfers of a pollutant between food chains and trophic levels affect the concentration of the pollutant in organisms. In cases where it has been possible to examine in some detail the distribution of pollutant concentrations among the trophic levels in a simple food chain, results have sometimes indicated a steady increase in concentration with increasing trophic level number. Table 1.1 shows concentrations of the pesticide DDT (plus the closely related compounds DDD and DDE) in the water and in various organisms taken from a Long Island, New York, salt marsh. The residue concentrations increase steadily from the plankton to the small fish to the larger fish and finally to the fish-eating birds. The total concentration factor from plankton to fish-eating birds is roughly 600. Observations such as this one led some scientists to believe that a common mechanism or explanation might underlie similar observations of increasing pollutant concentrations at higher trophic levels in some food chains, a phenomenon that they termed *food chain magnification.*

TABLE 1.1   DDT RESIDUES IN ORGANISMS TAKEN FROM A LONG ISLAND SALT MARSH

Organism	DDT Residues (ppm)[a]
Water	0.00005
Plankton	0.04
Silverside minnow	0.23
Sheephead minnow	0.94
Pickerel (predatory fish)	1.33
Needlefish (predatory fish)	2.07
Heron (feeds on small animals)	3.57
Tern (feeds on small animals)	3.91
Herring gull (scavenger)	6.00
Fish hawk (osprey) egg	13.8
Merganser (fish-eating duck)	22.8
Cormorant (feeds on larger fish)	26.4

[a] Parts per million (ppm) of total residues, DDT + DDD + DDE (all of which are toxic), on a wet weight, whole organism basis.
Source: Woodwell et al. (1967)

A logical explanation for food chain magnification is forthcoming from food chain theory if one assumes that certain pollutants ingested with an organism's food are not as effectively respired or excreted as is the remainder of the food. DDT would seem to be a likely candidate for such a pollutant because it is resistant to biological breakdown and tends to be stored in an organism's fatty tissues rather than being directly excreted with other waste materials that the organism is unable to utilize. Consider, for example, a case in which the ecological transfer efficiency of food between trophic levels is 20% but the transfer efficiency of DDT, caused by its resistance to respiration and excretion, is 60%. As a result, the steady-state concentration of DDT in a predator will be about three times greater than the DDT concentration in its food because the predator retains three times (60% versus 20%) as much of the DDT that it eats as it does of the remaining food. If this process is repeated through four trophic level transfers, the concentration of DDT in the fifth trophic level will be $3^4 = 81$ times greater than the DDT concentration in the first trophic level.

Although this reasoning is logical enough, the logicality of the reasoning by no means guarantees that biological magnification of the sort described is responsible for observations such as those presented in Table 1.1. Pollutant concentration trends of exactly this sort may be produced by mechanisms very different from food chain magnification. Only carefully designed experiments can sort out the possible causes of such concentration trends. In Chapter 10 we will examine one such experiment in some detail. The point here is not to argue the pros and cons of the theory of food chain magnification, but rather to show how some knowledge about the characteristics of food chains can lead to a logical hypothesis regarding pollutant effects that is worth testing with further study. Logical thinking and hypothesis testing of this sort, based on sound ecological principles, represent the best means for studying and solving water pollution problems.

FOOD WEBS

Now that we have developed a simple food chain model as a conceptual basis for examining ecological problems, it is best to back up a step and remind ourselves that the world is not really so simple. The feeding behavior of many animals is such that they cannot be assigned to a unique trophic level. Some animals, such as shrimp, will eat almost anything they can swallow, including plants, detritus, and other animals. Obviously, they cannot be assigned to one or even a few trophic levels. Other organisms may feed on one trophic level as juveniles, on a second trophic level at a later developmental stage, and on a third trophic level as adults. In such a case, one would have to treat each developmental stage of the species as a different organism in order to make unique trophic level assignments. Certainly the kind and quantity of available food influence the feeding habits of many organisms. Figure 1.5 indicates the complex feeding relationships found in a small stream community.

The pattern of lines indicating the feeding relationships in such a system form a sort of web, and the feeding pattern has therefore come to be known as a *food web*. A food web depicts a more complex system than that represented by a food chain. One may therefore ask whether any of the implications deduced from food chain theory are relevant to a world that seems to be much more complex. The answer to this question is "yes" for the following reasons.

First, even though it is true that many organisms feed on more than one trophic level, it is also true that many organisms show a preference for one kind of food or another and can be reasonably defined as belonging to primarily one or two trophic levels. Although there are exceptions to this rule, the exceptions are not so numerous as to warrant discarding the concept.

Second, and perhaps more important, is the fact that some implications of food chain theory do not even require one to assign specific organisms to specific trophic levels. The idea that a large percentage of a prey will be either respired or excreted by the predator is valid regardless of the identity of the prey and predator. Thus, food in an abstract sense is produced by plants and is passed through a series of feeding transfers from one trophic level to the next, with roughly 80% of the food being respired or excreted at each transfer. Therefore, on the average, the biomass of food that has passed through m such transfers is likely to be much less than the biomass of food that has passed through $m - 1$ transfers. If one thinks of trophic levels in terms of a series of food transfers, and not nec-

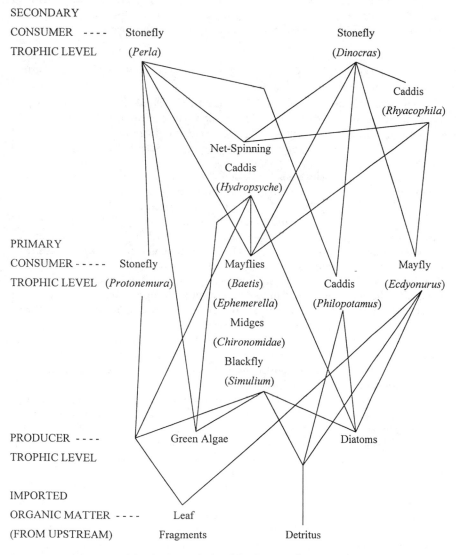

FIGURE 1.5   Diagram of the feeding relationships in a small stream community in South Wales. *Source*: Redrawn from Jones (1949). Reproduced with permission from Blackwell Science, Ltd.

essarily in terms of particular organisms eating particular organisms, the problem of assigning each organism to a unique trophic level disappears.

Despite this second point, it is in fact the case that the effect of pollutants on particular organisms is sometimes discussed with reference to an organism's position in the food chain. The relevance of food chain theory to such discussions is based on the first of the above two arguments, namely, that many organisms may be reasonably assigned to one or two trophic levels. For example, it is reasonable to assume that a pelican does not eat microscopic plants and animals, bacteria, protozoans, or even small fish; nor does a pelican eat whales, dolphins, or sea lions. Rather, a pelican eats fish of the size (approximately a few tens of centimeters) that can be conveniently scooped up in its mouth. Thus, a pelican would be assigned to roughly the fourth trophic level in a grazing food chain. Such reasoning, while not flawless, is unlikely to be wildly in error. Nevertheless, whenever food chain arguments are invoked, one should keep in mind that the theory has its shortcomings and that the feeding relationships of some organisms can be extremely complex.

FOOD WEBS AND ECOSYSTEM STABILITY

The complexity of aquatic food webs is associated with another controversy that is particularly relevant to the problem of water pollution. The controversy concerns the stability of ecosystems. To what extent will a biological community resist change when stressed by pollution, and will the system return to its original state if the pollution stress is removed?

In the 1950s and 1960s, a number of publications appeared in the literature suggesting that the complex interactions between organisms in a food web tended to stabilize the biological system. MacArthur (1955), for example, suggested that community stability might be roughly proportional to the logarithm of the number of links in the food web, a hypothesis that unfortunately has sometimes been accorded the status of a theorem. The hypothesis is based on conclusions derived from information theory, in which it is shown that such a logarithm provides a measure of the degree of organization or complexity. One then argues intuitively that the larger the number of links and pathways in the food web, the greater the ability of the system to damp down perturbations and hence the greater the chance that the system can absorb environmental shocks without falling apart. Examples of stable and very complex ecosystems include tropical rain forests and coral reefs.

A very elegant and understandable treatment of this issue has been provided by Robert May (1974). The somewhat counterintuitive result of May's analysis is that greater food web complexity per se does not impart greater stability to the ecosystem. In fact, the more complex the linkages in the food web, the more unstable the system is likely to become. As noted by May (1974, p. 75), "The greater the size and connectance of a web, the larger the number of characteristic modes of oscillation it possesses: since in general each mode is as likely to be unstable as to be stable (unless the increased complexity is of a highly special kind), the addition of more and more modes simply increases the chance for the total web to be unstable." The fact that tropical rain forests and coral reefs are highly complex systems may therefore be due more to the fact that these ecosystems exist in stable environments than to some inherent stabilizing characteristics conferred by the complex links in their food webs. May (1974) was careful not to rule out the latter possibility, and noted that some very special and mathematically atypical sorts of complexity might enhance ecosystem stability, but that it would be unwise to assume that, as a general rule, food web complexity implies stability. In fact, "If there is a generalization, it could be that stability [of the environment] permits complexity" (May 1974, p. 76).

An important point to bear in mind is that the foregoing discussion of stability and complexity pertains to complexity of food web linkages only. Other types of complexity may confer stability on ecosystems. For example, habitat complexity may provide refuges that allow some members of the biological community to survive adverse conditions. Similarly, complexity or diversity of the gene pool in a species may result in a subset of the population that is unusually resistant to a particular stress and hence better able to survive the impact of certain types of pollution. Similar considerations apply to the functionality of biological wastewater treatment systems. As we will see in Chapter 6, from a functional standpoint, a simple wastewater treatment system that relies almost entirely on bacteria to decompose the organic matter in sewage may be more easily perturbed by changes in the characteristics of the sewage than a system that relies on a more complex biological community. There are many types of complexity, and not all of them are inherently destabilizing. However, it appears from May's (1974) analysis that, as a general rule, increased complexity in food web linkages does not increase ecosystem stability.

REFERENCES

BERNER, R. A. 1994. Geocarb II: A revised model of atmospheric CO_2 over Phanerozoic time. *Amer. J. Sci.*, **294**, 56–91.

FALKOWSKI, P. G., R. T. BARBER, and V. SMETACEK. 1998. Biogeochemical controls and feedbacks on ocean primary production. *Science*, **281**, 200–206.

FIELD, C. B., M. J. BEHRENFELD, J. T. RANDERSON, and P. FALKOWSKI. 1998. Primary production of the biosphere: Integrating terrestrial and oceanic components. *Science*, **281**, 237–240.

FLAVIN, C., and N. LENSSEN. 1990. Beyond the Petroleum Age: Designing a Solar Economy. Worldwatch Institute, Washington, D.C. 65 pp.

JANNASCH, H. W., and C. O. WIRSEN. 1977. Microbial life in the deep sea. *Sci. Amer.*, **236**(6), 42–52.

JONES, J. R. E. 1949. A further ecological study of a calcareous stream in the 'Black Mountain' district of South Wales. *J. Anim. Ecol.*, **18**, 142–159.

LALLI, C. M., and T. R. PARSONS. 1993. *Biological Oceanography: An Introduction.* Pergamon Press, New York. 301 pp.

LEVINTON. J. S. 1982. *Marine Ecology*, Prentice-Hall, Englewood Cliffs, N.J. 526 pp.

LONGHURST, A. 1991. Role of the marine biosphere in the global carbon cycle. *Limnol. Oceanogr.*, **36**, 1507–1526.

MACARTHUR, R. H. 1955. Fluctuations of animal populations, and a measure of community stability. *Ecology*, **36**, 533–536.

MAY R. M. 1974. *Stability and Complexity in Model Ecosystems.* 2nd ed. Princeton Univ. Press, Princeton, N.J. 265 pp.

ODUM, E. P. 1971. *Fundamentals of Ecology.* 3rd ed. Saunders, Philadelphia. 574 pp.

RYTHER, J. H. 1969. Photosynthesis and fish production in the sea. *Science*, **166**, 72–76.

SCAVIA, D. 1988. On the role of bacteria in secondary production. *Limnol. Oceanogr.*, **33**, 1220–1224.

SHELDON, R. W., A. PRAKASH, and W. H. SUTCLIFFE, JR. 1972. The size distribution of particles in the ocean. *Limnol. Oceanogr.*, **17**, 327–340.

STRAYER, D. 1988. On the limits to secondary production. *Limnol. Oceanogr.*, **33**, 1217–1220.

WOODWELL, G. M., C. F. WURSTER, and P. A. ISAACSON. 1967. DDT residues in an east coast estuary: A case of biological concentration of a persistent insecticide. *Science*, **156**, 821–824.

QUESTIONS

1.1 As is apparent from Table 1.1, concentrations of toxic substances in water and aquatic organisms may be very low. Concentrations are frequently reported in parts per million (ppm) or even parts per billion (ppb). One part per million of substance X in water is 1 gram of X per million grams of water. Likewise, 1 part per billion of substance X in water is 1 gram of X per billion grams of water. A million grams is 10^6 grams or 1 metric ton. A metric ton is often written *tonne* and equals about 2,200 pounds. A billion grams is 10^9 grams or 1,000 tonnes. To get some feeling for the meaning of 1 ppm and 1 ppb, calculate the amount of time in days equal to 1 million seconds and the amount of time in years equal to 1 billion seconds.

1.2 The ecological efficiency in a food chain is 20%. Pollutant X is transferred up the food chain from one trophic level to the next with an efficiency of 60%. The concentration of X on trophic level 4 is 36 ppm. If the concentration of X on other trophic levels in the food chain is determined entirely by food chain magnification, what would you expect the concentration of X to be on

   a. Trophic level 2?

   b. Trophic level 5?

1.3 The Earth's atmosphere presently contains about 10^{21} grams of O_2. Assuming that this O_2 was produced by photosynthesis, how many grams of organic carbon (C) must have been produced to account for this much O_2? The total inventory of organic C on the Earth is currently estimated to be about 1.5×10^{22} grams (Berner, 1994). Assuming that this organic C was produced by photosynthesis, how much O_2 would have been associated with its production? How would you account for the discrepancy between this figure and the amount of O_2 currently in the atmosphere?

1.4 The net production of organic C by aquatic and terrestrial plants is currently estimated to be about 10^{17} grams per year on a global basis (Field et al., 1998). Heterotrophic respiration is estimated to convert about the same amount of organic C to CO_2 each year. The current imbalance between the production and consumption of organic matter is believed to be due to human activities—primarily the burning of fossil fuels. These activities are estimated to convert about 6×10^{15} grams of organic C to CO_2 each year (Flavin and Lenssen, 1990). About half of that C appears as additional CO_2 in the atmosphere each year. Assuming that the production of organic C amounts to 10^{17} grams per year and that consumption exceeds production by 6×10^{15} grams per year, what is the current ratio of production to consumption of organic C?

1.5 Suppose that a giant meteor hit the Earth, creating a nuclear winter that blocked out enough sunlight to completely shut down photosynthesis. Assuming that the respiration of heterotrophs continued to oxidize organic C at a rate of 10^{17} grams per year, how long would it take for heterotrophic respiration to consume all the O_2 in the atmosphere? Why is it unlikely that heterotrophic respiration would continue at its present rate for this length of time?

1.6 Current use of fossil fuels consumes about 6×10^{15} grams of C per year. Total fossil fuel reserves (coal, oil, and natural gas) are estimated to be about 4×10^{18} grams of C (Falkowski et al., 1998). If fossil fuel consumption continues at its present rate, how long will it take to consume all the fossil fuel deposits?

1.7 Current use of oil and natural gas amounts to about 3.4×10^{15} grams of carbon per year (Flavin and Lenssen, 1990). Total deposits of oil and natural gas are estimated to be about 5×10^{17} grams of C (Falkowski et al., 1998). If the consumption of oil and natural gas continues at its present rate, how long will it take to consume all the oil and natural gas deposits?

1.8 Sheldon et al. (1972) have argued that the concentrations of particles ranging from microscopic algae to tuna and whales varies by no more than a factor of 2–3 in the ocean. In other words, there is an ecological pyramid of biomass, but the biomass of microscopic algae is only two to three times greater than the biomass of tuna and whales. How can we account for this observation if the ecological efficiency of the food chains leading from algae to tuna and whales is only 20%? More recently, Longhurst (1991) has presented data concerning the relative size of predators and prey in a wide variety of marine food chains and covering a range in organism size from protozoans to whales. The median ratio of predator length to prey length in Longhurst's data set is about 10. Assuming that organism biomass scales as length cubed, the implication of Longhurst's analysis is that the biomass of a predator is roughly 1000 times greater than that of its prey. Assume that:

a. The biomass of individual predators is 1000 times greater than the biomass of their prey
b. Metabolic rate scales as individual biomass raised to the 0.75 power
c. Ecological efficiency is 20%

Given these assumptions, in a grazing food chain what would you predict the biomass on trophic level 6 to be relative to the biomass on trophic level 1? Is this result roughly consistent with the observations of Sheldon et al. (1972)?

Chapter Two

✧ ⌑ ✧ ⌑ ✧

Photosynthesis

1. Euphotic Zone
   a. 1% of Surface Light
2. Aphotic Zone
3. Phytoplankton
   a. Examples of Problems Caused by
      (1) red tides
      (2) excessive algal biomass
4. Light limitation of photosynthesis
   a. Light versus Depth—Exponential Decay
   b. Critical Depth
      (1) net community production
   c. Effects of Turbidity
5. Nutrient limitation
   a. Macronutrients—C, H, O, N, P, S, K, Mg, Ca
   b. Micronutrients—Fe, Mn, Cu, Zn, B, Si, Mo, Cl, V, Co, Na
   c. CO_2 Limitation?
   d. Si Limitation of Diatoms
   e. Mo
      (1) nitrate reductase
      (2) N fixation (competition between Mo and sulfate)
   f. Typical Nitrate and Phosphate Concentrations
6. Nutrient Enrichment Experiments
   a. Single and Multiple Nutrient Enrichments
      (1) complicating factors
   b. Long Island Bays
   c. Canadian Lakes
      (1) CO2 exchange—lake 227
      (2) PO_4^{3-} sedimentation with Fe^{3+}
         (a) Lakes 226, 304
7. N versus P Limitation
   a. N/P in Sewage
   b. N Fixation
      (1) availability of Fe and Mo in ocean
   c. P Sedimentation
      (1) $Fe^{3+} \leftrightarrow Fe^{2+}$
   d. Nutrient Cycling Within Mixed Layer
   e. Denitrification
8. Fe Limitation

A S THE ORGANISMS primarily responsible for the synthesis of organic matter from inorganic constituents, plants occupy a uniquely important position in aquatic food chains. Therefore, a study of the factors that control photosynthetic rates is a logical place to begin examining how changes in the environment may be expected to influence aquatic systems.

Aquatic plants are generally very different from the terrestrial plants with which most people are familiar. There are relatively few trees or grasses in aquatic systems. Along coastlines, mangrove and cypress trees are, to be sure, aquatic organisms, and benthic algae such as kelp are somewhat analogous to terrestrial grasses or shrubs. In shallow lakes or along coastlines where the water is sufficiently shallow for light to penetrate to the bottom effectively, benthic algae may be the most important primary producers in an aquatic ecosystem. However, since plants need light to carry out photosynthesis and since water absorbs light, the depth range within which rooted aquatic plants can survive is obviously limited. In systems where there is a high concentration of particulate materials suspended in the water, water transparency is further reduced because such particles scatter and absorb light. A similar effect is produced by certain dissolved organic substances such as tannic and humic acids, which also absorb light. In the clearest ocean water, only about 1% of surface light is transmitted to a depth of 100 m, and in clear coastal waters the 1% light level is typically reached at a depth of only 20 m. The net photosynthetic rate of most aquatic plants is close to zero at light intensities less than ~1% of surface light. The depth at which net photosynthesis equals zero is called the *compensation depth*. The limit of the *euphotic zone* (the region of the water column within which net photosynthesis is positive) equals the compensation depth. The euphotic zone is typically a few tens of meters in clear coastal waters and roughly 150 m or at most 200 m in the clearest open ocean water. These depth limits are reduced in waters containing significant concentrations of particulate or dissolved organic materials. Obviously, benthic algae and/or rooted aquatic plants are not to be found in parts of aquatic systems where the bottom lies below the euphotic zone. Since the depth below about 93% of the surface area of the world's oceans is greater than 180 m (Ryther, 1969), it is clear that benthic plants contribute nothing to primary production in most parts of the ocean. Furthermore, many lakes are sufficiently deep and/or turbid to prevent the development of benthic plants except in the immediate vicinity of the shoreline.

From the previous discussion, it is clear that aquatic plants found in the surface waters of the open ocean or of deep lakes must have the ability to float or drift about in the euphotic zone and must be able to derive all the essential nutrients for photosynthesis directly from the water. Seaweeds such as the *Sargassum* weed provide one example of such planktonic (drifting) plants. However, most such plants are microscopic or nearly so in size and are frequently unicellular. These tiny plants are called *phytoplankton* and may range in size from microscopic cells a micron or less in diameter to "giant" cells visible to the unaided eye with a diameter of as much as 2 mm. Some species of phytoplankton form colonies or long chains of cells that are visible to the unaided eye, although the individual cells are not. It is perhaps surprising to realize that such tiny organisms are responsible for the vast majority of aquatic primary production and that top-level carnivores such as sharks and whales depend almost entirely, although indirectly, on such minute organisms as a source of food. However, despite their essential role as primary producers in virtually all aquatic food chains, phytoplankton may create serious problems, particularly when their concentrations in the water exceed certain limits. For example, "red tides," which are an unwelcome but recurrent event in a number of coastal areas throughout the world, are caused by a population explosion of certain species of a class of phytoplankton called *dinoflagellates*. At such times the concentration of these dinoflagellates becomes sufficiently great to give the water a reddish color over distances of as much as several kilometers. Unfortunately, these dinoflagellates release neurotoxins into the water. During red tides these toxin concentrations may become high enough to kill great numbers of small fish, some of which frequently wash up on the beach to rot. Furthermore, the toxins may be concentrated by shellfish (which are seemingly unaffected) and may subsequently poison humans who eat the shellfish. In severe cases this poisoning may result in paralysis or even death. Although some phytoplankton population explosions are completely natural phenomena, there is no doubt that in many cases human activities have largely caused or greatly exacerbated conditions that lead to undesirably high phytoplankton concentrations. To avoid and/or correct conditions that stimulate excessive phytoplankton growth, it is obviously essential to first understand the factors that normally control phytoplankton growth rates, a problem to which we now turn.

LIGHT LIMITATION OF PHOTOSYNTHESIS

There is no doubt that light is the most important factor limiting photosynthetic rates in the world's oceans. Over 95% of the ocean's volume lies below the euphotic zone and is hence unable to support plant life. In addition, many lakes are sufficiently deep that the euphotic zone makes up only a small fraction of the lake's volume, and even shallow lakes may show pronounced seasonal patterns in plant production that are undoubtedly influenced in part by changes in insolation. Figure 2.1 shows the variation of light intensity with depth in a hypothetical body of water. Only visible light is graphed because the plant pigments that absorb light absorb only certain parts of the visible light spectrum. For example, chlorophyll absorbs predominantly blue and red light. In fact, not all parts of the

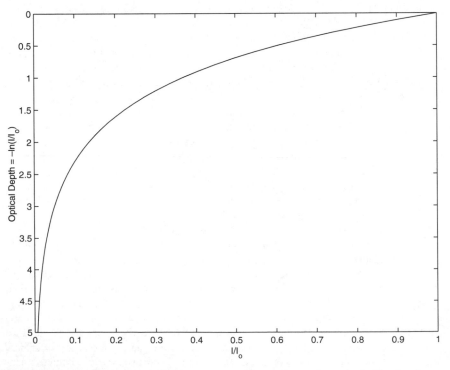

FIGURE 2.1   Variation of visible light intensity with depth in a hypothetical body of water.

visible light spectrum are equally useful for photosynthesis, but the total intensity of visible light does provide a convenient, if approximate, measure of the amount of light available for photosynthesis.

As indicated in Figure 2.1, light intensity drops off with depth in the water column in approximately an exponential fashion, that is, at a rate proportional to the intensity of the light. Photosynthetic rates tend to become saturated at high light intensities and may be inhibited by the ultraviolet (UV) light in direct sunlight. However, UV light is rapidly attenuated by water, and at depths where the irradiance (I) is less than about 30% of surface values (I_0), photosynthetic rates are almost directly proportional to light intensity. Thus, a graph of gross photosynthesis versus depth in a body of water might appear qualitatively very similar to the solid curve in Figure 2.2. Therefore, to the extent that light limits photosynthesis, phytoplankton growth rates should be highest near the surface, where light intensities are close to optimal. However, most phytoplankton cannot position themselves in the water column because they have very little locomotive ability, and for the most part are moved about by turbulence and currents in both the horizontal and vertical directions. At the surface, winds blowing over the water generate waves and turbulence that keep the water column well mixed and the phytoplankton concentration correspondingly uniform. If the depth of this mixed layer is shallow enough, the production of organic matter by the phytoplankton in the mixed layer will exceed losses of organic matter to plant and animal respiration. This difference between the production and consumption of organic matter by the plant and animal communities is called *net community production*. The dashed curve in Figure 2.2 shows the qualitative relationship between net community

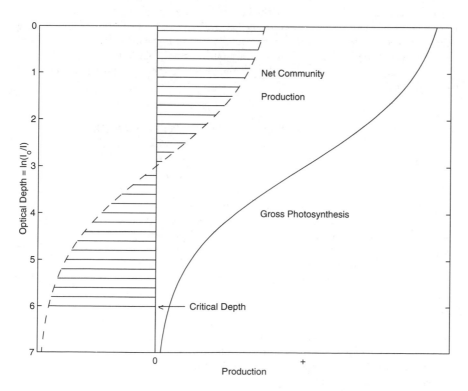

FIGURE 2.2   Relationship between gross photosynthesis, net community production, and depth in a well-mixed body of water. Average net community production is zero above the critical depth.

production and depth in a hypothetical body of water. When net community production is positive in the mixed layer, biomass may accumulate in this layer and/or may be exported to deeper parts of the system. Sinking of detritus is a common mechanism that exports organic matter from the mixed layer to deeper water. If the depth of the mixed layer equals a certain depth known as the *critical depth*, net community production is zero when averaged over the depth of the mixed layer (Smetacek and Passow, 1990). In Figure 2.2, net community production is positive above an optical depth ($= \ln(I_o/I)$ of about 3 and is negative below that depth. The hatched area of positive net community production exactly balances the hatched area of negative net community production above an optical depth of 6. The critical depth in this example therefore occurs at an optical depth of 6. If the mixed layer is deeper than the critical depth, there will be a net consumption of organic matter within the mixed layer, i.e., the average net community production within the mixed layer will be negative. Obviously, this situation cannot persist indefinitely in the absence of allochthonous (external) inputs of organic matter, or there would be no organic matter left.

Certain kinds of pollution associated with human activities may reduce the depth of the euphotic zone by increasing the turbidity of the water. The result is a decrease in net community production and sometimes a dramatic change in the composition of the phytoplankton community. For example, sediment runoff from construction sites may greatly diminish water clarity and therefore decrease the amount of light available for phytoplankton. In such cases, the phytoplankton population may become dominated by cyanobacteria (also referred to as *blue-green algae*), many species of which are able to maintain themselves near the surface of the water by means of special gas-filled vacuoles that give the plants a slightly positive buoyancy. Cyanobacteria may form floating surface scums or mats that are aesthetically objectionable, particularly if they wash ashore and rot along the water's edge. Furthermore, the cyanobacteria may be unpalatable to the natural community of herbivorous zooplankton, and through this and other feeding relationships the initial change in the composition of the phytoplankton community may impact the entire biological community.

NUTRIENT LIMITATION OF PHOTOSYNTHESIS

Excessive stimulation of algal production by the addition of essential nutrients to the euphotic zone is more often the cause of algal pollution problems than changes in water clarity. In fact, even in cases where cyanobacteria come to dominate a phytoplankton community, the principal cause of change is usually a shift in the relative input of certain nutrients to the euphotic zone rather than increased turbidity. The sensitivity of phytoplankton production and/or composition to changes in nutrient inputs arises because phytoplankton biomass in the euphotic zone is often controlled by the availability of certain nutrients that, in addition to light, are essential for growth. If the supply of these nutrients to a nutrient-limited phytoplankton community is increased, an increase in production and biomass roughly proportional to the increase in nutrient supply can be expected. Obviously, no immediate increase in phytoplankton production will occur if such nutrient additions are made to the water column below the euphotic zone (the so-called aphotic zone), where light is inadequate to support photosynthesis even if nutrients are abundant. However, production in the surface waters will subsequently increase if aphotic zone water is later mixed into the euphotic zone, a process that occurs regularly, albeit sometimes slowly, in all aquatic systems.

Usually only one or a few nutrients are limiting to the production of phytoplankton biomass in a given system at a given time, but the identification of these nutrients has proven to be a controversial problem. Table 2.1 lists chemical elements that are known to

TABLE 2.1 ESSENTIAL MACRO- AND MICRONUTRIENT
ELEMENTS FOR PLANTS

Essential Macronutrient Elements	Symbol	Essential Micronutrient or Trace Elements	Symbol
Oxygen	O	Iron	Fe
Carbon	C	Manganese	Mn
Nitrogen	N	Copper	Cu
Hydrogen	H	Zinc	Zn
Phosphorus	P	Boron	B
Sulfur	S	Silicon	Si
Potassium	K	Molybdenum	Mo
Magnesium	Mg	Chlorine	Cl
Calcium	Ca	Vanadium	V
		Cobalt	Co
		Sodium	Na

Source: Odum (1971), p. 127.

be essential for plants. Macronutrient elements are required in relatively large amounts compared to micronutrients, although for some elements the distinction is not clear-cut. Of the macronutrients, carbon, oxygen, and hydrogen are required in the largest amounts, since they are essential components of organic compounds such as carbohydrates, lipids, and proteins. Nitrogen (N) and phosphorus (P) are required in somewhat smaller amounts, with the atomic ratio of N to P averaging about 16 (range: 3–30) in phytoplankton and 30 (range: 10–70) in macroalgae (Ryther and Dunstan, 1971; Atkinson and Smith, 1983). N is an essential component of proteins, nucleic acids, and certain pigments (e.g., chlorophyll), and P is required to produce phospholipids and sugar phosphate bonds in molecules such as adenosine triphosphate (ATP). These high-energy phosphate bonds provide a convenient and essential storage unit for small amounts of energy derived primarily from the stepwise catabolism of organic molecules and may be used at any time by the organism as an energy source for metabolic processes. The remaining macronutrient elements are required in even smaller amounts, and their concentrations in both marine waters and freshwaters are more than adequate to supply the nutritional needs of aquatic plants. The micronutrients or trace elements are required in the smallest amounts, in many cases functioning as catalysts to speed up metabolic reactions. Such catalysts are not consumed or altered by the reactions they mediate, and therefore only small amounts are required by an organism.

Although there are some remarkable differences in the physiology of different species of algae, the elemental composition of most algae is remarkably constant with respect to C, N, and P. The constancy of their elemental composition reflects the fact that many algae are very similar in terms of the macromolecular composition of their organic matter. Phytoplankton tend to allocate about 50% of their C to protein, 35% to carbohydrate, and 15% to lipids. The corresponding percentages for macroalgae are 20%, 75%, and 5%. Because the C, N, and P composition of protein, carbohydrate, and lipid tends to be very similar among species, there is a corresponding similarity in the C:N:P elemental ratios in phytoplankton and likewise in macroalgae. The C:N:P ratio in many phytoplankton is roughly 106:16:1 on an atomic basis. Credit for this discovery is generally given to Alfred Redfield (1958), and atomic C:N:P ratios of roughly 106:16:1 are known as *Redfield ratios.*

In macroalgae, atomic C:N:P ratios are roughly 550:30:1 (Atkonson and Smith, 1983). Deviations by more than a factor of 2 from these ratios are uncommon.

Realizing the relative requirements of plants for the essential nutrients listed in Table 2.1, one may ask which of these elements is most likely to limit phytoplankton biomass. C, H, and O are needed in the largest amounts, but these elements are readily obtained from H_2O (H) and CO_2 (C and O). Obviously, there is no lack of H_2O in an aquatic environment. CO_2 is a gas that is found in the atmosphere and dissolves in water, reaching at equilibrium a concentration proportional to its concentration in the atmosphere. The chemistry of the oceans is such that there is invariably an abundance of inorganic C to support photosynthesis, although under certain conditions the supply of CO_2 may limit the production of some species (Riebesell et al., 1993). On the other hand, the inorganic C concentration in some freshwater lakes is extremely low, and it has sometimes been argued that photosynthesis in such systems might be limited by a lack of CO_2. However, Schindler (1974) has convincingly shown that the exchange of CO_2 between the atmosphere and the water is sufficiently rapid to provide adequate CO_2 for the development of large phytoplankton blooms over a time period of no more than a few weeks. In other words, the atmosphere acts as a CO_2 reservoir for the mixed layer of an aquatic system, and the flux of CO_2 from the atmosphere into the water may easily provide the CO_2 needed for photosynthesis even if the ambient CO_2 concentration in the water is low. We will examine more closely how Schindler arrived at this conclusion later in this chapter. For the moment, suffice it to say that CO_2 appears to limit phytoplankton biomass in few if any natural aquatic systems.

Table 2.2 lists the average concentrations of most of the remaining macro- and micronutrient elements in typical river water and seawater. Sulfur (as sulfate, SO_4^{2-}), potassium (Ka), magnesium (Mg), calcium (Ca), chlorine (Cl), and sodium (Na) are the elements that make up the principal salts in seawater. Their concentrations in seawater range from 10 millimolar (mM) to 0.56 molar (M). There is no evidence that a lack of these elements ever limits photosynthesis in the sea. The concentrations of the same ele-

TABLE 2.2   AVERAGE CONCENTRATIONS OF SELECTED ELEMENTS IN RIVER WATER AND SEAWATER (S = 35‰)[a]

Element	River Water	Seawater
B	1.67	416
Ca	332	10,300
Cl	226	546,000
Cu	0.003	0.00002
Co	0.024	0.004
Fe	0.716	0.001
K	38	10,200
Mg	128	53,200
Mn	0.149	0.0005
Mo	0.005	0.11
Na	391	468,000
S	116	28,200
Si	178	100
V	0.02	0.03
Zn	0.459	0.006

[a] Concentration units are μM.
Source: For seawater, Bruland (1983). For freshwater, Martin and Whitfield (1983), with the exception of S, Cl, and Na, which were taken from Riley and Chester (1971).

ments in river water are much lower and range from ~40 to 400 micromolar (μM). Again, however, there is no evidence that such concentrations are limiting to photosynthesis. The micronutrient elements are for the most part found at much lower concentrations than are any of the macronutrients. Although it is true that most of the micronutrients are required in very small amounts, there is evidence that the availability of certain micronutrients may limit algal biomass in some aquatic systems or at least limit the biomass of certain species of algae. For example, phytoplankton known as *diatoms* build an elaborate skeleton of silica that encloses the rest of the cell. Consequently, diatoms require much more silicate than other classes of phytoplankton, and there is fairly convincing evidence that changes in silicate concentrations may affect the abundance of certain diatoms in freshwater systems (Hutchinson, 1967, pp. 446–455). There is little evidence that silicate limits photosynthetic rates in the ocean, although the distribution of marine diatoms is undoubtedly influenced by silicate availability. Silicate concentrations in ocean surface waters can easily drop as low as 1–2 μM (Karl et al., 2000). The greatest abundance of marine diatoms is found in Antarctic seas, where silicate concentrations as high as 30–50 μM have been reported during the austral winter. Molybdenum (Mo) plays a role in the formation of the enzyme nitrate reductase and hence is required for the assimilation of N in the form of nitrate (NO_3^-). Molybdenum is also required for N fixation, the process by which certain plants and bacteria convert atmospheric nitrogen gas (N_2) into a form that can be used in primary production. There is evidence that the availability of Mo limits nitrate uptake and photosynthetic rates in Castle Lake, California (Axler et al., 1980), and evidence presented by Howarth and Cole (1985) indicates that Mo availability limits N fixation rates in the ocean. In the latter case, limitation may result from competitive interference with Mo uptake by sulfate, which has an effective radius and charge distribution nearly identical to those of molybdate. Similar competitive interference is presumably insignificant in freshwater because the sulfate concentration in freshwater is less than 1% of the sulfate concentration in seawater. Howarth and Cole's (1985) hypothesis has been challenged, however, by Paulsen et al. (1991), who found no evidence of Mo limitation of nitrogen fixation in North Carolina coastal waters.

N and P, the two essential macronutrients not yet discussed, are found in seawater below the euphotic zone in concentrations of about 20–40 and 1.3–2.5 μM, respectively. Typical aphotic zone concentrations of these elements in freshwater systems may be two or three times lower but can vary greatly from one system to another. However, when there is adequate light to support photosynthesis, it is not unusual for the concentrations of inorganic N and P in the euphotic zone of both freshwater and marine systems to be in the range 50–100 nanomolar (nM) (Edmondson 1972; Smith et al. 1986). These concentrations are several orders of magnitude smaller than the average concentrations of the other macronutrients listed in Table 2.1; and of those macronutrients, phytoplankton require N and P in much larger amounts than S, K, Mg, or Ca. Thus, based on observed concentrations, N and P are the macronutrients most likely to be limiting photosynthesis. However, because many of the micronutrients are present in extremely low concentrations in the euphotic zone, it is impossible to tell, based simply on measured nutrient concentrations, whether N, P, or one of the micronutrients is limiting photosynthetic rates. As a result, phytoplankton ecologists have resorted to a bioassay type of experiment to determine the identity of the nutrient(s) limiting phytoplankton production. These bioassay experiments are commonly referred to as *nutrient enrichment* experiments.

Nutrient Enrichment Experiments

The usual approach in a nutrient enrichment experiment is to fill a series of clear flasks with the water to be assayed and enrich some of the flasks with various nutrients to see whether these nutrient additions have any effect on phytoplankton production in the flasks. In some cases, the water is filtered first to remove the natural phytoplankton and other organisms and is then inoculated with a monoculture of a particular phytoplankton species. In other cases, the water is not filtered, so that the natural phytoplankton community becomes the test population. Following enrichment the flasks are incubated under appropriate light and temperature conditions, and the response of the phytoplankton is monitored for a period of time in both the enriched flask and a control flask, which receives no nutrient additions. Normally, the response of the phytoplankton is monitored by one of two methods. In the first method the phytoplankton biomass is determined, usually in terms of cell counts or chlorophyll concentration, and the effect of a given enrichment is measured in terms of the difference in phytoplankton biomass between the enriched flask and the control flask. One generally speaks of such a bioassay as being based on the *yield* (i.e., biomass) in the enriched flasks. In the second method the actual rate of photosynthesis in each flask is measured, usually after waiting for a fixed time interval (approximately a day to a week) after the nutrient enrichments. There are pros and cons to both methods. If, as is usually the case, the test population consists of the natural phytoplankton community, one often finds that the composition of the community in the flask changes after several days of incubation. This change results because some species evidently do not grow well under artificial conditions and because nutrient enrichments may not stimulate all species equally. Because the biomass in the enriched flasks generally peaks as much as several weeks following enrichment, one can argue that yield-type experiments may misrepresent the nutrient limitation characteristics of the natural phytoplankton assemblage. On the other hand, photosynthetic rates in enriched flasks may vary greatly with time following enrichment because there may be a lag in the response of phytoplankton to certain enrichments, whereas other enrichments produce a rapid increase in production that subsequently declines. For example, Menzel et al. (1963) were initially led to believe that Sargasso Sea water was iron-limited because, after a few days of incubation, photosynthetic rates in flasks enriched with N, P, and iron (Fe) were substantially higher than those in flasks that received only N and P additions. However, they later discovered that after about a week's incubation, flasks enriched with only N and P showed photosynthetic rates just as high as the flasks enriched with N, P, and Fe had shown after a few days. They concluded that N and P were the principal limiting nutrients and that addition of Fe simply speeded up the response of the phytoplankton to N and P additions. The effect of the added Fe is perhaps not surprising, since Menzel et al. (1963) added N in the form of nitrate, and Fe is required for the reduction of nitrate to a form that can be utilized in the photosynthetic production of organic N compounds. This example illustrates why nutrient enrichment experiments must be interpreted cautiously to avoid jumping to unwarranted conclusions.

Figure 2.3 illustrates the methodology and interpretation of so-called single nutrient and multiple-nutrient enrichment experiments. In a single-nutrient enrichment experiment, a series of experimental flasks receives an enrichment with only one nutrient. In a multiple-nutrient enrichment experiment, a series of experimental flasks is enriched with all essential plant nutrients except one. In both cases the productivity or yield in the experimental flasks is compared to that of the control flasks.

The rationale behind the interpretation of the results is as follows: If the water being studied contains an abundance of all essential nutrients except one, then addition of that single nutrient should greatly stimulate production. If addition of a single nutrient does not greatly stimulate production, then some other nutrient(s) is (are) limiting. Since several nutrients may be simultaneously limiting (i.e., there is very little of each of several nutrients in the water to support additional growth), it is impossible to say in the case where production is not stimulated whether the single nutrient added was one of the limiting

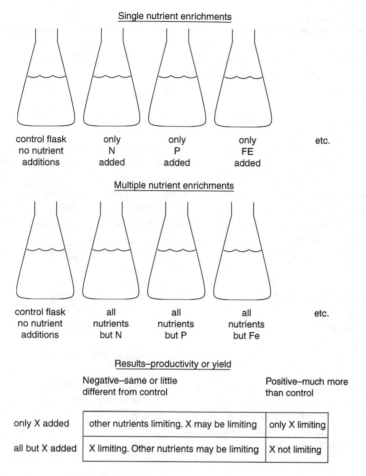

Single nutrient enrichments

| control flask no nutrient additions | only N added | only P added | only FE added | etc. |

Multiple nutrient enrichments

| control flask no nutrient additions | all nutrients but N | all nutrients but P | all nutrients but Fe | etc. |

Results–productivity or yield

	Negative–same or little different from control	Positive–much more than control
only X added	other nutrients limiting. X may be limiting	only X limiting
all but X added	X limiting. Other nutrients may be limiting	X not limiting

FIGURE 2.3   Single and multiple nutrient enrichment experiments and interpretation of results.

nutrients or not. We only know that there is at least one limiting nutrient other than the one tested.

In the case of multiple-nutrient enrichments, production little different from the control flask indicates that the nutrient omitted is limiting, since all other nutrients were added in the enrichment. It is impossible to tell without doing additional experiments whether other nutrients are simultaneously limiting. However, if production is much greater than that of the control flask, then the omitted nutrient must not have been limiting. Figure 2.3 summarizes the possible results and conclusions to be reached from the two types of nutrient enrichment experiments.

Unfortunately, the simple interpretability of the four outcomes indicated at the bottom of Figure 2.3 is not always encountered in practice. One obvious complicating factor is the variable composition of phytoplankton. For example, we previously noted that the N:P ratio by atoms in marine phytoplankton varies between roughly 3 and 30. Hence, a population initially with an N:P ratio of 10 might respond to N enrichment by assimilating additional N until the N:P ratio reached 20. However, the same population might respond to P enrichment by assimilating additional P until the N:P ratio was reduced to 5. In either case, the additional nutrient assimilation would probably result in an increase in photosynthetic rate and/or yield. Applying the simple logic in Figure 2.3 appropriate to a single nutrient enrichment, one might conclude in the first case (positive response to N addition) that only N was limiting and in the second case (positive response to P addition) that only P was limiting. These two conclusions are obviously inconsistent with one another.

Fortunately, changes in productivity and/or yield associated with shifts in cellular composition are unlikely to be large. For example, it is unlikely that yield, as measured by changes in cell numbers or chlorophyll, would change by more than a factor of 2 in response to shifts in N:P ratios. On the other hand, changes in yield by an order of magnitude or more are not uncommon in cases where a single nutrient is added to water that contains all other nutrients in abundance. It is therefore best to consider only large (e.g., order of magnitude) changes in yield or productivity as clear evidence of a positive response to nutrient addition, and Figure 2.3 accordingly defines a positive response as a productivity or yield much more than the control.

We have now seen that nutrient enrichment experiments may be complicated by the following factors:

1. Possible changes in the species composition of the culture in the enrichment flask over time
2. Variability over time in the response of the same population to different nutrient additions
3. Changes in cellular composition in response to nutrient additions

Such complications need not be serious problems if the enrichment experiments are designed carefully and if the enrichment flasks are monitored frequently. In fact, such "complications" may provide useful insights for the experimentalist (e.g., Menzel et al., 1963). Keeping in mind these observations, let us examine several cases in which nutrient enrichment experiments have been used to study nutrient limitation questions.

Long Island Bays

The release of organic waste from duck farms began to cause noticeable water pollution problems in certain bays and tributary streams along the southern side of Long Island as the Long Island duckling industry developed during the 1940s. The affected area is shown in Figure 2.4. Most apparent among these problems was the development of massive phytoplankton blooms in the waters of the tributary streams along which the duck farms were located. These blooms extended downstream from the duck farms and into the bays along the coast, where the demise of the once productive oyster and hard-shell clam fisheries closely coincided with the development of the phytoplankton blooms (Ryther and Dunstan, 1971). Oysters, which feed by filtering water through their gills, were unable to feed or respire effectively because their gills became coated with the phytoplankton cells (Wagner, 1971). Although clams thrived in the phytoplankton-rich waters, they were so contaminated with bacteria that they were often commercially unusable. Studies by the Woods Hole Oceanographic Institution between 1950 and 1955 showed that the algal concentrations in the bay water dropped off steadily with increasing distance from the mouths of the principal tributary streams in Moriches Bay, in a manner similar to what would be expected from dilution by coastal water. In other words, there was little indication that the dense phytoplankton populations in the bays were actively growing; rather, they had simply been washed into the bays from tributary streams and subsequently dispersed by tidal currents.

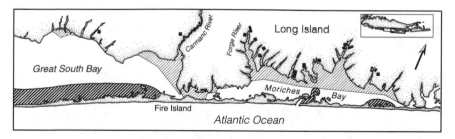

FIGURE 2.4   Shellfishing areas affected by duck farm wastes along the south shore of Long Island. Solid blocks along tributary streams indicate duck farms in 1966. Hatched regions indicate shellfishing areas. Lightly hatched shellfishing areas were closed to shellfishing due to pathogen contamination. (*Source*: Redrawn from U.S. Department of the Interior (1966).

Nutrient analyses performed on the bay and tributary stream waters showed that phosphate and phytoplankton concentrations were closely correlated, with the maximum phosphate concentrations in Moriches Bay of about 7 μM dropping off to about 0.25 μM at the eastern and western ends of the affected area. Similar analyses for N in the form of nitrate (NO_3^-), nitrite (NO_2^-), ammonium (NH_4^+), and uric acid (ducks excrete uric acid) revealed virtually no detectable N in any of these forms except in the immediate vicinity of the duck farms. It was tentatively concluded that N was the principal nutrient limiting phytoplankton production in the study area; that nitrogenous wastes from the duck farms were stimulating the massive algal blooms in the tributary streams; and that phosphate, being present in excess in the duck farm wastes, simply acted as a tracer of the water and phytoplankton from the tributary streams.

To see whether N was in fact the nutrient limiting phytoplankton biomass in the system, a series of nutrient enrichment experiments was performed on water samples taken from Moriches Bay, Great South Bay, and the Forge River, a Moriches Bay tributary on which several duck farms were located. Each water sample was filtered and then poured into three flasks, one of which served as a control and received no nutrient additions. The second flask was enriched with N and the third flask with P. No other nutrients were added. The experimental setup therefore conformed to the single-nutrient enrichment design shown in Figure 2.3. All three flasks at each station were inoculated with a pure culture of the phytoplanker *Nannochloris atomus*, the dominant algal species in the phytoplankton blooms, and were incubated for 1 week, at which time the number of cells in each flask was counted. Figure 2.5 shows the results of the experiments. Cell counts in the N-enriched flasks were an order of magnitude or more greater than the counts in the control flasks after 1-week incubations. Virtually no change in cell number occurred in the P-enriched flasks. The fact that counts in the control flasks increased by factors of about 2–4 indicated that the inoculum possessed a moderate potential for additional growth even without additional nutrients. The order-of-magnitude difference in cell numbers between the N-enriched flasks and the controls indicates that phytoplankton biomass in the system was limited by N, and only by N.

Canadian Experimental Lakes

During the 1960s and early 1970s, the Canadian government authorized a series of nutrient enrichment experiments on whole lakes in the Experimental Lakes Area of northeastern Ontario. The nutrient additions were made directly to the lakes rather than to water

Aquatic Pollution

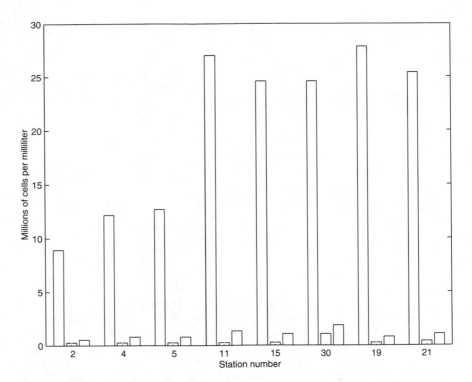

FIGURE 2.5   Cell counts after incubation for one week in nutrient enrichment flasks at indicated stations. At each station, the left-hand, middle, and right-hand vertical bars represent the N-enriched flask, the P-enriched flask, and the control flask, respectively. Stations 2, 4, and 5 were located in Great South Bay; station 11 was located between Great South Bay and Moriches Bay; stations 15, 19, and 21 were located in Moriches Bay; and station 30 was located in the Forges River. *Source:* Reprinted with permission from J. M. Ryther and W. M. Dunstan. 1971. Nitrogen, phosphorus, and eutrophication in the coastal marine environment. *Science,* **171,** 1008–1013. Copyright 1971 American Association for the Advancement of Science.

samples taken from the lakes to make the experiments as realistic as possible. Of particular concern was the importance of CO_2 exchange between the air and water. In laboratory flasks, no mechanism such as stirring or shaking is usually provided to simulate mixing effects and encourage gaseous exchange with the atmosphere. Thus, water taken from a lake with a low CO_2 content might be CO_2 limited if incubated in a flask in the laboratory. In the real world, CO_2 exchange with the atmosphere, stimulated by wind-generated turbulence and mixing, might be more than adequate to provide the phytoplankton population with CO_2.

As a test of the effectiveness of CO_2 exchange with the atmosphere, Lake 227, a lake having an extremely low dissolved CO_2 content, was enriched with N and P. Within weeks a massive bloom of phytoplankton had developed, with algal concentrations roughly 100 times greater than in other lakes of the area (Schindler, 1974). Gas exchange studies revealed that some of the CO_2 to support the phytoplankton bloom had come from the atmosphere. Considering the results of this experiment and the fact that most lakes (and the oceans) have a much higher CO_2 content than Lake 227, Schindler (1974) concluded that CO_2 probably does not limit phytoplankton biomass in most bodies of water.

To determine the nutrient that limited photosynthetic rates in the experimental lakes, multiple nutrient enrichments were made to a second lake, Lake 226. This lake consisted of two similar basins separated by a shallow neck. A divider consisting of a sea curtain made from vinyl reinforced with nylon was sealed into the sediments and fastened to the bedrock of the shallow neck (Schindler, 1974). The two basins were therefore physically separated. Beginning in May 1973, C and N were added to one basin and C, N, and P to the other basin. The phytoplankton population in the basin that received only C and N additions was little different from the population before nutrient additions, but a massive algal bloom developed in the basin that received added P. This experiment clearly showed P to be a limiting nutrient in the lake.

In a third series of experiments, P, N, and C additions were made in 1971 and 1972 to Lake 304. As expected, a large phytoplankton crop developed in response to the nutrient additions. In 1973, although additions of N and C continued, no P was added. As a result, the phytoplankton population declined dramatically, reaching typical pre-1971 levels. Table 2.3 shows the mean annual phytoplankton biomass in Lake 304, as measured by the mean chlorophyll *a* concentration in the water. Most of the P added to the lake in 1971 and 1972 was evidently trapped on the bottom of the lake, either in the form of inorganic solids or in microbial biomass, and was hence unavailable to support a phytoplankton bloom in 1973. This experiment showed convincingly not only that P was limiting photosynthetic rates in Lake 304, but also that little of the P added to the lake in one year was recycled in the next year. This latter result suggested that lakes having undesirably high phytoplankton populations might be effectively treated by reducing P inputs. In Chapter 4 we will see how well this idea has worked in practice.

Nitrogen versus Phosphorus Limitation

Ryther and Dunstan (1971) found convincing evidence that only N limited photosynthetic rates in the Long Island bays, whereas Schindler (1974) found P to be limiting in the Canadian experimental lakes. These results are rather typical of findings reported in freshwater and marine systems; that is, in freshwater systems, P is often found to be the principal limiting nutrient, and in marine systems, N is most commonly found to be limiting (e.g., Thomas, 1969). Why would P tend to be limiting in freshwater systems and N in marine systems?

A number of papers have been written on this subject, and Howarth (1988) has provided an excellent review and discussion of the issues. Which nutrient limits photosynthetic rates depends on the N:P ratio of the external inputs to the system and on the biogeochemical processes that alter the availability of N and P within the system. During the 1960s, domestic sewage typically had a N:P ratio of about 9 by atoms (Weibel, 1969),

TABLE 2.3   MEAN CHLOROPHYLL *a*
CONCENTRATIONS IN LAKE 304

Year	Chlorophyll a *(mg m^{-3})*
1969	6.5
1970	11.0
1971	21.5
1972	24.3
1973	8.9

Source: Data derived from figures reported by Schindler (1974).

which is about one-third to one-half of the values typically associated with benthic algae and phytoplankton, respectively. The use of polyphosphates as a component of laundry detergents undoubtedly accounted for the low N:P ratio of domestic wastewater at that time (see Chapter 6). However, it is clear from Ryther and Dunstan's (1971) work in the Long Island bays that wastewater from the duck farms also contained a low N:P ratio relative to phytoplankton needs. A consequence of the low N:P ratio in domestic wastewater and sewage during the 1960s was that systems artificially enriched with such water almost invariably become nitrogen limited. Ryther and Dunstan's (1971) results can be attributed not primarily to the fact that the systems contained seawater but rather to the fact that the water was enriched with wastewater having a low N:P ratio relative to the needs of the phytoplankton.

The biogeochemical processes that occur within many freshwater systems tend to create P limitation. One of the most important such process is N fixation, of which many species of cyanobacteria are capable. As a result, several experimental lakes deliberately enriched by Schindler (1977) with fertilizer containing a low N:P ratio developed large populations of cyanobacteria. The algae were able to make up for the N deficiency in the fertilizer by fixing atmospheric N. Thus, a mechanism exists for providing needed N from the atmosphere (N fixation) even if N is limiting in the external inputs to a lake. The atmosphere thus serves as a source both of C (by means of CO_2 exchange at the air-water interface) and of N. P has no such atmospheric reservoir, however, and all essential P must come from external inputs or from recycling within a lake.

In most parts of the ocean, on the other hand, N fixation by cyanobacteria appears to be comparatively unimportant. The principal reason appears to be the lower availability of Fe and perhaps Mo in the ocean compared to most fresh waters (Howarth et al., 1988a). Both metals are required for N fixation. The concentration of Fe in the ocean is orders of magnitude lower than the Fe concentration in typical freshwaters (Table 2.2). The idea that competition with sulfate lowers the availability of molybdate in marine waters is presently being debated (Paulsen et al., 1991). There is no doubt that at some times and in some places N-fixing cyanobacteria are an important source of allochthonous N in the ocean (Karl et al., 1997). However, in general, N fixation is much less important as a source of allochthonous N in the ocean than is the case in lakes (Howarth et al., 1988b).

An important factor in the P limitation question is the fact that the phosphate ion, PO_4^{3-}, forms insoluble compounds with several positive ions, including aluminum (Al^{3+}), calcium (Ca^{2+}), and iron (Fe^{3+}). In freshwater systems the most important mechanism of phosphate removal typically involves its adsorption to oxides and oxy-hydroxides of Fe such as Fe_2O_3 and FeOOH. The solid compound sinks to the bottom of a lake, effectively trapping phosphate in the sediments. This mechanism may have been partly responsible for the rapid recovery of Lake 304 in 1973. If the O_2 concentration in the water is very low, however, ferric iron, Fe^{3+}, is spontaneously converted to ferrous iron, Fe^{2+}. Ferrous iron does not effectively bind phosphate. Thus, if the bottom waters of a lake become anoxic, phosphate trapped in the sediments may be released and circulate back into the water. In his work on the Canadian experimental lakes, however, Schindler (1974, 1977) found that in some cases phosphate was efficiently trapped in the sediments even though the bottom waters were anoxic for several months. He reported that uptake of the phosphate by microorganisms in the bottom waters and subsequent sedimentation explained the failure of the phosphate to return to the water column. Nevertheless, the bottom waters of many lakes do remain oxidizing throughout the year, and in such cases the formation and sedimentation of ferric phosphate compounds may be a significant mechanism for removing phosphate from the water column. N does not form insoluble chemical precipitates, and there is no mechanism analogous to phosphate adsorption to ferric Fe to trap N in the sediments of either marine or freshwater systems.

Since the average depth of the ocean is roughly 4 km, particulate materials that sink out of the surface mixed layer are essentially lost from the euphotic zone. Upwelling of water from below the nutricline does occur at a slow rate in almost all parts of the ocean and at a rapid rate in certain upwelling areas. However, with the exception of shallow coastal regions, recycling of nutrients from below the nutricline and from the sediments is a very slow process in most parts of the ocean. Thus, recycling of nutrients within the surface mixed layer takes on special importance in most marine food chains. The principal mechanisms for recycling nutrients within the mixed layer are direct excretion (e.g., zooplankton excrete ammonia) and regeneration from detritus. Evidence to date suggests that both animal excretion and regeneration of nutrients from detritus tend to create N-limited conditions. Nutrient release from particulate detritus must be rapid if the nutrients are to be recycled before the particles sink out of the euphotic zone. P seems to be released more rapidly from detritus than N, presumably because phosphate ester bonds are more easily cleaved than the covalent bonds of organic N (Howarth, 1988). As a result, fecal material and sedimenting detritus tend to be enriched in N relative to P (Knauer et al., 1979; Lehman, 1984). For example, the studies of Knauer et al. (1979) revealed that material that settled into sediment traps near the base of the euphotic zone in the North Pacific subtropical gyre contained N and P in an atomic ratio of 29, about twice the Redfield ratio of 16 typically associated with oceanic particulate matter (Copin-Montegut and Copin-Montegut 1983). Corresponding ratios in coastal areas were 22 and 27 under upwelling and nonupwelling conditions, respectively. With respect to excretion, studies summarized by Lehman (1984) indicate that the soluble compounds released by zooplankton are enriched in P relative to N. Studies by Le Borgne (1982) clearly showed that the net growth efficiencies of zooplankton are higher for N than P, i.e., the zooplankton excrete a higher percentage of the P than of the N in the food they assimilate. Hence, biological processes that occur within the surface mixed layer tend to create N-limited conditions, and this fact accounts in part for the tendency of open-ocean systems to be N-limited.

Denitrification is the process by which nitrate N is reduced in a series of steps to N_2, the gas that makes up about 78% of the Earth's atmosphere. The conversion is mediated by certain species of bacteria that utilize the oxygen from the nitrate ion to oxidize organic matter under anoxic or nearly anoxic conditions. Denitrification obviously tends to create N-limited conditions because N_2 is unavailable to aquatic plants other than N-fixing cyanobacteria. Because of the requirement for low O_2 concentrations, denitrification is restricted to certain portions of aquatic habitats. Sediments are usually anoxic below the surface layer, particularly when the overlying water column is highly productive. According to Seitzinger (1988, p. 702), "During the mineralization of organic matter in sediments, a major portion of the mineralized nitrogen is lost from the ecosystem via denitrification" and "The loss of nitrogen via denitrification exceeds the input of nitrogen via N_2 fixation in almost all river, lake, and coastal marine ecosystems in which both processes have been measured."

Nor is denitrification confined to the sediments. Denitrification may occur in the water column when the O_2 concentrations drops below about 0.2 g m^{-3}, a condition that is quite common in some lakes and certain parts of the ocean. Furthermore, even though the concentration of O_2 in a bulk water sample may be well above 0.2 g m^{-3}, very low O_2 concentrations may exist in microzones associated with particles. Such microzones may be sites of denitrification. Studies summarized by Hattori (1983) indicate that the low-O_2 intermediate waters of the Eastern Tropical Pacific Ocean account for a major fraction of the denitrification that occurs in the water column of the ocean, and that substantial amounts of denitrification also occur in the low-O_2 intermediate waters of the Arabian Sea, in the bottom waters of the southwest African shelf, and perhaps in the Bay of Bengal. Following their formation, oceanic bottom waters typically do not return to the surface again for 500–1,000 years. It is reasonable to expect that denitrification would cause these subsurface waters to become depleted in N. The denitrification of bottom and intermediate waters is probably the major factor that causes subsequently upwelled marine surface waters to be N-limited.

The general picture that emerges from this analysis is that freshwater systems tend to be P-limited because of the sedimentation and burial of phosphate in the sediments, particularly in association with ferric oxides or oxy-hydroxides. This mechanism is of much less consequence in the marine P cycle because of the very low Fe concentration in seawater. The tendency of marine waters to be N-limited reflects the low Fe concentration in seawater, the fact that P is recycled more efficiently than N by biological processes in the euphotic zone, and the long time during which denitrification may act to deplete subsurface waters of fixed N.

While these generalizations are useful for understanding the role of nutrients in limiting photosynthetic rates in aquatic systems, it is important to realize that not all lakes and rivers are P-limited, nor are all marine waters N-limited. The magnitude and elemental composition of external nutrient inputs and the relative importance of various biogeochemical processes occurring within the system all combine to determine which nutrient or nutrients limit photosynthetic rates. The relative importance of these inputs and processes will vary from one system to another. Examples of exceptions to the general picture include the apparent limitation by Fe of photosynthetic rates in the Equatorial Pacific upwelling system, the offshore waters of the Antarctic, and the northeast Pacific subarctic gyre. In those areas the concentrations of phosphate and nitrate in the euphotic zone are well above the levels associated with nutrient limitation, and nutrient enrichment studies conducted by Martin and Fitzwater (1988), as well as *in situ* iron fertilization experiments (Kolber et al., 1994), have clearly shown that Fe is the single nutrient limiting primary production. The explanation seems to be that the Fe:N ratio of intermediate-depth ocean waters is low compared to the needs of phytoplankton (Martin and Gordon, 1988). In the absence of external Fe inputs, these waters become Fe-limited when upwelled into the euphotic zone, e.g., the Equatorial Pacific upwelling system (Landry et al., 1997). In coastal waters the input of Fe from land runoff and release from sediments is apparently more than adequate to supply the needed Fe, and in the large subtropical gyres that account for about 40% of the ocean's surface area, upwelling is so slow that atmospheric fallout of Fe from dust and rainfall is sufficient to provide the Fe required for photosynthesis. However, in the offshore waters of the Antarctic and northeast Pacific oceans, atmospheric inputs and lateral transport of Fe are insufficient to keep pace with the upward movement of N into the euphotic zone. As a result, phytoplankton strip the water of Fe long before the supply of N is exhausted.

REFERENCES

ATKINSON, M., and S. V. SMITH. 1983. C:N:P ratios of benthic marine plants. *Limnol. Oceanogr.*, 28, 568–574.

AXLER, R. P., R. M. GERSBER, and C. R. GOLDMAN. 1980. Stimulation of nitrate uptake and photosynthesis by molybdenum in Castle Lake, California. *Can. J. Fish. Aquat. Sci.*, 37, 707–712.

BRULAND, K. W. 1983. Trace elements in seawater. In J. P. Riley and R. Chester, Eds., *Chemical Oceanography*. Volume 8. Academic Press, New York. Pp. 157–220.

COPIN-MONTEGUT, C., and G. COPIN-MONTEGUT. 1983. Stoichiometry of carbon, nitrogen, and phosphorus in marine particulate matter. *Deep-Sea Res.*, 30, 31–46.

EDMONDSON, W. T. 1972. Nutrients and phytoplankton in Lake Washington. In. G. E. Likens, Ed., *Nutrients and Eutrophication*. American Society of Limnology and Oceanography, Lawrence, Kan. Pp. 172–193.

HATTORI, A. 1983. Denitrification and dissimilatory nitrate reduction. In E. J. Carpenter and D. G. Capone, Eds., *Nitrogen in the Marine Environment*. Academic Press, New York. Pp. 191–232.

HOWARTH, R. W. 1988. Nutrient limitation of net primary production in marine ecosystems. *Ann. Rev. Ecol.*, 19, 89–110.

HOWARTH, R. W., and J. J. COLE. 1985. Molybdenum availability, nitrogen limitation, and phytoplankton growth in natural waters. *Science*, **229**, 653–655.

HOWARTH, R. W., R. MARINO, and J. J. COLE. 1988a. Nitrogen fixation in freshwater, estuarine, and marine ecosystems. 2. Biogeochemical controls. *Limnol. Oceanogr.*, **33**, 688–701.

HOWARTH, R. W., R. MARINO, J. LANE, and J. J. COLE. 1988b. Nitrogen fixation in freshwater, estuarine, and marine ecosystems. 1. Rates and importance. *Limnol. Oceanogr.*, **33**, 669–687.

HUTCHINSON, G. E. 1967. *A Treatise on Limnology.* Volume 2. Wiley, New York. 1115 pp.

KARL, D. M., R. R. BIDIGARE, and R. M. LETELIER. 2000. Long-term changes in plankton community structure and productivity in the North Pacific subtropical gyre: The domain shift hypothesis. *Deep-Sea Res.* (in press).

KARL, D. M., R. LETELIER, L. TUPAS, J. DORE, J. CHRISTIAN, and D. HEBEL. 1997. The role of nitrogen fixation in biogeochemical cycling in the subtropical North Pacific Ocean. *Nature*, **388**, 533–538.

KNAUER, G. A., J. H. MARTIN, and K. W. BRULAND. 1979. Fluxes of particulate carbon, nitrogen, and phosphorus in the upper water column of the northeast Pacific. *Deep-Sea Res.*, **26A**, 97–108.

KOLBER, Z. S., R. T. BARBER, K. H. COALE, S. E. FITZWATER, R. M. GREENE, K. S. JOHNSON, S. LINDLEY, and P. G. FALKOWSKI. 1994. Iron limitation of phytoplankton photosynthesis in the equatorial Pacific Ocean. *Nature*, **371**, 145–149.

LE BORGNE, R. 1982. Zooplankton production in the eastern tropical Atlantic Ocean: Net growth efficiency and P:B in terms of carbon, nitrogen, and phosphorus. *Limnol. Oceanogr.*, **27**, 681–698.

LANDRY, M. R., R. T. BARBER, R. R. BIDIGARE, F. CHAI, K. H. COALE, H. G. DAM, M. R. LEWIS, S. T. LINDLEY, J. J. McCARTHY, M. R. ROMAN, D. K. STOECKER, P. G. VERITY, and J. R. WHITE. 1997. Iron and grazing constraints on primary production in the central equatorial Pacific: An EqPac synthesis. *Limnol. Oceanogr.*, **42**, 405–418.

LEHMAN, J. T. 1984. Grazing, nutrient release, and their impacts on the structure of phytoplankton communities. In D. G. Meyers and J. R. Strickland, Eds., *Trophic Interactions within Aquatic Ecosystems.* Westview, Boulder, Colo. Pp. 49–72.

MARTIN, J. H., and S. E. FITZWATER. 1988. Iron deficiency limits phytoplankton growth in the northeast Pacific subarctic. *Nature*, **331**, 341–343.

MARTIN, J. H., and R. M. GORDON. 1988. Northeast Pacific iron distributions in relation to phytoplankton productivity. *Deep-Sea Res.*, **35**, 177–196.

MARTIN, J.-M., and M. WHITFIELD. 1983. The significance of the river input of chemical elements to the ocean. In C. S. Wong, Ed., *Trace Metals in Sea Water.* Proc. NATO Advanced Research Institute, March 30–April 1, 1981, Erice, Italy. Plenum, New York. Pp. 265–296.

MENZEL, D. W., E. M. HULBERT, and J. H. RYTHER. 1963. The effects of enriching Sargasso Sea water on the production and species composition of the phytoplankton. *Deep-Sea Res.*, **10**, 209–219.

ODUM, E. P. 1971. *Fundamentals of Ecology.* 3rd ed. Saunders, Philadelphia. 574 pp.

PAULSEN, D. M., H. W. PAERL, and P. E. BISHOP. 1991. Evidence that molybdenum-dependent nitrogen fixation is not limited by high sulfate concentrations in marine environments. *Limnol. Oceanogr.*, **36**, 1325–1334.

REDFIELD, A. C. 1958. The biological control of chemical factors in the environment. *Am. Sci.*, **46**, 205—222.

RIEBESELL, U., D. A. WOLF-GLADROW, and V. SMETACEK. 1993. Carbon dioxide limitation of marine phytoplankton growth rates. *Nature*, **361**, 249–251.

RILEY, J. P., and R. CHESTER. 1971. *Introduction to Marine Chemistry.* Academic Press, New York. 465 pp.

RYTHER, J. H. 1969. Photosynthesis and fish production in the sea. *Science*, **166**, 72–76.

RYTHER, J. H., and W. M. DUNSTAN. 1971. Nitrogen, phosphorus, and eutrophication in the coastal marine environment. *Science*, **171**, 1008–1013.

SCHINDLER, D. W. 1974. Eutrophication and recovery in experimental lakes: Implications for lake management. *Science*, **184**, 897–899.

SCHINDLER, D. W. 1977. Evolution of phosphorus limitation in lakes. *Science*, **195**, 260–262.

SEITZINGER, S. P. 1988. Denitrification in freshwater and coastal marine ecosystems: Ecological and geochemical significance. *Limnol. Oceanogr.*, **33**, 702–724.

SMETACEK, V., and U. PASSOW. 1990. Spring bloom initiation and Sverdrup's critical-depth model. *Limnol. Oceanogr.*, **35**, 228–234.

SMITH, S. V., W. J. KIMMERER, and T. W. WALSH. 1986. Vertical flux and biogeochemical turnover regulate nutrient limitation of net organic production in the North Pacific Gyre. *Limnol. Oceanogr.*, **31**, 161–167.

THOMAS, W. H. 1969. Phytoplankton nutrient enrichment experiments off Baja California and in the eastern equatorial Pacific Ocean. *J. Fish. Res. Bd. Can.*, **26**, 1133–1145.

U.S. DEPARTMENT OF THE INTERIOR. 1966. Proceedings, Conference in the Matter of Pollution of the Navigable Waters of Moriches Bay and the Eastern Section of Great South Bay and their Tributaries, U.S. Interior Dept. of the Patchogue, N.Y. September. 496 pp.

WAGNER, R. H. 1971. *Environment and Man.* Norton, New York. 491 pp.

WEIBEL, S. R. 1969. Urban drainage as a factor in eutrophication. *In Eutrophication: Causes, Consequences, Correctives.* National Academy of Sciences, Washington, D.C. Pp. 383–403.

WEIBEL, S. R. 1969. Urban drainage as a factor in eutrophication. In Eutrophication: Causes, Consequences, Correctives. National Academy of Sciences, Washington, D.C. Pp. 383–403.

QUESTIONS

2.1 At the present time, the Earth's atmosphere contains about 7.5×10^{17} grams of inorganic C in the form of CO_2. The concentration of dissolved inorganic C in the ocean is about 28 grams per cubic meter. The volume of the ocean is about 1.4×10^{18} m^3. What is the ratio of the total amount of dissolved inorganic C in the ocean to the total amount of inorganic C in the atmosphere?

2.2 Assume that you work for the United Nations and that a group of scientists comes to you with a proposal to reduce the concentration of CO_2 in the atmosphere. The proposal is to fertilize the surface waters of the ocean to stimulate the fixation of about 7.5×10^{16} grams of C, i.e., about 10% of the CO_2 in the atmosphere. The scientists believe that the uptake of this much C from the surface waters of the ocean will cause an equivalent amount of inorganic C to enter the ocean from the atmosphere (recall Lake 227) and thereby reduce the concentration of CO_2 in the atmosphere by about 10%. They propose to fertilize with either N, P, or Fe. Assuming that the organic matter contains C, N, and P in the Redfield ratio, calculate the amount of N and P that would be required to stimulate the fixation of 7.5×10^{16} grams of carbon. Compare these amounts of N and P with the present global production of N and P for fertilizer use, 7×10^7 tonnes of N and 3.5×10^7 tonnes of P per year. How many years of fertilizer production would it take the world to produce enough N and P to stimulate the uptake of 7.5×10^{16} grams of C by marine phytoplankton? The ratio of C to Fe in marine phytoplankton is about 10^4 grams of C per gram of Fe. How many grams of Fe would be required to stimulate the uptake of 7.5×10^{16} grams of C? Global Fe production is presently about 5.5×10^{14} grams per year. How many days would it take the world to produce enough Fe to stimulate the uptake of 7.5×10^{16} grams of C by marine phytoplankton?

2.3 After some discussion, the scientists decide that fertilization with Fe would be the most practical way to stimulate the uptake of 7.5×10^{16} grams of C. They propose to fertilize the surface waters of the Southern Ocean over a wide area where studies have shown that production is Fe-limited. In the Southern Ocean, excess inorganic N is present in the surface waters at a concentration of about 20 μM. What concentration of Fe should be added to the surface waters to stimulate the uptake of this much inorganic N, assuming that the C/N ratio in the phytoplankton equals the Redfield ratio of 106/16 by atoms? Assume that this much Fe is mixed into the euphotic zone, which is 50 m deep. Over what area of the ocean would the Fe additions have to be made? How does this figure compare to the surface area of the ocean, which is 3.6×10^8 km^2? Suppose that the area of the Southern Ocean the scientists plan to fertilize is about the size of Alaska, which has an area of 1.5×10^6 km^2. How many times would an area the size of Alaska have to be fertilized, as described above, to stimulate an uptake of 7.5×10^{16} grams of C? Assume that a ship fertilizing the ocean with Fe can fertilize about 75 km^2 per day. How many ship-days would be required to fertilize an area the size of Alaska this many times?

2.4 A second group of scientists comes to you with another proposal for drawing down the CO_2 concentration in the atmosphere. These scientists points out that almost all of the organic matter that sinks into the aphotic zone of the ocean decomposes completely. As a result, the N and P incorporated into the organic matter are released as inorganic N and inorganic P, respectively, in deep-ocean water. They argue that pumping this deep ocean water to the surface would stimulate the synthesis of organic C by marine phytoplankton and cause an equivalent amount of CO_2 to enter the ocean from the atmosphere. What is the flaw in their reasoning?

Wastewater with an N/P ratio of 8 by atoms is being discharged into a lake. The N/P ratio of the algae in the lake is 16 by atoms. Assume that the wastewater is the only significant source of N and P for the lake and that either N or P limits algal biomass in the lake. What will happen to the algal biomass in the lake if

2.5 The N in the wastewater is reduced by a factor of 10?
a. No change
b. Reduced by a factor of 2
c. Reduced by a factor of 5
d. Reduced by a factor of 10

2.6 The P in the wastewater is reduced by a factor of 10?
a. No change
b. Reduced by a factor of 2
c. Reduced by a factor of 5
d. Reduced by a factor of 10

2.7 The N is the wastewater is increased by a factor of 10?
a. No change
b. Increased by a factor of 2
c. Increased by a factor of 5
d. Increased by a factor of 10

2.8 The P in the wastewater is increased by a factor of 10?
a. No change
b. Increased by a factor of 2
c. Increased by a factor of 5
d. Increased by a factor of 10

2.9 The optical depth is a dimensionless number defined as $\ln(I_0/I)$, where I is the irradiance at a particular depth and I_0 is the irradiance at the surface. The base of the euphotic zone is commonly taken to be the depth at which I equals 1% of I_0. What is the optical depth associated with the base of the euphotic zone?

Chapter Three

✧   ¤   ✧   ¤   ✧

Physical Factors Affecting Production

ALTHOUGH THE AVAILABILITY of light and nutrients most directly influences photosynthetic rates, purely physical processes such as upwelling, vertical mixing, and currents may greatly affect the availability of light and nutrients to aquatic plants. This chapter examines the physical properties of water relevant to these effects. Seasonal production cycles and the susceptibility of aquatic systems to cultural eutrophication stress are then examined using the physical and chemical concepts developed in Chapters 1 and 2. Estuaries are singled out for special consideration because of their importance in the life history of many aquatic organisms and because of their susceptibility to pollution stress.

PHYSICAL PROPERTIES OF WATER

One of the most peculiar properties of pure water is the fact that its density maximum occurs at a temperature (4°C) above the freezing point (0°C). Thus, either heating or cooling pure water initially at a temperature of 4°C causes its density to decrease. On the other hand, the density of typical seawater with a salinity of $35^0/_{00}$ steadily increases as the water is cooled, no matter what the initial temperature is. What is there about pure water that causes it to have a density maximum at 4°C?

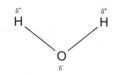

FIGURE 3.1   Simple chemical representation of a water molecule. Lines between the oxygen atom (O) and each hydrogen atom (H) represent electron-pair bonds, which hold the atoms in the molecule together. δ^+ and δ^- signs indicate partial polarization of charge.

First of all, water is, chemically speaking, a polar molecule. The fact that the oxygen atom in the water molecule is more electronegative than the hydrogen (H) atoms causes the distribution of negative electrons in the molecule to be shifted slightly toward the oxygen atom. As a result of this shift, the oxygen atom has a local partial negative charge, designated δ^-, and each of the two hydrogen atoms has a local partial positive charge, designated δ^+. Figure 3.1 shows a water molecule in its equilibrium configuration and the polarization of charge associated with that structure. Note that the molecule is not linear. The H-O-H angle is about 105°. In the liquid and solid phases, individual water molecules interact with one another. Since positive and negative charges attract each other, the molecules tend to become oriented with their positive and negative ends close together. Figure 3.2 shows the hexagonal arrangement of water molecules in ice I, the most common crystalline form of ice. The hexagonal pattern of the molecules actually extends in three dimensions, although only a single hexagonal unit is shown in Figure 3.2.

The bridge between two adjacent oxygen atoms formed by the H atom attached to one of the oxygen atoms is called a *hydrogen bond*. H bonds are indicated by dashed lines in Figure 3.2. H bonds are caused by the attraction of a positively charged hydrogen atom to a negatively charged atom, in this case oxygen. Although the strength of hydrogen bonds is small compared to that of normal electron-pair bonds,[1] it is the H bonds that largely serve to orient the water molecules in ice into structures such as ice I.[2] Each water mole-

---

[1]The H bonds in liquid water have a bond energy of about 4.5 kcal mol^{-1}. The covalent O-H bond in liquid water has a bond energy of 110 kcal mol^{-1}.
[2]We now know that the bonding between neighboring water molecules is augmented by some covalent bonding (Hellemans, 1999), but H bonds are the principal mechanism that orients neighboring water molecules.

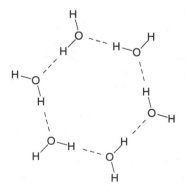

FIGURE 3.2   Hexagonal arrangement of water molecules in ice. Each H atom is joined by an electron-pair bond to the nearest O atom and by an H bond to the next closest O atom.

cule in the three-dimensional structure of ice I is H-bonded to the four nearest-neighbor water molecules. An important feature of this type of structure is its high degree of porosity. When ice melts, about 10% of the H bonds in the ice crystal are broken, and this breakage causes some of the structural units to collapse totally or partially. As these structural units collapse, the density of the water increases because some of the gaps in the original ice structure are filled. Thus, the density of liquid water at 0°C is greater than that of ice at 0°C, and ice therefore floats on water.

Although at any given time most of the water molecules in liquid water are H-bonded,[3] the H bonds break and reform about 10^{10} times per second. As the temperature of the water is raised further above 0°C, more H bonds break than form, and the structural units collapse even more. As a result, the density of the water is further increased as more gaps in the structure are filled in with smaller collapsed units or individual molecules. However, the increase in temperature is accompanied by another effect that tends to reduce the density. As the water is heated, the individual molecules and structural units move about at faster speeds, just as the translational speed of the atoms or molecules in a gas increases as the temperature of the gas rises. The higher speed of the molecules and structural units in the water causes the water to expand slightly and become less dense, just as a gas expands and becomes less dense when it is heated. Hot air balloon enthusiasts routinely utilize the latter phenomenon. As pure water is warmed from 0°C to 4°C, the density of the water increases because the effect of the collapsing structural units on the water's density more than offsets the reduction in density caused by the increased translational speed of the water molecules and structural units. However, the latter effect becomes more important above 4°C, and as a result, the density of water steadily decreases as the temperature of the water rises above 4°C. Hence, the maximum density of pure water occurs at 4°C.

Substances dissolved in water disrupt to a certain degree the arrangement of the structural units. If the dissolved substances themselves are ionic or polar, water molecules become oriented about them, with positive and negative ends in close proximity. Figure 3.3 shows how water molecules would be expected to orient themselves around a positive sodium ion and a negative chloride ion.

The concentration of dissolved ions in seawater with a salinity greater than about 24.7% is high enough to disrupt significantly the porous structure of pure water. As a result, the changes in structure associated with heating or cooling are never great enough to override the effect on the water's density of changes in the translational speed of the molecules and structural units in the seawater. Thus, the density of seawater with a salinity greater than 24.7% steadily decreases as the water is warmed above the freezing point.

WATER COLUMN STABILITY AND OVERTURNING

A column of water is resistant to vertical mixing (i.e., stable) if the density of the water increases steadily with increasing depth. In other words, the densest water is found on the bottom and the least dense water is found on the top. If the densest water is on the top, the water column is "top heavy" and will spontaneously mix vertically if slightly disturbed. In a freshwater system the temperature in a stable water column will decrease steadily with depth if the temperature of the water everywhere is greater than 4°C. In the ocean the temperature of a stable water column will almost always decrease with increasing depth. However, since changes in salinity also affect density (the higher the salinity, the higher the density), it is possible that in a stable water column the temperature may increase with

---

[3]Most of the molecules in liquid water are H-bonded even at 100°C, as evidenced by the high heat of vaporization of water.

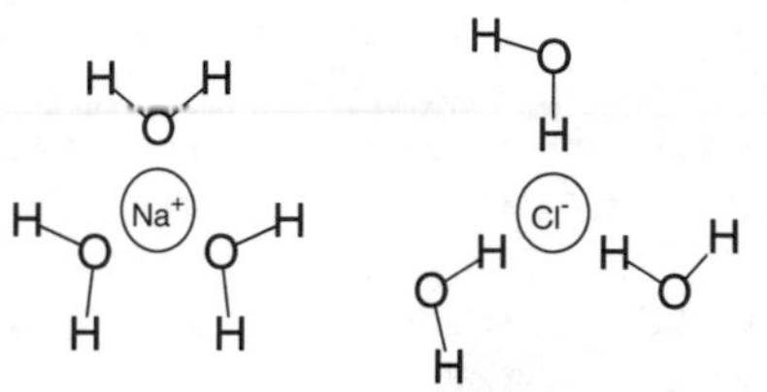

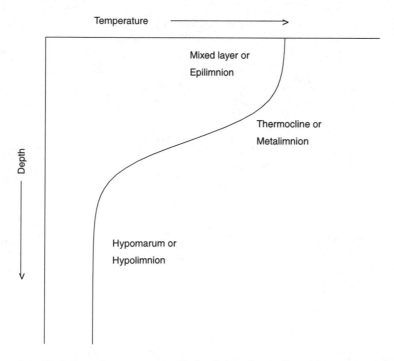

FIGURE 3.3 Qualitative arrangement of water molecules around a positive sodium (Na^+) ion and a negative chlorine (Cl^-) ion. Note that the negative (O) end of the water molecule is closest to the positive Na^+, while the positive (H) end of the water molecule is closest to the negative Cl^-.

depth if the salinity also increases. The water column will be stable if the effect of the salinity increase on the density more than offsets the effect of the temperature increase. The warm salt brines at the bottom of the Red Sea, for example, remain at the bottom of the water column because their high salinity makes them denser than the cooler overlying waters. In most parts of the ocean, however, the temperature, or more correctly the potential temperature, decreases steadily with increasing depth.

Figure 3.4 shows the characteristic variation of temperature with depth in a hypothetical body of water. The water column is being heated from above by radiant energy, and hence the temperature is highest at the surface. Waves and turbulence generated by winds have created a region of almost constant temperature near the surface. This region is called the *mixed layer* or, in strictly limnological work, the *epilimnion*. The depth of the

Temperature ⟶

Mixed layer or
Epilimnion

Thermocline or
Metalimnion

Depth

Hypomarum or
Hypolimnion

FIGURE 3.4 Variation of temperature with depth in a thermally stable marine water column or in a stable column of freshwater in which the temperature is everywhere >4°C.

mixed layer may vary greatly, depending on the strength of the winds and the stability of the water column. The mixed layer in a lake may be only 1 m or so deep during parts of the summer but perhaps 100 m or more deep at certain times during the fall or winter when the water column is destabilized. The region of relatively rapid temperature change below the mixed layer is called the *thermocline* or, in limnological work, the *metalimnion.* Below the thermocline is a region of relatively constant temperature referred to as the *hypolimnion* by limnologists. Oceanographers have coined no special term for the corresponding region of the ocean, although the word *hypomarum* would seem a reasonable choice. The decrease of temperature with depth shown in Figure 3.4 would stabilize any water column in the ocean and would stabilize any column of freshwater if the temperature everywhere were above 4°C. Figure 3.5 shows the variation of temperature with depth in a stable freshwater system in which the temperature everywhere is below 4°C. In this case the water column is being cooled from above, so that the lowest temperature is found at the surface. The water column is nevertheless stable because the density of freshwater increases with increasing temperature at temperatures below 4°C.

Now let us consider what happens to the temperature structure in a freshwater lake located in a temperate climate as the temperature of the atmosphere and the amount of radiant heating change seasonally. Figure 3.6 shows the sequence of temperature profiles that might be observed in the lake from the middle of summer until the middle of winter. As the atmosphere cools in the fall, heat fluxes from the surface waters of the warmer lake into the atmosphere, causing the temperature of the surface water to drop. As the surface water cools, it becomes denser and sinks. Consequently, the mixed layer becomes deeper and the thermocline begins to break down as shown. After sufficient cooling the water column becomes isothermal, and any further cooling of the surface water causes the water column to mix from top to bottom. This period of downward mixing of surface wa-

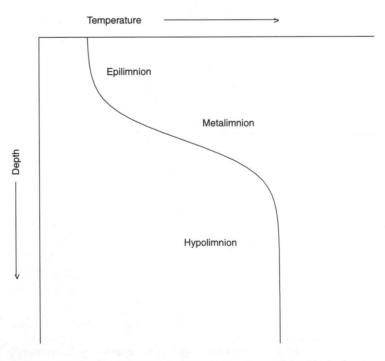

FIGURE 3.5   Variation of temperature with depth in a thermally stable freshwater system in which the temperature is everywhere <4°C.

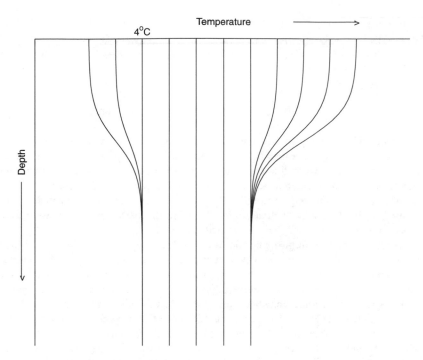

FIGURE 3.6   Temperature profiles in a hypothetical freshwater lake in a temperate climate from the end of summer (right-hand side) until the end of winter (left-hand side).

ters caused by surface cooling is called the *fall overturn*. If the temperature of the atmosphere becomes sufficiently low, the lake water will be cooled and will remain isothermal until the temperature of the water reaches 4°C. At that point any further cooling of the surface water causes the water column to stratify, with the coldest water being at the top and the warmest water at the bottom. If the lake eventually freezes, the water temperature near the surface will be 0°C, and the temperature of the bottom water will be somewhere between 0°C and 4°C.

Between the end of winter and the end of summer the sequence of temperature profiles follows a pattern similar to the mirror image of Figure 3.6. In the early spring, warming of the surface water causes it to become denser and hence induces downward mixing. The water column becomes isothermal at some temperature below 4°C and mixes to the bottom until the temperature reaches 4°C. This period of water column mixing is called the *spring overturn*. Once the water temperature has reached 4°C, any further warming at the surface causes the water column to stratify. The highest temperature is now found at the surface and the lowest temperature at the bottom. Thus, any further warming at the surface leads to a typical summer temperature profile such as that shown in Figure 3.4.

Bodies of water that have two overturning periods per year separated by periods of stratification are called *dimictic* (twice-mixing). Most lakes in temperate climates in which the summer water temperature is above 4°C and the winter water temperature is below 4°C are dimictic. Bodies of water that overturn only once per year are termed *monomictic* (once-mixing). Warm lakes in which the winter water temperature never drops below 4°C may be monomictic, as may be cold lakes in which the summer water temperature never rises above 4°C. In the former case, overturning occurs during the winter while the surface waters are being cooled; in the latter case, overturning occurs during the summer when the surface waters are being warmed. Temperate marine waters with a salinity above $24.7^0/_{00}$

never stratify thermally during the winter while the surface water is being cooled because the density of the water increases steadily all the way to the freezing point. In such waters overturning may occur throughout the winter, and the mixed layer at such times may be hundreds of meters deep.

The Importance of Overturning

Overturning of the water column serves two highly important functions. First, downward mixing of O_2-rich surface waters below the thermocline introduces O_2 into the bottom waters of aquatic systems. Without this mechanism of oxygen recharge the hypolimnions of many lakes would become anoxic, since the simple downward diffusion of gases through a stratified water column is quite slow. For the hypolimnion to remain oxygenated, the respiratory consumption of O_2 by organisms living below the thermocline must not be rapid enough to consume all the O_2 between overturning periods.

The O_2 concentration in the bottom waters of the deep ocean, at depths of several kilometers or more, is almost everywhere unaffected by winter overturning in the waters directly above, since overturning in the oceans rarely occurs to depths as great as several kilometers. Exceptions to this rule are found in the Atlantic Ocean near Greenland and in the Weddel Sea near Antarctica, where surface water temperatures drop as low as 1–2°C below zero during the winter. The formation of sea ice at such times leaves the surrounding waters enriched in dissolved salts because sea ice has a relatively low salt content. The high salinity and very low temperature of these surface waters cause them to sink (with some mixing) all the way to the bottom. These areas are the only parts of the ocean where bottom waters are formed. Deep-ocean currents transport these bottom waters to all the major ocean basins, where mixing and very slow upwelling gradually bring the water back to the surface after a period of approximately 500 years. Despite the long residence time of these bottom waters, only a very few areas of the ocean are anoxic at the bottom. The respiratory consumption of O_2 by organisms in the bottom waters of the ocean is very slow due to the cold temperatures (see the discussion of Q_{10} in Chapter 8) and the low rate of food supply. Consequently, the bottom waters remain oxygenated despite being out of contact with the atmosphere for hundreds of years. The only exceptions to this rule are found in places such as the Black Sea and the Cariaco Trench, where the fallout of organic matter is exceptionally high and/or the bottom waters are unusually stagnant.

The second important function of overturning is the recharging of surface waters with nutrients. Inorganic nutrients regenerated by excretion or detrital decay below the mixed layer diffuse at only a slow rate upward through the thermocline. Thus, in a lake at the end of the summer stratification period, one frequently finds very high nutrient concentrations in the hypolimnion and very low nutrient concentrations in the epilimnion. The low epilimnetic nutrient concentrations reflect uptake by the phytoplankton community. The breakdown of the thermocline and the mixing of epilimnetic and hypolimnetic waters mix regenerated nutrients into the surface waters. A similar process occurs in the sea. In aquatic systems where overturning is a weak phenomenon, biomass in the surface waters is low throughout the year due to the lack of an efficient mechanism for bringing nutrients from below the thermocline into the euphotic zone. The large oceanic subtropical gyres are illustrative of such systems.

SEASONAL PRODUCTION CYCLES

With this information one may explain, at least in a qualitative way, the seasonal production cycle found in many temperate aquatic systems. Table 3.1 lists the nutrient, light, and photosynthetic characteristics during each of the four seasons in a hypothetical temperate

TABLE 3.1 NUTRIENT, LIGHT, AND PHOTOSYNTHETIC CHARACTERISTICS DURING EACH OF THE FOUR SEASONS IN A TYPICAL TEMPERATE AQUATIC SYSTEM IN WHICH THE MIXED LAYER DOES NOT EXTEND TO THE BOTTOM

Season	Nutrients	Light	Photosynthesis
Summer	Low	High	Low and limited by lack of nutrients
Autumn	Increasing	Decreasing	An autumn bloom occurs and is terminated by decreasing light
Winter	High	Low	Low and limited by lack of light
Spring	Decreasing	Increasing	Vernal bloom occurs and is terminated by decreasing nutrient concentrations and grazing

system. During the summer the water column is highly stratified, with a shallow mixed layer whose depth is well above the critical depth. However, the concentration of nutrients in the mixed layer is quite low, and since no efficient mechanism exists for returning nutrients trapped below the thermocline to the mixed layer, photosynthetic rates are also low.

As the surface waters begin to cool in the fall, vertical mixing brings nutrient-rich waters into the euphotic zone. The surface light intensity is decreasing at this time, and the increasing depth of the mixed layer further reduces the average light intensity in the mixed layer. However, the bottom of the mixed layer is well above the critical depth, at least initially. With abundant nutrients and adequate light to support photosynthesis, a phytoplankton bloom frequently occurs at this time. The bloom is terminated when the declining surface light intensity and increasing mixed layer depth cause the bottom of the mixed layer to descend below the critical depth, or at least become close enough to the critical depth that net community production in the mixed layer is quite low.

During the winter nutrient levels in the mixed layer remain high, even if the water column stratifies, since the lack of light prevents any significant phytoplankton uptake of nutrients. Photosynthetic rates during this time are severely limited by the lack of light in the mixed layer.

Following the winter or spring overturning period, warming of surface waters causes the water column to stratify, and the depth of the mixed layer is therefore reduced. At the same time the average surface light intensity is increasing, and the shallow winter critical depth begins to deepen. The combination of increasing surface light intensity and decreasing mixed layer depth obviously increases the light available to phytoplankton in the mixed layer. By utilizing the high mixed layer nutrient concentrations left over from the fall and perhaps winter mixing periods, the phytoplankton population begins to multiply. A bloom develops when the mixed layer extends no deeper than the upper region of the euphotic zone, where photosynthetic rates are uniformly high (Smetacek and Passow, 1990). A combination of herbivore grazing and the exhaustion of nutrient reserves in the mixed layer usually terminates the spring phytoplankton bloom. Production during the rest of the summer remains at a low level because much of the nutrient reserve used to set off the spring bloom has been temporarily lost below the thermocline as detritus or is tied up in higher trophic level biomass.

TROPHIC STATUS

This admittedly simplified picture of photosynthetic seasonality in a temperate aquatic system underlines the importance of vertical mixing processes and water column stratification in determining the availability of light and nutrients for photosynthesis. Vertical mixing stimulates the recycling of nutrients from deep water, but at the same time reduces the amount of light available to the phytoplankton by increasing the mixed layer depth. How then do aquatic systems become highly productive (i.e., eutrophic) for more than the short periods of time typical of spring or fall blooms? The answer is that in shallow systems the mixed layer may extend all the way or most of the way to the bottom during much of the year. In such systems, recycling of nutrients is highly efficient because there is little or no part of the system into which detritus may sink and regenerated nutrients become trapped. Because the system is shallow, it is impossible for the mixed layer to extend to great depths, and in fact the entire water column may be in the euphotic zone during much of the year. By contrast, in a very deep system, recycling of nutrients from detritus that has fallen far below the euphotic zone is extremely inefficient. In fact, in most parts of the ocean and in some deep lakes, overturning of the water column may never extend to the bottom. Furthermore, in deep systems the mixed layer may extend to great depths during overturning periods, so that production is brought to a halt due to the lack of light in the mixed layer. Thus, barring unusual circumstances (e.g., sewage disposal, upwelling), deep aquatic systems are inherently less productive than shallow aquatic systems. Recalling that the eutrophication process involves a gradual reduction in the depth of an aquatic system, it should not be surprising to learn that annual production in a system undergoing eutrophication is an accelerating function of time. The rate of increase of production is small at first but becomes progressively larger over the years as the system both accumulates nutrients and recycles them more efficiently. Relatively deep, unproductive systems are often referred to as being *oligotrophic* (few nutrients), whereas highly productive, usually shallow systems are called *eutrophic* (many nutrients or nourishing). The term *mesotrophic* is sometimes applied to systems with intermediate characteristics. One should not get the impression that all aquatic systems are initially oligotrophic. A lake, for example, may be shallow from its inception and therefore tend toward an initial high rate of production. The present trophic status of a body of water is determined both by its original characteristics and by its history since formation. In all cases, however, the natural tendency of aquatic systems is to become shallower and more productive, i.e., to undergo eutrophication.

SUSCEPTIBILITY OF SYSTEMS TO O_2 DEPLETION

The aphotic zone of a body of water will become anoxic if the consumption of O_2 by biological or chemical processes exceeds the rate of resupply by vertical mixing and diffusion. Since virtually all aquatic organisms require O_2 for respiration, it is generally considered desirable for all parts of the water column to remain oxygenated. From our discussion of trophic status, it is obvious that depth is an important determinant of productivity in an aquatic system, and depth obviously influences the percentage of the water column that is impacted by vertical mixing and overturning. Consequently, depth plays an important role in determining the susceptibility of an aquatic system to O_2 depletion.

Deep, oligotrophic systems are the aquatic systems least likely to develop O_2 depletion problems. This conclusion is based on the following two considerations:

1. With the exception of allochthonous (external) inputs such as stream runoff, virtually all the organic C that is metabolized by aphotic zone organisms must have been produced in the euphotic zone. Consequently, the respiratory activity of aphotic zone organisms is directly related to the amount of productivity in the euphotic zone. Since oligotrophic systems are by definition unproductive systems, the amount of food available to aphotic zone organisms is small, so that the number of these organisms and their overall respiratory rate is also small.

2. In a deep aquatic system, the volume of water below the thermocline is large relative to the volume of the mixed layer. Therefore, the total amount of O_2 potentially available for respiration in a deep system is quite large relative to the small part of the system that is utilized for production. Thus, the productivity of the surface waters would have to be exceptionally high in order for respiration to reduce significantly the average deep-water O_2 concentration. If the deep waters are reoxygenated regularly by means of overturning, it is virtually impossible for a deep oligotrophic system to become anoxic at any depth.

Shallow systems that are mixed to the bottom at all times obviously do not develop seasonal O_2 depletion problems. If there is a sufficiently virogous mixing of the water column and exchange of gases with the atmosphere, the O_2 concentration in such systems is likely to remain near the saturation level at all times. If the wind dies down, however, so that O_2 exchange with the atmosphere is sluggish, the O_2 concentration in a highly productive, shallow system may drop to almost zero within a few days or even a few hours. The lowest concentrations are observed at night, when there is no photosynthetic production of oxygen. The shallow western basin of Lake Erie, with an average depth of only about 7.4 m, provides a good example of a shallow, productive system that may develop serious O_2 depletion problems after several days of calm weather. Aquaculture ponds, which are typically only about 1.0 m deep and receive large inputs of allochthonous organic matter in the form of feeds, are illustrative of the more extreme cases in which the respiratory activity of organisms in the system may strip the water of O_2 within a few hours on a calm night.

Seasonal O_2 depletion problems are undoubtedly most common in bodies of water that have somewhat intermediate depths, i.e., shallow enough to be highly productive, yet deep enough so that the mixed layer does not extend to the bottom except during overturning periods. In lakes of this sort the hypolimnion may be rather small compared to the eiplimnion. Hence, even a moderate amount of production in the epilimnion may result in enough food's being consumed and respired in the hypolimnion to reduce the hypolimnetic oxygen concentration to virtually zero between overturning events. The central basin of Lake Erie, with a mean depth of 18.5 m, is a good example of such a system. The disappearance from Lake Erie's central basin of certain fish species that normally inhabit cold, deep waters and that function efficiently only when oxygen concentrations are near saturation levels is undoubtedly explained in part by the periodic low O_2 concentrations that characterize the central basin's hypolimnion. We will study conditions in Lake Erie in more detail in Chapter 4. For the moment, suffice it to say that the development of low O_2 concentrations below the thermocline in any aquatic system is frequently associated with an undesirable change in the type and abundance of organisms living in the water.

Large-scale fish kills, in some cases involving hundreds of thousands or even millions of fish, are probably the most dramatic and most highly publicized results of O_2 depletion problems associated with eutrophication. Seasonal fluctuations in O_2 levels are not likely to be associated with the sudden killing of large numbers of organisms since seasonal declines in oxygen concentration are gradual. In such cases, organisms are more likely to be eliminated from the system due to their inability to function efficiently (e.g., to escape predators, obtain food, or reproduce) than to suffocation. It is possible, however, in extremely productive systems for O_2 levels to fluctuate from saturating or supersaturating conditions during the day to virtually zero at night. The aforementioned aquaculture ponds are examples of systems with the potential for this sort of behavior, but similar problems may develop in more natural systems seriously impacted by eutrophication. If the concentration of phytoplankton in the water is very dense, one can expect that the abundance of herbivores, primary carnivores, and higher trophic level organisms will also be high since organisms are naturally attracted to a source of food. If this situation develops in a fairly open system, and if the O_2 level does drop dangerously low at night, all motile organisms will rapidly try to leave the area, and for the most part will be successful as long as there are wide avenues of escape. Large-scale kills of aquatic organisms do, however, sometimes occur in bodies of water that have restricted escape routes. In such an isolated and highly eutrophic body of water, large numbers of organisms may be attracted by the abundance of food during the day when the O_2 concentration is high. At night, however, the respiration of all of these organisms may consume the O_2 in the water, and organisms that are unable to find their way out of the system will suffocate. A system must be highly eutrophic for such a situation to develop, but there is no doubt that large-scale kills of the sort described do occur from time to time in some systems.

ESTUARIES—A SPECIAL CASE

Estuaries are defined as semienclosed coastal bodies of water having a free connection with the open ocean, and within which seawater is measurably diluted by freshwater derived from land drainage (Lauff, 1967). Estuaries may be formed by the drowning of river valleys (Chesapeake Bay), by glacial scouring (Puget Sound), by the formation of barrier islands or sand spits (Pamlico Sound), or by tectonic processes (San Francisco Bay).

Regardless of their mode of formation, estuaries tend to be highly productive systems because of the nature of the estuarine circulation pattern that characterizes these systems. Since freshwater is less dense than saltwater of comparable temperature, there is a natural tendency in estuarine systems for the freshwater from land runoff to flow from the head to the mouth of the estuary along the surface, whereas seawater moves in and out with the tides along the bottom. Figure 3.7 depicts the general pattern of water movement in an estuary. As the curved arrows in Figure 3.7 indicate, there is invariably some upward mixing of seawater into the freshwater, so that some of the seawater that enters the estuary near the bottom flows back out near the surface. As a result, there is a net outflow of

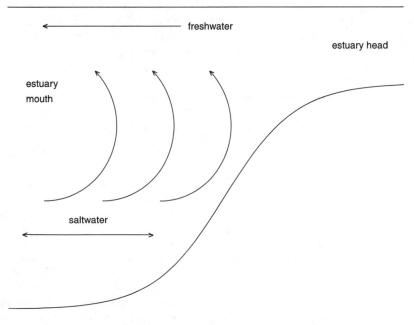

FIGURE 3.7   Simplified estuarine circulation pattern.

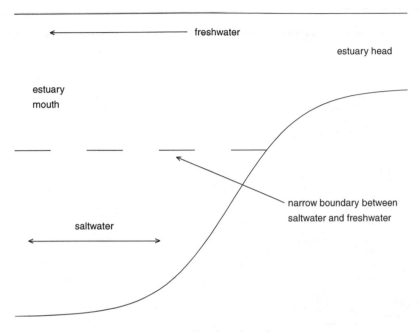

freshwater

estuary head

estuary
mouth

narrow boundary between
saltwater and freshwater

saltwater

FIGURE 3.8　Profile of a salt wedge estuary.

water (freshwater mixed with some saltwater) at the mouth of the estuary in the upper
water column and a net inflow of seawater in the lower water column. If the flux of freshwater into the estuary at the head is large compared to the tidal in-and-out flux of seawater, there is generally a sharp demarcation between the freshwater at the top of the water
column and the saltwater below. Due to the mixing of saltwater and freshwater, this sharp
transition region gradually blurs as one approaches the mouth of the estuary. Such an estuary is commonly referred to as a salt wedge estuary because of the shape of the saltwater "wedge" in the lower part of the water column when the estuary is viewed in profile
(Figure 3.8).

If the in-and-out flux of the tides is large compared to the flux of freshwater, then
freshwater and seawater tend to be thoroughly mixed together throughout the estuary, except in the immediate vicinity of the head. Such an estuary is commonly referred to as a
well-mixed estuary. Undoubtedly, many estuaries are best classified as having circulation
patterns intermediate between those of typical salt wedge and well-mixed estuaries. The
important point to bear in mind, however, is that there is a net outflow of water near the
surface and a net inflow near the bottom at the mouth of all estuaries, regardless of whether
the details of the circulation pattern correspond most closely to those of a salt-wedge, well-
mixed, or intermediate-type situation.

Because there is a net inflow and upward mixing of seawater at the bottom of an estuary, detritus that has sunk out of the mixed layer at the surface and regenerated nutrients from the deeper water are constantly being carried back into the estuary and mixed
up into the surface waters. Suspended organic matter that drifts out of the estuary on the
surface current, and that subsequently sinks offshore, or is eaten and then excreted offshore tends to be swept back into the estuary by the net influx of bottom water. Figure 3.9
depicts the cycling of nutrients and organic matter in an estuary as influenced by the estuarine circulation pattern. Thus, the physical circulation pattern in estuaries provides a
natural mechanism for recycling food and inorganic nutrients and thereby maintains a

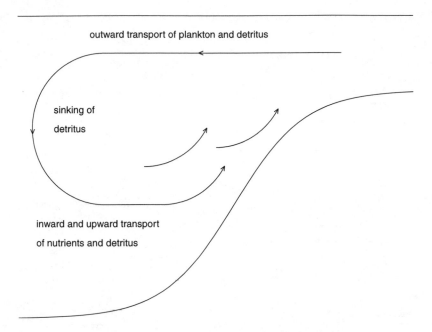

outward transport of plankton and detritus

sinking of

detritus

inward and upward transport

of nutrients and detritus

FIGURE 3.9   Cycling of nutrients and organic matter in a typical estuary.

high level of production in the system. Unfortunately, pollutants introduced into an estuary tend to be recycled by the same circulation mechanism. From this standpoint alone, estuaries are one of the last places one would choose to discharge pollutants since the pollutants will not be conveniently washed out to sea and dispersed. Rather, they tend to be recycled over and over within the estuarine system. Admittedly there will be leakage to the open ocean, and some pollutants will tend to be trapped in the sediments rather than recycled in the water column. However, as a general rule, the estuarine circulation pattern can be expected to exacerbate the impact of pollutants discharged into the estuary. Estuaries are particularly susceptible to problems associated with cultural eutrophication because the estuarine circulation pattern tends to recycle discharged nutrients and hence magnify their impact on production and biomass.

Estuaries are naturally eutrophic systems because of the high efficiency with which the estuarine circulation pattern recycles nutrients. Because of their high productivity, estuaries such as those found along the coastline of the United States account for some important coastal fisheries. Furthermore, estuaries serve as breeding and/or nursery grounds for many organisms that one usually associates with the open ocean such as sharks and whales. Finally, estuaries are traversed by a number of migratory species that breed either in freshwater (e.g., salmon) or in saltwater (e.g., the American eel). Contamination of estuaries by any sort of pollution may therefore have much graver consequences for aquatic systems than would perhaps be apparent from a casual examination of the abundance and kinds of organisms present in the estuary at any one time. Many large population and industrial centers have developed adjacent to estuaries because of the easy access both to the ocean and to inland river systems for water transportation. Unfortunately, the wastes from these large population/industrial centers have often been discharged carelessly into estuarine waters, with little or no awareness of the biological importance of the estuary or of the tendency of pollutants to be recycled within the system. The following two examples are particularly good illustrations of the role of physical processes in determining the impact of pollutant discharges on estuarine systems.

Escambia Bay

Escambia Bay is a well-mixed estuary on the Gulf of Mexico near Pensacola, Florida. The principal stream flowing into the bay is the Escambia River, although several other streams also contribute freshwater input. Figure 3.10 shows the principal features of the bay.

During the 1950s, Escambia Bay was an important commercial fishing area, with brown and white shrimp, oysters, and menhaden among the important species taken from

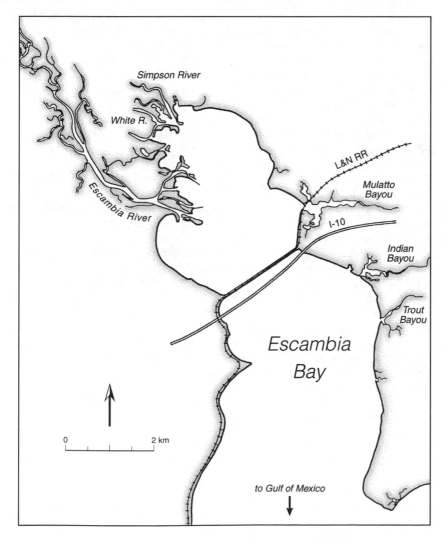

FIGURE 3.10 Escambia Bay and adjacent land areas.

its waters. By 1970, however, commercial fishing had disappeared from the bay, along with the benthic grass beds that had contributed both food and shelter for many of the organisms that had previously lived there. During this roughly 20-year period a number of industries had become established or expanded their operations around Escambia Bay, including a container plant, an electric power plant, and several chemical companies. All of these industries discharged wastes of one sort or another into the bay, including toxic chemicals such as polychlorinated biphenyls (Monsanto), sodium thiocyanate and acrylonitrile (American Cyanamid), heated wastewater (Gulf Power), and various organic wastes whose decomposition by microorganisms consumed oxygen in the bay.

The Louisville and Nashville railroad trestle across the bay (Figure 3.10) was supported by wooden pilings that had to be periodically replaced. However, no attempt was made to remove the old, slowly rotting pilings from beneath the trestle when replacement pilings were driven in at nearby positions. As a result, this transect across the bay gradually became cluttered with a great many wooden pilings, whose presence restricted tidal flushing and caused the northern part of the bay to become somewhat isolated from the rest of the system and relatively stagnant. The construction of an interstate (I-10) highway bridge across the bay adjacent to the railroad bridge further restricted circulation. Fill for the I-10 construction was taken from Mulatto Bayou, an extension of the bay on the eastern side and north of the railroad and highway bridges (Figure 3.10). Dredging for I-10 fill produced a number of deep holes in the bottom of the bayou. The circulation in Mulatto Bayou was sluggish even before this dredging and was further reduced by the restrictions to tidal mixing in the northern bay caused by the railroad and highway bridges. Not surprisingly, circulation in the deep holes dredged for the I-10 fill was extremely poor, and the water in the bottom of these holes frequently became anoxic. This condition was exacerbated by the construction of "dream homes" along Mulatto Bayou. The runoff from exposed ground on construction sites contained high nutrient concentrations, which stimulated the growth of phytoplankton in the bayou. Dredging of canals to serve the dream homes created additional pockets of water that were subject to very poor circulation. Mulatto Bayou became a highly eutrophic system, characterized by poor circulation and extreme fluctuations in oxygen levels.

As a result of these conditions, 12 fish kills involving over 1 million fish each occurred in and around Mulatto Bayou during 1971 alone (Figure 3.11). One kill involved over 5.5 million fish (Cubbison, 1973). These kills resulted from the literal suffocation of fish in the water during the night due to the lack of O_2. The O_2-consuming organic wastes discharged into the bay from industries and sewer outfalls, the runoff of nutrients into the bayou from construction sites, the poor circulation created in the northern bay and in Mulatto Bayou in particular, and the additional stress imposed on organisms by the discharge of toxic wastes undoubtedly all contributed to the situation that led to the fish kills. A spokesman for one of the industries suggested another explanation for the fish kills. He stated that the fish might have committed suicide, since the low nighttime O_2 levels in Mulatto Bayou would surely have caused a normal fish to avoid the bayou at night.

It is unlikely, however, that the millions of fish that died in Escambia Bay were trying to commit suicide, nor is it likely that many of them swam into Mulatto Bayou at night. The fish were undoubtedly attracted to the bay by the abundance of food and probably swam into Mulatto Bayou during the day, when O_2 levels in the euphotic zone were high. It is not difficult to imagine, from an examination of Figure 3.10, that fish might have found it difficult to escape quickly from Mulatto Bayou when the O_2 concentration rapidly dropped at night. The dredging of boat canals (not shown in Figure 3.10) along the bayou created additional blind alleys that compounded the escape problem. There is little doubt that the fish kills that occurred in Escambia Bay could have been largely prevented had some understanding of basic ecological principles been applied in the planning of developments in and around the bay.

Florida Wins Dubious Title: Nation's No. 1 Fish-Killer

By CHRISTOPHER CUBBESON
Of the Times Staff

Escambia Bay Was The Main Culprit In Boosting The
State's Ascendancy To The Fish-Kill Crown

For the schools of un-aware fish that swim through the inland and coastal waters of the United States, Florida is the most lethal spot to live.

Pollution from pesti-cides, fertilizers, man-ure, mining operations, improp-erly treated sewage and other sources killed an estimated 31.6-million Florida fish in 1971, mak-ing the state the nations's No. 1 fish-killer from pollution.

STATISTICS compiled by the U.S. Environmental Protection Agency (EPA) show that Florida waters killed almost twice as many fish from pollution as the next leading state, Texas, with a record 16.2-million fish reported killed in 1971, the last full year for which figures are available.

Florida now has led the nation in fish killed by pol-lution in two of the last three years.

The national total of 74.3-million fish killed was 81 percent higher than the previous high of 41-million reported in 1969 (federal records on fish kills date back to 1960). The 860

pollution-caused fish-kills were about a third higher than in 1970.

The overall figures do not represent all the fish killed by pollution, of course, because many small kills go unreported.

And the figures do not indicate whether the record total is due to better report-ing by a concerned public or to greater fish-kills.

Escambia Bay and envi-rons, often considered the most polluted waterway system in the state, was the main culprit in Florida's ascendancy to fish-kill supremacy.

Of 62 pollution-caused fish kills reported in Flor-ida in 1971, 41, or exactly two-thirds, occurred in or around Escambia Bay.

The figures, significantly, do not include the millions of fish killed by the Tampa Bay area's long siege of Red Tide in the summer of 1971. Red Tide is caused by a natural organism, not from pollution.

OHIO HAD more than twice as many reported kills as Florida – 134 – but 12 massive Florida kills of over 1-million each gave

the Sunshine State a huge margin over Ohio's 1971 total fish kill of 1.2 million.

All 12 massive Florida kills occurred in the Escambia Bay waterway system, including Amer-ica's largest pollution-caused kill of 1971: 5.5-million on Aug. 22, 1971. None of the big kills could be traced to a single source.

In addition to the munici-pal sewage that pours into the Escambia Bay system, paper mills and other heavy industries contribute heav-ily to the toxic environment for fish in the waters around Pensacola.

The largest kill outside the Escambia area occurred in Banana Lake in Lakeland, where an esti-mated 273,000 fish died March 29, 1971 from an overdose of sewage.

ELSEWHERE, two kills took place each in Avon Park, Sebring, Jacksonville and Lake Thonotosassa in Hillsborough County, where the largest pollution-caused U.S. fish-kill on record took place in 1969 – a staggering 26.5-million-victim debacle blamed on the discharge of wastes

from food-processing plants.

Nationally, municipal sewage systems were the leading killer of fish in 1971. Almost 30 percent of the fish killed nationally by pollution in 1971 liter-ally suffocated in sewage.

Sewage was the cause of 19.6 percent of the total fish kills in the southeast region, which includes Florida.

In Florida as the rest of the nation, more fish died in coastal waters than in lakes and rivers, with 44 of 62 kills occurring along the shore.

THE NUMBER of fish killed in estuarine waters like Escambia Bay rose from 44 percent to 77 per-cent from 1970 to 1971 in the nation, a fact that "could be of great national concern since estuaries serve as nursery grounds for many species of marine fish," the report states.

However, the report adds the large increase results from a series of massive kills principally in Escam-bia Bay and Galveston Bay in Texas and says that "interpretation as a national trend, therefore, is not in order."

FIGURE 3.11 Article that appeared in the February 21, 1973, edition of the *St. Peters-burg Times* page 1B. Reprinted with permission of the *St. Petersburg Times*.

Chesapeake Bay

Chesapeake Bay is the largest estuary in the United States. It was formed as a result of the rise in sea level following the last glaciation and is basically the drowned lower course of the Susquehanna River (Figure 3.12). The bay has an area of about 11,400 km^2, but its av-erage depth is only 6–7 m. The Susquehanna and Potomac rivers account for about 70% of the freshwater flow into the bay. Historically, Chesapeake Bay has been a highly pro-ductive ecosystem. It still supports several thousand full-time commercial seafood har-vesters and produces half of the U.S. catch of blue crabs and 20% if its oysters (Baker and Horton, 1990). Nevertheless, Chesapeake Bay is not well.

Between 1965 and 1985 the bay lost 80–90% of its benthic grass beds, which provided critical habitat for many fish and aquatic birds. During roughly the same period there were dramatic declines in the populations of fish such as striped bass, shad, yellow perch, alewife, blueback herring, and white perch. The oyster populations are estimated to be about 1%

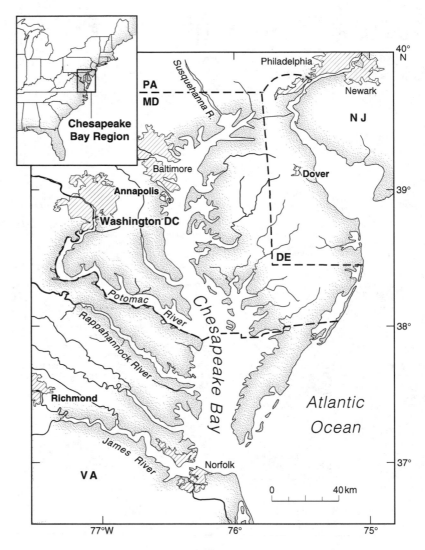

FIGURE 3.12   Chesapeake Bay and its major tributary streams.

of what they were in 1900 (Baker and Horton, 1990). Their decline has been attributed to a combination of overfishing, pollution, and disease.

Oxygen depletion in the bottom waters of Chesapeake Bay during the summer has been a common occurrence for many years (Newcombe and Horne, 1938), but some studies suggest that the problem has become much more severe in the last 20–30 years (Officer et al., 1984). The most obvious culprit has been eutrophication, and with this in mind, a serious effort to reduce nutrient loading to the bay began in 1987. The effort was part of a larger program to restore and protect the bay, formally set forth in mutual agreements between the Environmental Protection Agency (EPA) and the states of Virginia, Maryland, Pennsylvania, and the District of Columbia in 1983 and 1987 (CBP, 1999). The goals of the so-called Chesapeake Bay Program (CBP) were as follows:

1. Provide for the restoration and protection of the living resources, their habitats and ecological relationships.
2. Reduce and control point and nonpoint sources of pollution to attain the water quality conditions necessary to support the living resources of the bay.
3. Plan for and manage the adverse environmental effects of human population growth and land development in the Chesapeake Bay watershed.
4. Promote greater understanding among citizens about the Chesapeake Bay system, the problems facing it, and the policies and programs designed to help it and to foster individual responsibility for and stewardship of the bay's resources.
5. Provide increased opportunities for citizens to participate in decisions and programs affecting the bay.

What has the CBP accomplished to date? The answer in a few words is the following: "The Bay and it's rivers are doing better than they were 15 years ago, when the first Chesapeake Bay Agreement was signed, but we still have a ways to go before we reach our goals for a restored Chesapeake. The 'patient' has been stabilized and is showing signs of improvement, but it's not ready to go home yet" (CBP, 1999, www.chesapeakebay.net/pubs/shap1299.pdf, p. 1). What has been accomplished, and what remains to be done?

There has been roughly a 50% increase in the abundance of benthic grasses (Figure 3.13A), but the standing crop of these plants is still only about 60% of the CBP's interim

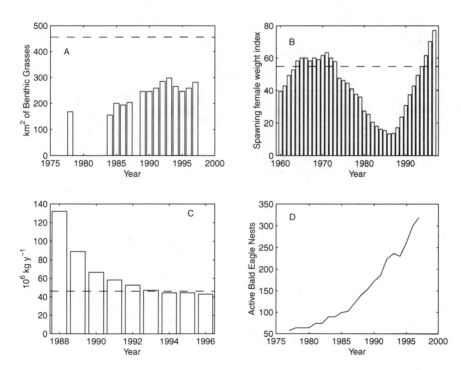

FIGURE 3.13   (A) Area of the bottom of Chesapeake Bay covered by benthic grasses. The dashed line is the interim goal of 460 km^2 established by the CBP. (B) Recovery of the Chesapeake Bay striped bass population. The dashed line is the restoration goal. (C) Industrial chemical discharges to Chesapeake Bay. The dashed line is the target reduction of 65% relative to 1988. (D) Recovery of the Chesapeake Bay bald eagle population. *Source:* CBP (1999). Reproduced with permission of the Chesapeake Bay Program.

goal of 460 km², about 4% of the bay's area. Striped bass have made a remarkable recovery since reaching their nadir in 1986–1987 (Figure 3.13B). Moratoria were imposed on striped bass fishing in Maryland and Delaware in 1985–1990 and in Virginia in 1989–1990. The stock was officially declared restored in January 1995. There has been about a 67% reduction in the discharges of industrial chemicals to the bay since 1988 (Figure 3.13C). In fact, the CBP year 2000 target figure of 65% was exceeded as early as 1994. Bald eagles in the bay area have recovered dramatically since the federal ban on DDT went into effect in January 1973 (Figure 3.13D) and are no longer considered endangered.

The nutrient loading picture is somewhat mixed. P loading has declined by almost 30% and is expected to meet the CBP year 2000 target level on schedule (Figure 3.14A). N loading, on the other hand, is not expected to reach the year 2000 target level, which would require a 20% reduction in overall loading from all sources. Furthermore, as noted by the CBP (1999, www.chesapeakebay.net/pubs/shap12.99.pdf, p. 2) "Maintaining reduced nutrient levels after 2000 will be a challenge due to [the] expected population growth in the region." Whereas striped bass and bald eagles seem to have made dramatic recoveries, the same cannot be said for blue crabs and oysters (Figure 3.14C,D). The latter two are both benthic organisms. What's the problem?

At least part of the problem can be traced directly to the fact that there continues to be a seasonal O_2 depletion problem in the bay. If the bottom waters do not literally become anoxic every year, they at least become hypoxic, and this condition imposes a seri-

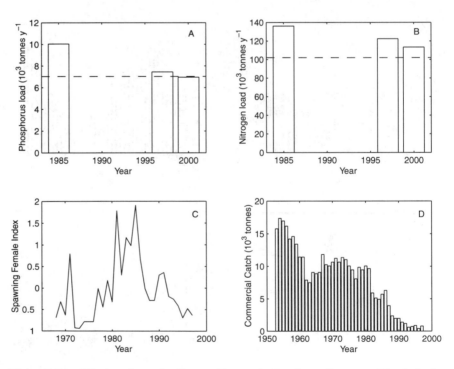

FIGURE 3.14   (A) phosphorus loading to Chesapeake Bay from all sources. The dashed line is the CBP goal of 7,000 tonnes y^{-1}. (B) N loading to Chesapeake Bay from all sources. The dashed line is the CBP goal and corresponds to a 20% reduction relative to 1985. (C) Status of the Chesapeake Bay's blue crab population. (D) Decline in the commercial catch of Chesapeake Bay oysters. *Source*: CBP (1999). Reproduced with permission of the Chesapeake Bay Program.

ous stress on benthic animals. Although much has been written about the role of allochthonous nutrient loading in causing hypoxic/anoxic conditions in the bay, there is a growing consensus that nutrient loading is not solely responsible for the problem. As noted by Smith et al. (1992, p. ix), "While the rate of water-column oxygen consumption appears relatively invariant from year to year, the rate of reaeration of lower-layer waters is strongly dependent upon seasonal stratification set up by spring runoff." In other words, the circulation in Chesapeake Bay generally follows the classical estuarine circulation paradigm. The upper water column consists of relatively freshwater derived from stream runoff, while the salinity of the bottom waters is much closer to that of seawater. The gradient in salinity from top to bottom creates a stable water column, and this stability is reinforced by a strong temperature gradient during the summer months. The temperature gradient does not change much from year to year, but the salinity gradient can change dramatically from wet years to dry years. When the water column is highly stratified during wet years, vertical mixing is minimal, and the bottom waters of the bay are physically isolated for several months. It is under these conditions that hypoxia/anoxia becomes most severe (Figure 3.15). In other words, interannual differences in seasonal bottom water O_2 depletion rates are not primarily due to year-to-year differences in allochthonous nutrient loading and concomitant bottom water respiration rates. Instead, they are due primarily to differences in the degree of physical isolation of the bottom waters of the bay from the atmosphere. The principal mechanism responsible for those differences is the amount of freshwater discharged to the bay and the resultant salinity gradient between the surface and bottom waters.

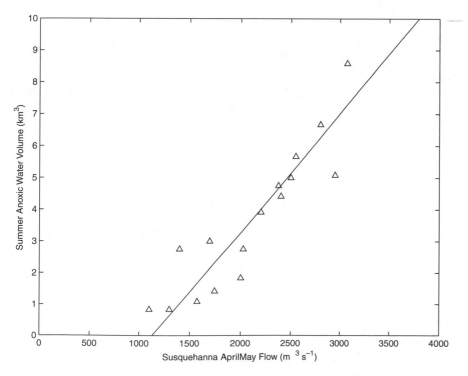

FIGURE 3.15   April–May flow of the Susquehanna River versus volume of anoxic bottom water during the summer in the Chesapeake Bay. *Source:* Flemer et al. (1983).

The importance of physical processes to hypoxia/anoxia in the Chesapeake Bay should not be taken to mean that nutrients are unimportant and can be ignored. As Boicourt has commented (1992, p. 36), "There is enough evidence for the deleterious effects of excess nutrient loading to the Bay on other grounds so that nutrient reduction would appear beneficial, for instance, for reversing the widespread loss of submerged aquatic vegetation." However, the importance of physical processes in this system suggests that focusing remedial efforts exclusively on allochthonous nutrient loading may prove to be futile. Is there any reason to believe that human activities have increased the amount of freshwater discharged to the Chesapeake Bay each year, and if so, is there any way to turn back the clock? The answer to the first question is certainly yes. Roughly 95% of the Chesapeake Bay watershed was forested 350 years ago (Figure 3.16). By the late 19th century, forests covered only about 40% of the watershed. It is worth pointing out here that while Chesapeake Bay is the largest estuary in the United States, its watershed is almost 15 times larger. This means that runoff from the watershed can have a large impact on water quality in the bay. This impact will be minimized to the extent that rain is able to percolate into the ground rather than flowing directly overland into the bay or tributary streams. However, urbanization is invariably associated with the creation of impervious surfaces such as streets, parking lots, and rooftops. The emergence of major urban population centers such as Baltimore and Washington, D.C., has undoubtedly increased the amount of freshwater that enters the Chesapeake Bay, and as noted by Baker and Horton (1990, p. 15), "Rainwater washing off urban pavement and other impervious surfaces can be shockingly polluted" (see Chapter 5). Obviously, one solution to this problem is to restore some of the natural mechanisms that trap and retain rain water. Planting trees is one step in the right direction, and the fact that forests now cover about 60% of the Chesapeake Bay watershed is encouraging (Figure 3.16). However, just planting trees may not be enough. In comment-

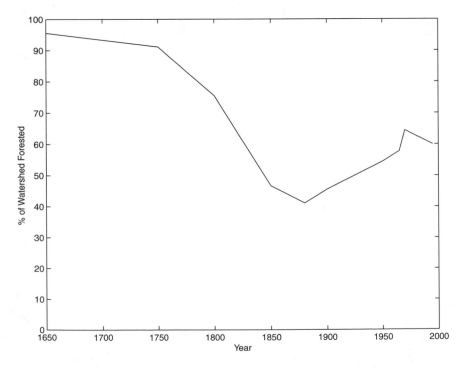

FIGURE 3.16   Percentage of the Chesapeake Bay watershed covered with forests. *Source:* CBP (1999). Reproduced with permission of the Chesapeake Bay Program.

ing on some well-intentioned but ineffective efforts to control runoff from construction sites, Baker and Horton (1990, p. 16) have commented, "There also appears to be too much reliance on the use of structural controls like filter fences and sediment basins to keep pollution from the water. Requiring that natural vegetation be left between development and waterways, thereby preventing pollution from occurring in the first place, may be more effective." In other words, a few trees and other vegetation planted in the right place can have a big impact. The CBP is certainly cognizant of this point, and to their credit, they have targeted the reestablishment of streamside (riparian) forest buffer zones as one means of improving water quality in the bay. The problem has been in implementation. Their year 2010 goal is to restore about 3200 km of streambank and shoreline riparian forest buffers. By 1998 they had reforested only 350 km (CBP, 1999).

Two other points are worth mentioning here. First, water quality problems are not infrequently the result of a complex set of interactions involving several forces. Much of the consumption of O_2 in Chesapeake Bay, for example, is due to bacteria in the water column as opposed to the sediments (Jonas, 1992). Why are there so many bacteria in the water column? A naïve answer might be that there are a lot of bacteria because there is a lot of organic matter for them to consume. That is certainly true, but who is eating the bacteria? Shellfish are filter feeders, and at the end of the 19th century there were enough oysters in the Chesapeake Bay to filter the entire volume of the bay in four days (Jonas, 1992). This kind of filtering activity would have had a very significant impact on the abundance of bacteria and other planktonic organisms. The oyster population now is only about 1% of its former size. The demise of the oysters has been due in no small part to overfishing. Would restoring the oyster population improve water quality in the bay? Perhaps, and experiments using a rafted oyster aquaculture approach are being used to explore this possibility (Jonas, 1992). The point is that a problem such as hypoxia/anoxia can be impacted by a variety of processes, some of which superficially have little to do with O_2 concentrations in the water. Overfishing and cutting down trees and other vegetation along stream banks are two examples.

The second point is that positive feedback loops sometimes conspire to exacerbate O_2 depletion problems. In the case of Chesapeake Bay, for example, denitrification is a potentially important mechanism for controlling the amount of nitrogen available to plants. Denitrification is a process that converts nitrate into N_2. Nitrate is a form of nitrogen readily utilized by almost all plants, including algae. N_2 is a gas that makes up about 80% of the atmosphere, and when nitrate is converted to N_2, much of the N is lost to the atmosphere. N released when organic matter decomposes is in the form of ammonium, and to undergo denitrification, it must be converted to nitrate. The transformation from ammonium to nitrate occurs only in the presence of O_2. In an anoxic or even hypoxic system, little or no nitrate is formed from ammonium. The result is that the N remains as ammonium, a form of nitrogen that, like nitrate, is readily utilized by plants. The point is that before the system goes anoxic (or hypoxic), there is a built-in mechanism for removing N and thereby limiting the amount of organic matter synthesized by plants. The less organic matter synthesized by plants, the less O_2 is later consumed when the organic matter decomposes. However, once the system goes anoxic or hypoxic, denitrification removes less N. More N is retained; more organic matter is produced; and more organic matter subsequently decomposes. The result is a vicious circle. It is believed to be one of the factors that has contributed to the hypoxia/anoxia problems in the Chesapeake Bay (Jonas, 1992).

REFERENCES

BAKER, W. C., and T. HORTON. 1990. Runoff and the Chesapeake Bay. *EPA Journal*, 16(6), 13–16.
BOICOURT, W. C. 1992. Influences of circulation processes on dissolved oxygen in the Chesapeake Bay. In D. E. Smith, M. Leffler, and G. Mackiernan, Eds., *Oxygen Dynamics in the Chesapeake Bay*. Maryland Sea Grant, College Park, Md. Pp. 7–59.

CHESAPEAKE BAY PROGRAM. 1999. Chesapeake 2000 Program. http://www.chesapeakebay.net/pubs/shap1299.pdf

CUBBISON, C. 1973. Florida Wins Dubious Title: Nation's No. 1 Fish Killer. *St. Petersburg Times,* Feb. 21, pp. B1 and B4.

FLEMER, D. A., G. B. MACKIERNAN, W. NEHLSEN, V. K. TIPPIE, R. B. BIGGS, D. BLAYLOCK, N. H. BURGER, L. C. DAVIDSON, D. HABERMAN, K. S. PRICE, and J. L. TAFT. 1983. *Chesapeake Bay; A Profile of Environmental Change.* U.S. Chesapeake Bay Program Report. Environmental Protection Agency. Washington, D.C.

HELLEMANS, A. 1999. Getting to the bottom of water. *Science,* **283,** 614–615.

JONAS, R. 1992. Microbial processes, organic matter and oxygen demand in the water column. In D. E. Smith, M. Leffler, and G. Mackiernan, Eds., *Oxygen Dynamics in the Chesapeake Bay.* Maryland Sea Grant, College Park, Md. Pp. 113–148.

LAUFF, G. H., Ed. 1967. *Estuaries.* American Association for the Advancement of Science, Washington, D.C. 757 pp.

NEWCOMBE, C. L., and W. A. HORNE. 1938. Oxygen-poor waters of the Chesapeake Bay. *Science,* **88,** 80–81.

OFFICER, C. B., R. B. BIGGS, J. L. TAFT, L. E. CRONIN, M. TYLER, and W. R. BOYNTON. 1984.Chesapeake Bay anoxia: Origin, development, and significance. *Science,* **223,** 22–45.

SMETACEK, V., and U. PASSOW. 1990. Spring bloom initiation and Sverdrup's critical-depth model. *Limnol. Oceanogr.,* **35,** 228–234.

SMITH, D. E., M. LEFFLER, and G. MACKIERNAN. 1992. Hypoxia studies in the Chesapeake Bay: An overview. In D. E. Smith, M. Leffler, and G. Mackiernan, Eds., *Oxygen Dynamics in the Chesapeake Bay.* Maryland Sea Grant, College Park, Md. Pp. vii–xvii.

QUESTIONS

3.1. A freshwater lake in Antarctica thaws in the austral summer, but the temperature never rises above 4°C.

    a. Would you expect the lake to be monomictic or dimictic?

    b. During what time(s) of the year would you expect overturning to occur?

3.2. The hypolimnion of a dimictic lake is 5 m thick. At the end of the spring overturning period, the concentration of dissolved O_2 in the hypolimnion is 10 g m^{-3}. The lake remains stratified for 90 days before the start of the fall overturn. What respiration rate (grams of O_2 per square meter per day) would be required to reduce the dissolved O_2 concentration in the hypolimnion to zero at the end of the 90-day stratification period? If respiration is adequately characterized by the equation $C_6H_{12}O_6 + 6O_2 \rightarrow 6CO_2 + 6H_2O$, what rate of organic C oxidation (grams of C per square meter per day) would have been required to support this much respiration? Assume that this organic C was supplied by fallout of organic matter produced in the surface waters of the lake and that the fallout represents 20% of net primary production in the euphotic zone of the lake during the 90-day period. What was the net photosynthetic rate in the surface waters of the lake (grams of C per square meter per day)? Maximum photosynthetic rates in temperate aquatic systems are roughly 5 g C m^{-2} d^{-1} during the summer months. Based on your answers to this question, is it possible that the euphotic zone of a temperate-latitude lake could export enough organic C to a 5-m-thick hypolimnion to strip the hypolimnion of O_2 during the summer stratification period?

3.3 Now consider a second lake where the net photosynthetic rate in the euphotic zone is the same as in question 3.2 but the hypolimnion is 25 m thick instead of 5 m thick. What would be the dissolved O_2 concentration in the hypolimnion of the second lake at the end of the 90-day stratification period?

3.4 Now let's apply similar logic to the ocean. The surface area of the ocean is about 3.6×10^{14} m^2, and net photosynthesis averages 1.3×10^{14} g C d^{-1}. The average depth of the ocean is about 3,800 m, and for purposes of this calculation, we can assume that the marine analogue of the hypolimnion, which we will call the *hypomarum*, has a thickness of about 3,600 m. Assume that the bottom water of the ocean has a dissolved O_2 concentration of 10 g m^{-3} when initially formed and that its residence time is 500 years. Assume furthermore that 10% of the organic C produced by marine algae is exported to the hypomarum. Assuming that all the organic C exported to the hypomarum is oxidized there, what would you expect the average dissolved O_2 concentration to be in the hypomarum? Now let's imagine that the formation of oceanic bottom water is slowed due to global warming and that the residence time of seawater in the hypomarum increases to 1,000 years. What would the average concentration of dissolved O_2 in the hypomarum become under these conditions?

3.5 If the hypomarum has an average thickness of 3,600 m and if the residence time of water in the hypomarum is 500 years, what is the average rate of upward movement of deep ocean water after it is formed? How does this compare to the upwelling rate of 1–3 m d^{-1} that moves water from depths of 50–100 m toward the surface in upwelling areas?

3.6 Suppose that you were a member of the Zoology Department at Florida State University during the 1970s, and you were called in to recommend remedial action to eliminate insofar as possible the occurrence of fish kills in Escambia Bay. What action(s) would you recommend to improve the circulation in the bay and in Mulatto Bayou in particular? What other actions would you recommend to reduce the probability of future fish kills?

Chapter Four

✧ ⌑ ✧ ⌑ ✧

Cultural Eutrophication—Case Studies

1. Problems Often Associated with Cultural Eutrophication
2. Lake Washington
   a. History of Eutrophication
      (1) combined sewer systems
         (a) stranded filth
      (2) septic tanks
      (3) *Oscillatoria*
   b. Water Clarity
      (1) Secchi disk and Secchi depth
   c. Annual Cycles of Nutrients and Chlorophyll
   d. Response to Sewage Diversion
      (1) nutrient limitation − P versus N
      (2) smelt → *Neomysis* → *Daphnia* → phytoplankton
   e. Seasonal O_2 Depletion
   f. Vollenweider Model
      (1) Lake Sammamish
3. Lake Erie
   a. Stresses and Modifications
      (1) habitat destruction
      (2) overfishing
      (3) anoxic bottom waters
      (4) toxic wastes
      (5) sewage
         (a) nutrients and pathogens
   b. Remediation
      (1) PCBs and mercury
      (2) P loading
         (a) laundry detergents
   c. Prospects
      (a) land runoff
      (b) exotic species
4 Kaneohe Bay
   a. Impact of Sewage Discharges on Coral Reefs
      (1) bubble algae, benthic filter feeders
   b. Urbanization of Portion of Watershed
   c. Response to Sewage Diversion
      (1) short- and long-term effects
   d. Overfishing

THIS CHAPTER is devoted to an in-depth study of three examples of cultural eutrophication. The examples include two freshwater lakes, Lake Washington and Lake Erie, and an estuary, Kaneohe Bay. Before beginning our study of these cases, it is appropriate to review some of the points covered in the first three chapters. In particular, what is cultural eutrophication, and what sorts of problems may be created or intensified by cultural eutrophication?

Eutrophication is a natural process that occurs in virtually all bodies of water. The gradual accumulation of nutrients and organic biomass, accompanied by an increase in production and a decrease in the average depth of the water column caused by sediment accumulation, constitutes the natural eutrophication process. Cultural eutrophication is simply the anthropogenic acceleration of eutrophication. This anthropogenic acceleration is often brought about by discharges of organic wastes and/or nutrients. What sorts of problems does cultural eutrophication cause?

Actually, cultural eutrophication may be beneficial to many aquatic systems. The deliberate fertilization of ponds or similar enclosed systems is a basic technique used in aquaculture to produce large crops of fish or shellfish. There is nothing inherently wrong with stimulating production. Cultural eutrophication may create problems, however, if the system of interest is not properly managed and, in particular, if the increased production levels and associated effects are incompatible with other uses of the system that are considered more important. The following effects are frequently encountered in cases of problem cultural eutrophication:

1. Species associated with eutrophic systems are sometimes less desirable than species characteristic of oligotrophic systems. Cyanobacteria, which are frequently associated with organic nutrient enrichment, are a class of organisms frequently associated with undesirable water quality conditions. Problems created by large cyanobacterial populations were discussed in Chapter 2. In Lake Erie some of the fish that became abundant during the period of greatest nutrient enrichment are commercially less valuable than species that were eliminated by overfishing and eutrophication. In Kaneohe Bay, a virtually worthless form of algae grew over and destroyed once-healthy coral reefs in some of the polluted sectors of the bay. From the human standpoint, such changes in the biota of polluted systems are certainly undesirable. Although cultural eutrophication almost invariably increases the productivity of a system, the value (monetary, aesthetic, scientific, etc.) of the organisms produced frequently declines.

2. O_2 concentrations in highly eutrophic systems generally fluctuate over a much wider range than is the case in oligotrophic or mesotrophic systems. Since it is impossible for some organisms to function efficiently unless the O_2 concentration in the water is near saturation, such organisms are often absent from eutrophic environments. A low O_2 concentration may be a seasonal phenomenon, as in the hypolimnions of eutrophic lakes, or a nocturnal phenomenon in the water columns of highly productive systems. Fish kills involving as many as several million fish are in some cases attributed largely to a rapid drop in O_2 concentration during the night in relatively confined and highly eutrophic bodies of water such as Mulatto Bayou. Large-scale fish kills and the elimination of desirable species as a result of oxygen depletion may constitute a serious eutrophication problem in some aquatic systems.

3. Excessive amounts of phytoplankton and/or macroscopic plants in the water create aesthetic problems and reduce the value of the body of water as a recreational resource. From a purely aesthetic standpoint, the crystal clear water characteristic of highly oligotrophic systems is most attractive for swimming or boating. High

phytoplankton concentrations cause the water to appear turbid and aesthetically unappealing. Macroscopic plants can literally cover the entire surface of eutrophic ponds or lakes and consequently make the water almost totally unfit for swimming or boating. In such highly eutrophic systems the large plant biomass is often not entirely grazed off by herbivores. The death and decomposition of large amounts of plant biomass may create a foul smell and a highly unaesthetic appearance. Such occurrences are clearly undesirable from the standpoint of recreational use of the body of water. It should be noted here that the various potential uses of a body of water, such as flood control, recreation, fish production, and irrigation, frequently conflict with one another. The high concentration of phytoplankton in a well-managed catfish pond, for example, would hardly be conducive to use of the pond for swimming; nor would the periodic raising and lowering of the water level in a reservoir used for flood control be desirable from the standpoint of recreational boating. The importance attached to the pros and cons of some eutrophication effects depends to a certain degree on the intended use of the body of water.

4. Because of the high concentration of organisms in a eutrophic system, competition for resources and predator pressure are often severe. This high degree of competition and predation, combined with sometimes severe chemical and/or physical stresses, makes the struggle for survival in eutrophic systems particularly keen. As a result, the diversity of organisms is frequently much lower in eutrophic than in oligotrophic systems. In other words, a smaller variety of organisms is able to survive in eutrophic systems. For certain purposes, it is desirable for the organisms in a system to be dominated by only a few species. Aquaculturists routinely design and manage systems that produce only one or a few species of fish or shellfish. Such systems, although highly productive, are rather monotonous and uninteresting from the standpoint of the naturalist-scientist or nature lover.

CASE STUDY 1. LAKE WASHINGTON

Lake Washington is located in the State of Washington adjacent to the city of Seattle and about 10 km east of Puget Sound (Figure 4.1). It was formed about 13,500 years ago as the result of glacial activity toward the end of the Pleistocene epoch. Colonization of the watershed by American settlers began in 1851. Forests were clear-cut, and some of the clear-cut land gradually became urbanized. However, large areas of the clear-cut land became reforested, with red alder (*Alnus rubra*) accounting for much of the initial regrowth. The hydrology of the lake was altered dramatically in 1916 as the result of a project to build a ship canal with locks connecting Lake Washington to Puget Sound. The level of the lake was lowered by about 3 m, and the Cedar River was diverted into the south end of the lake, where the outlet had been (Edmondson, 1993). Lake Washington presently has an area of about 88 km^2 and a mean depth of 33 m. Inflow comes primarily from the Cedar River to the south and from the Sammamish River to the north. Outflow occurs via two parallel locks to Puget Sound. The residence time of water in the lake (i.e., the ratio of the lake volume to the inflow rate of freshwater) is about 3.4 years. The lake is monomictic, with a minimum recorded winter water temperature of 4.6°C. Overturning generally occurs from the end of November until April or May.

History of Eutrophication

During the 1920s Lake Washington received increasing amounts of raw sewage from the growing Seattle metropolitan area. Concern over public health problems caused this sewage

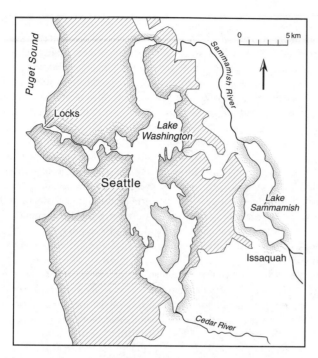

FIGURE 4.1 Lake Washington and surrounding area. Hatching indicates urbanized areas.

to be diverted to Puget Sound by 1936. Some communities used the lake as a source of drinking water until as late as 1965. Between 1941 and 1955, 10 secondary sewage treatment plants serving the Seattle metropolitan area were built, with outfalls entering Lake Washington. In 1941 treated sewage from about 10,000 persons was directly entering the lake, but by 1957 the population served by all 10 plants had risen to 64,300 (Edmondson, 1972). In 1959 an additional sewage treatment plant was built, with its outfall entering the Sammamish River. As of 1960, treated sewage from a population of about 70,000 people was being discharged into Lake Washington or its tributaries.

In addition to the treated sewage discharged into Lake Washington, additional sewage from the outflow of improperly operating septic tanks entered the lake from tributary streams. The actual amount of this septic tank overflow is not known, but in 1957 it was estimated to be equivalent to the treated sewage from a population of about 12,000 persons (Edmondson, 1972). Finally, sewage periodically entered Lake Washington from the combined sewer system of the city of Seattle. A combined sewer system consists of a single system of pipes for handling both street runoff and sanitary sewage. During dry weather all flow in the system consists of sanitary sewage, which in the case of Seattle was given secondary treatment before being discharged into Lake Washington. Secondary treatment, which is discussed in more detail in Chapter 6, usually consists of removal of at least 85% of the settleable and floatable solids and potentially O_2-consuming wastes and disinfection by means of chlorination. During storms the flow in a combined sewer system consists of both sanitary sewage and street runoff. If the street runoff is small the entire flow may be handled by the sewage treatment plant, but during heavy storms a portion of the combined flow is simply diverted from the treatment plant and discharged directly into the receiving body of water. This flow obviously contains raw sewage, both from the normal operation of the sanitary sewer system and from the *stranded filth* (Weibel, 1969), defined as sewage that is left in the pipes during normal low-flow operating conditions and scoured out during storm-flow conditions.[1] The amount of raw sewage entering Lake Washington from the Seattle combined sewer system during storms was estimated to be equivalent to the continuous input of sewage from a population of about 4,500 persons in 1957 (Edmondson, 1972). In summary, as of roughly 1960, sewage from a population of about 85,000–90,000 persons was entering Lake Washington, with about 15–20% of this sewage being partially decomposed sewage from septic tanks and combined sewer overflow and the remainder being secondarily treated sewage.

The effect of the steadily increasing enrichment of Lake Washington with sewage is a bit hard to document in quantitative terms, since only two significant studies had been performed on the lake prior to 1959, the first by Scheffer and Robinson (1939) in 1933 and the second by Comita and Anderson (1959) in 1950. Both studies were of a general limnological nature and were not directly concerned with the question of pollution. However, as a result of these studies, some information on water transparency, nutrient concentrations, and the abundance and characteristics of the phytoplankton community is available prior to the extensive studies conducted by Dr. W. T. Edmondson and colleagues beginning in 1955.

Scheffer and Robinson (1939) characterized the lake as "distinctly oligotrophic," although they noted several species of cyanobacteria and observed that *Anabaena lemmermanni*, a cyanobacterium, could be "abundant in early summer and very common again in fall." By 1950, when the studies of Comita and Anderson (1959) were performed, the lake had begun to show some definite signs of nutrient enrichment. For example, figures given by Edmondson (1969) indicate that winter phosphate concentrations were about 50% higher in 1950 than in 1933. In June 1955 a bloom of the cyanobacterium *Oscillatoria rubescens*, a species frequently associated with sewage pollution, was noticed in the lake, and a series of studies of water quality in the lake was begun shortly thereafter (Edmondson, 1969).

During this time public concern over conditions in Lake Washington had been growing, primarily as a result of the obvious reduction in summer water clarity that accompanied the steady increase in phytoplankton abundance during the 1950s. The concern over water clarity was based primarily on aesthetic considerations, since local residents used the lake extensively for swimming and boating. One simple measure of water clarity is the Secchi disk transparency or simply the Secchi depth.[2] A Secchi depth measurement is made by lowering a white disk into the water until it just disappears from sight. The disk is then lowered into the water a bit more and then raised until it just becomes visible. The mean of the two depths, disappearance depth and reappearance depth, is taken to be the Secchi depth (Hutchinson, 1957). In limnological and oceanographic work, the diameter of a Secchi disk is usually 20 cm and 30 cm, respectively (Hutchinson, 1957; Preisendorfer, 1986). Obviously, one would expect a Secchi disk to be visible at greater depths in clear water than in turbid water. Figure 4.2A shows the results of summer (July-August) Secchi depth measurements made in Lake Washington from 1950 to 1996. The mean Secchi depth in 1950 was 3.7 m, but it had declined to about 2 m by 1955 and had dropped to 1 m by 1963. Thus, almost a fourfold reduction in water clarity had occurred from 1950 to 1963. Suffice it to say that water with a Secchi depth of 1 m is not particularly enticing to swimmers.

---

[1]The diameter of the pipes in a combined sewer system must be large to handle flow during storm events. As a result, there is little scouring effect during dry-weather flow, and a nontrivial amount of the solids in the sanitary sewage may be stranded in the pipes.

[2]The disk described here was introduced by A. Secchi in 1865 on a cruise of the pontifical steam corvette *Immacolata Concezione* in the Mediterranean (Secchi, 1866). Variations on the diameter (10–100 cm) and coloring (e.g., alternating black and white quadrants) have been explored by various workers (see Hutchinson, 1957, pp. 401–402).

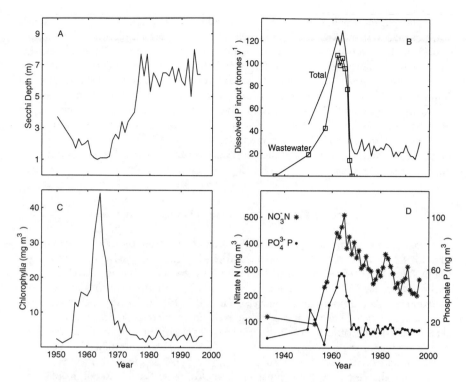

FIGURE 4.2 Changes in Lake Washington between 1933 and 1997. (A) Mean July–August Secchi depth. (B) Annual loading of dissolved P and wastewater P to the lake. (C) Mean July–August chlorophyll *a* in top 10 m. (D) Mean phosphate-P and nitrate-N during January–March. *Source of data*: Edmondson and Lehman (1981), Edmondson (1993), and Edmondson (pers. comm.).

Figure 4.2C–D shows that during the same 13-year period the average summer chlorophyll *a* concentration increased from less than 3 mg m^{-3} to almost 40 mg m^{-3}, and winter phosphate levels increased from 14 to 55 mg m^{-3}. From these graphs and from what we have learned in the first three chapters, it is reasonable to conclude that the high winter phosphate concentrations were assimilated by the phytoplankton during the spring and summer. Hence, the concentration of phytoplankton in the summer was roughly proportional to the concentration of phosphate in the winter. The fact that the chlorophyll *a* concentration increased almost 13-fold while winter phosphate levels increased only 4-fold probably reflects adaptation on the part of the phytoplankton community to the reduced water clarity. As the irradiance experienced by phytoplankton cells is reduced, the cells tend to increase their content of chlorophyll *a* and other light-absorbing pigments so as to capture as much of the available light as possible (Laws and Bannister, 1980). Certainly much of the decrease in summer water clarity can be attributed to the high summer phytoplankton population. A liter of water with a chlorophyll *a* concentration of 35 mg m^{-3} is visibly green. Thus the postulated sequence of events is as follows:

increased nutrient levels → increased summer phytoplankton population → reduced summer water clarity

Following similar but more quantitative and detailed reasoning, the city of Seattle voted in 1958 to spend $121 million to develop an effective sewage disposal system for the entire area (Edmondson, 1969). Although some of this money was spent to stop discharges of raw sewage into Puget Sound and for other improvements, the majority of the funds were spent to improve the water quality of Lake Washington. The plan developed by the Municipality of Metropolitan Seattle (METRO) called for diverting all of the secondary sewage discharged into Lake Washington to Puget Sound, where tidal mixing and dilution were expected to prevent the development of objectionably high phytoplankton populations. Even though the vote for this project took place in 1958, the first diversion did not occur until February 20, 1963, and the last sewage treatment plant diversion did not occur until February 1968.

Effects of Sewage Diversion

The effects of the sewage diversion on the lake can be seen from the data graphed in Figures 4.2–4.4. By the early 1970s, summer chlorophyll *a* concentrations had dropped to 5–6 mg m^{-3} and had declined further to 2–3 mg m^{-3} by the end of the decade. They remain at that level today. In 1950 *Oscillatoria* never accounted for more than 30% of the phytoplankton population at any time. However, from 1962 to 1967 *Oscillatoria* never accounted for less than 30% of the phytoplankton biomass, and sometimes it accounted for more than 90%. The contribution of *Oscillatoria* began to decline in 1968, and the genus has been absent from the lake since 1976. Winter phosphate concentrations dropped from about 55 mg m^{-3} in 1963–1965 to 10–15 mg m^{-3} by 1972 and in subsequent years have remained at that level, which is comparable to the concentration measured in 1950. Secchi depths increased from a low of 1 m just prior to sewage diversion to over 2 m between 1968 and 1970 and to an average of 3.4 m between 1971 and 1975. Thus, after a period of about 5–10 years, the lake in many respects had returned to a condition similar to the condition that existed before eutrophication became a serious problem. Secchi depths and the concentration of winter phosphate were comparable to the values reported in 1950, and *Oscillatoria* had disappeared from the lake. Considering the fact that the residence time of water in the lake is about 3.4 years, the time course of this response seems reasonable.

In 1976, however, the July-August Secchi depth increased to 5.6 m and has averaged almost 6.5 m in subsequent years. This dramatic increase in water clarity was apparently caused by an increase in the abundance of certain species of *Daphnia*, commonly known as *water fleas*. The *Daphnia* are herbivorous, and, "Phytoplankton become relatively scarce at times of *Daphnia* abundance" (Infante and Edmondson, 1985, p. 161). The increase in *Daphnia* abundance has been related to two factors, only one of which is associated with sewage diversion. The first important event was the decline of filamentous cyanobacteria of the genus *Oscillatoria* beginning in approximately 1973 (Figure 4.3). *Daphnia* cannot digest *Oscillatoria*, but the cyanobacteria accumulate in *Daphnia*'s feeding mechanism and must be ejected periodically. Inevitably, some edible algae are simultaneously ejected, and the act of ejection is metabolically costly (Edmondson and Litt, 1982). The presence of *Oscillatoria* therefore interfered with the ability of *Daphnia* to feed and grow. The increase of *Daphnia* is closely correlated with the decline of *Oscillatoria* and other cyanobacteria in the mid-1970s.

However, "Inhibition of feeding cannot explain the scarcity of *Daphnia* in 1933 and 1950, when *Oscillatoria* also was scarce" (Edmondson, 1994, p. 80). The second important factor contributing to the increase in the *Daphnia* population was a rapid decline in the abundance of the large carnivorous zooplankton *Neomysis mercedis* in the mid-1960s. The abundance of *Neomysis*, a *Daphnia* predator, would have made establishment of a sizable *Daphnia* population difficult prior to that time. The decline of *Neomysis* does not appear

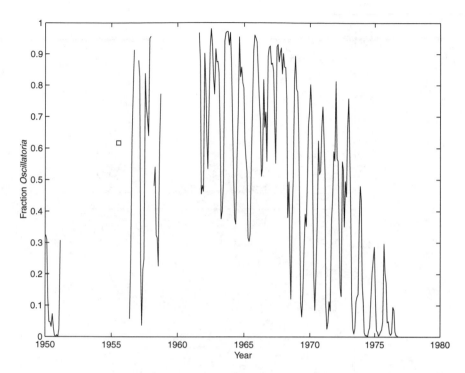

FIGURE 4.3   Fraction of total phytoplankton in Lake Washington composed of the genus *Oscillatoria*. Only one datum is available for 1955 (July). In that year, the first reported bloom of *Oscillatoria rubescens* occurred in the lake. *Source:* Data provided by Dr. W. T. Edmondson.

to have been related to diversion of the sewer outfalls, but instead may have been caused by an increase in the abundance of the long-fin smelt (*Spirinchus thaleichthys*), which is known to feed largely on *Neomysis* in Lake Washington. How the smelt was introduced into Lake Washington is unknown. It was first observed in 1960 (Edmondson, 1994). According to Edmondson (1993), conditions for spawning of the smelt improved following changes in a flood control program in the Cedar River. The smelt spawns primarily in the artificial channel of the Cedar River. For many years the spawning grounds were disrupted each summer when the middle part of the channel was dredged to permit water to flow faster, and gravel was thrown on the bank. However, in 1947 the dredge broke and was never replaced for lack of funds. Farther upstream, massive landslides sometimes occurred when high flow undercut the stream banks. Sediment that washed into the stream smothered the downstream spawning beds. To prevent such slides, the stream banks were stabilized with revetments and riprap (piles of large rocks) (Edmondson, 1994). The increase in smelt followed shortly thereafter.

The rapid decline of the phosphate concentration in Lake Washington as a result of the sewage diversion provides an interesting contrast to the more gradual and less dramatic change in nitrate levels (Figure 4.2D). During the winter overturning period the concentration of both nutrients in the lake is high, a characteristic of monomictic temperate lakes (Figure 4.4). The rapid decline of both elements during the summer stratification period attests to the efficiency with which phytoplankton can take up nutrients when there is adequate light to support photosynthesis. The winter increase in phosphate

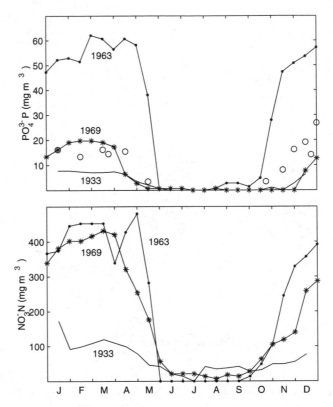

FIGURE 4.4   Seasonal changes in phosphate and nitrate in Lake Washington surface water. Solid lines are data from 1933. Solid lines and dots are data from 1963. Solid lines and asterisks are data from 1969. Circles are data from 1950. (*Source*: Redrawn from Edmondson (1972). Reproduced with permission.

was clearly much less pronounced in 1969 than in 1963, although still substantially higher than the winter buildup in 1933. The 1969 winter nitrate concentration, however, was quite comparable to the winter nitrate level in 1963, and by 1981–1984 winter nitrate-N concentrations were still averaging over 300 mg m^{-3}. At the present time, winter nitrate-N concentrations are 200–250 mg m^{-3}, about half of the peak values recorded in 1964–1965. Phosphate-P concentrations dropped by about a factor of 4 within a few years of sewage diversion and have remained at about 25% of their peak values ever since. What is the explanation for the relatively modest response of the winter nitrate concentrations to sewage diversion?

There are several components to the explanation. First, the sewage being discharged into Lake Washington contributed a minor percentage of the N input to the system. This condition is explained by the fact that the N/P ratio of the sewage was about 7 by atoms, whereas the N/P ratio in streams entering the lake was about 43. In the year (1962) before the sewage diversions began, sewage effluent contributed about 72% of the total P input to Lake Washington but only 35% of the total N (Edmondson and Lehman, 1981). The average winter phosphate and nitrate concentrations in Lake Washington from 1975 to 1980 were in fact 72% and 36%, respectively, lower than the corresponding 1962–1963 values. Thus, the short-term changes in the nutrient concentrations of the lake are in fact quite consistent with estimates of the contribution of the sewage effluent to nutrient loading just prior to the initiation of the diversion program.

A second component of the explanation concerns the N/P ratio in the streams entering the lake. Why was the N/P ratio in the streams so high? During the second half of the 20th century, there was much land clearing in the Lake Washington watershed associated with the growth of Seattle suburbs. Nitrate-N is more easily leached from exposed soils than is phosphate-P due to the tendency of phosphate to adsorb to soil particles and oxides and oxy-hydroxides of iron. Nitrate-N has relatively little tendency to bind to soil particles. Thus, the high N/P ratio in the stream inputs can be attributed in part to the differential leaching of N and P from exposed soils. However, in the case of the Lake Washington watershed, an additional factor was the fact that large areas of clear-cut land had been reforested by red alder. N-fixing bacteria colonize the roots of red alder. Hence, soil under stands of red alder is highly enriched in N. When areas reforested by red alder were cleared for urban development, the runoff from the exposed soils contained an unusually high concentration of N.

Nutrient Limitation

An examination of the phosphate and nitrate levels in Lake Washington during the development of spring blooms provides a good picture of the nutrient limitation situation in the lake. As the high winter nutrient concentrations are assimilated by the rapidly growing phytoplankton, the concentration of both nitrate and phosphate drops to a low level (Figure 4.4). Figure 4.5 shows plots of nitrate versus phosphate during the spring bloom period in each of five years. In 1933 and 1967 phosphate and nitrate were exhausted almost simultaneously, and the values therefore extrapolate to a point close to the origin. In 1968 and 1969 phosphate was exhausted first, and the values extrapolate to a point on the positive nitrate axis. However, in 1962, just before the sewage diversion began, nitrate was exhausted first, and the results therefore extrapolate to a point on the positive phosphate axis. The steady shift of the lines across Figure 4.5 from right to left during the period 1962–1969 may be interpreted as a shift in the system from nitrate limitation during sewage pollution to phosphate limitation following sewage diversion. The lake evidently became nitrate limited during the period of sewage pollution due to the low N/P ratio of the sewage and P limited following sewage diversion because of the high N/P ratio of the stream inputs. Recall that phytoplankton utilize N and P in ratios ranging from roughly 3 to 30 by atoms and that the Redfield ratio of about 16 is associated with optimal growth conditions. Judging from Figure 4.5, Lake Washington was more clearly phosphate limited in 1968 and 1969 than in 1933. This observation is consistent with the fact that the molar ratio of winter nitrate and phosphate concentrations increased from 36 in 1933 to 49 in 1969. Since that time, there has been a gradual decline in the ratio of nitrate to phosphate in the lake (Figure 4.6), from a peak molar ratio of almost 80 in 1972 to a mean of 36 from 1993 to 1996. This trend presumably reflects a reduction in land clearing and construction activities in the watershed as the urbanization of the Seattle suburbs reaches a more mature phase.

Oxygen Depletion

Despite the high productivity in Lake Washington during the period of sewage enrichment, severe O_2 depletion problems never occurred, primarily because of the large volume of the hypolimnion relative to the euphotic zone. The Secchi depths reported for the lake during the summer suggest that the euphotic zone probably did not extend much deeper

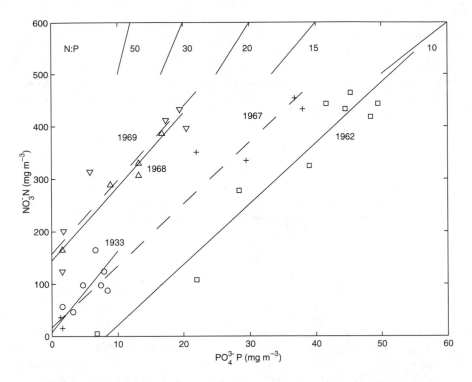

FIGURE 4.5   Correlation between surface values of phosphate and nitrate during the spring phytoplankton bloom when the concentrations of nutrients are decreasing. Circles, squares, plusses, triangles, and inverted triangles correspond to data collected in 1933, 1962, 1967, 1968, and 1969, respectively. The straight lines are regression lines fit to the data. N:P ratios are shown by the numbered lines radiating from the origin. *Source*: Redrawn from Edmondson (1970). Copyright 1970 by the American Association for the Advancement of Science. Reprinted with permission from W. T. Edmondson (1970). Phosphorus, nitrogen, and algal in Lake Washington after diversion of sewage. *Science*, **169**, 690–691. Copyright 1970 American Society for the Advancement of Sciences.

than about 3–6 m between 1950 and 1970.[3] The hypolimnion extends from roughly 20 m to the bottom (Lehman, 1988) and has a volume of 1.4×10^9 m^3, about 49% of the volume of the lake (Edmondson and Lehman, 1981). The summer stratification period lasts for about 200 days, and during that time there is a steady decline in the concentration of O_2 in the hypolimnion (Figure 4.7). This decline is due to the respiratory activity of organisms in the hypolimnion and to the fact that a stratified water column permits little downward movement of O_2 from the surface waters. In 1933 and 1951 the hypolimnetic O_2 depletion rates were less than 0.02 g m^{-3} d^{-1}, sufficient to reduce the O_2 concentration by less than 4 g m^{-3} during the stratification period. The hypolimnetic O_2 concentration at the end of the overturning period is typically 11–12 g m^{-3}. By 1957–1958 the O_2 depletion rates had increased by roughly a factor of 2. Although the average hypolimnetic O_2 concentration in those years never dropped below 4 g m^{-3}, the concentrations at 60 m were less than 1.0 g m^{-3} (Edmondson, 1993).

[3]The depth of the euphotic zone is about three times the Secchi depth.

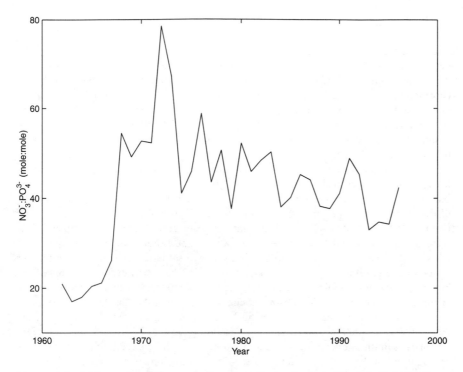

FIGURE 4.6   Molar ratio of nitrate to phosphate during January-March in Lake Washington based on data in Figure 4.2D.

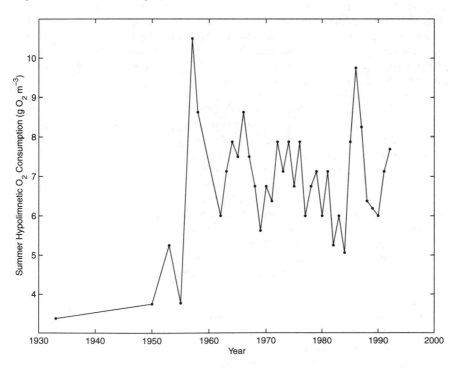

FIGURE 4.7   Depletion rate of O_2 below a depth of 20 m during the summer stratification period in Lake Washington. The duration of the stratification period averages 200 days, and the initial O_2 concentration averages 11.3 g O_2 m^{-3}. *Source:* Data provided by Dr. W. T. Edmondson.

What effect did diversion of sewage from the lake have on the pattern of seasonal O_2 depletion? From a statistical standpoint, there is a significant difference between the O_2 depletion rates measured from 1957 to 1964 and those measured from 1969 to 1997.[4] However, the difference in the mean rates is not very impressive: 0.030 g m^{-3} d^{-1} prior to sewage diversion versus 0.026 g m^{-3} d^{-1} after sewage diversion. Why did the seasonal O_2 depletion rates decline by less than 15% while the chlorophyll *a* concentrations dropped by more than a factor of 10?

There are several components to the explanation. First, as noted by Lehman (1988), the episode of sewage enrichment led to a higher phytoplankton biomass, but it did not increase the growth rate of the phytoplankton *in situ*. In fact, the gross photosynthetic rates of the phytoplankton per unit chlorophyll *a* were 2.3 times lower in 1964 than in 1975 (Lehman, 1988). Qualitatively, this is a common observation in aquatic ecosystems, i.e., there is a negative correlation between algal biomass and productivity per unit algal biomass. In the case of Lake Washington, this pattern to some extent undoubtedly reflects light limitation during the period of sewage enrichment. The water became so murky during the summer that the algae were in effect shading themselves (see below). An additional factor may have been the release of metabolic waste products by the algae that inhibited their own growth. The concentration of such waste products may be insignificant in dilute cultures but inhibitory when the concentration of algae is high.[5] A second consideration relates to the fate of the organic matter produced by the algae. Edmondson (personal communication) has noted that much of the *Oscillatoria* biomass may have been decomposed in the epilimnion of the lake during the period of sewage enrichment. Diatoms are more subject to sinking losses than other types of algae, and as noted by Lehman (1988, p. 1344), "A major portion of material fluxes . . . to the hypolimnion of Lake Washington probably comes with the spring diatom bloom." As the abundance of *Oscillatoria* declined during the 1970s (Figure 4.3), diatoms became a more important component of the phytoplankton community. Thus, based merely on changes in the composition of the phytoplankton community, it is reasonable to expect that a higher percentage of the organic matter produced in the euphotic zone was exported to the hypolimnion in the years following sewage diversion.

An additional factor that has undoubtedly influenced the export of organic matter to the hypolimnion has been the increase in the abundance of *Daphnia* beginning in 1977. Following the disappearance of *Oscillatoria*, *Daphnia* replaced another zooplankton, *Diaptomus*, as the dominant herbivore during the summer months. As noted by Lehman (1988, p. 1344), "*Daphnia* does not produce compacted fecal pellets as does *Diaptomus*." Consequently, one would expect that less organic matter was exported to the hypolimnion in the form of fecal pellets after *Daphnia* became dominant. This shift in the composition of the herbivore community would have tended to reduce hypolimnetic O_2 depletion rates.

The temporal pattern of O_2 depletion rates shown in Figure 4.7 appears to be nonrandom. There is a decline from 1964 to 1982, an increase from 1982 to 1985, and a decrease thereafter. As the foregoing discussion suggests, various factors have influenced this pattern, including nutrient loading to the lake and changes in the composition of the phytoplankton and zooplankton communities. No effort has been made to explain the pattern quantitatively in terms of the physics and biology of the system. Most important to the topic of eutrophication is the fact that the average rates of hypolimnetic O_2 depletion changed by less than 15% in response to a sewage diversion that cost more than $100 million. Had the major goal of the sewage diversion been to reduce the rate of seasonal O_2

[4]The probability that the mean values are the same is less than 2% based on a one-way analysis of variance.
[5]This observation has an analogue in the human experience. See D. L. Meadows, *The Limits to Growth*, Sussex University Press, 1973.

depletion in the hypolimnion of Lake Washington, the diversion might have been considered a failure. The relationship between nutrient loading and seasonal O_2 depletion is not as simple as one might expect. This lesson has important implications for bodies of water such as the Chesapeake Bay and Lake Erie. Fortunately, in the case of Lake Washington, water clarity was the principal factor that motivated voters to approve the expenditure of over $100 million to divert the sewage discharges, and the diversion did produce a dramatic improvement in water clarity.

Water Clarity

Since concern over water clarity was the principal motivation for diverting the sewage from Lake Washington, it seems appropriate to examine this issue in some detail. Figure 4.8 shows the empirical relationship between summer chlorophyll *a* concentrations and Secchi depths based on data summarized by Edmondson and Lehman (1981). It is apparent from this figure that Secchi depths in Lake Washington are very insensitive to chlorophyll *a* concentrations greater than about 15 mg m^{-3} and highly sensitive to chlorophyll *a* concentrations in the approximate range 1–10 mg m^{-3}. Based on Figure 4.8, it seems fair to say that a reduction of summer chlorophyll *a* concentrations by a factor of 2, for example, from prediversion values would have produced no discernable effect on water clarity in Lake Washington. Reduction of the chlorophyll *a* concentrations by at least a factor of 3–4 was necessary to produce a significant impact on water clarity. Just prior to the first

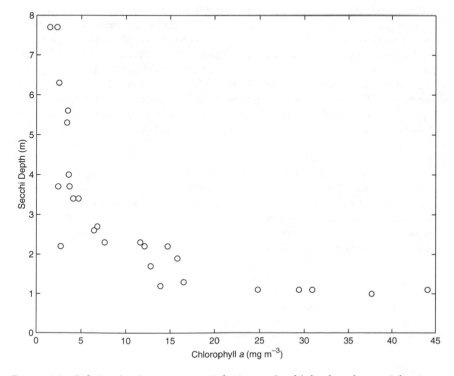

FIGURE 4.8   Relationship between mean July-August Secchi depth and mean July-August cylorophyll *a* in upper 10 m of the water column. *Source of data*: Edmondson and Lehman (1981).

sewage diversion, the ratio of total N to total P in all inputs to Lake Washington was about 15 on a molar basis, virtually identical to the Redfield ratio. It was therefore reasonable to assume, at least as a first approximation, that the reduction in phytoplankton biomass caused by a diversion of the sewer outfalls would be directly proportional to the reduction in P loading. Since wastewater accounted for 72% of the P loading at that time, the expected reduction would be a factor of $1/(1 - 0.72) = 3.6$, which is comparable to the estimated minimum reduction needed to improve water clarity significantly. The fact that the actual reduction in chlorophyll *a* was closer to a factor of 10 reflects changes in the composition of the zooplankton community (i.e., increased abundance of *Daphnia*) and (probably) a reduction in the chlorophyll *a* content of the phytoplankton in response to the improved water clarity. The former effect was not anticipated; the latter could have been predicted, at least in a qualitative way. In any case, at the time the sewage diversions were being considered, it was important that knowledgeable and respected scientists were able to say with some confidence that diverting the sewage would indeed improve water quality. Otherwise, it would have been difficult to justify spending over $100 million to finance the project.

Cyanobacteria

The disappearance of *Oscillatoria* from Lake Washington after 1976 and the general decline in the contribution of cyanobacteria to the phytoplankton community have been regarded as one of the successes of the sewage diversion. However, in 1988 there was a surprising bloom of another species of cyanobacteria, *Apanizomenon flos-aquae* (Edmondson, 1997). Although this species had been reported in the lake in earlier years, its peak abundance in 1988 was about three times higher than in any previous year. During 1988 it accounted for about half of the total cyanobacterial biomass in the lake and over 25–30% of the total phytoplankton biomass (Edmondson, 1997). Although the biomass of *A. flos-aquae* remained low in the next few years, the 1988 bloom raised concerns that conditions in Lake Washington might be changing in a way that would eventually shift the composition of the phytoplankton community toward cyanobacteria. Edmondson (1997) has provided a thoughtful discussion of the implications of the available information. Two noteworthy changes in water quality have been the gradual decline in nitrate/phosphate ratios (Figure 4.6) since 1972 and an increase in alkalinity dating back to at least the 1950s (Figure 4.9). Alkalinity is simply a measure of the acid-neutralizing capacity of a body of water, and in the case of Lake Washington it is for all intents and purposes equal to $[HCO_3^-] + 2[CO_3^{2-}] + [OH^-] - [H^+]$, where the brackets signify concentrations (Lehman, 1988). Edmondson (1994) has speculated that land development has been responsible for the increase in alkalinity. The interaction of water and air with soils and rocks leads to a series of chemical reactions known collectively as *weathering*. A simple example of a weathering reaction is the following:[6]

$$CO_2 + H_2O + MgSiO_3 \rightarrow Mg^{2+} + HCO_3^- + OH^- + SiO_2$$

In this particular reaction, alkalinity is increased by the addition of both hydroxide (OH^-) and bicarbonate (HCO_3^-) ions. Weathering and the associated addition of alkalinity to streams is a natural process that helps to buffer the pH of natural waters (see Chapter 15).

---

[6]Granite and basalt are aluminosilicate rocks that contain variable amounts of Mg, Ca, and other elements. In this equation, $MgSiO_3$ represents the magnesium silicate moiety of these rocks or soils derived from them.

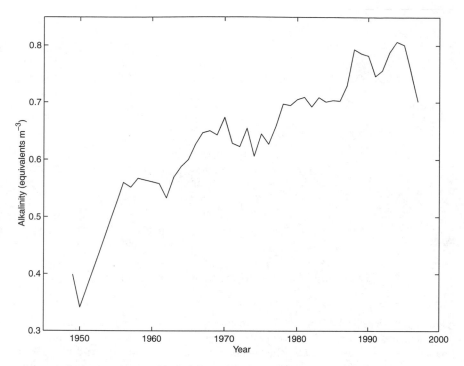

FIGURE 4.9   Mean alkalinity in Lake Washington measured from July 1 to August 20. Data provided by Dr. W. T. Edmondson.

The exposure of soils and rocks as a result of land-clearing activities accelerates the weathering process.

The extent to which the increase in alkalinity and the decrease in the nitrate/phosphate ratio may have contributed to the 1988 *A. flos-aquae* bloom is problematic. Edmondson (1997) compared a variety of physical and chemical characteristics of Lake Washington in 1988 versus 1963–1992. He concluded (p. 409), "The greatest differences between conditions in 1988 and other years were in alkalinity, mixing depth, wind velocity and solar radiation. Of these, wind velocity seems most likely as a critical factor in 1988." What does seem clear is that the chemistry of the lake has gradually been changing, and these changing conditions may allow cyanobacteria to bloom more frequently when other conditions (e.g., wind speed) are favorable. Eventually, the changes in chemistry may lead to a permanent shift in the species composition of the phytoplankton.

Theoretical Predictions

Predicting how an aquatic system such as Lake Washington will respond to a change in its nutrient loading is a problem that limnologists have discussed and debated for many years. Among the theoretical models most commonly applied to such problems are those developed by Vollenweider (1968, 1969, 1975, 1976). Pertinent equations can be derived by considering the simple model depicted in Figure 4.10. If one assumes that the system is in steady state, then the rate of nutrient addition to a lake must be balanced by losses due to sedimentation and washout. One assumes that per unit time a fraction f_w of the phos-

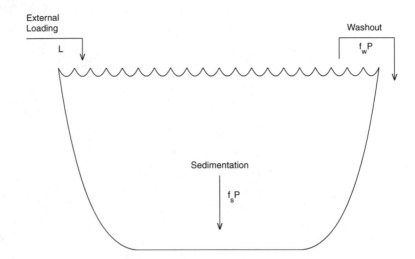

FIGURE 4.10   Vollenweider model of P fluxes in a lake.

phorus in the lake washes out and that a fraction f_s sinks to the bottom and is buried in the sediments. The decrease in P concentration per unit time due to washout and sedimentation is therefore $(f_w + f_s)P$. The loading rate L is commonly expressed per unit of lake surface area, and the increase in P concentration per unit time due to external inputs is therefore L/Z, where Z is the average depth of the lake. In the steady state $L/Z = (f_w + f_s)P$ or $P = L/[Z(f_w + f_s)]$. One therefore predicts that the average concentration of P in the lake will be directly proportional to the loading rate and inversely proportional to the depth of the lake and the sum of the fractional loss rates to sedimentation and washout. Pertinent data for Lake Washington are summarized in Table 4.1. Vollenweider (1976) somewhat arbitrarily defined the transitions from oligotrophy to mesotrophy and mesotrophy to eutrophy as occurring at total P concentrations of 10 and 20 mg m^{-3}, respectively. By these standards, Lake Washington was clearly eutrophic prior to the diversion of the sewer outfalls but became mesotrophic during the postdiversion period.

Toward the end of the Lake Washington sewage diversion effort, METRO appropri-

TABLE 4.1   CHARACTERISTICS OF LAKE WASHINGTON AND LAKE SAMMAMISH RELEVANT TO VOLLENWEIDER'S MODEL OF LAKE EUTROPHICATION

	Washington	*Sammamish*
Area (km^2)	8.6	19.8
Z (m)	3.9	17.7
f_s (per month)	.069	0.125
f_w (per month)	.024	0.133
L (mesotrophic $\rightarrow$ eutrophic) (g m^{-2} y^{-1})	.74	1.1
L (oligotrophic $\rightarrow$ mesotrophic) (g m^{-2} y^{-1})	.37	0.55
L (prediversion) (g m^{-2} y^{-1})	.71	1.02
L (postdiversion) (g m^{-2} y^{-1})	.64	0.67
P (prediversion) (mg m^{-3})	65	33
P (postdiversion) (mg m^{-3})	17.4	28.4

TABLE 4.2 CHARACTERISTICS OF LAKE SAMMAMISH RELEVANT
TO VOLLENWEIDER'S MODEL OF LAKE EUTROPHICATION
CORRECTED FOR EFFECTS OF INTERNAL P CYCLING

	Loading		*Total P (mg m^{-3})*	
	Allochthonous (g m^{-2} y^{-1})	*Autochthonous (g m^{-2} y^{-1})*	*Observed*	*Predicted*
Prediversion	1.02	0.79	33	33
1970–1975	0.67	0.79	26.7	26.6
1979–1984	0.59	0.34	18.5	17.0

ated an additional $3 million to initiate a similar wastewater diversion program for Lake Sammamish, which had begun to develop cultural eutrophication problems. The sewage effluent from the city of Issaquah and the waste from a dairy were both diverted in 1968. The lake was monitored from 1964 to 1966 to obtain baseline data and during 1970–1975 and 1979–1984 to determine the rate and extent of its recovery. The behavior of Lake Washington and Lake Sammamish and a comparison with the predictions of Vollenweider's model are summarized in Tables 4.1 and 4.2.

The response of Lake Washington was generally in accord with model predictions. The postdiversion P loading rate was about 14% below the loading rate associated with the transition from mesotrophy to eutrophy, and the postdiversion P concentration was about 13% below the corresponding P concentration of 20 mg m^{-3}. The behavior of Lake Sammamish, however, did not fit the model predictions. According to the model, the prediversion loading rate of the lake should have created mesotrophic conditions, but the P concentration was 33 mg m^{-3}, well above the concentration of 20 mg m^{-3} associated with the transition from mesotrophy to eutrophy. Furthermore, a reduction in the loading rate by 34% produced only a 19% drop in the average P concentration from 1970 to 1975. Associated with that decline in P concentration was a 24% reduction in summer chlorophyll *a* (6.6 to 5.0 mg m^{-3}) and almost no increase in Secchi depth (3.3 to 3.4 m). The explanation for the behavior of Lake Sammamish has been summarized in several papers by Welch and coworkers (Welch, 1985; Welch et al., 1980, 1986). Very simply, the model depicted in Figure 4.10 overlooks one important point. The bottom waters of Lake Sammamish became anoxic for several months during the summer stratification period, and during that time there was a considerable influx of P to the hypolimnion from the sediments. Table 4.2 is a revised assessment of the condition of Lake Sammamish before and after wastewater diversion. According to Welch et al. (1986), the internal cycling of P from the sediments of the lake increased the prediversion P loading rate by 77%, from 1.02 to 1.81 g m^{-2} y^{-1}. During the first few years following the wastewater diversion there was almost no change in internal P cycling, and the overall reduction in P loading (allochthonous plus autochthonous) was only about 19%, from 1.81 to 1.46 g m^{-2} y^{-1}. However, in subsequent years the internal recycling (autochthonous loading) eventually subsided in response to the reduction in external (allochthonous) loading. There was also a small additional reduction in external loading from 0.67 to 0.59 g m^{-2} y^{-1}. The net result was that during 1979–1984 the total P loading to Lake Sammamish was a little over half the prediversion loading rate. During that time the P concentration in the lake averaged 18.5 mg m^{-3}, 56% of the prediversion mean concentration. Thus, when the Vollenweider model was corrected for autochthonous loading, the predictions of the model were very consistent with the response of the lake. Of interest to local taxpayers was the fact that summer chlorophyll *a* concentrations during 1981–1984 averaged 2.5 mg m^{-3}, about 38% of the prediversion mean, and Secchi depths increased to 4.9 m. Furthermore, even by 1980 there had been a decline by almost a factor of 2 in the percentage of the phytoplankton accounted for by cyanobacteria. As a result, scums of cyanobacteria were no longer observed along the shoreline in late autumn (Welch et al., 1980).

The comparison of the responses of Lake Washington and Lake Sammamish to wastewater diversion provides some important lessons. First, why did the hypolimnion of Lake Sammamish become anoxic at a time when external P loading to the lake was only about 60% of the P loading to Lake Washington? The answer is that Lake Sammamish is shallower. The hypolimnion of Lake Washington accounts for almost half of the volume of the lake. The hypolimnion of Lake Sammamish is only about 28% of the lake's volume (Welch et al., 1986). Hence, there is only about half as much O_2 stored in the hypolimnion of Lake Sammamish compared to Lake Washington at the beginning of the summer stratification period. Just prior to sewage diversion, the O_2 depletion rate in the hypolimnion of Lake Washington was sufficient to reduce the O_2 concentration by about 6 g m^{-3} during the summer stratification period. If the hypolimnion of Lake Washington were only half as large, the reduction would have been twice as great, and anoxic conditions would have developed in the late summer. Another important point is that nutrient cycling in shallow systems is inherently more efficient than in deep systems. Thus, the same rate of external nutrient loading will tend to create more production in a shallow system than in a deep system. Hence, because of its shallower depth, Lake Sammamish is inherently more susceptible to seasonal O_2 depletion.

The delayed recovery of Lake Sammamish is also thought-provoking. The internal cycling of P did not subside for several years after the diversion of wastewater, and there continues to be substantial cycling of P. The most recent estimates indicate that autochthonous P inputs account for 35–40% of total P loading. If allochthonous P loading could be reduced sufficiently to prevent the lake's hypolimnion from becoming anoxic, there would be a dramatic reduction in total P loading and further improvement in water quality. Once hypolimnetic waters become anoxic, it is no longer possible to trap P in the sediments with Fe, and large amounts of P may therefore be released to the overlying waters. The transition from oxic to anoxic conditions in the hypolimnion of a lake, even for only a few weeks, is therefore a crucial event in the eutrophication process. It is a transition that is not easily reversed.

CASE STUDY 2. LAKE ERIE

Lake Erie has been turned into a cesspool. (Rene Dubois, cited in Cy Adler, 1973, p. 111)

Lake Erie represents the first large-scale warning that we are in danger of destroying the habitability of the Earth. Mankind is in an environmental crisis, and Lake Erie constitutes the biggest warning. (Barry Commoner, cited in Adler, 1973, p. 111)

You see, Lake Erie has died. The lake can no longer support organisms that require clean, oxygen-rich water. Much of this shallow body of water is a stinking mess—more reminiscent of a septic tank than the beautiful lake it once was. . . . No one in his right mind would eat a Lake Erie fish today. (Ehrlich, 1968, p. 39)

Once clear and filled with valuable fish and game, Lake Erie today is the embodiment of all that can go wrong in an aquatic system. The more desirable species of fish have disappeared; the once clear water is filled with excessive numbers of microorganisms; mats of filamentous algae at times cover whole square miles of lake surface; swimming is impossible in many places because of the quantity of untreated sewage in the water and the decaying vegetation covering the beaches; oil scums often cover harbors and coves, making boating and water sports unpleasant. (Wagner, 1974, pp. 128–129)

Despite some pollution, it should be obvious . . . that Lake Erie, with its enormous fish populations, its low bacterial count, and enormous quantities of clean, potable water, is far from dead. Millions of Americans living near the lake swim, fish and boat in Erie's waters each year. (Adler, 1973, p. 119)

The preceding statements indicate that there has been a certain amount of disagreement regarding the quality of Lake Erie's waters. Let us therefore begin this discussion by trying to make an intelligent evaluation of the condition of Lake Erie and of the changes that have occurred as a result of human activities.

Lake Erie is a glacially scoured depression that was filled with water from glacial runoff about 12,000 years ago. With an area of 25,820 km², it is the 13th largest lake in the world by area. Its mean depth of 18.5 m, however, makes it the shallowest of the Great Lakes. The western, central, and eastern basins of the lake are somewhat arbitrarily defined by lines drawn between Point Pelee (Bar Point) and Sandusky, Ohio, and between Long Point and Erie, Pennsylvania (Figure 4.11). The western basin is the shallowest, with a mean depth of about 7.4 m. The central and eastern basins have mean depths of 18.5 m and 24.4 m, respectively. The major inflow to Lake Erie comes from Lake Huron via the St. Clair and Detroit rivers, and the outflow is to Lake Ontario via Niagra Falls and the Welland Canal. The residence time of water in the lake is about 2.5 years. Lake Erie is dimictic and invariably freezes over in the winter.

The various ecological stresses that have been imposed on the biota of Lake Erie over approximately the last 200 years cannot all be classified as cultural eutrophication, although there is no doubt that cultural eutrophication has had a significant impact on the system. Unfortunately one sometimes gets the impression from newspaper articles or similar news sources that removing phosphates from detergents (for example) would solve all of Lake Erie's problems. The following summary of ecological stresses imposed on the lake will, it is hoped, provide a better perspective on the complex problems that have been created as a result of human activities in the Lake Erie watershed.

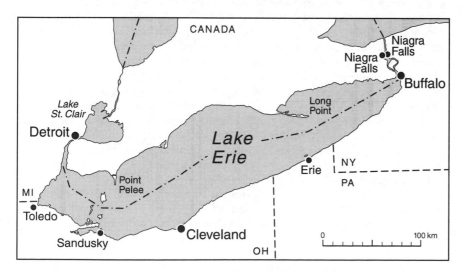

FIGURE 4.11   Lake Erie and environs.

Destruction of Fish Spawning and Nursery Grounds as a Result of Land Use Modifications

As of the 18th century, when the human population in the Lake Erie watershed was less than 0.1% of the present population, the land in the watershed supported large forests, primarily beech-birch, maple-hemlock, and oak-hickory associations (Regier and Hartman, 1973). Interspersed between the forests were savannas of grass and wild oats. The lake was bordered in many areas by marshes, the largest being the Great Black Swamp, which covered nearly 10,000 km² at the southwest corner of the lake. Because of the extensive vegetative cover, soil erosion was minimal, and streams flowing into the lake were generally clear. By roughly 1870 most of the forests had been cleared, the grasslands burned, and some of the swamps drained. Much of this modified land was given over to farming. Clay and silt eroded from exposed land and washed into streams and nearshore lake areas ruined the spawning grounds of many fish such as the walleye and lake whitefish. As a result of the increased sediment runoff, water quality in nursery marshes and bays declined, as did the biomass of aquatic vegetation, which provided food and shelter for the juvenile fish in these areas. By 1900 nearly all of the swamps, which had provided important nursery and spawning areas for the fish, had been drained. Furthermore, hundreds of milldams prevented or impeded fish such as walleye and sturgeon from reaching their traditional river spawning areas. Thus, the reproductive success of numerous species of fish was greatly reduced as a result of human activities that eliminated or seriously degraded the quality of traditional nursery and spawning areas.

Depletion of Fish Stocks Due to Overfishing

Although other factors undoubtedly contributed to the decline of certain species of fish in Lake Erie, the reduction of population numbers in many cases is strongly correlated with a known increase in fishing effort. Indeed, even in the absence of other ecological stresses, it seems likely that the number of certain valuable fish species would have been greatly reduced as a result of overfishing alone. The problem of regulating fishing efforts has been compounded by the fact that four U.S. states (Michigan, Ohio, Pennsylvania, and New York) and one Canadian province (Ontario) have jurisdiction in determining fishing practices. Unfortunately, fish are mobile creatures and do not recognize state or national boundaries. As a result, to a certain degree, several parties hold the fish resources of Lake Erie in joint or common possession. It is too often the case that when valuable resources are held in common ownership, the owners find it difficult to agree on a rational policy of utilization, the result being that an every-man-for-himself attitude prevails. Although relations between Canada and the United States are generally good, the so-called tragedy of the commons (Hardin, 1968) has undoubtedly influenced fishing policies in Lake Erie. The following is a summary of the changes that have occurred in the commercial fishing picture since roughly 1820.

The commercial fishery did not begin to develop around Lake Erie until after the War of 1812. Until that time, basically subsistence fishing was practiced in the tributary streams and nearshore waters of the lake. The failure of any significant commercial fishery to develop until roughly 1820 can be attributed to the following factors:

1. A lack of sophisticated methods for catching fish
2. A lack of reliable refrigeration methods for preserving fish
3. An inadequate transportation system for getting fish to market

During the period from roughly 1820 to 1890, a number of economic and technological changes resulted in the steady growth of the Lake Erie fishery at the rate of about 20% per year.

The development of reliable refrigeration methods during the mid-19th century eliminated the need for smoking and salt curing as a means of preservation and facilitated the transportation of fish without spoilage over greater distances to market. During the 1820s and 1830s the opening of canals greatly expanded the potential range of markets available to Lake Erie fishermen. The Erie Canal joining Buffalo with the Hudson River at Albany was completed in 1825; the Welland Canal connecting Lake Erie and Lake Ontario was opened in 1829; and the Erie–Ohio Canal joining the Ohio River with the Maumee River upstream from Toledo was completed in 1832. In the second half of the 19th century the development of an extensive railroad network eliminated the dependence of the fishery on canal transportation, reduced the transit time to markets, and further expanded the range of potential markets.

Prior to 1850, most Lake Erie fishing was done with hooks, seine nets, or small stationary gear such as traps and weirs. During the 1850s, fishermen began to use gill nets and pound nets in nearshore waters. During the next 20 years the use of these nets was extended further offshore, and by 1880 fishing gear was used throughout the lake (Regier and Hartmen, 1973). In 1905 the use of a large and very efficient gill net, the so-called bull net, was first introduced in the Lake Erie fishery. Due to its great efficiency the bull net became highly popular, but because of its nonselectivity it was eventually outlawed during the 1930s. The efficiency of gill nets was greatly increased in the early 1950s by the substitution of nylon for the traditional cotton or linen as a netting material. Furthermore, since nylon is much less subject to attack by microorganisms than is cotton or linen, it was no longer necessary for fishermen to rack and dry their nets every few days. As a result, the amount of fishing time per unit of gear was greatly increased.

Over the last 100 years, the annual commercial catch of fish from Lake Erie has remained remarkably constant at roughly 22 million kg per year, a figure that has equaled about 75% of the commercial catch from the other Great Lakes combined. However, the principal kinds of fish caught and their commercial value have varied greatly with time. In the early 19th century, large numbers of the generally preferred species of food and game fish inhabited the lake. In nearshore waters these species included smallmouth and largemouth bass, muskellunge, northern pike, and channel catfish. The important offshore species included lake herring (cisco), blue pike, lake whitefish, lake sturgeon, walleye, sauger, freshwater drum (sheepshead), white bass, and (primarily in the eastern end of the lake) lake trout. Of these species, the lake sturgeon and lake trout were among the first to suffer drastic population declines. The sturgeon was not valued commercially until the 1860s, when the knowledge of how to smoke it, render its oil, and make caviar from its eggs and isinglass from its air bladder was acquired from a European immigrant (Regier and Hartman, 1973). Prior to that time the lake sturgeon had been considered a nuisance fish, since its large size and external bony armor enabled it to tear nets set for smaller fish. To combat the lake sturgeon, Lake Erie fishermen had devised larger mesh and stronger nets specifically designed for its capture. However, they made no attempt to utilize the lake sturgeon they caught but simply disposed of the fish, often by burning piles of them on the beach. The effort to catch lake sturgeon intensified during the 1860s and 1870s after the fish's commercial value became known. Unfortunately, the lake sturgeon is a slow-growing fish that becomes sexually mature only at the age of 15 to 25 years and is rather easily taken with most types of fishing gear. As Figure 4.12 shows, the catch of lake sturgeon declined steadily after 1880 and by 1960 had reached a commercially insignificant level. How much of this decline can be attributed to overfishing and how much to other causes is problematic. The construction of milldams, for example, interfered with the ability of the sturgeon to reach river spawning grounds. However, there is no doubt that the slow growth rate and long mat-

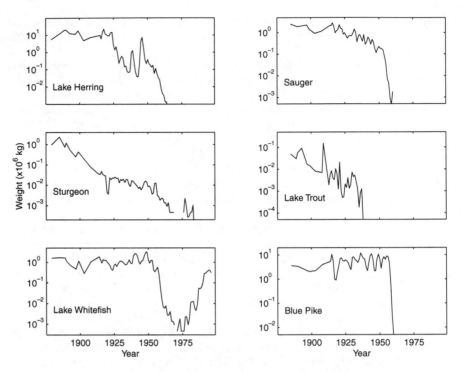

FIGURE 4.12   Commercial catch of six species of fish whose populations suffered serious declines in Lake Erie. *Source of data*: International Board of Inquiry (1943), Baldwin et al. (1979), and more recent catch statistics provided by the Great Lakes Fishery Commission.

uration period of the lake sturgeon, combined with its vulnerability to most types of fishing gear, made it a species particularly susceptible to overfishing effects.

As the lake sturgeon population declined during approximately the last 20 years of the 19th century, fishing pressure on other species intensified. As Figure 4.12 shows, the catch of lake trout began to drop precipitously in about 1880, and despite some pronounced peaks due to good year classes, the lake trout catch had become insignificant by 1930–1940. Although it is again impossible to evaluate accurately the impact of fishing pressure on the lake trout population, it seems likely that the irregular decline in annual catches between roughly 1880 and 1935 was due in no small part to overfishing. However, the elimination of native lake trout from Lake Erie after about 1940 was almost certainly brought about by deterioration of the environment as well as by overfishing.

During the period between 1890 and 1910, U.S. sports fishermen succeeded in having a number of species declared off limits to commercial fishermen. In Lake Erie the restricted species included smallmouth and largemouth bass, northern pike, and muskellunge. However, the Canadian government did not adopt similar restrictions, apparently because the large number of lakes in Ontario provided an adequate resource for Canadian anglers. Nevertheless, the U.S. restrictions provided these species with at least some refuge from commercial fishing pressure.

Figure 4.12 shows annual commercial catch statistics for several species of fish that suffered serious population declines. The whitefish, sauger, and blue pike catches all show precipitous declines beginning in about 1960. Of the species shown in Figure 4.12, only the lake whitefish population has recovered. Recalling the substitution of nylon for cotton or linen in the manufacture of gill nets during the early 1950s, one is tempted to attribute these rapid declines to overfishing pressure associated with the use of the more efficient nets. However, cultural eutrophication effects as well as other forms of pollution had become serious problems in Lake Erie by this time, and it seems unwise to attribute these declines entirely or even primarily to overfishing. A more realistic assessment might conclude that the increase in fishing pressure during this time may have triggered the rapid decline of populations that were already severely stressed from pollution and the elimination of many spawning and nursery grounds. The decline in the lake herring (cisco) catch began in about 1920 and continued rather steadily thereafter, with the exception of a few good years around 1945. It seems likely that some of this decline has been due to overfishing, although a quantitative assessment is again impossible.

Creation of Anoxic Bottom Water Conditions Due to Cultural Eutrophication

As the shallowest of the Great Lakes, Lake Erie has always been by far the most productive and, at the same time, the lake most likely to be adversely affected by cultural eutrophication. The human population in its watershed rose from about 3 million in 1900 to almost 15 million by 1980, a more rapid increase than in any other Great Lakes watershed. The major cities in the watershed are Detroit, Cleveland, and Erie. The sewage and other wastes from these population centers have been routinely discharged into Lake Erie. Between 1930 and 1965 the average N and P content of Lake Erie's water increased by about a factor of 3, as did the average biomass of phytoplankton. The concentration of all major ions in the water increased significantly during the first half of the 20th century, with chloride and sulfate, which are conspicuous components of wastewater, showing the greatest increases (Beeton, 1965). Although the average biomass of phytoplankton in the lake increased by a factor of 3, the rate of phytoplankton production increased by roughly a factor of 20 over the same time period, a consequence primarily of the longer duration of the spring and fall blooms (Regier and Hartman, 1973). The increase in biomass and in the productivity of the surface waters was reflected by an increase in the fallout of detritus into the hypolimnion and a concomitant increase in the respiration rate of the detritus food chain. Much of the increased O_2 demand was apparently associated with detritus that settled to the bottom of the lake and was decomposed by microorganisms living on or in the sediments. As of 1953, 28 days of calm weather were required for O_2 levels in the shallow western basin to drop below 3 ppm.[7] By 1963 only five days of similar weather were required to produce the same effect (Beeton, 1969). Thus, the sediment O_2 demand had increased by a factor of more than 5 during this 10-year period. Since the shallow western basin is often mixed to the bottom, O_2 depletion of the water column is not generally a problem, except during occasional periods of calm weather when the water column stratifies. However, the western basin was once an important breeding area for many fish. A female fish would typically deposit her fertilized eggs in the upper layer of sediment, where they would incubate prior to hatching. For the eggs to survive, the O_2 concentration in the immediate vicinity of the eggs must remain high. The high O_2 demand of the western basin's sediments was partially responsible for making this area poorly suited as a breeding ground from roughly 1960 to 1990. In addition, the accumulation of fine-grained clay and silt deposits on the bottom produced a compact sediment that inhibited

---

[7]The temperature of the bottom water in the western basin is about 19°C during the summer, and the saturation O_2 concentration in freshwater is about 8.6 ppm at that temperature. Concentrations below 3–4 ppm are considered dangerously low for fish by the EPA (EPA, 1986).

the circulation of oxygenated water within the sediment surface layer. Both factors—the high sediment O_2 demand and the low porosity of the sediments—created a sediment type that was virtually useless for fish reproduction.

The effect of nutrient enrichment on hypolimnetic O_2 levels has been most pronounced in Lake Erie's central basin, which is deep enough to stratify but has only a small hypolimnion. The average thickness of the central basin's hypolimnion is about 4.7 m (IJC, 1985). Low hypolimnetic O_2 concentrations were first noted in the central basin as early as 1929, and the O_2 demand of the hypolimnetic waters increased by a factor of 2 between 1930 and 1970. By the late 1950s, large areas of the hypolimnion were becoming anoxic for as much as several weeks during middle and late summer (Regier and Hartman, 1973). Lake trout, lake herring, lake whitefish, and members of the perch family including walleye, yellow perch, sauger, and blue pike are all typical cold-water fish that inhabit the hypolimnion during the summer and require O_2 concentrations at or near saturation to function effectively. Such fish were excluded from progressively larger areas of the central basin during the summer as low hypolimnetic O_2 conditions became more and more widespread and of longer duration. The eastern basin of the lake provided some refuge for these species, both because it received lower nutrient inputs and because its greater depth gave it a larger hypolimnion. However, by the 1960s, hypolimnetic O_2 levels in the eastern basin were found to be as low as 40–50% of saturation (about 4–5 ppm), a value approaching the critical level for many cold-water fish species (Beeton, 1965). It seems likely that the decline in commercial catches of cold-water fish such as the blue pike, lake whitefish, and cisco was due in part to the development of low O_2 concentrations first in the central basin and later in the eastern basin of Lake Erie as a result of cultural eutrophication.

Disposal of Toxic Wastes

About 500 toxic compounds have been identified in the Great Lakes. Some of these compounds have been discharged directly into the water; others have been introduced via stream runoff or fallout from the atmosphere (Table 4.3). Persistent biocides, toxic metals, and exotic organic chemicals have all been found in the flesh of Lake Erie fish. In many cases, the single most important source of these pollutants has been the Detroit River, which carries chemicals drained from the entire upper Great Lakes watershed. The impact of these substances on the health of the fish population has not been thoroughly studied, but the effect has undoubtedly not been beneficial.

The toxic substances of greatest concern have been polychlorinated biphenyls (PCBs) and mercury (Hg). PCBs are no longer manufactured or used in the United States, but they

TABLE 4.3   ESTIMATED ANNUAL LOADING OF SELECTED POLLUTANTS INTO LAKE ERIE

Chemical	Source of Chemical		Atmospheric Loading
	Detroit River	Lake Erie Tributaries	
Total PCB	512 (51%)	229 (23%)	257 (26%)
Total PAH	41,000 (79%)	— (0%)	10,870 (21%)
Mercury	2,050 (62%)	534 (16%)	728 (22%)
Lead	369,000 (67%)	53,400 (10%)	124,000 (23%)
Arsenic	102,000 (80%)	15,200 (12%)	9,909 (8%)
DDT	61 (47%)	31 (24%)	37 (29%)
Cadmium	7,170 (34%)	1,523 (7%)	12,329 (59%)

Source: Ohio Lake Erie Commission (1998).

are highly persistent and were used for many years in a variety of industrial applications. At present, sources of PCB contamination include leaking dumpsites and contaminated sediments, a legacy of poor waste management in the past. The toxicological properties of PCBs are discussed in some detail in Chapter 10. The 1978 Great Lakes Water Quality Agreement between the United States and Canada set an objective criterion of 0.1 ppm PCBs (whole organism, wet weight) for fish. However, a number of species of fish from the Great Lakes contain PCB concentrations that exceed this limit. Because of concern over PCB contamination, Ohio and Pennsylvania currently have fish consumption advisories for freshwater drum, walleye, carp, channel catfish, lake trout, rainbow trout, smallmouth bass, and coho salmon taken from Lake Erie.[8] In addition, Pennsylvania has issued fish consumption advisories for largemouth bass and northern pike, and Ohio has issued advisories for white perch and chinook salmon, again because of concern over PCB contamination. On the positive side, there is evidence that PCB concentrations in Lake Erie fish have been declining. In 1988, for example, fillets of walleye from western Lake Erie, with the exception of Maumee Bay, contained an average of 0.25 ppm PCBs. In 1992 the PCB levels at these same locations averaged 0.19 ppm (Ohio Lake Erie Commission, 1998).

Hg is a highly toxic metal that can cause genetic abnormalities and damage to the brain, kidneys, and liver (see Chapter 12). In 1970 the fishery in Lake Saint Clair was closed because of excessive levels of Hg in the fish, especially walleye; and with the exception of perch, similar action was taken for the commercial fishery in Lake Erie. Since that time, Hg use in the United States has declined by about 40%, and efforts to reduce Hg discharges from industrial operations have caused the Hg concentrations in Lake Erie fish to drop below federal guideline levels. Although a number of other toxic metals have been detected in fish taken from the Great Lakes, according to Shear (1984, p. 34), "The levels of metals . . . in whole fish samples are generally extremely low, well below any guidelines established for human consumption of fish flesh."

Other pollutants introduced into Lake Erie include a wide variety of organic compounds that enter the lake from atmospheric deposition or are discharged either directly into the lake or into tributary streams. For example, during 1969 more than 1,000 barrels of oil and grease per day were discharged into the Detroit River, the principal tributary of Lake Erie (IJC, 1970), and in 1973 Regier and Hartman reported that the Detroit River transported 6 billion liters of industrial and domestic waste water per day into the western basin of the lake. As is apparent from Table 4.3, the Detroit River remains a major source of many pollutants. The Cuyahoga River at Cleveland has carried such a high concentration of oil and other flammable industrial wastes that it actually caught fire in 1936, 1952, and 1969. The June 22, 1969, incident burned two railroad bridges to such an extent that the bridges were unusable. Runoff from farms, as well as from lawns and backyard gardens, has contributed sediment, persistent pesticides, and nutrients. The pesticides have been detected both in fish and in fish-eating birds (IJC, 1987). Fortunately, pesticide concentrations have been generally decreasing since the mid-1970s as a result of the greater awareness of their environmental impact and the use of alternative approaches to pest management in the United States. Interestingly, at the present time, the Great Lakes receive much of their DDT load from atmospheric deposition due to continued use of DDT as a pesticide in Third World countries in Central and South America (Ohio Lake Erie Commission, 1998).

Less progress has been made in the area of soil erosion and sedimentation. Sediment loading to Lake Erie is considered to be the primary nonpoint-source pollution problem in the lake (Ohio Lake Erie Commission, 1998). Efforts to reduce sediment loading have included the voluntary adoption of conservation tillage practices by farmers, the creation of stream vegetative filter strips, implementation of wetland and streambank restoration projects, stricter regulation of runoff from construction sites, and improvements in

---

[8]www.odh.state.oh.us/public/advisories/98fishadvisory1.html, www.health.state.pa.us/nr/fish-list.htm

stormwater sewer systems. The impact of these efforts has been estimated from a long-term study of sediment loading from the Maumee, Sandusky, and Cuyahoga rivers. The loading of sediment from these three rivers has averaged 1.5 million tonnes per year since 1982. Environmentalists are hoping that soil erosion control and conservation efforts such as those described above will reduce the sediment loading from these rivers to an average of no more than 0.5 million tonnes per year. However, in 1996 the three-year average sediment loading rate from these three rivers still stood at 1.5 million tonnes per year (Ohio Lake Erie Commission, 1998).

Contamination of Nearshore Areas with Sewage Wastes

Sewage, in some cases with inadequate treatment, has been discharged into Lake Erie or its tributaries for many years. Untreated sewage has also found its way into the lake as a result of overflows from combined sewer systems during storm runoff. As a result, bacterial counts in nearshore waters, particularly near large cities, have frequently been found to exceed limits considered safe for swimming. Although public health authorities have closed beaches, people sometimes continue to use the beaches despite posted warnings. Even though bacterial counts in offshore waters are rather uniformly low (see the quotation from Adler, 1973 on page 79), swimmers are interested in using nearshore rather than offshore waters. Since sewage is discharged at nearshore outfalls as a matter of convenience, it is the nearshore swimming areas that have become contaminated. The city of Cleveland has for many years drawn its water from an intake located 5 km offshore, and the bacterial counts in this water have usually been low enough to satisfy drinking water standards. Thus, even though it is correct to say that bacterial counts in most of Lake Erie's waters are low and that the water is potable, this generalization does not apply to nearshore areas that have become polluted by sewage discharges from nearby outfalls.

In addition to discharging pathogens into the water, sewage outfalls have introduced large amounts of nutrients, particularly N and P. The nutrients from the sewage have stimulated phytoplankton production, and the fallout of the organic matter produced by the phytoplankton has been a major cause of the hypolimnetic O_2 depletion problem. The increased phytoplankton concentrations also created an aesthetic problem that greatly reduced the recreational value of certain areas of the lake. The principal aesthetic problem was the formation of dense surface mats of algae, which have a tendency to drift ashore and rot on the beaches. The most offensive algal genus in this respect has been the green alga *Cladophora*, a macroscopic species that generally grows best attached to the bottom in shallow water. Although *Cladophora* has apparently inhabited Lake Erie for many years, its biomass did not begin to reach nuisance proportions until the 1950s. By the late 1950s, accumulations of *Cladophora* up to 0.5 m thick were being reported along the shoreline in some areas, and when these shoreline accumulations decomposed, local residents sometimes had to move out of their homes because of the odors produced by the decay process (IJC, 1975, p. 7). In addition to the *Cladophora* problem, cyanobacteria, which tend to form surface mats, began to dominate the fall phytoplankton bloom by about 1970 (Regier and Hartmen, 1973). The formation of these dense accumulations of algae was clearly undesirable from the standpoint of human recreational use of the lake.

Perhaps surprisingly, water transparency in Lake Erie during the 1970s was comparable to if not better than that of Lake Washington in the first few years after sewage diversion, Secchi depths in the central and eastern basins of the lake usually fall in the range 3–7 m (IJC, 1985). Historically, the murkiest water has been in the shallow western basin. In 1985 the IJC reported a mean Secchi depth of only 1.5 m in the western basin. Such poor water clarity reflects both the susceptibility of shallow aquatic systems to eutrophication and the generally high concentration of particulate material stirred up from the bottom of the western basin (Beeton, 1969).

Remedial Efforts

A massive cleanup of Lake Erie, initiated largely as the result of an agreement signed by Prime Minister Pierre Trudeau and President Richard Nixon in 1972, has been underway for some time. The cost of this program has run into billions of dollars. What has this cleanup accomplished to date, and what can the program ultimately be expected to achieve? Is it possible that the lake will be restored to a condition comparable to that existing in the early 1800s?

Toxic Substances

The technology is available to remove from wastewater or, even better, to recycle most of the toxic substances that have been discharged into Lake Erie. In some cases, the solution has been to eliminate the need for the toxic chemical, either by finding a suitable nontoxic substitute or by changing to an alternative technology that does not require the use of the toxic substance. In many cases, regulations regarding the discharge of toxic wastes have been established, and industries have been forced either to comply or to shut down (IJC, 1977, p. 40).

Figures 4.13–4.15 indicate some of the results that have been achieved as a result of the efforts to eliminate or at least reduce the discharge of toxic substances to Lake Erie.

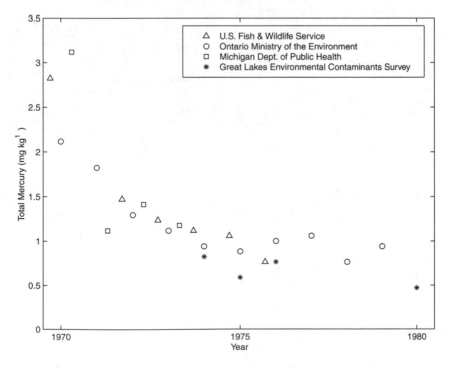

FIGURE 4.13   Hg levels in the fillets of walleyes taken from Lake St. Clair. *Source:* Great Lakes Water Quality Board. 1981. *Report on Great Lakes Water Quality to the International Joint Commission: Appendix—Great Lakes Surveillance.* Windsor, Ontario. November 1981, 174 pp.

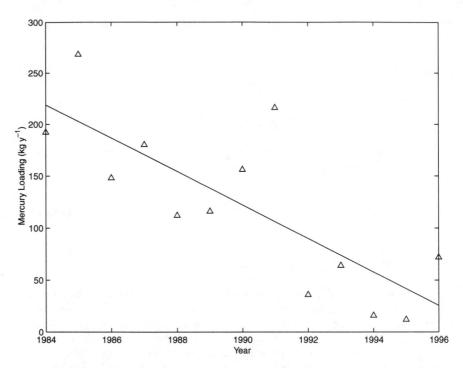

FIGURE 4.14   Hg loading to Lake Erie from point source discharges. *Source:* Ohio Lake Erie Commission (1998).

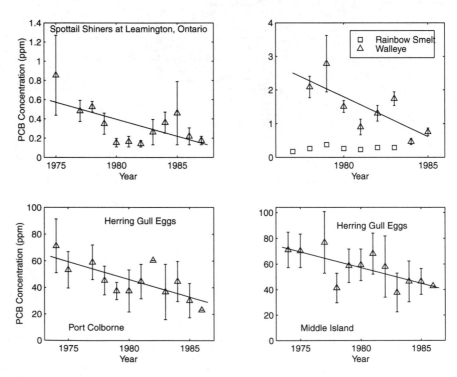

FIGURE 4.15   PCB levels in Lake Erie fish and herring gull eggs. *Source:* Great Lakes Water Quality Board. *1987 Report on Great Lakes Water Quality.* Windsor, Ontario, November 1987, 236 pp.

Hg pollution in the Great Lakes has been associated mainly with fish taken from Lake St. Clair and the western basin of Lake Erie. The principal cause of the contamination was a chlor-alkali plant operated at Sarnia, Ontario, by the Dow Chemical Company (see Chapter 12). Elimination of these discharges resulted in a rapid decline of the Hg content of Lake St. Clair fish (Figure 4.13). A similar pattern has been observed in fish sampled from Lake Erie. For example, Hg levels in the tissue of Lake Erie white bass and walleye declined from 1.19 to 0.21 ppm and from 0.55 to 0.31 ppm, respectively, between 1971 and 1977 (IJC, 1977). Historically, Hg was used in a variety of industrial applications (Chapter 12), and Hg concentrations in municipal sewage effluent were often elevated due to the number of industries tied into plants in the Lake Erie watershed. In recent years, these concentrations have declined dramatically (Figure 4.14) as industries have become more efficient at recovering and recycling Hg or have switched to alternative means of production that do not involve Hg.

Voluntary restrictions on the manufacture of PCBs in the United States went into effect in 1971, and all production of PCBs in the United States was voluntarily terminated in 1977. An EPA ban on the manufacture, processing, and distribution of PCBs in the United States took effect in July 1979. The result of these actions on the PCB levels in Lake Erie fish and fish-eating birds is shown in Figure 4.15. The pattern of decline is obvious, but the target goal of 0.1 ppm has still not been reached in some fish, hence the fish advisories noted above.

Eutrophication

The effort to reduce nutrient inputs to Lake Erie has been ongoing since approximately 1970. Controls have focused on P, which is believed to be the primary limiting nutrient in the lake. Figure 4.16 shows the record of P loading to Lake Erie from 1967 to 1994. The 1978 revisions to the 1972 Great Lakes Water Quality Agreement between the United States and Canada called for annual P loading of the lake to be reduced to 11,000 tonnes, and this target was achieved by roughly 1985. In recent years, annual total P loading has been averaging about 10,000 tonnes. Initial efforts to control P inputs focused on point sources such as municipal and industrial outfalls. The agreement required all such point sources that discharged more than 1 million gallons per day (mgd) to reduce the P content of their effluent to no more than 1.0 ppm. The annual P loading from these point sources was expected to be about 2,500 tonnes if the P concentration in the effluent was 1.0 ppm. The actual P loading rate from point sources dropped below 2,500 tonnes per year in 1982, although at that time the effluent from some sources was still above the 1.0 ppm goal.

At the present time, about 75% of the P loading to Lake Erie is from nonpoint sources such as agricultural runoff. The focus in recent years has therefore shifted from control of point sources to control of nonpoint sources. In fact, calculations have shown that if the effluent from all major point sources contained 1.0 ppm P, the total loading of P to Lake Erie would be 13,000 tonnes per year, or 2,000 tonnes per year above the goal established by the 1978 revisions. The 1983 Phosphorus Load Reduction Supplement to Annex 3 of the 1978 agreement therefore called for an additional reduction of 2,000 tonnes per year in the P loading of Lake Erie. About 70% of the additional reduction was to be achieved by erosion control, conservation tillage, and improved animal waste management in the area of Ohio west of Cleveland (IJC, 1987). For example, instead of plowing their fields each fall and spring, a practice that completely exposes the soil, farmers are leaving their fall stubble in the field (Ohio Lake Erie Commission, 1998). Figure 4.17 shows the dramatic increase in the use of conservation tillage in the Lake Erie watershed in recent years.

The impact of the P loading reduction is illustrated in part in Figures 4.18–4.21. Both total P and chlorophyll *a* concentrations in the central basin of the lake have declined by

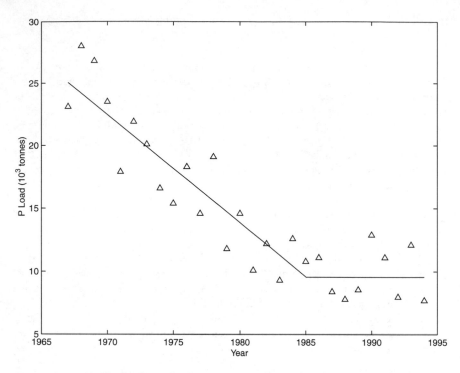

FIGURE 4.16   Total P loading of Lake Erie. *Source*: Data taken from IJC (1987), Dolan (1993), and Dolan and McGunagle (1998).

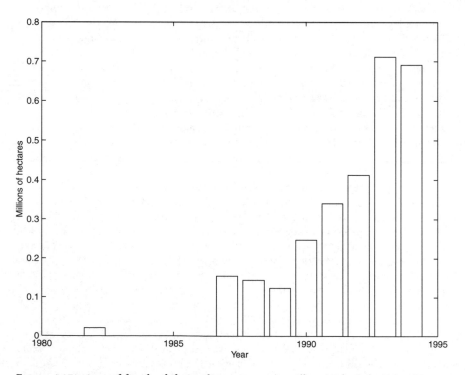

FIGURE 4.17   Area of farmland devoted to conservation tillage in the Lake Erie watershed. *Source*: Ohio Lake Erie Commission (1998).

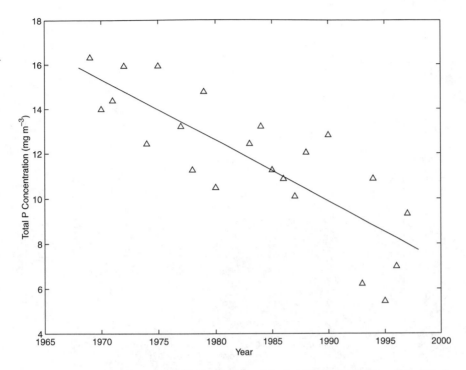

FIGURE 4.18 Total P concentrations in the central basin of Lake Erie reported by Charlton et al. (1998). Reproduced with permission from Backhuys Publishers, Leiden, the Netherlands.

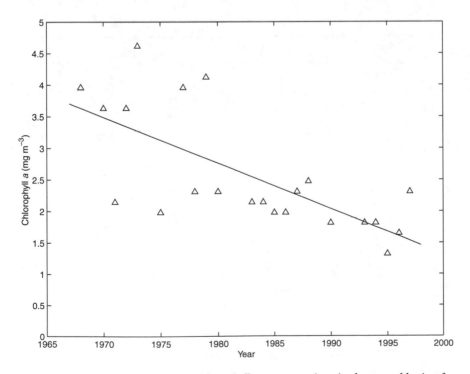

FIGURE 4.19 Average June-August chlorophyll *a* concentrations in the central basin of Lake Erie reported by Charlton et al. (1998). Reproduced with permission from Backhuys Publishers, Leiden, the Netherlands.

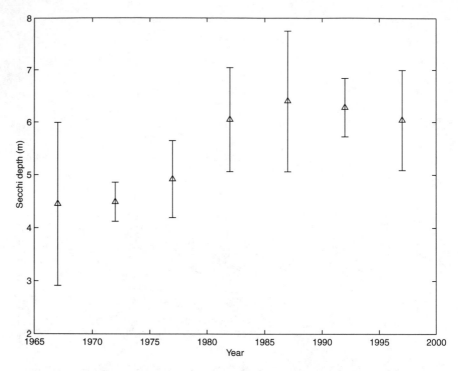

FIGURE 4.20   Means and standard deviations of June-August Secchi depths measured in the central basin of Lake Erie during consecutive five-year intervals based on data reported by Charlton et al. (1998).

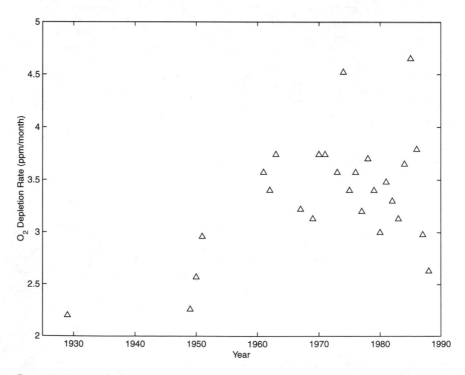

FIGURE 4.21 O_2 depletion rates in the hypolimnion of the central basin of Lake Erie. *Source of data*: IJC (1987, 1989).

about a factor of 2 since 1965 and now average about 8 and 2 mg m^{-3}, respectively. Based on Vollenweider's (1976) criteria, the central basin would now be classified as oligotrophic. Summer Secchi depths have increased from 4–5 m in about 1970 to 6–7 m at the present time. Diatoms and green algae rather than cyanobacteria are now the dominant forms of phytoplankton in the central portion of the lake (IJC, 1985, 1987). The O$_2$ depletion rate in the hypolimnion of the central basin peaked around 1980 and declined precipitously between 1986 and 1988 (Figure 4.21). Portions of the central basin hypolimnion become anoxic during the late summer if the O$_2$ depletion rates exceeds approximately 3.0 ppm per month (Hartman, 1973). In 1985 a period of late summer anoxia caused some 2,000 tonnes of P to be released from sediments in the central basin (IJC, 1987). This internal recycling of P from the sediments frustrated efforts to improve the O$_2$ regime in the central basin through control of allochthonous P inputs. In 1988 O$_2$ depletion rates in the hypolimnion of the central basin dropped substantially below 3.0 ppm per month for the first time in approximately 30 years. These rates should be no more than about 2.0 ppm per month to ensure that the hypolimnetic waters of the central basin provide a suitable habitat for cold stenotherms throughout the year. For such fish the O$_2$ concentration must remain above 3–4 ppm. Unfortunately, in 1990 the International Joint Commission discontinued the kind of systematic monitoring that produced the data shown in Figure 4.21. The data in Figure 4.21, as well as anecdotal information, indicate that there have been and continue to be significant interannual differences in the O$_2$ depletion rates. Much of this variability is evidently due to strictly physical factors, in particular differences in the thickness of the hypolimnion caused by weather conditions at the time the thermocline is established in the spring and year-to-year differences in lake level (D. Dolan, pers. comm.).

One informative biological indicator of the trophic status of Lake Erie is the condition of the mayfly (genus: *Hexagenia*) population. Mayflies are native to the Great Lakes and do well in shallow productive lakes with soft, organically rich sediments (Ohio Lake Erie Commission, 1998). Historically, the western basin of Lake Erie has been an ideal habitat for mayflies. The mayflies spend most of their life cycle in the sediments, where as nymphs they burrow in and feed for 1–2 years. In late June they swim to the surface, metamorphose, and take wing as adults. During the next 1–2 days they mate, drop their eggs into the water, and then die. The appearance of swarms of mayflies was a common summer event for many years in the western basin of Lake Erie, but in 1954 the mayflies failed to appear. The previous summer had been unusually hot and calm. With little wind mixing to oxygenate the water column, the decomposition of organic matter on the bottom of the lake stripped the water of O$_2$, and the mayfly nymphs suffocated. They did not reappear in significant numbers for 40 years. Figure 4.22 shows the dramatic increase in the number of mayfly nymphs in recent years. Historical data collected from 1930 to the early 1950s indicate that average nymph densities during that time were as high as 500 m^{-2}. The average density in 1997 was 404 m^{-2} (Ohio Lake Erie Commission, 1998).

Prospects for Lake Erie

Problems associated with the introduction of toxic substances into Lake Erie certainly appear to have diminished. For example, in the early 1970s the reproductive success of herring gulls in Lake Erie was reduced to nearly zero, apparently due to exposure to PCBs and other organochlorine substances (see Chapter 10). Bald eagles suffered a similar fate. Between 1969 and 1972 legislation was enacted to restrict or ban the use of PCBs, Hg, and the organochlorine pesticides dieldrin, heptachlor, DDT, and mirex within the Great Lakes basin. Virtually all use of DDT in the United States ceased in January 1973; production of mirex in the Great Lakes basin ended in 1976; and the manufacture of PCBs was voluntarily terminated by the Monsanto Corporation in 1977. In the second half of the 1970s,

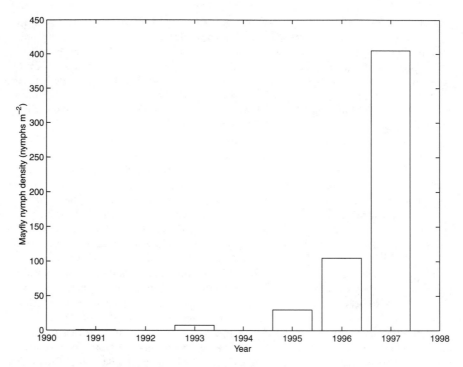

FIGURE 4.22   Density of mayfly nymphs in the sediments of the western basin of Lake Erie. *Source*: Ohio Lake Erie Commission (1998).

both the reproduction level and the number of adult herring gulls increased (Sonzogni and Swain, 1984). Bald eagles have made a similar dramatic recovery (Figure 4.23). The minimum reproductive rate needed for a stable bald eagle population is 0.7 eaglets fledged per nest. Bald eagles in the Lake Erie region have produced more than 0.7 eaglets per nest in 14 of the last 20 years, and reproduction exceeded 1.1 fledgling per nest in 4 of the 6 years between 1992 and 1997, inclusive (Ohio Lake Erie Commission, 1998).

The increase in the herring gull and bald eagle populations and the decrease in contaminant levels in fish have both been in the desired direction, but PCB levels in many species of fish remain disturbingly high. Rivers such as the Maumee, Black, Cuyahoga, and Ashtabula rivers are still significant sources of toxic organic chemicals, heavy metals, oil, and grease. Pollution control measures have improved water quality in these streams, but it is doubtful whether water quality in the former three will ever meet all of the objectives set by the 1978 Great Lakes Water Quality Agreement (IJC, 1982).

The effort to reduce P inputs to Lake Erie has obviously produced some positive results. The recovery of the mayfly population in the western basin of the lake and the fact that total P concentrations in the central basin are now less than 10 mg m^{-3} are both indications that the lake has made a remarkable recovery. Interestingly, the rates of hypolimnetic O_2 depletion reported in 1987 and 1988 were comparable to the rates observed around 1950, when the development of low O_2 conditions in the hypolimnion during the late summer was considered serious enough to have adversely impacted a number of species of fish. Within the context of this observation, it is noteworthy that the mayfly population in the western basin did not begin to recover until 1993. It would be interesting to know whether the hypolimnetic O_2 depletion rates in the central basin continued to decline between 1988 and 1993, but relevant data are lacking. Nonpoint source P control has been

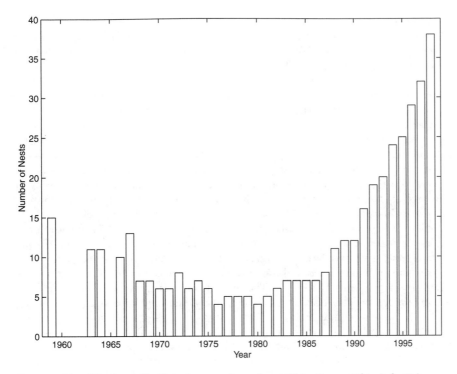

FIGURE 4.23.   Number of bald eagle nests throughout Ohio. *Source*: Ohio Lake Erie Commission 1998).

more difficult and expensive than point source P control. Despite the increase in conservation tillage, there has apparently been little change in the rate of sediment loading to the lake from land runoff. Almost 75% of existing P inputs come from nonpoint sources.

Between 1972 and 1979, New York, Indiana, Michigan, Minnesota, and Wisconsin all passed laws limiting the P content of laundry detergents to 0.5%, and Canada imposed a 2.2% limit in 1973. In 1990 legislation was passed that limited the P content of laundry detergents to 0.5% P in the counties of Ohio and Pennsylvania located in the Lake Erie watershed, a move long sought by environmentalists. Prior to these restrictions, detergent phosphates had accounted for about 50% of the P in municipal wastewater. In its 1989 report, the International Joint Commission recommended that Canada further reduce its detergent phosphate limit to 0.5%, because, "A further reduction in Canada would allow municipal treatment plants which remove phosphorus to achieve lower effluent concentrations with the same effort. In addition, a lower phosphorus limit would lower the effluent concentrations from those municipal and private plants that do not remove phosphorus, as well as reduce the phosphorus content in wastewater from combined sewer overflows and treatment plant bypasses" (IJC, 1989, p. 35). However, to this day, the limit on P in laundry detergents remains at 2.2% in Canada.

It is thought-provoking to realize that the P concentrations of 10–14 mg m^{-3} in Lake Erie during the 1980s were less than the postsewage diversion P concentration of 17 mg m^{-3} in Lake Washington (Table 4.1 and Figure 4.18). If 10 and 20 mg P m^{-3} are taken to be the concentrations associated with the oligotrophic/mesotrophic and mesotrophic/eutrophic transitions (Hern et al., 1981), respectively, then the central basin of Lake Erie during the 1980s was mesotrophic. It seems problematic, however, whether the very low late-summer O$_2$ concentrations in the central basin during that time should be associated with mesotrophy. It is certainly possible that the transition concentrations of 10 and 20 mg P m^{-3} are not the same for all lakes, and may be smaller in a relatively shallow lake such as the central basin of Lake Erie than in Lake Washington, which is almost 80% deeper and has a hypolimnion almost five times thicker (23 versus 4.7 m).

Contamination of nearshore waters with pathogens continues to be a problem in Lake Erie. Since 1984 the Ohio Department of Health has monitored bacterial indicator counts at 11 Lake Erie beaches. Advisories are issued when the counts at a beach imply an unacceptable health risk from water contact. Figure 4.24 shows the results of the monitoring program. No pattern of improvement is evident. In fact, the average number of days under advisement was higher (26 ± 10) during the last six years of the study than during the first six (19 ± 3). Since the recreational swimming season lasts for about 100 days, the implication is that these 11 beaches are under advisement 20–25% of the time during the swimming season. For comparison, individual beaches during the 1970s were routinely under advisement more than 50% of the time, and some were completely closed to the public due to pollution (Ohio Lake Erie Commission, 1998). Thus, there has been a definite improvement with respect to pathogen contamination since the 1970s. This improvement was achieved by upgrading sewage treatment plants, constructing water storage basins and retention tunnels, and connecting nonsewered communities and businesses to sanitary sewer systems. However, if "The goal is to have clean beaches all the time" (Ohio Lake Erie Commission, 1998, p. 12), the present condition still leaves much to be desired.

The technology is certainly available for sewage treatment plants to remove virtually all the pathogens from raw sewage. It seems likely, therefore, that the pathogen problem reflects nonpoint source pollution, the most important sources probably being combined

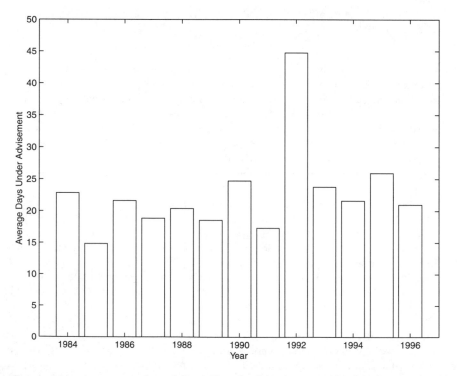

FIGURE 4.24   Average number of days of beach advisories at 11 Ohio beaches monitored by the Ohio epartment of Health along the south shore of Lake Erie. *Source*: Ohio Lake Erie Commssion 1998).

sewer system overflows and treatment plant bypasses. Correcting this problem will not be accomplished easily or cheaply. Construction and operation of facilities capable of treating stormwater runoff is an enormous task (see Chapter 5). The most cost-effective solution to the combined sewer system overflow problem is probably construction of separate sewer systems, but this task is also a major undertaking. Whether such decisions are made will undoubtedly depend very much on the public's perception of the environmental and public health costs associated with allowing untreated sewage to enter Lake Erie.

The fish population picture is rather sobering. The native populations of blue pike, sauger, and lake trout have been virtually eliminated from the lake; sturgeon, lake herring, and muskellunge are found only in small numbers; and the northern pike population has been greatly reduced. Human impact on the lake's natural fish community has taken four principal forms: overfishing, degradation of water quality, change or loss of habitat, and introduction of exotic species. Overfishing is a theoretically manageable problem, but implementation requires the cooperation and understanding of all concerned parties. To the extent that the United States and Canada remain good neighbors and recognize the errors of the past, there is reason to hope that Lake Erie fisheries will be managed in a more intelligent manner than was the case during the 19th century and the first half of the 20th century. The P control program has done much to improve the O_2 regime in the hypolimnion of the lake. This fact has undoubtedly been a factor in the recovery of the whitefish and walleye populations.

However, even with sensible fishing practices and improved water quality, it is highly unlikely that the pre-1800 distribution of fish will be restored in the lake. First, there is virtually no chance that many of the nursery and spawning grounds that were destroyed for various land developments, or that were covered with silt or dammed off, will be restored. Indeed, it is unlikely that many people would welcome the return of swamps, with their attendant mosquito populations. Second, a number of exotic species, introduced in many cases from the ballast water of ocean-going ships, have greatly altered the natural biological community of the Great Lakes. Of particular concern in Lake Erie are the alewife, rainbow smelt, sea lamprey, and zebra mussel. The alewife was first detected in Lake Erie during the 1870s (Roberts, 1990). It presumably entered the lake via the Welland Canal. The alewife is zooplanktivorous, and its consumption of herbivorous zooplankton may have contributed to the development of nuisance algal blooms in the lake (IJC, 1977). Rainbow smelt were initially introduced into Lake Michigan to provide forage for game fish. They first appeared in Lake Erie in 1931, increased greatly in number, and are now the most abundant pelagic fish. Like the alewife, rainbow smelt are largely zooplanktivorous, but yearling and older rainbow smelt feed to a certain degree on the young of other fish such as cisco, sauger, and blue pike. Any attempt to reintroduce large numbers of these fish (e.g., by seeding with fingerlings from hatcheries) would probably be inhibited by the now established large population of rainbow smelt. Other now abundant species such as the freshwater drum, carp, and goldfish would undoubtedly offer similar resistance to restocking attempts. The sea lamprey is a marine organism that reproduces and hatches in freshwater streams. However, in some parts of the world it has adapted to a completely freshwater life cycle and is an old inhabitant of the St. Lawrence River and Lake Ontario. The lamprey is an eel-like fish with a sucker-like mouth and sharp teeth that attaches itself to other fish, rasps a hole in the fish's body, and sucks the blood and body juices. A sea lamprey can kill a delicate fish such as the lake trout in as little as four hours. Niagra Falls blocked the access of sea lampreys to the upper Great Lakes until the Welland Ship Canal was completed in 1829. The lamprey was evidently slow to take advantage of this access route, however, because no lampreys were reported in Lake Erie until 1921 (Applegate and Moffett, 1955). The lamprey then moved on into Lake Huron and Lake Michigan, where it decimated the lake trout populations during the 1940s. During this time the sea lamprey never became a serious problem in Lake Erie, evidently because water quality conditions in the lake did not favor its propagation. However, as conditions in Lake Erie improved during the 1970s and 1980s, the population of sea lampreys in the lake began to increase. Even then, however, it was not considered a serious problem because its predation was spread over a large number of species, none of which was perceived as seriously impacted by its presence. Lamprey predation did become an issue, however, when attempts were made to introduce lake trout into the eastern basin of the lake. The lamprey seriously impeded these stocking efforts since lake trout seem to be its preferred prey in the Great Lakes. The problem was brought under control during the second half of the 1980s by applying chemical treatments to tributary streams where the lamprey offspring spend the first four years of their life as larvae in the sediments.

Zebra mussels are native to the Caspian Sea. Scientists believe that they were introduced to the Great Lakes when a transoceanic vessel discharged its ballast water into Lake St. Clair in 1986. The mussels multiplied rapidly, and by 1990 most hard substrates in Lake Erie had been colonized. Densities of zebra mussels as high as 200,000 m^{-2} have been reported in the western basin of the lake (Cooley, 1993), where mussel abundance is highest. The mussels are filter feeders and have a voracious appetite. In heavily infested portions of the western basin, scientists estimate that zebra mussels filter the entire water column 20 times a day (Cooley, 1993). Madenjian (1995) has estimated that they consume 26% of the primary production in western Lake Erie. Between 1988 and 1991, Secchi depths increased 77% and chlorophyll *a* concentrations decreased 60% in the western basin of the lake (Leach, 1994). Feeding by zebra mussels is believed to have been the primary cause of these changes. One of the principal economic impacts of the zebra mussel has been on the electric power industry. The mussels have colonized the cooling water intakes of power plants and restricted flow to such an extent that some plants have been forced to shut down temporarily (Roberts, 1990). The power industry estimates that the cost of cleaning and retrofitting cooling systems to discourage colonization by zebra mussels will be $2 billion over a period of 10 years (Roberts, 1990). Ecologically, there is concern that the zebra mussel will outcompete native shellfish and disrupt the lower food web, with possibly serious effects being transmitted up the food chain. For example, in the western basin the populations of crustaceans and microzooplankton are down 60% and 65%, respectively (Cooley, 1993). Such changes could have serious implications for zooplanktivorous fish such as smelt. There is convincing evidence that zebra mussels have outcompeted native unionid bivalves in the offshore waters of western Lake Erie. At some locations, native mussels have been completely replaced by zebra mussels (Schloesser and Nalepa, 1994). One of the major concerns has been that the zebra mussels would adversely impact walleyes since the mussels have colonized some of the walleye's prime spawning beds. Fortunately, the impact on walleye reproduction appears to have been small. In fact, walleye young-of-the-year survival was strong in both 1990 and 1991, when mussel abundance was highest (Fitzsimmons et al., 1995). Although zebra mussels have invaded all the Great Lakes, their number increased most dramatically in Lake Erie, particularly in the western basin. This fact probably reflects the more eutrophic condition of the lake and its western basin. There is some hope that the zebra mussels will simply eat themselves "out of house and home" (Roberts, 1990, pp. 1371–1372) and that the population will then crash or stabilize at a lower level. There is some evidence to support this scenario from the experience in Europe, where zebra mussels invaded from the Black and Caspian seas in the early 1800s. However, the ecosystem of the Great Lakes is different, and despite extensive study, even today, European biologists are "at a loss to explain why the zebra mussel crashed and then came back in force in Sweden but never quite recovered in mainland Europe" (Roberts, 1990, p. 1372). In 1991 a close relative of the zebra mussel, the quagga mussel, was discovered in Lake Erie. Unlike the zebra mussel, the quagga mussel thrives on soft substrates such as sand, silt, and even mud (Cooley, 1993). Although the quagga mussel has invaded all three basins of Lake Erie, it is of greatest concern in the central and eastern basins, where scientists believe it may be adversely affecting the deepwater amphipod *Diporeia* (C. Madenjian, pers. comm.). To date no definitive impacts on the fish community have been demonstrated.

At the present time, the goal of fisheries management in Lake Erie is not to reestablish the historical fish populations, but rather to manage sensibly the valued species that remain (e.g., walleye, yellow perch, white bass). In addition, some efforts have been made to introduce other valuable species that can coexist with the present populations. For example, Pacific coho salmon were introduced into the lake in 1968 to satisfy the needs of U.S. sports fishermen, but the impact on U.S. sport fishing was small. The salmon apparently migrated during the summer into Canadian waters, where the temperature was colder and the O_2 regime more favorable. Worse yet, the salmon preyed on smelt and other commercially important species, to the chagrin of Canadian fishermen. Recently, the salmon-stocking program was terminated. Both rainbow trout and lake trout have also been stocked into the lake, although in the latter case the stocking attempts have been significant only in the eastern basin. The trout, like the salmon, prey on smelt, and the stocking program is a bone of contention between the commercial and sport fishing interests. Ohio commercial fishermen are currently prohibited from catching walleye, sauger, sturgeon, cisco, trout, and salmon and from having gill nets in their possession. Canadian fishermen take almost all the commercial catch of smelt, white bass, whitefish, and walleye and 80% of the commercial catch of yellow perch. U.S. fishermen account for most of the commercial catch of carp and sheepshead. To a large extent, the restrictions on U.S. commercial fishermen have been imposed to protect the interests of U.S. sport fishermen, but in some cases they have also been aimed at maximizing the probability that a fish population will increase. The walleye population, for example, increased dramatically during the 1970s and is now a major target of U.S. sport fishermen.

Figure 4.25 shows the temporal pattern in the commercial catch of the six species of fish that presently contribute the most by weight to the commercial catch in Lake Erie. Of these six species, the carp and white bass are warm-water species and hence are little impacted by water quality conditions in the hypolimnion during late summer. Despite the collapse of cold-water fish populations such as lake trout and lake herring (Figure 4.12), it is obvious from Figure 4.26 that the total catch by weight has remained fairly constant since the latter part of the 19th century. The statement by Ehrlich at the beginning of this section that "Lake Erie has died" is therefore rather an overstatement of the case. However, it is obvious from an examination of Table 4.4 that there have been major changes in the composition of the commercial catch. While the weight of the catch has been relatively constant, the decline in species such as sauger and blue pike has greatly reduced its monetary value. At the present time, walleye and yellow perch are by far the most economically important species in the commercial catch.

CASE STUDY 3. KANEOHE BAY

The example of Kaneohe Bay provides an informative comparison with the cases of Lake Washington and Lake Erie. Kaneohe Bay was polluted with sewage for a period of about 30 years, but this stress was removed in 1978. Because the residence time of water in the bay is only a few weeks, the response of the system to this change was in some respects much more rapid than in the case of Lake Washington or Lake Erie. Nevertheless, several years were required for the bay to recover from some effects of the sewage enrichment, a thought-provoking fact when one considers that the bay is a relatively well-mixed and well-flushed system compared to Lake Washington or Lake Erie.

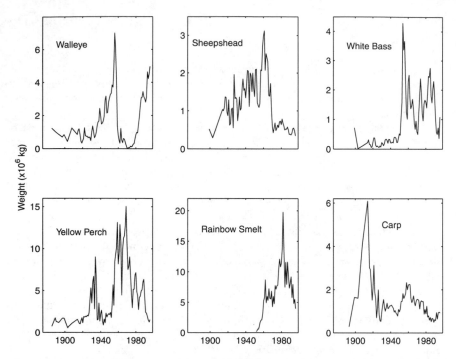

FIGURE 4.25   Historical commercial catches of the six most important species by weight in the present commercial fish catch from Lake Erie. *Source of data*: Baldwin et al. (1979) and subsequent catch summaries from the Great Lakes Fishery Commission.

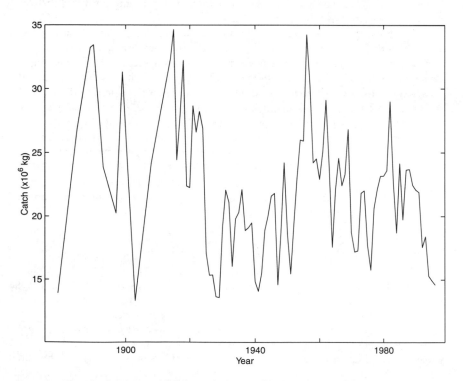

FIGURE 4.26   Total commercial fish catch from Lake Erie. *Source of data*: Baldwin et al. (1979) and subsequent catch summaries from the Great Lakes Fishery Commission.

TABLE 4.4 IMPORTANCE BY WEIGHT AND ECONOMIC VALUE OF SPECIES TO COMMERCIAL FISH CATCH IN LAKE ERIE IN SELECTED YEARS

1885		1918		1952		1996		
Species	*Wt*	*Species*	*Wt*	*Species*	*Wt*	*Species*	*Wt*	*value*
Cisco	11.5	Cisco	22.1	Blue pike	6.4	Walleye	5.0	12.6
Sauger	2.5	Carp	2.2	Walleye	3.3	Rainbow smelt	4.0	1.3
Sturgeon	2.4	Yellow perch	1.4	Sheepshead	2.1	Yellow perch	1.5	5.5
Whitefish	1.7	Sheepshead	1.4	Yellow perch	1.8	White bass	1.1	1.3
Yellow perch	0.7	Whitefish	1.2	Carp	1.5	Carp	0.9	0.1
Northern pike	0.1	Sauger	1.0	Whitefish	1.3	Sheepshead	0.4	0.1
Others	8.0	Others	2.9	Others	3.1	Others	1.9	1.7
Total	26.8	Total	32.2	Total	19.4	Total	14.6	22.6

Note: Weight (wt) expressed in millions of kilograms. Value expressed in millions of U.S. dollars.

Physical Setting

Kaneohe Bay is a subtropical embayment on the northeastern side of the island of Oahu in the Hawaiian Islands (Figure 4.27). The bay has an area of about 46 km^2 and a mean depth of about 6 m, although the average depth in the southeast sector is about 12–13 m. The salinity of the bay's waters normally falls in the range 33–35$^0/_{00}$, and water temperature varies seasonally between roughly 20°C and 27°C. (Bathen, 1968). A barrier reef that restricts circulation with the ocean extends along much of the bay's mouth, with two narrow channels, the Ship Channel and the Sampan Channel, providing access to the bay for boats. The southeastern sector of the bay is relatively isolated physically from the rest of the bay by Coconut Island and a system of shallow reefs. As a result, circulation in the southeastern sector is more sluggish than in the rest of the bay. However, the residence time of water in the southeastern sector is only about 24 days, and in the rest of the bay it is about 12 days (Sunn et al., 1975). These short residence times are due largely to tidal flushing through the channels and over the barrier reef.

The Kaneohe Bay watershed is only slightly larger in area than the bay itself and is bounded on its landward side by a series of steep cliffs rising 500–850 m above sea level. Runoff from the watershed normally drains into the bay by way of 11 small streams, but during heavy rains the streams may overflow their banks, and water may flow into the bay from virtually all shoreline areas (Banner, 1968). The northeast trade winds blow directly against the front of the bay about 70% of the year, and much rain is precipitated from the moist air as it rises against the cliffs at the landward margin of the watershed. The average annual rainfall in the watershed is estimated to be about 230 cm (Cox and Fan, 1973).

The Coral Reefs

Concern over pollution in Kaneohe Bay centered on the deterioration of the coral reef community. Although historical information on the reef community is rather sketchy, some idea of the characteristics of the coral reefs prior to their demise can be obtained by putting together bits and pieces of information. In a popular article written in 1915, MacKaye (1915, p. 135) commented that, "Probably no other one spot in the Territory of Hawaii

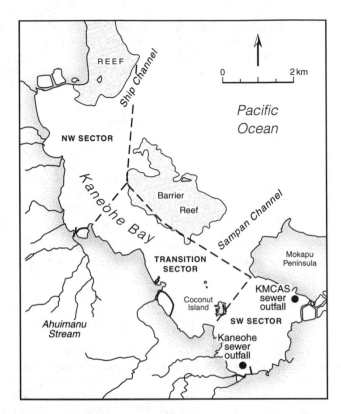

FIGURE 4.27   Kaneohe Bay and environs.

can show such a wonderful variety of corals as the waters of Kaneohe Bay". He noted in particular the beauty of the reefs in the southeastern part of the bay, the so-called coral gardens that were viewed by tourists from glass-bottom boats. In 1928, Edmondson stated that "Kaneohe Bay, on the windward coast, is recognized as one of the most favorable localities for the development of shallow water corals. Nearly all the reef-forming genera known in the Hawaiian Islands are represented . . . and many species grow luxuriantly" (p. 34). In 1946 Edmondson again commented that "Kaneohe Bay harbors a large population of marine animals. Here is one of the best exhibitions of living corals to be seen about the islands" (p. 5).

Banner (1974) provided a more detailed description of the coral reefs. He reported that, as of 1951, by far the dominant coral species in the lagoon area was *Porites compressa* vaughn, with *Montipora verrucosa* (Lamark) being of secondary importance. In areas exposed to greater wave action, the most abundant species included *Pocillopora meandrina* var. *nobilis* Verrill, *P. ligulata* dana, and the sturdier forms of *Porites*, such as *P. lobata* (Dana). Roughly 20 years later, Maragos (1972) reported a drastic change in the condition of the coral reefs. Based on a number of quantitative transects, he estimated that in the southeastern sector 99.9% of the original coral reefs had died or been destroyed, 87% had died or been destroyed in the transition sector, and 26% had died or been destroyed in the northwest sector. What caused this severe decline in a once-thriving coral reef community?

Urbanization of the Watershed

Most of the adverse effects on the coral reef system in Kaneohe Bay can be traced to problems associated with urbanization. To be sure, some corals were removed outright by dredging activities associated with construction of the Marine Corps Air Station on Mokapu Peninsula in 1939–1941 and, to a lesser extent, by private individuals seeking access for their boats across the reef flat. However, Banner (1974, p. 688) has commented that "By 1951 the reefs in most sections of the bay were again thriving, and some of the reefs cut to 3 m had a recolonization of usual bay corals." Evidently, had the environment of the bay remained favorable to the growth of corals, the recolonization of dredged areas noted by Banner in 1951 would presumably have continued. However, changes in the ecosystem caused by urbanization brought that recovery to a halt.

Changes in Land Runoff

Until roughly 1940, the Kaneohe Bay watershed was used primarily for agricultural purposes. The region's native Hawaiian population, which numbered only about 5,000 in 1830, cultivated taro as the staple of their diet (Chave and Maragos, 1973). The taro was grown in a system of terraces, into which stream runoff was diverted to provide irrigation water. Since most of the watershed's low floodplains were once covered with taro patches, runoff of sediment as well as overland water runoff into the bay was greatly reduced, as water diverted into the taro patches tended to percolate into the ground, depositing its sediment load on the taro patches. Overland runoff that reached the bay was of reduced intensity, and as a result had less erosive capability and transported less sediment than would otherwise have been the case. In short, there is little reason to believe that the agricultural practices of the native Hawaiians, who may have lived in the watershed for as long as 1,500 years, significantly increased the rate of land erosion and concomitant siltation of Kaneohe Bay. In fact, their extensive taro patch irrigation system may have reduced the rate of sediment runoff.

In the early 19th century, a variety of foreigners began to arrive in Hawaii following the early explorers. The diseases they brought with them devastated the native Hawaiian population, and by 1860 the population of the Kaneohe Bay region had dropped to less than 2,000 (Chave and Maragos, 1973). During the second half of the 19th century, the growing Hawaiian sugar cane industry began to import large numbers of Chinese, Japanese, and other ethnic groups to work in the cane fields, but the population of the Kaneohe Bay region did not reach 5,000 again until about 1940. During the influx of these nationalities, some changes in agricultural practices took place in the watershed. Rice replaced taro in some of the terraces, and small-scale commercial efforts were made to cultivate coffee, bananas, sugar cane, pineapple, and perhaps oranges (Cox and Fan, 1973). However, the major agricultural development affecting stream runoff appears to have been the utilization of the drier lands in the northern part of the bay's watershed for grazing cattle, sheep, horses, and goats. In some cases, overgrazing of these lands left persistent patches of bare ground. These patches, usually consisting of red volcanic clay, would "Dry to pavement-like hardness during dry weather and would wash away in heavy rains" (Banner, 1974, p. 687). During this time, Agassiz (1889) noted that some corals had been killed, evidently by sedimentation, in nearshore reef areas. However, the areal extent of this damage appears to have been small.

In 1913 a series of partial stream diversions was begun in the northern watershed to provide irrigation water for leeward Oahu. The tapping of several streams near their headwaters for this purpose reduced the annual stream runoff into the bay by about 42% (Chave and Maragos, 1973). At least some of the potential increase in sediment runoff from the effects of overgrazing and the reduction in the number of taro patches was undoubtedly counterbalanced by the reduced flux of stream runoff. Thus, sedimentation and freshwater runoff appear to have had only a limited effect on the coral reefs prior to at least 1925.

Since that time, several important changes in the watershed have greatly increased the runoff problem. The most significant of these changes has been the urbanization of the town of Kaneohe. The rapid rise in population began in roughly 1940 and has continued at a rate of about 4.8% per year since then. By 1990 the population of the watershed numbered about 55,000. Most of this population increase was confined to the southeastern sector's watershed, in which the town of Kaneohe is located.

Urbanization is inevitably associated with two processes that greatly increase runoff problems. First, the clearing of land for the construction of roads and buildings exposes, albeit temporarily, much bare soil to erosion from rainwater. In the case of Kaneohe, rather weak land clearing ordinances were passed to alleviate this problem, but in practice, even these weak laws were poorly enforced. Second, roads, parking lots, rooftops, and other such surfaces are impervious to water. Rainfall that lands on these areas has no chance to sink into the ground, as would be the case if the land were covered with vegetation. The percentage of land cover by impervious surfaces in business districts may approach 100%, and in residential areas it may easily be 50%. Thus, the amount of overland runoff from urban areas is much greater than in comparable rural areas (more on this subject in Chapter 5). Furthermore, runoff from impervious surfaces is more rapid than from land covered with vegetation since there is little to impede the flow of water over parking lots, streets, sidewalks, or rooftops. As a result, both the total runoff and the intensity of runoff are greater from urban areas than from rural areas. Because the erosive power of water increases with the intensity of water flow and because there is more total runoff in urban areas, exposed land is more rapidly eroded away in urban than in rural areas.

Freshwater runoff can adversely affect coral reefs because of both the stress of reduced salinity and the smothering effects of excessive sedimentation. Corals can tolerate salinities in the approximate range $27-40^0/_{00}$ (Shepard, 1973), and although salinities in Kaneohe Bay are generally well within this range, salinities of the surface waters may drop to much lower levels after particularly heavy rains. For example, on May 2–3, 1965, approximately 50–60 cm of rain fell on the Kaneohe Bay watershed; five days later, surface water salinities were still only $23-24^0/_{00}$ (Banner, 1968). As a result of this storm, corals in parts of the bay were killed to a depth of as much as 1.5 m. Although torrential rains of this sort undoubtedly occurred in the past in Kaneohe Bay, the effects of such rains have been exacerbated by the urbanization of the watershed. The high percentage of impervious surfaces in urban areas results in a greater volume of freshwater runoff and in a more rapid delivery of runoff waters to the bay.

With regard to sedimentation, it is noteworthy that Charles Darwin (1842) attributed the breaks in fringing reefs off the mouths of streams to the effects of sedimentation and not to the effect of reduced salinity. He reasoned that the freshwater would spread out on the surface, whereas the sediment particles would settle down on the corals. However, Dr. Robert Johannes (quoted by Banner and Bailey, 1970, p. 15) has noted that "The temporary production of turbid waters due to man's activities appears not to affect the surface reef community seriously unless siltation is so great that the biota is fairly evenly coated with sediments" but that "The exposure of reefs to brackish, silt-laden water associated with flood runoff seems to be a major cause of reef destruction historically." The combined stress of reduced salinity and heavy siltation may prove lethal in some cases where either stress taken separately would not be lethal.

There have been no direct measurements of sedimentation rates in Kaneohe Bay, but several indirect measurements provide some insight into the problem. Based on old bathymetric charts of the bay, Roy (1970) concluded that there had been very little change in the mean depth of the bay between 1882 and 1927, but a 1976 bathymetric survey conducted by Hollett (1977) indicated that the depth of the lagoonal area of the bay had decreased by about 1.0 m between 1927 and 1976. The shoaling was most pronounced in the southeastern sector of the bay, where the decrease in depth amounted to 1.6 m. According to Hollett, only about 27% of the bay's sediments were terrigenous, the remainder consisting of reef carbonate detritus (63%) and dredge spoils (11%). Hollett concluded that shoaling rates in the bay had increased substantially since 1927 and attributed the increase to higher stream sediment loads caused by urbanization and to extensive dredging and disposal activities. Based on water samples taken from Kamooalii Stream (the major stream contributing runoff to the southeastern sector) during runoff from a storm of about 6–7 cm of rainfall, Fan (1973) calculated that the stream discharged about 8,250 tonnes of sediment into the bay. The observed accumulation of terrigenous sediments on the bottom of Kaneohe Bay can easily be accounted for if sediment loadings of the magnitude calculated by Fan occur approximately eight or nine times per year. Given the present average depth of the bay (6 m) and the rate of sediment accumulation estimated by Hollett (1977) from his bathymetric studies, the bay will completely fill up with sediments in about another 300 years.

Sewage Disposal

The first sewage treatment plant built in the Kaneohe Bay watershed was constructed to serve the Kaneohe Marine Corps Air Station (KMCAS) on Mokapu Peninsula. The outfall from this plant was located near one corner of the southeastern sector of the bay, as indicated in Figure 4.27. Originally constructed as a primary treatment facility, the KMCAS plant was upgraded to secondary treatment in 1972. As of 1972, it was discharging about 4.1×10^3 m^3 of sewage per day into the bay. During dry weather, about 25% of this discharge was diverted to water the military golf course (Banner, 1974).

Until 1963, sewage from the nonmilitary population of the watershed was handled by private cesspools or septic tanks. In 1963, however, the Kaneohe municipal sewage treatment plant, a secondary treartment facility serving the population of Kaneohe Town, was put into operation. Its outfall, also located in the southeastern sector of the bay, is indicated in Figure 4.27. As of 1972, the Kaneohe municipal plant was discharging about 11.5×10^3 m^3 per day of treated sewage at this outfall. In 1970 a third sewage treatment plant serving a housing development in the Kahaluu Valley began discharging effluent into Ahuimanu Stream, which flows into the northwestern sector of the bay. The plant was designed as a tertiary treatment facility. As of 1972, it was discharging about 0.5×10^3 m^3 per day of effluent into Ahuimanu Stream. The remainder of the watershed is still served by cesspools.

As was the case with Lake Washington, the principal concern over sewage disposal into Kaneohe Bay centered on the nutrient enrichment problem. According to figures given by Sunn et al. (1975), about 74% of the N and 75% of the P that entered the bay came directly from the Kaneohe municipal and KMCAS sewage treatment plants, with the remainder attributed to stream runoff. The nutrient enrichment problem was aggravated by the fact that the sewage was discharged into the most stagnant part of the bay, so that mixing with low-nutrient open-ocean water was minimized, and the residence time of the discharged water was maximal. From a pollution standpoint, the southeastern sector was the worst part of the bay to choose for sewage disposal. The decisions to locate the outfalls in the southeastern sector were clearly made on the basis of convenience and expense rather than on the basis of ecological considerations.

Effects of Sewage Disposal

The sewage discharges into Kaneohe Bay affected the coral reef community in basically three ways. The first effect resulted from a reduction in water clarity caused by the increased abundance of phytoplankton, particularly in the southeastern sector. Hermatypic corals such as those found in Kaneohe Bay contain symbiotic algae, and these algae require light to carry out photosynthesis. The algal symbionts benefit the corals by producing and excreting organic compounds used by the corals as a major source of food. Furthermore, removal of the symbiotic algae has been shown to reduce coral calcification rates markedly (Shepard, 1973). A series of coral transplant experiments performed by Maragos (1972) showed that when healthy corals were transplanted to the southeastern sector of the bay they invariably died, and even when still alive, they grew slowly and erratically. The failure of these transplants appears to have been caused largely by the low water clarity in the southeastern sector of the bay.

The second effect of the sewage discharges was to create conditions favorable to the growth of filter-feeding organisms such as sponges, barnacles, tunicates, and zoanthids, which then overgrew many of the former coral reefs. Fig. 4.28 illustrates the food chain leading to these filter-feeding organisms. Had the principal signal perceived by the natural coral reef community been an elevation of inorganic nutrient concentrations, the effect on the reefs would have been a stimulation of photosynthesis (Kinsey and Domm 1974), as indicated in Fig. 4.28. However, because the effluent was not released directly onto the reefs, much of the nutrient content of the wastewater was assimilated by phytoplankton. Although nutrient concentrations in the water flowing over the reefs were elevated by the sewage discharges, the principal signal perceived by the reefs was an increase in the concentration of plankton rather than of nutrients. Hermatypic corals are also filter feeders, but they appear to derive most of their nutrition from their algal symbionts rather than from plankton. As a result, dense populations of plankton create a condition more favorable to the growth of noncoral filter feeders than to corals. Metabolic studies performed on a fringing reef in the southeastern sector of Kaneohe Bay reflect the takeover of the reef by organisms other than corals. These results are summarized in Table 4.5. Photosynthesis and calcification on the southeastern sector reef were both about 25% less than the rates characteristic of a typical coral reef, a result reflecting the reduced amount of light available to the reef community and the concomitant reduction in the activity of calcareous algae, which are among the principal calcifiers on healthy coral reefs. Respiration, on

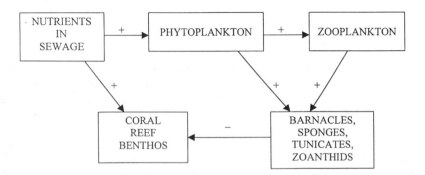

FIGURE 4.28   Effect of nutrients in sewage on production of organic matter by organisms in Kaneohe Bay. Positive effects are indicated by a +; negative effects, by a −. The direct effect of nutrients on production of organic matter by the coral reef benthic community is positive, but indirect effects, transmitted through the food chain, are negative.

TABLE 4.5 METABOLIC CHARACTERISTICS OF REEF FLAT
BENTHIC COMMUNITIES

	Production (P) $(g\ C\ m^{-2}\ d^{-1})$	Respiration (R) $(g\ C\ m^{-2}\ d^{-1})$	P/R	Calcification $(kg\ CaCO_3\ m^{-2}\ y^{-1})$
Typical coral reef	6.9	6.8	1.0	4.3
SE Kaneohe Bay reef	5.2	12.7	0.4	3.1

the other hand, was about 85% higher than that of a typical coral reef, a consequence of the greater abundance of plankton in the water column and hence the greater supply of food for benthic filter feeders. The more than twofold imbalance between photosynthesis and respiration is very uncharacteristic of a healthy coral reef. The behavior of the southeastern sector's fringing reef during the period of sewage enrichment contrasts sharply with the metabolic behavior of a reef enriched with inorganic nutrients. In the study of Kinsey and Domm (1974), for example, enrichment of an experimental reef with inorganic nutrients produced no change in respiration rates but caused a 27% increase in photosynthetic rates.

The third adverse effect of sewage discharges on the coral reefs was the stimulation of the alga *Dictyosphaeria cavernosa*, commonly known as *bubble algae*. This alga may establish itself within a coral head at the base of a frond and then grow outward, eventually enveloping the coral head and killing the coral. In the northwestern sector of the bay, bubble algae were grazed by certain fishes and other organisms (Banner, 1974) and consequently were prevented from enveloping coral heads. However, in the middle sector of the bay, the low abundance of *D. cavernosa* grazers allowed it to envelop and kill numerous coral heads. Maragos (1972) estimated that about 24% of the corals in the lagoon portion of the bay had been killed by this mechanism. A study reported by Smith et al. (1981) showed that bubble algae grew poorly, if at all, in offshore water unless the water was artificially enriched with nutrients. Thus, it seems likely that the proliferation of bubble algae in Kaneohe Bay resulted largely from the elevated nutrient concentrations in the water caused by sewage enrichment.

Response to Sewage Diversion

Between December 1977 and June 1978, both the Kaneohe Municipal and KMCAS sewer outfalls were diverted to a new outfall seaward of Mokapu Peninsula. Since the residence time of water in the bay is only about two weeks, it was expected that characteristics of the water column would rapidly respond to this reduction in nutrient inputs. Figure 4.29 summarizes the mean water column parameters in the southeastern sector and the remainder of the bay before and after the sewage diversions.

In the first year following sewage diversion, phytoplankton populations, as measured by chlorophyll *a* concentrations, decreased by 56% in the southeastern sector, and water clarity, as measured by Secchi depth, increased by 22%. In the remainder of the bay, chlorophyll *a* concentrations declined by 36%, and Secchi depths increased by 8%. Inorganic N and P concentrations declined by factors of 2–3 in all parts of the bay. The effect of the reduced plankton concentrations on the populations of benthic filter feeders was apparent within a matter of weeks, as colonies of these organisms began to disappear rapidly from previously colonized areas. Brock and Smith (1983) documented the decrease in these organisms by sampling at approximately 60-day intervals from June 1976 to August 1979.

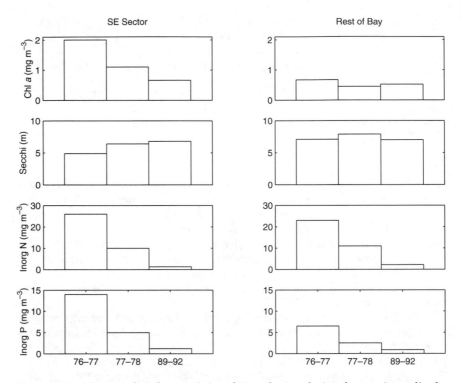

FIGURE 4.29   Water quality characteristics of Kaneohe Bay during the year immediately preceding sewage diversion (1976–1977), the year immediately following sewage diversion (1977–1978), and approximately 15 years following sewage diversion (1989–1992). *Source*: Data taken from Laws and Redalje (1982) and Laws and Allen (1996).

They discovered that the hard substratum benthic communities near the municipal sewer outfall were dominated by filter- and suspension-feeding organisms. Following the sewage diversion, the abundance of these organisms was 2.5–10 times lower on a fringing reef in the southeastern sector of the bay.

Although inorganic nutrient concentrations declined dramatically in the first year following sewage diversion, further decreases occurred in subsequent years. Within 10–15 years, the inorganic nutrient concentrations bordered on the limit of detection by traditional colorimetric methods (Figure 4.29). Laws and Allen (1996) speculate that these subsequent decreases reflected a gradual drawdown of the nutrients that accumulated in the bay's sediments during the period of sewage enrichment. Prior to the diversion of sewage, there was a rapid exchange of nutrients between the sediments and the water column, with a slight imbalance in favor of nutrient accumulation in the sediments (Smith et al., 1981). When the concentrations of particulate material and nutrients in the water column dropped by a factor of 2–3 shortly after the sewage diversion, a large imbalance developed between the influx and efflux of nutrients from the sediments, with the net flux being out of the sediments. This input of nutrients from the sediments persisted until the sediment nutrient reservoir was depleted and an equilibrium was reestablished between the flux of nutrients into and out of the sediments (Figure 4.30).

The increase in water clarity associated with the reduction in plankton created conditions more favorable to the growth of corals in all parts of the bay, and the decline of

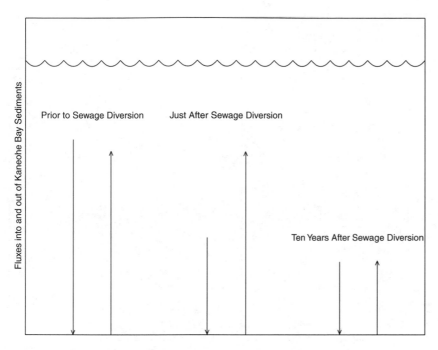

FIGURE 4.30   Fluxes of nutrients into and out of Kaneohe Bay sediments at various times before and after diversion of sewage discharges from the bay.

inorganic nutrient concentrations by a factor of 2–3 created conditions less favorable to the growth of bubble algae. The response of the corals to the sewage diversion was expected to be much slower than that of the plankton community because corals grow slowly even under optimal conditions (Shepard, 1973). However, studies summarized by Hunter and Evans (1995) revealed that the response of the corals was quite dramatic in the six years following the sewage diversion. The results are shown in Figure 4.31. On the average, live coral coverage in the lagoon almost doubled between 1971 and 1983. There was relatively little additional change in coral coverage between 1983 and 1990. Coverage by *D. cavernosa* declined by a factor of 4 between 1971 and 1983, but by 1990, *D. cavernosa* abundance at depths of 3 m or less was comparable to the values reported in 1971. The change in *D. cavernosa* coverage between 1983 and 1990 was unrelated to any obvious source of nutrients. Discharges from the Ahuimanu sewage treatment plant were diverted to the Mokapu Point ocean outfall in 1986, and the areas with the heaviest *D. cavernosa* coverage in 1990 were offshore regions not subject to the immediate impact of land runoff. A likely explanation for the increase in *D. cavernosa* between 1983 and 1990 is reduction in grazing pressure by herbivorous fish due to overfishing.

Prospects for Kaneohe Bay

The condition of the coral reefs in the central part of Kaneohe Bay only six years after the diversion of the sewer outfalls from the southeastern sector was a very encouraging discovery. Because corals grow slowly, the complete recovery process was expected to take 10–20 years. In the southeastern sector, which was most heavily impacted by the sewage

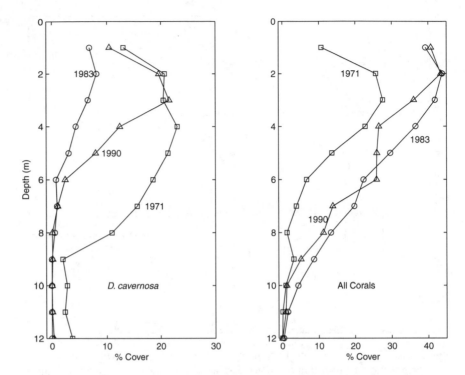

FIGURE 4.31   Percentage of cover of bottom by *Dictyosphaeria cavernosa* and corals along experimental transects in Kaneohe Bay. *Source:* Redrawn from Hunter and Evans (1995), p. 507. Reproduced with permission.

discharges, coral coverage at monitoring stations increased from virtually zero in 1971 to approximately 10% in 1983 and to 20% in 1990. *D. cavernosa*, on the other hand, has never become established in the southeastern sector. The fact that coral coverage in the other parts of the bay did not continue to increase between 1983 and 1990 may be related to the increase in *D. cavernosa* outside of the southeastern sector.

The resurgence of corals but not of *D. cavernosa* in the southeastern sector is a curious phenomenon. The amount of sediment that has washed into this sector has left in many areas a thick deposit of mud on the bottom. If these sediments are in the least stirred up, visibility near the bottom is reduced to virtually zero, so that "In diving, one would literally swim into the bottom" (Banner, 1974, p. 696). Coral larvae require a hard substrate for attachment, and such muddy substrates are therefore very unlikely places for corals to become established. The fact that coral coverage has increased dramatically in the southeastern sector indicates that despite the prevalence of muddy sediments, there are some hard substrates available for attachment. Why bubble algae have never become established in the southeastern sector remains something of a mystery but may be related to the physical circulation regime. Water movement in the southeastern sector is somewhat sluggish, and Hunter and Evans (1995) reported that the greatest coverage by bubble algae in the rest of the bay occurred near channel mouths where water movement was relatively brisk. The flow of water over and around benthic algae can greatly increase the efficiency of nutrient delivery to the plants and hence stimulate their growth.

With respect to the issue of nutrient limitation, it is noteworthy that stream runoff can introduce large amounts of nutrients as well as sediments into a body of water, and after heavy rains the perturbation caused by these nutrient inputs can be quite substantial. For example, from November 1987 to January 1988, rainfall in the southeastern sector watershed averaged 40 cm per month. The nutrients delivered to the bay in the runoff associated with these heavy rains stimulated a phytoplankton bloom that produced chlorophyll *a* concentrations in excess of 40 mg m^{-3} in the southeastern sector, almost four times the highest concentration measured during the year immediately preceding the sewage diversion (Taguchi and Laws, 1989). The amount of water and nutrients delivered to the southeastern sector as a result of these heavy rains would undoubtedly have been much less if the watershed were undeveloped rather than heavily urbanized. If urbanization were allowed to spread from Kaneohe northwestward through the watershed, it is quite probable that the fluxes of water, sediment, and nutrients to the central and northwestern sectors of the bay would increase. Such developments would undoubtedly have an adverse impact on the coral reefs.

The principal use of Kaneohe Bay is for recreation, primarily sailing and water skiing in the lagoon. For many years the local skipjack tuna fishery used the bay as a source of nehu, a small planktivorous anchovy used as baitfish. The diversion of sewage from the bay has been associated with a reduction in the abundance of plankton and hence of nehu. Public health problems associated with sewage disposal were noted in the past. At least one aquaculture operation was forced to move from the bay due to contamination of its shellfish with sewage pathogens, and water skiers and swimmers have suffered from time to time from infections that may have been caused by sewage-introduced organisms. Thus, diversion of sewage has been beneficial from the standpoint of some but not all important uses of Kaneohe Bay.

In several ways, the prospects for the coral reefs in Kaneohe Bay are good. Diversion of the sewage discharges from the bay has clearly been beneficial. Furthermore, the local government is now effectively enforcing land-clearing ordinances. Urbanization of the northern portion of the bay's watershed has been effectively blocked. The land will continue to be used for small-scale agricultural purposes.

At the present time, overfishing appears to be the principal threat to the coral reefs in Kaneohe Bay. Parrotfishes and surgeonfishes are the herbivorous fishes believed to provide the principal grazing control over *D. cavernosa*. As noted by Stimson et al. (1996), there is evidence that fishing pressure has significantly reduced the population of parrotfishes. The increase of *D. cavernosa* on the bay's barrier reef is most likely the result of a reduction in grazing pressure. Since the flow of water over the barrier reef comes entirely from the offshore ocean, any impact from land runoff is minimal. Regulation of recreational, commercial, and subsistence fishing is unfortunately confounded by a variety of social and political pressures. The long-term prospects for the coral reefs will probably be determined largely by their perceived value to the State of Hawaii. To the state government, the value of the reefs as a recreational resource and tourist attraction will be the most compelling reasons to protect them. Careful planning and control of activities in the bay will be necessary to provide that protection.

REFERENCES

Adler, C. A. 1973. *Ecological Fantasies*. Green Eagle Press, New York. 350 pp.

Agassiz, A. 1889. The coral reefs of the Hawaiian Islands. *Bull. Mus. Comp. Zool. Harvard College*, **17**(3), 121–170.

Applegate, V. C., and J. W. Moffett. 1955. The sea lamprey. *Sci. Amer.*, **192**(4), 36–41.

ATKINSON, M. J., B. CARLSON, and G. L. CROW. 1995. Coral growth in high-nutrient, low-pH seawater: A case study of corals cultured at the Waikiki Aquarium, Honolulu, Hawaii. *Coral Reefs*, **14**, 215–223.

BALDWIN, N. S., R. W. SAALFELD, M. A. ROSS, and H. J. BUETTNER. 1979. *Commercial Fish Production in the Great Lakes 1867–1977.* Great Lakes Fishery Commission Tech. Rept. No. 3. Ann Arbor, Mich. 187 pp.

BANNER, A. H. 1968. *A Fresh-Water "Kill" on the Coral Reefs of Hawaii.* Hawaii Inst. Mar. Biol. Tech. Rept. No. 15. 29 pp.

BANNER, A. H. 1974. Kaneohe Bay, Hawaii: Urban pollution and a coral reef ecosystem. *Proceedings of the Second International Coral Reef Symposium.* pp. 685–702. Great Barrier Reef Committee, Brisbane.

BANNER, A. H., and J. H. BAILEY. 1970. The Effects of Urban Pollution Upon a Coral Reef System. Hawaii Inst. Mar. Biol. Tech. Rept. No. 25. 66 pp.

BATHAN, K. H. 1968. *A Descriptive Study of the Physical Oceanography of Kaneohe Bay, Oahu, Hawaii.* Hawaii Inst. Mar. Biol. Tech. Rept. No. 14. 353 pp.

BEETON, A. M. 1965. Eutrophication of the St. Lawrence Great Lakes. *Limnol. Oceanogr.*, **10**, 240–254.

BEETON, A. M. 1969. Changes in the environment and biota of the Great Lakes. In *Eutrophication: Causes, Consequences, Correctives.* National Academy of Sciences, Washington, D.C. Pp 150–187.

BROCK, R. E., and S. V. SMITH. 1983. Response of coral reef cryptofaunal communities to food and space. *Coral Reefs*, **1**, 179–183.

CHARLTON, M. N., R. LE SAGE, and J. E. MILNE. 1998. *Lake Erie in Transition: The 1990s.* Contribution No. 98–241. National Water Research Institute, Burlington, Ontario, Canada.

CHAVE, E. H., and J. E. MARAGOS. 1973. A historical sketch of the Kaneohe Bay region. In *Atlas of Kaneohe Bay: A Reef Ecosystem under Stress.* University of Hawaii Sea Grant. University of Hawaii Press, Honolulu. Pp. 9–14.

COMITA, G. W., and G. C. ANDERSON. 1959. The seasonal development of a population of *Diaptomus ashlandi* marsh, and related phytoplankton cycles in Lake Washington. *Limnol. Oceanogr.*, **4**, 37–52.

COOLEY, J. 1993. Is Lake Erie changing—again? *Focus*, **18**, 5–6.

COX, D. C., and P. F. FAN. 1973. The Kaneohe Area. In *Estuarine Pollution in the State of Hawaii.* Vol. 2. *Kaneohe Bay Study.* Univ. of Hawaii Water Resources Res. Center Tech. Rept. No. 31. Pp. 7–25.

DARWIN, C. 1842. *The Structure and Distribution of Coral Reefs.* Smith, Elder, London. 214 pp. (Reprinted in 1962 by Univ. of California Press, Berkeley-Los Angeles.)

DOLAN, D. M. 1993. Point source loadings of phosphorus to Lake Erie: 1986–1990. *J. Great Lakes Res.*, **19**, 212–223.

DOLAN, D. M., and K. P. MCGUNAGLE. 1998. The effect of program cuts on Lake Erie total phosphorus loading estimates in the 1990s. Paper presented at the 41st Annual Conference on Great Lakes Research, McMaster Univ., Hamilton, Ontario, May 18–22.

EDMONDSON, C. H. 1928. Ecology of a Hawaiian coral reef. *B. P. Bishop Mus. Bull.*, **45**, 1–64.

EDMONDSON, C. H. 1946. *Reef and Shore Fauna of Hawaii*, 2nd ed. B. P. Bishop Museum, Special Publication Honolulu 22. 381 pp.

EDMONDSON, W. T. 1956. Artificial eutrophication of Lake Washington. *Limnol. Oceanogr.*, **1**, 47–53.

EDMONDSON, W. T. 1961. Changes in Lake Washington following an increase in the nutrient income. *Verh. Internat. Verein. Limnol.*, **14**, 167–175.

EDMONDSON, W. T. 1966. Changes in the oxygen deficit of Lake Washington. *Verh. Internat. Verein. Limnol.*, **16**, 153–158.

EDMONDSON, W. T. 1969. Eutrophication in North America. In *Eutrophication: Causes, Consequences, Correctives.* National Academy of Sciences, Washington, D.C. Pp. 124–149.

EDMONDSON, W. T. 1970. Phosphorus, nitrogen, and algae in Lake Washington after diversion of sewage. *Science*, **169**, 690–691.

EDMONDSON, W. T. 1972. Nutrients and phytoplankton in Lake Washington. In G. E. Likens Ed., *Nutrients and Eutrophication.* American Society of Limnology and Oceanography, Lawrence, Kan. Pp. 172–193.

EDMONDSON, W. T. 1993. Eutrophication effects on the food chains of lakes. *Mem. Ist. Ital. Idrobiol.*, **52**, 113–132.

EDMONDSON, W. T. 1994. Sixty years of Lake Washington: A curriculum vitae. *Lake and Reservoir Management*, **10**, 75–84.

EDMONDSON, W. T. 1997. *Aphanizomenon* in Lake Washington. *Arch. Hydrobiol.*, Suppl., **107**, 409–446.

EDMONDSON, W. T., and J. T. LEHMAN. 1981. The effects of changes in the nutrient income on the condition of Lake Washington. *Limnol. Oceanogr.*, **26**, 1–29.

EDMONDSON, W. T., and A. H. LITT. 1982. *Daphnia* in Lake Washington. *Limnol. Oceanogr.*, **27**, 272–293.

EHRLICH, P. 1968. *The Population Bomb.* Ballantine, New York. 223 pp.

EPA. 1986. *Quality Criteria for Water.* EPA 440/5–86–001. U.S. Government Printing Office, Washington, D.C.

EVANS, C. W., J. E. MARAGOS, and P. F. HOLTHUS. 1986. Reef corals in Kaneohe Bay. Six years before and after termination of sewage discharges (Oahu, Hawaiian Archipelago). In P. J. Jokiel, R. H. Richmond, and R. A. Rogers, Eds., *Coral Reef Population Biology.* Hawaii Institute Mar. Biol. Tech. Rept. No. 37. Pp. 76–90.

FAN, P. F. 1973. Sedimentation. *In Estuarine Pollution in the State of Hawaii.*, Vol. 2. *Kaneohe Bay Study.* Univ. of Hawaii Water Resources Res. Center Tech. Rept. 31. Pp. 229–265.

FITZSIMMONS, J. D., J. H. LEACH, S. J. NEPSZY, and V. W. CAIRNS. 1995. Impacts of zebra mussel on walleye (*Stizostedion vitreum*) reproduction in western Lake Erie. *Can. J. Fish. Aquat. Sci.*, **52**, 578–586.

HARDIN, G. 1968. The tragedy of the commons. *Science*, **162**, 1243–1245.

HARTMAN, W. L. 1973. *Effects of Exploitation, Environmental Changes, and New Species of the Fish Habitats and Resources of Lake Erie.* Great Lakes Fishery Commission Tech. Rept. No. 22. Ann Arbor, Mich. 43 pp.

HERN, S. C., V. W. LAMBOU, L. R. WILLIAMS, and W. D. TAYLOR. 1981. *Modifications of Models Predicting Trophic State of Lakes: Adjustment of Models to Account for the Biological Manifestations of Nutrients.* U.S. Environmental Protection Agency. Project Summary. EPA-600/S3-81-001. U.S. Government Printing Office, Washington, D.C.

HOLLETT, K. J. 1977. Shoaling of Kaneohe Bay, Oahu, Hawaii, in the period 1927 to 1976, based on bathymetric, sedimentological, and geographical studies. M.S. Thesis. Univ. of Hawaii, Honolulu. 145 pp.

HUNTER, C. L., and C. W. EVANS. 1995. Coral reefs in Kaneohe Bay, Hawaii: Two centuries of western influence and two decades of data. *Bull. Mar. Sci.*, **57**, 501–515.

HUTCHINSON, G. E. 1957. *A Treatise on Limnology.* Vol. 1. Wiley, New York. 1015 pp. Infante, A., and W. T. Edmondson. 1985. Edible phytoplankton and herbivorous zooplankton in Lake WAshington. *Arch. Hydrobiol. Beih.*, **21**, 161–171.

INTERNATIONAL BOARD OF INQUIRY FOR THE GREAT LAKES FISHERIES. 1943. *Report and Supplement.* U.S. Government Printing Office, Washington, D.C. 213 pp.

INTERNATIONAL JOINT COMMISSION. 1970. *Pollution of Lake Erie, Lake Ontario and the international section of the St. Lawrence River.* Information Canada, Ottawa. 105 pp.

INTERNATIONAL JOINT COMMISSION. 1975. *Cladophora in the Great Lakes.* Great Lakes Research Advisory Board, Windsor, Ontario. 179 pp.

INTERNATIONAL JOINT COMMISSION. 1977. *Great Lakes Water Quality.* 1977 Annual Report. Great Lakes Water Quality Board, Windsor, Ontario. 89 pp.

INTERNATIONAL JOINT COMMISSION. 1981. *Report on Great Lakes Water Quality.* Appendix. Great Lakes Surveillance. Windsor, Ontario. 174 pp.

INTERNATIONAL JOINT COMMISSION. 1982. 1982 *Report on Great Lakes Water Quality.* Windsor, Ontario. 153 pp.

INTERNATIONAL JOINT COMMISSION. 1985. *A Review of Trends in Lake Erie Water Quality with Emphasis on the 1978–1979 Intensive Survey.* Report to the Surveillance Work Group. Windsor, Ontario. 129 pp.

INTERNATIONAL JOINT COMMISSION. 1987. *1987 Report on Great Lakes Water Quality.* Windsor, Ontario. 236 pp.

INTERNATIONAL JOINT COMMISSION. 1989. *1989 Report on Great Lakes Water Quality.* Windsor, Ontario. 128 pp.

KINSEY, D. W. 1979. Carbon turnover and accumulation by coral reefs. Ph.D. dissertation. Univ. of Hawaii. 248 pp.

KINSEY, D. W., and A. DOMM. 1974. Effects of fertilization on a coral reef enrironment—primary production studies. *Proc. Second Int. Coral Reef Symp.*, **1**, 49–66.

LAWS, E. A., and C. ALLEN. 1996. Water quality in a subtropical embayment more than a decade after diversion of sewage discharges. *Pac. Sci.*, **50**, 194–210.

LAWS, E. A., and T. T. BANNISTER. 1980. Nutrient- and light-limited growth of *Thalassiosira fluviatilis* in continuous culture, with implications for phytoplankton growth in the oceans. *Limnol. Oceanogr.*, 25, 457–473.

LAWS, E. A., and D. G. REDALJE. 1982. Sewage diversion effects on the water column of a tropical estuary. *Mar. Environ. Res.*, 6, 265–279.

LEACH, J. H. 1994. Fluctuations in zebra mussel populations and ecological impacts in western Lake Erie, 1988–1993. *37th Conference of the International Association for Great Lakes Research and Estuarine Research Federation: Program and Abstracts.* International Assoc. for Great Lakes Research, Buffalo, N.Y. P. 166.

LEHMAN, J. T. 1988. Hypolimnetic metabolism in Lake Washington: Relative effects of nutrient load and food web structure on lake productivity. *Limnol. Oceanogr.*, 33, 1334–1347.

MACKAYE, A. I. 1915. Corals of Kaneohe Bay. In *Hawaiian Almanac and Annual* for 1916. B. P. Bishop Mus., Honolulu. Pp. 135–139.

MADENJIAN, C. P. 1995. Removal of algae by the zebra mussel (*Dreissena polymorpha*) population in western Lake Erie: A bioenergetics approach. *Can. J. Fish. Aquat. Sci.*, 52, 381–390.

MARAGOS, J. E. 1972. A study of the ecology of Hawaiian reef corals. Ph.D. dissertation. Univ. of Hawaii. 290 pp.

PREISENDORFER, R. W. 1986. Secchi disk science: Visual optics of natural waters. *Limnol. ceanogr.*, 31, 909–926

OHIO LAKE ERIE COMMISSION. 1998. *State of the Lake Report: Lake Erie Quality Index.* Toledo. 88 pp.

REGIER, H. A., and W. L. HARTMAN. 1973. Lake Erie's fish community: 150 years of cultural stress. *Science*, 180, 1248–1255.

ROBERTS, L. 1990. Zebra mussel invasion threatens U.S. waters. *Science*, 249, 1370–1372.

ROY, K. L. 1970. *Changes in Bathymetric Configuration, Kaneohe Bay, Oahu. 1882–1969.* Univ. of Hawaii, Hawaii Inst. of Geophysics Rept. No. 70-15. 226 pp.

SCHEFFER, V. B., and R. J. ROBINSON. 1939. A limnological study of Lake Washington. *Ecol. Monogr.*, 9, 95–143.

SCHLOESSER, D. W., and T. F. NALEPA. 1994. Dramatic decline of unionid bivalves in offshore waters of western Lake Erie after infestation by the zebra mussel, *Dreissena polymorpha. Can. J. Fish. Aquat. Sci.*, 51, 2234–2242.

SECCHI, A. 1866. Relazione della Esperienze Fatta a Bordo della Pontificia Pirocorvetta *L'Immacolata Concezione* per Determinare La transparenza del Mare [Reports on Experiments Made on Board the Papal Steam Sloop *L'Immacolata Concezione* to Determine the Transparency of the Sea]. *From* Cmdr. A. Cialdi, *Sul moto ondoso del mare e su le correnti di esso specialment auquelle littorali*, 2nd ed. Pp. 258–288. [Dept. of the Navy, Office of Chief of Naval Operations, ONI Transl. A-655, Op-923 M4B.]

SHEAR, H. 1984. Contaminants research and surveillance—a biological approach. In J. O. Nriagu, and M. S. Simmons, Eds., *Toxic Contaminants in the Great Lakes.* Wiley-Interscience, New York. Pp. 31–51.

SHEPARD, F. P. 1973. *Submarine Geology.* 3rd ed. Harper & Row, New York. 51 pp.

SMITH, S. V., W. J. KIMMERER, E. A. LAWS, R. E. BROCK, and T. W. WALSH. 1981. Kaneohe Bay sewage diversion experiment: Perspectives on ecosystem responses to nutrient perturbation. *Pac. Sci.*, 35, 279–395.

SONZOGNI, W. C., and W. R. SWAIN. 1984. Perspectives on human health concerns from Great Lakes contaminants. In J. O. Nriagu, and M. S. Simmons, Eds., *Toxic Contaminants in the Great Lakes.* Wiley-Interscience, New York. Pp. 1–29.

STIMSON, J., S. LARNED, and K. McDERMID. 1996. Seasonal growth of the coral reef macroalga *Dictyosphaeria cavernosa* (Forskål) Børgesen and the effects of nutrient availability, temperature and herbivory on growth rate. *J. Exp. Mar. Biol. Ecol.*, 196, 53–77.

SUNN, LOW, TOM, and HARA, INC. 1975. *Draft Kaneohe Bay Data Evaluation Report.* Honolulu.

TAGUCHI, S., and E. A. LAWS. 1989. Biomass and compositional characteristics of Kaneohe Bay phytoplankton inferred from regression analysis. *Pac. Sci.*, 43, 316–331.

VOLLENWEIDER, R. A. 1968. *Water Management Research. Scientific Fundamentals of the Eutrophication of Lakes and Flowing Waters with Particular Reference to Nitrogen and Phosphorus as Factors in Eutrophication.* Org. for Econ. Cooperation and Development Tech. Rept. DAS/CSI/68.27, 159 pp. Revised 1971. Paris.

VOLLENWEIDER, R. A. 1969. Moglichkeiten und Grenzen elementarer modelle der Stoffbilanz von Seen. *Arch. Hydrobiol.*, **66**, 1–36.

VOLLENWEIDER, R. A. 1975. Input-output models. *Schweiz. Z. Hydrologie*, **37**, 53–84.

VOLLENWEIDER, R. A. 1976. Advances in defining critical loading levels for phosphorus in lake eutrophication. *Mem. Ist. Ital. Idrobiol.*, **33**, 53–83.

WAGNER, R. H. 1974. *Environment and Man.* Norton, New York. 528 pp.

WEIBEL, S. R. 1969. Urban drainage as a factor in eutrophication. In *Eutrophication: Causes, Consequences, Correctives.* National Academy of Sciences, Washington, D.C. Pp. 383–403.

WELCH, E. B. 1985. The eventual recovery of Lake Sammamish following phosphorus diversion. *J. Water Pollut. Control. Fed.*, **57**, 977–978.

WELCH, E. B., C. A. ROCK, R. C. HOWE, and M. A. PERKINS. 1980. Lake Sammamish response to wastewater diversion and increasing urban runoff. *Water Res.*, **14**, 821–828.

WELCH, E. B., D. E. SPYRIDAKIS, J. I. SHUSTER, and R. R. HORNER. 1986. Declining lake sediment phosphorus release and oxygen deficit following wastewater diversion. *J. Water Pollut. Control. Fed.*, **58**, 92–96.

QUESTIONS

4.1. Studies of Lake Washington have show that the lake was P-limited during the 1930s and 1970s. However, during the 1960s, when treated sewage was being discharged into Lake Washington, the lake was N-limited. What is the explanation for the transition from P to N limitation and back to P limitation?

4.2. During the 1960s, some persons argued that the production of algae in lakes such as Lake Erie was limited by the availability of CO_2. To test this hypothesis, Canadian scientists fertilized an experimental lake with N and P. Within a few weeks, a massive algal bloom developed in the lake. The lake had a naturally very low concentration of inorganic C, and the C associated with the algal bloom was far more than could be accounted for by the inorganic C in the lake. How were the algae able to obtain enough CO_2 to produce such a huge algal bloom?

4.3. Seasonal anoxia was a serious problem in the central basin of Lake Erie for many years. How would you account for the fact that seasonal anoxia was never a problem in Lake Washington, although the P concentration in Lake Washington just prior to diversion of sewage discharges from the lake in the 1960s was about four times higher than the P concentration in Lake Erie at the peak of eutrophication?

4.4. Why do scientists think that the development of late summer anoxia in the central basin of Lake Erie had a greater impact on cold-water species of fish such as lake trout and lake herring than on warm-water species such as carp?

4.5. Why did the development of late summer anoxic conditions in the hypolimnion of the central basin of Lake Erie frustrate attempts to reduce nutrient loading to the lake?

4.6. During the time that sewage was being discharged into Kaneohe Bay, photosynthetic rates of benthic communities on perimeter zones of coral reefs in the vicinity of the discharges were depressed, and their respiration rates were much higher than normal. How would you account for these observations?

4.7. The graph on page 116 shows concentrations of nitrate and phosphate from late winter until early summer in a lake. The straight line is a linear regression fit to the data. The concentrations of both nutrients are high at the end of the winter and low by the beginning of summer. The drawdown of both nutrients is due to uptake by phytoplankton. Based on the pattern of nitrate and phosphate concentrations, which nutrient would you judge to be limiting phytoplankton biomass in the lake? Explain your reasoning.

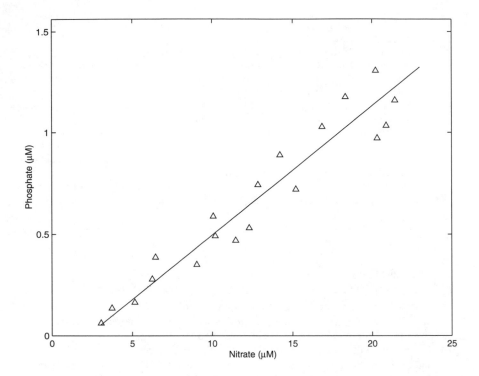

4.8. Between 1983 and 1990 there was an increase in the abundance of bubble algae on the Kaneohe Bay barrier reef. Why is it unlikely that this increase was caused by nutrients delivered by land runoff? What is a more likely cause of the increase in bubble algae?

4.9. The deterioration of the coral reefs in Kaneohe Bay during the 1960s has often been attributed to the nutrients introduced into the bay by sewer outfalls. Nevertheless, Atkinson et al. (1995) have reported that numerous species of corals maintained in an artificial reef system at the Waikiki Aquarium in Honolulu flourish in seawater containing higher concentrations of inorganic N and P than the concentrations reported in Kaneohe Bay during the peak of eutrophication. In the Waikiki Aquarium, seawater is pumped from a nearby seawater well and flows through the coral reef exhibit with a residence time of a few hours. How would you account for the different response of the Kaneohe Bay and Waikiki Aquarium coral reefs to nutrient loading?

4.10. Briefly explain why the central basin of Lake Erie is more susceptible to seasonal O_2 depletion than either the eastern or western basin.

Chapter Five

✧ ⌑ ✧ ⌑ ✧

Nonpoint Source Pollution

NONPOINT SOURCE POLLUTION is a term for a variety of water pollution problems caused by land runoff. The problems are associated with allochthonous inputs of sediment, nutrients, O_2-consuming wastes, pathogens, and toxic substances such as pesticides and heavy metals. The term *nonpoint source* indicates that the input does not occur at the end of a pipe or an artificial wastewater outfall. Instead, the input occurs in a more or less continuous manner along the shoreline and/or has a diffuse source. Some of the problems associated with nonpoint source pollution became apparent in our discussions of Lake Erie and Kaneohe Bay in Chapter 4. Peterson et al. (1985) have characterized nonpoint source pollution problems as among the most pervasive, persistent, and diverse water quality problems facing the United States.

Two important types of nonpoint source pollution are runoff from urban and agricultural lands. In both cases, the most significant water quality problem is frequently the introduction of sediment due to soil erosion. Following sedimentation, cultural eutrophication caused by allochthonous nutrient inputs is the most widespread nonpoint source pollution problem. Both soil erosion and eutrophication are natural phenomena and become water quality problems only when human activities accelerate them in an undesirable way. Pesticides are a problem associated more with runoff from agricultural land than with urban runoff. Problems caused by pathogens, heavy metals, oil, and O_2-consuming wastes are more frequently associated with urban runoff.

DEFINITIONS

Before going any further in our discussion of nonpoint source pollution, it is best to define a few terms that invariably arise in a discussion of land drainage problems. Land runoff may contain high concentrations of *sediments*. Some of these sediments may consist of relatively large particles such as pebbles and gravel that are too heavy to be suspended in the water column under normal flow conditions. Instead, these larger particles, along with some of the smaller particles such as sand and silt, are transported along the bottom of the stream or storm sewer. The flux of particles along a streambed is referred to as the stream's *bed load*. Typically, 20–40% of the sediment transported by a stream is bed load (Dunne and Leopold, 1978). The remainder consists of small particles such as sand, silt, and clay, many of which remain suspended in the water column rather than being transported along the streambed. The stream's suspended load is determined from the concentration of *suspended solids* (*SS*) in the water by simply filtering a known volume of the water and weighing the filter before and after filtering. The difference in weight is the weight of SS in the given volume of water. Normally, most of the material retained on the filter consists of inorganic sand, silt, and clay particles, although it is possible that in some cases significant amounts of biological material such as leaf fragments may also contribute. The total *sediment load* transported by a stream or storm sewer is defined as the sum of the bed load and the suspended load.

What about the O_2-consuming wastes in the runoff? How does one measure the O_2-consuming potential of a sample of water? There are two commonly used metrics, one called the *biochemical oxygen demand* (*BOD*) and the other called the *chemical oxygen demand* (*COD*). BOD is defined as the amount of O_2 consumed in a 300-ml sample of water incubated in a stoppered bottle in the dark at 20°C for five days. Since the sample is incubated in the dark, there is no possibility for photosynthesis to occur, so the O_2 concentration must either remain constant or decline. A decrease in O_2 concentration may be due to the respiratory activity of organisms and/or the simple chemical oxidation of substances that are unstable in the presence of oxygen, such as ferrous Fe or ammonia. Since both biological and strictly chemical processes may cause the decline in O_2 concentration, BOD should be understood to refer to biochemical oxygen demand rather than simply biological oxygen demand. The COD is the amount of O_2 consumed when the substances in water are oxidized by a strong chemical oxidant. The COD is measured by refluxing the water sample in a mixture of chromic and sulfuric acid for a period of two hours. This oxidation procedure almost invariably results in a larger consumption of O_2 than the standard BOD test since many organic substances that, because of their refractory nature, are not immediately available as food to aquatic microbes (e.g., cellulose), are readily oxidized by a boiling mixture of chromic and sulfuric acids. On the other hand, some compounds such as acetic acid that are utilized by aquatic microbes are not efficiently oxidized in the standard COD test. Because of the variability in the organic composition of different wastewaters, the ratio of COD to BOD may vary widely from one type of wastewater to another. Obviously, both BOD and COD are arbitrary measures of O_2-consuming potential. The value of the tests lies in the fact that, from experience, one can usually estimate the impact of a particular wastewater on the O_2 content of the receiving water from empirical correlations between either BOD or COD and the impact of the wastewater.

COMPOSITION OF LAND RUNOFF

With this introduction, let's turn our attention to Table 5.1, which lists concentrations of SS, BOD, COD, N, and P in urban runoff, agricultural runoff, raw sewage, and rainfall. The numbers for urban runoff and rainfall are taken from studies conducted in Cincin-

TABLE 5.1 AVERAGE CONCENTRATIONS OF VARIOUS
COMPONENTS OF URBAN RUNOFF, RAW SEWAGE, AND
RAINFALL (in mg L^{-1})

	Urban Runoff	Raw Sewage	Rainfall
SS	227	200	13
COD	111	350	16
BOD	17	200	No data
Total N	3.1	40	1.3
Total phosphate-P	0.36	10	0.08

Source: Weibel (1969).

nati, Ohio, with the urban runoff numbers coming from a residential–light commercial area (Weibel, 1969). The numbers are therefore not to be considered a national or world-wide norm because the characteristics of both urban runoff and rainfall can vary greatly from one time and place to another. For example, Abernathy (1981) has reported that urban runoff may contain COD levels as high as 1,000 mg L^{-1}, SS up to 2,000 mg L^{-1}, and total P as high as 15 mg L^{-1}. However, it is probably fair to say that the order of magnitude of the urban runoff and rainfall numbers in Table 5.1 is representative of many areas. The characteristics of domestic raw sewage are much less variable than those of urban runoff and rainfall, and the numbers for raw sewage are typical of the raw sewage entering many sewage treatment plants in the United States.

Although urban runoff obviously contains water derived from rainfall, it is clear from Table 5.1 that the concentration of most pollutants in urban runoff is much higher than that in rainfall. The implication is either that rainfall is a minor source of the pollutants or that the urban watershed retains rainwater more efficiently than the pollutants present in the rainwater. The latter explanation seems to account in part for the elevated concentrations of COD and nutrients in urban runoff, but there is no doubt that the watershed itself is a significant source of SS and, at least in commercial areas, of COD and nutrients as well (Randall et al., 1981).

Federal law requires that most sewage treatment plants remove at least 85% of the BOD and SS from raw sewage and keep the BOD and SS of the plant effluent below 30 mg L^{-1}. Given this information, Table 5.1 implies that urban runoff may contain BOD and SS at concentrations comparable to or much higher than the concentrations in treated sewage. The concentrations of N and P in urban runoff are usually much lower than those found in either raw sewage or treated sewage, but they are still several times higher than the N and P concentrations in rainfall. The contribution of the watershed to the SS, BOD, COD, N, and P in urban runoff reflects the results of soil erosion, the leaching of nutrients from exposed soils and detritus, and the transportation by the runoff of various accumulated wastes in the watershed during dry weather.

The daily discharge of sewage from a sewage treatment plant is a very predictable number, but land runoff, which is strongly influenced by rainfall, is irregular and highly unpredictable (weather reporters notwithstanding). Acknowledging that the concentrations of certain substances in urban runoff are of the same order of magnitude as the concentrations of the same substances in sewage, one may therefore ask whether urban runoff might not create pollution problems similar to those created by sewage. To answer this question, one must have some idea of the total flux of substances delivered to a body of water by urban runoff versus sewage. Thus, both concentrations and flow rates must be taken into account. Table 5.2 lists fluxes of SS, BOD, COD, N, and P delivered by raw sewage and urban runoff in the Cincinnati watershed studied by Weibel (1969) and estimated to

TABLE 5.2 FLUXES (kg ha^{-1} y^{-1}) OF VARIOUS CONSTITUENTS FROM URBAN RUNOFF,[a] AGRICULTURAL LAND,[b] AND FORESTED LAND[c] COMPARED TO RAW SEWAGE[d]

Land Use	SS	BOD	COD	Total N	Total Phosphate-P
Urban runoff	641	47	310	8.8	1.1
Cropland					
Cultivated	7,940	5–10	40–80	15.5	1.4
Noncultivated[e]	2,040	5–10	40–80	16	0.18
Pastureland[e]	2,270	10–15	40–80	20	0.15
Rangeland[f]	2,720	2–3	10–15	3	0.5
Forest	410	2–3	10–15	0.5–3.5	0.01–0.15
Raw sewage	683	683	1,194	137	35

[a] The numbers are taken from a residential-light commercial watershed in Cincinnati, Ohio. *Source:* Weibel (1969).
[b] The figures for SS are averages for the United States in the year 1992. *Source:* www.nhq.nrcs.usda.gov/NRI/tables/1992/table.
[c] *Source:* Cooper (1969), Likens and Borman (1974), Meyer et al. (1981), and Fraser et al. (1995).
[d] The population density of Cincinnati at the time of the study was about 2,500 persons per km^2.
[e] N and P fluxes taken from Pionke et al. (1996) and http://pswmrl.arsup.psu.edu/cris/nitrogencris.html.
[f] N and P fluxes estimated from data in Helgesen et al. (1995).

wash off various types of agricultural land and forested land in the United States. Recalling that secondary sewage treatment should remove at least 85% of the SS and BOD from raw sewage, Table 5.2 implies that urban runoff can contribute about half as much BOD and six to seven times as much SS to the receiving body of water as treated sewage. Most sewage treatment plants remove only about 30% or so of the N and P in raw sewage, and Table 5.2 therefore implies that, in general, treated sewage is a far more important source of nutrient loading than is urban runoff. Whether the nutrient loading from urban runoff is sufficient to create eutrophication problems depends on the characteristics of the receiving body of water. The winter phosphate-P concentrations in Lake Washington at the height of eutrophication were only about 0.06 mg L^{-1}. The phosphate-P concentrations in the urban runoff characterized by Weibel (1969) were 0.36 mg L^{-1}. Clearly, the nutrient levels in urban runoff can be more than enough to cause eutrophication problems if the runoff is not sufficiently diluted by the receiving body of water.

The loss of topsoil from agricultural land is considered to be a serious problem in the United States, and the data in Table 5.2 indicate why. The areal loss of topsoil from cultivated cropland is more than 10 times the flux of SS from urban runoff and raw sewage and roughly 20 times the flux of soil from forested land. Losses of topsoil from cultivated cropland declined by roughly 20% between 1982 and 1992 as a result of the implementation of practices such as no-till agriculture, but the losses are still enormous. If all the topsoil eroded each year from cropland in the United States were placed on a football field, it would form a pile about 200 miles high. About 2 billion tonnes of topsoil are eroded from cropland in the United States each year. Of that total, water erosion accounts for 55% and wind erosion for 45%. Five states—Texas, Minnesota, Iowa, Montana, and Kansas—account for 40% of the total erosion (www.nhq.nrcs.usda.gov/NRI).

The fluxes of nutrients from agricultural land are closely connected to fertilizer use. Fertilizer use in the United States has been increasing at the rate of about 1% per year for the last 20 years and currently stands at about 7.0 million tonnes of N and 1.1 million tonnes of P per year (www.fertilizer.org/stats.htm). Cultivated cropland in the United States is about 1.5×10^8 ha (www.nhq.nrcs.usda.gov/NRI). The flux figures for cultivated crop-land in Table 5.2 were calculated assuming that 35% of the N and 20% of the P applied as fertilizer to cultivated cropland appear in streams via either overland flow or ground-water seepage (M. Ver, pers. comm.). The areal fluxes so calculated are about 4–10 times higher than the figures for forested watersheds and 25–75% higher than the figures for urban runoff.

TYPES OF SEWER SYSTEMS

Given this state of affairs, one may logically ask, what should be done with land runoff? For many years the response of most communities in the United States was to do little or nothing. Until approximately 1980, flooding was considered to be the principal danger associated with land runoff, and systems intended to handle land runoff were designed to minimize the potential damage from flooding. Little or no attention was paid to the pollution potential of the runoff. However, as discharges from point sources of pollution were reduced during the 1970s, it became apparent that significant water pollution problems were being caused by nonpoint sources. Lake Erie is a case in point. Another example that may have caught the attention of the federal government is the Occoquan Reservoir in northern Virginia, which supplies drinking water for about 650,000 persons in the Virginia suburbs of Washington, D.C. Water quality in the reservoir began to deteriorate noticeably during the 1960s, and a program of point source pollution control in the watershed was initiated. Despite progress in reducing the point sources of pollution, water quality in the reservoir continued to decline, and extensive monitoring revealed that urban runoff was the cause of the problem. Urbanization of the watershed combined with the successful effort to reduce point source pollution had created a situation in which urban runoff was the primary source of the pollutants in the reservoir. This and similar incidents made it clear to public authorities that urban runoff could cause serious water quality problems. The extent of the problem was apparent in a 1983 EPA report to Congress, which summarized the results of 30 studies conducted under the auspices of its Nationwide Urban Runoff Program.

The crux of the problem was and still is the fact that in many communities land runoff is simply routed to the nearest convenient watercourse and discharged without treatment. The conduit for routing the water from streets and parking lots to the discharge point is called the *storm sewer*. In most cases, this system of pipes is separate and distinct from the sanitary sewer system, which routes domestic and other point source wastewater to the sewage treatment plant. According to the EPA (1983), about 46% of the U.S. population was served by separate storm and sanitary sewer systems in 1980. This figure is expected to increase to about 70% by the year 2000. Separate sewer systems obviously present no opportunity for treatment of land runoff. However, it is much more difficult to design a system to treat land runoff than sanitary sewage because during storms the former system must be able to handle a potentially enormous volume of runoff in a short time. The flux of sanitary sewage is more predictable and less variable, although leaks in the sanitary sewer system can lead to substantial increases in flow during storms due to infiltration of the sewer lines by groundwater.

The wastewater from approximately 20% of the U.S. population is handled by so-called *combined sewer* systems in which a single system of pipes transports both the land runoff and sanitary sewage. During dry weather flow, the sanitary sewage is routed to an interceptor point, where it is diverted to the sewage treatment plant rather than flowing to the stormwater outlet. The interceptor system is deliberately designed to divert 1.5–5 times the normal dry-weather sewage flow to the treatment plant. Hence, runoff from small storms is given the same treatment as the sanitary sewage. When the flow in the pipes exceeds the capacity of the interceptor system, however, the additional flow is simply routed to the stormwater outlet. Stormwater runoff can easily amount to 50–200 times the dry-weather flow of sanitary sewage (Weibel, 1969). Since the sanitary sewage and stormwater are mixed in the combined sewer system, almost all the sanitary sewage is discharged untreated with the stormwater during moderate to heavy rains. According to Weibel (1969), if one ignores the contribution of so-called stranded filth, the amount of raw sewage that escapes untreated during storms amounts to about 3% of the total annual flux of raw sewage. The advantages of a combined sewer system are lower cost and treatment of runoff of small storms. The major disadvantage is that a combined system allows raw sewage to escape when stormwater flow is greater than 1.5–5 times the dry-weather flow. The negative consequences of allowing 3% of the raw sewage to escape untreated are generally considered to outweigh the positive effects resulting from the treatment of runoff from small storms.

CORRECTIVES MEASURES

Land runoff has been viewed, not illogically, by some people as a misplaced resource. If runoff is intercepted from impervious surfaces such as rooftops and parking lots, it is potentially usable for a variety of purposes, including cooling, watering lawns and gardens, and recharging groundwater. There is no doubt that water is put to a great many uses that do not require it to be potable (drinkable). If the quality of runoff from a particular area is inconsistent with some uses, it is possible that only a moderate amount of treatment would be required to upgrade the quality of the runoff sufficiently. Such considerations have been little applied in the United States but may receive more attention in the future, particularly in areas where water conservation is critical (see Chapter 16). The following discussion summarizes a few approaches that have been proposed or implemented for dealing with land runoff.

Use of Settling Basins

A settling basin is simply a large basin through which runoff is routed. In some cases, the settling basin is constructed by merely deepening and greatly widening the channel of an existing stream near its mouth. In other cases, the settling basin is an excavated pit or trench located adjacent to a parking lot, shopping mall, housing development, or highway. In the past, such basins were designed primarily to reduce the intensity of runoff from storms, and they are referred to as *detention* or *retention basins*. If some or most of the water intercepted by these basins is percolated into the ground, the basins are referred to as *infiltration* or *recharge basins*. In all cases, the speed of the runoff water is reduced because of the enlarged cross-sectional area of the channel across the settling basin. Because of the reduced speed of the water as it crosses the basin, some of the sediments sink to the bottom before the water flows through the exit. Since many of the pollutants in urban runoff are associated with suspended solids, settling of the suspended solids will usually remove a significant portion of the BOD, nutrients, hydrocarbons, metals, and pesticides from the water. During dry-flow conditions, the accumulated sediments are periodically dredged out and trucked away to be used for landfill.

In practice, such settling basins sometimes have little impact on pollutant fluxes associated with land runoff. During dry-flow conditions, for example, nutrient concentrations in the water column of the settling basin may reach high levels due to recycling from accumulated sediment, and during even a small storm these nutrients can be rapidly flushed from the basin. Although detention basins may serve a useful purpose in reducing the intensity of storm runoff, the residence time of water in the basins during storms is often too short to allow a significant portion of the suspended solids to settle out. In fact, sediment accumulated on the bottom during dry-flow conditions is more than likely to be resuspended and overflow with the runoff water. Furthermore, a significant fraction (about 25% or more) of the SS may consist of silt and clay-sized particles (Harriss and Turner, 1974) that stay in suspension much longer than the residence of water in the settling basin. Such small particles are not amenable to removal by settling even in large settling basins unless chemical flocculants are added to speed up the settling process. Evans et al. (1968), for example, found that completely quiescent settling for less than 1 hour removed negligible amounts of SS, BOD, and nutrients from urban runoff, and Whipple and Hunter (1980) found that even 16 hours of settling removed only about 50% of the SS and P and less than 40% of the BOD from urban runoff. Extending the settling time to 32 hours increased the removal efficiency of the SS to almost 70%, but there was little improvement in the removal of P and BOD (Whipple and Hunter, 1980). Since conditions during storm runoff in a settling basin are far from quiescent and since the retention time of storm runoff in typical settling basins is no more than a few hours, it should be obvious why settling basins have sometimes been ineffective in handling land runoff problems.

Ground Recharge Basins

If land area is available and if the soil is sufficiently porous, it may be possible to route land runoff to recharge basins where the water has a chance to percolate into the soil rather than running off the land. Such basins may be classified as either dry or wet, the latter containing a permanent pool of water below the level of stormwater containment. In areas where the water table is sufficiently low, percolation of runoff may also be accomplished with the use of dry wells, which are pits or trenches in porous soil backfilled with rock. A system of recharge basins has been used for many years in certain parts of Long Island, New York City, for example, where the sandy subsoil is favorable to rapid percolation. Some of the recharge basins are actually underground pits; others are fenced-off open basins. Runoff from roads, construction sites, and some industrial operations are routed to the recharge basins (Weibel, 1969). From time to time, recharge basins must be dredged to remove accumulated sediments, but in principle, one tries to locate the basins so that runoff from impervious surfaces has little chance to traverse erodible soils. Hence, the sediment load is minimized. From this standpoint, a system of many small pits is clearly preferable to one large pit.

An important concern with the use of recharge basins is the possibility that pollutants in the runoff will contaminate groundwater supplies. This possibility is minimal to the extent that the pollutants are associated with SS, since percolation of the water through the soil removes virtually 100% of the SS. Pathogens derived from animal wastes are also removed with virtually 100% efficiency by the percolation process (see Chapter 6). Groundwater may become contaminated, however, with certain dissolved substances that are not effectively adsorbed by soil particles. Whether the presence of such substances in land runoff precludes percolation of the water depends on the concentrations of the substances in the runoff and the extent to which they are diluted by the water in the underground aquifer. In the case of Long Island, for example, cadmium (Cd) concentrations in the groundwater near an aircraft manufacturing company were found to be over 100 times acceptable levels during the early 1950s (Chapter 12). The company responsible had been using Cd to anodize aluminum (Al) and other metals and was disposing of its wastewater in a recharge basin. While runoff from streets and parking lots is unlikely to contain pollutant concentrations comparable to those in industrial wastewaters, one should be aware that some pollutants are not effectively removed by percolation and that the possibility of groundwater contamination does exist. Some studies or estimates of the characteristics of runoff from existing or proposed sites may therefore be useful in determining whether groundwater recharge basins are a reasonable mechanism for treating the runoff.

Storage in Underflow Tunnels or Tanks

Overflow from combined sewer systems is sometimes diverted into underflow tunnels or tanks until it can be treated. The tunnels or tanks are placed underground, and the overflow from storm events is pumped out during dry weather and routed to a treatment plant. If the capacity of the underground reservoir system is large enough, the runoff from virtually all storms can be treated in this way. The lower the capacity of the reservoir system, the more frequent will be the release of untreated runoff.

Berlin, Germany, for example, has a system of underground tanks to handle overflow from the combined sewer system that serves a third of West Berlin. The system is able to handle small overflows completely. In the case of large overflows, water passes through the tank system, receives nominal sedimentation, and is discharged to local streams. With the system in operation, about 85% of the total waste handled by the combined sewer system receives treatment. Before the tanks were installed, only 55% of the waste was treated (Cohrs, 1962).

The most ambitious treatment system for combined sewer overflow is undoubtedly the tunnel and reservoir plan (TARP) developed by the city of Chicago. TARP consists of a system of huge tunnels, reservoirs, pumping stations, and treatment facilities. The system was designed in two phases, the first intended primarily for pollution control and the second aimed at flood control from large storms. Phase I includes approximately 175 km of tunnels, into which combined sewer overflow is routed. The tunnels are 2.4–10 m in diameter and are located 45–90 m below street level. Runoff is pumped out of the underflow tunnel system to the treatment plants during dry weather (Figure 5.1). Sediment and sludge are removed from the tunnels with a floating hydraulic dredge and pumped to the same treatment plants. The treated wastewater is pumped into the ground through deep groundwater recharge wells located near the treatment plants. When completed in November 2002, the Phase I system will be capable of handling about 85% of the pollution caused by combined sewer overflows from a 970 km^2 area. Phase II, which is targeted for completion in 2013, will extend the tunnel system to a total length of 210 km and will add three storage reservoirs with a combined capacity of 59 million m^3 for combined sewer overflow. The principal reservoir, with a capacity of 40 million m^3, will be located in the McCook-Summit area. When completed, it will be the largest combined sewage reservoir in the world. A second reservoir, the Thornton reservoir, will be located at the southern extension of the Calumet system (Figure 5.1). It will have a capacity of 18 million m^3 for combined sewage and an additional 12 million m^3 for flood relief for areas serviced by separate sewer systems. A third reservoir, the O'Hare reservoir, has a capacity of 1.3 million m^3 and was substantially complete in 1998. When completed, TARP is expected to eliminate over 99% of the release of SS and BOD from combined sewer overflows in the Chicago area. As of January 1, 2000, 150 km of Phase I tunnels had been completed and were in operation, an additional 13 km were under construction and scheduled for completion in 2002; awards for construction of the remaining 12 km were expected to be made in the first half of 2000. The cost of the work completed to date or underway is $2.2 billion, with 75% of the bill paid by the EPA and the remainder by local authorities. The total cost of Phase 1 is expected to be $2.4 billion, with operating costs of about $9 million per year. The Corps of Engineers will pay 75% of the cost of Phase II, which is scheduled for completion in 2013. Construction of the entire TARP system is expected to cost $3.1

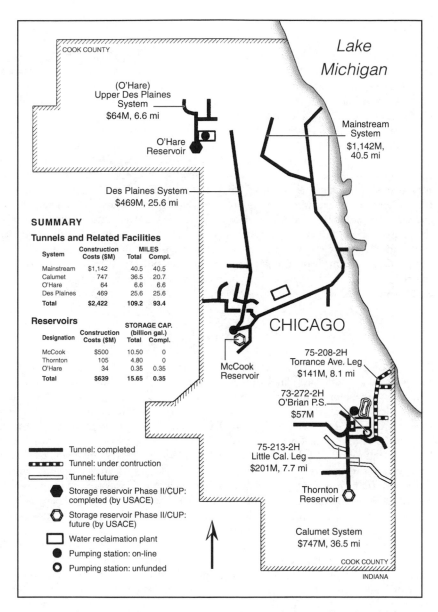

FIGURE 5.1   Schematic of the TARP to store and treat urban runoff in the Chicago area. *Source.* Metropolitan Water Reclamation District of Greater Chicago Engineering Department.

billion, with annual operating costs of $13.6 million (City of Chicago, 1999; Di Vita, 1990). The combined TARP system will virtually eliminate backflows of combined sewage into Lake Michigan, the source of Chicago's drinking water, and end basement flooding of about 200,000 homes. The flood relief benefits alone are estimated to be $96 million per year (City of Chicago, 1999).

Minimizing Runoff

Urbanization is a process that invariably increases the percentage of impervious surfaces in a watershed. As a result, less rainwater percolates into the ground and more flows overland. For example, it has been estimated that if a well-managed wooded area is converted to quarter-acre housing, the runoff from a storm of 5 cm would increase from 6 to 18 mm (Whipple et al., 1983). With careful planning, however, it is possible to reduce the amount of overland water flow. For example, one can intercept runoff from rooftops and percolate it directly into the ground with the use of dry wells or similar devices. Where practical, the use of porous paving material for sidewalks, parking lots, and streets can significantly reduce runoff from these surfaces. The usual impervious concrete or asphalt may be replaced with gravel or crushed rock. In some cases, an even better solution is to use modular pavement, which consists of impervious material, usually concrete, interspersed with void areas filled with porous materials such as sod, sand, or gravel. In suburban areas where the slope of the land is not too great, the traditional curb and gutter system may be replaced with gently sloping grassed channels or swales. In the past, relatively little use was made of such preventive techniques, but as awareness of the problems caused by urban runoff increases, these methods will likely receive more attention, particularly in the planning of new developments.

Summary

The foregoing examples give some idea of the variety of methods that have been used to deal with land runoff. There are basically two approaches to the problem. The first is to minimize the amount of overland runoff by intercepting the water as close to the source as possible. This goal is achieved most effectively by using porous paving materials wherever practical, but it may also be accomplished by intercepting runoff in small detention or recharge basins before it has had a chance to traverse exposed soils. The principal pollutant in land runoff is almost always suspended sediment. The amount of soil erosion at construction sites can be minimized by planning developments, so that as little land as possible is laid bare at one time, and by building interceptor basins around the construction site to permit percolation of runoff water. This first approach requires a relatively large number of individually small preventive measures, and a good deal of planning by public authorities and cooperation by developers.

The second approach is to intercept the runoff in the storm sewer system and give it some form of treatment before it is released. This approach is exemplified by the Berlin underflow tank system and by TARP. The principal advantage of the first system is that soil erosion is minimized, and because of its low SS concentration, the runoff may be percolated without treatment or utilized for purposes that do not require water of potable quality. The principal advantage of the second system is that its operation and maintenance usually fall under the authority of a single government entity. Consequently, its performance is more likely to be monitored and maintained than is the performance of the many components of the first system, which are likely to be separately owned and operated by numerous persons and agencies. The principal disadvantage of the second system is its high cost. The price tag reflects the fact that the system must be able to handle very large volumes of water in a relatively short time. The $3.1 billion cost of TARP, for example, works out to about $824 per person served by the system. Similar projects for the control of runoff into Kaneohe Bay have been estimated to cost $100–$1,300 per resident of Oahu (Gray and Lau, 1973). Obviously, a rather aroused public is required to finance such projects. The fact that the EPA and the Corps of Engineers agreed to pay 75% of the cost of TARP undoubtedly enhanced the image of the project in the eyes of local taxpayers.

A CASE STUDY—LAKE JACKSON, FLORIDA

Lake Jackson is a shallow lake (mean depth about 2–3 m) with an area of about 20 km^2 located just north of Tallahassee, Florida (Figure 5.2). Because of the shallowness of Lake Jackson, almost all of its benthos lies in the euphotic zone. As a result, benthic algae and grasses contribute substantially to the overall primary production of the lake. There is no doubt that the lake is highly productive, but this eutrophic condition by itself does not appear to have led to any undesirable effects. Water clarity in most parts of Lake Jackson is good, with Secchi depths of about 2–3 m where it is possible to make a Secchi measurement.[1] Until the early 1970s, there were no reports of nuisance algal blooms involving either benthic algae or phytoplankton in any part of the lake. During the 1960s, Lake Jackson gained national prominence due to the large number of trophy-sized largemouth bass that were taken from its waters. It has undoubtedly been a valuable recreational resource to the city of Tallahassee, since it is used extensively for fishing, pleasure boating, and picnicking at parks along its shores.

Unfortunately, the northward expansion of the city of Tallahassee during the 1960s led to increasing urbanization of the lake's southern watershed, particularly the subwater-

---

[1]The bottom is visible from the surface in many locations.

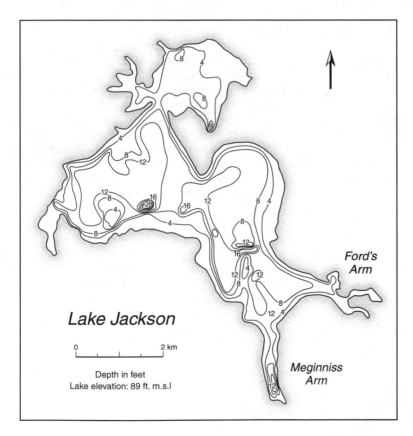

FIGURE 5.2   Bathymetry of Lake Jackson. *Source:* Redrawn from Harriss and Turner (1974). Reproduced with permission.

TABLE 5.3   MEAN VALUES OF SELECTED PARAMETERS OVER A
THREE-YEAR PERIOD IN LAKE JACKSON

	Inner Megginnis Arm	North Midlake[a]
Secchi depth (m)	0.52	2.9
SS (mg per liter)	61.3	4.0
Primary production (mg C m^{-3} h^{-1})	139	32.4

[a] Deep hole in central part of lake (see Figure 5.2).
Source: Harriss and Turner (1974).

sheds draining into Megginnis Arm and Ford's Arm (Figure 5.2). Associated with this urbanization was the construction in the southern watershed of two shopping malls in 1968 and 1969 and an interstate highway link in 1972. In the early 1970s, the effect of this urbanization became apparent to those using the lake, as large amounts of fine-grained sediment washed into both Megginnis Arm and Ford's Arm after every rain. Fortunately, domestic sewage from northern Tallahassee was pumped to a sewage treatment plant south of the city rather than being discharged into Lake Jackson, but the nutrient loading from urban runoff alone was sufficient to create water quality problems in the lake. Massive blooms of rooted aquatic plants began to appear during the summer in both southern arms of the lake, creating an unsightly appearance and impeding boat traffic. In addition, cyanobacterial blooms characteristic of hypereutrophic waters began to occur in the same areas. As a result of these observations, the Florida Game and Freshwater Fish Commission initiated a study of Lake Jackson to determine how serious the eutrophication problem had become and to decide what corrective measures needed to be taken. This study lasted for three years, from July 1971 to June 1974.

As a result of this study, the deterioration of water quality in the southern part of the lake caused by the effects of urban runoff from a rapidly developing watershed was documented in quantitative terms. Table 5.3 summarizes a few of the results reported in the study. The concentration of SS and the rate of primary production at the inner Megginnis Arm station were about 20 and 4.3 times greater, respectively, than the same parameters at the north midlake station. The mean Secchi depth of 0.5 m at the inner Megginnis Arm station reflects the combined effects of a high algal biomass and high concentration of SS. It may be compared to the mean Secchi depth of about 1 m in Lake Washington at the height of eutrophication in 1963–1965 and to the mean Secchi depth of 2.9 m at the north midlake station in Lake Jackson.

To quantify the effects of urban runoff on water quality in the southern part of Lake Jackson, a study was initiated comparing the runoff characteristics of two subwatersheds. The two subwatersheds, which we will henceforth refer to as the *forested watershed* and the *urban watershed*, were of comparable area, topography, and soil type.[2] The principal difference between the two was the use of the land. The forested watershed, which drained into the northern part of the lake, consisted in large part of an old estate with limited public access. About 52% of the forested watershed was covered by mixed pine-hardwood forest, with the remainder being in light agricultural use or in old fields. The urban watershed, which drained into the head of Megginnis Arm, was used primarily for residential housing (67%). About 13% of the land was occupied by commercial developments, 12% was forested, and 9% was used for agricultural purposes or was in old fields (Harriss and Turner, 1974). Figure 5.3 depicts the experimental watersheds and their relationship to Lake Jackson.

[2] Area: 6.3 km²—forested, 7.2 km²—urban; topography: 4.4% average slope—forested, 4.2% average slope—urban; soil type: sandy loam, well drained over loamy subsoil.

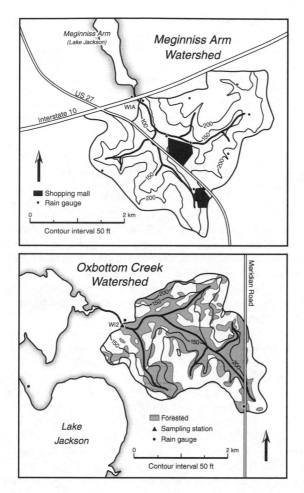

FIGURE 5.3   Maps of the urban watershed (Megginnis arm watershed) and the forested watershed (Oxbottom Creek watershed). *Source*: Redrawn from Harriss and Turner (1974). Reproduced with permission.

During the course of the watershed study (August 1973–April 1974), water samples were taken by automatic sampling devices from the streams draining the experimental watersheds. During storms, the urban watershed stream was sampled at 20- or 30-minute intervals, and the forested watershed stream was sampled at 1-hour intervals. The U.S. Geological Survey measured stream flow rates. During the study period, 13 storms were sampled in the urban watershed and 8 storms were sampled in the forested watershed.

Table 5.4 lists mean concentrations of selected constituents of runoff from the urban and forested watersheds based on this sampling. In the urban watershed, SS and nutrient concentrations were both significantly higher in storm runoff than in base flow. In the forested watershed, SS concentrations were over twice as high in storm runoff as in base flow, but nutrient concentrations were little different. The most dramatic comparison in Table 5.4 is the 30-fold increase in SS concentrations in urban stormwater runoff versus urban base flow. This large difference is undoubtedly attributable to the many areas of exposed land that had been cleared for building construction purposes and for the construction of an interstate highway (I-10). In the case of the I-10 construction, the contractor

TABLE 5.4   MEAN CONCENTRATIONS OF SS, INORGANIC N, AND INORGANIC P IN DRY WEATHER RUNOFF AND STORM RUNOFF FROM THE FORESTED AND URBAN WATERSHEDS (mg L^{-1})

	Baseflow (Dry Weather)		Stormwater	
	Urban	Forested	Urban	Forested
SS	10	12	299	34
Inorganic N	0.17	0.10	0.29	0.12
Inorganic P	0.06	0.13	0.12	0.10

Source: Turner and Burton (1975).

had signed an agreement to clear only small areas of land at a time. In fact, after construction had begun, the entire right-of-way for the highway was bulldozed all at once, leaving an enormous area of bare land that contributed large amounts of sediment to storm runoff for many months before highway construction was completed and the ground finally sodded. The failure of local authorities to enforce the signed contract or to penalize the construction company in any way is a good example of the futility of antipollution laws that are not properly enforced.

In this particular case, the Department of Transportation, which had contracted for the work, was persuaded to install settling basins downstream of the I-10 construction in an attempt to reduce sediment loading to Megginnis Arm. As noted earlier, such sediment traps are futile in moderate or heavy storms, since the water flushes through the traps so rapidly that there is virtually no chance for settling, and in fact accumulated sediments may be scoured out. It is true that the traps in this case retained some of the larger sediment particles that washed into them during base flow or during light storms. As a result, it was necessary to dredge the traps periodically. Incredibly, the sediment dredged from the traps was initially deposited next to the stream on the downstream side of the trap, where it was well situated to wash back into the stream with the next storm and be transported down to Megginnis Arm. This situation was later corrected after notification of local authorities.

Table 5.5 lists annual discharges of water, SS, and nutrients into Lake Jackson from the urban and forested watersheds estimated from sampling during the study period. The discharge of water was about 1.5 times greater from the urban watershed than from the forested watershed. The relatively large area of the urban watershed covered by impervious surfaces undoubtedly caused this difference. As a result, a greater percentage of the

TABLE 5.5   RAINFALL AND ESTIMATED ANNUAL LOADINGS OF WATER, SS, AND NUTRIENTS FROM THE URBAN AND FORESTED WATERSHEDS

	Urban Watershed	Forested Watershed
Rainfall (cm y^{-1})	249	254
Total runoff (cm y^{-1})	48.3	31.2
SS (kg ha^{-1} y^{-1})	2376	40
Inorganic P (kg ha^{-1} y^{-1})	0.18	0.15
Inorganic N (kg ha^{-1} y^{-1})	0.41	0.16

Source: Turner et al. (1977).

rainfall that fell on the urban watershed ran off rather than percolating into the ground. The discharge of SS was almost 60 times greater from the urban watershed than from the forested watershed, a result caused by both the greater amount of water runoff and the higher SS concentration in the water from the urban watershed. Inorganic P loading from the urban watershed was only 20% greater than from the forested watershed, but inorganic N loading from the urban watershed was about 150% higher. The greater sensitivity of inorganic N loading to urbanization probably reflects the fact that nitrate N, unlike phosphate P, does not bind to soil particles. As a result, nitrate is more likely to be flushed from exposed soils than phosphate.

Finally, Table 5.6 compares the importance of storm runoff versus base flow in the materials balance of Lake Jackson. The comparison has been made by calculating the number of days of base flow that would be required to deliver to the lake the amount of the given substance delivered by an average storm. The importance of storm events in the materials balance of the lake is apparent from Table 5.6. If an average storm occurs about once per week, it is clear from Table 5.6 that storm runoff loadings are of at least comparable importance and in most cases of greater importance than base flow loadings for all the constituents considered. Therefore, any realistic attempt to obtain a materials balance for such a system, whether the watershed is forested or urban, must be based on sampling during both storms and dry weather. It is not adequate to sample only when the sun is shining. The most dramatic figure in Table 5.6 is the number of years of sediment base flow loading to which an average storm event was equivalent in the urban watershed. This study dramatically illustrates the tremendous perturbation to the sediment loading characteristics of a watershed that can be caused by urbanization and land clearing.

A bit more can be learned about the characteristics of urban runoff by examining the time course of pollutant concentrations and runoff during storm events. Figure 5.4, for example, shows water discharge and the concentrations of SS, silicate, and dissolved inorganic P in the runoff from both the urban and forested watersheds during a storm on November 21, 1973. The rain itself was characterized by two intensity peaks that are mimicked by the water discharge from the urban watershed. This close correspondence between rainfall intensity and stream discharge is characteristic of watersheds covered by a large percentage of impervious surfaces. Runoff from such watersheds is rapid. In this case, the runoff from the urban watershed was almost complete six hours after the beginning of the storm, whereas storm runoff from the forested watershed was barely underway by that time. The time lag in water discharge from the forested watershed reflects the fact that much of the early rainfall was absorbed by vegetation and soil, and the flow of overland runoff was delayed and impeded by the vegetative cover. In fact, much of the early runoff from the forested watershed probably consisted of subsurface flow in the form of percolating groundwater rather than overland flow. This subsurface flow transports no sediment

TABLE 5.6 DAYS OF BASEFLOW REQUIRED TO DELIVER AS MUCH OF THE INDICATED CONSTITUENT AS WAS DELIVERED BY AN AVERAGE STORM DURING THE LAKE JACKSON WATERSHED STUDY

	Urban Watershed	*Forested Watershed*
Water	19.3	6.2
SS	3859 (10.6 years)	21
Inorganic N	30.0	8.9
Inorganic P	21.8	5.8

Source: Turner and Burton (1975).

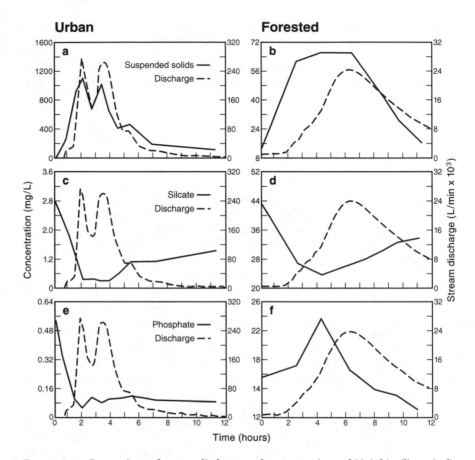

FIGURE 5.4   Comparison of stream discharge and concentrations of SS (a,b), silicate (c,d), and phosphate (e,f) versus time in the two experimental watersheds for the storm of November 21, 1973. Note the differences in horizontal (time) and vertical (discharge and concentration) scales between watersheds. *Source:* R. Turner, personal communication.

and creates no water pollution problems. Stream discharge from the forested watershed in no way reflects the two-peak intensity pattern evident in the urban runoff because the runoff has been delayed and the rainfall intensity characteristics consequently smeared out or integrated in the actual stream discharge. The relative importance of subsurface flow is much smaller in the case of the urban watershed than the forested watershed due to the high percentage of impervious surfaces in urban areas.

It is clear from Figure 5.4 that the concentrations of all three constituents shown in the figure tend to peak prior to the peak in stream discharge. This phenomenon is referred to as the *first flush effect* and arises because nutrients and particulate materials that have accumulated on the ground since the last storm wash into the storm sewer system with the early runoff. After this first flush, concentrations tend to drop because most of the easily eroded or flushed materials are already gone. This explanation is somewhat more applicable to silicate and phosphate than to SS, since the peak concentration of the former two clearly occurred before the peak in water discharge. SS concentrations, on the other hand, were at or very near their maximum at approximately the same time as the peak in water discharge. The concentration of SS was thus more closely correlated with stream flow than were the concentrations of the two dissolved constituents. This result should not be particularly surprising considering the fact that sediment erosion is closely correlated with flow intensity, whereas simple chemical dissolution is not. The greater intensity of runoff from the urban watershed is apparent in Figure 5.4. The peak discharge from the urban watershed was about 10 times greater than the peak discharge from the forested watershed. This difference reflects both the greater volume of water that ran off the urban watershed and the fact that the runoff was confined to a shorter time interval. The greater intensity of runoff from the urban watershed means that the urban runoff has a much greater erosive potential than the runoff from the forested watershed.

The first flush phenomenon has potential utility in the treatment of urban runoff if in fact a large percentage of the discharge of certain pollutants is associated with a relatively small percentage of the total runoff. In other words, if the early runoff from a storm is grossly contaminated and if subsequent runoff is relatively clean, then one may be able to eliminate much of the discharge of the pollutant from the watershed by retaining and treating only the first portion of the runoff. Figure 5.5 illustrates the first flush effect in the case of a hypothetical storm in which the initial pollutant concentration is 10 times (Figure 5.5C) and 100 times (Figure 5.5D) the final concentration. In the first case, the first 20% of the runoff accounts for almost 50% of the pollutant discharge. In the second case, the first 20% of the runoff accounts for 70% of the pollutant discharge. The magnitude of the first flush effect depends on the nature of the pollutant, the time since the pre-

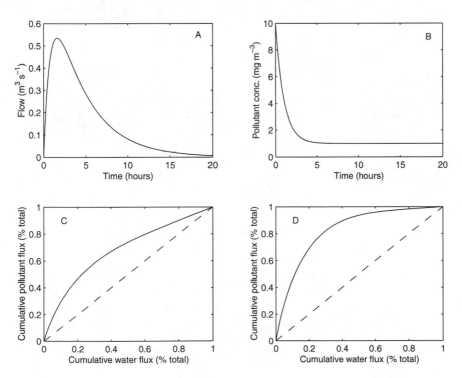

FIGURE 5.5   (A) Stream flow versus time during a hypothetical rainfall event, (B) concentration of a pollutant during the same event, (C) the cumulative flux of that pollutant versus cumulative water flux given the concentration time course shown in panel B, and (D) cumulative pollutant flux with a concentration time course similar to that shown in panel B but with the initial concentration 100 times the final concentration. In panels C and D, the dashed line is the cumulative flux if the concentration of the pollutant is constant.

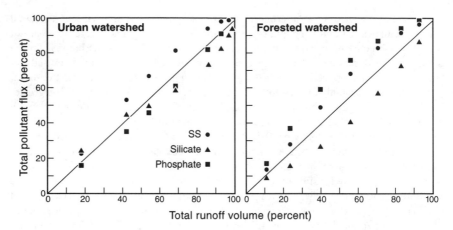

FIGURE 5.6   Percentage of total runoff of SS and nutrients versus percentage of total runoff of water calculated from the data in Figure 5.4.

vious storm, the degree and type of developments in the watershed, and perhaps a number of other factors. When the data shown in Figure 5.4, for example, are plotted in the same manner as the data in Figure 5.5, it becomes apparent that in the case of this particular storm, the percentage of SS efflux exceeded the percentage of water runoff by at most about 10% at any time (Figure 5.6). Furthermore, the percentage of silicate efflux lagged behind the percentage of water runoff because silicate concentrations tended to increase after the peak in water runoff. Thus, while it is generally true that the early runoff from a storm contains the highest concentration of many contaminants, it is by no means the case that retention and treatment of only the early runoff can be consistently relied upon to eliminate a major portion of the pollution caused by land runoff. Treatment of the early runoff is certainly a step in the right direction, but it is only a step.

Correctives

As a result of the three-year study of Lake Jackson, Harriss and Turner (1974) made the following recommendations to local authorities to minimize the damaging effects of urban runoff on the lake:

1. Land clearing, site preparation, and utility installations should be confined, as far as possible, to the dry season of the year, when the possibility of heavy rains washing away large amounts of exposed sediment is minimal.[3]
2. Developments requiring drastic disturbances of the subsoil, such as deep grading for level parking lots and slab foundations, should be prohibited in the watershed or, where permitted, extensive stabilization measures should be required.
3. Developers should be required to provide facilities to percolate any additional storm runoff generated by land use change, where disposal of such additional runoff would otherwise be to the primary drainage system in the watershed.
4. Impact zoning principles should be adopted for all future land development in the watershed. In other words, the impact of certain types of real estate developments on the environment should be considered in planning the growth of the city.

[3]Rainfall in the Tallahassee area is highly seasonal.

5. Early storm runoff from large shopping malls and other heavily used, paved parking surfaces should be percolated or treated before disposal, since such runoff was found to be grossly contaminated with heavy metals, hydrocarbons, and nutrients.

6. Porous paving materials should be considered for the paving of storm drainage ditches, road shoulders, residential driveways, roadways, and parking areas near the lake and in any areas where the benefits caused by reduced runoff from the use of such porous paving materials are judged to exceed any possible liabilities.

7. Extant and future sedimentation and nutrient control ordinances should be effectively enforced, as was not the case with the I-10 land clearing.

8. The effect of small and large ponds, including settling basins, on nutrient and SS loading to the lake from selected watersheds should be determined, in the hope that certain types of ponds, including, for example, ponds containing marsh plants or similar vegetation or a system of artificial baffles to retard flow, might prove effective in reducing stream inputs of SS and nutrients.

Most of these recommendations are nothing more than commonsense measures that any municipality should consider if urban runoff is perceived to be a problem. The last suggestion is a bit more extreme. As it turns out, however, water quality conditions in Lake Jackson became of such concern that the City of Tallahassee finally decided to implement the last suggestion. A facility to treat urban runoff into the head of Megginnis Arm was completed in 1983 and is shown schematically in Figure 5.7. Runoff initially enters a detention pond with a capacity of 163 million liters. The water then passes through a sand filter with a surface area of 1.8 ha. The filtrate from the sand filter is routed to a triple box culvert running underneath I-10, and from there enters a diversion impoundment, where it is joined with direct runoff from I-10. The combined runoff then enters a 2.5 ha artificial marsh. The marsh has an average depth of 0.5 m and is divided into three cells. The first, second, and third cells are planted with the macrophytes *Thypha*, *Scirpus*, and *Pontederia*, respectively.

The detention basin is intended to settle out suspended solids, and with a capacity of 163 million liters it can contain all the runoff from small and moderate storms. The sand filter is intended to remove particulate materials that do not settle out in the detention basin. Much of the material collected by the sand filter consists of silt- and clay-sized particles that settle very slowly even under quiescent conditions. The artificial marsh is intended primarily to remove dissolved nutrients, which the aquatic macrophytes and associated epiphytes are expected to assimilate. A concrete spillway built into the detention pond dam and a stoplog weir at the diversion impoundment leading into the marsh permit runoff that exceeds the capacity of the system components to bypass the sand filter and marsh, respectively. LaRock (1988) studied the performance of the treatment facility from 1983–1987, and as might be expected, the ability of the system to treat the runoff from a particular storm was found to depend very much on the amount of runoff and the antecedent weather conditions. It is sufficient for our purposes to examine the performance of the system only during the storm of June 11, 1985, because this analysis clearly reveals both the strengths and the weaknesses of the treatment facility.

The storm in question actually consisted of three storm events lasting for nearly one week. The total amount of runoff entering the head of the detention basin was 315 million liters. Of that amount, 155 million liters passed through the detention basin and sand filter, 108 million liters flowed over the detention basin dam and bypassed the sand filter, and 52 million liters remained in the detention basin. The 155 + 108 = 263 million liters that flowed through the triple box culvert were joined in the diversion impoundment by an additional 1,080 million liters of runoff from I-10. Of the 263 + 1,080 = 1,343 million liters of water that passed through the diversion impoundment, 223 million liters were ac-

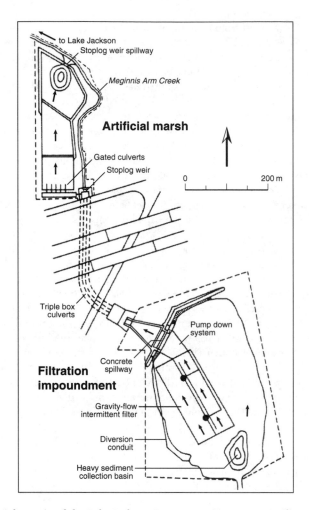

FIGURE 5.7   Schematic of the Lake Jackson Stormwater Treatment Facility. *Source:* Redrawn from LaRock (1988). Reproduced with permission.

tually routed through the artificial marsh, the remaining $1{,}343 - 223 = 1{,}120$ million liters being bypassed to Megginnis Arm.

The facility did a rather good job of removing pollutants from the water that it treated (Table 5.7). The combination of the detention basin and sand filter reduced SS and total P concentrations by at least 97% and total N concentrations by 92%. The artificial marsh reduced SS concentrations by about 73% and total N and P concentrations by 36% and 60%, respectively. It is obvious that the system is capable of effecting a dramatic reduction in SS and nutrient concentrations when not overloaded.

One obvious problem with the system is the fact that a large percentage of the runoff from I-10 received no treatment at all, and much of the remaining I-10 runoff passed only through the artificial marsh. The magnitude of the runoff from I-10 was evidently not appreciated at the time the facility was built. As the drainage system is designed, a large percentage of the runoff from I-10 cannot reach the head of the detention basin and instead is routed directly to the diversion chamber. However, even if all the I-10 runoff were routed to the head of the detention pond, it is obvious, in this case at least, that most of the wa-

TABLE 5.7   PERCENTAGE CHANGE IN FLUX OF VARIOUS
CONSTITUENTS THROUGH DETENTION BASIN/SAND FILTER
AND ARTIFICIAL MARSH DURING STORM OF JUNE 11, 1985

	Detention Basin/Sand Filter	*Artificial Marsh*
Inorganic solids	−99.0	−76.4
Organic solids	−98.1	−69.1
Total N	−91.6	−36.4
Total P	−97.0	−60.2
Ca	91.7	−5.3
Mg	221.1	−22.1

Source: LaRock (1988).

ter would have bypassed both the sand filter and the artificial marsh. This observation underscores the fact that land runoff treatment systems must have the capacity to retain very large volumes of water if they are to give adequate treatment to the runoff from large storms. One major flaw in the Lake Jackson stormwater treatment facility is the fact that there is no adequate mechanism for removing water from the detention basin between storm events. As a result, the capacity of the detention basin is sometimes much less than 163 million liters. The detention basin actually contains a dewatering system that pumps water out of the detention basin and sprays it over the surface of the sand filter when the water level in the detention basin drops below the level of the sand filter. However, use of this dewatering system was discontinued after only a few weeks of use due to the high electrical costs associated with pumping the water. Subsequently, the storage capacity of the detention basin was regulated by manually operating the drain valve in the spillway. This procedure, however, causes the water to bypass the sand filter and in practice has not provided adequate storage capacity for subsequent storms. Some bypass of the sand filter has occurred during 60% of all storm events (LaRock, 1988).

A second problem associated with the inefficient removal of water from the detention basin has been the accumulation of ammonia in the water that remains in the detention basin after storms. The ammonia accumulates as the result of the breakdown of organic matter trapped in the detention basin. The design of the sand filter is such that lateral water movement can occur from the detention basin through the sides of the filter. Ammonia-rich water therefore enters the sand filter and percolates down into the limestone that underlies the sand. As the water level is lowered, air enters the sand filter and the ammonia is oxidized to nitric acid by chemosynthetic bacteria. The nitric acid is then neutralized by the Ca and Mg ions in the limestone. The result is dissolution of the limestone and export of significant amounts of Ca and Mg with the water that leaves the sand filter (Table 5.7). If this situation is allowed to continue, the sand filter will someday disappear into a sinkhole.

A third problem, and one that was anticipated, is clogging of the sand filter. The combination of the detention basin and sand filter does an excellent job of removing SS (Table 5.7), and much of the particulate material retained by the sand filter consists of clay- and silt-sized particles, which are too small to be settled efficiently in the detention basin. As the clay and silt accumulate on the sand filter, its effective porosity is greatly reduced and the flow rate of water through the filter is slowed. According to LaRock (1988), the flow rate of water through the sand filter was reduced by about 70% during the first three years of operation of the sand filter. This problem could be corrected by periodically replacing roughly the upper 25 cm of the material in the sand filter. Serious clogging of the sand filter should not occur if a reasonable schedule of maintenance is followed.

The performance of the Lake Jackson stormwater treatment facility illustrates both the potential and the problems of systems designed to treat urban runoff. The capacity of the system must be large to handle the amount of water that flows off urban areas during large storms. It is often difficult to identify sufficient areas of land for treating urban runoff in existing urban settings. The problem is undoubtedly solved most economically if the treatment facilities are incorporated at the planning stage into the design of future urban developments. Since public awareness of the problems caused by urban runoff is a relatively recent phenomenon, most storm sewer systems have not been designed in a way that easily accommodates large treatment facilities. Chicago has solved the problem with its underflow tunnel system, but at considerable expense.

The Lake Jackson facility does an excellent job of treating the water that passes through it. Flaws in the design, however, and inadequate maintenance have resulted in the bypass of at least some runoff during about 60% of storm events. The artificial marsh is a creative idea, but at higher latitudes the biological uptake of nutrients by marsh plants would cease during the winter months. It seems fair to say that one important component of any plan to minimize pollution from urban runoff should be to minimize both the amount of runoff that must be treated and the concentration of pollutants in the water. In this regard, the recommendations made by Harriss and Turner (1974) should be given serious consideration.

REFERENCES

ABERNATHY, A. R. 1981. *Oxygen-Consuming Organics in Nonpoint Source Runoff.* EPA-600/S3-81-033. Envirnmental Protection Agency, Washington, D.C.

CITY OF CHICAGO. 1999. *TARP Status Report as of January 1, 1999.* Metropolitan Water Reclemation district of Greater Chicago. 12 pp.

COHRS, A. 1962. Storm water tanks in the combined sewerage system of Berlin. *Gas Wasserfach.*, 103, 947–952.

COOPER, C. F. 1969. Nutrient output from managed forests. In *Eutrophication: Causes, Consequences, Correctives.* National Academy of Sciences, Washington, D.C. Pp. 446–463.

DI VITA, L. R. 1990. *TARP Status Report as of 9/14/90.* Metropolitan Water Reclamation District of Greater Chicago, Chicago.

DUNNE, T., and L. LEOPOLD. 1978. *Water in Environmental Planning.* W. H. Freeman, San Francisco. 818 pp.

EPA. 1983. *The 1982 Needs Survey. Conveyance, Treatment, and Control of Municipal Wastewater Combined Sewer Overflows, and Stormwater Runoff.* Summary of Technical Data. EPA/43019-83-002. U.S. Government Printing Office, Washington, D.C.

EVANS, F. L., E. E. GELDREICH, S. R. WEIBEL, and G. G. ROBECK. 1968. Treatment of urban stormwater runoff. *J. Water Pollut. Control Fed.*, 40, R162–R170.

FRASER, R. H., M. V. WARREN, and P. K. BARTEN. 1995. Comparative evaluation of land cover data sources for erosion prediction. *Water Resources Bull.*, 31, 991–1000.

GRAY, B., and L. S. LAU. 1973. Alternative methods for sewage disposal. In *Estuarine Pollution in the State of Hawaii*, Vol. 2. Univ. of Hawaii Water Resources Research Center Tech. Rept. No. 31. Pp. 343–435.

HARRISS, R. C., and R. R. TURNER. 1974. *Job Completion Report, Lake Jackson Investigations.* Florida State University Marine Laboratory, Tallahassee, Fla. 231 pp.

HELGESEN, J. O., R. B. ZELT, and J. K. STAMER. 1995. Nitrogen and phosphorus in water as related to environmental setting in Nebraska. *Water Resources Bull.*, 30, 809–822.

LAROCK, P. 1988. *Evaluation of the Lake Jackson Stormwater Treatment Facility.* Final Report to Florida Dept. of Environmental Regulation, Tallahassee, FL. 246 pp.

LILTENS, G. E., and F. H. BORMANN. 1974. Linkages between terrestrial and aquatic ecosystems. *Bioscience* 24, 447–456.

MEYER, J. L., G. E. LIKENS, AND J. SLOANE. 1981. Phosphorus, nitrogen, and organic carbon flux in a headwater stream. *Arch. Hydrobiol.* 91, 28–44.

PETERSON, S. A., W. E. MILLER, J. C. GREENE, and C. A. CALLAHAN. 1985. Use of bioassays to determine potential toxicity effects of environmental pollutants. In *Perspectives on Nonpoint Source Pollution*. EPA 440/5-85-001. Environmental Protection Agency, Washington, D.C. Pp. 38–45.

PIONKE, H. B., W. J. GBUREK, A. N. SHARPLEY, and R. R. SCHNABEL. 1996. Flow and nutrient export patterns for an agricultural hill-land watershed. *Water Resources Res.* 32, 1795–1804.

RANDALL, C. W., T. J. GRIZZARD, D. R. HELSEL, and D. M. GRIFFIN, JR. 1981. A comparison of pollutant mass loading in precipitation and runoff in urban areas. In *Proceedings, 2nd International Conference on Urban Storm Drainage.* IAWPR, Univ. of Illinois, Urbana. Pp. 2, 29–38.

TURNER, R. R., and T. M. Burton. 1975. The effects of land use on stormwater quality and nutrient and suspended solids, exports from three north Florida watersheds. Proc. Stormwater Management Workshop. Feb. 26–27, Orlando.

TURNER, R. R., T. M. BURTON, and R. C. HARRISS. 1977. "Lake Jackson watershed study." In D. L. Correll, Ed., *Watershed Research in Eastern North America. A Workshop to Compare Results.* Vol. 1. Chesapeake Bay Center for Environmental Studies. Edgewater, Md. Pp. 19–32, 211–224, 323–342, 471–485.

WEIBEL, S. R. 1969. Urban drainage as a factor in eutrophication. In *Eutrophication: Causes, Consequences, Correctives.* National Academy of Sciences, Washington, D.C. Pp. 383–403.

WHIPPLE, W., N. S. GRIGG, T. GRIZZARD, C. W. RANDALL, R. P. SHUBINSKI, and L. SCOTT TUCKER. 1983. *Stormwater Management in Urbanizing Areas.* Prentice-Hall, Englewood Cliffs, N.J. 234 pp.

WHIPPLE, W., and J. V. HUNTER. 1980. *Settleability of Urban Runoff Pollution.* Rep. Water Resources Research Institute, Rutgers Univ., New Brunswick, N.J.

QUESTIONS

5.1. Suppose that you are contracted to design a system to minimize the flux of oil and heavy metals from a very large parking lot to a nearby stream during storm events. Given the amount of land available for your treatment system, it is clear that you cannot realistically expect to retain and treat all of the runoff from storms if the rainfall exceeds about 2 cm. Given this constraint, how could you take advantage of the first flush effect to minimize the release of pollutants to the stream?

5.2. What is it about the nature of the pipes/conduits in a combined sewer system that causes so much particulate material to become stranded filth during low-flow conditions?

5.3. Suppose that a billionaire philanthropist decides to build a utopian city on a large area of land he owns. You are contracted to design two wastewater treatment systems, one to handle the sanitary sewage and the other to handle runoff from the city streets during rain events. The annual rainfall in the area is about 100 cm. Which wastewater treatment system would you expect to have the higher capital cost and why?

5.4 A contractor is hired to build a condominium on a parcel of land with an area of 4 ha. In an effort to minimize runoff from the construction site, the contractor builds a sedimentation basin with a depth of 3 m and an area of 400 m^2. At the same time, a nearby town decides to build a sedimentation basin to intercept runoff from the town that is polluting a recreational lake. The town occupies an area of 250 ha. About 50% of the town consists of impervious surfaces such as streets, parking lots, and rooftops. The town builds a sedimentation basin with a depth of 3 m and an area of 4,000 m^2. Which sedimentation basin do you think will do a better job of minimizing nonpoint source pollution and why?

5.5 City A is located in a place where there is a rainfall event every week. City B is located in a place where there is a rainfall event every month. In which city is the first flush effect likely to be more important and why?

Chapter Six

✧ ⌑ ✧ ⌑ ✧

Sewage Treatment

> The toilet area shall be outside the camp. Each man must have a spade as part of his equipment; after every bowel movement he must dig a hole with the spade and cover the excrement. (Deuteronomy 23: 12,13)

RECOGNITION of the need to dispose of sewage properly has existed for a long time. However, in the United States, the first comprehensive sanitary sewer system was not constructed until 1855 (in Chicago), and the first sewage treatment facility was not completed until 1886 (in New York City). Problems associated with inadequate treatment or improper disposal of sewage led to the proliferation of sewage treatment plants in the United States during the 20th century. At the present time, the wastewater generated by about 75% of the U.S. population is processed by municipal sewage treatment plants (NAS, 1996), commonly referred to as *publicly owned treatment works* (*POTWs*). The remainder of the population is served primarily by individual household septic systems. Since 1972, U.S. governmental agencies have spent more than $72 billion upgrading the nation's sewage systems (Marinelli, 1990). At the present time, there are approximately 19,400 municipal POTWs in the United States, and they treat roughly 50 billion m^3 of raw sewage per year (NAS, 1996).

As we have seen in the previous few chapters, the discharge of sewage, whether treated or not, can create serious water pollution problems in the receiving body of water. The high concentrations of SS and nutrients, as well as the BOD of raw sewage, create a great potential for causing cultural eutrophication problems. The fact that raw sewage also contains a high concentration of pathogens is cause for concern from a public health standpoint as well. Standard sewage treatment removes most of the SS, BOD, and pathogens from the raw sewage, but less than half of the N and P. Hence, even treated sewage can be an environmental problem from the standpoint of nutrient enrichment. In this chapter, we focus primarily on the methods used to remove SS and BOD from wastewater, but we also consider methods for dealing with the nutrients, either by removing them or by putting them to practical use. We consider the problem of sewage pathogens in detail in Chapter 7.

PRIMARY, SECONDARY, AND TERTIARY TREATMENT

The basic features of most sewage treatment plants are similar, although they may differ greatly in the details of their operations. The differences reflect the variable characteristics of sanitary sewage and, in the case of combined sewer systems, the need for the plant to process both sanitary sewage and street runoff during storms. Although the characteristics of domestic sewage are rather well defined, the composition of industrial sewage is highly variable. Some industrial wastes contain toxic organic substances, pesticides, and/or heavy metals. As pollution laws have become stricter, the release of such substances by industries has declined, but sewage plants that receive the effluent from such operations must still be able to handle wastewater with characteristics rather different from those of typical domestic sewage. Cesspool pumping is another factor that can create problems for sewage treatment plants. Once a cesspool has been pumped out, its contents are injected directly into the sanitary sewer line, sometimes only a short distance from the sewage treatment plant. The contents of one or a few cesspools may have characteristics quite different from those of the average wastewater entering the plant. Given these considerations, it is perhaps surprising that the general design of most POTWs is so similar.

Figure 6.1 illustrates the basic features of a secondary POTW. When the raw sewage first arrives at the POTW, it usually passes through a screening device to remove large objects that have entered the sewer pipes and might damage components of the POTW if they are allowed to pass through. The screening device may consist of nothing more than a series of metal bars spaced 5–10 cm apart. The trash that accumulates on these racks is removed periodically and deposited in a landfill. An alternative to the trash rack is a mechanical device called a *comminuter*, which pulverizes any solid objects in the raw sewage. Once the sewage has passed through the screening device or comminuter, it flows into a *grit chamber*, where particles roughly the size of sand grains or larger settle out. Removal of this coarse grit is intended to protect the pumps in the rest of the system. With large objects and grit thus removed, the sewage is pumped to a *primary settling tank* or *primary clarifier*. Much of the flow through the rest of the system is by gravity.

The primary clarifier is nothing more than a large quiescent tank in which settlable

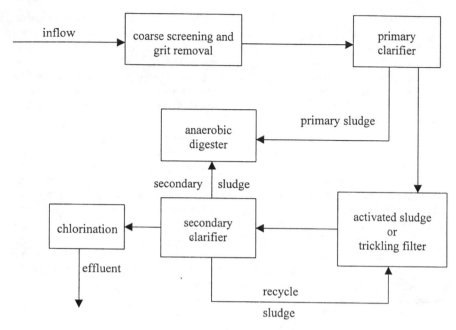

FIGURE 6.1   Diagram of flow through a secondary sewage treatment plant.

or floatable solids are removed from the sewage. The tank may be circular or rectangular. In the former case, the sewage is introduced at the center and slowly flows to the edge, where it exits under a baffle. In the latter case, the sewage is introduced at one end and exits at the other. Solid materials that float to the surface during the residence time of the sewage in the tank[1] are removed by a mechanical skimming device that pushes floating objects to a hopper, from which they are pumped to the anaerobic digester. The bottom of the tank is inclined, and a similar mechanical device moves the sludge that accumulates on the bottom toward the deepest part of the tank, where it is drawn off and pumped to the anaerobic digester. The material skimmed from the surface and pumped from the bottom of the primary clarifier is called the *primary sludge*. Treatment involving no more than removal of this primary sludge is termed *primary sewage treatment*. Federal law requires that a POTW operating with nothing but primary treatment remove at least 30% of the BOD and SS from the sewage.

If the POTW is a secondary treatment plant, the effluent from the primary clarifier flows into a second tank where biological processes are used to remove much of the BOD in the effluent. The basic idea is to allow organisms living in the tank to consume the organic substances in the primary effluent. In this process, the consumer organisms grow and multiply, but at the same time, a large percentage of the organic substances is respired. As a result, the amount of potentially oxidizable material in the waste is greatly reduced. This phase of the treatment will be described in detail later on. The effluent from this treatment process consists in part of the organisms that have been feeding off the organic wastes. Most of the biomass of these organisms can be removed by running the effluent into a secondary settling tank or clarifier. The organisms' biomass settles to the bottom in the form of a floc or sludge and is removed in a manner similar to that employed in the pri-

---

[1]The residence time is usually a few hours.

mary clarifier. Normally, no attempt is made to skim the surface of the secondary clarifier, since little if any of the secondary sludge floats to the surface. A portion of the secondary sludge is pumped to the anaerobic digester, while the remainder is recycled.

The entire treatment process just described is called *secondary sewage treatment*. According to federal law, secondary treatment should remove at least 85% of the SS and BOD from the raw sewage and reduce the SS and BOD concentrations to less than 30 mg/L. Primary sewage treatment typically removes 40–75% of the SS and any BOD associated with the SS. BOD removal efficiencies associated with primary treatment are usually in the range 30–40%. Regardless of whether the POTW is a primary or secondary facility, the liquid effluent is usually not released to the environment until it has been disinfected, usually by chlorination, to reduce the concentration of pathogens to an acceptably low level. With this overview of a typical secondary POTW, let's take a closer look at several of the steps in the treatment process.

Secondary Treatment for BOD Removal

By far the most commonly used methods for removing BOD at the secondary level are the so-called activated sludge and trickling filter processes. Both methods use primarily biological means to reduce BOD.

Trickling Filters

A trickling filter usually consists of a cylindrical tank 2–3 m in depth filled with rocks. The rocks typically have a diameter of 2–10 cm, and should be loosely packed and as nearly spherical in shape as possible to facilitate the downward flow of sewage and the upward flow of air through the system. The effluent from the primary clarifier is sprinkled over the surface of the bed of rocks by one or more rotating arms. Usually a given area of the bed is sprinkled approximately every 30 seconds, but faster application rates are possible and sometimes necessary.[2]

As the sewage trickles down through the bed of rocks, bacteria and fungi that grow on the surface of the rocks in a glutinous film ingest the dissolved and suspended organic substances in the water. The bacteria and fungi are, in turn, fed on by a variety of higher trophic level organisms, including protozoa, rotifers, nematodes, worms, and insects. Since the composition of the sewage changes (due to the feeding of the bacteria and fungi) as it trickles from the top to the bottom of the trickling filter, there is a tendency for the species composition of the food chain to change from top to bottom. Because the biological community living in the trickling filter consists of several trophic levels, the organic substances assimilated from the sewage by the bacteria and fungi may pass through several food chain transfers, and a large percentage of the organics is typically respired. About 70% of the organic wastes in the primary clarifier effluent are oxidized in the trickling filter. Of the remainder, 15–20% are converted to organism biomass. The film of organic biomass that develops on the trickling filter rocks periodically sloughs off and is carried by the water to the bottom of the filter, where it is drained off with the liquid waste. Part of the trickling filter effluent is routed to the secondary clarifier, where settling of the sludge removes additional BOD from the water. However, part of the effluent is normally recycled back to the trickling filter to maintain the population of decomposer organisms in the filter and to further reduce the BOD.

Bacteria grow by simple cell division, and given adequate food, their populations increase exponentially with time. Biomass production during this so-called logarithmic phase

[2]For example, if the plant is being forced to treat more sewage than it was designed to handle.

of growth is relatively efficient, i.e., respiration losses are relatively small. The operator of a POTW, however, is not interested in producing large numbers of bacteria. Instead, the operator is interested in having the bacteria and other organisms burn up as much of their food as possible, or in other words, to grow inefficiently. As the concentration of bacterial food in the water declines below a certain level, both the efficiency of bacterial biomass production and the rate of biomass production begin to decline. Realizing this fact, the POTW operator attempts to operate the trickling filter so that the organisms in the system grow rapidly enough to consume most of the organic substances in the water but slowly enough to keep biomass production efficiency as low as possible. In a trickling filter, the concentration of food is obviously highest at the top of the filter and lowest at the bottom. Therefore, the microorganisms near the top grow rather rapidly and efficiently, whereas organisms near the bottom grow more slowly and oxidize much of their food. The POTW operator can control the growth rate and growth efficiency of the organisms in the system to a degree by adjusting the amount of trickling filter effluent that is recycled. He is constrained, of course, by the fact that the net outflow of water from the trickling filter (gross minus recycled flow) must equal the inflow of water to the plant.

Activated Sludge

An activated sludge system usually consists of a long rectangular tank or series of tanks 3–5 m deep. The effluent from the primary clarifier is introduced at one end of the tank and exits at the other end. The residence time of water in the system is usually about four to eight hours. During its passage through the tank(s), the sewage is vigorously mixed from below with bubbles of air that keep the contents of the tank in a state of great turbulence. This vigorous mixing and aeration are necessary to prevent the O_2 concentration in the water from dropping below 2–3 mg/L. Otherwise the respiratory activity of the organisms in the tank is likely to be slowed.

Because of the extreme buffeting that goes on in an activated sludge tank, only very small aquatic organisms are suited to the environment. Bacteria are the principal decomposer organisms in the tank, with fungi being of strictly secondary importance unless the system is not operating properly. Protozoan ciliates and flagellates, small rotifers, and nematode worms usually complete the food chain, although other classes of organisms such as insects may occasionally become established. Most of the bacteria, protozoa, and other organisms in the system are found aggregated together in flocculent masses that collectively are called the *activated sludge*. Because the food chain in an activated sludge system is usually shorter than the food chain in a trickling filter, a smaller percentage of the organics in the inflow are oxidized and a larger percentage are incorporated into sludge biomass. Typically, about 50% of the organics from the primary clarifier effluent are oxidized in the tank, and the volume of secondary sludge may be as much as twice that produced by a trickling filter. This secondary sludge is removed from the effluent in the secondary clarifier. A small part of the secondary sludge is usually pumped to the anaerobic digester; the remainder is recycled to the activated sludge tank(s). This recycling of sludge is essential with the activated sludge process, since the primary clarifier effluent normally produces only about 5% of the sludge required for its treatment (Warren, 1971, p. 358). As is the case with a trickling filter, the operator of an activated sludge POTW may control to a certain extent the age and hence the biomass production efficiency of the sludge by adjusting the proportion of the sludge that is recycled. Although one pass through the activated sludge system may take only four to eight hours, recycling increases the effective age of the sludge to three to four days. Manipulations of the system are constrained by the total volume of sewage per unit time that the plant must handle and by economic factors. For example, additional recycling of sludge will normally require additional aeration and

TABLE 6.1 PERCENT REMOVAL OF BOD AND SS USING VARIOUS
TREATMENT METHODS

Process	% BOD Removal	% SS Removal
Septic tank	25–65	40–75
Primary treatment	30–40	40–75
Primary + trickling filter	80–90 or more	80–90 or more
Primary + activated sludge	85–96 or more	85–96 or more

Source: Klein (1966).

therefore increase the cost of treatment. The objective of the treatment process is to oxidize as much of the waste organics as possible and therefore to minimize the amount of
secondary sludge that must be handled by the anaerobic digester.

Pros and Cons of Trickling Filter and Activated Sludge Treatment

Table 6.1 provides a comparison of the efficiency with which various sewage treatment
systems remove SS and BOD. As the table shows, efficient trickling filter and activated
sludge POTWs are both capable of achieving the 85% removal of BOD and SS required
by federal law. What, then, are the relative merits of these two types of treatment processes?

There is virtually no difference in the efficiency with which chlorination kills
pathogens in the effluent from the two types of treatment. However, activated sludge plants
tend to achieve higher SS and BOD removals than trickling filter plants. It is noteworthy
that the difference between 90% and 95% BOD removal is a factor of 2 in the amount of
BOD released by the plant. Thus, activated sludge plants tend to be significantly more successful at accomplishing what they are designed to do, namely, removing SS and BOD.
However, the trickling filter system produces less secondary sludge than the activated sludge
system, and trickling filter biological communities seem to be less perturbed by sudden
changes in the character of the sewage than are activated sludge communities. This greater
resilience of the trickling filter communities probably reflects the greater variety of organisms that inhabit trickling filters. For example, if certain bacterial species in a trickling
filter are killed due to a sudden change in the characteristics of the sewage, other bacterial species already present in the trickling filter community may move in to fill the vacant
niche. However, in an activated sludge system there are likely to be fewer types of decomposer organisms, and if some of these are killed, there are relatively fewer species to move
in and take their place. Because of the greater resilience of trickling filter communities to
perturbations, the cost of aerating the activated sludge tanks,[3] and in particular the cost
of disposing of the additional sludge produced by activated sludge plants, trickling filters
have proven to be generally less expensive to operate than activated sludge systems. On the
other hand, trickling filters require more land area to build and are more costly to construct. Furthermore, trickling filters tend to attract flies[4] and to have a more serious odor
problem than activated sludge systems.

Although the foregoing discussion makes it clear that there are pros and cons to both

---

[3]In recent years, however, more efficient mechanisms for aerating activated sludge tanks have been
developed. The strategy has been to reduce the size of the air bubbles and hence increase the contact
area between air and water. The result has been roughly a twofold reduction in aeration costs.
[4]Recall that insects are one component of the trickling filter community.

trickling filter and activated sludge systems, present federal regulations mandating 85% removal of BOD and SS have tended to make activated sludge systems more popular because it is generally easier to achieve 85% BOD and SS removal with an activated sludge system than with a trickling filter. Trickling filters may be preferred, however, in cases where large fluctuations in sewage characteristics are anticipated. For example, one might prefer to build a trickling filter plant to treat sewage containing wastewater from industrial operations where accidents or shutdowns might lead to a sudden change in the characteristics of the sewage. On the other hand, sewage that is largely of domestic origin has rather predictable and constant characteristics. Judging from experience, fluctuations in the characteristics of domestic sewage usually do not create unmanageable problems in activated sludge POTWs.

The Anaerobic Digester

What happens to the sludge that is collected from the primary and secondary clarifiers? This organic biomass is putrescible and therefore cannot be simply deposited at a convenient landfill. Although it is theoretically possible to oxidize this sludge completely by burning it in a furnace, only a minority of POTWs in the United States incinerate their sludge. In the past, some POTWs disposed of their sludge in the ocean, but this practice came to an end when federal legislation prohibiting ocean dumping of sludge was enacted in 1981 (Clark et al., 1984). The principal alternatives to incineration are either deposition in a sanitary landfill or some form of land application. In either case, the putrescible nature of the sludge must be eliminated before it is disposed of. This task is accomplished by pumping the sludge to a closed, usually cylindrical tank, where it provides food for a special class of microorganisms that carry on their metabolic activities in the absence of O_2, i.e., anaerobically. The products of the anaerobic digester are digested sludge, a supernatant fluid, and gases. The digested sludge is relatively stable and inoffensive compared to the primary and secondary sludges. Furthermore, its volume is only about one-third that of the primary and secondary sludges. This reduction in volume is achieved both by the conversion of some of the organics in the raw sludge to gases, primarily methane and carbon dioxide, by anaerobic catabolism, and by dewatering the sludge to reduce its moisture content. The raw sludge typically has a moisture content of 94–99% by weight, and most of its volume is due to water.

The anaerobic digester is invariably heated to speed up the metabolic rate of the microbes. Operating temperatures are usually in the range 27–35°C. The methane gas produced in the digester is more than adequate to provide fuel to heat the digester to the desired temperature. The excess methane produced in the digester is burned as a waste product. The supernatant fluid from the digester retains some objectionable properties (e.g., odor) and is usually recycled through the secondary treatment process. Thus, the only product of the digester that requires disposal is the stabilized sludge, which is either trucked off to a sanitary landfill or applied to the land as an organic fertilizer and soil conditioner.

The availability of suitable land to accommodate the sludge produced by the anaerobic digester can be a limiting factor in the use of this technique to treat the primary and secondary sludge. Under such conditions, the POTW may resort to burning as a means of disposal. Because of its high water content, the sludge must first be dewatered, and the organic matter is then burned in a furnace. A small amount of inorganic ash remains, but the volume of this ash is far less than the amount of stabilized sludge produced by anaerobic digestion. About 16% of the sewage sludge produced in the United States is incinerated (NAS, 1996).

Tertiary Treatment

Tertiary treatment, strictly speaking, consists of any additional processing of secondary treatment effluents undertaken to improve their quality. In some countries, tertiary treatment means additional removal of SS and BOD, but in the United States, tertiary treatment usually means nutrient removal (Ellis, 1980). From some of our previous case studies, it is clear that the nutrient concentrations in raw sewage are sufficiently high to stimulate objectionable algal growth in receiving waters unless the effluent is greatly diluted in the receiving system. To get some idea of the nutrient problem, let's compare the amount of O_2 required to oxidize the algal biomass that could be produced using the N in raw sewage versus the typical BOD of 200 mg/L found in raw sewage. Taking the total N of raw sewage to be 40 mg/L and assuming the oxidation of typical phytoplankton biomass to proceed via the equation (Riley and Chester, 1971, p. 173)

$$(CH_2O)_{106}(NH_3)_{16} H_3PO_4 + 138O_2 \rightarrow 106CO_2 + 122H_2O + 16NH_3 + H_3PO_4$$

one concludes that it would take

$$\frac{40 \text{ mg N/L}}{14 \text{ mg N/mg atom N}} \frac{138 \text{ mmoles } O_2}{16 \text{ mg atoms N}} \frac{32 \text{ mg } O_2}{\text{mmole } O_2} = 789 \text{ mg/L } O_2$$

to oxidize the phytoplankton biomass produced from the N in raw sewage. Let's assume that a typical secondary sewage treatment plant removes 30% of the total N from the raw sewage and discharges the remaining 70% in the form of ammonia, and that the ammonia is subsequently assimilated by phytoplankton and converted into algal biomass. The decomposition of that algal biomass would be expected to consume $(0.7)(789) = 552$ mg/L O_2, which is about 2.8 times the BOD of the raw sewage. The depletion of O_2 in the hypolimnion of Lake Erie's central basin has been attributed largely to this effect, that is, the loading of a system with nutrients that are incorporated into organic biomass whose ultimate decomposition consumes O_2.

It should be clear from our discussion of primary and secondary sewage treatment that neither process is directly aimed at removing nutrients. In fact, the oxidation of the organic substances in sewage releases nutrients rather than removing them. However, the sludge that is removed in the primary and secondary clarifiers does contain nutrients. The nutrients are bound primarily in the organic biomass that makes up most of the sludge. If POTWs operated so as to try to maximize BOD removal via sludge production rather than via oxidation, then nutrient removal would be more efficient. However, since finding suitable sites to dispose of sludge is often a problem, POTWs prefer to remove as much BOD as possible through oxidation. Therefore, nutrient removal efficiencies are generally rather low. Primary treatment plants and conventional activated sludge plants remove 5–15% and 30–50%, respectively, of the N and P from raw sewage. Under special conditions, P removal efficiencies as high as 70–90% have been reported in some POTWs, but N removal using conventional secondary treatment methods is rarely more than 30—40% efficient.

Phosphorus Removal

What methods are available to a POTW operator for removing P? Conceptually, P removal is very straightforward because inorganic P forms insoluble compounds with Ca, Al, and Fe under appropriate conditions. A common procedure for removing phosphate consists of simply adding lime (CaO), aluminum sulfate (alum), sodium aluminate, or ferric salts to the effluent from the secondary clarifier. The phosphate precipitate that forms in the effluent must then be allowed to settle out before the effluent is discharged. Since phosphate is of commercial value in fertilizer, the precipitate may be chemically processed to recover the phosphate. Since the solubility of the phosphate precipitates is pH-dependent, some adjustment of effluent pH may be necessary before addition of the cation (Al^{3+}, Ca^{2+}, or Fe^{3+}) and again after separation of the precipitate to bring the pH of the final effluent back to a suitable value. In the case of Ca^{2+}, for example, the pH must be raised above 10 (highly basic) to obtain phosphate removals in the 90–95% range. Phosphate removal efficiencies by this chemical precipitation technique are typically in the range 88–95%, regardless of whether Al, Fe, or Ca is used as the precipitating agent.

A variation on the strict chemical precipitation technique is to add the precipitating agent at some stage in the primary or secondary treatment process. The chemical precipitate is then collected along with the primary or secondary sludge. Addition of alum or coagulating chemicals to the primary clarifier promotes the removal of organic substances as well as phosphate. However, this approach obviously produces a larger volume of sludge and also requires greater chemical additions.

A variety of other methods, including sorption, ion exchange, reverse osmosis, and distillation, are capable of removing 90% or more of the P from sewage. The popularity of chemical precipitation as a tertiary treatment method for P removal stems from its low cost and from the fact that its removal efficiency compares favorably with that of alternative methods. The actual cost depends very much on the degree of P removal. Miller (1972), for example, has estimated the cost of tertiary treatment achieving 85% P removal by chemical addition to be only about 15–30% higher than that obtained with conventional secondary treatment, and 90% P removal can be achieved at a cost of roughly 2 cents per person per day at POTWs serving more than 10,000 persons.[5] However, 95% P removal can cost as much as or more than conventional secondary treatment (Miller, 1972). The difference between 85% and 95% P removal is a factor of 3 in the effluent P concentration. These considerations help to explain the rationale for restricting the use of phosphates in laundry detergents in the Lake Erie watershed. If raw sewage contains 10 mg/L total P, reducing that concentration to less than 1 mg L^{-1} would require more than 90% P removal. If 40% of the P in the raw sewage originates from laundry detergents, then eliminating the P in the detergents would mean that an effluent P concentration of 1 mg L^{-1} could be achieved with 83% P removal. The difference between 83% and 90% P removal may seem small, but it can be significant in terms of tertiary treatment costs. Although 95% P removal is theoretically possible, P removal efficiencies at tertiary POTWs are seldom that high in practice. Hall et al. (1999), for example, cite a P removal efficiency of 84% at a POTW in Regina, Canada.

Nitrogen Removal

There are a variety of possible ways to remove N from sewage. Secondary treatment involves vigorous oxidation of the organic wastes in sewage. At the pH of typical sewage, the inorganic N produced by this process initially appears primarily in the form of ammonium (NH_4^+). The ammonium ion, however, can be converted to ammonia gas, NH_3, by shedding a proton (H^+). Therefore, a common method for removing N is to convert NH_4^+ to NH_3 and then drive off the NH_3 gas by vigorous aeration. The equilibrium between NH_4^+ and NH_3 is strongly pH dependent, and at neutral pH over 99% of an ammonium-ammonia mixture is in the form of NH_4^+ at equilibrium. However, if the pH is raised above 11, over 98% of the mixture is NH_3 at equilibrium. The procedure in the *ammonia stripping process* is therefore to raise the pH of the secondary effluent artificially to 11 or

---

[5]Based on Lee and Jones (1986) and an inflation factor of 2 (Grogan, 1998).

more, vigorously aerate the effluent, and then bring the pH back down to a suitable level. Since P removal by means of Ca^{2+} addition also requires a high pH, it is possible to combine the two procedures in the same step. A disadvantage of the ammonia stripping technique is the high solubility of NH_3 in water. As a result, a very large volume of air is required to strip the NH_3 from a small volume of effluent. About 3,000 m^3 of air is required to strip 95% of the NH_3 from 1 m^3 of effluent (Rohlich and Uttormark, 1972). Furthermore, ammonia stripping efficiency is reduced in cold weather because the solubility of NH_3 in water increases as the temperature decreases. To deal with this problem, special countercurrent, low-pressure airflow stripping towers have been designed to remove the NH_3 from the effluent. Such a system has achieved about 90% NH_3 removal at a POTW near Lake Tahoe (Slechta and Culp, 1967).

A second technique for removing N is to convert it to nitrogen gas, N_2. There are two common procedures for accomplishing this task, one biological and the other chemical. In the biological procedure, the inorganic N is first oxidized to nitrate by prolonged, vigorous aeration. The nitrate is then converted to N_2 by denitrification, which proceeds in an anaerobic environment and is mediated by a special class of microorganisms that are able to utilize the O_2 from NO_3^- to oxidize organic substrates. Usually the organic content of secondary effluent is sufficiently low that an additional C source such as acetone, alcohol, or acetic acid must be added to provide sufficient food for the denitrifying microbes. In practice, some NO_3^- is converted to nitrous oxide gas (N_2O), but both N_2O and N_2 may be stripped by aeration.

The chemical procedure is called *breakpoint chlorination* and involves addition of Cl to the effluent until the total dissolved residual Cl reaches a minimum (the breakpoint) and virtually all the ammonia N has disappeared. The overall chemical reaction is

$$3Cl_2 + 2NH_3 \rightarrow N_2 + 6HCl$$

The efficiency of the process is pH-dependent, and production of nitrate and chloramines may become significant if reaction conditions are not properly maintained. Ammonia removal efficiencies of 95–99% are achievable with breakpoint chlorination if the pH is maintained in the range 6.5–7.5 (Pressley et al., 1972), but computer control and sophisticated monitoring of reaction conditions are necessary if the process is to work efficiently (Fertik and Sharpe, 1980).

N removal from sewage has been tried using a variety of other techniques, including ion exchange, electrodialysis, reverse osmosis, electrochemical treatment, and distillation. All of these methods, as well as ammonia stripping and conversion to N_2, are capable of achieving over 90% N removal. Ammonia stripping and denitrification are by far the most cost competitive. The cost of ammonia stripping is little different from that of P removal by chemical precipitation; and as pointed out earlier, the two methods may be conveniently combined if Ca^{2+} is used to precipitate the phosphate. In activated sludge systems, substantial N removal can be achieved via denitrification by simply routing secondary effluent to a quiescent (unaerated) portion of the activated sludge tanks. As is the case with P, the cost of N removal escalates rapidly as removal efficiency approaches 100%. Removing 95% of the N and P from sewage can cost as much as or more than conventional secondary treatment. Removing 85% of the N and P from sewage may increase the cost of treatment by no more than 20–30% above the cost of secondary treatment.

Cost of Conventional Sewage Treatment

How high is the cost of conventional sewage treatment? Table 6.2 lists estimated costs of building and operating POTWs providing various degrees of sewage treatment. The ratio

TABLE 6.2 ESTIMATED CAPITAL COST AND OPERATING COST
OF A 60 mLd and 180 mLd POTW AS OF 2000

Degree of Treatment	Capital Cost (Millions of Dollars)		Operating Cost (Cents per 1,000 L)	
	60 mLd	*360 mLd*	*60 mLd*	*360 mLd*
Primary	11.6	45.2	16.5	10.0
Secondary	23.8	95.1	34.9	23.7
Tertiary—general Tertiary—potable	26.9	114.1	47.6	37.1
Tertiary—potable water	132.1	366.1	171.2	97.0

Note: Capital and operating costs were multiplied by factors of 5 and 3, respectively, to adjust for inflation (Grogan, 1998).
Source: Stephen and Weinberger (1968).

of sewage flow to the number of persons served by a POTW is about 600 L per person per day (NAS, 1996). The operating costs for 60 and 360 million liter per day (mLd) secondary sewage treatment plants are about 21 and 14 cents per person per day, respectively. If the construction of the plants were financed by selling 20-year municipal bonds at 7% interest, the cost of building the secondary treatment plants would be 6.1 and 4.0 cents per person per day for the 60 and 360 mLd plants, respectively. This latter calculation assumes that the cost of financing the plants is equally divided among 100,000 taxpayers (=60 mLd/600 L day^{-1}) and 600,000 taxpayers (=360 mLd/600 L day^{-1}) in the case of the 60 mLd and 360 mLd plants, respectively. Thus, the total cost (capital plus operating cost) of secondary POTWs would be 27 and 18 cents per person per day for the 60 mLd and 360 mLd plants, respectively. These results, along with similar total cost calculations for other degrees of sewage treatment, are listed in Table 6.3. Note that increasing the degree of treatment from primary to secondary roughly doubles the cost to the taxpayer. No matter what the degree of treatment, most of the cost is associated with operating the plant. Although a typical plant may cost tens of millions of dollars to build, its lifetime is relatively long, typically 20 years or longer. As a result, most of the cost of building and operating a POTW is associated with operations. The cost of tertiary treatment depends on the nature and degree of the tertiary treatment. Upgrading secondary effluent to a quality suitable for irrigation (see below) increases overall costs by about 50%. Upgrading secondary effluent

TABLE 6.3 ESTIMATED TOTAL COST TO THE TAXPAYER
OF THE POTWs LISTED IN TABLE 6.2

	Total Cost to the Taxpayer (Cents per Person per Day)	
	60 mLd Plant	*360 mLd Plant*
Primary	13	8
Secondary	27	18
Tertiary—general irrigation supply	35	27
Tertiary—potable water supply	136	74

to the point where it could theoretically be used to recharge drinking water supplies raises the overall treatment cost by roughly a factor of 4.

LAND APPLICATION OF SEWAGE

Until now, our discussion of sewage treatment has implied that sewage effluent is an undesirable, problem-creating substance. A little thought will convince one, however, that sewage effluent is, at least potentially, a valuable resource. This conclusion stems from the following reasoning:

1. The high nutrient concentrations in sewage obviously stimulate plant growth. Although excessive plant growth in aquatic systems is undesirable, sewage could be applied at a controlled rate to fields or forests to stimulate the growth of valued crops, just as high-nutrient fertilizer is presently applied to achieve the same purpose. Use of sewage to irrigate crops would eliminate the need to discharge the sewage into aquatic systems that might be adversely affected by the high nutrient levels in the sewage, reduce the dependence of the crop grower on synthetic fertilizers, and allow part of the sewage irrigation water to percolate down through the soil into the underground water table and thus to recharge groundwater supplies. This last observation is directly related to point 2.
2. In some parts of the world, freshwater is in scarce supply or is rapidly becoming scarce. Discharging sewage at controlled rates to land areas rather than to the nearest convenient watercourse could be used to recharge groundwater, since under suitable conditions a large percentage of the effluent could be expected to percolate into the underground water table. In any case, use of wastewater for purposes such as irrigation that do not require water of potable quality can be an effective way to increase drinking water supplies. As noted in Section 13550 of the California Water Code, "The use of potable domestic water for nonpotable uses, including, but not limited to, cemeteries, golf courses, parks, highway landscaped areas, and industrial and irrigation uses, is a waste or an unreasonable use of the water within the meaning of . . . the California Constitution if reclaimed water is available."

Thus, sewage effluent is potentially useful both for growing crops and for recharging groundwater. However, several problems associated with such applications are immediately suggested.

1. The effluent from a conventional secondary POTW invariably contains some pathogens. Obviously, no one would want to eat crops or drink groundwater that had become contaminated with sewage pathogens.
2. High concentrations of nitrate in drinking water may create health problems of two sorts. Nitrate breaks down in saliva and in the digestive tracts of humans and animals into nitrite. Nitrite may subsequently combine with amines to form nitrosamines, a class of compounds shown to cause cancer in laboratory animals (Smith, 1978; Tannenbaum et al., 1978). In addition, nitrite may be absorbed into the blood, where it converts hemoglobin to methemoglobin. The latter pigment is unable to transport O_2, and in severe cases of methemoglobinemia the victim may literally suffocate. EPA water quality standards with respect to nitrate have been based largely on the methemoglobinemia problem. The disease, which is largely confined to infants less than 3 months old, is fatal in about 7–8% of reported cases (EPA, 1986). Approximately 2,500 cases of infant methemoglobinemia have been reported in Europe and North America since 1945. Older children and adults are

less affected because (a) the stomach pH of infants is higher than that of adults[6] and (b) infant gastrointestinal illnesses may permit reduction of nitrate to nitrite to occur in the intestinal tract.

Methemoglobinemia is associated with nitrate-N concentrations in excess of 10 mg/L, although it is known that many infants have consumed water containing higher nitrate concentrations without developing the disease. To be on the safe side, the EPA has established an upper limit of 10 mg/L nitrate-N for public water supplies. Since the total N concentration in raw sewage is about 40 mg/L and since only about 30–40% of this is removed by conventional secondary treatment, it is possible that continuous recharging of groundwater with sewage effluent could produce dangerously high nitrate levels in the groundwater.

3. Continuous irrigation of land with water containing dissolved salts can lead to an increase in the salt content of the soil because some of the irrigation water inevitably evaporates, leaving its salt content behind. If the accumulation of salt is sufficiently great, the soil may become unfit for growing crops. This problem would be most severe in arid climates where evaporation rates are high. In addition to the dissolved salts that are abundant in aquatic systems, sewage effluent may also contain small amounts of industrial wastes, including heavy metals such as Hg, Pb, Cu, Cd, and Zn. Accumulation of elements such as Cu and Zn in the soil could poison the soil for plant growth, and uptake of elements such as Pb, Hg, and Cd could render crops unfit for human consumption. A gradual deterioration in soil quality due to the accumulation of salts and heavy metals might be difficult to detect for many years, since crop production would be stimulated at the same time due to the high nutrient concentrations in the effluent. Once soil quality had become sufficiently degraded to affect crop production noticeably, it might be difficult to restore soil quality, at least in a reasonably short time.

4. Spraying of sewage for irrigation purposes obviously introduces many small droplets of water into the air. The smaller of these droplets may be transported for considerable distances by the wind. Studies conducted at conventional POTWs have shown that aerosols injected into the air from trickling filter or activated sludge operations may transport viable pathogens as much as several hundred meters downwind from POTWs (Hickey and Reist, 1975). Spray irrigation systems would undoubtedly inject many more aerosols into the air than do trickling filters or activated sludge tanks. Thus, spray irrigation operations utilizing sewage effluent might create significant health problems for persons living immediately downwind from the irrigation systems.

Results of Spray Irrigation Studies

Realizing the potential benefits as well as the potential problems of land application of sewage effluent, a number of scientists have set out to determine whether land application of sewage effluent is in fact a safe and more beneficial means of disposing of sewage effluent than the traditional method of discharging the effluent into the nearest body of water. A very thorough study was carried out at Pennsylvania State University beginning in 1963 (Kardos, 1970; Kardos et al., 1974). Interest in land application of sewage was stimulated both by the results of that study and by the passage in 1972 of the Clean Water Act, which proposed a zero discharge concept that encouraged the philosophy of reuse and recovery. A long-term study of sewage spray irrigation began near Tallahassee, Florida, in

---

[6]This higher pH permits the growth of bacteria that reduce nitrate to nitrite.

TABLE 6.4   CROP YIELDS FROM CONTROL PLOTS
(COMMERCIAL FERTILIZER) AND PLOTS SPRAYED WITH
SECONDARY SEWAGE EFFLUENT

	Control Plot	*2.5 cm/wk Effluent*	*5 cm/wk Effluent*
Alfalfa hay, tonnes/ha	5.2	10.5	12.1
Corn silage, tonnes/ha	6.9	8.7	9.6
Corn, bushels/ha	156	283	274
Oats, bushels/ha	117	298	179

1971 (Payne and Overman, 1987; Allhands and Overman, 1989; Allhands et al., 1995) and reports summarizing work carried out at numerous other locations throughout the United States began to appear during the 1970s and 1980s (Gilbert et al., 1976; Hershaft and Truett, 1981; Majeti and Clark, 1981; Page et al., 1984; Shuval et al., 1984; Sheikh et al., 1990). At the present time, the treated wastewater from a population of about 6 million persons is reclaimed and reused in the United States. About 70% of the reclaimed water is used for irrigation (NAS, 1996). Virtually all the studies conducted to date have led to the same conclusion, namely, that land application of sewage effluent is a safe and practical way to dispose of sewage wastewater if reasonable precautions are exercised. The principal limiting factors are the amount of land required and accumulation of nitrate in groundwater supplies. For our purposes, it is sufficient to examine in some detail the results of the Pennsylvania State study and then to summarize briefly the results of some of the more recent work.

The Pennsylvania State studies were carried out near State College, Pennsylvania, where a series of experimental land plots were established on which various crops were grown using either recommended amounts of commercial fertilizer to stimulate growth (the control plot) or various application rates of secondary sewage effluent. The effluent was applied to the experimental crops from April to November each year for a period of five years beginning in 1963. Other plots consisting of mixed hardwood trees and grassy areas were sprinkled throughout the year. Table 6.4 compares some of the crop yields from the control and experimental plots in 1965. Irrigation with 2.5 cm wk^{-1} of effluent roughly doubled the yields of alfalfa, hay, corn, and oats and increased the yield of corn silage by about 26%. Increasing the irrigation rate further to 5 cm/week produced little significant additional change in the crop yields.

The results of sprinkling forested areas with sewage effluent were somewhat mixed. The growth of red pines was actually retarded by the irrigation, perhaps because red pines grow better in dry soil[7] and perhaps because of the relatively high boron content of the sewage effluent. Boron (B) is a common component in detergents; and B, although an essential plant nutrient, may be toxic to plants if its concentration is in the approximate range 0.3–0.4 mg L^{-1} or higher (Allen, 1973). However, other trees responded favorably to effluent sprinkling, although in most cases the growth response was rather small (Kardos, 1970).

To determine the effect of the spray irrigation program on groundwater quality, a number of sampling wells were dug to various depths and at various locations in and around the experimental area. There was no evidence of fecal pathogens in the groundwater. Similar results have been reported by Bouwer et al. (1974), Gilbert et al. (1976), and Hershaft and Truett (1981). Percolation of treated wastewater through roughly 1 m of well-

[7]Control plots in the forested areas received neither fertilizer nor well water.

aerated soil appears to be an extremely efficient mechanism for removing sewage pathogens. Such pathogens simply do not survive well in soil. According to Allen (1973, p. 39), "Almost all die within two weeks." For example, a study of wastewater irrigation in California reported a 100,000-fold reduction in viruses in soil after only 10 days (NAS, 1996). It is true that the transmission of infectious diseases has been linked to the irrigation of food crops with untreated sewage or treated wastewater of questionable quality. However, an epidemiological review of disease transmission from irrigation with reclaimed water by Shuval (1990) concluded that only untreated wastewater had been implicated in the transmission of infectious disease. Except for the use of untreated sewage or primary sewage effluent on sewage farms in the late 19th century, there has been no documented case of infectious disease resulting from the use of reclaimed wastewater in the United States (EPA, 1992).

The effect of sewage irrigation on the nutrient concentrations of groundwater presents a somewhat different picture. In the Pennsylvania State study, P removal by the soil-vegetation complex was generally quite efficient, although not as impressive as the removal of pathogens. On the plots that received 5 cm of effluent per week throughout the year, P removal was about 92% at the 120-cm depth level during the third year of application. After the first year or so, most of the P removal in the forested areas was evidently due to adsorption of the P to soil particles, since much of the P taken up by the vegetation during the spring and summer was returned to the soil in the fall via dead leaves and other litter. In the plots where crops were grown, P removal was more efficient because much of the P taken up by the crops was permanently removed when the crops were harvested. By the third year of the study, P removal via harvested crops ranged from 22% in corn silage irrigated with 5 cm/wk to 63% in red clover irrigated with 2.5 cm wk^{-1} (Kardos, 1970). With the additional P removal due to adsorption to soil particles, P removal at the 120-cm depth level was 99% in the plots where crops were grown.

N removal by the soil-vegetation complex was not as efficient as P removal, primarily because N in the form of nitrate does not adsorb effectively to soil particles.[8] In fact, soil water nitrate levels in a white-spruce area sprayed throughout the year with 5 cm wk^{-1} of sewage effluent were actually higher by the third year of the study than the nitrate levels in the effluent. The recycling of N utilized by vegetation during the spring and summer and then returned via litter in the fall presumably caused this increase in nitrate. The other forested plots (red pine, oak) that received similar sprinkling did remove "a large percentage of the nitrate added to the soil by the effluent" (Kardos 1970, p. 15). However, on forested plots that were sprinkled with 10 cm/wk of effluent, soil water nitrate levels approached the EPA guideline for potable water of 10 mg L^{-1} nitrate-N.

N removal in the plots planted with crops was more efficient because of the harvesting of the crops. In fact, calculated removal efficiencies based on the N content of the crops sometimes exceeded 100%, presumably because the crops were utilizing N in the soil in addition to that added with the effluent. Calculated N removal efficiencies were 60–90% for wheat, 65–127% for corn (grain), and 105–210% for corn silage (ear, stalk, and leaves). After three years of irrigation, soil water samples at a depth of 120 cm in the crop plots were found to have a nitrate-N concentration of about 6 mg L^{-1}, and groundwater samples from an unspecified depth had a nitrate-N concentration of 3 mg L^{-1}. The highest soil water nitrate levels were usually found in the early spring, when the ground had thawed but when the plant root systems were not yet fully active. Kardos (1970) points out that perennial grasses are most effective in removing N from the soil in the early spring because their root systems are already fully established, whereas the roots of annual grasses

---

[8]NH_4^+ does effectively adsorb to soil particles, but NH_4^+ is spontaneously oxidized to nitrate in the presence of even small amounts of O_2.

and most crops are just beginning to develop. At latitudes where there is a long period of the year when plant root systems are inactive, it may be necessary to set aside additional land areas for sewage irrigation during the winter to prevent the buildup of unacceptably high soil nitrate levels by late winter–early spring. Such measures would obviously not be necessary in tropical latitudes where root systems are active throughout the year.

Results reported from the Tallahassee study have been qualitatively similar to the Pennsylvania State results but have also revealed some interesting temporal trends. The Tallahassee work has involved irrigation with secondary treated effluent on approximately 700 ha of sandy soil south of the city of Tallahassee. Since 1981, application rates have been 4.5–6.0 cm/wk. Corn, soybeans, rye, canola, and coastal Bermuda grass have all been grown on the experimental plots. The soil pH, which was initially acidic (5.3), increased to an apparently steady-state value of 7.0 (neutral) in the upper 1.5 m after about six to seven years of irrigation and reached a value of 6.0 at a depth of 7.6 m (Allhands and Overman, 1989; Allhands et al., 1995). The organic content of roughly the upper 50 cm of the soil increased by about a factor of 2 after eight years of irrigation and has stabilized at about 1% near the surface (Allhands et al., 1995). Virtually all the P in the effluent was removed in the upper 1.0 m of soil during the first six to seven years of irrigation. The P content of the soil surface layer increased during that time and eventually stabilized at about 90 mg kg^{-1} (Allhands et al., 1995). N removal was initially less impressive, in part because supplemental N was added to both the corn and coastal Bermuda grass plots during the growing season. Fertilizer was added because percolation of the sewage effluent through the soil was so rapid that the crop roots were unable to obtain adequate N from the sewage effluent. The corn and coastal Bermuda grass removed 25% and 47%, respectively, of the total applied N (fertilizer N plus sewage effluent N). By fine-tuning the sewage effluent application schedules and using a mixture of Bermuda grass and rye, Allhands et al. (1995) were able to boost the N removal efficiency to 65%. Because the N removal efficiencies were well below 100%, nitrate concentrations in groundwater increased for several years and had reached an average of about 18 mg L^{-1} by 1988, virtually identical to the N content of the applied wastewater. Subsequent upgrading of the Tallahassee POTW reduced the N content of the effluent to 9.3 mg L^{-1} (Allhands et al., 1995). Incidentally, yields of corn, Bermuda grass, and rye have averaged about 320 bushels ha^{-1}, 7.4 tonnes ha^{-1}, and 4.3 tonnes ha^{-1}, respectively (Allhands and Overman, 1989; Allhands et al., 1995).

The buildup of salts in the soil because of repeated irrigation with sewage was not a problem in either the Pennsylvania State or Tallahassee studies. However, it was obvious to scientists in the former case that harvesting of crops was in most cases removing only a small percentage of the sewage salts. Kardos (1970) states that in 1965 an alfalfa plot removed 118% of the irrigation water K, 19% of the Ca, 11% of the Mg, and 0.4% of the Na. These figures suggest that significant accumulations of Ca, Mg, and Na might be expected in alfalfa fields that are irrigated year after year with sewage effluent. In the Tallahassee study, Ca and Mg concentrations were as much as 10-fold or more higher in the upper 1 m of sewage-irrigated plots after six or seven years of irrigation, but Na and K concentrations were relatively unchanged. Heukelekian (1957) reported that salt accumulation was not a problem in sewage-irrigated fields in Israel because rainfall tended to leach out accumulated salts. The irrigation systems used in Israel, however, typically called for application rates of 50–75 cm of effluent per year, a somewhat lower rate than the 2.5–6.0 cm wk^{-1} application rates used in the Pennsylvania State and Tallahassee studies. Whether the gradual accumulation of salts in the soil would eventually create a problem is certainly a legitimate question. Rodhe (1962), for example, reported that soils at sewage farms that had been in operation for over 100 years near both Paris and Berlin had shown marked decreases in productivity, and stated that the cause had been traced to an accumulation of Cu and Zn in the soils. Hershaft and Truett (1981), however, reported no evidence of metal accumulation in soils at six sewage irrigation projects in the western United States, one of which had been in operation for over 30 years. None of the six facilities received industrial wastewater with a high concentration of metals.

As noted by the NAS (1996), even under ideal conditions, plants remove less than 10% of the salts present in irrigation water. To sustain soil quality, salts must therefore be leached from the root zone. In areas where irrigation is practiced only during the dry season, rainfall is usually sufficient to leach salts to an acceptable level. This may not be the case in arid regions. In such areas, an undesirable accumulation of salts can be avoided by applying more irrigation water than can be used by the crop. The quantity of water in excess of the amount required by the crop is called the *leaching requirement* (NAS, 1996).

Until now, we have discussed harvesting of crops from irrigated plots as a means of removing nutrients, but it should be obvious that such crops themselves are potentially useful as sources, for example, of food, fuel, and fibers. If the crops were to be used for food, one would obviously be concerned over possible contamination with pathogens. There is also the possibility that such crops might concentrate toxic substances such as heavy metals from the sewage effluent.

As far as pathogens are concerned, it is important to keep in mind that enteric pathogens do not survive well in soil and that proper cooking will kill virtually any pathogen. Furthermore, there is no evidence that germs can be assimilated through a plant's root system and transported, for example, to fruits and seeds. A number of studies with sewage sludge (see the later discussion) have clearly indicated that plants can pick up toxic substances from the soil (Sopper and Kerr, 1981; Dowdy et al., 1984), but such effects are not always observed (Robson and Sommers, 1982), and it appears that uptake can be minimized through plant breeding (Hinesly et al., 1984). The best safeguard against contamination with toxic substances is common sense and careful monitoring of crops.

The EPA published guidelines for agricultural use of wastewater in 1992, and by that time, at least 19 states had set regulations or guidelines for the use of reclaimed water on food crops (NAS, 1996). A summary of the EPA guidelines is presented in Table 6.5. Although the EPA guidelines recommend secondary treatment in all cases, the guidelines are not binding, and some states permit the use of primary effluent to irrigate food crops that are commercially processed as long as requirements are met to protect workers and nearby residents and to prevent water pollution (NAS, 1996). In Israel, where water conservation is critical and where sewage irrigation is common, it is acceptable to use primary treated sewage to irrigate both vegetables and fruits destined for human consumption, with the proviso that the vegetables must be cooked and the fruits peeled before eating or, in the case of fruits that are not peeled before eating, that irrigation be stopped one month before harvesting (Heukelekian, 1957). At the present time, Israel recycles the sewage from

TABLE 6.5   SUMMARY OF EPA GUIDELINES FOR THE AGRICULTURAL USE OF RECLAIMED WASTEWATER

Type of Reuse	Treatment Recommended	Water Quality
Food crops not commercially processed	Secondary filtration Disinfection	<2.2 fecal coliform/100 mL 1 mg L^{-1} Cl_2 residual after 30 min contact time (minimum) Turbidity ≤2 NTU BOD ≤10 mg L^{-1}
Food crops commercially processed, including orchards and vineyards and nonfood crops such as pasture, fodder, fiber, and seed	Secondary disinfection	≤200 fecal coliform/100 mL 1 mg/L Cl_2 residual after 30 min contact time (minimum) BOD ≤30 mg L^{-1}

about two-thirds of its population, primarily for agricultural use. Israel's goal is to recycle 80% of its wastewater (Shuval et al., 1984).

Until now, we have discussed the pathogen problem from the standpoint of contamination of groundwater and crops. However, studies by Shuval et al. (1984) have shown that sprinkler irrigation results in aerosolization of about 0.1–1.0% of the effluent, and pathogens may be dispersed as much as several hundred meters in the resultant aerosols (Majeti and Clark, 1981). This aerial transport of pathogens could create public health problems for persons living or working downwind from the spray irrigation operation.

Perhaps the most noteworthy study of this problem was carried out by Shuval et al. (1984) in Israel. During a four-year period they compared the incidence of 12 illnesses associated with enteric pathogens at 11 kibbutzim (cooperative agricultural settlements) where lagoon-treated sewage effluent was used to irrigate crops during two of the four years. The incidence of an additional group of 12 illnesses not associated with enteric pathogens was monitored to check for any changes in the health of the roughly 3,000 people in the study unrelated to the sewage spray irrigation. No effort was made to disinfect the sewage effluent, and based on the reported coliform and enteric virus concentrations, the effluent was grossly contaminated with pathogens. Shuval et al. (1984) found that the incidence of enteric diseases was 32–112% higher among children less than five years old during the periods of wastewater irrigation, but the corresponding excess risk in other age groups appeared to be no more than 10%. Shuval et al. (1984) concluded that any health problems associated with the spray irrigation system would very likely have been negligible if the sewage had been disinfected using standard procedures.

A somewhat different approach to land application of sewage effluent was taken at the Flushing Meadows project, where the Salt River bed west of Phoenix, Arizona, was used for ground recharge of secondary sewage effluent (Bouwer et al., 1974). The objective of this project was to apply sewage effluent at as high a rate as possible to experimental plots on the river bed without degrading soil quality while at the same time producing groundwater that could subsequently be used for unrestricted irrigation and recreation. The application rates averaged about 90 m yr^{-1}, or about 30–35 times the application rates used in the Pennsylvania State and Tallahassee studies. The effluent was applied to the plots by simply flooding the plots to various depths with effluent and then allowing the water to percolate into the ground. The soil was allowed to dry for variable periods of time between floodings. Some experimental plots contained no vegetation, but others were planted with various grasses or rice. No attempt was made to harvest the vegetation. The following is a brief summary of the experimental results:

1. Percolation of the effluent through the sand and gravel of the riverbed removed virtually 100% of the SS, BOD, and fecal coliform bacteria.
2. With alternate flooding and drying periods of two weeks each, it was possible to remove about 30% of the N from the effluent. The mechanism appeared to be adsorption of NH_4^+ to soil particles during flooding, conversion of the NH_4^+ to NO_3^- during subsequent drying periods as the soil became partially aerobic, and finally, denitrification in anaerobic parts of the soil during the drying period. This sequence of steps is very sensitive to O_2 levels. The conversion of NH_4^+ to NO_3^- requires O_2, but the conversion of NO_3^- to N_2 gas can occur only in the almost complete absence of O_2. Despite this problem, laboratory studies indicated that N removal efficiencies as high as 90% might be achieved by further modifications in the system.
3. Phosphate removal of about 90% was achieved after about 100 m of underground travel. The mechanism of removal presumably involved a combination of precipitation and adsorption to soil particles.
4. Cu and Zn concentrations were reduced by about 80% and Hg concentrations by about 35%, but Cd and Pb concentrations were virtually unchanged. None of these

metals was present in the renovated water at concentrations that would be objectionable for irrigation or recreational use.

Limiting Factors

The conclusion that emerges from the studies of land application of sewage effluent is that the two most important factors limiting the feasibility of this mode of disposal are the availability of suitable land and the accumulation of nitrate in groundwater. If the effluent is applied at a rate of 6 cm/wk, for example, and if the per capita sewage flow is about 600 L day^{-1}, then a city with a population of 100,000 persons would need about 7 km^2 to dispose of its sewage effluent. Additional land areas would probably be needed in places where plant root systems are inactive during the winter. Large areas of land suitable for irrigation are apt to be scarce near large population centers, and the cost of such land is not likely to be low. The Flushing Meadows project showed that much higher application rates are possible if the soil is sufficiently permeable, but had the soil consisted of clay rather than sand, it would have been impossible to sustain such high irrigation rates.

One of the most serious limiting factors in the land application of sewage, regardless of whether the effluent is used to grow crops or not, is the accumulation of nitrate in groundwater. The Pennsylvania State study made it clear that under appropriate conditions, a major portion of the N in the effluent may be assimilated by crops and removed at harvest time. In the case of the Tallahassee work, however, the percolation rate of the effluent through the soil was so rapid that even after fine-tuning, the crop root systems removed only about two-thirds of the N from the wastewater. Furthermore, crop root systems will obviously assimilate none of the N in temperate latitudes during the winter months. An alternative to growing crops to remove N is the approach used at the Flushing Meadows project, in which high average application rates and alternate flooding and drying periods resulted in N removal via denitrification. Regardless of which approach is used, it is clear that careful planning and monitoring of sampling wells are necessary if groundwater nitrate levels are to be maintained below 10 mg L^{-1}.

It is thought-provoking to realize that the United States produces about 134 million m^3 of wastewater per day, about 26% of current water use for crop irrigation (518 million m^3 day^{-1}). Only about 25% of U.S. cropland is used to grow food crops (NAS, 1996). About 70% of U.S. cropland is used to grow hay and feed for livestock. In other words, all the wastewater produced in the United States could be used to irrigate hay and feed crops, a scenario that poses virtually no risk to humans. The fact is that only about 3% of the wastewater produced in the United States is reclaimed, and only about 70% of that water is used for irrigation. In most parts of the country, it is cheaper to pump irrigation water out of the ground than to pump it from the nearest POTW. For example, in a recent study in the Tampa, Florida, area, the cost of supplying reclaimed wastewater for agricultural irrigation was estimated to be 18–24 cents per cubic meter. The cost to the farmer of pumping water from the local aquifer was 2.5–4.0 cents per cubic meter. However, exceptions to this pattern do exist. In Southern California, for example, the wholesale cost of freshwater is roughly 40 cents per cubic meter and can exceed 65 cents per cubic meter (NAS, 1996). In such cases, there is an economic incentive to reclaim wastewater. One might assume that the nutrients in sewage effluent would add to its economic value for irrigating crops. In fact, fertilizer accounts for only about 6% of the production expenses of farming in the United States (NAS, 1996). The nutrients in sewage effluent therefore have little positive impact on the economics of its use for irrigating crops. Furthermore, when crop nutrient needs are not in phase with irrigation demands, the high nutrient content of sewage effluent may become a liability. Overfertilization can cause excessive growth, reduce crop yield, and encourage weed growth (NAS, 1996). In some areas, however, pollution problems associated with the discharge of wastewater and/or competing demands for potable water may dictate that wastewater be recycled. This has been the case, for example, with the wastewater produced by Hawaiian sugar mills (see Chapter 9).

Use of Sewage Sludge

In addition to the liquid effluent from a POTW, the solid waste (sludge) may also be put to good use in land application. Typical sludge from POTWs contains about 3.2% N, 1.8% P, and 0.3% K (Abron-Robinson et al., 1981). Because of its low K content the sludge is not an ideal fertilizer, but it is an excellent source of N and P. Because the nutrients are bound in organic matter, they are slowly released following land application, and experience has shown that a single application of sludge can provide the N and P requirements of terrestrial plants for as long as 3–5 years (Sopper and Seaker, 1984). At the present time, POTWs in the United States generate about 8 million tonnes of sludge per year (Marinelli, 1990). If this sludge were used as fertilizer to supply the N needs of nonleguminous plants, an application rate of about 23 tonnes ha^{-1} yr^{-1} would be required (Lewicke, 1972). Thus, the fertilizer N requirements of about 0.35 million ha of cropland could be supplied with sewage sludge. The market value of the sludge as a source of N, P, and K would be about $200 million per year (Abron-Robinson et al., 1981). While these figures may sound impressive, the fact is that all the sewage sludge produced in the United States could supply only 1–2% of the annual fertilizer N required for crop production (Dowdy et al., 1984; NAS, 1996). Furthermore, municipalities would have to bear the cost of transporting the sludge if land application were to be cost competitive with commercial fertilizer (Abron-Robinson et al, 1981).

Sewage sludge will therefore not significantly change U.S. reliance on commercial fertilizer for growing crops. Several factors, however, do make land application of sewage sludge very appealing. First, legislation prohibiting ocean dumping of sludge was enacted in 1981 (Clark et al., 1984). Second, in many parts of the United States, sanitary landfills for disposal of sludge have been or are becoming rapidly filled. Third, incinerators used to burn sludge have been phased out in some parts of the United States due to air pollution problems (Sopper and Kerr, 1981). Finally, strip mining in 31 states has laid bare approximately 2.0 million ha of land, less than half of which has been properly reclaimed. About 0.25 million ha of land are disturbed each year as a result of strip mining activity (Sopper and Kerr, 1981). The Federal Surface Mining Control and Reclamation Act of 1977 requires backfilling and restoration of strip-mined land and maintenance for periods of either 5 or 10 years, depending on whether annual rainfall in the area is greater or less than 66 cm, respectively. Restoring strip-mined land has proven not to be an easy task, but research has shown that the job is greatly facilitated by applying sewage sludge to the land.

Thus, there is considerable motivation for applying sludge to the land. The EPA first promulgated criteria for land application of sewage sludge to cropland in 1979, and final standards for the use and disposal of sewage sludge were promulgated in 1993 (EPA, 1993). The final standards are commonly referred to as the "Part 503 Sludge Rule." The Sludge Rule puts sludge in one of two categories. Class A sludge is safe for direct contact and can be used in an unrestricted manner. Restrictions apply to the use of Class B sludge.

During the 1980s, a number of studies addressed the feasibility of land application of sludge as a method of disposal. The concerns were similar to those associated with land application of the liquid effluent: pathogens, nitrate in groundwater, and heavy metals. The studies conducted to date, some of which have lasted as long as 5–15 years, have revealed no evidence of contamination of groundwater with fecal pathogens or nitrate (Hinesly et al., 1984; Sopper and Seaker, 1984). According to Page et al. (1984), groundwater monitoring for nitrate is unnecessary where sludge N application rates do not exceed fertilizer N recommendations. The lack of a nitrate problem is probably associated with the slow release of the N from the organic matter in the sludge. Crop yields on land fertilized with sludge often exceed those obtained with commercial fertilizer (Hinesly et al., 1984; Sopper and Seaker, 1984). The lack of a pathogen problem in groundwater is not surprising considering the results obtained in spray irrigation studies, but reasonable precautions to avoid pathogen exposure must be taken by persons directly involved in applying the sludge to the land (Clark et al., 1984). The principal concern is the presence of parasite ova and cysts, which tend to become concentrated in the sludge during the process of sewage treatment. Helminth ova are resistant to the environmental factors that reduce the concentrations of bacteria and viruses in sludge, and the concentration of such ova is one of the principal criteria for determining whether sludge falls into Class A or Class B. Heavy metals are sometimes found at much higher concentrations in municipal sludge than in typical agricultural soils. The uptake of these metals by crops and/or their migration into groundwater appear to be the most serious factors limiting land application of sludge. Dowdy et al. (1984), for example, reported that Cd and Zn concentrations were elevated in corn silage produced on sludge-fertilized land, but there was no evidence of elevated metal concentrations in the tissues of animals fed the corn silage. Sopper and Kerr (1981) found that metal concentrations were higher than usual in vegetation grown on strip-mined land restored with sewage sludge, but they felt that the metal concentrations were below problematic levels. However, Pb concentrations in groundwater exceeded standards for potable water, and in a later study, Sopper and Seaker (1984) reported that both Pb and Cr concentrations periodically exceeded drinking water standards in groundwater below sludge-fertilized land. A National Academy of Sciences report concluded that Cd was the inorganic chemical of greatest concern in sewage sludge and noted that Cd could be absorbed by crops and reach "levels that are dangerous to humans if a high percentage of the consumer's diet is derived from crops grown on cadmium-contaminated soil over an extended time period" (NAS, 1996, pp. 109–110).

The most dramatic success of land application of sewage sludge has undoubtedly been the restoration of strip-mined land. Establishing vegetation on such land is extremely difficult due to the lack of nutrients, low organic matter content, low pH, low water retention, and toxic levels of metals in the soil (Sopper and Kerr, 1981). The acidity of the soil results from the oxidation of sulfides such as pyrite (FeS_2) to produce sulfuric acid when the sulfides are exposed to air and water. The acidic nature of the soil mobilizes metals and makes retention of certain essential nutrients difficult. Restoration of the land is accomplished by first grading and tilling. Lime is then applied to neutralize the acid temporarily. Sludge is then applied at typically 100–250 tonnes ha^{-1}, and the ground is seeded with vegetation. Plants then grow rapidly, and the soil pH stabilizes at a neutral value. Vegetative cover generally increases over the course of the next few years. In several documented cases, strip-mined land that had resisted previous attempts at restoration recovered dramatically following application of sewage sludge (Sopper and Kerr, 1981; Hinkle, 1984; Sopper and Seaker, 1984). In the study of Hinkle (1984), progress was slowed by several years of drought, and there was no improvement in the acidity of a stream draining the disturbed area. However, about 90% vegetative cover was established on the land after seven years, and heavy metal concentrations generally declined in the drainage stream.

One important consideration in the land application of sludge is the fact that sludge can be transported economically much greater distances than wastewater. For example, New York City's treated sludge is currently shipped to northeastern Texas and eastern Colorado for cropland application. Boston ships a portion of its sludge in the form of heat-dried pellets to Florida for application to cropland and pastures, and some of the sludge generated in the Los Angeles basin is transported by truck for cropland application in Yuma, Arizona (NAS, 1996).

The consensus that emerges from this analysis is that land application of sewage sludge is a very practical means of disposal. This conclusion undoubtedly accounts in part for the fact that 36% of municipal sewage sludge is now applied to the land in the United States. However, except for the reclamation of strip-mined land, land application of sludge appears to be little more than a convenient means of disposal. Sludge is a source of nutrients and a good soil conditioner, but as already noted, fertilizer accounts for only a small percentage of the cost of crop production in the United States. Furthermore, the nutrient content of sludge is not as consistent as that of commercial fertilizer. The negative side of land application of sludge is the presence of potentially toxic substances and pathogens in the sludge. The chief cause for concern appears to be the accumulation of Cd and possibly other toxic metals in the soil, groundwater, or vegetation. To date, however, there has been no evidence of a problem with phytotoxicity, and to the extent that crops are not to be marketed for food, there is little reason for concern over contamination of terrestrial food chains (Abron-Robinson et al., 1981). Furthermore, there is no evidence that animals fed crops grown on sludge-amended soils accumulate objectionable concentrations of heavy metals (Dowdy et al., 1984; Hinesly et al., 1984). Groundwater contamination with heavy metals is perhaps the most sensitive problem associated with land application of sewage sludge, and it clearly seems advisable to monitor groundwater in areas where sludge is being applied. Adjustments of sludge application rates may well be sufficient to keep groundwater metal concentrations within acceptable limits. The use of sewage sludge to restore strip-mined land is probably the most beneficial use of the sludge. Given the present and future anticipated extent of strip mining activity, it seems likely that a large percentage of the sludge generated in the United States could be used for this purpose for many years.

UNCONVENTIONAL SEWAGE TREATMENT

There are several alternatives to conventional sewage treatment as it has been described in this chapter. For example, one can forgo biological systems for removing BOD and rely instead on largely physical and chemical methods to treat the sewage. Kreissl and Lewis (1981), for example, describe a treatment process involving chemical clarification with hydrated lime, nitrification of the effluent from the clarifier, removal of additional particulate material with a dual-media filter, and removal of dissolved organics with activated C columns. The complete system achieved 97%, 93%, 80%, and 32% removal of SS, BOD, phosphate, and total N, respectively. The normal operation and maintenance of the plant, however, was more time-consuming and complex than that of conventional treatment, and the staff required for the hybrid system was about three times that of an extended aeration system designed to achieve comparable removals of SS and BOD. One advantage of the physical/chemical system over conventional secondary treatment was the much greater removal of phosphate, but it does not appear that physical/chemical treatment is any more cost effective than advanced biological treatment for tertiary nutrient removal (U.S. Army, 1972). The chief appeal of physical/chemical treatment systems is the fact that they can be packaged in small units and brought on line rapidly, which may provide special appeal under certain conditions (Kreissl and Lewis, 1981).

Another alternative to conventional treatment is natural systems of microorganisms and higher plants. There are basically three types of such systems. The first consists of artificial marshes analogous to the marsh used by Tallahassee to treat runoff into Lake Jackson. The second involves artificial rock marshes, in which wastewater flows laterally underground through a bed of rocks within which plants are rooted. The third is sometimes referred to as *solar aquatics* and involves growing microorganisms, plants, and in some cases animals in artificial enclosures illuminated with natural sunlight. At the present time, there are an estimated 150 artificial marsh and solar aquatic systems treating municipal sewage in the United States.

Probably the best-known artificial marsh sewage treatment system in the United States is operated by the town of Arcata, California. Sewage from the town's 15,000 residents is first routed to a primary treatment facility and then to 20 ha of oxidation ponds containing microalgae. The effluent from the oxidation ponds is routed to two 1.0-ha artificial marshes planted with bulrush and cattails, and then into 18 ha of additional marshes constructed in part by the California Coastal Conservancy to help restore fish, shellfish, waterfowl, and other wildlife in the area. The residence time of water in the treatment system is about two months (Stewart, 1990), and by the time the effluent flows into Humboldt Bay, it is cleaner than the water in the bay.

Rock marsh systems are variations on the more common artificial marshes and are capable of producing effluent that satisfies secondary treatment standards. Wastewater is routed through lined trenches filled with rocks. Wastes in the sewage are broken down by microorganisms growing on the rocks and assimilated by plants rooted in the rocks. One of the largest rock marsh systems in the United States is operated by the town of Denham Springs, Louisiana. Wastewater from the town's 20,000 residents is partially treated in two 16-ha ponds and then routed through three 2-ha rock marshes. The rock marshes are planted primarily with ornamental flowers and are designed to treat 11 mLd. Compared to the cost of a secondary POTW, the rock marshes saved the town of Denham Springs about $1 million in construction costs and are saving about $60,000 per year in maintenance costs (Marinelli, 1990).

Solar aquatic treatment systems have been pioneered by scientists at the New Alchemy Institute on Cape Cod, Massachusetts, and frequently involve growing plants and animals in greenhouse environments. Microalgae are grown in vertically oriented cylindrical tanks; macrophytes are grown hydroponically. Fish and aquatic invertebrates are often grown in the microalgal tanks. Solar aquatic systems involve more manipulation and control of the environment than artificial marshes and are capable of achieving much higher wastewater treatment rates per unit area. In fact, solar aquatic systems require no more land area than conventional sewage treatment and can provide advanced wastewater treatment at about two-thirds the cost of conventional secondary treatment systems (Marinelli, 1990).

Whether artificial marsh and solar aquatic sewage treatment systems are the wave of the future remains to be seen. They are certainly less expensive to build and operate than conventional treatment systems. For example, the Arcata, California, treatment system cost only a little over $500,000 to build, or about $33 per capita. Much larger secondary treatment plants, which should be cheaper based on economies of scale, actually cost five to seven times as much per capita to build.[9] Furthermore, the artificial marsh and solar aquatic systems produce little or no sludge, achieve a fair degree of tertiary nutrient removal, and have even been shown capable of removing pollutants such as heavy metals, pesticides, and industrial toxins. The artificial marsh systems, however, require considerably more area than a conventional POTW. The two-month residence time of wastewater in the Arcata artificial marsh, for example, is several orders of magnitude longer than the residence time of water in a conventional POTW. Solar aquatic systems require about the same amount of land as conventional POTWs, but at this time, solar aquatic systems have been less thoroughly tested than the other systems, and a different kind of expertise is needed to operate them. However, despite some uncertainties and limitations, the interest these nonconventional forms of treatment have generated in recent years suggests that they will be given careful consideration in the future and may one day largely replace conventional sewage treatment systems.

---

[9]Given the figures in Table 6.3, the cost of a 60-mLd secondary treatment plant is about $23.8 million. Assuming that this plant services 100,000 persons, the per capita cost is $238. The corresponding cost of a 360-mLd plant is $159 per capita.

DETERGENT PHOSPHATES

The use of phosphates in laundry and dishwashing detergents is pertinent to our discussion of the treatment and use of sewage, because historically as much as 50% or more of the P in municipal wastewaters has come from the phosphates used in these detergents. P in the form of sodium tripolyphosphate (STP) was used as a component of laundry and dishwashing detergents from the time detergents were first introduced on the market in the 1930s until roughly 1985. Initially the amount of P released to the environment as a result of detergent use was quite small because the public was rather slow to switch from the traditional soap flakes or soap powders for cleaning clothes and dishes. However, detergents had largely replaced soap products for such cleaning purposes by the 1950s, and by 1971 it was estimated that about 30–40% of the P entering the aquatic environment was coming from the phosphate in the wastewater from detergent washing operations (Grundy, 1971). Since there is reason to believe that many freshwater systems are P-limited (Schindler, 1974), it occurred to many ecologists that removal of P from detergents might be an easy and cheap way to reduce or perhaps eliminate cultural eutrophication problems in some freshwater systems. As a result, pressure was brought on detergent manufacturers to find a substitute for STP in their products, and a number of new detergent products containing no P rapidly appeared on the market. The president of Procter and Gamble was led to comment, "We recognize that the public wants phosphates out of laundry detergents, and we intend to take them out. Our job is to make certain that we remove them as rapidly as we can do so in a thoroughly reasonable manner. This we are doing" (Morgens, 1970, p. 1). A glance at the detergent section in many grocery stores in the United States today will reveal that many detergents indeed contain little no phosphorus, and virtually all contain less than 0.5% P.[10] If it is possible to produce detergents without phosphorus, why was STP added to detergents in the first place?

To answer this question, we must understand what role STP played in laundry detergents. A detergent contains two important components, a surfactant and a builder, which together often account for over half of the weight of the product. A filler, typically sodium sulfate, and moisture account for most of the remaining weight of the detergent (Layman, 1984). Surfactants consist primarily of alkylsulfonates, fatty alcohol ethoxylates, and fatty alcohol sulfates. All of these compounds are bipolar, and because of this characteristic, they are able to penetrate between dirt and fabric. One side of the surfactant molecule is attracted to the fabric; the other side is attracted to the dirt. Surfactants also help to lift off grease by lowering surface tension. The principal purpose of the builder is to soften the water by removing Ca and Mg ions. Otherwise, these ions combine with the surfactant to form a gummy precipitate or curd that cannot be effectively removed with the rinse water. The curd settles onto the clothes, forming a dull film, and may build up in the washer, ultimately clogging the washing machine. STP acts as a builder by chemically sequestering Ca and Mg so that they do not have a chance to combine with the surfactant. STP is an excellent water softener. Phosphates also peptize and help suspend certain kinds of dirt, aid in germ killing, buffer the pH of the water to aid in the removal of fatty soils, and increase the effectiveness of the surfactant by breaking up surfactant polymeric units.

How soft must water be before a surfactant alone (e.g., soap) will do an adequate job of washing clothes? According to Duthie (1972), about 3 grams of P is required to neutralize the hardness contributed by perspiration, soil, food, and previous rinsing of the clothes. If any Ca and Mg are present in the water, then additional P is required. Figure 6.2 shows the approximate amount of P that would be required to neutralize all the hardness in a washload as a function of water hardness. The implication of Figure 6.2 is that

---

[10]A product containing 0.5% P in the form of STP contains about 2% STP.

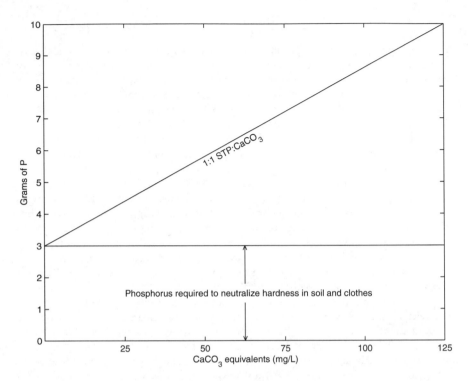

FIGURE 6.2   P required to neutralize total hardness from both washwater and clothes in a 64-L capacity washer. *Source:* Redrawn from Duthie (1972). Reproduced with permission.

a certain amount of P, or of some other builder, is required to neutralize hardness even in water with virtually no Ca and Mg. The challenge for detergent manufacturers was to find a suitable substitute for STP as a water-softening agent.

The search for a substitute turned out to be more difficult than detergent manufacturers initially expected. The principal alternatives to STP are sodium citrate, nitrilotriacetate (NTA), zeolites such as sodium aluminosilicates, and sodium carbonate. Sodium citrate and NTA are sequestering builders, and in this respect are similar to STP. Zeolites soften water through ion exchange. Sodium carbonate softens water by combining with Ca and Mg ions to form a granular precipitate that is easily flushed from the washing machine during the rinse cycle. During the late 1960s, it appeared that the STP in many detergents would be replaced by NTA, but this alternative, upon which detergent manufacturers placed great hopes and in which they invested much money, was squelched by the U.S. Surgeon General on December 18, 1970, as a result of tests performed by the National Institute of Environmental Health Science (NIEHS). The tests indicated that NTA interacted with heavy metals such as Cd and Hg in such a way as to increase the transmission of these metals across the placental barrier to the fetus, thus increasing the likelihood of birth defects. Ruckelshaus and Steinfeld (1970, p. 1) admitted that the doses of NTA and toxic metals used in the NIEHS animal tests were "considerably higher than would ordinarily be encountered by the human population" and that, "there is no evidence at this time to indicate that anyone has been or is being harmed by the combination of NTA with metals in the environment." Furthermore, NTA is normally degraded in sewage treatment plants (Swisher et al, 1967) and in aerobic aquatic systems, and therefore would not be present to interact with the metals anyway. However, it is not clear how effectively NTA may be degraded in anaerobic systems, so that it is possible, for example, that disposal of anaerobically digested sewage sludges might contaminate certain systems with NTA. Procter and Gamble did some limited test marketing of NTA as a builder for detergents in Indiana and New York, but concern over a possible legal ban on NTA caused the company not to pursue the matter further. At the present time, almost all detergents marketed in the United States are built with sodium carbonate, sodium citrate, or a combination of sodium carbonate and sodium aluminosilicates. In contrast to the United States, both Canada and Finland have permitted the substitution of NTA for STP in detergents. In Japan, over 90% of detergents are built with zeolites.

Impact of Detergent Phosphate Reductions

The impact of detergent phosphate reductions lies primarily in two areas, one financial and the other ecological. First of all, what has been the financial cost of eliminating or reducing the P content in many laundry detergents? Given the fact that detergent phosphates never accounted for more than 50% of the P in typical municipal wastewater and that the wastewater from most households is routed through sewage treatment plants in the United States, would it not make more sense to remove the P at the sewage treatment plant? The answer is that where P enrichment of receiving waters is perceived to be a problem, it certainly makes sense to remove the P from raw sewage. This approach, for example, is being taken in the Lake Erie watershed, and in Sweden the wastewater from about 80% of the urban population is given tertiary treatment to remove nutrients (Layman, 1984). However, one should keep in mind that marine waters receive about half of the sewage effluent from the U.S. population (Ryther and Dunstan, 1971), and a substantial body of data now indicates that N rather than P is the principal limiting nutrient in many marine waters. Given this state of affairs, it might be more cost effective to remove the N from the sewage than the P.

If P removal is considered to be a cost-effective way to control cultural eutrophication, then P removal from point source discharges may be the most cost-effective way to deal with the problem. However, the lower the P concentration in the raw sewage, the easier it is for tertiary treatment to reduce the P concentration in the effluent to a specified level. Ecologically, the most effective strategy is to remove the P from detergents and also provide tertiary P removal at sewage treatment plants. P removal from detergents alone is ecologically a much less effective strategy, since on the average, detergents never accounted for more than 30–40% of P inputs to aquatic systems in the United States. Lee and Jones (1986) argue that removal of P from laundry detergents will have no discernible effect on water quality unless detergent P accounts for at least 20% of the P load to a body of water.

The cost of substituting a builder like NTA for STP in laundry detergents increases the cost to the consumer by about 40 cents per kilogram.[11] At the present time, per capita use of detergents amounts to about 5 kg yr^{-1} (Layman, 1984). Hence, the incremental cost of replacing STP with NTA is $2 per person per year. Ninety percent removal of P via tertiary treatment costs about $7 per person per year.[12] The advantage of spending the additional $2 to remove the P from laundry detergents stems from the fact that 90% P removal reduces the P content of typical raw sewage from 10 to 1 ppm. In the Great Lakes basin, 1 ppm P is the standard for treated sewage, and POTWs need to operate with some margin of error. In some cases, even 1 ppm P in sewage effluent causes problems (see below). In other words, reducing the P concentration from 10 to 1 ppm is not really adequate in

---

[11]Based on Hammond (1971) and an inflation factor of 5 (Grogan, 1998).
[12]Based on Lee and Jones (1986) and an inflation factor of 2 (Grogan, 1998).

many cases. It is technically possible to achieve as much as a 95% reduction in P, but the additional cost of consistently achieving 95% P removal is much more than $2 per person per year. The cheaper and technologically simpler solution has been to remove the P from laundry detergents. In this way, the concentration of P in the raw sewage is reduced by an average of 30–40%, and 90% P removal reduces the effluent P concentration to 0.6–0.7 ppm. That provides the POTW with the margin of error it needs.

Judging from the case studies presented in this book, it seems unlikely that P removal from laundry detergents by itself would effect much improvement in the trophic status of many receiving waters. A 30–40% reduction in the P input to Lake Washington or Lake Erie would have produced little improvement in water quality. A very interesting example of what can and cannot be achieved by reducing the P content of laundry detergents comes from Onondaga Lake, located just north of Syracuse, New York. About 20% of the inflow to the lake is municipal wastewater, and the lake has been characterized as "the most polluted lake in the United States" (Effler and Hennigan, 1996, p. 1). One of the serious problems with the lake is cultural eutrophication. In an effort to deal with this problem, on July 1, 1971, Syracuse enacted legislation requiring that detergent P concentrations not exceed 8.7%. The State of New York passed similar legislation in January 1972 (Murphy, 1973) and then further reduced the P limit to 0.5% in July 1973. The result was about a fivefold reduction in the concentration of P in the sewage entering the lake, from 11.5 mg L^{-1} in 1970 to 2.3 mg L^{-1} in 1974–1975 (Figure 6.3). Likewise, between 1970 and 1972, the concentration of P in the epilimnion of the lake dropped from 730 to 110 μg L^{-1} (Figure 6.4). This change was certainly desirable, but 110 μg L^{-1} is far above the mesotrophic → eutrophic transition at 20 μg L^{-1}. The lake continued to experience serious eutrophication problems, including low O_2 concentrations and poor water clarity. In an effort to fur-

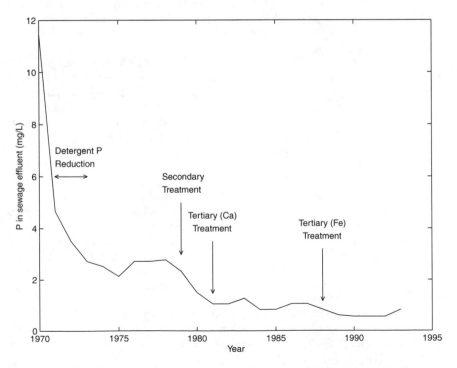

FIGURE 6.3   Total P concentrations in sewage effluent entering Onondaga Lake from 1971 to 1993. *Source:* Effler et al. (1996). Reproduced with permission.

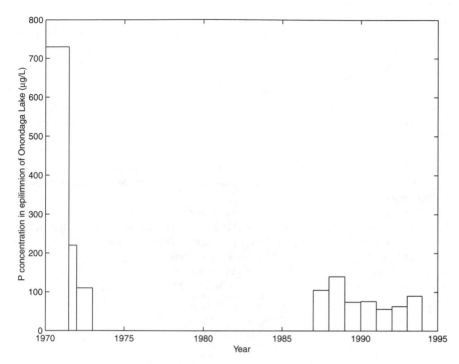

FIGURE 6.4   Total P concentrations in the epilimnion of Onondaga Lake from 1970 to 1972 and from 1987 to 1993. *Sources:* Murphy (1973) and Effler et al. (1996).

ther improve conditions in Onondaga Lake, Onondaga County upgraded the local sewage treatment plant (METRO) from primary to secondary in 1979. The effect of the upgrade was to reduce the P concentration of the sewage effluent from 2.3 to 1.5 mg L^{-1} (Figure 6.3). This was still above the standard of 1.0 mg L^{-1} for POTWs in the Great Lakes watershed,[13] and tertiary treatment using Ca^{2+} to precipitate P was introduced in 1981. This upgrade reduced the P concentration of the METRO effluent to an average of 1.0 mg L^{-1} from 1982 to 1987. In this case, the choice of Ca^{2+} to precipitate phosphate was motivated by the cheap availability of Ca-rich waste from a local soda ash (sodium carbonate) chemical plant. However, the soda ash plant closed in 1986, and in subsequent years tertiary removal of P was achieved by precipitation with Fe^{3+}. Between 1989 and 1993 the P concentration in the METRO effluent averaged 0.6 mg L^{-1}, and the P concentration in the epilimnion of Onondaga Lake averaged 72 μg L^{-1} (Figure 6.4).

Several lessons can be learned from the data in Figures 6.3 and 6.4. First, the most dramatic changes in the P concentrations in the sewage effluent and in the epilimnion of Onondaga Lake were achieved by virtually eliminating P from laundry detergents and not by upgrading to secondary and tertiary sewage treatment. The implication of Figure 6.3 is that laundry detergents accounted for roughly 80% of the P in the wastewater entering the lake. This is an unusually high percentage. The rapid response of Lake Onondaga was not unexpected. The residence time of water in the lake is about 95 days (Effler and Hennigan, 1996). Despite the fact that the concentration of P in the METRO wastewater was ultimately reduced by about 95%, Lake Onondaga has continued to experience serious eutrophication problems (Effler and Hennigan, 1996). The explanation stems from the fact

[13]Onondaga Lake drains into Lake Ontario via the Seneca/Oswego river system.

that wastewater accounts for about 20% of the inflow to the lake. This is an extremely high percentage. The result is that even after removal of 95% of the P from the METRO effluent, the areal loading rate of P is still 7.8 g m^{-2} y^{-1}. This is more than 10 times the post-diversion areal loading rates of P to Lake Washington and Lake Sammamish (see Table 4.1). Furthermore, the mean depth of Onondaga Lake is only about 11 m, making it very susceptible to eutrophication problems. Realistically, the only way to overcome the eutrophication problems experienced by Onondaga Lake is to find some other place to discharge the METRO wastewater. Land application comes to mind, but the discharge from the METRO plant averages 2.7 × 10^5 m^3 d^{-1}. At an application rate of 5 cm wk^{-1}, almost 38 km^2 of land would be required to dispose of this much effluent. Finding this much land suitable for wastewater irrigation near a city of 450,000 is very problematic, and the cost of pumping wastewater over long distances becomes prohibitive. The future for Onondaga Lake is grim.

The net result of all the commotion over detergent phosphates has been that most laundry detergents contain no P, and those that do contain less than 0.5% P. However, with a few exceptions, cultural eutrophication problems in most aquatic systems are unlikely to be solved by removal of detergent phosphates alone. Tertiary sewage treatment or land application of sewage will be needed to eliminate most of the point source inputs, and in some cases, such as Lake Erie, control of nonpoint source inputs will probably be necessary as well.

REFERENCES

ABRON-ROBINSON, L. A., C. LUE-HING, E. J. MARTIN, and D. W. LAKE. 1981. *Production of Non-Food-Chain Crops with Sewage Sludge.* EPA-600/S2-199. U.S. EPA, Cincinnati, Ohio.

ALLEN, J. 1973. Sewage farming. *Environment,* 15(3), 36–41.

ALLHANDS, M. N., S. A. ALLICK, A. R. OVERMAN, W. G. LESEMAN, and W. VIDAK. 1995. Municipal water reuse at Tallahassee, Florida. *Trans. American Soc. Agricultural Eng.,* 38(2), 411–418.

ALLHANDS, M. N., and A. R. OVERMAN. 1989. *Effects of Municipal Effluent Irrigation on Agricultural Production and Environmental Quality.* Report for Wastewater Operations, Water and Sewer Department of Tallahassee, Florida. 377 pp.

BOUWER, H., J. C. LANCE, and M. S. RIGGS. 1974. High-rate land treatment. *J. Water Pollut. Contrl Fed.,* 46, 834–859.

CLARK, C. S., H. S. BJORNSON, C. C. LINNEMANN, JR., and P. S. GARTSIDE. 1984. *Evaluation of Health Risks Associated with Wastewater Treatment and Sludge Composting.* EPA-600/S1-84-014. U.S. EPA, Research Triangle Park, N.C.

DONOVAN, J. F., and J. E. BATES. 1980. *Guidelines for water reuse.* EPA-600/8-80-036. U.S. EPA, Cincinnati, Ohio.

DOWDY, R. H., R. D. GOODRICH, W. E. LARSON, B. J. BRAY, and D. E. PAMP. 1984. *Effects of Sewage Sludge on Corn Silage and Animal Products.* EPA-600/S2-84-075. U.S. EPA, Cincinnati, Ohio.

DUTHIE, K. V. 1972. Detergents: Nutrient considerations and total assessment. In G. E. Likens, Ed., *Nutrients and Eutrophication.* American Society of Limnology and Oceanography. Lawrence, Kan. Pp. 205–216.

EFFLER, S. W., C. M. BROOKS, and K. A. WHITEHEAD. 1996. Domestic waste inputs of nitrogen and phosphorus to Onondaga Lake, and water quality implications. *Lake and Reserv. Manage.,* 12, 127–140.

EFFLER, S. W., and R. D. HENNIGAN. 1996. Onondaga Lake, New York: Legacy of pollution. *Lake and Reserv. Manage.,* 12, 1–13.

ELIASSEN, R., and G. TCHOBANOGLOUS. 1969. Removal of nitrogen and phosphorus from waste water. *Environ. Sci. Technol.,* 3, 536–541.

ELLIS, K. V. 1980. The tertiary treatment of sewages. *Effluent and Water Treatment J.,* 20(9), 422–430; 20(11), 527–537.

EPA. 1972. *Water Quality Criteria.* EPA-R3-73-003. Washington, D.C. 594 pp.

EPA. 1986. *Quality Criteria for Water.* EPA 440/5-86-001. Washington, D.C.

EPA. 1992. *Manual Guidelines for Water Reuse.* EPA 625/R-92/004. Washington, D.C.

EPA. 1993. Standards for the use and disposal of sewage sludge, Final Rule, 40 CFR Part 257, 403 and 503. Federal Register **58**(32), 9248–9415.

FERTIK, H. A., and R. SHARPE. 1980. Optimizing the computer control of breakpoint chlorination. *ISA Trans.*, **19**(3), 3–17.

GILBERT, R. G., R. C. RICE, H. BOUWER, C. P. GERBA, C. WALLIS, and J. L. MELNICK. 1976. Wastewater renovation and reuse: virus removal by soil filtration. *Science*, **192**, 1004–1005.

GROGAN, T. 1998. Builders' construction cost indexes. *Engineering News-Record*, **241**(23), 30.

GRUNDY, R. D. 1971. Strategies for control of man-made eutrophication. *Environ. Sci. Tech.*, **5**, 1184–1190.

HALL, R. I., P. R. LEAVITT, R. QUINLAN, A. S. DIXIT, and J. P SMOL. 1999. Effects of agriculture, urbanization, and climate on water quality in the northern Great Plains. *Limnol. Oceanogr.*, **44**, 739–756.

HAMMOND, A. L. 1971. Phosphate replacements: Problems with the washday miracle. *Science*, **172**, 361–363.

HERSHAFT, A., and J. B. TRUETT. 1981. *Long-Term Effects of Slow-Rate Land Application of Municipal Wastewater*. EPA-600/S7-81-152. U.S. EPA, Washington, DC.

HEUKELEKIAN, H. 1957. Utilization of sewage for crop irrigation in Israel. *J. Water Pollut. Contr., Fed.*, **29**, 868–874.

HICKEY, J. L. S., and P. C. REIST. 1975. Health significance of airborne microorganisms from wastewater treatment processes. *J. Water Pollut. Contr. Fed.*, **47**, 2741–2773.

HINESLY, T. D., L. G. HANSEN, D. J. BRAY, and K. E. REDBORG. 1984. *Long-Term Use of Sewage Sludge on Agricultural and Disturbed Lands*. EPA-600/S2-84-128. U.S. EPA, Cincinnati, Ohio.

HINKLE, K. R. 1984. *Reclamation of Toxic Mine Waste Utilizing Sewage Sludge: Contrary Creek Demonstration Project*. Addendum report. EPA-600/S2-84-016. U.S. EPA, Cincinnati, Ohio.

INTERNATIONAL JOINT COMMISSION. 1977. *Annual Report of the Research Advisory Board*. Windsor, Ontario 45 pp.

KARDOS, L. T. 1970. A new prospect. *Environment*, **12**(2), 10–27.

KARDOS, L. T., W. E. SOPPER, E. A. MYERS, R. R. PARIZEK, and J. B. NESBITT. 1974. *Renovation of Secondary Effluent for Reuse as a Water Resource*. EPA-660/2-74-016. Washington, D.C.

KLEIN, L. 1966. *River Pollution. 3. Control*. Butterworth, Washington, D.C. 484 pp.

KREISSL, J. F., and R. F. LEWIS. 1981. *Demonstration Physical Chemical Sewage Treatment Plant Utilizing Biological Nitrification*. EPA-600/S2-81-173. Cincinnati, Ohio.

LAYMAN, P. L. 1984. Brisk detergent activity changes picture for chemical suppliers. *Chem. and Eng. News*, **62**, 17–20, 31–49.

LEE, G. F., and R. A. JONES. 1986. Detergent phosphate bans and eutrophication. *Environ. Sci. Technol.*, **20**, 330–331.

LEWICKE, C. K. 1972. Recycling sludge and sewage effluent by land disposal. *Environ. Sci. Technol.*, **6**, 871–873.

MAJETI, V. A., and C. S. CLARK. 1981. *Potential Health Effects from Viable Emissions and Toxins Associated with Wastewater Treatment Plants and Land Application Sites*. U.S. Gov. Printing Office, EPA-600/S1-81-006. Cincinnati, Ohio.

MARINELLI, J. 1990. After the flush. The next generation. *Garbage*, **2**, 24–35.

MILLER, S. S. 1972. Debugging physical-chemical treatment. *Environ. Sci. Technol.*, **6**, 984–985.

MORGENS, H. J. 1970. The issue of phosphates in detergents. A statement to shareholders of Procter & Gamble Company, Cincinnati, October 13.

MURPHY, C. B., JR. 1973. Effect of restricted use of phosphate-based detergents on Onondaga Lake. *Science*, **182**, 379–381.

NATIONAL ACADEMY OF SCIENCES. 1996. *Use of Reclaimed Water and Sludge in Food Crop Production*. National Research Council, Washington, D.C. 178 pp.

PAGE, A. L., T. L. GLEASON, III, J. E. SMITH, JR., I. K. ISKANDAR, and L. E. SOMMERS. 1984. *Utilization of Municipal Wastewater and Sludge on Land: Proceedings of the 1983 Workshop*. EP-600/S9-84-003. U.S. EPA, Washington, D.C.

PAYNE, J. F., and A. R. OVERMAN. 1987. *Performance and Long-Term Effects of a Wastewater Spray Irrigation System in Tallahassee, Florida*. Report for Wastewater Division Underground Utilities City of Tallahassee. 329 pp.

PRESSLEY, T. A., D. F. BISHOP, and S. G. ROAN. 1972. Ammonia-nitrogen removal by breakpoint chlorination. *Environ. Sci. Technol.*, **6**, 622–628.

RILEY, J. P., and R. CHESTER. 1971. *Introduction to Marine Chemistry*. Academic Press, New York. 465 pp.

ROBSON, C. M., and L. E. SOMMERS. 1982. *Spreading Lagooned Sewage Sludge on Farmland: A Case History* EPA-600/S2-82-019. U.S. EPA, Cincinnati, Ohio.

RODHE, G. 1962. The effects of trace elements on the exhaustion of sewage irrigated land. *Water Pollution Abstracts*, **36**, 421(2063).

ROHLICH, G. A., and P. D. UTTORMARK. 1972. Wastewater treatment and eutrophication. In G. E. Likens, Ed., *Nutrients and Eutrophication*. American Society of Limnology and Oceanography, Lawrence, Kan. Pp. 231–245.

RUCKELHAUS, W. D., and J. L. STEINFELD. 1970. Statement on NTA. Joint press release of the Environmental Protection Agency and the Surgeon General, December 18.

RYTHER, J. H., and W. M. DUNSTAN. 1971. Nitrogen, phosphorus, and eutrophication in the coastal marine environment. *Science*, **171**, 1008–1013.

SAWYER, C. N. 1960. *Chemistry for Sanitary Engineers*. McGraw-Hill, New York. 367 pp.

SCHINDLER, D. W. 1974. Eutrophication and recovery in experimental lakes: Implications for lake management. *Science*, **164**, 897–899.

SHEIKH, B., R. P. CORT, W. R. KIRKPATRICK, R. S. JHAQUES, and T. ASANO. 1990. Monterey wastewater reclamation study by agriculture. *J. Water Pollut. Contr. Fed.*, **62**, 216–226.

SHUVAL, H. I. 1990. *Wastewater Irrigation in Developing Countries: Health and Technical Solutions*. Summary of World Bank Tech. Rept. 51. Washington, D.C. World Bank,

SHUVAL, H. I., B. FATTAL, and Y. WAX. 1984. *Retrospective Epidemiological Study of Disease Associated with Wastewater Utilization*. EPA-600/S1-84-006. U.S. EPA, Research Triangle Park, N.C.

SLECHTA, A. F., and G. L. CULP. 1967. Water reclamation studies at the south Lake Tahoe public utility district. *J. Water Pollut. Contrl Fed.*, **39**, 787–813.

SMITH, R. J. 1978. Ever so cautiously, the FDA moves toward a ban on nitrites. *Science*, **201**, 887–891.

SOPPER, W. E., and S. N. KERR. 1981. *Revegetating Strip-Mined Land with Municipal Sewage Sludge*. EPA-600/S2-81-182. Cincinnati, Ohio.

SOPPER, W. E., and E. M. SEAKER. 1984. *Strip Mine Reclamation with Municipal Sludge*. EPA-600/S2-84-035. Cincinnati, Ohio.

STEWART, D. 1990. Flushed with pride in Arcata, California. *Smithsonian*, **21**(1), 174–180.

STEPHAN, D. G., and L. W. WEINBERGER. 1968. Wastewater reuse—has it "arrived?" *J. Water Pollut. Control Fed.*, **40**, 529–539.

SWISHER, R. D., M. M. CRUTCHFIELD, and D. W. CALDWELL. 1967. Biodegradation of nitrilotriacetate in activated sludge. *Environ. Sci. Technol.*, **10**, 820–827.

TANNENBAUM, S. R., D. FETT, V. R. YOUNG, P. D. LUND, and W. R. BRUCE. 1978. Nitrite and nitrate are formed by endogenous synthesis in the human intestine. *Science*, **200**, 1487–1489.

U. S. ARMY CORPS OF ENGINEERS. 1972. *Regional Wastewater Management Systems for the Chicago Metropolitan Area*. Summary report and technical appendix. Office of the Chief of Engineers, Dept. of the Army, Washington, D.C.

WARREN, C. E. 1971. *Biology and Water Pollution Control*. Saunders, Philadelphia. 434 pp.

QUESTIONS

6.1 Given the fact that photosynthesis produces O_2, why do the inorganic nutrients in sewage effluent sometimes lead to O_2 depletion problems?

6.2 Federal law requires that primary POTWs remove at least 30% of the BOD and 30% of the SS from the raw wastewater. Why do you suppose primary treatment plants have had a more difficult time meeting the requirement for 30% BOD removal?

6.3 The principal mechanisms used to remove N from wastewater are ammonia stripping and denitrification. Which of these two methods is more likely to lead to eutrophication problems caused by high concentrations of inorganic N in rainwater? Explain your reasoning.

6.4 At the present time, only 3% of the wastewater produced in the United States is recycled, but 36% of the municipal sewage sludge is applied to the land. How would you account for the much greater percentage of municipal sludge that is used for land application purposes?

6.5 The City and County of Honolulu operates two POTWs that discharge a total of roughly 350 mLd of primary treated sewage into Mamala Bay, a coastal indentation on the south side of the island of Oahu. Assume, for the sake of argument, that because of concerns over cultural eutrophication, the EPA orders the city to remove most of the inorganic N in the effluent. Assume that the city is considering two methods of inorganic N removal, ammonia stripping and denitrification. If you were a consultant to the city, what form of inorganic N removal would you recommend and why?

6.6 The U.S. Navy operates a sewage treatment plant that discharges roughly 25 mLd of effluent into the mouth of Pearl Harbor. The sewage receives secondary treatment via the activated sludge process. The plant removes over 95% of the BOD from the raw wastewater but only 30% of the N and P. Because of concerns over eutrophication of Pearl Harbor, the EPA orders the Navy to use tertiary treatment to remove most of the inorganic N from its effluent. The Navy is considering using either ammonia stripping or denitrification to remove the inorganic N. If you were a consultant to the Navy, what form of inorganic N removal would you recommend and why?

6.7 The effluent from a wastewater treatment plant is being discharged into a lake, and for that reason, the lake is experiencing serious cultural eutrophication problems. The city operating the plant is considering two forms of tertiary nutrient removal; the first would reduce the N concentration in the effluent to 1 ppm, and the second would reduce the P concentration in the effluent to 1 ppm. Both options cost exactly the same amount of money. The city can afford only one option. If you were a consultant to the city, which option would you recommend and why?

6.8 Imagine that you are a lawyer and that Phil N. Thropist, an aging millionaire, asks you to write his will. He wants to leave his entire 25 km^2 estate to nearby city X, with the stipulation that the land be used as a sewage farm and be irrigated with 5 cm wk^{-1} of sewage effluent from city X. The population of city X is currently 200,000. How many square kilometers of Phil's estate do you estimate will be needed to accommodate the sewage from city X? How large can the population of city X become before Phil's estate will no longer be able to accommodate all of the wastewater at an irrigation rate of 5 cm wk^{-1}?

6.9 The Suds family lives in Honolulu, where the public water supply comes entirely from groundwater stored in basaltic aquifers below the island of Oahu. The Suds family wash their clothes using Ivory soap flakes and are very pleased with the results. The Suds family later moves to Miami, where the public water supply comes from a limestone aquifer. They are upset to find out that their clothes no longer appear to be clean when washed with Ivory soap flakes. They suspect that the washing machine in their Miami home is defective and consider buying a new one. Can you suggest an alternative explanation for the problem the family is having washing their clothes in Miami? Explain your reasoning.

Chapter Seven

✧ ¤ ✧ ¤ ✧

Pathogens in Natural Waters

1. Sources of Pathogens
   a. Permanent and Temporary Carriers
2. Types of Pathogens and Their Detection
   a. Culture Methods
   b. Viable-But-Nonculturable Condition
   c. Bacteria
      (a) *Shigella, Salmonella, E. coli, Campylobacter, V. cholera, Leptospira, Legionella*
   d. Protozoans
      (a) *Cryptosporidium, Giardia, Naegleria, E. histolytica*
   e. Viruses
      (a) infectious hepatitis, Norwalk, polio
         (1) Salk and Sabin vaccines
   f. Helminths
      (a) *Ascaris, Schistosoma*, tapeworm,
3. Tests for Pathogens
   a. Indicator Organisms
      (a) coliforms, fecal coliforms, *E. coli, Enterococcus*
4. Treatment of Public Water Supplies
   a. Suspended Solids Removal, Filtration, Chlorination
   b. Alternatives to Chlorination
      (1) ultraviolet light, ozone
   c. Impact of Treatment

A WIDE VARIETY of human pathogens may be found in the excrement from humans as well as from other animals. Most human pathogens can be classified as either viruses, protozoans, helminths (intestinal worms), or bacteria. Both raw sanitary sewage and land runoff contain pathogenic organisms, and virtually every sizable body of water contains some pathogens. However, it does not follow that bathing in a stream, drinking water from a lake, or swimming in the ocean will automatically cause a person to become ill. One must first come in contact with pathogens, the pathogens must gain entry to the body, and the dose of pathogens must be sufficiently great to overcome the body's natural defense mechanisms. Under special circumstances, an infection can develop from a single virus, protozoan, or helminth, but the minimum infective dose for bacteria is between 100 and 100 million, depending on the species (Majeti and Clark, 1981; NAS, 1996).

It follows that neither recreational waters nor public water supplies need be absolutely free of pathogens to be nominally safe, but the higher the concentration of pathogens, the greater the probability that health problems will develop. The fact that the concentrations of pathogens in recreational waters and public water supplies have sometimes been far above safe limits is well documented in the historical record of waterborne disease outbreaks. Even in an advanced nation such as the United States, where sanitary procedures and waste disposal practices are far superior to those used in Second and Third World nations, waterborne disease occurrences are not uncommon. Lippy and Waltrip (1985), for example, have estimated that between 1946 and 1980, there were over 1,300 waterborne disease outbreaks in the United States and these outbreaks caused approximately 350,000 cases of illness.

It is certainly possible to treat water so that the concentration of pathogens is reduced to a safe level, but careful treatment of drinking water and monitoring of recreational water quality have become common only in the last 100 years or so, due in no small part to the recognition and acceptance of the germ theory of disease as developed by Louis Pasteur and Robert Koch during the 19th century (Cohen, 1994). There remain, however, practical and scientific limitations on the extent to which water is treated or monitored, even in advanced nations. In their analysis of waterborne disease outbreaks in the United States, for example, Lippy and Waltrip (1985, p. 74) commented that "The glaring deficiencies were that disinfection was not in place where it was needed and not properly operated where it was in place." In this chapter we examine the nature of the waterborne pathogen problem and the methodologies available for treatment and monitoring.

SOURCES OF PATHOGENS

Pathogens found in human excrement, whether urine or fecal material, come from persons who are presently infected by the disease organism. That such persons are infected by a pathogen does not necessarily imply that they feel or show any signs of disease. In other words, a person infected by a pathogen may be either symptomatic or asymptomatic. Some persons, following infection by a particular disease organism, may harbor the pathogen in their bodies for the rest of their lives without showing any ill effects. Such persons are referred to as *permanent carriers* of the pathogen. In most populations, the percentage of persons who are permanent carriers of a particular pathogen is quite small. For example, roughly 2–4% of the persons who recover from typhoid fever become permanent carriers of *Salmonella typhi*, the bacterium that causes the disease (Frobisher et al., 1969). In some cases, a permanent carrier may be completely unaware that he or she has ever had the disease caused by the organism carried, since some infections may be so mild as to pass almost without notice.

In the case of some pathogens there are no truly permanent carriers, but persons who are presently sick or recovering from a disease may excrete enormous numbers of the causative disease organisms in their feces or urine. Such persons are termed *temporary carriers* of the pathogen. Animals also become infected by organisms, some of which are pathogenic to humans. Thus, animals may be carriers of human pathogens, just as humans are. For this reason, both sanitary sewage and land runoff may contaminate water supplies with human pathogens.

TYPES OF PATHOGENS AND THEIR DETECTION

Table 7.1 lists the causative agents of waterborne disease outbreaks in the United States during the 10-year period 1987–1996. An *outbreak* is here defined as an incident in which at least two cases of acute infectious disease or one case of an acute intoxicating illness[1] occurred and originated from a common source. This definition is important to bear in

[1]Infectious diseases are associated with microbiological agents. Intoxicating illnesses are caused by chemical agents.

TABLE 7.1 CAUSATIVE AGENTS OF WATERBORNE DISEASE OUTBREAKS IN THE UNITED STATES, 1987–1996

Cause	Outbreaks	% of Total	Cases	% of Total
Bacteria				
Campylobacter jejuni	3	1.0	223	0.0
Cyanobacteria	1	0.3	21	0.0
Escherichia coli	11	3.6	576	0.1
Legionella	4	1.3	30	0.0
Leptospira	2	0.7	14	0.0
Plesiomonas shigelloides	1	0.3	60	0.0
Pseudomonas aeruginosa	49	16.1	985	0.2
Salmonella	2	0.7	628	0.1
Shigella	28	9.2	4,726	1.0
Vibrio cholerae	1	0.3	11	0.0
Protozoans				
Cryptosporidium parvum	23	7.5	429,553	91.5
Giardia lamblia	33	10.8	4,111	0.9
Naegleria spp.	16	5.2	16	0.0
Viruses				
Adenovirus	1	0.3	595	0.1
Hepatitis A	5	1.6	101	0.0
Norwalk	4	1.3	6,294	1.3
Unidentified virus	6	2.0	289	0.1
Helminths				
Schistosoma	5	1.6	161	0.0
Chemical				
Nitrite	5	1.6	14	0.0
Other chemicals	19	6.2	538	0.1
GI	86	28.2	20,408	4.3

Source: Various *Morbidity and Mortality Weekly Surveillance Summaries*.

mind. During the 1990s, for example, leptospirosis cases in the United States averaged about 55 per year. However, between 1987 and 1996 there were only two reported outbreaks of leptospirosis involving a total of 14 people. In other words, most cases of leptospirosis were isolated incidents involving one person. Thus, in some cases, the number of persons who became ill from a particular pathogen as a result of drinking contaminated water or from recreational water use may substantially exceed the number of cases associated with waterborne disease outbreaks as defined by the Centers for Disease Control.

Bearing this caveat in mind, let's take a look at the data. Waterborne disease outbreaks have been averaging about 31 per year in the United States since 1980. That figure is down from an average of about 37 per year from 1946 to 1980 (Lippy and Waltrip, 1985). The number of persons affected by these outbreaks can vary greatly from year to year. During the period 1987–1996, by far the greatest number of cases occurred in Milwaukee, Wisconsin, in 1993. During April of that year, over 400,000 persons became ill as a result of ingesting oocysts (eggs) of the parasite *Cryptosporidium parvum*. Excluding this atypical event, the number of cases associated with waterborne disease outbreaks in the United States has been averaging about 7,000 per year since 1980. That compares to an average of about 10,000 cases per year from 1946 to 1980 (Lippy and Waltrip, 1985). Between 1971 and 1985, about half of the 485 waterborne disease outbreaks reported in the United States were due to unknown sources and were simply reported as outbreaks of acute gastrointestinal illness (AGI). Between 1987 and 1996, 28% of the outbreaks were still listed as AGI.

In the case of microbiological agents, two factors have confounded the problem of identifying the causative agent. First, the incubation period preceding the development of overt symptoms of illness ranges from about one day to several weeks. As a result, outbreaks are normally not recognized less than one to two weeks following exposure. In the meantime, the causative agent may have died or been flushed out of the water. Second, the isolation, culture, and identification of microbiological agents from natural waters are not trivial tasks. In the case of bacteria, detection normally relies on the appearance of visible colonies of cells after an incubation period of one or more days on a defined growth medium. The presence of viruses is more troublesome because they are too small to be seen under a light microscope and are not self-propagating organisms. Because viruses are obligatory intracellular parasites, they can be isolated and cultured only in the presence of living cells in the form of live animals, embryonated eggs, or, more commonly, *in vitro* cell cultures. The presence of viruses is evidenced by the characteristic lesions that appear in the culture when the virus multiplies. Most identifications of microbiological agents have been based on samples of stools or blood from infected persons (Lippy and Waltrip, 1985).

Many waterborne pathogens have the ability to survive for long periods of time in water by entering a dormant state. In this dormant condition, pathogenic bacteria may enter what is called a *viable-but-nonculturable (VBNC)* condition. A cell in the VBNC condition is defined as, "a cell which can be demonstrated to be metabolically active, while being incapable of undergoing the sustained cellular division required for growth in or on a medium normally supporting growth of that cell" (Oliver, 1993, p. 240). Recent reports have indicated that VBNC bacteria may retain the potential to be pathogenic, even though many of their metabolic functions are stopped (Colwell et al., 1985, 1990; Rahman et al., 1994, 1996). The implication of these reports is that traditional culturing techniques may fail to detect virulent pathogens. An alternative approach to pathogen detection relies on the use of molecular biological techniques to detect DNA from specific pathogens. The methodology makes use of the polymerase chain reaction (PCR) to amplify the signal from the DNA of a specific microorganism (Bej et al., 1991, 1994; Chary et al., 1993; Spierings et al., 1993; Way et al., 1993; Kong et al., 1995; Saruta et al., 1995; Kwang et al., 1996; Martineau et al., 1998). It is likely that pathogen detection using molecular biological methods will eventually replace culture techniques (Fleisher, 1990). Unfortunately, the fact that a sample of water contains DNA from a particular pathogen does not prove that there are living pathogens in the water, and even if the pathogens are alive, there is no proof that they are virulent. The expectation is that molecular biological methods will be refined to the point where they can be used to determine viability and virulence, but at the moment, assays designed to determine viability and virulence are still being developed.

Bacterial Pathogens

About 16% of the waterborne disease outbreaks reported between 1987 and 1996 were attributed to the bacterial pathogen *Pseudomonas aeruginosa*. All of the *P. aeruginosa* outbreaks were associated with the use of recreational waters. *P. aeruginosa* is an opportunistic pathogen that under suitable conditions can infect wounds as well as the urinary and respiratory tracts. *P. aeruginosa* infections can lead to pneumonia, endocarditis, and meningitis. The bacterium is found in both humans and animals and in both soil and water. Infected persons may shed *P. aeruginosa* in their feces. *P. aeruginosa* is probably best known for its effect on persons suffering from cystic fibrosis. Cystic fibrosis is an incurable and progressive disease associated with accumulation of mucus in the lungs, which then become susceptible to infection by pathogens such as *P. aeruginosa*. Fortunately, the waterborne outbreaks of *P. aeruginosa* infections have involved less severe effects. Illnesses have included dermatitis, folliculitis (inflammation of hair follicles), otitis (ear inflammation), and conjunctivitis (eye inflammation).

The bacteria that ranked first and third in terms of number of cases from 1987 to 1996 were *Shigella* and *Salmonella*, respectively. The outbreaks associated with *Salmonella* and *Shigella* involved both drinking water and recreational waters. The diseases associated with these two bacterial genera are commonly referred to as *salmonellosis* and *shigellosis*, respectively. Salmonellosis usually involves acute gastroenteritis with diarrhea and stomach cramps. Fever, nausea, and vomiting may also occur. Under normal conditions, about 1–4% of human beings excrete *Salmonellae*. The excretion percentage is much higher during salmonellosis outbreaks. Over several hundred serotypes of *Salmonella* are known to be pathogenic to humans. Typhoid fever, which is caused by *Salmonella typhi*, is probably the best known of the salmonelloses. *S. typhi* and other *Salmonellae* apparently pass through the stomach but multiply in the intestines. Subsequently, *S. typhi* appears in the blood and may establish itself in the periosteum (membranes covering bones), liver, gallbladder, bone marrow, spleen, or kidneys. *S. typhi* may cause meningitis or pneumonia. In cases of death, it is usually these latter complications that are responsible for death rather than the original intestinal infection (Frobisher et al., 1969). Between 1920 and 1945, an average of 37 deaths per year due to waterborne diseases were reported in the United States, and most were caused by typhoid fever. Since that time, the number of deaths associated with waterborne diseases has averaged only one per year, and the number of typhoid fever cases has declined to an average of about 400 per year (Figure 7.1). Most cases of typhoid fever are not associated with contaminated water.

Shigellae are bacillus-type bacteria, and shigellosis is therefore sometimes called *bacillary dysentery*. The disease is spread via the feces of a carrier, whereas *S. typhi* is excreted both in feces and in urine. The percentage of persons excreting *Shigellae* is about 0.3–3%. Unlike *Salmonellae*, *Shigellae* are not commonly found in animals.

Survival of *Salmonellae* in water is variable and is enhanced by low temperatures and

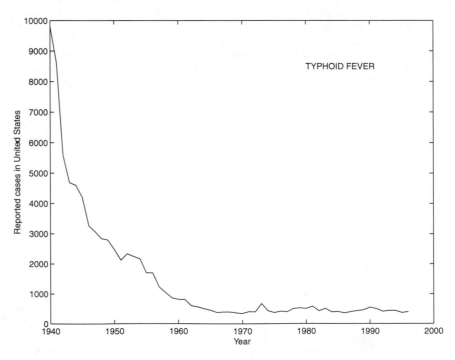

FIGURE 7.1   Rate of occurrence of typhoid fever in the United States from 1940 to 1996. *Source*: Various issues of *Morbidity and Mortality Weekly Reports*.

high nutrient levels. For example, *Salmonellae* discharged into the Red River at Fargo, North Dakota, and Moorhead, Minnesota, during January 1966 were detected 120 km downstream, four days' flow time from the point of discharge (Geldreich, 1972). *S. typhi* can survive for several weeks in river water and for 39 days in ice cream (Frobisher et al., 1969). *Shigellae*, on the other hand, do not survive for long periods either in feces or in sewage. As with *Salmonellae*, their survival in aquatic systems is enhanced by low temperatures. With the exception of typhoid fever, the reported incidence of salmonellosis has been increasing in recent years, and the disease is now about twice as common as shigellosis in the United States (Figure 7.2). Only a small percentage of both salmonellosis and shigellosis cases are caused by contaminated water.

Several other types of waterborne bacterial pathogens are worthy of mention, either for historical or other reasons. *Escherichia coli* is found in the intestines of all warm-blooded animals, including humans. In fact, human feces may consist of as much as 5–50% *E. coli*. Certain serotypes of *E. coli*, referred to as *enteropathogenic E. coli*, are pathogenic in the sense that they may cause diarrhea. Infants are particularly susceptible to this form of diarrhea. *E. coli* has attracted much attention in recent years as a result of the emergence of a particularly pathogenic strain designated *E. coli* O157:H7. The strain first appeared in the United States in 1982, when it caused severe diarrhea in persons who ate contaminated hamburgers at a national fast-food chain (Swinbanks, 1996). The pathogen struck again in 1993, when it sickened about 500 persons who ate contaminated hamburgers at Jack-in-the-Box restuarants in the Pacific Northwest. Four children died as a result of the incident. In Japan in 1996, *E. coli* O157:H7 sickened thousands of people and killed several school children who ate uncooked radish sprouts, and later in the same year it sickened

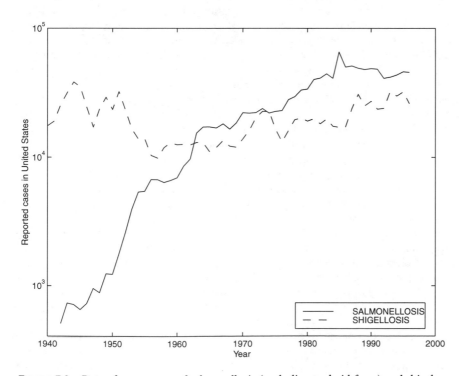

FIGURE 7.2   Rate of occurrence of salmonellosis (excluding typhoid fever) and shigellosis in the United States from 1940 to 1996. *Source*: Various issues of *Morbidity and Mortality Weekly Reports*.

more than 60 Americans who consumed fruit drinks containing unpasteurized apple juice, killing one child (Grady, 1996). In August 1999, more than 600 persons were infected with a pathogenic strain of *E. coli* at a fair in New York. Several persons died of the infection. Rain runoff apparently washed the *E. coli* from cow manure at a nearby barn into the fair's underground water supply. On average, *E. coli* O157:H7 is now believed to be responsible for about 20,000 cases of food-related illnesses and 250 deaths in the United States each year (Kluger, 1998). Waterborne-disease outbreaks account for only about 50–60 *E. coli*-related illnesses each year in the United States, but some of those are fatal. In June 1998, a two-year-old girl contracted *E. coli* in a pool at a Georgia water park. She developed kidney complications and died several weeks later (Kluger, 1998).

Most strains of *E. coli* are benign. In aquatic systems contaminated with human feces, the proportion of enteropathogenic *E. coli* is probably less than 1% (Geldreich, 1972), and the percentage of persons excreting enteropathogenic *E. coli* is no more than 1–10%. The survival time of *E. coli* in water is influenced by a great many factors and is usually much shorter in seawater than in freshwater (Fujioka et al., 1981). Multiplication of *E. coli* in aquatic systems is believed to be rare, but it may occur in heavily polluted warm water containing high concentrations of bacterial nutrients. In most cases, the presence of *E. coli* in aquatic systems can be taken as evidence of recent fecal pollution.

Campylobacteriosis is among the most common bacterial infections of humans throughout the world. The disease may be caused by any of several members of the bacterial genus *Campylobacter*. Although in the United States campylobacteriosis outbreaks are uncommon, in most parts of the world *Campylobacter* is more commonly isolated from fecal specimens of diarrheal patients than either *Salmonella* or *Shigella*. The number of reported *Campylobacter* infections in England exceeds those of *Salmonella* and *Shigella* combined (Blaser, 1990). The most common symptoms of the disease are diarrhea, malaise, fever, and abdominal pain. Symptoms may last for periods ranging from one day to a week or longer. In the United States, the peak incidence of campylobacteriosis is in children under 1 year old and persons in the 15–29 age group. The disease is an important cause of acute diarrhea suffered by travelers who visit developing areas. In endemic regions, *Campylobacter* infections are frequently symptomatic in early life but tend to be asymptomatic later on (Blaser, 1990). The implication is that the body develops a certain resistance to the symptoms of the disease as a result of repeated infections.

Cholera is a serious and acute intestinal disease caused by the bacterium *Vibrio cholera*. The disease is characterized by diarrhea, vomiting, suppression of urination, rapid dehydration, fall in blood pressure, subnormal temperature, and complete collapse. Death may occur within a few hours of onset unless treatment is started in time. The treatment, which involves infusion of water and electrolytes into the veins, produces rapid relief from the symptoms of the disease (Frobisher et al., 1969). For example, in a recent case in Louisiana, a 67-year-old woman suffering from cholera was admitted to the intensive care unit of a hospital with hypotension and bradycardia. She was resuscitated after administration of approximately 22 L (6 gal) of fluids over a 24-hour period (Gergatz and McFarland, 1989).

Cholera is transmitted via contaminated feces. Roughly 2–10% or more of the human population are healthy carriers, and convalescent carriers may shed vibrios intermittently for as long as 4–15 months. Outbreaks of cholera have usually been associated with contaminated water supplies. The disease apparently first appeared in the Orient, the earliest recorded account of an epidemic being a 1563 medical report from India. Cholera did not spread to other parts of the world until the 19th century, when it first appeared in Europe. European epidemics were eventually brought under control through the protection and treatment of public water supplies. In the United States, reports of cholera are rare but have been increasing at an accelerating rate. The number of cases averaged 2, 7, and 32 per year during the 1970s, 1980s, and 1990s, respectively. Although cholera is still rare in the United States, cholera epidemics are quite common in other parts of the world. Since 1871 there have been seven cholera pandemics, the most recent of which began in 1961 and continues to the present day (Mekalanos and Sadoff, 1994). Approximately two epidemics per year now occur in the endemic area of Bangladesh (Islam et al., 1990), and during the late 1980s outbreaks were reported in Thailand, India, Singapore, West Africa, and Tanzania (Killewo et al., 1989; Tabtieng et al., 1989; Fule et al., 1990; Goh et al., 1990; Murthy et al., 1990; St. Louis et al., 1990). A 1985 epidemic in Bangladesh resulted in 12,194 registered cholera cases and 51 deaths (Siddique et al., 1989), and there have been nearly a million cholera cases in Latin America following an outbreak in Peru in 1991 (Mekalanos and Sadoff, 1994). Kenya has been experiencing a major cholera outbreak since 1997, with 22,432 cases and 1,237 deaths as of January 1999. (http://www.who.int/emc/outbreak_news/). Because of the severity of the symptoms associated with cholera and the potential of the disease to be fatal, outbreaks of cholera can have very serious consequences. For the same reasons, there is speculation that *V. cholerae* might be used as a biological warfare agent by governments that do not subscribe to the accords of the 1972 Bacteriological (Biological) and Toxin Weapons Convention.

Leptospirosis and tularemia are diseases caused by bacteria of the genus *Leptospira* and the bacterium *Pasteurella tularensis*, respectively. In both cases, the bacteria enter the bloodstream through skin abrasions or mucous membranes. Leptospiral bacteria may cause acute infections involving the kidneys, liver, and central nervous system. Tularemia is characterized by chills and fever, swollen lymph nodes, and a generally prostrate condition. Both pathogens tend to be transmitted to humans by animals. *Leptospira* are excreted in urine but are not normally found in feces. Transmission to humans usually involves contact with water in which infected animals have urinated. In the United States, leptospirosis outbreaks are confined almost exclusively to the summer recreational period and are associated with swimming in polluted waters. Leptospirosis is often severe and may be fatal. However, only about 55 cases per year are reported in the United States (Figure 7.3).

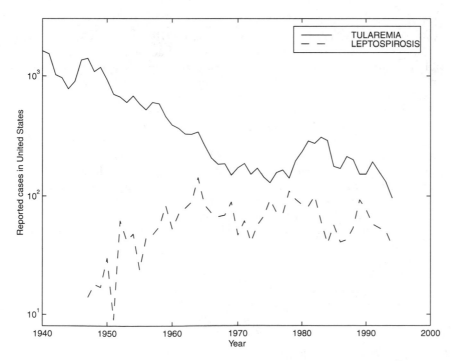

FIGURE 7.3   Rate of occurrence of leptospirosis and tularemia in the United States from 1940 to 1996. *Source*: Various issues of *Morbidity and Mortality Weekly Reports*.

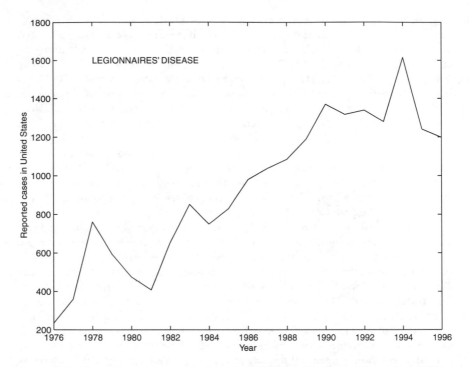

FIGURE 7.4   Rate of occurrence of legionnaires' disease in the United States from 1976 to 1996. *Source*: Various issues of *Morbidity and Mortality Weekly Reports.*

Most cases of tularemia occur in the hunting season as a result of contact with wood ticks or infected animals. However, the disease may also be spread through water contaminated with the urine, feces, or dead bodies of infected animals, and a number of outbreaks of tularemia have been associated with drinking contaminated water (Geldreich, 1972). Reported cases of tularemia in the United States declined rapidly from 1940 to 1980 and have averaged about 150–200 cases per year since then (Figure 7.3).

Legionella is a genus of bacterium that was not discovered until 1976, when it caused an outbreak of pneumonia and 29 deaths at an American Legion convention. Since that time, a number of species of *Legionella* have been discovered. The species of greatest public health concern is *L. pneumophila*, which causes Legionnaires' disease and Pontiac fever. Infection by *L. pneumophila* leads to a gradual onset of flu-like symptoms and, in the case of Legionnaires' disease, to severe pneumonia that is unresponsive to penicillins or aminoglycosides.[2] Legionnaires' disease may spread beyond the lungs and, in particular, into the gastrointestinal tract and central nervous system. Hence, it may lead to diarrhea, nausea, disorientation, and confusion. Pontiac fever produces the same flu-like symptoms as Legionnaires' disease, but the infection does not spread beyond the lungs and does not lead to pneumonia. Reported cases of Legionnaires' disease in the United States increased steadily between 1976 and 1990 and since then have averaged about 1,300 cases per year (Figure 7.4). Most of the cases are not associated with contaminated water.

---

[2]Effective treatment usually involves administration of erythromycin.

Protozoan Pathogens

Numerous protozoan species normally inhabit the intestinal tracts of warm-blooded animals, including humans. Many of these protozoan species are completely harmless to humans and are routinely found in the feces of both healthy and sick persons. However, some protozoans are pathogenic. In the United States, the most important of the enteropathogenic protozoans are *Cryptosporidium parvum*, *Giardia lamblia*, and *Naegleria* spp. *Cryptosporidium* was first discovered in the gastric mucosa of mice in 1907. For many years it was thought to be a rare, opportunistic animal pathogen. However, in 1976 *C. parvum* was isolated from the intestinal mucosa of a three-year-old girl from rural Tennessee who had been suffering from severe gastroenteritis for two weeks (Flanigan and Soave, 1993). Since then, *C. parvum* has been recognized as a ubiquitous human pathogen. Symptoms of *C. parvum* infection include prolonged watery diarrhea, which is life-threatening in immunosuppressed persons. *C. parvum* produces eggs called *oocysts* that are not readily killed by chlorination and can be effectively removed from public water supplies only by filtration (Nash, 1993). The oocysts remain dormant until they are swallowed. Digestive juices dissolve the thick walls of the oocysts in the gastrointestinal tract, releasing the protozoans, which then attach themselves to the intestinal walls, where they mature and produce more oocysts. While attached to the intestinal walls, they presumably release the toxin that causes diarrhea and vomiting. A person infected with *C. parvum* may excrete as many as 100 million oocysts per day (Nash, 1993). Most cases of cryptosporidiosis are associated with contaminated water, but transmission has also been reported via person-to-person contact in child-care centers. One of the disturbing aspects of waterborne *C. parvum* outbreaks is that tests routinely performed by water purification plants to detect biological contaminants fail to detect *C. parvum* (Berkelman et al., 1994). In the 1993 Milwaukee incident, for example, the municipal water supply met both state and federal standards for quality.

G. lamblia causes a disease known as *giardiasis*. The most common symptom of the disease is diarrhea, but giardiasis may also be associated with weakness, weight loss, abdominal cramps, nausea, greasy stools, abdominal distention, flatulence, vomiting, belching, and fever. Anthony van Leeuwenhoek first described *Giardia* in 1681, incidentally from a sample of his own stools. Intestinal disorders caused by *G. lamblia* were not documented in the United States until 1966 (AWWA, 1985), but of the waterborne disease outbreaks in which a causative agent has been identified, *G. lamblia* has accounted for more outbreaks of gastroenteritis in recent years than any other single pathogen (Table 7.1). According to Hill (1990), roughly 4% of persons in the United States are carriers, but rates may be as high as 16% in certain areas.

G. lamblia occurs in two forms, a free-living or trophozoite form and an encapsulated cyst form. Both forms may be excreted by an infected organism, but the cysts are more resistant and tend to be the dominant forms in natural waters. Furthermore, the trophozoite stage is not resistant to the initial stages of digestion, and the cyst form is therefore believed to be the cause of infections. Once the cysts reach the small intestine, they undergo an excystation process and are transformed into the active trophozoite state, which reproduces and infects the host.

Giardia cysts are associated primarily with surface waters and have been reported in groundwater only in cases involving sewage contamination (AWWA, 1985). Cold temperatures enhance the survival of the cysts in water, and the cold environment of mountain streams may allow them to remain viable for up to two months. Virtually all the outbreaks in the United States have been reported in mountainous areas, where the frequent use of easily contaminated surface waters by hikers and campers has undoubtedly contributed to the prevalence of the disease. Although most mammals are susceptible to *G. lamblia* infection, beavers, because of their aquatic habitat, are believed to be significant sources of *G. lamblia* in surface waters. Because *G. lamblia* cysts are resistant to chlorination, filtration appears to be the most practical and effective method of eliminating *G. lamblia* from public water supplies (Ongerth et al., 1989). Boiling is an obvious option for hikers and campers.

One of the more frightening waterborne pathogens is the amoeba *Naegleria*. It is normally found in soil, but it is capable of invading humans. All reported waterborne disease outbreaks due to *Naegleria* have been associated with contaminated recreational waters. *Naegleria* has a flagellate stage that enters the host via either the mouth or the nose. It is then believed to invade the mucosal lining of the nasal passages and to migrate up the olfactory nerves into the brain and central nervous system. It causes an inflammation of the brain and meninges known as *amebic meningoencephalitis*. If untreated, the infection leads to rapid death. Amphotericin B is an effective treatment for amebic meningoencephalitis, but the relative rarity of the disease and its rapid progression often result in death because of late or incorrect diagnosis.

Entamoeba histolytica is a protozoan that causes a disease of the large intestine called *amebiasis*. Although of little public health significance in the United States, amebiasis causes millions of cases of diarrhea worldwide and is estimated to account for 100,000 deaths each year (Cohen, 1995). The largest center of research on amebiasis is in Mexico, where the disease causes 1,200 deaths annually. A recent survey of almost 70,000 Mexican blood samples revealed that 8.4% showed signs of prior infection with a virulent form of *E. histolytica* (Cohen, 1995). The amebiasis syndrome may range from mild abdominal discomfort involving diarrhea alternating with constipation to a more severe chronic dysentery referred to as *amebic dysentery*. Using tissue-dissolving enzymes, *E. histolytica* burrows into the intestinal lining and occasionally ruptures the intestine. The organism is eliminated in the feces, often in the form of cysts, which may remain viable for many days in water. The concentration of cysts in sewage is usually quite low, typically one to five per liter, and these low densities are greatly reduced in receiving bodies of water by dilution, settling, and death of the cysts. Incidents of amebiasis in the United States have generally involved defects in plumbing systems, such as cross-connections between sewer lines and water supply pipes, or leaky sewer and/or water lines.

VIRAL PATHOGENS

Over 130 different viral pathogens are excreted in human feces and urine. Table 7.2 lists the major categories of these viruses and their associated diseases. According to the data in Table 7.1, viral pathogens accounted for about 5% of waterborne disease outbreaks in the United States between 1987 and 1996. However, because of the difficulty in detecting and identifying viruses, there is reason to suspect that viruses may also have caused a significant portion of the 28% of unexplained waterborne disease outbreaks. There is circumstantial evidence to support this viewpoint. For example, a study of infant diarrheal cases in Houston, Texas, between 1964 and 1967 revealed that Group A coxsackie viruses were present 3.7 times more often in diarrheal than in nondiarrheal infants (Geldreich, 1972). In developing countries, rotaviruses cause an estimated 870,000 deaths per year and are detected in 20–70% of fecal specimens from children hospitalized with acute diarrhea (Glass et al., 1996). They are now considered to be the most common cause of severe dehydrating diarrhea in children worldwide. In the United States, rotaviruses are associated with 3% of all hospitalizations of children less than five years old (Glass et al., 1996).

A practical consideration in the case of viruses is the fact that they are more resistant to standard water treatment procedures than the commonly used indicator bacteria (see the later discussion). Consequently, public water supplies and recreational waters that appear safe based on standard assays may in fact contain dangerous concentrations of viruses. Based on these as well as other considerations, Cabelli (1983) was led to conclude that the

TABLE 7.2   HUMAN ENTERIC VIRUSES
THAT MAY BE PRESENT IN WATER

Virus group	Number of Types	Disease or Symptom
Enteroviruses		
Poliovirus	3	Paralysis, meningitis, fever
Echovirus	34	Meningitis, respiratory disease, rash, fever, gastroenteritis
Coxsackievirus A	24	Herpangina, respiratory disease, meningitis, fever, hand, foot, and mouth disease
Coxsackievirus B	6	Myocarditis, congenital heart anomalies, rash, fever, meningitis, respiratory disease, pleurodynia
New enterovirus types 68–71	4	Meningitis, encephalitis, respiratory disease, rash, acute hemorrhagic conjunctivitis, fever
Hepatitis A (enterovirus 72)	1	Infectious hepatitis
Norwalk virus	2	Epidemic vomiting and diarrhea, fever
Rotavairus	4	Gastroenteritis, diarrhea
Reovirus	3	Not clearly established
Adenovirus	47	Respiratory disease, conjunctivitis, gastroenteritis
Parvovirus		
Adeno-associated virus	3	Associated with respiratory disease of children, but etiology not clearly established
?	?	Acute infectious nonbacterial gastroenteritis

Source: Rao and Melnick (1986) and R. Fujioka (personal communication).

cases of gastroenteritis he documented in a study of swimming-related illnesses in the United States and Egypt were probably caused by human rotaviruses and/or parvo-like viruses, although no attempt was made to isolate and identify the etiologic agent(s).

In the United States, the viruses most commonly linked with waterborne disease outbreaks are the infectious hepatitis (hepatitis A) virus and the Norwalk virus. The infectious hepatitis virus is excreted in feces of infected persons. The disease is associated with a marked inflammation of the liver. Symptoms include malaise, transient fever, myalgia, nausea, anorexia, and abdominal pain. The fact that this virus accounts for about 30% of the waterborne disease outbreaks linked to viruses probably reflects its unusually high resistance to chlorination. The issue of resistance was dramatically illustrated during an outbreak of infectious hepatitis that occurred in Delhi, India, during 1955–1956. In that outbreak over 20,000 clinical cases of infectious hepatitis were diagnosed. The cause of the outbreak was contamination of the public water supply with raw sewage. Although the water was treated by filtration and chlorination, the treatment process was evidently inadequate to kill off sufficient numbers of the hepatitis A viruses. However, the fact that there was no change in the incidence of typhoid fever or dysentery during the infectious hepatitis epidemic suggests that the treatment process had been adequate to reduce the concentrations of other pathogens to an acceptable level. Indian officials stated that during the period of contamination *E. coli* counts in the treated water did not exceed two counts per 100 mL (Dennis, 1959), a level that the Indian officials considered acceptable for drinking water. According to Dennis (1959), analysis of the raw water entering the treatment plant during November 1955 indicated that as much as 50% of the water was sewage during the peak of contamination. Indian officials were aware that the raw water was contaminated and raised the level of residual chlorine in the treated water in order to reduce the pathogen concentrations to a safe level (Fox, 1976). They evidently succeeded in the case of all pathogens except hepatitis A, an indication of its unusually high resistance to chlorination. The incident emphasizes the problem of using the concentration of a single indicator organism (*E. coli.*) as a criterion of water quality. The survival time in water and the susceptibility of different pathogens to standard water treatment methods vary greatly.

The Norwalk virus was unknown until 1969, when it was detected after an epidemic of gastroenteritis in Norwalk, Ohio (Kapikian et al., 1972). At the present time, it is estimated that in a typical year (i.e., excluding 1993), almost 65% of nonbacterial gastroenteritis in the United States is due to the Norwalk and Norwalk-like viruses. The virus is shed in the feces of infected persons, and contaminated food is the most common means of transmission. Raw oysters, cake frosting, and salads have been implicated in several outbreaks. Shellfish are particularly susceptible to contamination, since they can readily filter viral particles from the water. The virus can also be transmitted directly via water and by person-to-person contact. About 180,000 cases of gastroenteritis are attributed to the Norwalk and Norwalk-like viruses each year in the United States. Fortunately, there have been no known deaths associated with the virus. Symptoms of infection include nausea, vomiting, diarrhea, and abdominal cramps. The incubation period is about 36 hours, and the duration of symptoms is 12–60 hours. The most common therapy is fluid replacement.

Only one other waterborne disease outbreak could be traced to a specific virus during the period 1987–1996. That outbreak occurred in June 1991 in North Carolina, where 595 persons developed pharyngitis after swimming in a pond at a camp. The outbreak was traced to an adenovirus. Adenoviruses are a family of nonenveloped viruses that contain a linear double-stranded DNA genome. There are two genera in the family. One genus, *Aviadenovirus*, contains five groups and a total of 22 species that infect birds. The other genus, *Mastadenovirus*, contains 10 groups and a total of 103 species that infect mammals. Within the *Mastadenovirus* is a group of 47 species that infect humans. The human adenoviruses commonly cause infections of the eyes and upper respiratory and gastrointestinal tracts. The 1991 North Carolina outbreak was caused by human adenovirus type 3.

Poliomyelitis is not a significant health problem in the United States today, but historically it has taken on much greater significance. In recent years there have been fewer than 10 reported cases per year of poliomyelitis in the United States (Figure 7.5). Between 1945 and 1955 the frequency of occurrence was over 1,000 times greater. The disease is marked by inflammation of nerve cells in the anterior horns of the spinal cord, and is characterized by fever, motor paralysis, and atrophy of skeletal muscles, often with permanent disability and deformity. Poliomyelitis is transmitted primarily by person-to-person contact, and since 1946 only a single outbreak in the United States has been attributed to waterborne transmission. In that case, a 1952 epidemic involving defects in the water distribution system in Huskerville, Nebraska, a total of 16 persons developed poliomyelitis. As indicated in Figure 7.5, the incidence of clinical poliomyelitis has greatly declined since the introduction of the Salk vaccine (1954–1955) and the Sabin vaccine (1961–1962). The Sabin vaccine has proven particularly effective because it involves the oral administration of live but nonvirulent polioviruses. The live viruses replicate in the guts of vaccinated persons for several weeks, and as a result, individuals who have been vaccinated spread the virus among their contacts, who become similarly immunized. The Salk vaccine does not provide as strong or long-lasting immunity as the Sabin vaccine, and because it involves the administration of inactive viruses, there is no chance for immunity to be transferred from vaccinated to nonvaccinated persons. The negative aspect of the Sabin vaccine is that in approximately 1 in 3 million vaccinations it causes a case of clinical poliomyelitis.

In 1985 the Pan American Health Organization began an effort to eradicate polio from the Americas. The immunization program was successful. The last confirmed case of polio in the Americas from the wild virus occurred in 1991 (Olshansky et al., 1997). In

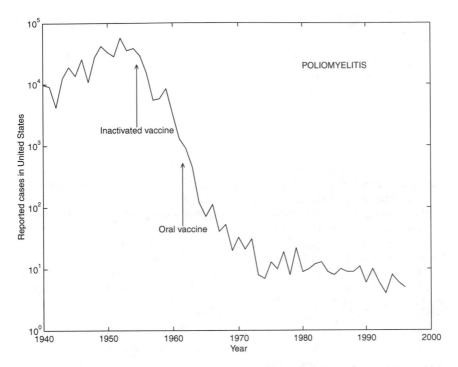

FIGURE 7.5   Rate of occurrence of poliomyelitis in the United States from 1940 to 1996. *Source*: Various issues of *Morbidity and Mortality Weekly Reports.*

1988 the World Health Organization (WHO) began a similar immunization program in Asia and Africa, and by 1998 had succeeded in reducing the number of polio cases by 90% on those two continents (Schlein, 1998). The WHO hopes to eradicate the disease by the year 2000. If that effort is successful, polio may join smallpox as the second major disease to be eradicated worldwide by immunization.

The problem of clinical polio cases caused by immunization unfortunately still remains. The 5–10 cases of poliomyelitis reported in the United States each year are all caused by the Sabin vaccine. Recently, the U.S. government began making available a so-called enhanced inactivated polio vaccine (E-IPV), which is similar to the original Salk vaccine but rivals the Sabin vaccine in effectiveness and, because the viruses are inactivated, cannot cause poliomyelitis. It is unlikely, however, that there will be a complete switch to E-IPV because such a switch would eliminate the benefits derived from transfer of immunization to unvaccinated members of the population. In some disadvantaged areas of the United States, less than 50% of young children are vaccinated against poliomyelitis, in contrast to the national average of 95% (Roberts, 1988). Since the greatest probability of developing poliomyelitis is associated with the first few administrations of the Sabin vaccine, a compromise solution has been to administer the E-IPV vaccine at ages 2 and 4 months and the Sabin vaccine at age 12–18 months and 4–6 years (Roberts, 1988; Connaught Laboratories, 1997). The Sabin vaccine continues to be used in the WHO eradication program because the vaccine is taken orally[3] and hence can be administered by an untrained volunteer. The Sabin vaccine is also inexpensive, about 8 cents per dose (Schlein, 1998).

Why does the United States continue to vaccinate children for poliomyelitis if the wild poliovirus has been eradicated from the country? The answer, as noted by Schlein (1998,

---

[3]The E-IPV vaccine is administered by injection.

p. 168), is the realization that as long as even one pocket of the poliovirus remains on Earth, "The poliovirus is one day's journey from any spot on the globe." Hence the United States has been spending about $230 million annually to immunize children against a disease whose wild type has been absent for more than 20 years. It is estimated that eradicating polio will save $1.5 billion each year in immunization, treatment, and rehabilitation costs on a worldwide basis (Schlein, 1998).

Helminths

A number of parasitic intestinal helminths (worms) may be found in sewage. Most of them enter the human body through the mouth, although a few gain entry through the skin. Contamination of drinking water with these worms can be effectively prevented with modern water treatment methods. However, swimming or wading in sewage-polluted waters or use of raw sewage to fertilize crops may also lead to outbreaks of intestinal worm infections. Such incidents are of little public health significance in the United States today, but they are much more important in some other parts of the world.

Perhaps the best-known helminth is the beef tapeworm, *Taenia saginata*. The adult tapeworm lives in the intestinal tract and may discharge as many as 1 million eggs per day in the feces of an infected person. The eggs are frequently found in sewage but usually at low concentrations, on the order of one or two eggs per 100 mL. The eggs remain viable for hundreds of days if kept cool and moist. Humans occasionally become infected by swallowing water containing the eggs. Symptoms of tapeworm infection include abdominal pain, digestive disturbance, and weight loss. Geldreich (1972) cites a 50% incidence of beef tapeworm infection in East Africa, but the percentage of carriers in the United States is less than 1%.

Ascariasis is an infection of the small intestine caused by the roundworm *Ascaris lumbricoides*. The disease produces variable symptoms and is more common among young children. In the intestine the worm may produce as many as 200,000 embryonated eggs per day. After being excreted with the feces, the eggs may remain viable in water or soil for as much as several months. *A. lumbricoides* has been found in 2% of the feces of sewage plant workers and in 16% of the feces of farm workers who were involved in cultivating sewage-irrigated crops (Geldreich, 1972).

Certain species of the blood fluke genus *Schistosoma* spend part of their life cycle in humans, where they may cause a debilitating infection known as *schistosomiasis*. The worm enters the body through the skin, and from there may be transported by the circulatory system to the lungs, liver, and intestines. The organism is excreted in both urine and fecal material. Completion of the organism's life cycle requires it to infect snails, and a given species of *Schistosoma* infects only one species of snail (Mahmoud, 1990). It is impossible for the infection to be transmitted in the United States because of the absence of the appropriate snail intermediate hosts. Nevertheless, it is estimated that as many as 400,000 persons in the United States are infected with *Schistosoma* (Mahmoud, 1990). Most of these persons are immigrants who were infected before entering the United States. The disease is endemic to certain areas in Africa, South America, and Asia. Worldwide it is estimated that over 200 million persons are carriers of *Schistosoma*, and schistosomiasis is estimated to cause about 8 million deaths per year (Olshansky et al., 1997).

Penetration of the skin by larval *Schistosomae* may cause a skin rash known as *swimmer's itch*. The five schistosomiasis outbreaks reported in the United States between 1987 and 1996 all involved polluted recreational waters, and in each case the symptom of the infection was dermatitis. The more serious symptoms of the disease, which generally do not develop until four to eight weeks later, include fever, chills, sweating, and headaches. Chronic cases are associated with fatigue, abdominal pain, and intermittent diarrhea or dysentery. The liver and intestines are the organs most commonly affected.

TESTS FOR PATHOGENS

The preceding discussion suggests one of the serious problems associated with testing for pathogens in water supplies. There are simply too many pathogens to make testing for all or even most of them practical. Since the survival of different enteric pathogens in natural waters and their resistance to standard water treatment methods vary greatly from one species to another, pathogen concentrations are likely to show considerable temporal and spatial variations that are poorly correlated with each other. Consequently, monitoring the concentration of only one or a few pathogens would not necessarily provide a good measure of disease probability. In addition, the cost of setting up and operating a water quality laboratory capable of testing municipal water supplies and sewage treatment plant effluents for a large variety of organisms would be prohibitive for most communities. Viral pathogens pose a particularly difficult detection problem because they are invisible under the light microscope and can be propagated only in living cells. Furthermore, although it is possible to estimate the concentrations of many viruses using sophisticated culture methods, present techniques are capable of demonstrating only the presence or absence of the infectious hepatitis virus and the parvo-like viruses. For a number of years, the general approach to checking water for pathogens has therefore been to look for an indicator organism whose presence can be taken as evidence of sewage pollution. Ideally such an indicator organism should satisfy the following criteria:

1. The organism is proportional in abundance to the number of pathogens in the water, and survives at least as long as the pathogens outside the intestinal tract.
2. The organism is easily detectable and is present in greater quantity than any potential pathogen.
3. The organism is absent from the aquatic environment unless the water has been polluted with sewage or animal excrement.

Because human enteric pathogens are likely to be found in the intestinal tracts of only a small percentage of the population, the concentration of none of these pathogens is particularly high in sewage under normal circumstances. Because the concentrations in sewage are likely to be greatly reduced in receiving waters by dilution, such pathogens would not make good indicator organisms of sewage pollution. What is needed is an indicator organism that is likely to be detected by a scientist who takes a sample from only a small part of a natural body of water. Such an indicator organism must be consistently present in high concentrations in raw sewage.

The search for a suitable indicator organism has not met with unqualified success. In the first half of the 20th century, many water quality standards were written in terms of the concentration of coliform bacteria because it was believed that all coliform bacteria were of fecal origin. We now know, however, that the total coliform population includes four genera from the family *Enterobacteriaceae* and some species from the genus *Aeromonas*, and of the various coliform species, only *E. coli* of the family *Enterobacteriaceae* is found exclusively in feces (Cabelli, 1983). The use of total coliforms as an indicator of fecal pathogens received a severe jolt in 1941, when the Illinois Supreme Court overruled a lower court conviction of the director of the Illinois department of public welfare for failing to take action to correct contamination of the drinking water supply in a state hospital. The contamination resulted in a typhoid fever epidemic that produced many deaths. Contamination was suggested by the numbers of total coliforms in the water. However in *People v. Bowen* (1941), 376 Ill. 317, 33 N.E. (2d) 587, the court ruled as follows:

> It appears from the record that coli aerogenes or colon bacillus may be friendly or inimical, and that the mere presence of the colon bacillus in water proves exactly nothing as far as typhoid fever is concerned. The tests seem to have been made by a method of broth fermentation, and

determined nothing more than the presence or absence of some kind of colon bacillus. It further appears that this type of bacillus is present in the air one breathes, in milk, on fruits and practically everywhere.

Despite this ruling, federal and many state water quality guidelines were still written in terms of total coliforms during the early 1970s. For example, the 1972 EPA water quality criteria state that a well-operated water treatment plant "Can be expected to meet a value of 1 total coliform per 100 ml with proper chlorination practice" (EPA, 1972, p. 58).

As realization emerged that total coliform counts could be a very misleading indicator of fecal pollution, guidelines were rewritten in terms of fecal coliforms. The fecal coliforms are distinguished operationally from other coliforms on the basis of their ability to grow on a defined medium at 44.5°C. The test for total coliforms can be conducted using the same growth medium, but the incubation temperature is 35°C (Wolf, 1972). Unfortunately, the assay for fecal coliforms does not select exclusively for *E. coli*, but also includes thermotolerant members of the *Enterobacteriaceae* genus *Klebsiella*. *Klebsiella* is infrequently found in human feces and, when present, is usually a minor portion of the coliform population (Cabelli, 1983). Furthermore, there are substantial nonfecal sources of *Klebsiella*, including the thermotolerant biotype. Hence the designation *fecal* coliform is something of a misnomer. Nevertheless, the use of fecal coliform counts was regarded as a significant improvement over total coliform counts as an indicator of water quality, and fecal coliform counts largely replaced or at least supplemented total coliform counts in water quality criteria during the 1970s. For example, the 1972 EPA guidelines for public water supplies stated that the geometric mean fecal coliform and total coliform counts should not exceed 2,000 per 100 mL and 20,000 per 100 mL, respectively (EPA, 1972, p. 58).

The empirical evidence for setting these standards was surprisingly slim. At the time the 1972 EPA water quality criteria were established, the only data available from freshwater stream pollution studies on the correlation between pathogen occurrence and fecal coliform concentrations were for *Salmonella*. The data are summarized in Table 7.3. Taken at face value, Table 7.3 indicates that most water supplies containing the public water supply standard of 2,000 fecal coliforms per 100 mL would be contaminated with *Salmonella*. Although treatment processes were expected to reduce the fecal coliform counts to 1 or less per 100 mL, it is unclear from the table whether even 1 count per 100 mL is a safe level. Indeed, the 1955–1956 infectious hepatitis epidemic in Delhi, India, dramatically showed that coliform counts of less than 2 per 100 mL could be associated with a very unsafe level of at least one enteric pathogen.

Realizing the paucity of empirical evidence to support its recommended water quality standards with respect to fecal coliforms, the EPA authorized a systematic study of both freshwater and marine recreational waters. The study of marine waters lasted from 1972 to 1978 and involved beaches in New York City, Boston, and Lake Pontchartrain in the United States and Alexandria, Egypt. The freshwater studies lasted from 1978 to 1982 and

TABLE 7.3 FECAL COLIFORM ABUNDANCES AND *SALMONELLA* OCCURRENCE

Fecal Coliform Density Range (Counts/100 mL)	% of Samples Containing Salmonella
1–200	31.7
201–1,000	83.0
1,001–2,000	88.5
Over 2,000	97.6

Source: EPA (1972, p. 57).

were conducted at Keystone Lake, Oklahoma, and Lake Erie near Erie, Pennsylvania. In both the marine and freshwater studies, weekend beachgoers were interrogated 7–10 days after their visits to the beach to determine whether they had experienced any health problems that might be related to enteric pathogens in the water. The beachgoers were divided into a swimming and a nonswimming group, depending on whether they had immersed their heads in the water while swimming. The difference in the incidence of gastrointestinal (GI) illness symptoms between the swimmers and nonswimmers was correlated with the abundance of various possible indicator organisms in the water. The GI illness symptoms included vomiting, diarrhea, stomachache, and nausea.

The results of the study proved quite revealing. Not unexpectedly, there was a rather poor correlation between the incidence of GI illness among swimmers and total coliform concentrations. Surprisingly, there was an even poorer correlation between fecal coliform concentrations and GI illnesses. Of the concentrations of indicator species tested in the study, the concentration of enterococci bacteria was by far best correlated with the frequency of GI illnesses in the study of marine bathing beaches (Figure 7.6). Although *E. coli* counts were positively correlated with GI illnesses, the correlation coefficient was not statistically significant (Figure 7.6). In the freshwater beach studies, the occurrence of GI illnesses again showed no significant correlation with fecal coliforms concentrations but did correlate significantly with the abundances of both enterococci and *E. coli*.

An important, and perhaps surprising, implication of the study was that fecal coliform concentrations are almost worthless as indicators of the probable occurrence of GI illnesses associated with waterborne pathogens. Evidently, for natural waters at least, the

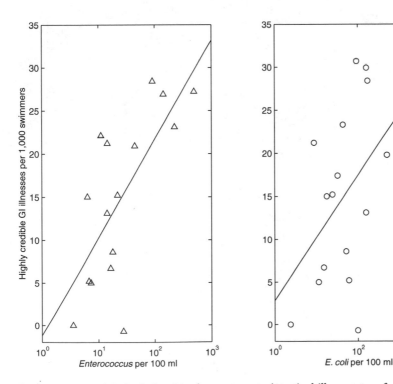

FIGURE 7.6   Empirical relationships between gastrointestinal illness rates of swimmers at marine beaches and the concentrations of *Enterococcus* (left) and *E. coli* (right). *Source*: Cabelli et al. (1983).

presence of thermotolerant *Klebsiella* genera in the fecal coliform assay thoroughly confounds the implications of the fecal coliform count regarding public health. The poor correlation between *E. coli* counts and GI illnesses in marine waters probably reflects differences in the survival rates of *E. coli* and the etiological agent(s) responsible for causing the health problems. With respect to this point, it has been reported that enterococci survive better than *E. coli* in salt water (Cabelli, 1983). However, enterococci are usually found at lower densities in human feces and sewage effluents than *E. coli* (Dufour, 1984), and this fact may compromise their usefulness as indicator organisms in marine waters (see the following discussion). Nevertheless, their concentrations were rather well correlated with the incidence of GI illnesses in both the marine and freshwater studies.

It was obvious from the 1972–1982 studies that the relative survival rates of enterococci and the etiologic agent(s) responsible for the GI illnesses differed in marine waters and freshwaters. This fact is apparent when the regression lines relating the frequency of GI illnesses and enterococci concentrations are plotted on the same graph (Figure 7.7). The concentration of enterococci is lower in marine waters than in freshwaters at the same incidence of GI illnesses. The cause of this offset is believed to be the fact that enterococci die off more rapidly in marine than waters, whereas the die-off rate of the etiologic agent(s) is similar in both water types (Dufour, 1984).

Given the regression lines in Figure 7.7, what is an acceptable concentration of enterococci in bathing waters? The answer to this question has proven to be controversial. The average swimming-associated illness rates recorded in the 1972–1982 EPA studies were 15 and 6 GI illnesses per 1,000 swimmers in marine waters and freshwaters, respectively. The associated geometric mean enterococcus concentrations were 25 and 20 per 100 mL, re-

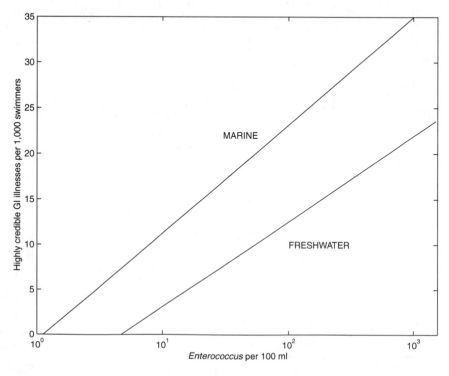

FIGURE 7.7   Empirical relationships between gastrointestinal illness rates of swimmers at marine and freshwater beaches and the concentrations of *Entercoccus*. *Source*: Cabelli (1983) and Dufour (1984).

spectively. On the assumption that six GI illnesses per 1,000 swimmers represented an acceptable illness rate, the EPA (1984) proposed that the criterion for fresh recreational waters be set at 20 enterococci per 100 mL. Based on the regression line in Figure 7.6, six GI illnesses per 1,000 swimmers in marine waters would be associated with three enterococci per 100 mL. Rationalizing that comparable public health standards should be applied to both marine waters and freshwaters, the EPA (1984) proposed a criterion of three enterococci per 100 mL for marine waters. A number of public health officials, however, objected to the marine criterion, in part because laboratory methods for quantifying enterococci at such low densities were thought to be undependable. In response to public comments, the EPA (1986) revised the marine and freshwater criteria to 35 and 33 enterococci per 100 mL, respectively. The rationale was that the previously recommended fecal coliform criterion for both marine and fresh recreational waters had been 200 per 100 mL. Based on the data collected in the 1972–1982 study, enterococci concentrations of 35 and 33 per 100 mL would be associated with a fecal coliform concentration of 200 per 100 mL in marine waters and freshwaters, respectively. These enterococci concentrations correspond to GI illness rates of about 16 and 8 per 1,000 swimmers in marine waters and freshwaters, respectively. However, some public health officials consider 16 GI illnesses per 1,000 swimmers unacceptably high. Furthermore, since the EPA study revealed no significant correlation between fecal coliform counts and GI illnesses in either marine waters or freshwaters, it is unclear why the fact that states had historically accepted a fecal coliform count of 200 per 100 mL in recreational waters should have any bearing on the water quality standard with respect to enterococci. It seems fair to say that these standards will continue to be debated and discussed in the years ahead.

TREATMENT OF PUBLIC WATER SUPPLIES

Surface waters are frequently treated to remove objectionable substances if they are to be used for drinking purposes. Groundwater is much less likely to require treatment, although aquifers sometimes become contaminated and require treatment prior to use (see Chapter 16). Often the treatment is designed primarily to ensure that the treated water does not contain a dangerous concentration of pathogens. However, the treatment process may also involve removal of suspended solids or dissolved organics, and in some cases the water is softened by precipitating out Ca and Mg ions. The actual treatment process may differ greatly from one municipality to another, depending on the characteristics of the public water supply. Where treatment does occur, one or more of the following three steps is usually a part of the process.

Removal of Suspended Solids

Perhaps the simplest form of water treatment is removal of suspended solids. This may be accomplished by merely drawing the water from the downstream end of a large reservoir where the residence time of the water is weeks or months. This long residence time in a quiescent system allows many of the suspended particulates to settle to the bottom. An alternative is to route the water through a special sedimentation basin where alum, a compound containing Al and sulfate ions, is added to abet sedimentation of particulate and colloidal substances. In most natural waters, the alum forms a positively charged precipitate that attracts colloidal substances, which are negatively charged at normal pH levels. The combination of positively and negatively charged substances tends to conglomerate into a larger and larger composite particle, or floc. The flocculation process is speeded up by slow agitation. The large floc particles gradually settle out, taking with them a large portion of the bacterial population. The efficiency of bacterial removal in the sedimentation step can vary greatly, but a well-designed flocculation and sedimentation system may remove 90–95% of the bacterial population.

Filtration

In some cases, sedimentation and flocculation do not remove a sufficiently large percentage of objectionable substances from the water. In such cases, a filtration step, either alone or following sedimentation/flocculation, is part of the water treatment process. Filtration is necessary, for example, to remove most of the *G. lamblia* cysts from raw surface waters. The filtration unit consists of layers of sand built up over layers of gravel or some other coarse material. The filter bed is typically several meters deep (Figure 7.8). The system is usually referred to simply as a *sand filter*, although the designation *dual-media filter* is sometimes applied to this or similar filtration systems. Particulate matter strained out on the surface of the sand provides food for microorganisms, which grow to form a film over the sand surface. This biological filter helps to strain out both colloidal and dissolved organic matter. In the early days of water treatment, there was usually no treatment step preceding filtration. As a result, the water was percolated rather slowly through the filters in order to bring about effective treatment. Bacterial removal by these slow sand filters could be as high as 99.99% (McKinney, 1962). With modern pretreatment methods, it is possi-

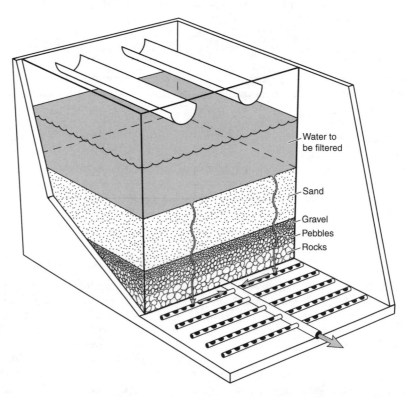

FIGURE 7.8   Diagram of a rapid sand filter. The filter is usually backwashed every one to two days. The two troughs above the sand carry away washwater when the filter is backwashed. *Source*: Redrawn from Frobisher et al. (1969).

ble to run water through the sand filters at a much higher rate without impairing water quality. Ongerth et al. (1989), for example, indicate that a well-functioning flocculation/filtration system can reduce the concentration of *G. lamblia* cysts by a factor of 10–100.

Chlorination

In many cases, surface waters destined for drinking are chlorinated to kill pathogens, and sewage effluent is usually disinfected by the same procedure. When chlorine gas (Cl_2) is added to water, the gas reacts with the water to form hydrochloric acid (HCl) and hypochlorous acid (HOCl) according to the reaction

$$Cl_2 + H_2O \rightarrow HCl + HOCl$$

The HOCl is actually the chemical agent primarily responsible for killing microorganisms. It oxidizes key reducing enzyme systems and prevents normal respiration, and at high Cl concentrations it denatures sufficient proteins to destroy cells completely. Viruses are generally more resistant to chlorination than bacteria because killing viruses requires denaturation of the viral nucleoprotein. About twice the free Cl residual is required to kill viruses as is needed to kill bacteria (McKinney, 1962). If the raw water supply contains suspended particles, chlorination following rather than preceding removal of these particles increases the effectiveness of the Cl kill because a large percentage of the pathogens are removed in the sedimentation and/or filtration step(s) and because most of the particles that might otherwise react with Cl and/or provide refuges for pathogens have been removed.

Normally, sufficient Cl is added to the water to maintain a residual Cl concentration in the water until it reaches the user. This residual Cl level prevents reestablishment of pathogen populations in the water while the water stands in storage. If the residual Cl level is too high, however, customers are likely to complain about the water's taste. Certainly no one expects tap water to taste as if it had been drawn from a swimming pool. The degree of chlorination therefore represents something of a compromise between a high degree of microorganism kill, on the one hand, and water taste and treatment cost considerations, on the other hand.

Alternatives to Chlorination

Chlorination is not the only mechanism for killing waterborne pathogens, and in recent years much interest has developed in identifying suitable alternatives to chlorination for this purpose. The principal reasons are the potential formation of carcinogenic chlorinated organic compounds during the treatment process; the resistance of certain pathogens such as viruses, *G. lamblia*, and *C. parvum* to chlorination; and the toxicity of residual Cl to nontarget aquatic organisms (Scheible and Bassell, 1981). Possible alternatives include UV radiation, ozone, and chlorine dioxide (ClO_2), and a number of studies during the 1980s addressed the feasability of these methods (Fluegge et al., 1981; Roberts et al., 1981; Stover et al., 1981; Carlson et al., 1985; Scheible et al., 1986). Neither UV light nor ozone is associated with the formation of chlorinated organics, and in studies with sewage treatment plant effluent, disinfection with ClO_2 produced less than 10% as many chlorinated organics as Cl_2 (Roberts et al., 1981).

With respect to pathogen killing efficiency, ClO_2 inactivates viruses more effectively than Cl, but at least for typical sewage treatment plant effluent, the cost of ClO_2 treatment would be two to five times as great as that of disinfection with Cl_2 (Roberts et al., 1981). The studies of Stover et al. (1981) showed that satisfactory reduction of total coliform concentrations in treated sewage effluent could be achieved with ozone, but Scheible and Bassell's (1981) cost analysis indicated that disinfection of secondary sewage treatment plant effluent with ozone would cost at least twice as much as disinfection with Cl_2. For sewage treatment plants that handle flow rates of 40–400 million L d^{-1}, UV light can reduce fecal coliform concentrations to less than 200 per 100 mL at a cost that is only about 25% higher than that of chlorination. For small plants that handle on the order of 4 million L d^{-1}, UV light disinfection may actually be 30% cheaper than chlorination (Scheible and Bassell, 1981). *G. lamblia* cysts do, however, survive UV light treatment substantially better than *E. coli*. Based on the studies of Scheible and Bassell (1981), Scheible et al. (1986), and Carlson et al. (1985), UV light doses that reduce fecal coliform concentrations by a factor of 10^3–10^4 would reduce the concentration of *G. lamblia* cysts by only a factor of 5. In a study of virus inactivation, Fluegge et al. (1981) found that Cl, ozone, and UV light treatments of sewage effluent all produced comparable reductions in virus isolation rates, the median reduction being 75%. At the present time, therefore, it does not seem that either UV light or ozone offers any advantage over Cl in terms of killing pathogens. However, UV light treatment is more or less cost competitive with chlorination and is not associated with the production of chlorinated organics. UV light treatment may therefore become more common as new wastewater treatment plants are built and old ones upgraded. Cl will probably continue to be used to treat public water supplies because of the need to discourage the growth of pathogens in the distribution system between the waterworks and the consumer. Maintaining a residual Cl concentration in the distribution system is the most effective way to accomplish this task.

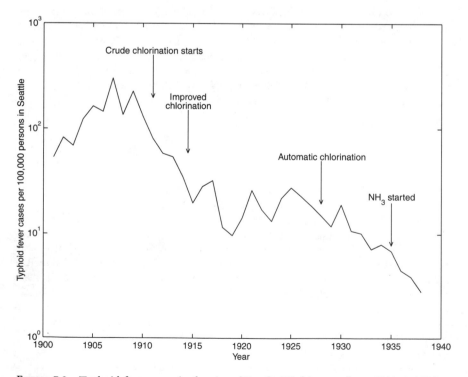

FIGURE 7.9   Typhoid fever cases in the city of Seattle, Washington, from 1901 to 1938. *Source*: Green et al. (1944).

Impact of Treatment

Numerous case studies could be cited to illustrate the impact of treatment of water supplies on public health. The following example comes from the city of Seattle. Toward the end of the 19th century, Seattle decided to obtain its permanent water supply from the Cedar River (Green et al., 1944). Public health problems associated with the use of that water soon became apparent. Between 1901 and 1910 there were almost 2,500 cases of typhoid fever in Seattle, resulting in 418 deaths. The city began chlorinating its water in a crude way in 1911 and upgraded the chlorination process in 1914–1915. The result was a dramatic reduction in typhoid fever cases to an average of about 60 cases per year between 1915 and 1925. In 1928 the chlorination process was automated, and in the following six years typhoid fever cases dropped to about 40 per year. In 1935 an ammonia-Cl treatment process was introduced, and by 1938 the number of typhoid fever cases had dropped to 11. No deaths were reported from typhoid fever in either 1937 or 1938. Figure 7.9 shows the pattern of typhoid fever cases from 1901 to 1938. The treatment process reduced the number of typhoid fever cases by more than a factor of 20 and reduced the number of typhoid fever related deaths from more than 40 per year to virtually zero. As noted by Green et al. (1944, p. 17), "The decline in cases of enteric disease which follows the practice of chlorinating city water supplies can be substantiated by the records of hundreds of cities and towns. The Seattle record demonstrates equally forcibly the potential hazards of the supply without adequate treatment. The fact that there is imminent danger of water-borne infection at all times should lead to the greatest care in the maintenance, operation and control of the treatment facilities."

REFERENCES

AMERICAN WATER WORKS ASSOCIATION. 1985. *Giardia lamblia in Water Supplies—Detection, Occurrence, and Removal.* American Water Works Association, Denver, CO. 242 pp.

BEJ, A. K., J. L. DiCESARE, L. HAFF, and R. ATLAS. 1991. Detection of *Escherichia coli* and *Shigella. spp.* in water by using the polymerase chain reaction and gene probes for *uidA. Appl. Environ. Microbiol.* 57(4), 1013–1017.

BEJ, A. K., M. H. MAHBUBANI, M. J. BOYCE, and R. ATLAS. 1994. Detection of *Salmonella* spp. in oysters by PCR. *Appl. Environ. Microbiol.* 60(1), 368–373.

BERKELMAN, R. L., R. T. BRYAN, M. T. OSTERHOLM, J. W. LeDuc, and J. M. HUGHES. 1994. Infectious disease surveillance: A crumbling foundation. *Science,* 264, 368–370.

BLASER, M. J. 1990. *Campylobacter* species. In G. L. Mandell, R. G. Douglas, Jr., and J. E. Bennett, Eds., *Principles and Practice of Infectious Diseases,* 3rd ed. Churchill Livingstone, New York. Pp. 1649–1658.

CABELLI, V. J. 1983. *Health Effects Criteria for Marine Recreational Waters.* EPA-600/1-80-031. Research Triangle Park, NC. 98 pp.

CARLSON, D. A., R. W. SEABLOOM, F. B. DEWALLE, T. F. WETZLER, J. ENGESET, R. BUTLER, S. WANGSUPHACHART, and S. WANG. 1985. *Ultraviolet Disinfection of Water for Small Water Supplies.* EPA/600/S2-85/092. Cincinnati, OH.

CHARY, P., R. PRASAD, A. K. CHOPRA, and J. W. PETERSON. 1993. Location of the enterotoxin gene from *Salmonella typhimurium* and characterization of the gene products. *FEMS Microbiol. Lett.,* 111, 87–92.

COHEN, J. 1994. Fulfilling Koch's postulates. *Science,* 266, 1647.

COHEN, J. 1995. A stubborn amoeba takes center stage. *Science,* 267, 822–824.

COLWELL, R. R., P. R. BRAYTON, D. J. GRIMES, D. B. ROSZAK, S. A. HUQ, and L. M. PALMER. 1985. Viable, but non-culturable *Vibrio cholerae* and related pathogens in the environment: Implications for release of genetically engineered microorganisms. *Bio/Technol.,* 3, 817–820.

COLWELL, R. R., M. L. TAMPLIN, P. R. BRAYTON, A. L. GAUZENS, B. D. TALL, D. HARRINGTON, M. M. LEVINE, S. HALL, S. A. HUQ, and D. A. SACK. 1990. Environmental aspects of *V. cholerae* in transmission of cholerae, In R. B. Sack and Y. Zinnaka, Eds., *Advances in Research on Cholera and Related Diarrheas.* K.T.K. Scientific Publishers, Tokyo. Pp. 327–343.

CONNAUGHT LABORATORIES. 1997. *What You Need to Know About Polio and Its Prevention.* Pasteur Mérieux Connaught, Swiftwater, Pa.

DENNIS, J. M. 1959. Infectious hepatitis epidemic in Delhi, India. *J. Amer. Water Works Assoc.*, 51, 1288–1296.

DUFOUR, A. P. 1984. *Health Effects Criteria for Fresh Recreational Waters.* EPA-600/1-84-004. Research Triangle Park, NC. 33 pp.

EPA. 1972. *Water Quality Criteria.* EPA-R3-73-033. Washington, DC. 594 pp.

EPA. 1984. Water quality criteria: Request for comments. Environmental Protection Agency. *Federal Register*, 49, 21987–21988.

EPA. 1986. Bacteriological ambient water quality criteria: Availability. Environmental Protection Agency. *Federal Register*, 51, 8012–8016.

FLANIGAN, T. P., and R. SOAVE. 1993. Cryptosporidiosis. *Prog. Clin. Parasitol.*, 1–20.

FLEISHER, J. 1990. Conducting recreational water quality surveys: Some problems and suggested remedies. *Marine Pollut Bull*, 21(12), 562–567.

FLUEGGE, R. A., T. G. METCALF, and C. WALLIS. 1981. *Virus Inactivation in Wastewater Effluents by Chlorine, Ozone, and Ultraviolet Light.* EPA-600/S2-81-088. Cincinnati, OH.

FOX, J. P. 1976. Human-associated viruses in water. In G. Berg, H. L. Bodily, E. H. Lennette, J. L. Melnick, and T. G. Metcalf, Eds., *Viruses in Water.* American Public Health Association, Washington, DC. Pp. 39–49.

FROBISHER, M., L. SOMMERMEYER, and R. FUERST. 1969. *Microbiology in Health and Disease*, 12th ed. Saunders, Philadelphia. 549 pp.

FUJIOKA, R. S., H. H. HASHIMOTO, E. B. SIWAK, and R. H. F. YOUNG. 1981. Effect of sunlight on survival of indicator bacteria in seawater. *Appl. Environ. Microbiol.*, 41, 690–696.

FULE, R. P., R. M. POWER, S. MENON, S. H. BASUTKAR, and A. M. SAOJI. 1990. Cholera epidemic in Solapur during July-August, 1988. *Indian J. Med. Res.*, 91, 24–26.

GELDREICH, E. E. 1972. Water-borne pathogens. In R. Mitchel, Ed., *Water Pollution Microbiology.* Wiley-Interscience, New York. Pp. 207–241.

GERGATZ, S. J., and L. M. MCFARLAND. 1989. Cholera on the Louisiana Gulf coast: Historical notes and case report. *J. Louisiana State Med. Soc.*, 141(110), 29–34.

GLASS, R. I., J. R. GENTSCH, and B. IVANOFF. 1996. New lessons for rotavirus vaccines. *Science*, 272, 46–48.

GOH, K. T., S. H. TEO, S. LAM, and M. K. LING. 1990. Person-to-person transmission of cholera in a psychiatric hospital. *J. Infect.*, 20(3), 193–200.

GRADY, D. 1996. Quick-change pathogens gain an evolutionary edge. *Science*, 274, 1081.

GREEN, C. E., B. L. GRONDAL, and A. WOLMAN. 1944. *Report on the Water Supply and the Cedar River Watershed of the City of Seattle, Washington.* Cedar River Watershed Commission. Seattle, WA. 79 pp.

HILL, D. R. 1990. *Giardia lamblia.* In G. L. Mandell, R. G. Douglas, Jr., and J. E. Bennett, Eds., *Principles and Practice of Infectious Diseases*, 3rd ed. Churchill Livingstone, New York. Pp. 2110–2115.

ISLAM, M. S., B. S. DRASAR, and D. J. BRADLEY. 1990. Long-term persistence of toxigenic *Vibrio cholera* 01 in the mucilaginous sheath of a blue-green alga, *Anabaena variabilis. J. Trop. Med. Hyg.*, 93(2), 133–139.

KAPIKIAN, A. Z., R. G. WYATT, R. DOLIN, T. S. THORNHILL, A. R. KALICA, and R. M. CHANOCK. 1972. Visualization by immune electron microscopy of a 27 nm particle associated with acute infectious nonbacterial gastroenteritis. *J. Virol.*, 10, 1075–1081.

KAPLAN, J. E., G. W. GARY, R. C. BARON, N. SINGH, L. B. SCHONBERGER, R. FELDMAN, and H. B. GREENBERG. 1982. Epidemiology of Norwalk gastroenteritis and the role of Norwalk virus in outbreaks of acute nonbacterial gastroenteritis. *Ann. Intern. Med.*, 96, 756–761.

KILLEWO, J. Z., D. M. AMSI, and F. S. MHALU. 1989. An investigation of a cholera epidemic in Butiama village of the Mara. *J. Diarrheal Dis. Res.*, 7(1–2), 13–17.

KLUGER, J. 1998. Anatomy of an outbreak. *Time*, 152(5), 57–62.

KONG, R. Y. C., W. F. DUNG, L. L. P. VRIJMOED, and R. S. S. WU. 1995. Co-detection of three species of waterborne bacteria by multiplex PCR. *Marine Pollut. Bull.*, 31, 317–324.

KWANG, J., E. T. LITTLEDIKE, and J. E. KEEN. 1996. Use of the polymerase chain reaction for *Salmonella* detection. *Lett. Appl. Microbiol.*, 22, 46–51.

LIPPY, E. C., and S. C. WALTRIP. 1985. Waterborne disease outbreaks—1946–1980: A thirty-five year perspective. In *Giardia lamblia in Water Supplies—Detection, Occurrence, and Removal.* American Water Works Association, Denver, CO. Pp. 67–74.

MAHMOUD, A. F. 1990. Schistosomiasis. In G. L. Mandell, R. G. Douglas, Jr., and J. E. Bennett, Eds., *Principles and Practice of Infectious Diseases*, 3rd ed. Churchill Livingstone, New York. Pp. 2145–2151.

MAJETI, V. A., and C. S. CLARK. 1981. *Potential Health Effects from Viable Emissions and Toxins Associated with Wastewater Treatment Plants and Land Application Sites.* EPA-600/S1-81-006. Cincinnati, OH.

MARTINEAU, F., F. J. PICARD, P. H. ROY, M. OUELLETTE, and M. G BERGERON. 1998. Species-specific and ubiquitous-DNA-based assays for rapid identification of *Staphylococcus aureus. J. Clin. Microbiol.*, 36(3), 618–623.

MEKALANOS, J. J., and J. C. SADOFF. 1994. Cholera vaccines: Fighting an ancient scourge. *Science*, 265, 1387–1389.

McKINNEY, R. E. 1962. *Microbiology for Sanitary Engineers.* McGraw-Hill, New York. 293 pp.

MURTHY, G. V., A. GOSWAMI, S. NARAYANAN, and S. AMAR. 1990. Effect of educational intervention on defecation habits in an Indian urban slum. *J. Trop. Med. Hyg.*, 93(3), 189–193.

NASH, M. 1993. The waterworks flu. *Time*, 141(16), 41.

NATIONAL ACADEMY OF SCIENCES. 1996. *Use of Reclaimed Water and Sludge in Food Crop Production. National Research Council.* Washington, D.C. 178 pp.

OLIVER, J. D. 1993. Formation of viable but nonculturable cells. In S. Kjelleberg, Ed., *Starvation in Bacteria.* Plenum, New York. Pp. 239–272.

OLSHANSKY, S. J., B. CARNES, R. G. ROGERS, and L. SMITH. 1997. *Infectious Diseases—New and Ancient Threats to World Health.* Population Reference Bureau, Washington, DC. 52 pp.

ONGERTH, J. E., J. RIGGS, and J. COOK. 1989. *A Study of Water Treatment Practices for the Removal of Giardia lamblia cysts.* American Water Works Association, Denver, CO. 56 pp.

RAHMAN, I., M. SHAHAMAT, M. A. R. CHOWDHURY, and R. R. COLWELL. 1996. Potential virulence of viable but nonculturable *Shigella dysenteriae* type 1. *Appl. Environ. Microbiol.*, 62, 115–120.

RAHMAN, I., M. SHAHAMAT, P. A. KIRCHMAN, E. RUSSEK-COHEN, and R. R. COLWELL. 1994. Methionine uptake and cytopathogenicity of viable but nonculturable *Shigella dysenteriae* Type 1. *Appl. Environ. Microbiol.*, 60, 3573–3578.

RAO, V. C., and J. L. MELNICK. 1986. *Environmental Virology.* American Society of Microbiology, 88 pp.

ROBERTS, L. 1988. Change in polio strategy? *Science*, 240, 1145.

ROBERTS, P. V., E. MARCO AIETA, J. D. BERG, and B. M. CHOW. 1981. *Chlorine Dioxide for Wastewater Disinfection: A Feasibility Evaluation.* EPA-600/S2-81-092. Cincinnati, OH.

SARUTA, K., S. HOSHINA, and K. MACHIDA. 1995. Genetic identification of *Staphylococcus aureus* by polymerase chain reaction using single-base-pair mismatch in 16s ribosomal RNA gene. *Microbiol. Immunol.*, 39(11), 839–844.

SCHLEIN, L. 1998. Hunting down the last of the poliovirus. *Science*, 279, 168.

SPIERINGS, G. , C. OCKHUIJSEN, H. HOFSTRA, and J. TOMMASSEN. 1993. Polymerase chain reaction for the specific detection of *Escherichia coli/Shigella. Res. Microbiol.*, 144, 557–564.

SWINBANKS, D. 1996. Outbreak of *E. coli* infection in Japan renews concerns. *Nature*, 382, 290.

SCHEIBLE, O. K., and C. D. BASSELL. 1981. *Ultraviolet Disinfection of a Secondary Wastewater Treatment Plant Effluent.* EPA-600/S2-81-152. Cincinnati, OH.

SCHEIBLE, O. K., M. C. CASEY, and A. FORNDRAN. 1986. *Ultraviolet Disinfection of Wastewaters from Secondary Effluent and Combined Sewer Overflows.* EPA/600/S2-86/005. Cincinnati, OH.

SIDDIQUE, A. K., Q. ISLAM, K. AKRAM, Y. MAZUMDER, A. MITRA, and A. EUSOF. 1989. Cholera epidemics and natural disasters; where is the link? *Trop. Geogr. Med.*, 41(4), 377–382.

ST. LOUIS, M. E., J. D. PORTER, A. HELAL, K. DRAME, N. HARGRETT-BEAN, J. G. WELLS, and R. V. TAUXE. 1990. Epidemic cholera in west Africa: The role of food handling and high risk. *Am. J. Epidemiol.*, 131(4), 719–728.

STOVER, E. L., R. N. JARNIS, and J. P. LONG. 1981. *High-level Ozone Disinfection of Municipal Wastewater Effluents.* EPA-600/S2-81-040. Cincinnati, OH.

TABTIENG, R. S., S. WATTANASRI, P. ECHEVERRIA, J. SERIWATANA, L. BODHIDATTA, A. CHATKAEOMORAKOT, and B. ROWE. 1989. An epidemic of *Vibrio cholerae* el tor Inaba resistant to several antibiotics with a conjugative group C plasmid coding for type II dihydrofolate reductase in Thailand. *Am. J. Trop. Med. Hyg.*, 41(6), 680–686.

WAY, J. S., K. L. JOSEPHSON, S. D. PILLAI, M. ABBASZADEGAN, C. P. GERBA, and I. L. PEPPER. 1993. Specific detection of *Salmonella* spp. by multiplex polymerase chain reaction. *Appl. Environ. Microbiol.,* **59**(5), 1473–1479.

WOLF, H. W. 1972. The coliform count as a measure of water quality. In R. Mitchell, Ed., *Water Pollution Microbiology.* Wiley-Interscience, New York. Pp. 207–241.

QUESTIONS

7.1 You and a friend go hiking up a stream. Your friend cuts his hand on some sharp rocks. He washes the cut in the stream before applying a Band-Aid. Several weeks later, your friend comes down with a serious infection of the kidney and liver. What organism is the likely cause of his illness?

7.2 You go hiking with a friend in the Rocky Mountains. Your friend fills his canteen from a stream and adds some Cl tablets to disinfect the water. A few days later, he develops a serious case of diarrhea. He goes to the doctor and is told that he does not have a viral infection. What would you suspect to be the cause of his diarrhea? What would have been a more effective way to disinfect the water?

7.3 A friend of yours goes swimming in a lake. About a week later, he develops an annoying case of dermatitis. Which of the following pathogens might be the cause of the dermatitis?

    a. *Naegleria* and *Leptospira*

    b. Cryptosporidium and *Giardia*

    c. *E. coli* and *Campylobacter*

    d. *Pseudomonas* and *Schistosoma*

7.4 The city of Potus is considering switching from Cl to UV light to disinfect both its sewage effluent and to treat its public water supply. If you were hired as a consultant by the Potus City Council, would you recommend that they switch from Cl to UV light (a) to disinfect their sewage effluent and (b) to treat their public water supply? Explain your reasoning.

7.5 The outbreak of infectious hepatitis in Delhi, India, in 1955–1956 and the outbreak of cryptosporidiosis in Milwaukee in 1993 illustrate several problems associated with the treatment and monitoring of public water supplies. What are those problems, and how might they be solved in the future?

Chapter Eight

✧　◻　✧　◻　✧

Toxicology

> We assumed [10 or 15 years ago] we knew what the bad pollutants are, at what levels they cause adverse environmental or public health threats. We [assumed] we knew how to drive them down to a no-effect level, how to measure them, how to control them at a reasonable cost. All you needed was an enforcement presence that was sufficiently strong. [Except for the need for enforcement,] all those assumptions were wrong. (Statement by EPA administrator William Ruckelhaus, quoted by M. Sun, *Science*, **227**, 496–497 [1985])

Toxicology is the quantitative study of the effects of harmful substances or stressful conditions on organisms. This rather broad field is broken down into three major divisions: economic, forensic, and environmental toxicology. *Economic toxicology* is concerned with the deliberate use of toxic chemicals to produce harmful effects on target organisms such as bacteria, parasites, and insects. The obvious applications of economic toxicology are in medicine, agriculture, and forestry management, where it is frequently desirable to eliminate or at least control the numbers of various infectious organisms, parasites, and pests (see Chapter 10). *Forensic toxicology* is concerned with the medical and legal aspects of the adverse effects of harmful chemicals and stressful conditions on humans. The medical aspects of forensic toxicology concern the diagnosis and treatment of the adverse effects of toxic chemicals and stressful conditions. The legal aspects concern the cause-and-effect relationships between exposure to harmful chemicals or conditions and the effect of this exposure on human health. *Environmental toxicology* is concerned with the incidental exposure of plants and animals, including humans, to pollutant chemicals and unnatural environmental stresses. Environmental toxicology is related directly to the subject of water pollution because it is environmental toxicological studies that reveal the quantitative relationships between, for example, the concentrations of chemicals found in the water or in aquatic organisms and the effect of these chemicals on aquatic organisms and on persons who drink the water or consume the organisms.

THE ROLE OF TOXICOLOGY IN WATER QUALITY MANAGEMENT

The role of toxicology in water quality management is controversial, because not everyone agrees, and indeed many people disagree, about what sorts of information are needed to manage water quality and how that information should be applied in setting water quality standards. Much information is available, for example, on short-term exposures that produced fatal effects. While the results of such studies are certainly relevant to water quality management, we must also be concerned about sublethal effects resulting from long-term exposure to much lower concentrations of a toxic substance. How do we extrapolate from the results of short-term studies concerned with fatal effects when we are trying to set water quality standards that will protect organisms, including humans, from subtle sublethal effects resulting from long-term exposure?

A second consideration is the question of who or what we are trying to protect. Attitudes generally differ, depending on whether we are considering humans or other organisms. In the latter case, we may not be concerned about protecting every organism in an ecosystem, but where do we draw the line? Do we try to ensure the long-term survival of every species, or is it acceptable for some species to become extinct? In the case of humans, do we set standards that would protect every individual in the population, or are we willing to sacrifice a few subsistence fishermen who eat a great many fish from a contaminated stream or lake or a few individuals who happen to live close to a nuclear power plant?

A third consideration is whether or not there is a threshold level associated with exposure to a particular toxic substance or stressful condition. In other words, is there a level of exposure below which there is no adverse effect? For many toxic substances and stressful conditions we believe that there is a threshold or no adverse effect level, but determining whether such a threshold exists and where it lies may be a difficult task, since subtle adverse effects are inherently hard to detect. In other cases, we assume that there is no threshold and must then decide what level of adverse effects we are willing to tolerate. It is often assumed, for example, that there is no threshold associated with the probability of developing cancer as a result of exposure to carcinogenic chemicals. Is it acceptable then for the concentrations of certain chemicals in drinking water to cause 100 cases of cancer per year in a nation with a population of 275 million people? This is the sort of question that public health authorities must answer in the case of any substance or condition for which there is presumed to be no threshold level.

A major concern in toxicology today is the relevance of studies with animals, particularly rats and mice, to human health. It has been argued, for example, that the physiology of rats and mice is sufficiently different from that of humans that chemicals that are harmful to such animals when administered in large doses over the lifetime of the organism may pose no threat to humans. This argument may have merit. The fact that we often rely on studies carried out with experimental animals when setting environmental standards for the protection of human health reflects a more general problem, namely, that of setting environmental standards in the absence of directly relevant information. The fact is that we often do not have information on the adverse effects of substances or conditions on humans, and because it is unacceptable to carry out controlled experiments on human guinea pigs, we must rely instead on other types of information, usually the results of experimental studies with animals. Whether rats and mice are the best human analog for such purposes is certainly debatable. Rats and mice are easy and relatively inexpensive to maintain in captivity, but monkeys, for example, are physiologically more closely related to humans. In some cases, we do have some knowledge of the effects on human exposure, the ongoing study of 120,321 persons resident in Hiroshima and Nagasaki when the atomic bombs were dropped in August 1945 being a case in point. Of those persons, 91,228 were exposed to radiation from the bombs. Incidents such as this are the exception rather than the rule, however, and in all such cases the degree of exposure is uncontrolled and in some cases it is difficult to estimate.

A final concern in water quality management is the problem of estimating effects when more than one toxic substance is present in the water. If there is no interaction between the toxic substances, then the toxic substance present in the greatest relative amount will determine the toxicity of the mixture, the presence of the other toxic substances being of no consequence. It is certainly possible, however, that there will be some interaction with respect to toxic effects. Since relatively very little information exists on the toxicity of mixtures of toxic substances, establishing water quality standards is difficult in situations where two or more toxic substances are likely to be present at the same time. This situation may often exist in the vicinity of discharges of industrial wastewater. How are government agencies charged with protecting the environment and human health to decide under what conditions such water is safe?

The foregoing discussion gives some idea of the complex problems that surround the use of toxicological information in setting water quality standards. Certainly not everyone agrees on how these issues should be resolved, but practical considerations have forced at least a working resolution, subject to change and further debate. In the following sections of this chapter, we review the toxicological considerations that come into play when setting water quality standards and where appropriate examine how the EPA has used toxicological information to establish water quality criteria for the United States. The use of EPA water quality criteria for illustrative purposes is not to be construed as an endorsement of EPA policy but rather as an example of how a major industrialized nation has dealt with the problem of protecting the quality of its waters.

KINDS OF TOXICITY

In a general sense, studies of toxic effects may be assigned to one of two categories. The first category involves studies of the general overall effects of a compound or stress on an organism. Such overall effects are often described as the result of either acute or chronic exposure. Acutely toxic effects are the result of short-term exposure, usually no more than a few days. Chronic effects are the result of exposure for a relatively long time interval, usually 10% of an organism's life span or longer, and generally involve a lingering or continuous stimulus.

The effects of acute and chronic exposure may be either lethal or sublethal. Falling off a building is an example of an acute stress. It may be fatal, but it need not be. It may produce sublethal damage, such as broken bones, which the body repairs within a short time. It may also produce sublethal adverse effects, such as damage to the brain or spinal cord, which the body is unable to repair. The effect of eating fish with a high concentration of mercury for a period of many years is an example of a chronic stress. The effect may be fatal; indeed, some people have died from such exposure. The effect may also be sublethal and manifest itself in the form of brain damage or lack of coordination. The Mad Hatter in Lewis Carroll's *Alice's Adventures in Wonderland* is an example of a person suffering from chronic exposure to mercury. Mercury was used for years in making felt for hats, a practice banned in the United States since 1941.

The second category of toxicity studies involves experiments designed to evaluate in detail specific kinds of toxicity. The tests may be designed, for example, to study the tendency of a toxicant to cause abnormal development of the fetus (teratogenic tests), to affect the reproductive capacity of an organism, to cause mutations, to produce tumors, to cause cancer, to affect the photosynthetic rates of plants, and so forth. These specific tests are necessary because many of the important effects of toxicants on organisms, particularly at the sublethal level, do not become apparent in standard tests to evaluate overall effects.

Sublethal Effects

Concentrations of a toxic substance that are insufficient to kill organisms outright may nevertheless have a devastating effect on the population of the same organism over a sufficient period of time as a result of sublethal effects. Most sublethal effects may be classified as modifications or interferences with reproduction, development or growth, and behavior. The aim of setting water quality standards in the United States is to eliminate or at least minimize the sublethal effects of pollutants. Unfortunately, the task of determining the level of stress that causes no significant sublethal effects has proven much more difficult than was once imagined, in part because of the large number of potentially toxic substances and because of the large number of possible sublethal effects. In some cases, simply designing an experiment to study certain types of sublethal effects is no trivial matter. A few examples will illustrate the variety and complexity of sublethal effects.

Reproduction

Obviously, any pollutant that interferes with or blocks the reproduction of a species may completely eliminate that species without having any apparent effect on the adult members of the population. For example, DDE, which is a metabolite of DDT,[1] was implicated as the cause of the reproductive failure of certain fish-eating birds during the 1960s and 1970s. Certain toxicological tests designed to examine the specific effects of DDE on the reproduction of various birds showed convincingly that high levels of DDE in the feed of female birds lowered the level of estrogen in the birds and caused them to lay thin-shelled eggs by inhibiting the Ca pump in the oviduct membrane, preventing adequate Ca from

---

[1]DDE and DDT stand for dichlorodiphenyltrichloroethene and dichlorodiphenyltrichloroethane, respectively.

TABLE 8.1 REPRODUCTION AND EGGSHELL THICKNESS DATA FROM PENNED MALLARD DUCKS MAINTAINED FOR TWO SEASONS ON FEED CONTAINING DDE

DDE Added to Feed (ppm)	No. of 14-Day Ducklings per Hen (% of Control Results)		Shell Thickness (% of Control Results)	
	Year 1	Year 2	Year 1	Year 2
10	51	24	92	89
40	35	20	87	86

Source: Heath et al. (1969).

reaching the inside of the oviduct where the shell is formed. Table 8.1 illustrates the results of one such study.

In this particular experiment, which lasted for two years, test ducks began receiving feed containing either 10 or 40 parts per million (ppm) DDE several weeks before the onset of the first laying season. A control group of ducks received feed containing no added DDE. The ducks fed 10 or 40 ppm DDE produced 49–80% fewer 14-day ducklings per hen than the control group and produced eggshells that were about 10% thinner than the control eggshells.

That fish-eating birds might have encountered food containing as high as 10 ppm DDE during the 1960s is confirmed by the data in Table 8.2. In 1969 northern anchovies, which account for about 92% of the food consumed by brown pelicans in the Southern California bight (Anderson and Gress, 1983), were found to contain an average of 3.24 ppm DDE on a fresh weight basis. On a dry weight basis, this concentration is equivalent to about 9.7 ppm[2] and is therefore virtually identical to the concentration of 10 ppm dry weight that Table 8.1 indicates reduced the reproduction of mallard ducks by 50–80%. The sharp drop in anchovy DDE content between 1969 and 1970 coincides with the time that the Montrose Chemical Company, located in Los Angeles County and the sole manufacturer of DDT in the United States, began disposing of its liquid wastes in a sanitary landfill rather than discharging the wastes into the municipal sewer system.[3] The DDE content of the anchovies then remained constant for about three years but dropped sharply between 1972 and 1973, when the EPA restriction on DDT use in the United States went into effect.[4] The EPA restriction resulted in no small part from the examination of toxicological results such as those shown in Table 8.1, which convincingly demonstrated the adverse effects of DDE on the reproduction of aquatic birds.

The correlation between DDT use and the reproductive success of the Southern California brown pelican is illustrated in Figure 8.1. In 1969 and 1970 only nine fledglings were produced from a total of 1,852 nesting attempts, and the thickness of intact eggshells averaged about 0.40 mm. In 1971 and 1972, following the diversion of Montrose's liquid waste to a sanitary landfill, reproduction improved to 249 fledglings out of 1,161 nesting attempts, and eggshell thickness increased to about 0.45 mm. In 1973–1975, following the restriction on DDT use in the United States, reproductive success improved further to 1,575

[2]The difference between the dry weight and wet weight concentration is the moisture content of the food. If two-thirds of the weight of an anchovy is water, then the DDE concentration per unit dry weight is three times larger than the concentration per unit wet weight.
[3]Use of the sanitary landfill began in April 1970.
[4]The restriction became effective in January 1973.

TABLE 8.2   GEOMETRIC MEAN RESIDUES OF DDE IN
ANCHOVIES OFF THE SOUTHERN CALIFORNIA COAST

Year	Anchovy Whole Bodies (ppm, Fresh Weight)
1969	3.24
1970	0.84
1971	0.87
1972	0.74
1973	0.18
1974	0.12
1975[a]	—

[a] Data for 1975 from Anacapa and Santa Cruz islands only.
Source: Anderson et al. (1975, 1977).

fledglings out of 2,175 nesting attempts, and intact eggshell thickness averaged 0.50 mm. The most recent data have shown almost no evidence of crushed eggs due to thin eggshells, and by 1986–1987, eggshell thickness was averaging about 0 54 mm, within 5% of the pre-1947 norm of 0.57 mm (Anderson et al., 1975).

The number of brown pelicans in the Southern California bight colonies increased dramatically during the 1970s, largely due to the increase in reproductive success of the

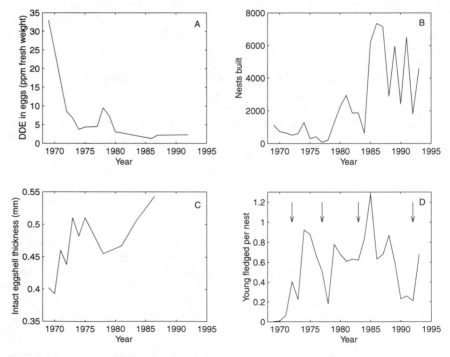

FIGURE 8.1   Recent history of DDE concentrations in southern California brown pelican eggs, eggshell thickness, and pelican reproductive success as reported by Anderson et al. (1975, 1977), Gustafson (1990), and Gress (1995). Vertical arrows in panel D indicate years during which El Niño events produced unusually warm water conditions off the southern California coast.

adult birds. Since 1985 the pelicans have built an average of 5,000 nests per year, up from an average of about 1,200 per year during the previous decade. However, since 1984, the fledgling rates of these colonies have averaged only about 0.63 per nest compared to an average of 1.0 per nest at similar colonies in the Gulf of California (Anderson and Gress, 1983). It has been estimated that perhaps 25% of the adult Southern California population in 1980 was the result of immigration from colonies in southern Baja California or the Gulf of California (ibid.). Had the disastrous reproductive failures in the years immediately preceding the EPA's restriction on DDT been allowed to continue for much longer, the Southern California brown pelican probably would have become extinct. Although some of the population decline during the late 1960s may have been due to adult mortality from direct poisoning (ibid.), the major cause of the decline was reproductive failure. Fortunately, careful toxicological studies of the sublethal effects of DDE on bird reproduction had been carried out and were presented to the EPA at the time of the 1972 DDT hearings. These studies did much to influence the EPA decision to restrict DDT use in the United States.

In recent years, reproductive rates of the Southern California bight brown pelicans have been strongly influenced by fluctuations in the local availability of anchovies (Anderson et al., 1980, 1982), whose abundance typically declines during El Niño years (Figure 8.1D). However, the fact that fledgling rates in these colonies are below the average rates observed at colonies outside the Southern California bight suggests that the former colonies may still be under some unnatural stress. The most recent analyses of Gress (1995) have shown that Southern California brown pelican eggs still contain measurable concentrations of both DDE and PCBs, roughly 2 and 1 ppm, respectively, on a fresh weight basis. The latter are believed to be at least as potent as DDE with respect to their adverse effects on reproduction (see Chapter 10).

Development and Growth

Interference with the normal physiological processes of an organism may adversely affect its ability to grow and develop without directly killing the organism. Nevertheless, because of the pressure of competition and predation, a population of organisms whose development or growth has been retarded may be rapidly eliminated from an ecosystem. For example, studies have shown that exposure of certain species of marine phytoplankton to Hg concentrations as low as 0.5 parts per billion (ppb) can cause more than a 50% reduction in photosynthesis. Figure 8.2 illustrates some of the experimental results obtained with the diatom *Nitzschia delicatissima*. If the presence of a toxin in the water significantly reduces the photosynthetic rate of a particular species, it is possible that the species will be grazed to extinction by herbivores without being directly killed by the toxin. Similarly, phytoplankton division rates and/or photosynthetic rates have been found to be reduced at PCB concentrations as low as 0.1 to 10 ppb (Mosser et al., 1972a; Fisher and Wurster, 1973; Harding and Phillips, 1978). The effect of PCBs on phytoplankton differs markedly among species, some species being much more sensitive than others. As a result, the species composition of a phytoplankton community may be greatly altered due to the presence of PCBs in the water at concentrations as low as 0.1 ppb (Mosser et al., 1972b; Fisher et al., 1974; Fisher, 1975). Studies conducted in a natural estuarine marsh showed that exposure to as little as 1.0 ppb PCBs reduced phytoplankton production and caused a shift in the size structure of the phytoplankton community toward smaller cells (O'Connors et al., 1978). This shift in the mean cell size of the phytoplankton could effectively increase the number of trophic levels between the primary producers and commercially useful fish (Ryther, 1969) and/or could divert more production toward various gelatinous predators such as jellyfish (Greve and Parsons, 1977). Thus, the sublethal effects of a toxin on organisms at the base of the food chain can have a profound effect on the structure of the entire food

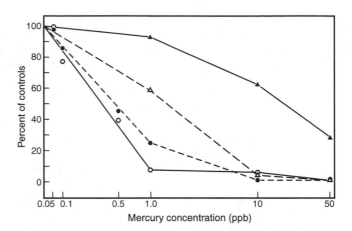

FIGURE 8.2.   Photosynthesis by the marine diatom *Nitzschia delicatissima* after 24 hours of exposure to the following mercurials: ▲, diphenyl mercury; △, phenylmercuric acetate; ●, methylmercury dicyandiamide; ○, MEMMI (a compound containing methylmercury). *Source*: Redrawn from Harriss et al. (1970). Reprinted with permission from R. C. Harriss, D. B. White, and R. B. MacFarlane. 1970. Mercury compounds reduce photosynthesis by plankton. *Science*, **170**, 736–737. Copyright 1970 American Association for the Advancement of Science.

chain and may significantly affect the productivity of the system with respect to commercially useful organisms.

One of the best-documented effects of a stress on the development and growth of aquatic organisms is that of temperature. All organisms grow most efficiently over a limited temperature range. Temperatures above or below the optimum range lead to reduced growth rates or to reduced growth efficiency. Figure 8.3 illustrates in a qualitative way how temperature can influence the growth rate and efficiency of an organism. In this figure, A_c is the amount of food consumed by the organism and A_g is the amount of that food converted into biomass (growth). A_w is the amount of food that is excreted (waste), and A_a, A_d, and A_s represent the amounts of food that are respired to support the activity of the organism (A_a), the degradation and metabolism of consumed food (A_d), and standard or basal metabolism (A_s). It is clear from Figure 8.3 that there is some temperature at which growth, A_g, is at a maximum because the rate of food consumption, A_c, peaks at a certain temperature, whereas the basal respiration rate, A_s, rises exponentially with temperature. This dependence of A_c and A_s on temperature is characteristic of the behavior that might be expected from a poikilothermic (cold-blooded) organism. The amount of food converted to biomass is positively correlated with temperature only as long as the difference between A_c and A_w increases more rapidly than A_s.

Figure 8.3 illustrates one of the difficult problems encountered in converting toxicological information into water quality standards. In all cases shown, it is clear that a small deviation in temperature from the optimal value will result in some decrease in growth, but this effect by itself will not be lethal. However, the effect of a change in temperature on growth is obviously dependent on the availability of food to the organism. A change in temperature that would cause only a modest reduction in growth when the food supply is unlimited could be lethal when food availability is low. Furthermore, although a modest reduction in growth might not by itself be lethal, in a highly competitive system, even a small reduction in growth could lead to the elimination of a species through the combined effects of competition and predation. Similarly, the ecological impact of a given

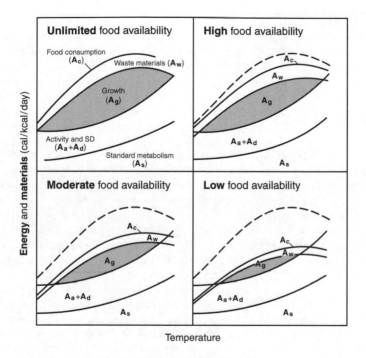

FIGURE 8.3   Effect of temperature and food availability on food consumption (A_c) and the partitioning of consumed food for various purposes by a poikilothermic animal. The curves are qualitative and purely theoretical. *Source*: BIOLOGY AND WATER POLLUTION CONTROL by Charles E. Warren, copyright © 1971 by Saunders College Publishing, reproduced by permission of the publisher.

temperature stress might be far greater in a system in which the organisms were already subjected to other forms of stress, such as the presence of toxic chemicals, than in a system in which no other stresses existed. Thus, translating apparently straightforward toxicological information into water quality standards can be quite difficult.

Behavior

Modification of behavior patterns involves a wide variety of effects, such as alterations in migratory behavior, learning ability, feeding behavior, predator avoidance, and so forth. A few examples will illustrate some of the problems that pollutants may cause as a result of behavioral modification.

Salmon migrate from the ocean back to the freshwater stream where they were born in order to spawn. Although the mechanism that these fish use to find their way from the open ocean back to the general vicinity of their home stream is not well understood, experiments have shown clearly that salmon use their sense of smell to guide their migration upstream once they have entered freshwater. The chemical substances that guide this migration have been shown to include certain volatile organic compounds (Hasler and Larsen, 1971). Unfortunately, a number of streams once characterized by large salmon runs have become seriously polluted with industrial wastes, particularly in the northeastern United States. In many of these streams, salmon runs are now no longer observed or involve insignificant numbers of fish. There is good reason to believe that the presence in these streams of industrial wastes, including a variety of organic substances, has altered the characteristic smell of the water. As a result, the fish find it difficult to identify their home stream. Thus, the presence of chemicals in the water at concentrations well below lethal levels may completely disrupt the migratory patterns of the salmon and, through this behavioral modification, effectively eliminate the species.

Seasonal changes in water temperature can lead to a variety of natural behavioral responses in aquatic organisms. These responses include the onset of reproductive cycles and migratory behavior. The discharge of heated wastewater from large electric power plants can completely disrupt these temperature cues and therefore seriously interfere with natural behavior patterns. For example, menhaden found in the coastal waters of New Jersey during the summer months normally migrate to warmer waters off the North Carolina coast during the winter. At Barnegat Bay, New Jersey, the Jersey Central Power and Light Company operates a nuclear power plant that discharges cooling water at a temperature about 14°C above ambient levels. During the winter, large numbers of menhaden are often found in the waters near the Oyster Creek outfall, either because they are attracted naturally to the warm water in the winter or because, finding themselves in the warmer waters near the outfall, they do not receive the proper temperature cue to migrate south. In this case, the migratory behavior of the menhaden is altered completely by the presence of the heated effluent water, which by itself has no adverse physiological effect on the fish. In fact, during the winter months, the temperature of the effluent water is undoubtedly much closer to the optimum temperature for the menhaden than is the temperature of the surrounding water. Unfortunately, the water temperature near the outfall can drop quickly back to ambient levels if the power plant is shut down. Such a shutdown occurred on January 28, 1972, and as a result 100,000–200,000 menhaden died of thermal shock (Clark and Brownell, 1973). In a similar incident in Long Island Sound, approximately 10,000 bluefish that had been attracted to the heated plume from the Northport power plant died from thermal shock when winds and tidal currents caused the plume to shift its position suddenly on January 17, 1972 (Silverman, 1972). Similar incidents can be anticipated wherever the heated wastewater from power plants attracts aquatic organisms during the colder months of the year.

DETERMINATION OF TOXICITY

One can perhaps imagine a set of water quality criteria based on essentially two types of standards, the first a concentration or stress level that should rarely or never be exceeded and the second a concentration or stress level whose long-term average value should not be exceeded. In both cases, one would presumably be interested in protecting organisms from any form of stress, whether lethal or sublethal. In order to examine the adverse effects of long-term exposure, one must obviously keep the test organisms alive for a long time. For practical reasons, then, the study of chronic toxicity almost always has involved the study of sublethal effects, and ideally, the period of exposure has been the natural lifetime of the organism. Studies of shorter duration are acceptable where a long-term study is impractical, but in the United States at least, the duration of exposure must be 10% or more of an organism's natural lifetime to qualify as a chronic toxicity study. Acute toxicity studies in the United States generally involve exposure to a stress for 48 to 96 hours. The length of exposure for purposes of determining acute toxicity is arbitrary, but for practical purposes, the line needs to be drawn somewhere in order to facilitate comparison between studies. In practice, acute toxicity studies have generally involved the observation of lethal effects, i.e., the organisms are dead after 48–96 hours. There is no compelling reason why this should be the case, for as noted, the real concern should be sublethal effects resulting from acute exposure. Therefore, some allowance must be made for the fact that acute toxicity studies have generally involved lethal stresses and that the goal of water quality management is to protect organisms from both lethal and sublethal effects.

Regardless of the type of toxicity test being carried out, some allowance must be made for the natural variability within a species to the toxic substance or stress being studied. Because of the natural variability among members of a given species, any sample of that population can be expected to contain individuals that differ in their tolerance to a given stress. How then can the response of the species to the stress be quantified? Should one use the response of the most sensitive individual or the least sensitive individual, or should one use the mean or median response? In the United States at least, the tendency has been to use the median response of a sample of individuals of a given species as a measure of the tolerance of that species to a stress. Whether one considers this policy conservative or liberal depends to a certain extent on one's frame of mind. For example, suppose one were interested in knowing how long a particular species could withstand a given stress before the stress became fatal. A polluter might argue that the time required to kill all the members of the species should be termed the *survival time*, and that the median survival time is unduly conservative because half of the members of the species were still alive at the end of the experiment. An environmentalist might argue that the time during which all members of the species survived should be termed the *survival time*, and that the median survival time was unduly liberal because half of the members of the species were dead at the end of the experiment. The argument in favor of using the median response is in part statistical. Median tolerance levels are far more reproducible from one subgroup of a species to the next than are the tolerance levels of the most sensitive and least sensitive members of a subgroup. Furthermore, in the United States, water quality criteria are ultimately based on the tolerance levels of the 5% of the genera that are most sensitive to a stress. In practice, establishing water quality criteria in this way protects virtually all members of most genera, not just the 50% who happen to be most tolerant.

In the following sections, we examine in some detail how acute and chronic toxicity studies are carried out and then how the information obtained from those studies is incorporated into water quality criteria.

Acute Toxicity Determination

As previously noted, acute toxicity studies have generally involved studies of lethal toxicity. Typically, organisms are placed in a series of aquaria, each containing water with a known concentration of a toxic substance. The range of concentrations is chosen based on experience and/or educated guesswork to vary from concentrations too low to have any effect on survival to concentrations expected to kill all the organisms during the period of exposure. The survival of the organisms is noted after either 48 or 96 hours. The percentage survival of the test organisms is then plotted against the concentration of the toxic substance. By graphical or other means, the toxicant concentration corresponding to 50% survival is determined. This toxicant concentration is called the *median tolerance limit* (*TLm*). In some cases, the abbreviations *TL50* (tolerance limit corresponding to 50% survival) or *LC50* (lethal concentration corresponding to 50% mortality) are used to designate the same parameter (EPA, 1985b). In a more general sense, the abbreviation EC50 is often used to designate the concentration that produces a specified effect on 50% of the organisms after a certain period of exposure. In all cases, the symbol should be preceded by a designation of the associated period of time. Thus, a 96-hour TLm would be the concentration of a toxicant associated with 50% survival after 96 hours.

Figure 8.4 illustrates the calculations of a 96-hour TLm. The data are taken from a study of the toxicity of copper to bluegill fish (Trama, 1954). The fish were divided into

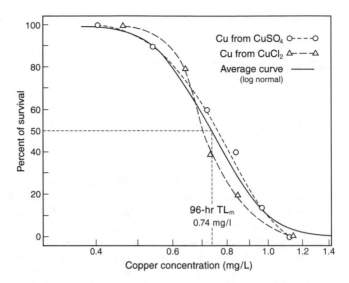

FIGURE 8.4   Percentage survival of bluegill fish after 96 hours versus Cu concentration. The smooth curve is a log-normal curve adjusted to give the best fit to the combined CuSO$_4$ and CuCl$_2$ data. The intersection of this curve with the horizontal line at 50% survival gives the TLm after 96 hours, 0.74 mg L^{-1}. *Source*: Redrawn from Trama (1954). Reproduced by permission.

two groups and exposed to various concentrations of either copper sulfate (CuSO$_4$) or copper chloride (CuCl$_2$) for a period of 96 hours. Percentage survival was then plotted against time. The two data sets were very similar, and a smooth curve was therefore drawn through the combined results. The intersection of this curve with the horizontal line at 50% survival corresponds to the TLm, which in this case was 0.74 mg L^{-1}.

Chronic Toxicity Determination

In some respects, chronic toxicity studies are conducted in the same way as acute toxicity studies, but the former require much longer periods of time. Ideally, studies would last throughout the life cycle of the organism, and where such studies are feasible, this approach is the one recommended by the EPA (1985a). Partial life cycle studies are acceptable for fish that require more than a year to reach sexual maturity, and early life stage toxicity tests may be used to establish water quality criteria if complete or partial life cycle tests with a given species are unavailable. In the latter case, exposure should begin shortly after fertilization and should last for 28 to 32 days through early juvenile development, or 60 days posthatch for salmonids. According to the EPA (1985a) guidelines, these chronic toxicity studies should be conducted in flow-through systems, and toxicant concentrations must be measured in the test solutions. In the case of daphnids,[5] renewal systems may be substituted for flow-through systems. Obviously, such tests involve considerably more time and effort than the 48- or 96-hour bioassays conducted for acute toxicity determination.

In the case of lethal effects, the goal of chronic toxicity studies is to estimate the stress level that would kill half of the test organisms after a period of exposure roughly equal to

---

[5]Commonly referred to as *water fleas*.

their natural lifetime. Since in practice it is often impractical to monitor the test organisms for such a long time, survival may be monitored for shorter time periods and the results extrapolated to time infinity. This approach leads to two important toxicological concepts, median survival times and incipient lethal levels.

Median Survival Times

The median survival time is the time half of the organisms in a random sample from a test population are able to survive a given level of stress. Figure 8.5 illustrates how median survival times are calculated. The data are taken from a study of the toxicity of low O_2 levels to juvenile brook trout (Shepard, 1955). The trout were conditioned for a period of time in water containing 10.5 mg L^{-1} dissolved O_2 and were then placed in experimental aquaria having lower concentrations of dissolved O_2. The survival of the trout was then monitored over a period of 2,000 minutes. As each fish died, it was removed from its aquarium and its time of death recorded. A total of 10 fish were studied at each O_2 level. The results were plotted on semilog paper, and the data fit with straight lines. At the two highest O_2 levels, over half of the fish survived for the entire 2,000 minutes, and no effort was therefore made to estimate median survival times at these O_2 concentrations. However, at the eight lowest O_2 levels, over half of the fish died during the study period, and estimates of median survival time were made from the intercept of the regression lines fit to the data with the horizontal line at 50% mortality.

Incipient Lethal Levels

Once median survival times have been determined at a series of stress levels, the median survival times are plotted against the stress level and the results extrapolated to an infinite survival time, assuming that the data permit such an extrapolation. The hypothetical concentration of the toxicant corresponding to an infinite median survival time is called the *incipient lethal level.* Figure 8.6 illustrates how incipient lethal levels are calculated. The data are taken from a study of the toxicity of Zn to rainbow trout (Lloyd, 1960). Median

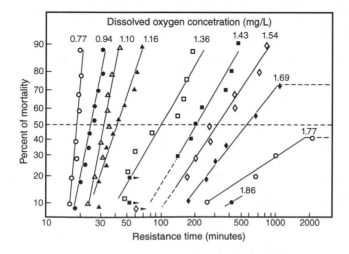

FIGURE 8.5   Percentage mortality versus resistance time for juvenile brook trout exposed to the indicated concentrations of dissolved O_2 after being conditioned to 10.5 mg L^{-1} O_2. *Source*: Redrawn from Shepard (1955). Reproduced with permission.

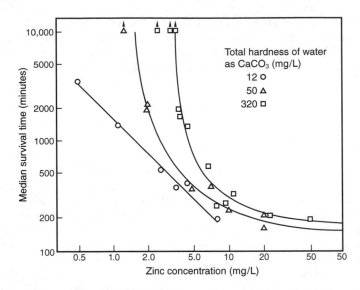

FIGURE 8.6   Median survival time of rainbow trout as a function of Zn concentration at three levels of water hardness. *Source*: Redrawn from Lloyd (1960). Reproduced by permission.

survival times were calculated as a function of Zn concentration at three different levels of water hardness. Note that both axes are plotted on a logarithmic scale. At water hardness levels of 50 and 320 mg L^{-1}, the data show distinct curvature, and a function of the form $(C - C_i)^n(T - T_i) = K$ was fit to the data. In this equation, C is the concentration of Zn corresponding to median survival time T. As T approaches infinity, C approaches C_i, and C_i is therefore interpreted as the incipient lethal level. The values of C_i estimated from this curve fitting were 3.5 and 1.5 mg L^{-1} at water hardness levels of 320 and 50 mg L^{-1}, respectively. At a water hardness of 12 mg L^{-1}, it was impossible to estimate an incipient lethal level because the data were linear over the range of concentrations studied. All that may be said is that the incipient lethal level at 12 mg L^{-1} hardness is less than 0.5 mg L^{-1} Zn.

It should be obvious that the incipient lethal level is a hypothetical concept because no organism survives for an infinite time, regardless of the level of stress. However, the incipient lethal level does provide a standard measure of the concentration or stress level that produces no adverse effects on survival.

Sublethal Effects

As already noted, there is a very wide variety of sublethal effects that could be examined in chronic toxicity studies. Ideally, the procedure is to expose the test organisms to a given level of stress for their natural lifetime and to monitor their behavior, development and growth, and reproduction during the period of exposure. The results are compared to those from a control group, which is exposed to no stress. The treatment, i.e., exposure to the given concentration of toxicant or level of stress, is considered to have adversely affected the organisms if a standard statistical test indicates that the performance of the stressed organisms was significantly worse than that of the control group. "Worse" in this context means, for example, that the treatment group produced fewer viable offspring, grew more slowly, or developed or behaved in an aberrant way.

Results for a given species or genus are generally reported as *mean chronic values*, meaning that they are the average of the levels of stress that produced adverse effects on

TABLE 8.3   CHRONIC TOXICITY OF FOUR COMPOUNDS TO
DAPHNIA MAGNA

	Chronic Values (ppb)		
Compound	Survival of Adults	Reproduction	Mean Chronic Value
1,1,2-Trichloroethene	32,000	18,000	24,000
Dieldrin	100	32	57
Pentachlorophenol	180	320	240
3,4-Dichloroaniline	56	5.6	18

Source: Adema (1978).

the organisms in chronic toxicity bioassays. Since there are many effects that might be examined, and since species will invariably be more sensitive with respect to some kinds of effects than others, the particular effects that were studied can influence to some extent the mean chronic value for the species. For example, Table 8.3 lists the values of four pollutants that have been shown to adversely affect reproduction and the survival of adult *Daphnia magna* in chronic toxicity studies (Adema, 1978). In the case of reproduction, the reported values are the smallest concentrations that had a statistically significant adverse effect on reproduction in paired comparisons with a control group subject to no stress. The difference in the chronic values that affected reproduction and survival is as much as a factor of 10. Obviously, one would reach rather different conclusions if one estimated chronic levels of these pollutants for *D. magna* solely on the basis of reproduction or solely on the basis of adult survival. Standard EPA practice is to average the reported chronic values by calculating their geometric mean, which is the nth root of n numbers. In this case, since two chronic values are reported for each toxicant, the geometric mean is the square root of the product of the two numbers. The geometric means so calculated are listed in the last column of the table. It is noteworthy that the mean chronic value of 18 ppb for 3,4-dichloroaniline is over three times the concentration reported to adversely affect reproduction.

WATER QUALITY STANDARDS

Once acute and chronic toxicity information is available for a sufficient number and variety of aquatic organisms, the process of establishing water quality criteria can begin. The rationale and procedures for determining these water quality guidelines in the United States are described in guidelines promulgated by the EPA (1985a). In establishing these guidelines, the EPA has not attempted to suggest standards that would make the water safe for every organism. The rationale behind the guidelines is summarized as follows:

> Aquatic communities can tolerate some stress and occasional adverse effects on a few species, and so total protection of all the species all of the time is not necessary. Rather, the Guidelines attempt to provide a reasonable and adequate amount of protection with only a small possibility of considerable overprotection or underprotection. Within these constraints, it seems appropriate to err on the side of overprotection. (EPA, 1980a, p. 79342)

Acute Effects

Within the context of this statement, it is now useful to examine the minimum database that the EPA feels is needed to determine acutely toxic concentrations. The rationale behind establishing a minimum data base is that "Results of acute and chronic toxicity tests with a reasonable number and variety of aquatic animals are necessary so that data available for tested species can be considered a useful indication of the sensitivies of the numerous untested species" (ibid., p. 79343).

The minimum data base differs for freshwater and marine organisms and is stipulated as follows in the EPA (1985b) guidelines. The criterion in freshwater must be based on acute tests with freshwater animals in at least eight different families, including all of the following categories: (1) salmonid fish, (2) nonsalmonid fish, (3) a third vertebrate family, (4) planktonic crustaceans, (5) benthic crustaceans, (6) insects, (7) a family not included among vertebrates or insects, and (8) a family in any order of insect or any phylum not already represented. For marine organisms, the criterion should be based on acute tests with saltwater animals in at least eight different families subject to the following five constraints: (1) at least two different vertebrate families are included, (2) at least one species is from a family not included among the vertebrates and insects, (3) either the Mysidae[6] or Penaeidae[7] family or both are included, (4) there are representatives from at least three other families not included among the vertebrates, and (5) at least one other family is represented.

If studies have been carried out on an appropriate number and variety of organisms, it may be possible to establish what the EPA (1985b) refers to as the *final acute value* of the given toxicant, as long as the duration of exposure to the toxicant was appropriate. The guidelines require that results be based on 48–96 hour exposures, the recommended duration depending on the type of organism and the nature of the physiological response. In some cases, acute values are based on TLm's, but EC50 values are used for effects such as decreased shell deposition in oysters. For each species for which satisfactory studies with a particular toxicant have been performed, the acute concentration is calculated as the geometric mean of the reported values. When available, the results of flow-through tests in which the toxicant concentrations were measured are used to determine the geometric mean. When no such data exist, the geometric mean is calculated from all available acute values, including flow-through tests in which the toxicant concentration was not measured and static and renewal tests in which the initial total toxicant concentration was specified.

Once mean acute concentrations have been determined for the appropriate number and variety of species, the species are grouped by genera, and genus mean acute values are calculated as the geometric mean of all the species mean acute values for each genus. The genus mean acute values are then ranked from lowest to highest, and a graphical technique is used to estimate the concentration of the toxicant that would exert an acutely toxic effect on no more than 5% of the species. The genus ranks are converted to cumulative probabilities by dividing each rank by $N + 1$, where N is the number of genera in the list. The four genera with cumulative probabilities closest to 0.05 are then used to determine the final acute value. A plot is made of the logarithm of the genus mean acute values for the four genera against the square root of the cumulative probability, and a model II geometric mean least squares regression line is fit to the data. The concentration of the toxicant corresponding to a cumulative probability of 0.05 on the regression line is taken to be the final acute value.

Table 8.4 and Figure 8.7 illustrate how this procedure would be used to establish a final acute value for the pesticide dieldrin in saltwater at the time the EPA water quality criteria for dieldrin were promulgated. In this case, acute values were available for a total of 21 species, the range of values being about a factor of 70. The least sensitive species was the grass shrimp *Palaemonetes vulgaris*, with a TLm of 50.0 ppb, and the most sensitive species was the pink shrimp *Penaeus duorarum*, with a TLm of 0.7 ppb. Table 8.4 includes

---

[6]Crustaceans commonly referred to as *opposum shrimps* and resembling miniature crayfish.
[7]Certain shrimp.

TABLE 8.4   MEAN ACUTE VALUES OF DIELDRIN IN SALTWATER

Species	Species Mean Acute Value (ppb)	Genus Mean Acute Value (ppb)
Sphaeroides maculatus, northern puffer	34.0	34.0
Crassostrea virginica, eastern oyster	31.2	31.2
Mugil cephalus, striped mullet	23.0	23.0
Palaemonetes vulgaris, grass shrimp	50.0	
		20.7
Palaemonetes vulgaris, grass shrimp	8.6	
Morone saxatilis, striped bass	19.7	19.7
Pagurus longicarpus, hermit crab	18.0	18.0
Gassterosteus aculatus, threespine stickleback	14.2	14.2
Palaemon macrodactylus, Korean shrimp	10.8	10.8
Cyprinodon variegatus, sheepshead minnow	10.0	10.0
Crangon septemspinosa, sand shrimp	7.0	7.0
Fundulus heteroclitus, mummichog	8.9	
		6.7
Fundulus majalis, striped killifish	5.0	
Thalassoma bifasciatum, bluehead	6.0	6.0
Menidia menidia, Atlantic silverside	5.0	5.0
Mysidopsis bahia, mysid shrimp	4.5	4.5
Micrometrus minimus, dwarf perch	3.5	3.5
Cyamatogaster aggregata, shiner perch	2.3	2.3
Oncorhynchus tshawytscha, chinook salmon	1.5	1.5
Anguilla rostrata, American eel	0.9	0.9
Panaeus duorarum, pink shrimp	0.7	0.7

Source: EPA (1980b).

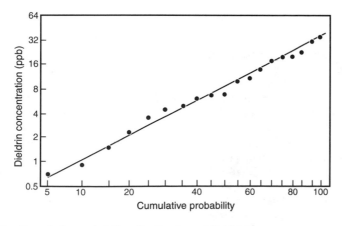

FIGURE 8.7   Cumulative probability distribution of dieldrin genus mean acute values based on the data in Table 8.4. The straight line is a least squares fit to the four lowest data points.

species from a total of 19 genera, and the cumulative probabilities associated with the four most sensitive genera are therefore 0.05, 0.10, 0.15, and 0.20. The genus mean acute values and the least squares regression line are shown in Figure 8.7. The regression line actually gives an excellent fit to the entire data set, although it was fit to only the four lowest genus mean acute values. According to the regression line, the dieldrin concentration corresponding to a cumulative probability of 5% is 0.63 ppb, and according to the 1986 EPA guidelines, this concentration would therefore become the final acute value for dieldrin in saltwater. Based on the data available for dieldrin, it is expected that if a large random sample of saltwater genera were taken, only 5% would have TLm values less than 0.63 ppb. In 1980, when the dieldrin criteria were being determined (EPA, 1980b), the EPA was using a somewhat different procedure for estimating the final acute value, and the final acute value determined at that time was 0.71 ppb. According to the regression line in Figure 8.7, this concentration would be associated with a cumulative probability of 6%. The difference does not seem worth arguing about.

Chronic Toxicity

If data on a sufficient number and diversity of organisms are available, a final chronic value for a particular toxicant may be calculated in the same way that final acute values are determined. In practice, however, there are seldom sufficient data to allow a direct graphical estimation of the toxicant concentration that would exert a chronic stress on no more than 5% of the species in the system. In such cases, an acute toxicity standard is established based on an adequate amount of short-term toxicity tests, and an average acute/chronic toxicity ratio is then calculated based on a smaller amount of information. The rationale for this procedure is that for a given pollutant the acute/chronic ratio is likely to be more constant between species than is the chronic or sublethal stress level itself. Hence, less information is required to estimate the acute/chronic ratio. The chronic toxicity standard is established by dividing the acute toxicity standard by the so-called final acute/chronic ratio. The EPA considers this procedure acceptable if acute/chronic ratios are available for at least three species and (1) at least one of the species is a fish, (2) at least one is an invertebrate, and (3) at least one is an acutely sensitive freshwater species or saltwater species when the ratio is being used to establish freshwater or marine criteria, respectively.

Acute/Chronic Ratios

The final acute/chronic ratio is considered to be a measure of the ratio of the concentration of the toxic substance associated with acutely toxic effects to the concentration associated with chronic toxicity effects. According to the methodology recommended by the EPA (1985a), the final acute/chronic ratio may be calculated in one of four ways. If no major trend is apparent among the acute/chronic ratios for the different species and the species mean acute/chronic ratios lie within a factor of 10 for a number of species, the final acute/chronic ratio is the geometric mean of all the species acute/chronic ratios available for both freshwater and saltwater species. If the species mean acute/chronic ratios seem to be correlated with the species mean acute values, the final acute/chronic ratio should be taken to be the acute/chronic ratio of species whose acute values lie close to the final acute value. In the case of acute tests conducted on metals and possibly other substances with embryos and larvae of barnacles, bivalve mollusks, sea urchins, lobsters, crabs, shrimp, and abalones, the acute/chronic ratio is assumed to be 2. The rationale is that chronic tests are very difficult to conduct with such species, and the sensitivities of embryos and larvae would likely determine the results of life-cycle tests. Assuming the acute/chronic ratio to

TABLE 8.5   ACUTE/CHRONIC RATIOS FOR DIELDRIN TOXICITY
TO THREE SPECIES OF AQUATIC ANIMALS

Species	Acute Value (ppb)	Chronic Value (ppb)	Ratio
Salmo gairdneri, rainbow trout	2.5	0.22	11
Poecilia reticulata, guppy	4.1	0.45	9.1
Mysidopsis bahia, mysid shrimp	4.5	0.73	6.2

Source: EPA (1980b).

be 2 causes the final chronic value to equal the criterion maximum concentration (see the later discussion). Finally, if the most appropriate species mean acute/chronic ratios are less than 2, acclimation to the stress probably occurred during the chronic test (see the later discussion). In such cases, the final acute/chronic ratio is assumed to be 2. The final chronic value for the toxicant is then calculated by dividing the final acute value by the final acute/chronic ratio.

An example of the calculation of a final acute/chronic ratio is shown in Table 8.5 for the pesticide dieldrin. Chronic values for this pesticide were available for only four species in 1980, when the guidelines for dieldrin were established (EPA, 1980b), and therefore it was necessary to use acute/chronic ratios to establish the final chronic value. Acute toxicity values were available in only three of the four cases in which chronic effects were studied, but the three species satisfied the criteria for calculating acute/chronic ratios in both freshwater and saltwater. Since the acute/chronic ratios for the three species differed by less than a factor of 2, it was appropriate to calculate the final acute/chronic ratio for dieldrin by taking the geometric mean of the three ratios, which is $[(11)(9.1)(6.2)]^{1/3} = 8.5$. Final acute values for dieldrin in freshwater (EPA, 1996a) and saltwater were 0.48 and 0.71 ppb, respectively. Hence, the final chronic values for dieldrin in freshwater and saltwater were $0.48/8.5 = 0.056$ ppb and $0.71/8.5 = 0.084$ ppb, respectively.

Toxicity to Plants

The EPA guidelines for calculating acute and chronic concentrations of a toxicant clearly pertain to aquatic animals. Nevertheless, aquatic plants may be affected adversely by the presence of toxicants in the water and sometimes at exceedingly low concentrations (e.g., Figure 8.2). According to EPA guidelines (EPA, 1986), a substance is toxic to a plant at a given concentration if the growth of the plant is decreased in a 96-hour or longer experiment with an alga or a chronic test with a vascular plant. Growth may be measured in a variety of ways, including photosynthetic rate, change in dry weight, or change in chlorophyll concentration. Table 8.6, for example, provides a summary of data on the toxicity of dieldrin to aquatic plants. In this particular case, the concentrations that reduced plant growth were substantially higher than both the final acute and final chronic values for animals and therefore had no influence on the final water quality guidelines for dieldrin. However, because plants are the chief producers of organic matter in most aquatic food chains, it is critical that the effects of toxicants on aquatic plants be assessed. Reductions in photosynthetic rates and/or changes in the composition of the plant community may cause repercussions throughout the entire food web. The final plant value is the lowest concentration of the toxicant that reduces plant growth in an appropriate experiment. Based on the data in Table 8.6, the final plant value for dieldrin was set at 100 ppb in freshwater and 950 ppb in saltwater (EPA, 1980b).

TABLE 8.6 CONCENTRATIONS OF DIELDRIN THAT REDUCE GROWTH IN AQUATIC PLANTS

Species	Effects	Concentration (ppb)
Scenedesmus quadricaudata	22% reduction in biomass in 10 days	100
Navicula seminulum	50% reduction in growth in 5 days	12,800
Wolffla papulifera	Reduced population growth in 12 days	10,000
Agmenellum quadruplication	Reduced growth rate	950

Source: EPA (1980b).

THE TWO-NUMBER CRITERION

In the United States, EPA water quality guidelines with respect to toxic substances are based on a two-number criterion (EPA, 1986). The first number is the one-hour average concentration that is not to be exceeded more than once every three years. This number is referred to as the *criterion maximum concentration*. The second number is the four-day average concentration that is not to be exceeded more than once every three years. This number is referred to as the *criterion continuous concentration*. The rationale for using a two-number criterion is that organisms can tolerate brief exposures to toxicant concentrations substantially higher than can be tolerated over a longer period. This fact is evident in Figure 8.6. There is invariably a negative correlation between the concentration of a toxicant that causes a particular adverse effect and the duration of exposure. The recommended exceedence frequency of three years is the EPA's best scientific judgment of the average amount of time it would take an unstressed system to recover from a pollution event in which exposure to a toxicant exceeded the criterion.

Under the present system (EPA, 1996a), the two numbers in the criterion are calculated from the final acute value, the final chronic value, and the final plant value. The three values for dieldrin in freshwater and saltwater are shown in Table 8.7. The criterion maximum concentration is equated to half the final acute value. Division by 2 in this case to some extent corrects for the fact that much of the acute toxicity information is based on observations of lethal effects, whereas the real concern is protection of organisms from sublethal stresses. The criterion continuous concentration is the smaller of the final chronic value and the final plant value.

TABLE 8.7 THE THREE TOXICANT CONCENTRATIONS USED TO ESTABLISH WATER QUALITY GUIDELINES FOR DIELDRIN IN FRESHWATER AND SALTWATER

	Value in Freshwater (ppb)	Value in Saltwater (ppb)
Final acute value	0.48	0.71
Final chronic value	0.056	0.084
Final plant value	100	950
Criterion maximum concentration	0.24	0.355
Criterion continuous concentration	0.056	0.084

COMPLICATING FACTORS

Interactions with Harmless Substances or Conditions

The toxicity of a particular substance to aquatic organisms may sometimes depend strongly on the concentrations of other substances or on environmental conditions that by themselves are harmless. For example, in aquatic systems, a continuous interconversion takes place between ammonium ions (NH_4^+) and un-ionized ammonia (NH_3). The equation describing this interchange can be written $NH_3 + H^+ \leftrightarrow NH_4^+$. Although NH_4^+ is nontoxic, NH_3 is highly toxic. The NH_3 96-hour median tolerance limits for a variety of freshwater fish lie in the range 0.08–4.6 ppm. The equilibrium distribution of NH_3 and NH_4^+ is a function of the hydrogen ion concentration (H^+) in the water, and the equilibrium shifts to NH_3 as the concentration or activity of H^+ is reduced. In water quality studies the H^+ activity commonly is reported in terms of pH, where pH is the negative of the common logarithm of the H^+ activity.[8] At steady state the concentrations of NH_3 and NH_4^+ are equal at a pH of 9.3, and the $[NH_3]/[NH_4^+]$ ratio increases by a factor of 10 whenever the pH increases by one unit. Thus water that was virtually harmless at a pH of 7 might become extremely toxic if the pH were raised to 9. For this season, it is of relatively little use in water quality work to assay merely for $NH_3 + NH_4^+$ (by standard procedures) without simultaneously measuring the pH.

Temperature is another factor that can greatly influence the toxicity of a substance without itself causing any significant stress. In general terms, the performance of an organism is positively correlated with the temperature below the optimum temperature for the organism and then drops rapidly as the temperature increases further (Figure 8.8). The rate of increase of the respiration rate or metabolic rate of an organism is usually expressed in terms of the organism's Q_{10}, which is the change in respiration rate or metabolic rate per 10°C increase in temperature. For temperature conformers (i.e., cold-blooded organisms), metabolic rates typically increase by a factor of 2 or 3 for every 10°C rise in temperature as long as the temperature is not too close to the lethal temperature for the species. In other words, typical Q_{10} values for cold-blooded organisms are 2–3. Obviously, any increase in respiration rate implies an increased demand for O_2, and as a result, the TLm for O_2 can be expected to be positively correlated with temperature. This fact in part underlies the tendency of fish kills caused by inadequate O_2 concentrations to occur during the summer, when water temperatures are highest. An increase in the respiration rate of a fish obviously implies an increase in the rate at which water is pumped over the gills. As a result, the intensity of exposure of the fish to any toxic substance in the water is increased. The effects of a given toxicant concentration on the organism may therefore become apparent more rapidly in warm water than in cold water.

Figure 8.6 provides another clear example of the interaction between otherwise harmless water quality characteristics and toxicant effects. The toxicity of Zn to rainbow trout is clearly correlated in a negative way with water hardness. Similar correlations between metal toxicity and water hardness have been noted for a variety of toxic metals. The mechanism responsible for the correlation is not well understood, and considering the wide range of metals for which the effect has been observed, it is unlikely that exactly the same mechanism is operative in all cases. Water hardness, which is a measure of the concentration of Ca and Mg ions in the water, tends to be correlated with a number of other water quality characteristics, such as pH and alkalinity. Therefore, it is possible that at least in some cases, the observed negative correlation between metal toxicity and water hardness actually reflects a cause-and-effect relationship with one of these other covariates.

---

[8]Most natural freshwaters have a pH in the range 6–8. The pH of seawater is about 8.1.

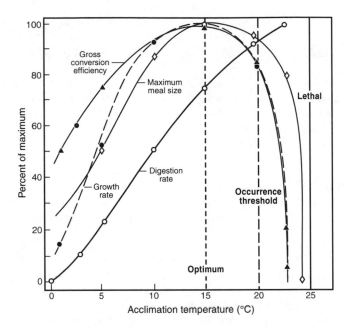

FIGURE 8.8   Performance of sockeye salmon as a function of acclimation temperature. Redrawn from Brett, *Amer. Zool.*, **11**, 99–113 (1971).

Incorporation into Water Quality Guidelines

If there is enough relevant information, it is possible to incorporate these interactions between toxicant effects and water quality characteristics into water quality criteria. In the United States, for example, the criterion maximum concentration and the criterion continuous concentration are sometimes expressed as explicit functions of certain water quality characteristics. In the case of NH_3, for example, toxicity has been shown to be negatively correlated with pH and, below 20–25°C, with temperature. The EPA water quality criteria for NH_3 therefore are expressed in terms of mathematical equations that relate the criterion maximum and continuous concentrations of NH_3 to temperature and pH (EPA, 1986).

   In the case of heavy metals, the criteria often are expressed as functions of water hardness. Figure 8.9 shows some of the data that were used to determine the freshwater criterion maximum concentration in the case of Cd (EPA, 1984, 1999). Pertinent data relating Cd toxicity to water hardness were available for five species, and the logarithms of the Cd 96-hour TLm values were plotted against water hardness. The slopes of the five regression lines fit to the data were similar, the average being 1.128. The implication is that Cd 96-hour TLm values can be expected to be proportional to water hardness raised to the 1.128 power. Freshwater 96-hour TLm data reported in the literature were then normalized to the same water hardness by assuming this dependence on water hardness, and the normalized 96-hour TLm values were then plotted as in Figure 8.7. The normalized final acute value was then calculated in the manner previously described. However, this procedure resulted in a normalized final acute value greater than the normalized species mean acute values of four salmonids, which were the most sensitive species among the 50 species included in the study. Since the salmonids are a commercially and recreationally important group of fish, the normalized final acute value was arbitrarily set to the normalized species mean acute value of rainbow trout, the most sensitive salmonid for which results of flow-through tests were available. The normalized freshwater criterion maximum concentration

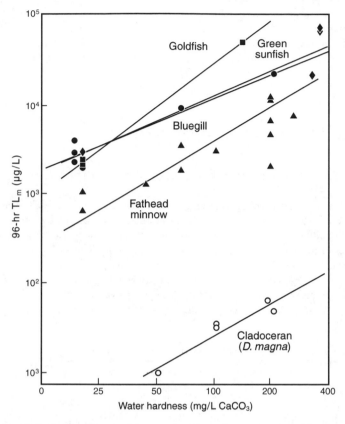

FIGURE 8.9   Cd freshwater TLm values as a function of water hardness for five species. *Source*: Data taken from EPA (1984)

was then set equal to half of this normalized final acute value. The criterion maximum concentration was then determined from the equation

$$\text{Criterion maximum concentration (ppb)} = 0.025 \, \text{Hd}^{1.128}$$

where Hd is water hardness in milligrams per liter. The value so calculated is the concentration of the so-called total recoverable fraction of the Cd (EPA, 1997) and is operationally determined by measuring the Cd concentration in an acidified sample of water. For purposes of establishing water quality criteria, this concentration is now multiplied by a conversion factor (cf) to determine the concentration of the dissolved metal (EPA, 1997). In the case of Cd, the conversion factor is a function of the water hardness and is calculated from the equation $cf = 1.137 - 0.0418 \log_e (\text{Hd})$. The complete equation for the Cd criterion maximum concentration expressed as dissolved Cd is then (EPA, 1999)

$$\text{Criterion maximum concentration (ppb)} = (0.025)(\text{cf})\text{Hd}^{1.128}$$

An analogous equation for the final chronic value may be calculated by simply dividing the equation for the final acute value by the final acute/chronic ratio. However, if there is evidence that there is a difference in the functional dependence of chronic toxicity and acute toxicity on water quality characteristics such as temperature and hardness, then the final chronic equation may be determined independently of the final acute equation. In the case of Cd, for example, chronic toxicity appears to be less sensitive to water hardness than acute toxicity, and therefore a final freshwater chronic equation was developed solely from chronic toxicity studies performed with a total of 16 freshwater species. The final chronic equation for dissolved Cd in freshwater is

$$\text{Final chronic value (ppb)} = (0.066)(\text{cf})\text{Hd}^{0.785}$$

where cf in this case equals $1.1017 - 0.0418 \log_e(\text{Hd})$.

Conditioning and Acclimation

The toxicity of a particular stress to an aquatic organism can depend very much on the previous life history of the organism. This fact reflects the ability of many organisms to acclimate or adapt to changes in the environment and in particular to the presence of toxic substances in the water. Figure 8.10 provides a good example of the effects of adaptation on the toxic effects of temperature stress. The data are taken from a study of both lethal and sublethal temperature stresses on young sockeye salmon. Two features of the graph are of particular interest. First, sublethal effects on spawning and growth become apparent outside a much narrower temperature range than is the case for lethal effects. For ex-

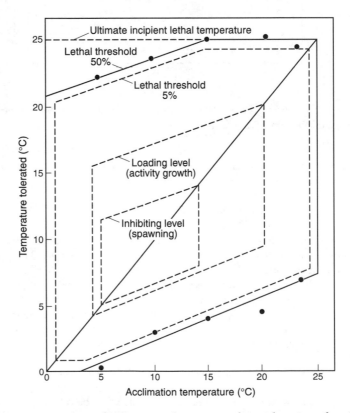

FIGURE 8.10   Temperature tolerance zones for young sockeye salmon as a function of acclimation temperature. Redrawn from Brett in C. M. Tarzwell, *Biological Problems of Water Pollution*, U.S. Department of Health, Education, and Welfare (1960).

ample, the fish do not begin to die outright until the temperature exceeds 21–25°C or falls below 5–7°C. These observations simply illustrate and reemphasize the importance of distinguishing sublethal effects from lethal effects. Second, the range of temperatures within which the fish can effectively carry out certain activities is shifted to higher temperatures as the acclimation temperature is increased. For example, a young sockeye salmon acclimated to a temperature of 20°C would evidently die if placed in water at 4°C, but a similar fish acclimated at 10°C would not die if placed in the same water. There is, however, a limit to the adaptive capabilities of aquatic organisms. For example, it is apparent from Figure 8.10 that young sockeye salmon cannot spawn outside the approximate temperature range 5–14°C, regardless of the acclimation temperature.

In some cases, exposure of an organism to a stress may actually increase the sensitivity of the organism to that stress, but in many cases, prolonged exposure to a sublethal level of stress may actually desensitize or acclimate the organism to the stress. For example, many fish are capable of acclimation to low O_2 levels as well as to high and low temperatures. Some fish have been shown to acclimate to ammonia, cyanide, pH, phenol, synthetic detergents, and Zn (Degens et al., 1950; Bucksteeg, et al., 1955; Neil, 1957; Lloyd, 1960; Lemke and Mount, 1963; Jordan and Lloyd, 1964; Edward and Brown, 1967; Lloyd and Orr, 1969). In organisms such as bacteria that can multiply rapidly, adaptation may actually involve genetic changes. Mutant strains may by chance be more resistant to a particular stress than the other members of the population and over time will gradually dominate the population. Genetic selection of this sort is believed to have been responsible for the emergence of pesticide-resistant strains of about 200 different insect pests. Among organisms that reproduce more slowly, genetic selection would take much longer, and acclimation in the short term must therefore involve other mechanisms such as biochemical, physiological, or behavioral adjustments. The processes involved in these adjustments are in general not well understood, but probably involve changes in the functioning of hormones and/or enzymes and the response characteristics of the organism's nervous system.

In a general sense, organisms that adapt to a stress may be classified as being either *conformers* or *regulators* (Warren, 1971). For example, cold-blooded animals are conformers with respect to temperature because their internal body temperature is correlated strongly with the temperature of the environment. Warm-blooded animals are regulators with respect to temperature because their internal temperature is relatively independent of the temperature of the environment. Of course, for both conformers and regulators there is an optimum internal state at which the organism functions most effectively. Acclimation among regulators usually produces a change in the range of the external temperature within which the organism can effectively control its internal state, but it does not change the optimum state of the organism. Acclimation among conformers may involve a change in the organism's optimal internal state and usually involves a change in the range of internal variability within which the organism can function efficiently.

Should the possible effects of acclimation be considered when setting water quality guidelines? The attitude of the EPA is that acclimation effects should in general not be considered and that toxicity data obtained with acclimated organisms should not be used in deriving water quality guidelines. The rationale for this attitude is that "Acclimated organisms are the exception rather than the norm. Rarely, if ever, can acclimation be depended on to protect organisms in a field situation because concentrations often fluctuate and motile organisms do not stay in one location very long" (EPA, 1980a, p. 79364). This attitude seems reasonable to the extent that acclimation in toxicological studies is avoidable and tends to desensitize organisms to a stress. However, conditioning may sometimes sensitize rather than desensitize an organism to a stress, and in the case of temperature stress, it is impossible to avoid the issue of acclimation in experimental studies. Certainly one should be aware of the potentially confounding effects of acclimation in toxicological work.

Interactions Between Toxic Substances

The discussion to this point has considered the effect of a toxicant as if it were the only stress-producing substance. What effect can be anticipated, however, if two or more toxic substances are present in the water at the same time? For example, according to the present EPA guidelines, the four-day average concentrations of Pb and Zn in saltwater should not exceed 5.6 and 86 ppm, respectively, more than once every three years. Now suppose that a particular body of saltwater is found consistently to contain 4 ppm of Pb and 60 ppm of Zn. Would this water be likely to exert a chronic stress that would in some sense violate the intentions of the EPA guidelines?

The answer to this question is that the effect of two or more toxic substances on the organisms in the water will depend on the manner in which the toxicants interact. The following example illustrates the various ways in which two toxicants might interact. Suppose that the 96-hour TLm for a particular organism is 1 and 10 ppb for toxicants A and B, respectively. Now suppose that toxicants A and B are added to the water in a flow-through acute toxicity experiment so that the concentrations of A and B are in fact 1 and 10 ppb, respectively. The test organisms are placed in the water, and their survival is monitored for 96 hours. Any of the following outcomes is possible:

1. Exactly half of the organisms are dead after 96 hours. This result is exactly what one would have expected if one had added 1 ppb A but no B to the water or 10 ppb B but no A. In this case, one says that there has been *no interaction* between A and B.
2. Fewer than half of the organisms are dead after 96 hours. In this case, the toxicity of the mixture is apparently less than the toxicity of a solution containing only 1 ppb A or only 10 ppb B, and the interaction between A and B is said to be *antagonistic.*
3. If more than half of the organisms are dead after 96 hours, there are several possible interpretations:
    a. If the percentage of dead organisms is exactly what would have been expected if the water contained only 2 ppb A or only 20 ppb B, then the interaction of A and B is said to be *strictly additive.* In other words, 1 ppb A produces the same toxicity as 10 ppb B, and when the two substances are present together, the toxicity of the mixture can be calculated by assuming that every 10 ppb of B are equivalent to 1 ppb of A.
    b. If the fraction of organisms that are dead after 96 hours is greater than in case 3a, then the interaction is said to be *supra-additive.*
    c. If the percentage of organisms dead after 96 hours is greater than 50% but less than in the case 3a, then the interaction is said to be *infra-additive.*

It is useful in discussing these interactions to speak in terms of *toxicity units* rather than actual *concentrations.* In this hypothetical example, one acute toxicity unit of A would be 1 ppb of A and one acute toxicity unit of B would be 10 ppb of B. The possible interactions between A and B may then be indicated graphically as in Figure 8.11.

How should one take account of these interactions in setting water quality guidelines? One obvious practical problem is that not much is known about the interactions of toxic substances (Madia, 1998). In fact, a review of current EPA water quality criteria (EPA, 1987, 1999) shows that in many cases, insufficient information is available to establish numerical guidelines for toxicants considered separately. The present guidelines make no allowance for interactions between toxicants. In response to criticisms concerning this policy, the EPA has stated, "Synergism and antagonism are possible between numerous combination [sic] of two or more pollutants, and some data indicate that such interactions are not only

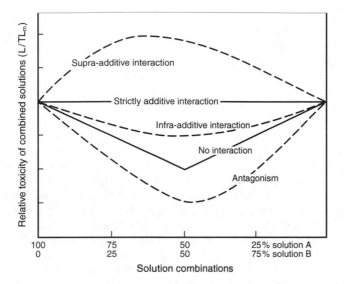

FIGURE 8.11   Possible interactions between two hypothetical toxicants, A and B. *Source*: *Biology and Water Pollution Control* by Charles E. Warren, copyright © 1971 by Saunders College Publishing, reproduced by permission of the publisher.

species specific, but also vary with the ratios and absolute concentrations of the pollutants and the life stage of the species" (EPA, 1980a, p. 79358).

In other words, not much is known, the interactions may be very complicated, and given these facts, it is very hard to estimate the toxicity of a solution containing more than one toxic substance. To the extent that the effects of two or more toxicants can be normalized by expressing concentrations in terms of toxicity units, Figure 8.11 shows that in two cases, calculation of the toxicity of a mixture of toxicants is straightforward. First, if there is no interaction between the toxicants, then the toxicity of the mixture would be determined by the toxicant present in the greatest number of toxicity units. For example, if a solution contained 0.75 toxicity unit of A and 0.5 toxicity unit of B, then if A and B do not interact, the toxicity of the solution would be equal to that of a solution containing only 0.75 toxic unit of A. Similarly, if the interaction were strictly additive, the toxicity of the mixture would be equal to that of a solution containing 1.25 toxicity units. In other words, if one assumes that there is no interaction or that the interactions are strictly additive, then the toxicity of a mixture of toxicants can easily be calculated. However, if the interactions are supra-additive, infra-additive, or antagonistic, then one must have detailed information about the nature of the interaction to calculate the toxicity of the mixture.

It is apparent from Figure 8.11 that the assumption of no interaction is conservative only if the true interactions are antagonistic. The assumption of no interaction would underestimate the true interactions if the true interactions were in any sense additive. The assumption of strictly additive interactions would appear to be conservative in all cases except the case in which the true interactions were supra-additive. This latter conclusion, however, is valid only to the extent that a given toxic effect can be associated uniquely with a certain number of toxicity units. Figure 8.11, for example, implicitly assumes that X toxicity units of A have the same effect as X toxicity units of B for all values of X. In other words, there is a universal curve that relates toxic effect to toxicity units. The normalization procedure ensures that the effects are the same when $X = 1$, but it does not follow automatically that the effects are the same for all values of X. For example, suppose that

2 acute toxicity units of A kill 75% of a given species in 96 hours and that 2 acute toxicity units of B kill 95% of the same species in 96 hours. Now suppose that 85% of the test organisms die after 96 hours of exposure in water containing 1 toxicity unit of both A and B. The interactions are clearly additive in some sense, since more than 50% of the test organisms died after 96 hours. However, the interactions appear to be supra-additive relative to toxicant A and infra-additive relative to toxicant B. Strictly speaking, then, one cannot apply the assumption of strictly additive interactions unless one feels confident that the curves relating toxic effects and toxicity units are the same or very similar for all toxicants of concern.

There is no doubt that some toxicants do not interact in a strictly additive manner. For example, the toxicity of Cd is decreased by other metal ions, particularly low concentrations of Cu and Zn (Parizek, 1957; Gunn et al., 1963a, 1936b; Hill et al., 1963; Bunn and Matrone, 1966; Ferm and Carpenter, 1967; Parizek et al., 1969), and it is known that the cholinesterase-inhibiting pesticides EPN (*O*-ethyl-*O*-paranitrophenyl phenylphosphonothioate) and malathion interact in a supra-additive manner. The TLm of a mixture containing equal numbers of toxicity units of EPN and malathion is only 1/12th the TLm of the individual compounds (Frawley, 1965). The effect is due to the ability of EPN to inhibit the enzyme system that is responsible for detoxifying malathion after it has been converted to malaoxon. In contrast, chlorinated hydrocarbons such as DDT that induce some of the liver microsomial enzyme systems may interact antagonistically with chemicals that are detoxified by the same enzyme systems (e.g., hexobarbital). Although admittedly it is illogical to expect the toxicity of poisons with different toxicological properties and different concentration-response curves to be strictly additive, "Nevertheless, the method has been found to work empirically. . . . If, for the present, the poisons can be regarded as agents producing stress . . . , each of which produces a degree of shock with resulting nonspecific effects, it might be considered reasonable that summation of the overall stress is possible" (Brown, 1968, p. 731).

Application of the strictly additive interaction assumption to water quality guidelines was made in the 1973 EPA water quality criteria, where it was noted, "The system of adding different toxicants in this way is based on the premise that their lethal actions are additive. Unlikely as it seems, this simple rule has been found to govern the combined lethal action of many pairs and mixtures of quite dissimilar toxicants, such as copper and ammonia, and zinc and phenol in the laboratory. . . . The rule holds true in field studies (Herbert, 1965; Sprague et al., 1965). . . . The method of addition is useful and reasonably accurate for predicting thresholds of lethal effects in mixtures" (EPA, 1973, p. 122).

This statement notwithstanding, the assumption of strictly additive interactions was abandoned in 1980 (EPA, 1980a, pp. 79318–79379) and replaced with the assumption of no interactions. Since the latter assumption causes one to underestimate the toxicity of a mixture whenever the interactions are in any sense additive, adoption of this policy seems rather inconsistent with the expressed intention of the EPA "to err on the side of overprotection" (EPA, 1980a, p. 79342). If a general guideline in establishing water quality criteria is to err on the side of overprotection and if detailed information on the interactions between combinations of toxicants is not available, then the assumption of strictly additive interactions would seem preferable to the assumption of no interactions at all.

PUBLIC HEALTH

In the previous discussion, we considered the use of toxicological studies for purposes of setting water quality criteria for the protection of marine and freshwater organisms. An equally if not more important use of toxicology is to establish water quality criteria for the protection of the general public. In the United States at least, such criteria are established by different methods, depending on whether the toxicants are believed to be carcinogenic or not.

Noncarcinogenic Effects

In the case of noncarcinogenic substances, it is generally assumed that there is a threshold dose below which the substance exerts no adverse effects on humans. It is obvious that such a threshold must exist in the case of essential trace metals, which are required in small amounts by all organisms but become toxic if administered in large doses. In all cases, however, the existence and value of the threshold must be established based on experiments with animals or experience with humans. Water quality criteria are then established based on the estimated thresholds.

One can imagine several types of experimental or observational data that might be used to estimate the threshold doses for noncarcinogenic substances. First, there may have been unfortunate incidents in which humans were exposed to toxic concentrations of a particular substance. If the dose received by the humans is known or can be estimated and the effect on the persons documented, then the data are useful for establishing an upper bound on the threshold dose. In some cases there actually may have been experimental feeding studies with human volunteers, and in such cases it may be possible to obtain an even better estimate of the threshold dose. In the absence of information directly related to humans, there may be data obtained from experimental feeding studies with animals. These data may be used to estimate the effects on humans, but the uncertainty factor is greater in the absence of any direct observations on humans.

The duration of exposure must also be considered. Water quality criteria are based on an assumed lifetime exposure, but experimental observations, whether on humans or animals, do not always cover the lifetime of the organisms. If the only data available are from acute rather than chronic observational or experimental studies, then the uncertainty factor used in establishing water quality criteria must be increased.

Finally, one can imagine several general types of effects or responses associated with exposure to a toxicant. The first type of response would be no observed adverse effect. The dose associated with no observed adverse effect is referred to as the *no observed adverse effect level* (*NOAEL*) and is theoretically a lower bound to the threshold dose that causes some adverse effect. The second type of response would be an *observed adverse effect*, and the *lowest observed adverse effect level* (*LOAEL*) is theoretically an upper bound to the threshold dose. Adverse effects are defined by the EPA as "any effects which result in functional impairment and/or pathological lesions which may affect the performance of the whole organism, or which reduce an organism's ability to respond to an additional challenge" (EPA, 1980a, p. 79353). It is also possible that exposure to a toxicant will produce a response that is not an adverse effect according to the above definition. For example, rather than passing through the digestive tract and being excreted, the toxicant may be assimilated and then detoxified or broken down by the organism. Therefore, the organism has been affected by the toxicant to the extent that it has diverted some of its metabolic energy to detoxifying or breaking down the toxicant, but should this response be regarded as an adverse effect? The job of classifying such responses is obviously somewhat subjective, and is becoming increasingly difficult as more sophisticated testing protocols are developed and more subtle responses are identified. One result has been that some responses are classified simply as observed effects. One therefore finds in the EPA literature, for example, reference to *no observed effect levels* (*NOELs*) and *lowest observed effect levels* (*LOELs*). It is up to public health administrators, given information on these various types of responses or effects and after applying appropriate uncertainty factors, to determine what is called the *reference dose* (*RfD*), which is the maximum acceptable rate of con-

TABLE 8.8   UNCERTAINTY FACTORS FOR DERIVING CRITERIA
FOR THRESHOLD EFFECTS OF TOXICANTS FROM NOEL,
NOAEL, LOEL, AND/OR LOAEL DATA

Nature of experimental results	Uncertainty Factor
Valid experimental results from studies on prolonged ingestion by humans with no indication of carcinogenicity	10
Experimental results of studies of human ingestion not available or scanty (e.g., acute exposure only), with valid results of long-term feeding studies on experimental animnals, or in the absence of human studies, valid animal studies on one or more species; no indication of carcinogenicity	100
No long-term or acute human data; scanty results on experimental animals with no indication of carcinogenicity	1,000
Additional judgmental uncertainty factor to be applied to LOAEL data	1–10

Source: National Academy of Sciences (1977).

sumption of the toxic substance per unit body weight. Table 8.8 lists the uncertainty factors used in the United States.

Once an RfD has been established, the human health criterion (HHC) for the toxicant is calculated from the equation

$$HHC = \frac{(RfD)(BW)}{WC + (FC)(PBCF)} \tag{8.1}$$

where HHC is the concentration of the toxic substance in the water ($\mu g\ kg^{-1}$), RfD is the reference dose ($\mu g\ kg^{-1}\ d^{-1}$), BW is body weight (kg), WC is the water consumption rate ($kg\ d^{-1}$), FC is the fish and shellfish consumption rate ($kg\ d^{-1}$), and PBCF the practical bioconcentration factor (dimensionless). For adults, BW is assumed to be 70 kg and WC is assumed to be 2 kg (2 L) per day. The product (RfD)(BW) is often referred to as the *acceptable daily intake (ADI)*. Average fish and shellfish consumption rates in the United States are as follows (EPA, 1996b):

Fresh and estuarine fish and shellfish	$0.0056\ kg\ d^{-1}$
Marine fish and shellfish	$0.0121\ kg\ d^{-1}$
Total	$0.0177\ kg\ d^{-1}$

Application to Cadmium

Application of the foregoing methodology is illustrated in the following example of Cd. In this case, there are a number of examples of humans who were exposed to levels that clearly caused adverse effects. The best-documented example concerns people living in the Jintsu River basin of Japan. The health effects caused by their consumption of Cd, commonly referred to as *itai-itai disease*, included severe pain in the back, joints, and lower abdomen, development of a waddling or duck-like gait, kidney lesions, proteinuria, glycosuria, and loss of Ca from the bones, leading in some cases to multiple bone fractures. Approximately 200 persons were afflicted with itai-itai disease, and of these, nearly 100 died. In the area of Japan where itai-itai disease occurred (see Chapter 12), about 85% of the Cd intake was due to the consumption of Cd-contaminated rice (Muramatsu, 1974). A sensitive indicator of the disease was tubular proteinuria, the incidence of which rose above that observed in control populations when the Cd concentration in the rice exceeded about 0.45 ppm (Nogawa et al., 1978). Japanese in the area were known to consume about 430 g of rice per day (Friberg et al., 1974). Therefore, an adverse effect level for Japanese equals $(430)(0.45)/(0.85) = 230$ μg d^{-1}. For Americans and Western Europeans, who are about 32% larger than Japanese, the corresponding adverse effect level would be about 300 μg d^{-1}. Since only about 5% of ingested Cd is actually absorbed by the body (EPA, 1980c), the figure of 300 μg d^{-1} ingested corresponds to only about 15 μg d^{-1} absorbed.

Similar conclusions have been reached from other studies. For example, a study of workers who inhaled Cd dust concluded that effects began to appear when the concentration of Cd in the air equaled 21 μg m^{-3}. If one assumes that the average person inhales about 10 m^3 of air per day, that lung retention of Cd is about 25% efficient (EPA, 1980c), and that the workers were exposed to the contaminated air five days per week, the calculated level of absorption corresponding to the appearance of effects was $(21)(10)(0.25)(5/7) = 38$ μg d^{-1}. Similarly, the Working Group of Experts for the Commission of European Communities has estimated that the threshold level of Cd exposure by ingestion is about 200 μg d^{-1} (Comm. Eur. Communities, 1978), which would correspond to absorption of 10 μg d^{-1}.

Applying the uncertainty factor of 10 in Table 8.8 appropriate for studies on prolonged ingestion by humans, one would conclude that an ADI for adults (assuming 5% absorption efficiency) would be $10/((10)(0.05)) = 20$ μg d^{-1} to $38/((10)(0.05)) = 76$ μg d^{-1} for Cd. It is apparent from Table 8.9 that for the average American, the rate of Cd intake in food exceeds the lower of these two figures and equals about 40% of the higher figure. It is also apparent that drinking water contributes a very minor fraction of the total Cd absorbed. In fact, the average drinking water in the United States contains only 1.3 ppb Cd. Only 3 of a total of 969 community water supplies studied by the EPA contained Cd in excess of 10 ppb (EPA, 1980c).

How much Cd would an average person absorb from drinking water and consuming fish and shellfish if all water contained the existing drinking water standard of 5 ppb Cd (EPA, 1991)? *Bioconcentration factors* (*BCFs*) for Cd in the edible parts of commercially important fish and shellfish are shown in Table 8.10. The BCFs are consistently higher for the five bivalve mollusks than for the other two species, the geometric mean BCFs being 763 and 11, respectively. Bivalve mollusks account for only 0.6 g d^{-1} of the 17.7 g d^{-1} of fish and shellfish consumed by the average American. It therefore seems appropriate to weight the PBCF according to the relative amounts of the different kinds of fish and shellfish. The weighted PBCF is therefore $[(0.6)(763) + (17.1)(11)]/17.7 = 36.5$. The daily intake of Cd from drinking water and the consumption of fish and shellfish if all water contained 5 ppb (i.e., 5 μg kg^{-1}) Cd would therefore be $5[2 + (0.0177)(36.5)] = 13.2$ μg d^{-1}. Of this figure, only 24% comes from the consumption of fish and shellfish. For the average American, then, the consumption of fish and shellfish makes a minor contribu-

TABLE 8.9 AVERAGE DAILY INTAKE AND ABSORPTION OF CADMIUM BY AMERICANS

Exposure Source	Exposure	Cd Intake (μg)	% Absorbed	Absorption (μg d^{-1})
Air, ambient	0.03 μg m^{-3}	0.6	25	0.15
Air, smoking (one pack)	3.0 μg/pack	3.0	25	0.75
Food	—	30.0	5	1.50
Drinking water	1.3 μg L^{-1}	2.6	5	0.13

Source: EPA (1980c).

TABLE 8.10 CADMIUM GEOMETRIC MEAN BCFs FOR EDIBLE
PARTS OF CONSUMED AQUATIC SPECIES

Species	Mean BCF
Bivalve molluscs	
Corbicula fluminea, asiatic clam	2570
Mytilus edulis, blue mussel	186
Argopecten irradians, bay scallop	2040
Crassostrea virginica, eastern oyster	1660
Mya arenaria, soft-shell clam	160
Other fish and shellfish	
Salmo gairdneri, brook trout	21.5
Carcinus maenas, green crab	5.9

Source: EPA (1984).

tion to Cd intake. This conclusion may not apply, however, to persons who consume large amounts of bivalve mollusks.

In considering the effects of Cd on human health, the EPA (1980c, p. C-66) commented, "It appears that a water criterion needs to be no more stringent than the existing primary drinking water standard (10 μg/L) to provide ample protection of human health." This conclusion is thought-provoking. If a drinking water supply actually contained 10 μg L^{-1} of Cd, the intake of Cd from drinking water alone would be 20 μg d^{-1}, which equals the lower bound of the ADIs for Cd estimated from the observed adverse effects of Cd on human health. If the estimated intake of 30 μg d^{-1} from food (Table 8.9) is added to this figure, the total is 2.5 times the lower bound. The EPA's conclusion seems to have been based largely on the fact that "Water constitutes only a relatively minor portion of man's daily cadmium intake" (EPA, 1980c, p. C-66). This fact arises, however, not because average water contains 10 μg L^{-1} Cd, but because average water contains almost eight times less Cd. The decision to lower the drinking water standard to 5 μg L^{-1} Cd (EPA, 1991) was a step in the right direction.

Carcinogenic Effects

The whole issue of carcinogenic effects has become quite controversial because of the suggestion that faulty experimental and analytical methods may have been used to determine whether a substance is carcinogenic and, if so, the relationship between dose and response. Although in some cases evidence of carcinogenicity has been based on unintentional human exposure, in many cases toxicological studies relative to cancer have been conducted with rats or mice. In the United States, the standard procedure has been to conduct initial feeding experiments with the test chemical and, on the basis of these preliminary experiments, to determine the *maximum tolerated dose (MTD)*. The MTD is the highest dose of the chemical that does not literally kill the animals, although if repeated over a two-week period, the MTD "usually leads to a noticeable but tolerated reduction in weight" (Abelson, 1987, p. 473). Chronic feeding studies are then conducted at doses equal to or less than the MTD over the course of the animal's lifetime. The experimental doses vastly exceed those to which humans are likely to be exposed.

If an abnormal incidence of cancer occurs in the experimental animals compared to a control group that is not exposed to the chemical, the excess incidence of cancer is plotted as a function of dose and an analytical function is fit to the data. The analytical function has the form (Crump, 1984)

$$A(d) = 1 - \exp(-q_1 d - q_2 d^2 - \cdots - q_k d^k) \tag{8.2}$$

where $A(d)$ is the extra risk over background associated with a dose d and the q_i are constants to be determined by a least squares curve-fitting procedure. An important characteristic of Equation 8.2 is the fact that at low doses it reduces to the form

$$A(d) = q_1 d \tag{8.3}$$

and hence conforms to the assumption that there is no threshold dose below which the chemical exerts no carcinogenic effects. The levels of exposure for most humans are likely to fall in the range where, according to the model, $A(d)$ can for all intents and purposes be assumed to be directly proportional to d. The cancer risks at these low dose levels represent considerable extrapolations of the experimental data.

So what's wrong with these studies? Actually, this approach was probably a reasonable one 20 years ago because medical understanding of the etiology of cancer was relatively crude compared to that of the present time, and it undoubtedly seemed wise to error on the side of safety. As our understanding of the mechanisms that cause cancer has improved, however, this approach has drawn much criticism. Abelson (1990a, p. 1497), for example, has commented, "The standard carcinogen tests that use rodents are an obsolescent relic of the ignorance of past decades."

The problems with the rodent tests are several. First, the rodents in question are pure strains that have been inbred for numerous generations. This inbreeding was intended to produce rats and mice with a more or less well-defined genotype, but it often leads to genetic impairments. In the experimental mice, for example, there is a high natural incidence of liver tumors, and "The usual response of these animals to massive doses of a chemical is to develop an even higher incidence of liver tumors" (Abelson, 1987, p. 473). This increase in liver tumors among naturally tumorigenic mice is of doubtful relevance to humans because primary liver cancer is rare in humans with the exception of alcoholics and persons who have suffered from hepatitis (Abelson, 1987).

Second, recent evidence indicates that inbreeding has in fact led to substantial genetic drift in the rodents used in risk assessment experiments (Abelson, 1995; Festing, 1997). In most tests conducted by the EPA, one or more of three strains of rodents have been used: Sprague-Dawley (SD) rats, Fischer (F-344) rats, and B6C3F1 mice. Abelson (1995) points out, for example, that at the Merck Research Laboratory in the 1970s, the survival rate at age two years of SD rats used as controls was 58%. In the 1980s the corresponding survival rate was 44%, and in the 1990s it had dropped to 24%. The incidence of tumors in control rodents has also changed. The number of liver tumors, for example, in control B6C3F1 mice increased from an average of 32% in 1980 to roughly 50% in 1984 (Abelson, 1995).

Third, it has been customary for the companies responsible for breeding the rodents to give them as much food as they want to eat. This practice is referred to as feeding the animals *ad libitum* (*ad lib*). The result is that the animals overeat, and their health is impaired (Festing, 1997). A recent experiment involving SD rats compared the longevity of control rats fed *ad lib* with diet-restricted rats fed 65% of the *ad lib* amounts. At maturity the *ad lib* male rats weighed 60% more than the diet-restricted males. Only 7% of the *ad lib* males survived to age two years, while 72% of the diet-restricted males survived longer than two years (Abelson, 1995).

A major criticism of the rodent studies is the fact that prolonged feeding at the MTD tends to cause mitogenesis (induced cell division), and it is now recognized that nongeno-toxic chemicals such as saccharin can be carcinogenic at high doses simply because they cause mitogenesis. The reason is that a dividing cell is much more at risk of mutating than a quiescent cell (Ames and Gold, 1990). Since there is believed to be a threshold dose below which such nonmutagenic toxins exert no effect on cell division, "At the low doses of most human exposures (where cell-killing and mitogenesis do not occur), the hazards [of nonmutagenic toxins] may be much lower than is commonly assumed and often will be zero" (Ames and Gold, 1990, p. 971).

Given the above information, one should not be surprised to learn that about 58% of all chemicals tested chronically at the MTD are judged to be carcinogens. Although much more emphasis has been given to testing synthetic chemicals than naturally occurring compounds, the percentage of synthetic chemicals judged to be carcinogenic on the basis of these tests is not much different from the corresponding percentage of naturally occurring compounds, 61% versus 48%, respectively (Ames and Gold, 1990). Of the natural chemicals that have been tested, about two-thirds are natural pesticides produced by plants to defend themselves. According to Ames and Gold (1990), these natural pesticides account for over 99.99% of the pesticides consumed by persons in the United States and, based on the rodent bioassays, 27 out of 52, or about 52% of the natural pesticides tested are carcinogens. These 27 naturally occurring pesticides are found in 57 different foods, including apples, bananas, carrots, celery, coffee, lettuce, orange juice, peas, potatoes, and tomatoes (Abelson, 1990a). A cup of coffee, for example, contains about 10 mg of rodent carcinogens, primarily caffeic acid, catechol, furfural, hydrogen peroxide, and hydro-quinone. The implication of these revelations is that the rodent bioassays probably have been very misleading concerning the carcinogenic threat associated with numerous chemicals, both synthetic and natural. Diets rich in fruits and vegetables, for example, tend to reduce the incidence of cancer in humans. This observation seems inconsistent with the results of the rodent bioassays.

Although not all scientists agree that the rodent bioassays have been misleading (Cogliano et al., 1991; Rall, 1991; Weinstein, 1991), there does seem to be a consensus that a better understanding of the mechanisms that cause cancer is needed. For example, we need to know the hormonal determinants of breast cancer, the viral determinants of cervical cancer, and the dietary determinants of stomach and colon cancers (Ames and Gold, 1990). Unleaded gasoline has been implicated as a carcinogen because it causes kidney tumors in male rats, but we now know that branched-chain hydrocarbons, which are key components of unleaded gas, interfere with the mechanism for excreting a low-molecular-weight protein by the male rat. Research conducted at the Chemical Industry Institute of Toxicology indicates that this interference may be the cause of kidney cancer in male rats exposed to gasoline. A similar mechanism does not exist in female rats, in male or female mice, or in humans (Abelson, 1990b). Similarly, we now know that "Saccharin's ability to induce bladder tumors in male rats is solely due to the proliferative effects that high doses have on the bladder lining" (Marx, 1990, p. 744). Clearly, it will be impossible to make an informed judgment about the carcinogenic threat of various chemicals to humans until the mechanisms responsible for causing cancer in humans are more thoroughly understood (Madia, 1998).

PROTECTION OF WILDLIFE

Recently, the EPA has published procedures for the determination of water quality criteria for the protection of wildlife that depend on fish for food either directly or indirectly (EPA, 1995). The criteria were developed as part of an effort to establish various water quality criteria for the Great Lakes. The rationale behind the so-called wildlife criteria is straightforward. The criteria are intended to protect fish-eating animals and animals further up the food chain from being adversely affected by the presence of toxic substances in the prey they consume and the water they drink. The approach used in establishing the criteria is similar to the approach used to determine human health criteria, but as noted by the EPA (1995, p. 19), the approach "focuses on endpoints related to reproduction and population survival rather than the survival of individual members of a species." At least in the case of the Great Lakes, the analysis focused on biocumulative chemicals (BCCs) because "These are the chemicals of greatest concern to the higher trophic level wildlife species feeding from the aquatic food web" (EPA, 1995, p. 19). The food chains leading to three representative avian species (bald eagle, kingfisher, and herring gull) and two representative mammal species (mink and otter) were chosen for analysis. A wildlife value was determined for each species, and an avian and a mammal value was determined from the geometric mean of the three avian species and two mammal species, respectively. The lower of the mammalian and avian values became the Great Lakes Wildlife Criterion.

Wildlife Criteria (WC) are not listed in national EPA water quality criteria documents, in part because the development of these criteria is relatively new and in part because the criteria are specific to particular aquatic ecosystems. The WC for the Great Lakes are thought-provoking. In the case of Hg, for example, the WC is 1.3 ng L^{-1}. This compares to the national human health criterion for Hg of 50 ng L^{-1} (see Chapter 12). Why is the WC almost 40 times lower than the human health criterion? The answer is that only a small fraction of the diet of most Americans comes from fish. Animals that prey on fish almost exclusively or are predators on such animals are obviously at much greater risk from the standpoint of exposure to toxic substances.

COMMENTARY

At the present time, there are approximately 75,000 chemicals in use or being distributed through the environment, and an additional 500–1,000 are added each year (Postel, 1987). Although not all of these chemicals are toxic, merely keeping up-to-date information on the possible toxicity of so many compounds is a prodigious task. The discussion in this chapter has indicated how involved the process of setting water quality guidelines is for even a single chemical. Large amounts of the right kind of information are needed. In its 1999 update to *Quality Criteria for Water*, the EPA listed water quality criteria for 120 priority toxic chemicals. In only 20 of those cases was there sufficient information to determine a criterion maximum concentration and criterion continuous concentration in both saltwater and freshwater. This fact dramatically illustrates the nature of the problem that confronts governmental agencies charged with the responsibility of controlling water pollution. On the one hand, there is an increasingly urgent demand to establish water quality guidelines for the large number of toxic chemicals that pollute both groundwater and surface water. On the other hand, there is a need to develop intelligent and legally defensible techniques for establishing those guidelines. Ultimately, government agencies must decide in a somewhat arbitrary manner how much of what kind of toxicological information is necessary and sufficient to establish water quality criteria. Establishing the rules of the game is obviously a critical first step, and in the case of carcinogenic substances, there is disagreement over what the rules should be. Even when the rules have been established, there often remain information gaps that must be filled before criteria can be recommended. Invariably there will remain disagreement over whether the methodology required to establish the guidelines is too simple or too cumbersome, too strict or too lax, and too detailed or too ambiguous. There will remain differences of opinion as to what types of toxicological data are useful and acceptable and how best to utilize those data in developing guidelines. The wisdom and collaboration of many conscientious scientists and administrators will be needed if the legal efforts to control the quality of water are to keep pace with the growing threat of pollution.

REFERENCES

ABELSON, P. H. 1987. Cancer phobia. *Science*, **237**, 473.

ABELSON, P. H. 1990a. Incorporation of new science into risk assessment. *Science*, **250**, 1497.

ABELSON, P. H. 1990b. Testing for carcinogens with rodents. *Science*, **249**, 1357.

ABELSON, P. 1995. Flaws in risk assessments. *Science* **270**, 215.

ADEMA, D. M. M. 1978. *Daphnia magna* as a test animal in acute and chronic toxicity tests. *Hydrobiology*, **59**, 125–134.

AMES, B. N., and L. S. GOLD. 1990. Too many rodent carcinogens: Mitogenesis increases mutagenesis. *Science*, **249**, 970–971.

ANDERSON, D. W., and F. GRESS. 1983. Status of a northern population of California brown pelicans. *Condor*, **85**, 79–88.

ANDERSON, D. W., F. GRESS, and K. F. MAIS. 1982. Brown pelicans: Influence of food supply on reproduction. *Oikos*, **39**, 23–31.

ANDERSON, D. W., F. GRESS, K. MAIS, and P. R. KELLY. 1980. Brown pelicans as anchovy stock indicators and their relationships to commercial fishing. *Calif. Coop. Oceanic Fish. Invest. Rep.*, **21**, 54–61.

ANDERSON, D. W., J. R. JEHL, R. W. RISEBROUGH, L. A. WOODS, L. R. DEWEESE, and W. G. EDGECOMB. 1975. Brown pelicans: Improved reproduction off the southern California coast. *Science*, **190**, 806–808.

ANDERSON, D. W., R. M. JUREK, and J. O. KEITH. 1977. The status of brown pelicans at Anacapa Island in 1975. *Calif. Fish Game*, **63**, 4–10.

BRETT, J. R. 1960. Thermal requirements of fish—three decades of study. In C. M. Tarzwell, Ed., *Biological Problems of Water Pollution*. U.S. Department of Health, Education, and Welfare, Robert A. Taft Sanitary Engineering Center, Cincinnati. Pp. 110–117.

BROWN, V. M. 1968. The calculation of the acute toxicity of mixtures of poisons to rainbow trout. *Water Res.*, **2**, 723–733.

BUCKSTEEG, W., H. THIELE, and K. STOLTZEL. 1955. Die Beeinflussung von Fischen durch Giftstoffe aus Abwassern. *Vom Wasser (Jahrburch fur Wasser-Chemie)*, **22**, 194–211.

BUNN, C. R., and G. MATRONE. 1966. *In vitro* interaction of cadmium, copper, zinc, and iron in the mouse and rat. *J. Nutr.*, **90**, 395–399.

CLARK, J., and W. BROWNELL. 1973. *Electric Power Plants in the Coastal Zone: Environmental Issues*. American Littoral Society Special Publication No. 7, Highlands, N.J.

COGLIANO, V. J., W. H. FARLAND, P. W. PREUSS, J. A. WILTSE, L. R. RHOMBERG, C. W. CHEN, M. J. MASS, S. NOSNOW, P. D. WHITE, J. C. PARKER, and S. M. WUERTHELE. 1991. Carcinogens and human health: Part 3. *Science*, **251**, 606–607.

COMMISSION OF THE EUROPEAN COMMUNITIES. 1978. *Criteria (Dose/Effect Relationships) for Cadmium*. Report of the working group of experts prepared for the Commission of the European Communities, Directorate-General for Social Affairs, Health and Safety Directorate. Pergamon Press, London.

CRUMP, K. S. 1984. An improved procedure for low-dose carcinogenic risk assessment from animal data. *J. Environ. Pathol. Toxicol.* **5**, 339–348.

DEGENS, P. N., H. VAN DER ZEE, J. D. KOMMER, and A. H. KAMPHUIS. 1950. Synthetic detergents and sewage processing. Part 5. The effect of synthetic detergents on certain water fauna. *J. Inst. Sewage Purification*, **1950**(1), 63–68.

EDWARD, R. W., and V. M. BROWN. 1967. Pollution and fisheries: A progress report. *Water Pollut. Contr. (J. Inst. Water Pollut. Contr.)*, **66**, 63–78.

EPA. 1973. *Water Quality Criteria 1972*. Environmental Protection Agency. EPA/R3/73/033. Washington, DC. 594 pp.

EPA. 1980a. *Water Quality Criteria Documents: Availability.* Environmental Protection Agency. *Federal Register*, **45**(231), 79318–79379.

EPA. 1980b. *Ambient Water Quality Criteria for Aldrin/Dieldrin.* Environmental Protection Agency. EPA 440/5-80-019. Washington, DC.

EPA. 1980c. *Ambient Water Quality Criteria for Cadmium.* Environmental Protection Agency. EPA 440/5-80-025. Washington, DC.

EPA. 1984. *Ambient Water Quality Criteria for Cadmium—1984.* Environmental Protection Agency. EPA 440/5-84-032. Washington, DC.

EPA. 1985a. *Guidelines for Deriving Numerical National Water Quality Criteria for the Protection of Aquatic Organisms and Their Uses.* PB85-227049. Washington, DC.

EPA. 1985b. *Methods for Measuring the Acute Toxicity of Effluents to Freshwater and Marine Organisms.* EPA/600/4-85/013. Washington, DC.

EPA. 1986. *Quality Criteria for Water.* Environmental Protection Agency. EPA 440/5-86-001. Washington, DC.

EPA. 1987. *Update #2 to Quality Criteria for Water 1986.* May 1, 1987. U.S. Environmental Protection Agency, Office of Water Regulations and Standards, Criteria and Standards Division, Washington, DC.

EPA. 1991. National Primary Drinking Water Standards. *Federal Register*, **56**, 3526.

EPA. 1995. Final Water Quality Guidance for the Great Lakes System; Final Rule. *Federal Register*, **60**(56), 15365–15425.

EPA. 1996a. *1995 Updates: Water Quality Criteria Documents for the Protection of Aquatic Life in Ambient Water.* EPA-820-B-96-001. Washington, DC.

EPA. 1996b. *Exposure Factors Handbook. Volume II. Food Ingestion Factors.* Office of Research and Development, Environmental Protection Agency, National Center for Environmental Assessment. EPA 600/P-95/002 Bb. Washington, DC.

EPA. 1997. Water Quality Standards; Establishment of Numeric Criteria for Priority Toxic Pollutants for the State of California; Proposed Rule. 40 CFR, Part 131. *Federal Register*, **62**, 42159–42208.

EPA. 1999. *National Recommended Water Quality Criteria—Correction.* EPA 822-Z-99-001. Washington, DC.

FERM, V. H., and S. J. CARPENTER. 1967. Teratogenic effect of cadmium and its inhibition by zinc. *Nature*, **216**, 1123.

FESTING, M. F. W. 1997. Fat rats and carcinogenesis screening. *Nature*, **388**, 321–322.

FISHER, N. S. 1975. Chlorinated hydrocarbon pollutants and photosynthesis of marine phytoplankton: A reassessment. *Science*, **189**, 463–464.

FISHER, N. S., E. J. CARPENTER, C. C. REMSEN, and C. F. WURSTER. 1974. Effects of PCB on interspecific competition in natural and gnotobiotic phytoplankton communities in continuous and batch cultures. *Microbiol. Ecol.*, **1**, 39–50.

FISHER, N. S., and C. F. WURSTER. 1973. Individual and combined effects of temperature and polychlorinated biphenyls on the growth of three species of phytoplankton. *Environ. Pollut.*, **5**, 205–212.

FRAWLEY, J. P. 1965. Synergism and antagonism. In C. D. Chichester, Ed., *Research in Pesticides.* Academic Press, New York. Pp. 69–83.

FRIBERG, L., M. PISCATOR, G. F. NORDBERG, and T. KJELLSTROM. 1974. Cadmium in the Environment. 2nd ed. Chemical Rubber Company Press, Cleveland, OH.

GRESS, F. 1995. *Organochlorines, Eggshell Thinning, and Productivity Relationships in Brown Pelicans Breeding in the Southern California Bight.* Ph.D. dissertation, University of California, Davis.

GREVE, W., and T. R. PARSONS. 1977. Photosynthesis and fish production: Hypothetical effects of climatic change and pollution. *Helgol. Wiss. Meeresunters*, **30**, 666–672.

GUNN, S. A., T. C. GOULD, and W. A. D. ANDERSON. 1963a. Cadmium-induced interstitial cell tumors in rats and mice and their prevention by zinc. *J. Natl. Cancer Inst.*, **31**, 745–759.

GUNN, S. A., T. C. GOULD, and W. A. D. ANDERSON. 1963b. The selective injurious response of testicular and epididymal blood vessels to cadmium and its prevention by zinc. *Am. J.Pathol.*, **42**, 685–702.

GUSTAFSON, J. R. 1990. Five-year Status Report for California brown pelican. California Department of Fish and Game. *Nongame Bird and Mammal Sec. Rep. 90-4* (draft).

HARDING, L. W., JR., and J. H. PHILLIPS, JR. 1978. Polychlorinated biphenyls (PCB) effects on marine phytoplankton photosynthesis and cell division. *Mar. Biol.*, **49**, 93–101.

HARRISS, R. C., D. B. WHITE, and R. B. MACFARLANE. 1970. Mercury compounds reduce photosynthesis by plankton. *Science,* 170, 736–737.

HASLER, A. D., and J. A. LARSEN. 1971. The homing salmon. In J. R. Moore, Ed., *Oceanography. Readings from Scientific American.* Freeman, San Francisco. Pp. 253–256.

HEATH, R. G., J. W. SPANN, and J. F. KREITZER. 1969. Marked DDE impairment on mallard reproduction in controlled studies. *Nature,* 224, 47–48.

HERBERT, D. W. M. 1965. Pollution and fisheries. In G. T. Goodman, R. W. Edwards, and J. M. Lambert, Eds., *Ecology and the Industrial Society.* Blackwell Scientific Publications, Oxford. Pp. 173–195.

HILL, C. H., G. MATRONE, W. L. PAYNE, and C. W. BARBER. 1963. *In vivo* interactions of cadmium with copper, zinc, and iron. *J. Nutr.,* 80, 227–235.

JORDAN, D. H. M., and R. LLOYD. 1964. The resistance of rainbow trout (Salmo gairdnerii Richardson) and roach (*Rutilus rutilus* (L.)) to alkaline solutions. *Int. J. Air Water Pollut.,* 8, 405–409.

LEMKE, A. E., and D. I. MOUNT. 1963. Some effects of alkyl benzene sulfonate on the bluegill, *Lepomis macrochirus. Trans Am. Fish. Soc.,* 92, 372–378.

LLOYD, D., and L. D. ORR. 1969. The diuretic response by rainbow trout to sublethal concentrations of ammonia. *Water Res.,* 3, 335–344.

LLOYD, R. 1960. The toxicity of zinc sulfate to rainbow trout. *Annals Appl. Biol.,* 48, 84–94.

MADIA, W. J. 1998. A call for more science in EPA regulations. *Science,* 282, 45.

MARX, J. 1990. Animal carcinogen testing challenged. *Science,* 250, 743–745.

MOSSER. J. L., N. S. FISHER, T. TENG, and C. F. WURSTER. 1972a. Polychlorinated byphenyls: Toxicity to certain phytoplankters. *Science,* 175, 191–192.

MOSSER, J. L., N. S. FISHER, and C. F. WURSTER. 1972b. Polychlorinated biphenyls and DDT alter species composition in mixed cultures of algae. *Science,* 176, 533–535.

MURAMATSU, S. 1974. Research about cadmium pollution in the Kakehashi River basin. *Jap. J. Public Health,* 21, 299–308.

NATIONAL ACADEMY OF SCIENCES. 1977. *Drinking Water and Health.* National Academy of Sciences, Washington, DC.

NEIL, J. H. 1957. Some effects of potassium cyanide on speckled trout (*Salvelinus fontinalis*). *Ontario Industrial Waste Conference,* 4, 74–96.

NOGAWA, K., A. ISHIZAKI, and S. KAWANO. 1978. Statistical observations of the dose–response relationships of cadmium based on epidemiological studies in the Kakehashi River basin.*Environ. Res.,* 15, 189–198.

O'CONNORS, H. B., JR., C. F. WURSTER, C. D. POWERS, D. C. BIGGS, and R. C. ROWLAND. 1978. Polychlorinated biphenyls may alter marine trophic pathways by reducing phytoplanktonsize and production. *Science,* 201 737–739.

PARIZEK, J. 1957. The destructive effect of cadmium ion on testicular tissue and its prevention by zinc. *J. Endocrin.,* 15, 56–63.

PARIZEK, J., I. BENES, J. KALOUSKO, A. BABICKY, and J. LENER. 1969. Metabolic interrelationships of trace elements. Effects of zinc salts on survival of rats intoxicated by cadmium. *Physl. Bohem.,* 18, 89–94.

POSTEL, S. 1987. *Defusing the Toxics Threat: Controlling Pesticides and Industrial Waste.* Worldwatch Paper 79. Worldwatch Institute, Washington, DC. 69 pp.

RALL, D. P. 1991. Carcinogens and human health: Part 2. *Science,* 251, 10–11.

RYTHER, J. H. 1969. Photosynthesis and fish production in the sea. *Science,* 166, 72–76.

SHEPARD, M. P. 1955. Resistance and tolerance of young speckled trout (*Salvelinus fontinalis*) to oxygen lack, with special reference to low oxygen acclimation. *J. Fish. Res. Bd. Canada,* 12, 387–446.

SILVERMAN, M. J. 1972. Tragedy of Northport. *Underwater Naturalist,* 7(2), 15–18.

SPRAGUE, J. B., P. F. ELSON, and R. L. SAUNDERS. 1965. Sublethal copper-zinc pollution in a salmon river: A field and laboratory study. *Air Water Pollut.,* 9(9), 531–543.

SUN, M. 1985. Legislative paralysis on the environment. *Science,* 227, 496–497.

TRAMA, F. B. 1954. The acute toxicity of copper to the common bluegill (*Lepomis macrochirus* Rafinesque). *Acad. Nat. Sci. Phila., Notulae Naturae,* 257. 13 pp.

WARREN, C. E. 1971. *Biology and Water Pollution.* W. B. Saunders. Philadelphia. 434 pp.

WEINSTEIN, I. B. 1991. Mitogenesis is only one factor in carcinogenesis. *Science,* 251, 387–388.

WEIR, P. A., and C. H. HINE. 1970. Effects of various metals on behavior of conditioned goldfish. *Arch. Environ. Health,* 20, 45–51.

QUESTIONS

8.1 Classify the following examples according to whether they illustrate the effects of acute toxicity (A) or chronic toxicity (C).

a. A two-year-child survived the bombing of Hiroshima. Twenty years later, he is diagnosed as being sterile as a result of the radiation does he received when the bomb was dropped.

b. A woman works for a company that manufactures pipettes. Her job requires her to place a pipette on an instrument, push a lever, allow Hg to rise in the pipette, and put a calibration mark at the point to which the Hg rises. After 20 years of performing this task, she suffers from neurological and personality disorders due to her exposure to Hg.

c. A woman living in the Jintsu River basin of Japan eats Cd-contaminated rice for 55 years. Before her 60th birthday she dies from itai-itai disease.

d. A person buys a bottle of vegetable oil and uses the oil to make some brownies. The oil is seriously contaminated with PCBs, and the person dies a few days after consuming three of the brownies.

8.2 L'eau Calcaire College is located in a part of the country where the water is hard. The biology lab at the school includes numerous aquaria with fish. The water distribution system at the university consists of copper pipes connected with Pb-based solder. The middle-aged faculty in the biology department organize a jogging club to help them get into shape. Subsequently, they complain to the department chairman about the hard water in the shower room because they are unable to get their soap to lather. The chairman responds by installing a water softener for the whole biology building. Shortly after the water softener is installed, all the fish in the aquaria die. How would you account for the death of the fish?

8.3 Assuming that people drink 2 L (2 kg) of water each day and eat 17.7 grams (0.0177 kg) of fish and shellfish caught from the same or similar water each day, the maximum acceptable concentration of a noncarcinogenic substance in the water for the protection of human health is calculated from the equation

$$X = \frac{ADI}{2 + (0.0177)(PBCF)}$$

where X is the maximum acceptable concentration (milligrams per liter) of the substance in the water, ADI is the acceptable daily intake of the substance (milligrams per day), and PBCF is the practical bioconcentration factor for the substance. Suppose that for substance Z the ADI is 181 mg d^{-1} and PBCF is 5,000. What is the maximum acceptable concentration of Z in the water for the protection of human health? Express your answer as milligrams of Z per liter of water.

8.4 Complete the following table by calculating the number of chronic toxicity units for toxic substances A, B, C, and D.

Toxic Substance	Concentration in Water	96-Hour TLm	Acute/Chronic Ratio	Chronic Toxicity Units
A	5	500	10	
B	10	125	5	
C	15	750	10	
D	18	600	10	

8.5 Assume that we discover that the water in question 8.4 is exerting a chronic stress on species Y. Which of the following may be true (more than one answer may be correct)?
   a. The interaction between A, B, C, and D is antagonistic.
   b. The interaction between A, B, C, and D is infra-additive.
   c. The interaction between A, B, C, and D is strictly additive.
   d. The interaction between A, B, C, and D is supra-additive.
   e. There is no interaction between A, B, C, and D.

8.6 Assume that we discover that the water in question 8.5 is not exerting a chronic stress on species Y. Which of the following may be true (more than one answer may be correct)?
   a. The interaction between A, B, C, and D is antagonistic.
   b. The interaction between A, B, C, and D is infra-additive.
   c. The interaction between A, B, C, and D is strictly additive.
   d. The interaction between A, B, C, and D is supra-additive.
   e. There is no interaction between A, B, C, and D.

8.7 The graph below shows the concentration of pollutant X versus median survival time of a certain species of fish. Judging from this graph, what would you estimate the incipient lethal level of pollutant X to be for this species of fish?

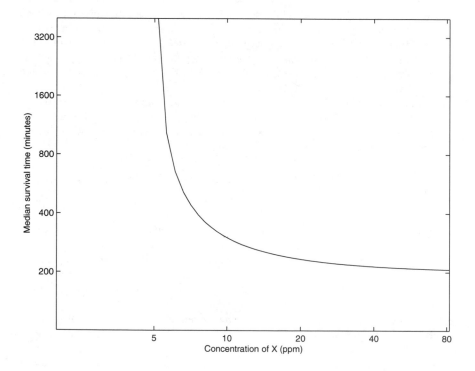

8.8 The graph below shows the cumulative probability that a given concentration of pollutant X will equal or exceed the mean acute values of the genera in a natural aquatic system. Based on the graph, what would you expect the EPA final acute value to be for pollutant X?

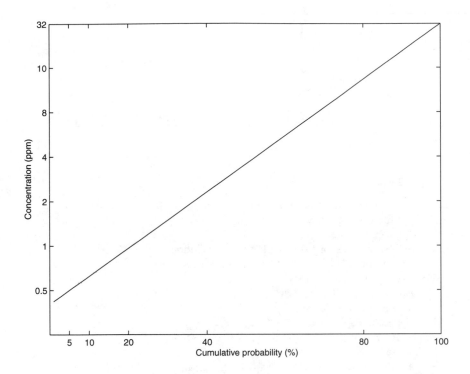

8.9 In the table below are listed the final acute, final chronic, and final plant values for pollutant X. Based on the data in the table, what would the EPA criterion maximum concentration and criterion continuous concentration be for pollutant X?

Final acute value	500 ppm
Final chronic value	20 ppm
Final plant value	50 ppm

Chapter Nine

✧ ◻ ✧ ◻ ✧

Industrial Pollution

1. Characteristics of Industrial Wastewaters
   a. BOD, SS, Toxic Substances
      (1) the oxygen sag
2. Innovative Strategies for Reducing Industrial Pollution
   a. Recycling and Modification of Production Process
3. Hawaiian Sugar Cane Industry
   a. Field and Factory Operations
      (1) need for cane cleaning
   b. Response to EPA Study
      (1) use of mill wastewater to irrigate cane
      (2) use of bagasse as fuel
      (3) irrigation modifications: furrow → drip irrigation
   b. Recent Developments
      (1) effort to bypass cane cleaning operation
      (2) one-year growth cycle
      (3) particle board from bagasse
4. Pulp and Paper Industry
   a Chemical and Mechanical Pulping
      (1) kraft and sulfite
   b. Bleaching Methods
      (1) Cl_2, ClO_2, oxygen
   c. Characteristics of Untreated Effluents
      (1) SS, BOD, toxic compounds
   d. Treatment Methods
      (1) use of lagoons with long residence times
   e. Buckeye (Procter and Gamble) Cellulose
      (1) comparison of Fenholloway and Econfina rivers
         (a) anoxic conditions in Fenholloway
         (b) impact of effluent on marsh areas and Gulf of Mexico
      (2) impact of clarifier and bio-oxidation pond
   f. Recent Developments
      (1) phase-out of chlorine bleaching
      (2) paper recycling

WASTEWATERS DISCHARGED by industrial operations are in some cases among the worst sources of water pollution. Although the nature of the pollutants associated with these wastewaters differs greatly from one industry to another, in almost all cases the problems are caused by one or a combination of the following conditions in the wastewater:

1. High BOD
2. High concentration of SS
3. Presence of toxic substances

Indeed, the BOD and SS in some industrial wastewaters may be a factor of 10 or more higher than the BOD and SS in raw sewage. It is not hard to imagine that severe O_2 depletion and/or turbidity and sedimentation problems could result from the discharge of such wastewater. The high concentration of toxicants in some industrial effluents can be lethal or at least stressful to organisms in the receiving waters, and may be concentrated by fish or shellfish to such an extent that they are unfit for human consumption. Efforts to reduce or eliminate the toxicant problem have tended to focus on three general solutions. The first is to eliminate the need for the toxicant in the manufacturing process. The second is to recover and recycle the toxicant before the wastewaters are released. The third is to remove the toxicant from the wastewater and dispose of it in an environmentally acceptable manner.

This chapter begins with a general examination of the problems caused by the release of BOD, SS, and toxic substances with industrial wastewaters and the nature of corrective measures taken to alleviate these problems. This general discussion lays the groundwork for a more detailed examination of the wastewater problems associated with specific industries. The case studies that follow provide good examples of the types of problems that are caused by industrial wastewaters and illustrate rather well some of the approaches that have been tried, in all cases not successfully, to alleviate those problems. The remainder of the chapter is devoted to an in-depth study of two industries with significant wastewater problems, the Hawaiian sugar cane industry and the pulp and paper industry.

THE OXYGEN SAG

The wastewater from industrial operations such as pulp mills, sugar refineries, and some food processing plants may easily have a BOD as high as several thousand parts per million or more. By contrast, raw sewage, with a typical BOD of only about 200 ppm, seems like the nectar of the gods. If industries discharged only small amounts of such wastewater, the effect of these enormous BODs on receiving waters would be small in many cases, since the effluent would be greatly diluted in the receiving system. Unfortunately just the opposite has too often been the case. For example, as of approximately 1980, a typical 500-tonne day^{-1} sulfite process pulp mill would have produced about 100 million L day^{-1} of wastewater with a BOD of about 1,000 ppm (McCubbin, 1983). The discharge of wastewater alone from such a plant is equal to the volume of sanitary sewage produced by a city of about 167,000 persons and the BOD loading from a city of about 830,000 persons.

If such large amounts of BOD are discharged into a system that is not vigorously mixed, there is an excellent chance that the O_2 concentration will drop below the level necessary to maintain the natural aquatic community. Water initially saturated with O_2 at 20°C, for example, and containing 1,000 ppm BOD would become completely anoxic in about one hour if placed in a closed container in the dark. Obviously, such wastewater must be greatly diluted or the receiving waters vigorously aerated to prevent serious O_2 depletion.

Figure 9.1 illustrates qualitatively how the dissolved O_2 concentration may vary with distance in a river downstream from an industry discharging wastewater with a high BOD. The oxidation of the organic substances in the wastewater by microbes initially consumes O_2 faster than the exchange of O_2 between the stream and atmosphere can resupply O_2. The result is that the O_2 concentration in the water drops rapidly downstream from the wastewater outfall. Due to the metabolic activity of the microbes, the concentration of oxidizable organic wastes steadily decreases with increasing distance from the outfall. Since

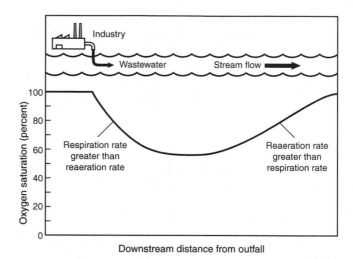

FIGURE 9.1   Qualitative variation of dissolved O_2 concentration resulting from BOD discharge into a stream. A characteristic oxygen sag curve.

the respiration rate of the microbes is positively correlated with the concentration of their food supply, the rate of microbial O_2 consumption also steadily decreases with distance from the outfall. The net influx of O_2 from the atmosphere to the stream is proportional to the difference between the O_2 concentration at 100% saturation and the actual concentration in the water. With increasing distance from the outfall, the net rate of O_2 influx eventually exceeds the microbial respiration rate, and the O_2 concentration in the water begins to rise. At a sufficient distance downstream from the outfall, the O_2 concentration is back to virtually 100% saturation. The characteristic decline and subsequent rise in O_2 concentration downstream from an outfall discharging wastewater with a high BOD is usually referred to as an *oxygen sag*. The phenomenon was first treated theoretically by Streeter and Phelps (1925), who developed an analytical expression to describe the characteristics of the oxygen sag curve.

Of particular importance are two variations of the oxygen sag curve illustrated in Figure 9.2. If the BOD loading from an outfall is sufficiently great, the O_2 concentration in a stream may drop to virtually zero and remain at that level for a considerable distance downstream from the outfall. In extreme cases, the zone of anoxic water may extend many kilometers downstream before the combined effects of reaeration and declining respiration rates allow the O_2 concentration to increase again (Livingston, 1975). Such anoxic zones result in virtually a complete displacement of the stream's natural fauna. The second variation concerns the case in which two or more industries are discharging high BOD wastewater into a stream at locations along the stream sufficiently close to one another that the oxygen sag produced by one industry's wastes is still apparent at the point downstream where the next industry is discharging wastewater. As a result, the combined effect of two or more oxygen sags may have a much more deleterious effect on water quality than would any one of the contributing oxygen sags taken by itself. Obviously it is possible to reduce O_2 concentrations to virtually zero by a series of several relatively small but closely spaced BOD discharges along the course of a stream.

What can be done to solve the oxygen sag problem? Obviously, one solution is to oxidize the organic matter in the wastewater before the water is discharged. This approach is the one adopted in conventional sewage treatment plants. If SS are also a problem, they can also be removed with conventional treatment. The problem with using nothing more than conventional wastewater treatment methods to remove the BOD and SS from in-

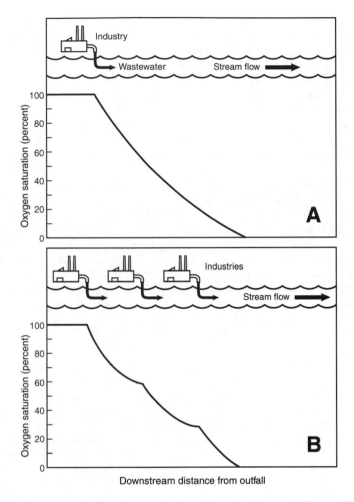

FIGURE 9.2 Oxygen sag leading to anoxia caused by a single large discharge of BOD (A) or a series of smaller but closely spaced BOD discharges (B).

dustrial wastewaters is twofold. First, conventional treatment may do a poor job of reducing the concentrations of toxicants in the wastewater. Second, the concentrations of BOD and SS are sometimes so high that the cost of treatment becomes a serious obstacle. A more practical alternative may be to reduce the amounts of toxic substances, BOD, and SS released from the industrial process in the first place. This objective may be accomplished by altering the industrial process or by recycling. These latter options may be not only more cost effective but also more environmentally appealing in terms of both resource conservation and pollution.

INNOVATIVE STRATEGIES FOR REDUCING INDUSTRIAL POLLUTION

How much does it cost to dispose of industrial wastes properly? This question is rather difficult to answer, in part because many industries did not dispose of their wastes properly in the past. In the United States, about 95% of industrial hazardous waste is disposed of on the site where it is generated (Postel, 1987), and in a disturbing number of cases, it would appear that the principal concern has been to simply get rid of the waste, without much concern about potential environmental impacts (see Chapter 16). Furthermore, in cases where environmentally acceptable disposal practices are followed, the costs do not always reflect manufacturing strategies designed to minimize waste production. Waste management costs at DuPont, the largest chemical producer in the United States, currently exceed $100 million per year (Postel 1987). Companies like DuPont, realizing the high cost of waste disposal, have begun to look for ways to minimize waste production and in particular to replace waste disposal with resource recovery (Iranpour et al., 1999). There now seems to be a consensus that this approach is the real key to solving the industrial waste problem.

Table 9.1 summarizes a few of the innovative techniques that have been used by companies to minimize waste production. In the United States, the Minnesota Mining and Manufacturing Company (3M) has probably had the longest commitment to waste reduction of any major corporation. Its "Pollution Prevention Pays" program, initiated in 1975, has reportedly reduced waste production by a factor of 2 and has saved 3M over $300 million. In some cases, one of the most effective ways to reduce pollution is to recycle secondary materials rather than manufacturing products from virgin resources. Table 9.2 summarizes the results of recycling in the production of aluminum, steel, paper, and glass. Not only does recyling of manufactured products often substantially reduce water pollution, it may also greatly reduce energy costs, water use, and air pollution. The impact of such recycling efforts on some industries has been substantial. For example, by 1983 the

TABLE 9.1   SUCCESSFUL INNOVATIVE TECHNIQUES FOR MINIMIZING INDUSTRIAL WASTE

Company/Location	Products	Strategy and Effect
Astra Sodertalje, Sweden	Pharmaceuticals	Improved in-plant recycling and substitution of water for solvents cut toxic wastes by half
Borden Chemical California, USA	Resins; adhesives	Altered rinsing and other operating procedures cut organic chemicals in wastewater by 93%; sludge disposal costs reduced by $49,000/yr
Cleo Wrap Tennessee, USA	Gift wrapping paper	Substitution of water-based for solvent-based ink virtually eliminated hazardous waste, saving $35,000/yr
Duphar Amsterdam, the Netherlands	Pesticides	New manufacturing process cut toxic waste per unit of one chemical produced by factor of 20
DuPont Barranquilla, Colombia	Pesticides	New equipment to recover chemical used in making a fungicide reclaims materials valued at $50,000 annually; waste discharges were cut by 95%
DuPont Valencia, Venezuela	Paint, finishes	New solvent recovery unit eliminated disposal of solvent wastes, saving $200,000/yr
3M Minnesota, USA	Varied	Companywide 12-year pollution prevention effort has halved waste generation, yielding total savings of $300 million
Pioneer Metal Finishing New Jersey, USA	Electroplated metal	New treatment system design cut water use by 96% and sludge production by 20%; annual net savings of $52,500; investment paid back in three years

Source: Postel (1987).

TABLE 9.2 ENVIRONMENTAL BENEFITS DERIVED FROM
SUBSTITUTING RECYCLED MATERIALS FOR VIRGIN RESOURCES

	Product			
Percent reduction of:	*Aluminum*	*Steel*	*Paper*	*Glass*
Energy use	90–97	47–74	23–74	4–32
Air pollution	95	85	74	20
Water pollution	97	76	35	—
Mining wastes	—	97	—	80
Water use	—	40	58	50

Source: Pollock (1987).

world's industrial market economies produced 30% of their raw steel in electric arc furnaces, which rely exclusively on scrap metal as a feedstock (Pollock, 1987). In the United States, more than half of the 300 billion aluminum cans sold since 1981 have been recycled, and the aluminum industry used 22% less energy to produce a kilogram of aluminum in 1984 than in 1972 (Pollock, 1987). Building a paper mill designed to use recycled paper rather than raw wood pulp is 50–80% cheaper and reduces the demand for valuable forest resources. According to Pollock (1987, p. 22), "Simply recovering the print run of a Sunday edition of the *New York Times* would leave 75,000 trees standing." In 1996, 63% of newsprint was recovered in the United States, up from 43% in 1990 (*Christian Science Monitor*, 1997). Overall, about 45% of the paper used by Americans is now recovered for domestic recycling, export, and other uses, an increase of more than 50% since 1990 (AFPA, 1999b). In the United States, approximately 200 paper mills now use recycled paper exclusively. These mills use less than half as much water to produce a given amount of paper product as do conventional pulp mills.

While some progress toward reducing industrial waste has obviously been made, there is much room for improvement. Japan, which seems to have gone further than any major industrial country toward reducing and reusing industrial waste, recycled more than half of its estimated 220 million tons of industrial waste in 1983 (Postel, 1987). In the United States, on the other hand, only 4% of hazardous waste was recycled in 1981. The U.S. Congressional Budget Office has estimated that 80% of waste solvents and 50% of the metals in industrial wastewater could be recovered (Postel, 1987). In so-called industrial ecoparks, wastes generated by one company are often made available as feedstock to another company, the idea being that "one company's sludge is another's manna" (Kaiser, 1999, p. 686). For example, at a Danish industrial park, flare gas from an oil refinery heats other factories; a power plant sends gypsum (produced by scrubbing SO_2 from flue gas) to a drywall factory; and a biotech firm's fermentation waste is shipped to farmers for use as fertilizer (Kaiser, 1999). Such recycling of waste would obviously do much to reduce the amount of toxicants released to the environment by some industries. As population pressures increase the demand for clean water, serious efforts to reduce the amount of water used by and polluted by industry will likely become more mandatory than optional. The following examples serve to illustrate this point.

THE HAWAIIAN SUGAR CANE INDUSTRY

Cultivation of sugar cane in Hawaii dates back to at least 1835, when the first successful plantation was started on the island of Kauai. In 1837 the first harvest yielded 2 tonnes of

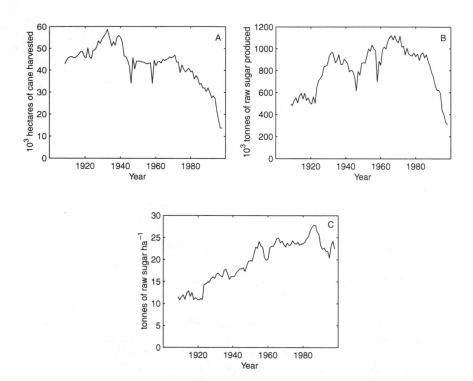

FIGURE 9.3   (A) Area of sugar cane land harvested each year, (B) tonnes of raw sugar produced, and (C) raw sugar produced per hectare by Hawaiian sugar cane industry. *Source*: USDA (1999).

raw sugar, which sold for $200. During the next 40 years the industry gradually expanded, and by 1876 sugar planters were growing cane on all four of the main Hawaiian islands— Kauai, Maui, Oahu, and Hawaii. Raw sugar production in 1876 equaled about 12,000 tonnes. During that year, a Treaty of Reciprocity was signed between the United States and the Kingdom of Hawaii. The United States received a coaling station at Pearl Harbor. In return, the 2-cent-per-pound duty fee on raw sugar imported into the United States was waived on Hawaiian-produced raw sugar (HSPA, 1990). As a result of this competitive advantage over foreign sugar growers, the Hawaiian sugar cane industry rapidly expanded, and many land areas that otherwise would have been marginally profitable for cane growing were converted to sugar cane cultivation. By 1898 raw sugar production equaled 205,000 tonnes, and by 1908 production had increased to more than 500,000 tonnes. Figure 9.3 shows how raw sugar production and the area of land devoted to cane cultivation have varied since that time.

Sugar production per unit area (yield) roughly doubled between 1910 and 1950 due to improvements in cane growing and harvesting and in sugar extraction techniques. Since 1950, yields have remained relatively constant at about 20–25 tonnes of raw sugar per hectare. The industry reached its peak in approximately 1966, when raw sugar production exceeded 1.1 million tonnes. Both the amount of land planted in cane and raw sugar production declined during the 1980s as a result of foreign competition and capture of the liquid sweetener market by high-fructose corn syrup. The decline of the industry continued during the 1990s, and the islands of Oahu and Hawaii saw their final sugar cane harvests in 1996. Today sugar cane is grown only on Kauai and Maui. While the sugar cane industry in the state has declined in recent years, both the tourist industry and military

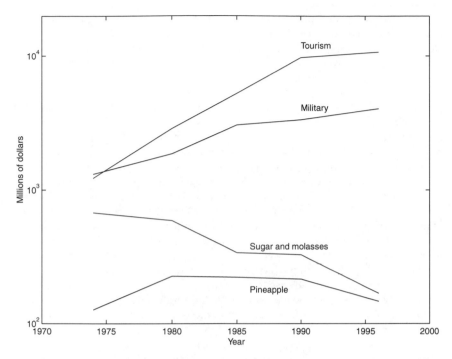

FIGURE 9.4   Direct income to Hawaii from major export industries. *Source*: USDA (1999).

operations have grown steadily throughout the 20th century. Although the sugar cane industry was a major source of income to the State of Hawaii for much of the 20th century,
the industry's economic importance has declined considerably in recent years (Figure 9.4).
It now ranks far behind tourism and military operations as a source of income to the state.

Until the early 1970s, it was the custom of Hawaiian sugar cane companies to discharge their wastewater into the ocean. This wastewater consisted primarily of processing
plant effluent but also included excess irrigation water, called *tailwater*, and storm runoff
from the cane fields. Although some companies treated their wastewater by means of settling basins or similar devices, other companies discharged their wastewater with no treatment whatsoever. During the cane-harvesting season, roughly from February through
November, a cane-processing mill would typically discharge about 20–40 million L day^{-1}
of wastewater into the ocean. As of 1966 there were a total of 26 sugar mills in the state:
6 on Kauai, 4 on Oahu, 4 on Maui, and 12 on Hawaii. Ultimately, it became apparent to
state officials that the discharges from the mills were creating conditions in the coastal
ocean that were in direct conflict with the tourist promotional campaign that described
Hawaii's coastal waters as blue and sparkling clear. Because of the increasing importance
of the tourist industry to the state, changes in the cane industry's waste disposal practices
were called for. In June 1963, a presidential board visited the state to survey the extent of
the pollution problem. As a result of the recommendations of this board and the request
of the Hawaii Department of Health, an in-depth study of the Hawaiian sugar cane industry wastewater problem was initiated by the Federal Water Pollution Control Administration (FWPCA).[1] The study, ultimately published by the EPA, lasted from November
1966 to September 1968. We will henceforth refer to this study as the *EPA study*.

[1]The FWCPA became the Federal Water Quality Administration (FWQA) on April 30, 1970. The FWQA
was absorbed by the EPA on December 2, 1970.

As a part of the EPA study, three sugar mills in the state were singled out for intensive study: the McBryde Sugar Co. mill on Kauai, the Pioneer Co. mill on Maui, and the Honokaa Sugar Co. mill on Hawaii. Of the three mills, the McBryde and Pioneer mills both subjected their wastewater to primary and secondary clarification before releasing it. The Honokaa wastewater received no treatment prior to discharge.

Table 9.3 summarizes the characteristics of the wastewater from the three mills as reported by the EPA (1971). The numbers in Table 9.3 are simply mean values of the given parameters measured at the three mills. The McBryde and Pioneer mill values were taken from the analyses of the washwater prior to clarification; the numbers would therefore be comparable to the Honokaa values, since Honokaa wastewater received no treatment.

The concentrations of total N and total P in the mill wastewater were both comparable to the values for raw sewage. The SS, BOD, and COD were much higher in the mill wastewater than in raw sewage. In fact, the SS concentration in the mill wastewater was over 30 times the SS concentration in raw sewage. Given the typical mill wastewater discharge rate of 20–40 million L day^{-1} and assuming a per capita raw sewage discharge of 600 L day^{-1}, the loading of SS in the untreated wastewater from a single sugar cane mill would be equivalent to the loading of SS in the raw sewage from a population of 1–2 million persons. This calculation provides an indication of the magnitude of the water pollution problem created by discharges of untreated cane mill wastes. Fecal coliform counts in the mill wastewater were about a factor of 1,000 lower than fecal coliform counts in raw sewage (e.g., Gunderson, 1973) but were far in excess of levels considered safe for water contact. Although the fecal coliform counts may in part have reflected the presence of nonfecal thermotolerant biotypes of *Klebsiella* (Chapter 7), the high counts raised the concern that the wastewater contained significant concentrations of fecal pathogens. What was the source of all the SS, BOD, nutrients, and fecal coliforms in the cane mill wastewater? To answer this question, one must understand something about the way sugar cane is grown, harvested, and processed to make raw sugar in Hawaii.

Sugar Cane Production: Field Operations

Fields in which sugar cane is to be planted are first graded to control runoff and then deep-plowed to a depth of as much as 0.5 m. Short sections of freshly cut 8- to 10-month-old cane stalks are then planted in the furrows, given an initial dose of fertilizer, and covered with dirt. The planting is done by a mechanical planter rather than manually. The cane is

TABLE 9.3 MEAN CONCENTRATIONS OF SELECTED CONSTITUENTS OF THE WASTEWATER FOR THE McBRYDE, PIONEER, AND HONOKAA MILLS AND TYPICAL RAW SEWAGE

Constituent	Mill Wastewater[a]	Raw Sewage
Fecal coliforms	53,000/10 mL	~10^6/100 mL
SS	6,900 ppm	200 ppm
Settleable solids	6,600 ppm	—
BOD	755 ppm	200 ppm
COD	2,033 ppm	350 ppm
Total N	39 ppm	40 ppm
Total P	15 ppm	10 ppm

[a] The values reported for the three mills have been averaged. In the McBryde and Pioneer effluents, the values averaged pertain to the washwater prior to clarification.
Source: EPA (1971) and Weibel (1969).

TABLE 9.4 MEAN CONCENTRATIONS OF SELECTED
CONSTITUENTS IN TAILWATER FROM THE McBRYDE
PLANTATION

Constituent	Concentration
Fecal coliforms	1,934/100 mL
SS	843 ppm
Settleable solids	576 ppm
COD	91 ppm
Total N	10 ppm
Total P	5.5 ppm

Source: EPA (1971).

heavily fertilized during the first year of growth with N, P, and K. During the second year no fertilizer is applied, so that the cane will store sugar rather than produce additional plant foliage. In Hawaii, herbicides are used to control weeds, with three or four applications per crop being typical treatment rates. The heaviest applications are made during the first six months of growth.

Prior to 1876, sugar cane in Hawaii was grown only in those areas where there was adequate rainfall to support cane growth. After passage of the 1876 Reciprocity Treaty, the industry rapidly expanded its plantings into areas where irrigation was required, and in recent years about 60% of Hawaiian sugar cane fields have been irrigated. The principal sources of pollution associated with field operations were the tailwater and stormwater runoff from the fields. The average concentrations of selected constituents in the tailwater from five different fields on the McBryde plantation are listed in Table 9.4. These values are considerably lower than the mean values for mill wastewater but still high enough to cause pollution problems. The SS concentration, for example, is over four times the typical SS concentration in raw sewage, and the fecal coliform count is still well above values considered safe for water contact.

Sugar Cane Production: Harvesting

Prior to World War II, sugar cane in Hawaii was cut by hand and transported to the mill by flumes. After World War II, however, the cost of labor increased to such an extent that mechanical harvesting became more economical than manual harvesting. Unfortunately, conventional cane cutters proved generally unsatisfactory in Hawaii due to the thickness of the cane growth, the rocky soil, and the furrowed and sometimes hilly terrain. A modified mechanical harvesting procedure was therefore adopted. Just prior to harvesting, the fields are burned to remove excess foliage. The succulent cane stalk is virtually unaffected by this burning, but much of the relatively dry leafy material is removed. A conventional tractor with a modified push rake in front then snaps off the cane stalks near ground level. The root structure is left intact to allow growth of a second, third, or even fourth crop from shoots that grow from the old roots. The snapped-off stalks of cane are raked into windrows by bulldozers, and the cane is then lifted into 40-ton-load cane haulers by cranes equipped with special grabs. The cane haulers then transport the cane to the mill. This harvesting procedure accounts for many of the undesirable characteristics of cane mill wastewater because rocks and dirt as well as fecal material[2] are invariably loaded onto the cane haulers along with the cane. This material must be removed at the mill.

[2]The fields are inhabited by cane rats, mongooses, and other animals.

Sugar Cane Production: Factory Operations

A flow diagram of factory operations in a typical sugar cane mill is shown in Figure 9.5. The sugar cane factory can be divided into basically three sectors: the cane preparation plant, the milling plant, and the boiling plant. The cane is brought from the field to the preparation plant, where the first step in processing consists of removal of the dirt and rocks. The cane is loaded onto a steeply inclined conveyer belt fitted with widely spaced prongs that hold the cane on the belt as it moves upward but presumably allow rocks to

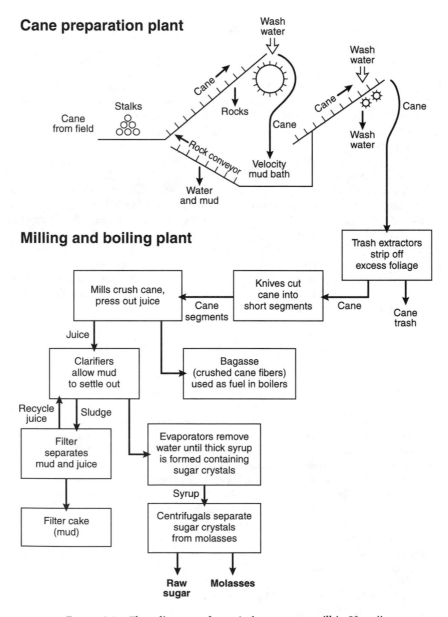

FIGURE 9.5   Flow diagram of a typical sugar cane mill in Hawaii.

roll back down to the bottom, where they are collected in a hopper. The rocks are trucked off to a landfill. Once over the top of the inclined plane, the cane travels through a floatation bath. The cane floats on the surface of the water while rocks sink to the bottom and some dirt is washed from the cane. Following the floatation bath, the cane is sprayed with jets of water to remove additional dirt. The cane stalks next pass over a series of oppositely spinning rollers that strip off any excess foliage. Finally, a series of revolving knives cut the cane stalks into short segments in preparation for milling.

Wastewater from the cane preparation plant contains high concentrations of SS and fecal coliforms from the dirt washed from the cane. The preparation plant wastewater also contains the cane trash, which consists primarily of excess foliage and broken stalks.

In the milling plant, high-pressure rollers, usually operated as a tandem of three per mill, squeeze the juice from the stalk segments. Usually a series of three to five mills are used to extract the juice. The stalk tissue that remains after this milling has the appearance of coarse sawdust and is called *bagasse*. In the past, at least some of this bagasse was discharged with the mill wastewater into the ocean. At the present time, however, all the bagasse is burned as fuel in the plant, and on the island of Hawaii, excess energy was sold to the local electric utility. Drying and pelletizing the bagasse further increases its energy yield. The bagasse has therefore become an asset rather than a liability.

In the boiling plant, lime is added to the cane juice to help prevent fermentation and to aid in clarification. The juice is then pumped into a clarifier similar in design to the secondary clarifier in a sewage treatment plant. The supernatant juice taken from the top of the clarifier is pumped directly to the evaporators, while the muddy sludge from the bottom is pumped to a cylindrical filtration unit. Normally, the juice that passes through the filtration unit is recycled to the clarifier. The sludge that is separated out is continuously scraped off the rotating filter and is referred to as *filter cake*.

In the evaporators, the water in the clarified juice is gradually evaporated off until a thick syrup consisting of about 50% sugar remains. This thick syrup is then pumped to the vacuum pans, where a final evaporation step results in the precipitation of sugar crystals from the syrup. The mixture of sugar crystals and syrup is then centrifuged in a rotary filter. The crystals of sugar adhere to the inside of the spinning filter, while the liquid syrup passes through. The raw sugar is then scraped from the inside of the centrifuge. The syrup collected from the first vacuum pass is called *A molasses*. This syrup still has a high sugar content and is evaporated down further in a second vacuum pan to yield additional raw sugar. The syrup from the second centrifuge step is called *B molasses*, and has a rather low sugar content. Usually no attempt is made to obtain sugar from the B molasses. The chief waste products from the boiling mill are the filter cake from the clarification step and the heated wastewater from the boilers.

Survey of Water Pollution Problems

The 1966–1968 EPA study revealed that a variety of pollution problems were created when wastewater from cane factory operations was discharged to the ocean. Plumes of brown water discolored by sediment from the wastewater were visible along the coastline as much as 3 km or more from the outfall. Secchi depths taken near the outfalls were sometimes no more than 5–10 cm. Total coliform counts exceeded 1,000/100 mL for distances of up to 3–5 km from mill outfalls. In some cases, surfers were noted in the areas of high total coliform counts.

Mills that discharged cane trash and bagasse created an additional water pollution problem. The trash and bagasse tended to float on the surface in mats as much as 1.5 km long and 50 m wide. These mats were unsightly, created a hazard to navigation, interfered with fishing, and, given an onshore wind, washed up along the shoreline, creating an unsightly mess. If the cane trash and bagasse did not wash up on the shore, it ultimately sank to the bottom.

Benthic surveys revealed that the accumulation of sediment and sludge from the mill discharges had killed or severely disrupted the natural coral reef communities up to as much as 1.5 km from the outfalls. Off the Honokaa mill outfall, sludge deposits as much as 3 m deep were found.[3] Killing of corals was presumably due to direct smothering by sludge and/or the high turbidity of the water near the outfalls. As was the case in Kaneohe Bay (Chapter 4), sponges and benthic algae had colonized many areas of dead coral.

Fish populations were reduced in both number and diversity near the outfalls. This change in the fish populations was believed to be caused in part by the turbidity of the water, since most of the local species use visual means for locating prey, and in part by the demise of the natural coral communities that normally would have provided food and shelter for the fish. Toxicity bioassays conducted with four local fish species revealed that tilapia, mullet, and aholehole (flagtail fish) could survive in almost 100% mill wastewater for 96 hours if the wastewater was adequately aerated. However, nehu (an anchovy) had a 48-hour TLm of about 6–10% mill wastewater. Because of the wide variety of fish on a typical undisturbed coral reef, it seems safe to say that a more thorough study of acute toxicity effects on fish would have been required to determine whether water quality near the outfalls might have been lethal for a significant portion of the natural fish population.

Despite the high BOD and COD of the mill wastewater, O_2 levels in the ocean near the outfalls were virtually identical to those in control areas. In both control areas and discharge areas, bottom water O_2 levels were typically 80–90% of saturation, whereas those in the remainder of the water column were usually 90% of saturation or higher. These results indicate the effectiveness of ocean mixing processes in dispersing the effluent and in promoting O_2 exchange with the atmosphere. It is likely that cane mill effluent discharged into a confined body of water would create serious oxygen depletion problems (Officer and Ryther, 1977), but in Hawaii no mill wastewater discharges were made to estuaries with restricted circulation patterns.

Response to the EPA Survey

The EPA study revealed that cane mill wastewater discharges to the ocean were creating conditions that violated eight different Hawaii state water quality guidelines. These guidelines pertained to the following conditions:

1. Objectionable sludge or bottom deposits (sediment, cane trash, and bagasse)
2. Floating debris (cane trash, bagasse)
3. Objectionable color or turbidity (sediment)
4. Pathogens (suggested by the presence of fecal coliforms)
5. Total nutrient concentrations (total N and total P in the effluent)

To eliminate water pollution problems associated with cane mill operations, the EPA stated that the following three policies should be followed:

1. No cane trash or bagasse should be discharged by any mills in the state.
2. Runoff of tailwater and stormwater should be minimized through the use of ponds and embankments for trapping water and through storage and reuse of irrigation water.

---

[3]Recall that the Honokaa wastewater received no treatment.

3. All wastewater discharged to the ocean should be treated to minimize the discharge of suspended solids.

Relevant to the second of these recommendations was a study of field runoff from the Waialua Sugar Company's fields on Oahu, which revealed that during most storms there was virtually no runoff from the fields (EPA, 1971). The fields were graded so that runoff was directed to ponds or against embankments at the end of the furrows. The fields could retain all the runoff from a 5-cm rain, and in one rain of 15 cm, only 1 cm of water ran off the fields. In fact, the Waialua cane fields retained significantly more rainfall than a control area of undeveloped land. Figure 9.6 shows the cumulative runoff from the Waialua cane fields and the control area during the 15-cm rain on February 1, 1969. Despite the much smaller total runoff from the cane fields, the higher concentration of SS in the cane runoff resulted in a total discharge of SS from the cane fields that was little different from the SS discharged from the control area. The results do indicate, however, that the use of ponds and embankments for retaining field runoff can at least reduce runoff to the point where soil erosion from the fields is no worse than soil erosion from undeveloped land.

The State of Hawaii followed through on the first two recommendations of the EPA panel. No cane mills in the state are allowed to discharge cane trash or bagasse, and all plantations must minimize field runoff by using methods similar to those practiced at Waialua. With respect to the third recommendation, the state was more stringent than the EPA panel. No sugar cane mills on the islands of Kauai, Oahu, and Maui were allowed to

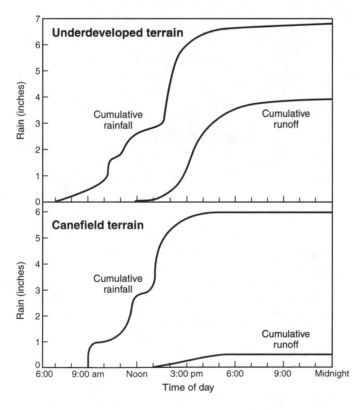

FIGURE 9.6 Cumulative runoff from the Waialua cane fields (bottom) and control area (top) versus cumulative rainfall for the storm of February 1, 1969. *Source*: Redrawn from EPA (1971).

discharge any mill wastewater. On the island of Hawaii, an exemption to this restriction was granted to mills along the Hamakua coast, where pumping wastewater back to the fields was regarded as impractical because of the steep terrain and the fact that irrigation requirements were minimal or nonexistent due to the abundant natural rainfall, which averages 180–380 cm yr^{-1} along the Hamakua coast (Grigg, 1985). Hamakua coast cane mills were, however, required to treat their wastewater with the use of hydroseparators and settling ponds. This treatment process reduced the discharge of sediment by roughly a factor of 30 between 1978 and 1982 (Grigg, 1985). The procedure was effective in removing much of the soil from the wastewater because of the high percentage of settleable SS (Table 9.3). On the other hand, the EPA study revealed that fecal coliform counts were not reduced to satisfactory levels through settling, and at the Pioneer mill fecal coliform counts actually increased by a factor of 10 between the mill washwater and the final settling pond effluent. This observation undoubtedly influenced the state's decision to require no wastewater discharges at most mills.

Present Status of the Industry

The response of the State of Hawaii and the Hawaiian sugar cane industry to the problems caused by wastewater discharges associated with the industry illustrate rather well the nature of solutions to such problems and, to some extent, the practical limitations associated with implementing those solutions. The complete ban on discharges at most mills led to a policy of recycling and reuse that was beneficial to the industry. The bagasse that was previously discharged to the ocean is now burned as a fuel. The use of bagasse as a fuel significantly reduced the energy costs of the industry. Because Hamakua coast plantations did not pump water for irrigation purposes, their energy costs were substantially lower than those of plantations that irrigated. As a result, they were able to generate and sell electricity to the Hawaiian Electric Company. These sales further reduced their operating costs.

Discharging topsoil and nutrients into the ocean is an obvious waste of a valuable resource. Along the Hamakua coast alone, the loss of topsoil amounted to 200,000–225,000 tonnes yr^{-1} just prior to the implementation of measures to treat mill wastewater (Grigg, 1985). Recycling of mill wastewater and minimization of irrigation water runoff largely eliminated such losses. Mill wastewater is treated to separate the water from the sediment. The water is then pumped back to the fields for irrigation. The topsoil is either returned to the fields or in some cases applied to low-elevation plots of land that would otherwise be unsuitable for cane cultivation. On Oahu, the Waialua Sugar Company used the latter approach to expand its cane production acreage.

Irrigation with mill wastewater throughout the two-year growth cycle of the cane reduces the sugar yield because the high nutrient levels in the wastewater cause the cane to store less sugar during the second year of growth.[4] Related to this problem have been the studies conducted by Lau et al. (1975) on the possibility of using secondary sewage effluent to irrigate sugar cane in Hawaii. They found that irrigation with secondary sewage effluent throughout the two-year cane-growing period reduced sugar production by about 6% compared to that of control fields that received standard fertilizer applications during the first year but no fertilizer during the second year. Sugar yields increased by about 6%, however, in fields irrigated with secondary sewage during the first year of growth and with groundwater or surface water during the second year of growth. Since the total N and total P concentrations in mill wastewater and raw sewage are comparable, it seems likely that a similar increase in sugar production might be achieved on fields irrigated with mill wastewater if the wastewater were applied during only the first year of growth. However, this

[4]Instead, more foliage is produced.

approach would obviously require that about twice as much cane be irrigated with wastewater, and the additional cost of pumping the wastewater to higher-elevation fields might more than offset the net revenue associated with the gain in sugar yield.

The industry has experienced difficult times since 1975 due to the rapid expansion in the use of high-fructose corn syrup in the liquid sweetener market. This competitive pressure forced an overall improvement in the economic efficiency of Hawaii's sugar cane industry, so that the real cost of production declined by about 25% from 1981 to 1986 (HSPA, 1990). One change of particular relevance to the issue of water pollution was a major conversion to drip rather than furrow irrigation on irrigated cane fields. Drip irrigation is now used on about 85% of irrigated fields. This change in methodology led to much more efficient use of irrigation water and eliminated tailwater runoff. Eventually, economic pressures forced all but the most efficient plantations out of business. The industry now produces about 30% as much raw sugar as it did during its heyday.

The plantations that remain in the industry have continued to explore innovative technologies for improving profitability and reducing pollution. Recently, the Hawaiian Commercial & Sugar Company (HC&S) on Maui began experimental studies with a special cane harvester that cuts the cane in the field, strips off the excess foliage, and chops the cane into short segments that can be fed directly into the milling operation (Tanji, 1999). Use of the new harvester eliminates the need to burn the fields prior to harvesting. Furthermore, because the new harvester picks up virtually no dirt and rocks, the cane stalks can bypass the cane cleaning operation, the major source of wastewater at conventional mills. HC&S is also exploring the possibility of using its bagasse to produce a particle board made with a nonformaldehyde binder. Finally, the company is conducting tests with over 600 varieties of cane with the goal of identifying a strain that can be harvested in one year.

Hawaiian sugar yields are among the highest in the world, about 22 tonnes $ha^{-1} yr^{-1}$ in 1998. This figure compares to annual yields of 10.3, 7.9, and 8.2 tonnes ha^{-1} for raw sugar produced from sugar cane in Florida, Louisiana, and Texas, respectively. The profitability of the Hawaiian industry, however, is reduced by the high cost of labor in Hawaii. Hawaii's cane field workers have the highest standard of living of any agricultural workers in the world (HSPA, 1990). Their daily earnings (including benefits) currently average more than $130 (DBEDT, 1997). High labor costs compromise the industry's competitive position with producers in countries such as India, Brazil, and Cuba, which are major sugar producers with much lower labor costs.

The profitability of the industry became a factor in the issue of water pollution during the 1980s, when an effort was made by Hamakua coast mills to obtain a relaxation of the standards applied to their wastewater discharges. The Hamakua coast mills collectively lost about $40,000 in 1981 and 1982 from sugar production (Grigg, 1985), and although other sources of income including revenues from the sale of molasses and electricity largely offset this loss, it was argued that without some economic relief the plantations would be forced to close. Studies conducted by Grigg (1985) indicated that the approximately 30-fold reduction in SS discharged by the mills had in fact produced only slight changes in the zones of measurable impact near the mill outfalls. The areas of coral cover, for example, had increased by only 4–8%. The EPA (1989), however, noted that the costs of wastewater treatment amounted to only 2–3% of operating costs for Hamakua coast mills, and that the tenuous financial position of the mills[5] would not be substantially altered by reducing the level of wastewater treatment. The EPA further noted that turbidity and concentrations of nitrate plus nitrite, Cu, Hg, Pb, As, and Mn exceeded Hawaii water quality standards and EPA criteria within designated zones of mixing around one or another of the mill outfalls. The levels of Cu and Hg exceeded EPA criterion maximum concentration limits. Based on these observations and other considerations, the EPA (1989) decided

---

[5]In 1988 the mills showed a profit of $1.6 million, which was 2.5% of their operating costs.

not to approve a reduction in the level of wastewater treatment for Hamakua coast mills. While it is true that the Hamakua mills closed a few years later, their demise was not due to the cost of wastewater treatment.

THE PULP AND PAPER INDUSTRY

By almost any metric, the pulp and paper industry is one of the largest industries in the world. It is certainly one of the largest producers of wastewater. The industry has more than doubled in size during the last 30 years (Figure 9.7A). Total capacity in 1998 was about 215 million tonnes of pulp per year (FAO, 1999). Actual production for the top seven producing nations averaged about 85% of capacity (FAO, 1999). Although the percentage of paper and paperboard products accounted for by recycling has increased from 18% to 36% since 1969 (Figure 9.7B), the production of pulp from raw wood has continued to increase. The United States and Canada are the major producers of wood pulp. Their combined capacity in 1998 totaled 91.5 million tonnes, and production equaled 81.4 million tonnes of pulp.

To appreciate the size of the industry, consider the fact that the pulp and paper industry in the United States and Canada employs about 937,000 persons[6] (AFPA, 1999a) and is the single largest industrial employer in Canada (McCubbin, 1983). In the United States, the industry is the largest discharger of conventional pollutants subject to national effluent standards (GAO, 1987). Annual production of paper and paperboard in the United States amounts to about 87 million tonnes, with a market value of more than $160 billion, 2.2% of the gross domestic product (AFPA, 1999a). Canadian production is about 19 million tonnes, with a market value of $30.5 billion, about 5.6% of Canada's gross domestic product. The industry is a major contributor to the U.S. economy, with a total of 345 mills in 44 states, primarily in the southeastern and western sections of the country (AFPA, 1999a). In Canada, pulp, paper, and forest products account for the largest dollar value of foreign exports, exceeding the value of such other exports as minerals, petroleum, and agricultural products (McCubbin, 1983). Obviously, the pulp and paper industry is a major industry in North America.

As a general statement, it is fair to say that the pulp and paper industry uses a lot of water. The amount of water used varies greatly, however, depending on the pulping procedure and the desired characteristics of the pulp. Water requirements for high-quality stationary paper may be as much as 30 times those of dark, unbleached products such as cardboard (McCubbin, 1983). Packaging and newsprint account for about 60% of paper and paperboard products (Roberts, 1996). These are relatively low-quality products that require the least amount of water to produce. Paper destined for printing or writing accounts for about 30% of the demand (FAO, 1999), but only a small fraction of writing-quality paper is fine stationary. Due to conservation efforts, water requirements for the industry have decreased by about a factor of 2 during the last 20 years but still average about 70 m^3 per tonne of pulp (CPPA, 1997). With an annual output of roughly 81.4 million tonnes, the combined U.S. and Canadian pulp and paper industries discharge about 5.7 km^3 of wastewater per year, a figure that exceeds the annual flow of the Colorado River. Without treatment, the discharge of BOD associated with that wastewater would equal about 60% of the BOD associated with the raw sanitary sewage produced by the entire population of the United States and Canada. Obviously, the task of adequately treating this wastewater before releasing it to the environment is a major undertaking. Since the characteristics of pulp and paper mill effluent depend very much on the procedure used to

[6]This figure includes persons employed in the manufacture of allied products such as containers and boxes.

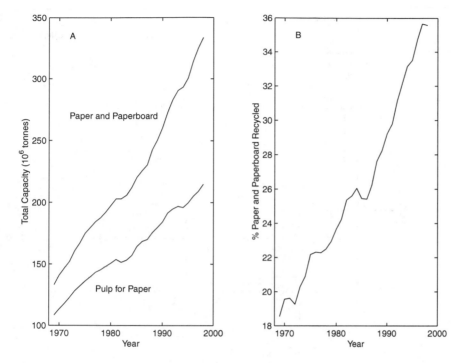

FIGURE 9.7 (A) Capacity of world pulp and paper industry and (B) percentage of paper and paperboard products accounted for by recycling since 1969. *Source*: FAO (1999).

make the pulp and the nature of the paper products, it is appropriate here to examine in some detail how pulp and paper are produced.

Steps in the Production of Paper

A simplified diagram of typical pulp and paper processes is shown in Figure 9.8. There are basically four steps to the procedure. First, the bark is separated from the rest of the wood in the so-called woodroom. Debarking may be accomplished with the use of rotating drums, which use friction to remove the bark; hydraulic jets; or mechanical debarkers, which use rotating knives to strip off the bark. The bark may account for as much as 15% of the weight of the raw wood. In the past, the bark was regarded as waste and its disposal was a major problem. In virtually all modern mills, however, the bark is utilized as a source of fuel in special bark-burning boilers. The costs associated with preparing the bark and burning it in this manner are more than offset by the elimination of bark disposal costs and the reduced requirements of the mill for fossil fuels to provide energy. Stone ground-wood mills use the debarked logs directly to make pulp, but all other pulping procedures require that the logs be cut into small chips, generally 12–18 mm in size (McCubbin, 1983). The debarking and chipping operations may be bypassed to the extent that the mill receives wood chips from an external source.

The next step is the production of the pulp itself. This process consists of partially breaking down the cellulose and lignin that make up the cell walls of the wood tissue to produce a soft, spongy mass of pulp. Pulping may be accomplished by either mechanical or chemical means or some combination of the two.

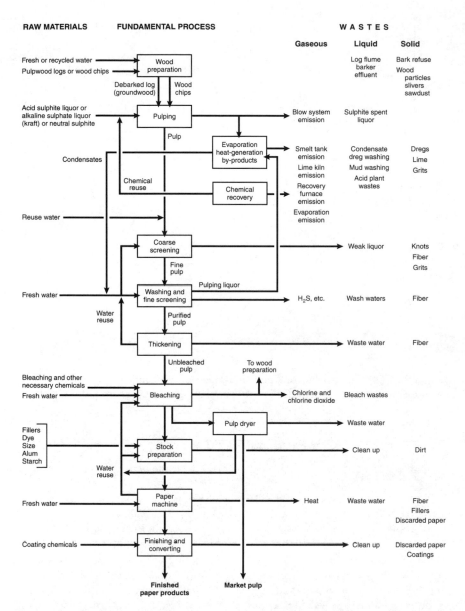

FIGURE 9.8   Diagram of typical pulp and paper processes. *Source*: Redrawn from McCubbin (1983). Reproduced with the permission of the Minister of Public Works and Government Services Canada, 2000.

Once the pulp has been produced, it may be treated in a variety of ways to give the paper particular qualities. In many cases, the pulp is bleached. Most of the color in the pulp is due to the lignin in the wood, and the bleaching process involves oxidation of the lignin with either chlorine or oxygen. Reclaimed paper or rag stock may be added to give the pulp a desired texture. Fillers such as clay, talc, and diatomite are usually added to give the paper increased opacity, brightness, and smoothness. In many cases, the pulp is sized to provide wet strength, improved printability, and resistance of the paper surface to damage due to the pull of the ink in printing. Dyes may be added to impart color. Finally, the pulp goes to the paper machines, where it is formed into paper and dried. Although there are several different types of paper machines, all consist of a wet-end or forming section, a press section, and a dryer section. The pulp is formed into paper on a wire or plastic mesh. The press section removes much of the moisture from the paper and gives it a smoother surface. The dryer section removes the remaining moisture and prepares the paper for reeling.

This general procedure is followed in virtually all pulp and paper mills that use raw wood as feedstock. The chief factor that determines the quantity and characteristics of the mill's wastewater is the method used to produce the pulp. When the debarked wood enters the pulping department, it consists primarily of cellulosic fiber (45%), lignin (20–30%), and hemicellulose[7] (25%). Chemical pulping removes most of the lignin and hemicellulose; mechanical pulping allows much of the nonfibrous material to remain in the pulp.

Mechanical pulping is used mainly to produce pulp for newsprint and currently accounts for 10% of world wood pulp production (FAO, 1999). The earliest form of mechanical pulping was stone groundwood, a process in which logs were forced into contact with a revolving grindstone in the presence of water. The wastewater from this process contains a relatively small amount of dissolved organic substances leached from the wood and a great many small wood fibers. Until the mid-1970s, newsprint was made from a mixture of about 75% groundwood pulp and 25% chemical (sulfite) pulp. In recent years, however, an increasing percentage of mechanical pulp has been produced with refiners, which use rotating discs to break down wood chips into pulp. The pulp produced with refiners is stronger than groundwood pulp and superior for most purposes (McCubbin, 1983). A variation on this theme is the production of thermomechanical pulp (TMP), in which the wood chips are softened by steam under pressure and then defibered in a disc refiner. TMP currently accounts for 15% of world wood pulp production (FAO, 1999). With the use of TMP, it is possible to reduce or even eliminate the need for chemical pulp in the production of newsprint. Both stone groundwood and TMP convert over 90% of the wood into pulp.

There are basically two ways of producing pulp by chemical means: the sulfite process and the sulfate or kraft process. Both methods consist of cooking the wood chips in a pressure cooker to which various chemicals have been added to help separate the wood fibers and break down the lignin. In most cases, about 95% of the lignin is removed in chemical pulping, and only about 35–60% of the material in the wood chips is converted to pulp. Chemical pulping presently accounts for 71% of world wood pulp production (FAO, 1999).

In the sulfite process, sulfurous acid (H_2SO_3) is added along with Ca, Mg, or some other base to cook the wood. The cooking solution is acidic. Lignin compounds are removed by sulfonation and hydrolysis reactions that form soluble ligno-sulfonates, and these account for roughly half of the dissolved organic by-products in the cooking solution. Sugars derived from hemicellulose and cellulose, and sugarsulfonic and aldonic acids derived from the bisulfite substitution and oxidation of sugars, account for most of the remaining dissolved organics in the spent sulfite liquor. While it is technically possible to recycle and burn much of the organic matter in the sulfite liquor, historically most sulfite mills have made no effort to recover these chemicals. The corrosive nature of the liquor and the lack of an economic incentive have been primarily responsible for this attitude. The dissolved organics leached from the wood chips and produced in the cooking process account for much of the BOD and toxicity of sulfite mill effluent. The BOD of untreated sulfite mill effluent is typically about 1,000 ppm if no effort is made to recover the organics produced in the cooking process (McCubbin, 1983).

In the kraft or sulfate pulping process, wood chips are cooked in a basic medium containing sodium hydroxide (NaOH) and sodium sulfide (Na_2S). The pulp produced by the

[7]Hemicellulose is a portion of the wood fiber consisting of sugar-like substances intimately associated with cellulose in the fiber wall.

kraft process has high strength properties relative to sulfite pulp, and carbohydrates that tend to degrade in the acidic sulfite medium are largely stable in the kraft process. Kraft mills are able to handle a greater variety of wood types than sulfite mills because extracts of certain wood species that lead to problems with pitch in acid sulfite pulping are dissolved or dispersed in the basic kraft liquor (McCubbin, 1983). Historically, the chief disadvantage of the kraft process was the fact that a relatively large amount of residual condensed lignin remains in the pulp. This residual lignin gives kraft pulp a dark color that is much harder to remove by bleaching than the color associated with the lignin in sulfite pulp. Until 1946, in fact, most chemical pulping was done with the sulfite process because it was impossible to produce bleached kraft pulp of comparable whiteness without impairing the strength of the product. In that year, however, two Canadians worked out a bleaching process using chlorine dioxide, which allowed the bleaching of kraft pulp to a whiteness previously thought impossible without loss of strength. Since that time, the kraft process has become the preferred chemical pulping method for most purposes because of the difficulty of recovering extracted organics from spent sulfite liquor and the limited availability of wood species that could be efficiently pulped by the sulfite process.[8] In the United States and Canada, for example, less than 3% of chemical wood pulp is currently produced by the sulfite process (FAO, 1999).

From the standpoint of water pollution, a key feature of the kraft process is the fact that most of the organic matter that is not converted to pulp is burned, and about 95% of the cooking chemicals are recycled. The procedure is to evaporate down the cooking liquor to about 65% solids and then burn the concentrated liquor in a recovery furnace (McCubbin, 1983). Combustion of the organic matter provides heat for steam generation, and under the conditions in the furnace, sodium sulfate is reduced to Na_2S. In subsequent steps, NaOH and CaO are recovered from calcium hydroxide ($Ca(OH)_2$) and sodium carbonate ($Na_2 CO_3$) via the overall reaction

$$Ca(OH)_2 + Na_2CO_3 \rightarrow 2NaOH + CaO + CO_2$$

Chemical recovery is economically essential for the kraft process. Because most of the organics that are not converted to pulp are burned, the BOD of untreated kraft mill effluent is only 25–35% that of untreated sulfite mill effluent. Kraft mill effluent may contain small amounts of hydrogen sulfide (H_2S) and various mercaptans such as methyl mercaptan (CH_3SH), but these highly toxic compounds are unstable in the presence of O_2, and under oxidizing conditions kraft mill effluent is about 100 times less toxic than sulfite mill effluent.

A hybrid of the standard mechanical and chemical pulping methods is semichemical pulping, in which wood chips are first treated chemically in a digester followed by a mechanical defibrating stage in a disc refiner. About 60–80% of the material in the wood chips can be converted to pulp with this procedure. The two major semichemical pulping methods are the neutral sulfite and kraft semichemical techniques. In the former case the cooking medium is neutralized, usually with the addition of sodium carbonate. Semichemical pulping currently accounts for 4% of world wood pulp production (FAO, 1999).

Objectionable Characteristics of Pulp and Paper Mill Effluent

Until roughly the mid-1960s, little effort was made to treat pulp and paper mill wastewater in order to reduce its undesirable characteristics. Since that time, growing environmental concern has caused the industry to make a serious effort to lessen the adverse impacts of

[8]Species usually pulped by the sulfite process are spruce, balsam, western hemlock, birch, and poplar.

its wastewater discharges both through modifications of the production process and by the installation of wastewater treatment systems. In this section, we briefly review the undesirable characteristics of pulp and paper mill wastewater and the environmental impacts associated with its release.

Suspended Solids

Table 9.5 lists typical characteristics of untreated pulp and paper mill effluents circa 1980. The SS concentrations range from about half to 2.5 times the SS of raw sanitary sewage. These SS may exert directly toxic effects on certain organisms. Sprague and McLeese (1968b), for example, report that pulp and paper mill SS may plug the gills of fish, leading in extreme cases to suffocation. Lower SS concentrations can produce sublethal and chronic stresses by the same mechanism.

Historically, the principal adverse impact of pulp and paper mill SS on aquatic communities occurred when the SS settled to the bottom of the receiving system. Virtually all the pulp and paper SS consists of small chips of wood that are not converted to pulp. Unless currents scour the bottom, these small wood chips can accumulate to such a degree that the natural benthic community is completely destroyed. The impact of this destruction could extend beyond the benthos, since many water column organisms depend on the benthos for food and shelter, and fish often utilize the benthos as a spawning and nursery area. When the natural benthos is covered by pulp mill sludge, it ceases to play its usual role in the aquatic ecosystem. Until wastewater treatment systems were installed at pulp and paper mills, the discharge of SS with the mill wastewater often had devastating effects on the benthic community in the receiving waters. Waldichuk (1962), for example, described the environmental impact of a pulp and paper mill that had been discharging wastewater at the head of Cousins Inlet, British Columbia, since 1913. The effluent from the mill consisted of a mixture of wastewater from the production of sulfite and kraft pulp, newsprint, and specialty paper. The solids in the wastewater had settled to the bottom of the inlet, due in part to the flocculating effect of saltwater on the solids. Decomposition of this precipitated sludge had produced anoxic conditions both in the sediments and in the bottom waters of the inlet. No living animals were found in the area of sludge deposits, and the odor of H_2S was apparent when a sediment sample was brought to the surface. Periodic dredging of this accumulated sludge to keep channels clear stirred the toxic H_2S into the water column. At least one fish kill was attributed to such dredging effects (Hourston and Herlinveaux, 1957).

The classic study of the impact of the SS in pulp mill effluent on benthic communities is that of Pearson and Rosenberg (1976). They studied the impacts of two mills. The first was a Swedish sulfite pulp mill that had operated from 1890 to 1966. By the time of its closure, it was discharging about 55 million L of wastewater and 25 tonnes of SS per

TABLE 9.5   TYPICAL CHARACTERISTICS OF PULP AND PAPER
MILL EFFLUENTS CIRCA 1980

Pulping Method	Flow (m^3 tonne^{-1})	SS (ppm)	BOD (ppm)
Unbleached kraft	30	500	350
Bleached kraft	150	300	250
Newsprint mill (inc. sulfite)	50	400	1,000
Sulfite without recovery	200	100	950
Neutral sulfite semichemical	30	350	500

Source: McCubbin (1983).

day about 5 km upstream from the mouth of a river that flowed into the Saltkällefjord. The second mill was a Scottish mill that began operating in 1966. The effluent from this mill was discharged directly into the Annat Narrows through a diffuser pipe at a depth of 5–9 m. During the study period, 1967–1973, the mill was discharging 10–14 tonnes of suspended solids per day, with the exception of 1970–1971, when the amount rose to a peak of 30 tonnes d^{-1}.

The impacts of the two mills on the benthic communities in the Saltkällefjord and Annat Narrows are summarized in Figures 9.9 and 9.10. Figures 9.9A and 9.9B show plots of numbers of species versus numbers of individual benthic macrofauna in selected years in the two discharge areas. The figures dramatically illustrate the changes in species diversity associated with the mill discharges. Very few species were present when the systems were stressed. In both cases, the benthic communities were dominated by two polychaete worms, *Capitella capitata* and *Scolelepis fuliginosa*. The Saltkällefjord benthic community recovered dramatically after cessation of the pulp mill discharges, and by 1974 about 60 benthic macrofauna species were present. Over the same time period, the diversity of the Annat Narrows benthic community plummeted in response to the loading of SS from the Scottish mill. Figure 9.10 diagrammatically shows the faunal and sedimentary changes associated with the increased loading of SS from the pulp mills. At the extreme right the sediments are completely anoxic, and no macrofauna are present. Under somewhat less extreme conditions, the macrofauna are dominated by the polychaetes *Capitella* and *Scolelepis*. As organic loading is further reduced, other species appear, including bivalve mollusks, brittle stars, a sea urchin, and finally a Norway lobster (*Nephrops*).

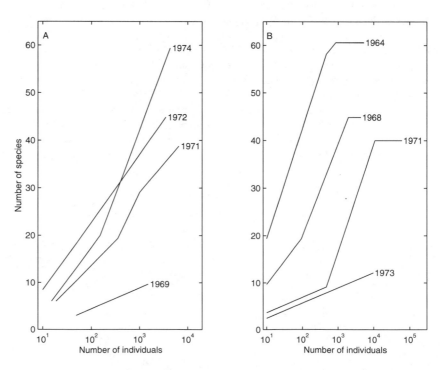

FIGURE 9.9   Depiction of species diversity based on plots of numbers of individual macrofauna versus number of species in pulp mill discharge area in (A) Saltkällefjord (Sweden) and (B) Annat Narrows (Scotland). *Source*: Pearson and Rosenberg (1976). Reproduced with permission.

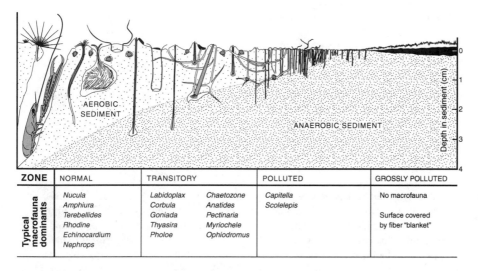

ZONE	NORMAL	TRANSITORY		POLLUTED	GROSSLY POLLUTED
Typical macrofauna dominants	*Nucula* *Amphiura* *Terebellides* *Rhodine* *Echinocardium* *Nephrops*	*Labidoplax* *Corbula* *Goniada* *Thyasira* *Pholoe*	*Chaetozone* *Anatides* *Pectinaria* *Myriochele* *Ophiodromus*	*Capitella* *Scolelepis*	No macrofauna Surface covered by fiber "blanket"

FIGURE 9.10   Diagrammatic representation of faunal and sedimentary changes under increasing organic loading from pulp mill effluent. From right to left are seen a fiber blanket, burrows of polychaete worms, bivalves, brittle stars, a sea urchin, and finally a Norway lobster. Redrawn from Pearson and Rosenberg (1976). Reproduced with permission.

Dissolved Organics

The dissolved organics in untreated pulp and paper mill effluent created several types of problems. First, due to the presence of wood sugars and other readily metabolized organic substrates, the dissolved organics contributed a substantial amount of BOD. In fact, dissolved organics account for about 90% of the BOD in the untreated effluent from a typical pulp and paper mill, and the latter can be as much as five times the BOD of raw sewage (Table 9.5). Hence there was considerable potential for anoxia to develop if such effluent were discharged into receiving waters that were not well mixed.

Second, if the effluent were discharged into freshwater, objectionable growths of slime-forming fungi of the genus *Sphaerotilus* could develop. These slime fungi were frequently reported downstream of sulfite pulp mills (Waldichuk, 1962). Overgrowth of fish eggs by this slime could reduce egg survival by as much as several orders of magnitude, apparently because the *Sphaerotilus* slime inhibited the exchange of gases between the eggs and the surrounding water.

Some of the dissolved organics in untreated pulp and paper mill effluent have surface-active properties that can lead to foaming in the receiving system, particularly if the water is turbulent. Although foaming is generally suppressed in saltwater, high concentrations of either kraft or sulfite effluent could produce considerable foaming even in seawater (Waldichuk, 1962). Foaming is of relatively little importance from a toxicological standpoint, but there is no doubt that water covered with foam is aesthetically unappealing. Surface-active agents may also retard the exchange of O_2 between the water and air, and in this way compound the O_2 depletion problem caused by the high BOD of the effluent. From the standpoint of wastewater treatment, one interesting benefit of the foaming caused by surface-active compounds is the restriction of heat exchange between the water and atmosphere. In Canada, for example, the foam blanket on lagoon-type secondary treatment systems helps prevent heat loss to the atmosphere during the winter months and is therefore important to the maintenance of an effective biological treatment process (McCubbin, 1983).

Both sulfite and kraft mill effluents are highly colored. Sulfite effluent has a characteristic amber color; kraft effluent is typically deep brown. These colors result primarily from the kinds and amounts of lignin-derived substances found in the wastewater. As with foaming, there is nothing particularly toxic about these colors or the compounds that produce them, but the appearance of such water is rather unappealing to most persons. The color does provide a sensitive tracer of the effluent.

Toxic Substances

Untreated pulp and paper mill wastewater contains a variety of toxic chemicals. These chemicals originate from three main sources: cooking chemicals and their by-products, dissolved organics derived from the nonfibrous material broken down in chemical pulping, and chlorinated organics produced in the bleaching process if chlorine is used to bleach the pulp. The most toxic cooking chemicals are the H_2S and mercaptans associated with kraft pulping, but fortunately these are unstable in the presence of O_2. Acids and bases used in the pulping and bleaching operations create wastewaters that are often highly acidic or alkaline. Kraft pulping effluent is basic; sulfite process wastewater is acidic. Unless neutralized, these wastewaters can be toxic to aquatic organisms. For purposes of biological treatment, pH values in the range 6.5–8.0 are desirable (McCubbin, 1983).

The various dissolved organic substances in pulp and paper mill wastewater are the source of toxicity that is most difficult to eliminate. Toxicology studies have often involved simply diluting raw effluent with either freshwater or seawater and testing the effect of the diluted effluent on various organisms. In most cases, no effort is made to determine the component or components of the effluent that are causing the toxic effects. For both fish and shellfish, such studies have indicated that the incipient lethal level of untreated sulfite effluent is about 1 part per thousand.[9] Jones et al. (1956), for example, found that the 30-day TLm for chinook and coho salmon exposed to sulfite effluent was about 1–2 parts per thousand, and Wilber (1969) reported that most oysters died within 2–29 days if exposed to sulfite effluent diluted with seawater to concentrations of 0.67–10 parts per thousand. Similar toxicity studies have indicated that untreated kraft mill effluent is 10–100 times less toxic than sulfite mill effluent. Sprague and McLeese (1968a), for example, found that for juvenile Atlantic salmon the incipient lethal level of neutralized bleached kraft mill effluent was about 12–15% (120–150 parts per thousand) and that larval lobster survival was little affected by concentrations less than 10%. Galstoff et al. (1947) observed no lethal effects on oysters exposed for up to 30 days to 0.1 part per thousand of untreated kraft mill effluent.

Sublethal effects have unfortunately been observed at concentrations several orders of magnitude below the concentrations associated with lethal toxicity. Chronic effects on oyster growth and reproduction, for example, have been detected at untreated sulfite effluent concentrations of 3–16 ppm. In other words, such effluent would have to be diluted by a factor of about 1 million to make it safe in terms of chronic stresses, at least for oysters. To dilute all the U.S. and Canadian pulp and paper mill effluent by even a factor of 1,000 would require more than all of the freshwater that flows into the ocean from U.S. and Canadian river systems. Obviously, dilution is not a practical answer to the toxicity problem.

One practical approach to detoxifying pulp and paper mill effluent was suggested by experiments conducted by Sprague and McLeese (1968b). They stored fresh bleached kraft

[9]In other words, 1 mL of effluent diluted to 1 L.

mill effluent (BKME) under quiescent conditions for periods of time ranging up to two weeks and then tested its toxicity on juvenile salmon and lobster larvae. During the storage period, the O_2 level in the effluent dropped to 1–2 ppm due to microbial consumption of the organics in the effluent. No effort was made to aerate the effluent. After two weeks of storage, the BKME had lost virtually all of its acute toxicity for juvenile salmon. After seven days of exposure to 100% two-week-old BKME, not a single fish died. However, quiescent storage produced no significant change in the toxicity of the BKME toward lobster larvae. These results suggested to Sprague and McLeese that the acutely toxic effects on lobster larvae and juvenile salmon were being caused by different substances in the BKME. After two weeks, microbial activity had apparently degraded the components toxic to the juvenile salmon, but substances toxic to the lobster larvae remained in the effluent.

To speed up microbial degradation of the BKME, Sprague and McLeese set up a biological oxidation system in which fresh BKME was seeded with a culture of microbes, primarily bacteria and protozoa, that had been conditioned to grow on BKME. The BKME was then stirred to effect aeration. During the first week of bio-oxidation the O_2 level in the BKME dropped to 1 ppm, an indication that aeration was not keeping pace with the microbial respiration rate. During the second week, however, the O_2 level recovered to 8 ppm. After two weeks of this treatment, the toxicity of the BKME to lobster larvae was tested. The median survival time of the lobster larvae in 100% BKME after two weeks of bio-oxidation was about 36 hours compared to a median survival time of only about 2 hours in fresh BKME. If the concentration of the two-week bio-oxidized BKME was diluted by about a factor of 2, the median survival time of the lobster larvae exceeded seven days.[10]

Wastewater Treatment

By the mid-1960s, it was becoming clear that simply discharging untreated pulp and paper mill wastewater was not environmentally acceptable and that some form of wastewater treatment would be necessary. The issue became a legal matter in 1972, when both the U.S. and Canadian governments passed legislation that controlled the discharge of pollutants into aquatic systems by industrial operations. The U.S. Clean Water Act, for example, requires industrial facilities to have permits for effluent discharges, and permit limits specify the types and amounts of pollutants allowed in the effluent over time. The Canadian government has passed legislation specific to pulp and paper liquid effluent discharges (EPA, 1972).

The strategy adopted in the treatment of pulp and paper mill effluent is similar to that used in the conventional treatment of sanitary sewage, but with some important modifications to deal with the toxicity problem. Normally, some form of pretreatment is used to remove grit and debris and to bring the pH to an acceptable value, usually in the range 6.5–8.0. The primary treatment process then consists of removal of SS and is accomplished in a settling basin or clarifier similar to those used at conventional sewage treatment plants. Wastewaters that contain only a small concentration of SS may bypass the primary clarifier and go directly to secondary treatment.[11] This strategy minimizes the amount of water that must be handled by the primary clarifier. Normally, 80–95% of the settleable portion of the suspended solids is removed in the primary clarifier. Flocculants are sometimes added to abet settling if a large fraction of the SS are colloidal or very fine material

---

[10]The survival experiments were terminated after seven days, and at that time fewer than 50% of the lobster larvae had died.

[11]Bleach plant effluents from new kraft mills, for example, contain very few SS.

that does not settle effectively by gravity. Roughly 10% of the BOD is removed along with the primary sludge. After dewatering, the sludge may be burned as a source of fuel. In some cases, it is deposited in a landfill.

The secondary treatment step is intended to reduce both the BOD and the toxicity of the effluent. Where land area is limited, some mills have opted for activated sludge-type secondary treatment because BOD reductions of 70–95% can be achieved in a relatively small space. The problem with this type of treatment is that the residence time of water in the system is not long enough to allow microorganisms to detoxify the wastewater to an acceptable degree. As a result, many pulp and paper mills use bio-oxidation lagoons in which the residence time of the water is typically 4–10 days (McCubbin, 1983). The lagoons are aerated to ensure that there is an adequate concentration of O_2 to support aerobic catabolism of the organics in the wastewater. In addition to being more efficient in reducing the toxicity of the effluent, bio-oxidation lagoons produce very little sludge relative to activated sludge systems. Much of the particulate material settles to the bottom of the lagoon and is consumed by auto-oxidation (McCubbin, 1983). The lagoons are also better able to absorb shock loads of concentrated effluent without appreciable change in treatment efficiency and are less costly to operate than activated sludge systems.

Tertiary treatment is possible to further reduce BOD and SS concentrations, to lower the toxicity of the effluent, or to mitigate color, odor, and taste problems. Activated carbon absorption, for example, can be used to eliminate most of the dissolved organics.

A thought-provoking point made by McCubbin (1983) is that the design of the wastewater outfall can be comparable in importance to the treatment process in mitigating the environmental impact of the effluent. It is important to try to dilute the wastewater as quickly as possible and to take steps to reduce or eliminate visible foam and color in the receiving system. A mill located near the mouth of a river, for example, might do well to build an outfall as much as several kilometers long to the ocean rather than discharging directly into the river.

A CASE STUDY: THE BUCKEYE CELLULOSE CORPORATION PULP AND PAPER MILL AT PERRY, FLORIDA

In 1954 the Buckeye Cellulose Corporation[12] began operating a kraft process pulp and paper mill at Perry, Florida, a small town in the Florida panhandle. As of approximately 1990, the mill was producing about 1,250 tonnes d^{-1} of pulp[13] and discharging wastewater at a rate of about 210 million L d^{-1} into the Fenholloway River at a point about 20 km upstream from the mouth of the river (Figure 9.11). To legally accommodate this wastewater, the State of Florida declared the Fenholloway an "industrial river."

Prior to the operation of the pulp mill, the Fenholloway River had supported an abundant fish population, as evidenced by the presence of a fish camp near the present outfall. A survey of the Fenholloway River by Beck (1954) just prior to the opening of the pulp mill revealed that the river supported a healthy fauna of invertebrates and fish.

Unfortunately, the discharge of wastewater from the pulp mill displaced virtually all the natural fauna from the stream. Although this displacement in no sense violated the designation of the stream as an industrial river, the company's right to pollute was strictly confined to the Fenholloway River and did not extend into the Gulf of Mexico. As a result

---

[12]The Buckeye Cellulose Corporation was a subsidiary of Procter and Gamble. The company's name was later changed to Procter and Gamble Cellulose.
[13]Current production is about 965 tonnes d^{-1}.

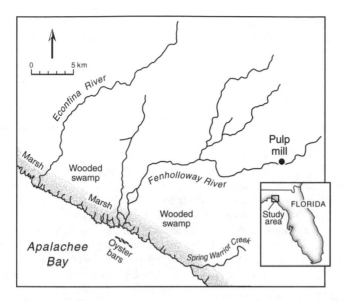

FIGURE 9.11   Apalachee Bay drainage area showing the Econfina and Fenholloway rivers and the Buckeye Cellulose pulp mill. *Source:* Redrawn from Livingston (1975). Reproduced by permission.

of preliminary studies conducted by several investigators in Apalachee Bay (**Figure 9.11**), it became apparent that the effects of the pulp mill pollution extended out into the bay and along the coast for as much as several kilometers. As a result of this discovery, the company initiated efforts to minimize the impact of its wastewater on the river and bay. In conjunction with this effort, an extensive two-year investigation (1972–1974) of the bay and river was conducted by Dr. Robert Livingston and his students (Livingston, 1975; Zimmerman and Livingston, 1976). Follow-up studies have continued to the present time (Livingston, 1984a, 1984b; Livingston et al., 1998). As a result, a rather clear picture of the impact of the pulp mill effluent on the Fenholloway River and Apalachee Bay is available.

In the 1972–1974 study conducted by Livingston, the Econfina River and its estuary were used as a control system to be compared with the Fenholloway River and its estuary. The two streams have similar lengths and discharge rates and drain the same swamp area. Oyster bars are found off the mouths of both rivers.

As a result of the mill discharges, water quality in the Econfina and Fenholloway rivers was found to be dramatically different. Below the mill outfall, O_2 concentrations at all depths in the Fenholloway River were virtually zero until the river flowed into Apalachee Bay. By comparison, O_2 concentrations in the Econfina River were about 6–8 ppm throughout the water column. The bottom of the Fenholloway River was covered with wood fibers and chips, and the odor of H_2S was evident in the air above the stream. Water clarity, as measured by Secchi depth in the Fenholloway River, was significantly lower than in the Econfina River due to the greater turbidity and color of the Fenholloway's water.[14] Lower Secchi depths associated with greater turbidity and water color were also apparent in the Fenholloway estuary compared to the Econfina estuary.

Given this state of affairs, it is not surprising to learn that significant differences were found in the biological communities inhabiting the two systems. The anoxic conditions in the Fenholloway River precluded the existence of most natural inhabitants of the marsh

[14]Recall that pulp and paper mill effluent is strongly colored.

ecosystem. The most obvious animals in the Fenholloway River were a few turtles and alligators. Sampling for fish in the Fenholloway River, not surprisingly, revealed no permanent fish assemblages, and a total of only three species were taken from the river during the study period. Most of these fish were topminnows, which could have moved into the river from adjacent marsh areas. By comparison, control areas yielded a total of 30 species of fish.

A study of fish populations was also conducted in marsh areas adjacent to the two rivers. A total of 48 species of fish were taken from the Econfina marsh area versus only 6 species from Fenholloway marshes. In addition, over 80 times as many fish were taken from Econfina marsh areas as were taken from Fenholloway marsh areas. Although the Fenholloway marshes supported a seemingly healthy stand of the usual marsh vegetation, none of the usual marsh invertebrates (e.g., fiddler crabs) was found.

In the Fenholloway estuary, several extensive fish kills were noticed during the study period. No such kills were observed anywhere in the Econfina system. The Econfina nearshore benthos was covered by extensive beds of benthic macrophytes, but these plants were largely absent from the Fenholloway nearshore area. Instead, the Fenholloway nearshore benthos was characterized by mud flats and only scattered clusters of benthic plants. No living oysters were found on the oyster bars off the mouth of the Fenholloway.

Moving away from the mouth of the Fenholloway, the effects of the pulp mill were apparent for the greatest distance in a southeasterly direction. The effects were evident in comparisons of both invertebrate and fish populations sampled at comparably positioned stations at various distances from the mouths of the Fenholloway and Econfina rivers. At the Fenholloway stations, significantly fewer numbers of invertebrates were taken in trawls (Hooks, 1973), and fish populations were significantly reduced both in terms of numbers of species and numbers of individuals taken each month. These effects were most acute in the area within 2–3 km from the mouth of the Fenholloway River.

In summary, the pulp and paper mill effluent had virtually eliminated all natural fauna from the Fenholloway River, had greatly reduced the biomass of plants and animals in the Fenholloway estuary, had eliminated many of the natural fauna from marsh areas adjacent to the Fenholloway, and had significantly reduced the numbers and kinds of organisms found in Apalachee Bay within a distance of roughly 2–3 km from the mouth of the Fenholloway. Within Apalachee Bay, the areal extent of the effluent's impact was determined by both wind-driven and tidal currents, and extended both west and southeast from the mouth of the Fenholloway.

In an effort to reduce the deleterious effects of its effluent on the Fenholloway River and Apalachee Bay, Buckeye Cellulose installed a primary clarifier to remove SS and a bio-oxidation pond with a capacity of about 1 billion L in January 1974. Given the rate of effluent discharge from the mill, the residence time of the wastewater in the oxidation pond was expected to be about five days.

The effects of the wastewater treatment system during the first few years of its operation have been reported in a study by Livingston (1977). In the Fenholloway River itself, observations made within a month or two after initiation of water treatment showed that the accumulated wood fibers and chips had largely disappeared from the river bed and that natural species of fish and other animals (including porpoises) were beginning to reappear in the river (P. LaRock, pers. comm.). Livingston's (1977) study, conducted from 1974 through 1976, revealed that both water quality and turbidity had improved markedly in the Fenholloway River. Some recolonization by sea grasses was noted in the Fenholloway estuary, and in offshore areas there was an increase in the number of species of benthic macrophytes, although the biomass of benthic macrophytes remained lower than in the Econfina control area. Fishes in Fenholloway marsh areas increased significantly both in numbers of species and in biomass, at times exceeding the number of species and individuals found in control areas. Fishes in nearshore Fenholloway coastal areas showed little change in biomass, but there was an increase in species numbers.

Toxicity studies conducted with pinfish showed that the fish suffered no mortality when placed in water containing up to 50% oxidation pond effluent, but the TLm for untreated mill effluent was about 10%. Growth and food conversion efficiency for these fish were adversely affected in water containing only 0.01–0.1% untreated effluent, but about 10 times as much oxidation pond effluent was required to produce the same effect.

Based on these reports, it appears that the Fenholloway River system made a significant but incomplete recovery following installation of Buckeye Cellulose's initial wastewater treatment system. The treatment facility was upgraded in 1980 with the installation of an additional oxidation pond with a capacity of about 310 million L between the primary clarifier and original oxidation pond. The overall system now removes 90% of the BOD and SS from the mill's wastewater. Effluent BOD and SS concentrations are 19 and 27 ppm, respectively. The O_2 concentration in the treated wastewater averages 7.3 ppm, but the BOD in the effluent is sufficient to produce a considerable oxygen sag in the Fenholloway River. The minimum O_2 concentration downstream from the mill averages 0.7 ppm (C. Henry, pers. comm.). In follow-up studies, Livingston et al. (1998) reported that Secchi depths were about 2.5 times greater within the estuary and at the mouth of the Econfina River compared to the Fenholloway River. Furthermore, sediments in nearshore areas affected by the discharge from the Fenholloway River were found to contain a higher percentage of silt and clay than sediments near the mouth of the Econfina River. The differences in light transmission and sediment composition were found to account for much of the variability in the distribution of benthic macrophytes. Mesocosm experiments showed that mill effluent in direct contact with several species of sea grass adversely affected growth at concentrations of 1–2%.

In one sense, the Buckeye Cellulose case study represents a worst-case scenario because the natural flow of the Fenholloway River is small compared to the discharge from the mill. Even in the rainy season about 60% of the water in the river consists of mill effluent, and the river is dry upstream of the mill about 35% of the time. Thus, in this case, dilution of the mill's wastewater by river water is of minor significance in reducing the adverse effects of the effluent. The partial recovery of the Fenholloway River after January 1974 is thus an encouraging sign that the harmful effects of pulp and paper mill effluents can be greatly reduced by a rather straightforward treatment process. However, the fact that adverse effects are still discernible in the Fenholloway River and nearshore areas of Apalachee Bay underscores the fact that reducing the impact of such a large amount of wastewater to a negligible level is a difficult task.

COMMENTARY

Great progress has been made in reducing the adverse effects of pulp and paper mill effluents on aquatic systems since roughly 1970. The industry, whether by choice or by force, has taken the matter of wastewater treatment seriously. In the United States, for example, the industry spent $6.95 billion to buy and operate water pollution control equipment between 1979 and 1998, an average of almost $350 million per year (GAO, 1987; AFPA, 1999a). McCubbin (1991) has estimated that the BOD of pulp mill effluent in the United States declined by a factor of 5–10 between 1980 and 1990. In Canada, the pulp and paper industry has invested more than $5 billion in pollution prevention technology since 1990. Between 1975 and 1997, water use at Canadian mills declined from an average of 156 to 70 m^3 per tonne of pulp, effluent BOD concentrations declined from 370 to 32 ppm, and effluent SS concentrations dropped from 177 to 53 ppm (CPPA, 1997).

Despite the success that has been achieved with secondary treatment, some troublesome issues remain. The use of chlorine to bleach pulp, for example, produces a wide variety of chlorinated organics, a number of which are known to contribute to the toxicity of pulp and paper mill effluent (Wallin and Condren, 1981). Dioxins and furans are particularly toxic examples. Organizations such as Greenpeace have advocated the elimination of chlorine-bleached paper products by switching to oxygen-based bleaching methods or simply using unbleached paper products. Concern over this issue produced economic incentives that caused the industry to greatly reduce its reliance on chlorine bleaching methods. Some pulp and paper companies in Canada are estimated to have lost millions of dollars of business to competitors who used chlorine-free bleaching methods (Bohn, 1991). Since 1975 the discharge of chlorinated organics by Canadian pulp and paper mills has declined by 75%, and the release of dioxins and furans has dropped by a factor of 100 (CPPA, 1997). Much of this reduction has been accomplished through conversion to oxygen-based bleaching methods (ozone or peroxide) and a 92% reduction in the use of elemental chlorine in bleaching (CPPA, 1997). In the United States, the release of dioxin from pulp and paper mills declined by 70–85% from 1988 to 1991 (Focus, 1993b).

A major legal development on the dioxin front occurred in 1997, when the U.S. Congress passed the Cluster Rule. The Cluster Rule specifically concerns the pulp and paper industry and addresses issues related to both air and water pollution. The American Forest and Pulp Association has estimated that compliance with the Cluster Rule will cost the U.S. industry $2.6 billion in capital expenses and $273 million per year in operating costs (AFPA, 1999a). The water-related portion of the Cluster Rule targeted dioxin in part because of the results of a lengthy study conducted by the EPA during the middle- to late 1990s. The study revealed that not only is dioxin a carcinogen, but it also appears to be a reproductive toxin and endocrine disrupter, even at very low doses (AFPA, 1999a). Legal motivation came from a court-imposed consent agreement that required the EPA to enact rules controlling dioxins and furans in the effluent from bleached pulp mills. The EPA then extended its mission to cover all pulp, paper, and paperboard producers and revised the rules to include conventional pollutants as well (AFPA, 1999a). Interestingly, the final Cluster Rule endorsed the industry-backed proposal to cut dioxin pollution using bleaching technologies that relied on chlorine dioxide rather than pure Cl_2. The original Cluster Rule proposal would have required the use of oxygen-based bleaching methods. The industry has contended that dioxin concentrations in mill effluent are below the limit of detection (AFPA, 1999a).

Another major issue in the pulp and paper industry is recycling. In Canada, the Pulp and Paper Research Institute has recently embarked on an $88 million research program with government and industry suppliers to develop technologies needed to launch closed-cycle production of paper and paperboard products (CPPA, 1997). According to Pollock (1987), paper recycling programs in 9 of the world's 11 largest paper-consuming nations saved more than 4,000 km^2 of trees in 1984; in addition, the use of recycled paper rather than raw wood considerably reduces the pollution and demands for energy and water associated with the production of paper products (Table 9.2). As indicated in Figure 9.7B, there has been a twofold increase in the percentage of paper and paperboard products accounted for by recycling since 1969. There is, however, a practical limit on the extent to which paper products can be recycled. The process of paper recycling involves blending the paper with water and stirring the mixture until the individual paper fibers are broken down. This process inevitably causes a reduction in the average length of the fibers, and that reduction translates into a decrease in strength. Ultimately, this loss of strength limits the number of times paper products can be recycled. Recycled fibers are well suited for products such as tissue paper, which does not have to be particularly strong. As noted by Klungness (1999 personal communication), "You want bulky, absorbent sheets, with lots of capillary action—that's exactly what you have with recycled fibers." At the present time, however, most paper recycling involves newsprint (used to make newspapers) and fiberboard (used to make corrugated containers). In the United States almost 5 million tonnes of newsprint is recycled each year, primarily to make more newsprint. And each year, more than 18 million tonnes of corrugated material (56% of the wastepaper used in U.S. mills) is recycled, primarily into more corrugated containers (Devitt, 1999).

Recognition of the environmental benefits associated with recycling paper products and of the problems possibly caused by chlorine bleaching has translated into some noteworthy consumer actions. In Ontario, for example, the Domtar Specialty Fine Papers company is now producing Cl-free paper made from 100% recycled paper. The white paper, called Comet or Infinity 100, is acid-free, which translates into a longer life when stored in archives (Focus, 1993a).[15] McDonald's fast food chain now uses carryout bags made of 100% recycled brown paper and oxygen-bleached coffee filters in its restaurants. The wrap for the Big Mac is also made with unbleached paper (EDF, 1991). Such changes reflect an increasing public awareness of environmental issues. This awareness will very likely lead to even greater reliance on recycled paper and less use of chlorine bleaching in the future.

REFERENCES

AMERICAN FOREST & PAPER ASSOCIATION. 1999a. *Pulp and Paper 1999 North American Factbook.* Miller Freeman, Inc. San Francisco.

AMERICAN FOREST & PAPER ASSOCIATION. 1999b. U.S. paper recycling: an environmental success. http://205.197.9.134/recycling/afandpa_recycletoday.html. 8/19/99.

BECK, W. M., JR. 1954. *A Stream Quality Survey of the Fenholloway River.* Report, Bureau of Sanitary Engineering, Florida State Board of Health, Jacksonville, FL. 3 pp (unpublished).

BOHN, G. 1991. Minister says polluters risk losing sales. *Vancouver Sun.* June 4, p. C1.

ENVIRONMENTAL DEFENSE FUND. 1991. A deeper shade of green: Cutting McDonald's waste. *EDF Letter,* **23**(3), 7.

CANADIAN PULP AND PAPER ASSOCIATION. 1997. *Canadian Pulp and Paper Association Annual Review.* http://open.doors.cppa.ca/english/index.htm.

CHRISTIAN SCIENCE MONITOR. 1997. Americans recycle more paper despite low demand. October 20. http://csmonitor.com/durable/1997/10/20/feat/feat.2.html.

DBEDT, 1997. *Hawaii Data Book.* Hawaii Department of Business, Economic Development and Tourism, Honolulu, HI.

DEVITT, T. 1999. Sensible recycling, or stupendous waste. http://whyfiles.news.wisc.edu/063recycle/paper.html.

EPA. 1971. *Hawaii Sugar Industry Waste Study.* Environmental Protection Agency, San Francisco. 106 pp.

EPA. 1972. *Pulp and Paper Effluent Regulations.* Environmental Protection Agency, EPA 1-WP-72-1, Water Pollution Control Directorate, November, 1971.

EPA. 1989. *Report on the Evaluation of Wastewater Discharges from Raw Cane Sugar Mills on the Hilo-Hamakua Coast of the Island of Hawaii.* Environmental Protection Agency, Washington, DC.

FAO. 1999. *Pulp and Paper Capacities.* Food and Agriculture Organization of the United Nations. Report of the FAO Survey of World Pulp and Paper Capacities 1998–2003. Rome.

FOCUS. 1993a. Briefs. *Focus,* **18**(2), July–August, p. 14.

FOCUS. 1993b. Briefs. *Focus,* **18**(3), November–December, p. 7.

GALSTOFF, P. S., W. A. CHIPMAN, J. B. ENGLE, and H. N. CALDERWOOD. 1947. Ecological and physiological studies of the effect of sulphate pulp mill wastes on oysters in the York River, Virginia. *Fish. Bull.,* **51**, 59–186.

GAO. 1987. *Water Pollution: Application of National Cleanup Standards to the Pulp and Paper Industry.* National Technical Information Service PB87-193231. U.S. Department of Commerce. Washington, DC. 36 pp.

GRIGG, R. W. 1985. *Hamakua Coast Sugar Mill Ocean Discharges Before and After EPA Compliance.* University of Hawaii Sea Grant Technical Report UNIHI-SEAGRANT-TR-85-02. U.H. Sea Grant College Program. Honolulu, HI. 25 pp.

GUNDERSEN, K. R. 1973. Microbiology. In *Estuarine Pollution in the State of Hawaii,* Vol. 2. Water Resources Research Center Technical Report No. 31. University of Hawaii, Honolulu. Pp. 235–341.

[15]Interestingly, in April 1996, Environment Canada charged Domtar with 30 violations related to the release of SS and failure to meet release reporting requirements at its Red Rock, Ontario, mill.

HOOKS, T. A. 1973. *An Analysis and Comparison of the Benthic Invertebrate Communities in the Fenholloway and Econfina Estuaries of Apalachee Bay, Florida.* M.S. thesis, Florida State University, Tallahassee.

HOURSTON, A. S., and R. H. HERLINVEAUX. 1957. A "mass mortality" of fish in Alberni Harbour, British Columbia. *Fish. Res. Bd. Can., Pac. Prog. Rep. 109.*

HSPA. 1990. *Hawaiian Sugar Manual 1990.* Hawaiian Sugar Planters Association, Aiea, HI. 27 pp.

IRANPOUR, R., M. STENSTROM, G. TCHOBANOGLOUS, D. MILLER, J. WRIGHT, and M. VOSSOUGHI. 1999. Environmental engineering: Energy value of replacing waste disposal with resource recovery. *Science*, **285**, 706–710.

JONES, B. F., C. E. WARREN, C. E. BOND, and P. DOUDOROFF. 1956. Avoidance reactions of salmonid fishes to pulp mill effluents. *Sewage Indust. Wastes*, **28**, 1403–1413.

KAISER, J. 1999. In this Danish industrial park, nothing goes to waste. *Science*, **285**, 686.

KLUNGNESS, J. 1999, verbal interview.

LAU, L. S., P. C. EKERN, P. C. S. LOH, R. H. F. YOUNG, N. C. B. BURBANK, and G. L. DUGAN. 1975. *Recycling of Sewage Effluent by Irrigation: A Field Study on Oahu.* Final progress report for August 1971 to June 1975. University of Hawaii Water Resources Research Center Technical Report No. 94. Honolulu, HI. 151 pp.

LIVINGSTON, R. J. 1975. Impact of kraft pulp-mill effluents on estuarine and coastal fishes in Apalachee Bay, Florida, USA. *Mar. Biol.*, **32**, 19–48.

LIVINGSTON, R. J. 1977. *Recovery of the Fenholloway Drainage System: Analysis of Water Quality, Benthic Macrophytes, and Fishes in Apalachee Bay (1971–76) following Implementation of a Waste Control Program.* Final report to the Buckeye Cellulose Corp., Perry, FL. 18 pp. + figures.

LIVINGSTON, R. J. 1984a. Trophic response of fishes to habitat variability in coastal seagrass systems. *Ecology*, **65**, 1258–1275.

LIVINGSTON, R. J. 1984b. The relationship of physical factors and biological response in coastal seagrass meadows. Proceedings of a Seagrass Symposium, Estuarine Research Foundation. *Estuaries*, **7**, 377–390.

LIVINGSTON, R. J., S. E. MCGLYNN, and X. NIU. 1998. Factors controlling seagrass growth in a gulf coastal system: Water and sediment quality and light. *Aquatic Botany*, **60**, 135–159.

MCCUBBIN, N. 1983. *The Basic Technology of the Pulp and Paper Industry and Its Environmental Protection Practices.* EPA 6-EP-83-1. Environmental Protection Service. Environment Canada, Ottawa. 204 pp.

MCCUBBIN, N. 1991. Paper versus polystyrene: Environmental impact. *Science*, **252**, 1361.

OFFICER, C. B., and J. H. RYTHER. 1977. Secondary sewage treatment versus ocean outfalls: An assessment. *Science*, **197**, 1056–1060.

PEARSON, T. H., and R. ROSENBERG. 1976. A comparative study of the effects on the marine environment of wastes from cellulose industries in Scotland and Sweden. *Ambio*, **5**(1), 77–79.

POLLOCK, C. 1987. *Mining Urban Wastes: The Potential for Recycling.* Worldwatch Paper 76. Worldwatch Institute, Washington, DC. 58 pp.

POSTEL, S. 1987. *Defusing the Toxics Threat: Controlling Pesticides and Industrial Waste.* Worldwatch Paper 79. Worldwatch Institute, Washington, DC. 69 pp.

ROBERTS, J. C. 1996. *The Chemistry of Paper.* The Royal Society of Chemistry, Letchworth, United Kingdom.

SPRAGUE, J. B., and D. W. MCLEESE. 1968a. Toxicity of kraft pulp mill effluent for larval and adult lobster and juvenile salmon. *Water Res.*, **2**, 753–760.

SPRAGUE, J. B., and D. W. MCLEESE. 1968b. Different toxic mechanisms in kraft pulp mill effluent for two aquatic animals. *Water Res.*, **2**, 761–765.

STREETER, H. W., and E. B. PHELPS. 1925. *A Study of the Pollution and Natural Purification of the Ohio River. III. Factors Concerned in the Phenomena of Oxidation and Re-aeration.* U.S. Public Health Service Bulletin 146. Cincinnati, OH.

TANJI, E. 1999. Company hopes to grow sugar without cane fires. *Honolulu Advertiser*, August 31, p. A1.

USDA. 1999. *Hawaii Sugarcane Acreage and Production.* Hawaii Agricultural Statistics Service. U.S. Department of Agriculture, Honolulu.

WALDICHUK, M. 1962. Some water pollution problems connected with the disposal of pulp mill wastes. *Can. Fish Culturist*, **31**, 3–34.

WALLIN, B. K., and A. J. CONDREN. 1981. *Fate of Toxic and Nonconventional Pollutants in Wastewater Treatment Systems within the Pulp, Paper, and Paperboard industry.* EPA-600/S2-81-158. Environmental Protection Agency, Cincinnati, OH.

WILBER, C. G. 1969. *The Biological Aspects of Water Pollution*. Charles C. Thomas, Springfield, IL. 296 pp.

WEIBEL, S. R. 1969. Urban drainage as a factor in eutrophication. In *Eutrophication, Causes, Consequences, Correctives*. National Academy of Sciences, Washington, DC. Pp 383–403.

ZIMMERMAN, M. S., and R. J. LIVINGSTON. 1976. Effects of kraft-mill effluents on benthic macrophyte assemblages in a shallow-bay system (Apalachee Bay, North Florida, USA). *Mar. Biol.*, **34**, 297–312.

QUESTIONS

9.1 How would you account for the fact that O_2 depletion was never a problem in the vicinity of Hawaiian sugar mill discharges, even in cases where the wastewater contained BOD at concentrations three to four times higher than that of raw sewage?

9.2 Why are sugar yields from Hawaiian sugar cane fields irrigated with mill wastewater somewhat less than from fields irrigated with groundwater?

9.3 Recycling is a strategy used more and more often by industries to reduce or eliminate the amount of waste they discharge to the environment. Cite one example each from the Hawaiian sugar cane industry and from the pulp and paper industry that illustrate the use of recycling to reduce or eliminate waste discharges.

9.4 Modification of the production process has often been used by industries to reduce or eliminate the production of waste. Cite one example each from the Hawaiian sugar cane industry and from the pulp and paper industry that illustrate how modifications of the production process (exclusive of recycling) have been used to reduce or eliminate waste production.

9.5 During the 20th century, both the sulfite and kraft processes were used to produce wood pulp chemically. The sulfite method was more commonly used during the first half of the century, and the kraft process was more commonly used during the second half of the century. What is the explanation for the transition from the sulfite to the kraft process? In particular, what considerations led pulp mills to prefer the sulfite process during the first half of the century and the kraft process during the last 50 years?

9.6 Pulp mills routinely treat their wastewater in large lagoons in which the residence time of the wastewater is one to two weeks. Why do pulp mills opt for this mode of treatment as opposed to something akin to activated sludge or trickling filters, in which the residence time of the wastewater is less than one day and the amount of land devoted to treatment is far less?

9.7 What is the principal reason that pulp mills have recently begun to switch from chlorine to oxygen as a means of bleaching their pulp?

9.8 Some pulp mills operate entirely with recycled paper products rather than raw wood as feedstock. This approach dramatically reduces water pollution and energy use. What limits the extent to which pulp and paper products can be recycled in this way? In other words, why do we not just recycle pulp and paper products *ad infinitum* and give the trees a break?

Chapter Ten

✧   ¤   ✧   ¤   ✧

Pesticides

1. Introduction
   a. Why Do We Use Pesticides?
   b. Problems with Pesticides
      (1) resistance and impacts on nontarget species
2. Classification of Pesticides
   a. Herbicides, Insecticides, Fungicides, etc.
   b. Chlorinated Organics, Organophosphates, Carbamates, Pyrethroids
   c. Modes of Action
3. Pesticide Use
   a. Public Health
      (1) use of DDT to control malaria
   b. Agriculture
      (1) green revolution
   c. Forestry
4. Effects on Nontarget Species
   a. Salmon in New Brunswick
   b. Western Grebes on Clear Lake, California
5. Exaggerated and/or Erroneous Charges
   a. Destruction of Sea Trout—Laguna Madre, Texas
   b. DDT Impacts on Phytoplankton Photosynthesis
   c. DDT Effects on Trout Reproduction
   d. DDT Causes Cancer
6. Pesticide Persistence/Food Chain Magnification
   a. Southern California Brown Pelicans
   b. Hamelink Study
7. Effects on Birds
   a. Field Studies
      (1) eggshell thinning
   b. Laboratory Studies
      (1) effects on estrogen levels, carbonic anhydrase, Ca-ATPase
      (2) DDE vs DDT
8. Pest Resistance
   a. Mechanisms
      (1) detoxification and target site modification
   b. Costs
9. Alternatives to Use of Conventional Chemical Pesticides
   a. Biological Control
      (1) natural predators and/or parasites
      (2) pathogens and natural toxins
   b. Genetic Control

I am deeply disturbed and sadly disillusioned by the actions of numerous educators. For years they have made almost a fetish of the "scientific method" of investigation, and have stressed that method of analysis in their classes and in their writings. I had presumed that they also followed their own advice, but it is now obvious that often they do not.

Despite dozens of scientific studies proving that DDT is broken down rather quickly by the environment, many so-called "ecologists" are heard telling the public that exactly the opposite is true.

Why do they do this, when it is certain that they will be later exposed as being either ignorant of the facts or being deliberately untruthful? Perhaps the reason is simply that they are either ignorant of the facts or untruthful. . . .

If the emphasis in our classrooms and in the public media can be shifted toward more concern for the truth, no matter whose sacred cows get gored, then we'll be on the way back to the proper role of scientists in the American scene.

To begin with we must discount the erroneous beliefs that have been perpetuated by certain scientists and conservationists concerning DDT and attempts to inform the public as to the truth about those topics. DDT is *not* terribly persistent (under environmental conditions). People in the U.S. only ingest about 0.0005 parts per million of DDT [0.5 μg DDT per kilogram of body weight] daily (and excrete all excess DDT from their body). A diet containing thousands of times the amount we ingest daily [still] does *not* produce cancer, sterility, mutations, or other undesirable effects, even in experimental mice, birds, and fish.

So-called "substitutes for DDT" are [usually] more toxic than DDT to most organisms, and are much more costly. Substitutes for DDT are exterminating honeybees, and we will not have them around to pollinate the crops and orchards and "set" the crops. DDT is not widespread in the environment . . . not even in the United States or in the states that have used it most heavily.

DDT has *increased* the number of birds in the U.S., rather than causing declines, and the recent British government report stated that DDT has not been responsible for any decline of bird populations in that country, either. DDT probably causes wild birds and animals to be healthier if they are exposed to moderate doses of it, because it reduces tumors, inhibits cancer, . . . eliminates insect-transmitted diseases, and induces the production of enzymes that destroy harmful substances in the body. (Excerpt of a letter written by Dr. J. Gordon Edwards, Professor of Entomology, San Jose State College and published in the Feb. 12, 1974, edition of the *Oregon State University Barometer*. Italics and words in brackets were added later at Dr. Edwards' request.)

SINCE 1945 chemical pesticides in one form or another have become a significant form of pest control throughout much of the world. Literally thousands of different commercial pesticides are available on the market. At the present time, however, there is a growing consensus that conventional chemical pesticides are by no means the panacea they were once considered to be, and that in many cases their indiscriminate use has created far more problems than it has solved. This realization has come gradually and not without controversy, as the quotation from Dr. Edwards suggests.

Undoubtedly the most notorious of the conventional chemical pesticides has been DDT, and this chapter is devoted in part to a case study of its use and misuse. DDT was chosen as an example because the impact of its use on the environment has been thoroughly studied, because the problems that have beset its use are in many ways typical of the adverse effects associated with pesticide use in general, and because the emotional nature of the debate over DDT provides a revealing picture of the way emotions and prejudices may cloud rational discussions of environmental issues.

DDT (*dichlorodiphenyltrichloroethane*) was first synthesized in 1877 but did not come into use as a pesticide until 1942. It was used on a broad scale during World War II by the U.S. Army, both to stop typhus fever epidemics that had broken out in Italy and to eradicate malaria in the Mediterranean basin (Jukes, 1974). DDT was later used extensively in many tropical countries by the World Health Organization (WHO) to control diseases such as malaria, plague, typhus fever, yellow fever, sleeping sickness, and river blindness (Jukes, 1974). DDT has been used to control agricultural pests on many crops throughout the world,[1] and in forestry management to control pests such as the spruce budworm and the Dutch elm disease vector (a beetle). Today, however, use of DDT is banned in developed countries such as the United States, and its use in other countries is limited largely to malaria control. Similar fates have befallen the use of a number of other pesticides. What is it about some pesticides that causes them to rapidly lose popularity or to be banned following a period of widespread use?

First, pesticides often affect nontarget species, i.e., organisms for which they are not intended. Determining the magnitude and weighing the importance of these side effects is a critical problem in judging the merits of pesticide use. A country with a serious public health problem, for example, may well feel that pesticide use to reduce the incidence of disease(s) is an acceptable policy even if certain species of wildlife are greatly reduced in number or even wiped out because of their susceptibility to pesticide poisoning.

Second, target pests, particularly many species of insects, have shown a remarkable ability to develop resistance to pesticides. Since many pesticides are synthetic compounds that do not occur naturally in the environment, there has been no reason, until recently, for organisms with a natural resistance to pesticides to have any better chance of survival than otherwise similar organisms lacking such resistance. Obviously, this picture has changed in recent years. The success of many pesticides, particularly many of those introduced since 1945, rests on the fact that pest populations have not been, at least initially, resistant. In other words, pest populations have not tended to be dominated by individuals that by chance were immune or resistant to the pesticide's effects. As pesticide use continues, pest populations tend to be dominated more and more by naturally resistant or immune individuals, since susceptible individuals are killed. Thus, the pesticide user may be forced to switch constantly from one pesticide to another in order to keep one step ahead of the pest's genetic adaptability, or may be required to use larger and larger applications of the same pesticide in order to bring about the same reduction in pest numbers. These two problems, the killing of nontarget species and the development of resistance in target species, underlie many of the difficulties that have been encountered in pesticide use.

---

[1]For example, cotton, tobacco, soybeans, and potatoes.

CLASSIFICATION OF PESTICIDES

Pesticides may be classified in a variety of ways based on physical state, target species, purpose of application, or chemical nature. When classified according to target species, the most common pesticides may be broadly defined as being herbicides, insecticides, or fungicides, depending on whether they are designed to kill plants, insects, or fungi, respectively. These three categories of pesticides account for 57%, 14%, and 8% by weight, respectively, of conventional pesticide use in the United States (EPA, 1999).

When classified according to chemical nature, most synthetic chemical pesticides used since 1945 fall into one of the following categories: chlorinated organics, organophosphates, carbamates, and pyrethroids (Brattstein et al., 1986). In the United States, for example, the two most widely used herbicides are atrazine and metolachlor, both of which are chlorinated organics (EPA, 1999). In malaria control, the most commonly used pesticides are DDT and malathion. The former is a chlorinated organic; the latter is an organic phosphate compound.

The chlorinated organics consist of Cl atoms attached to organic moieties. They began to be marketed as pesticides during the early 1940s. Many are rather stable compounds that tend to accumulate in lipid tissue. Because of their persistence, organochlorine pesticides tend to be associated with biomagnification and food chain transfer problems. Some, such as DDT and toxaphene, are highly toxic to fish. Some are also highly toxic to birds.

The chlorinated organics may be broken down into subgroups based on similarities in chemical structure. For example, the formulas for DDT, DDD, DDE, and methoxychlor are shown in Figure 10.1. The similarity between these chlorinated hydrocarbons is obvious. DDE is not itself a pesticide, but is a relatively stable metabolite of DDT and has been responsible for many of the adverse effects associated with DDT use. The chemical similarity of these compounds is of more than academic importance, since pest resistance often closely follows such group similarities (Metcalf, 1981; Erickson et al., 1985). Thus, pests that have developed resistance to DDT are likely to show resistance to DDD and methoxychlor as well. In the same vein, it is interesting to note that the chemically similar chlorinated organics chlordane, heptachlor, aldrin, and dieldrin (Figure 10.2) have all been banned by the EPA on the grounds that they are carcinogenic.[2] Historically, some of the most notorious synthetic chlorinated organic pesticides have been chlorinated hydrocarbons.

Organophosphorus pesticides began to appear on the market in approximately 1944. As their name implies, these compounds consist of one or more phosphate groups attached to an organic moiety. The chemical formulas for two such compounds, malathion and parathion, are shown in Figure 10.3. Organophosphorus pesticides are much less stable than organochlorine compounds, and their use is therefore not associated with biomagnification and food chain transfer problems. However, they are general biocides that are

---

[2]Cancellation of most uses of heptachlor was announced in 1978, with the last cancellation going into effect on July 1, 1983. The Velsicol Chemical Company, the sole manufacturer of heptachlor in the United States, voluntarily stopped selling heptachlor and chlordane in August 1987. Use of chlordane was officially prohibited on April 15, 1988. Use of aldrin and dieldrin was suspended in August 1974.

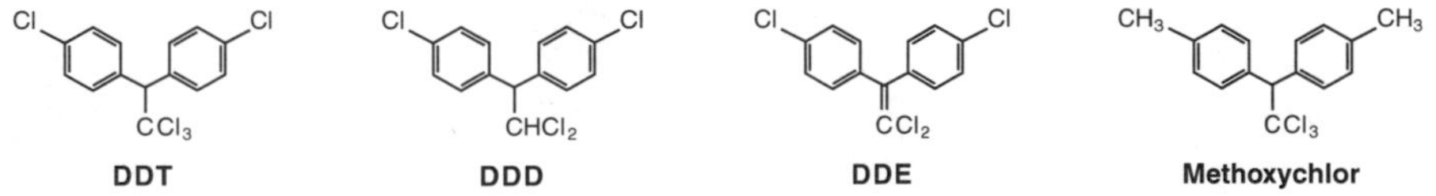

FIGURE 10.1 Structures of DDT (dichlorodiphenyltrichloroethane), DDD (dichlorodiphenyldichloroethane), DDE (dichlorodiphenyldichloroethene), and methoxychlor.

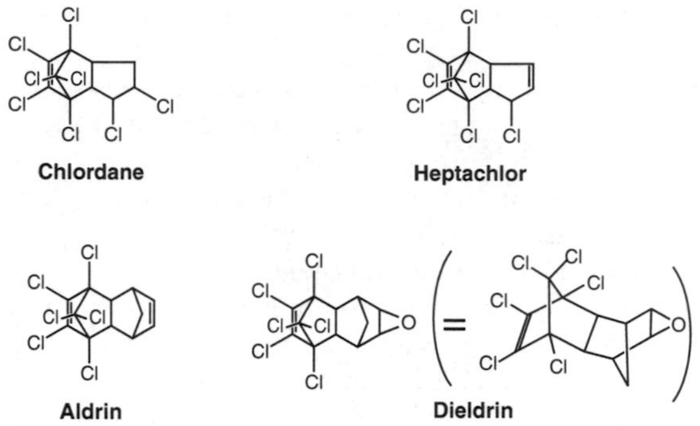

FIGURE 10.2   Structures of chlordane, heptachlor, aldrin, and dieldrin.

toxic to nearly all animals. They are highly toxic to bees and to natural insect parasites and predators. Parathion and related compounds are also highly toxic to humans, and the majority of pesticide poisonings throughout the world are the result of the improper use of these organophosphorus pesticides. Others, such as malathion, are much less toxic to humans and are readily available in garden store formulations. The cost of pest control with organophosphorus compounds is usually much greater than the cost of comparable control with chlorinated organics, the difference being as much as a factor of 5 or more (Metcalf, 1981).

Carbamate pesticides did not become available until 1956. They are all derivatives of carbamic acid, NH_2CO_2H. Examples such as aldicarb and carbofuran are shown in Figure 10.4. Like the organophosphorus pesticides, carbamates are biodegradable and nonpersistent. They are not particularly toxic to fish, but they are highly toxic to birds and bees (Metcalf, 1981).

The pyrethroid pesticides are synthetic relatives of the natural pyrethrin esters obtained from *Chrysanthemum* flowers. The rapid and selective insecticidal action of the natural pyrethrins combined with their high cost stimulated an effort to develop synthetic derivatives, the first of which was produced in 1949 (Metcalf, 1981). Examples of two such compounds, allethrin and dimethrin, are shown in Figure 10.5. Compared to the organochlorine, organophosphorus, and carbamate pesticides, the application rates of pyrethroids required to achieve a given level of pest control are lower, and in general, the pyrethroids are less toxic to animals. The pyrethroids, however, are highly toxic to beneficial insects and are extremely toxic to fish (Metcalf, 1981).

Mode of Action

The mode of action of pesticides depends on both the pesticide and the pest. Herbicides, for example, kill weeds by interfering with growth, respiration, or photosynthesis. Growth

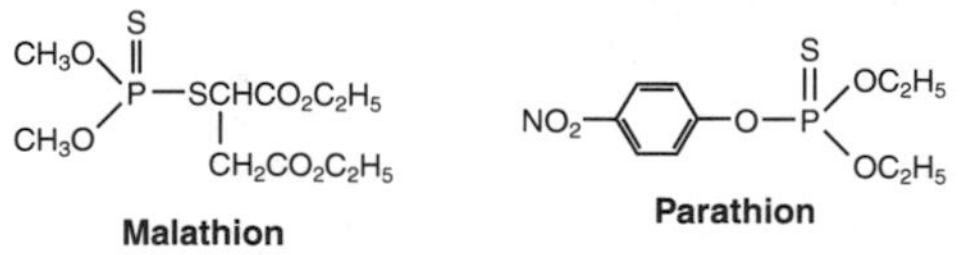

FIGURE 10.3   Structures of two organophosphorus pesticides, malathion and parathion.

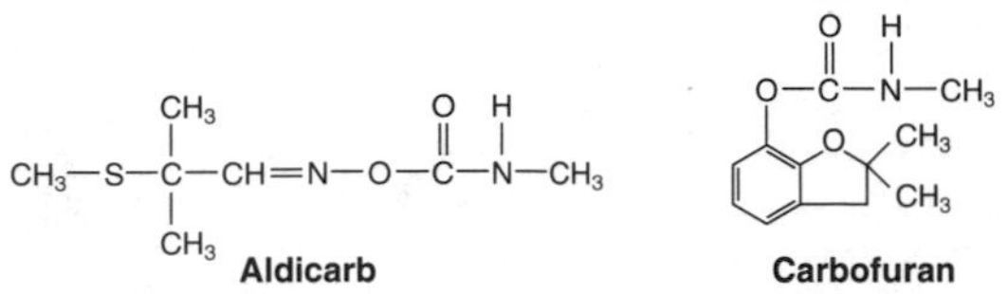

FIGURE 10.4   Structures of two carbamates, aldicarb and carbofuran.

inhibitors interfere with the natural function of plant hormones such as auxins and gibberellins. Herbicides can inhibit respiration and photosynthesis in a variety of ways, for example, by reacting with enzyme systems or uncoupling oxidative phosphorylation. Atrazine interferes with photosynthesis by inhibiting the Hill reaction in photosystem II (Plimmer, 1980). Metolachlor is a germination inhibitor.

Among the insecticides, it is known that many chlorinated organics affect the insect's nervous system. DDT, for example, affects peripheral sensory organs and causes violent trains of afferent impulses that lead to hyperactivity and convulsions. The paralysis and death that follow are believed to result from metabolic exhaustion or from elaboration of a naturally occurring neurotoxin (Metcalf, 1981). Chlordane and its relatives (Figure 10.2), however, affect the ganglia of the central nervous system rather than the peripheral nerves. Symptoms of poisoning include hypersensitivity, hyperactivity, convulsions, prostration, and death. Fish and shellfish are extremely sensitive to chlorinated hydrocarbons, but death comes from suffocation due to interference with O_2 uptake at the gills rather than from effects on the nervous system (Rudd, 1964). Chlorinated hydrocarbons tend to be stored and concentrated in an organism's fatty tissues, and reach an equilibrium concentration that is determined by the balance between pesticide intake and excretion. They do not accumulate indefinitely. When pesticide exposure is removed, concentrations gradually return to normal.

Organophosphates and carbamates both inhibit the action of the enzyme cholinesterase, which normally hydrolyzes acetylcholine at nerve junctions in the body. Nerve impulses are conducted across these junctions by acetylcholine. If the acetylcholine is not subsequently hydrolyzed by cholinesterase, electrical impulses can fire away continuously. Such repeated firing of impulses can lead to uncontrolled, rapid twitching of some muscles, paralyzed breathing, convulsions, and in extreme cases death (GTI, 1999). Symptoms of poisoning in animals include lacrimation, vertigo, muscular weakness, tremors, and labored respiration (Rudd, 1964). Animals usually die from depression of activity in the respiratory control center of the brain (i.e., they suffocate). There is very little residual accumulation of organic phosphates in tissue, although additive effects are possible if exposure is repeated before cholinesterase levels have returned to normal.

Pyrethroids interfere with the transmission of nerve impulses, the sodium channel being their primary target site (Brattsten et al., 1986). They readily penetrate the insect cuticle, and it is believed that the effects on the insect are associated with a specific stereochemical interaction with a biological receptor. Paralyzed insects exhibit a characteristic vacuolization of the nerve tissues (Metcalf, 1981).

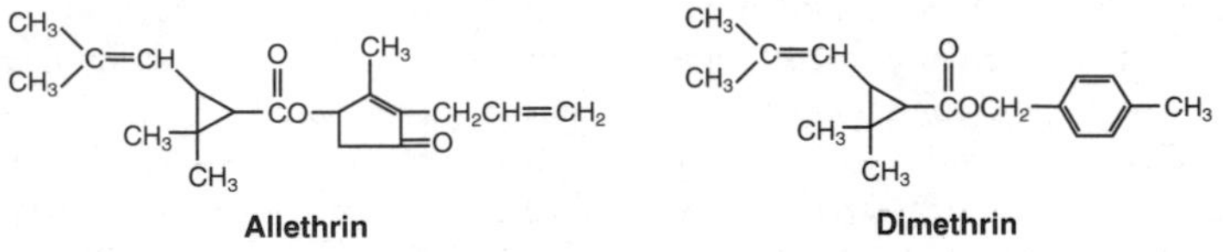

FIGURE 10.5   Structures of two pyrethroids, allethrin and dimethrin.

PESTICIDE USE

Most large-scale pesticide use is undertaken for reasons of public health, agricultural production, or forestry management. In the following sections, we review briefly the history of pesticide use in these three areas.

Public Health

When infectious microorganisms or parasites are transmitted to humans by means of other carrier animals, it is often easier to attack the host animal or vector than the microorganism itself. The mosquito provides an excellent example of such a vector. Mosquitoes transmit more than a dozen important diseases, including malaria, yellow fever, encephalitis, filariasis, and dengue fever. In the past, DDT was the principal pesticide used to control such diseases, but DDT use is now confined almost exclusively to control of malaria. The fact that other pesticides have been substituted for DDT in most disease control programs reflects some of the problems associated with pesticide use in general. Even in the case of malaria control, where DDT is still used, the control program has been less than an unqualified success. The history of DDT use in malaria control is a good example of the benefits, costs, and limitations of pesticide use in general.

Use of DDT to Control Malaria

Sixty of the 380 species of *Anophelies* mosquitoes carry the protozoan parasite that causes malaria. The parasite finds its way into the mosquito's salivary gland, from which it is injected into the next victim bitten by the mosquito. Once inside a human victim, the parasite reproduces in enormous numbers in the red blood cells and, in the case of the most malignant genus, often attacks the brain. Around 1900 there were about 250 million cases of malaria each year worldwide, about 1% of which were fatal (WHO, 1988). Although most persons who become infected survive, the effects of the disease, ranging from lethargy to total incapacitation, are so widespread in the tropics that "malaria has been blamed for impeding the development of entire nations" (Marshall, 1990, p. 399). Survivors, often severely debilitated, are subject to reinfection, although persons do develop resistance as a result of repeated infections. Most persons who die from malaria are young children.

The mosquitoes that transmit the disease tend to rest on the interior walls of homes during the day and to attack sleeping humans at night. Effective control of malaria was initially achieved by simply spraying the interior walls of homes with DDT. To be effective when used in this manner, a pesticide must be persistent, since spray teams cannot be expected to treat homes at frequent intervals. Also, the pesticide must be safe to humans.

Both the success and failure of DDT spraying for malaria control are well illustrated by the experience of Sri Lanka. There were about 2.8 million cases of malaria in Sri Lanka in 1946. DDT spraying for malaria control began the following year, and the incidence of malaria rapidly dropped to fewer than 40,000 clinical cases by 1954 (Figure 10.6). In the next few years the government reduced the extent and frequency of DDT spraying, presumably for economic reasons. The spraying interval was extended from six weeks to six months, and by 1955 it was confined to development projects and settlements. The result was a resurgence of malaria in 1957, but the number of cases dropped rapidly in subsequent years with the recommencement of the spraying program. By 1963 there were only 17 reported cases of malaria in Sri Lanka, 11 of which were imported (Edirisinghe, 1988).

Perhaps feeling that malaria had effectively been eradicated from the country and perhaps partially in response to the publication of Rachel Carson's *Silent Spring* in 1962, the government of Sri Lanka reduced the DDT spraying program in 1963 and terminated it

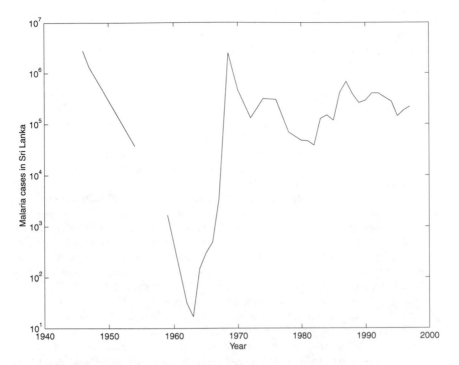

FIGURE 10.6   Incidence of malaria in Sri Lanka from 1946 to 1997. *Source*: Jukes (1974), Fonseka and Mendis (1987), Edirisinghe (1988), *Weekly Epidemiological Record* (13 August 1999).

altogether in 1964. The result was a rapid increase in the incidence of malaria until an epidemic of 2.5 million cases occurred in 1968–1969 (Jukes, 1974). Spraying with DDT was recommenced on a four-month cycle in 1969, but tests conducted at that time indicated that the mosquito population was developing resistance to the pesticide. The spraying program nevertheless reduced the incidence of malaria to 130,000 cases by 1972. Tests conducted in 1974, however, indicated that resistance within the mosquito population had increased to an alarming level, and there were over 300,000 clinical cases of malaria that year. In response to this development and perhaps also as a result of the U.S. ban on DDT use in 1972, the government began substituting malathion for DDT in the spray control program in 1974, and DDT use was ultimately phased out entirely. Because malathion is less persistent than DDT, it must be applied more frequently than DDT to achieve equivalent pest control. Largely because of this fact, the malathion spraying program cost more per unit area than the DDT spraying program. Because of this increased cost, the government of Sri Lanka opted for less than total coverage spraying, the result being protection with malathion of only a portion of the population at risk (Edirisinghe, 1988). Since 1978 the incidence of clinical malaria in Sri Lanka has been about 250,000 cases per year. This figure is certainly an improvement over the 1 million or more cases reported in 1946 and 1968, when there was no systematic spraying program to control the mosquito population, but it is sufficiently high to make malaria a significant public health problem in Sri Lanka.

There was euphoria in Sri Lanka in the early 1960s when malaria seemed to have been all but eradicated. The euphoria was short-lived. The impact of DDT on nontarget species, the development of resistance within the mosquito population, and the increased cost of alternative pesticides all conspired to reduce the effectiveness of the control program. This scenario is a reflection of the world situation as well. According to Gardner et al. (1998), at the present time 300–500 million persons are infected with malaria each year, and 1.5–2.7% of those infections lead to death. More than 90% of the infections occur in Africa, and most of the deaths occur among children in sub-Saharan Africa. These figures are no better than the 250 million cases and 2.5 million deaths per year that occurred around 1900. Malaria is still a major public health problem in more than 90 countries inhabited by 40% of the world's population. Treatment for the disease involves administration of quinine-related drugs, usually chloroquine, but the most lethal strain of the protozoan parasite that causes malaria, *Plasmodium falciparum*, is now genetically resistant to chloroquine (Marshall, 1990). In the late 1980s health authorities began switching to a new compound, mefloquine, but by 1990 there was evidence from Asia and Africa of resistance of the parasite to mefloquine as well (Marshall, 1990). Recently, there have been reports that the effectiveness of chloroquine can be improved when the drug is administered with chlorpheniramine, a cheap, easily available antihistamine (Bagla, 1997). In the last 10–15 years, considerable progress has been made on the development of a malaria vaccine (Hoffman et al., 1991), but none is yet available. However, in 1997 researchers at the Walter Reed Army Institute developed a candidate vaccine based on a protein that appears on the surface of *P. falciparum*. The vaccine protected six of seven persons from infection after they had been bitten repeatedly by mosquitoes carrying live parasites (Marshall, 1997). An effort is currently underway to sequence the genome of the malaria parasite (Kaiser, 1996b). Mapping of the genome should lead to more informed efforts to develop vaccines and drugs.

It is fair to say that the use of pesticides such as DDT to prevent diseases such as malaria has saved tens of millions of lives and prevented hundreds of millions of illnesses since 1945. It is also fair to say that despite the use of pesticides in public health, some diseases, such as malaria and river blindness, remain major public health problems in some parts of the world (Walsh, 1986). The continued use of pesticides in these disease-control programs largely reflects our inability to identify a more satisfactory alternative. The most effective defense against yellow fever, for example, which is transmitted by mosquito vectors, is vaccination, not pesticide spraying for mosquito control.

Agriculture

The use of pesticides in agriculture expanded greatly after World War II in response both to the greater availability of pesticides and to concerted efforts to increase the world's food supply. Basic elements of the so-called Green Revolution have been the development of high-yield agricultural crops through genetic manipulations, increased use of water and fertilizers, and modifications of traditional agricultural practices to control crop pests (Walsh, 1991). For example, farmers reduced their use of mechanical tilling to crontrol weeds and increased their use of chemical herbicides. This change in crop husbandry allowed them to plant row crops at more closely spaced intervals and hence increase yields per unit area. Similarly, control of insect pests through crop rotation was in some cases abandoned in favor of planting the same crop year after year. Such changes in agricultural practices allowed farmers to achieve greater yields of the desired crops, particularly rice and wheat, and turned chronic agricultural shortages into surpluses. There is no question that from the standpoint of crop production the Green Revolution has been a tremendous success.

This success, however, has not been achieved without some cost, and part of the cost has been a great increase in the use of chemical pesticides. In India, for example, pesticide use increased from about 2,000 tons annually in the 1950s to more than 80,000 tons in the mid-1980s (Postel, 1987). DDT and benzene hexachloride (BHC), both of which have been banned in the United States and much of Europe, account for about 75% of pesticide use in India (Postel, 1987). The variety and extent of pesticide use in agriculture during the Green Revolution are reflected in Table 10.1, which lists chlorinated hydrocarbon insecticides used on agricultural crops during the 1960s. In the United States, pesticide use in agriculture continued to increase until approximately 1980, when annual use peaked at roughly 370,000 tonnes (Figure 10.7). Current use is about 350,000 tonnes. Interestingly, the decline in agricultural pesticide use in the United States since 1980 reflects to a large extent the substitution of more biologically effective pesticides for some of the earlier synthetic pesticides, such as those listed in Table 10.1. For example, essentially the same agricultural pest control that was achieved by applying DDT at a rate of 2 kg ha^{-1} in 1945 is now achieved by applying pyrethroids and aldicarb at rates of 0.1 kg ha^{-1} and 0.05 kg ha^{-1}, respectively (Pimentel et al., 1991).

While it is true that agricultural production has increased greatly as a result of the Green Revolution, it is remarkable that the percentage of crops lost to pests has not declined as a result of the tremendous increase in pesticide use. For example, between 1942 and 1951, U.S. agricultural losses to pests were estimated to be about 31%. In 1974 and 1986 the percentage loss was estimated to be 33% and 37%, respectively (Pimentel et al., 1991). The fact that heavy reliance on pesticides has not reduced the percentage of crops lost to pests, and that the percentage loss in the United States has actually increased, has suggested to many persons that such heavy use of pesticides may not be the most effective way to control crop losses to pests.

TABLE 10.1   AREAS AND MAJOR CROPS ON WHICH CHLORINATED HYDROCARBON INSECTICIDES HAVE BEEN USED, EXCLUDING THE UNITED STATES, CANADA, MAINLAND CHINA, AND THE FORMER SOVIET UNION

Crop	Areas of Major Significance in Use	Insecticides Used
Cotton	Mexico, Nicaragua, Egypt, Sudan, Brazil, Guatemala, Colombia, Australia, Turkey, Uganda, Ivory Coast, El Salvador	Various, but mainly DDT
Rice	Japan, India, Indonesia, Cambodia, Colombia, Spain, Venezuela, Taiwan, Brazil	BHC, endrin, DDT, toxaphene, aldrin
Other cereals (maize, small grains, sorghum, etc.)	India, United Kingdom, Mexico, Upper Volta, Niger, Turkey, France, Colombia, Chile, Spain, Japan, Argentina, Greece, East Africa	BHC, DDT
Vegetables (excluding potatoes)	India, Japan, Mexico, Spain, Chile, United Kingdom, Thailand, South Africa	BHC, DDT
Onions, tomatoes, chilies, cabbage	Japan, Italy, Spain, France, Portugal	Cyclodienes
Potatoes and sweet potatoes	France, Spain, Brazil, Peru, Colombia, United Kingdom, Japan, Greece, Taiwan, Mexico, Australia	Cyclodienes, DDT
Sugar beets	Belgium, Italy, France, Spain, Greece, Chile, Turkey	Aldrin, heptachlor
Sugar cane	Mexico, Australia, India, Brazil, Ivory Coast, Taiwan, South Africa, Pakistan	BHC/lindane, endrin, dieldrin, aldrin, heptachlor
Tobacco	Mexico, Australia, South and East Africa, Japan Italy, Colombia, Greece, Spain	DDT, BHC, cyclodienes
Oil seeds soybeans, groundnuts, sunflower, etc.)	India, Japan, Argentina, Colombia, Nicaragua, Brazil, France, Venezuela	DDT, toxaphene
Crops in general	Asia and Africa	Dieldrin, BHC

Source: Ling et al. (1972).

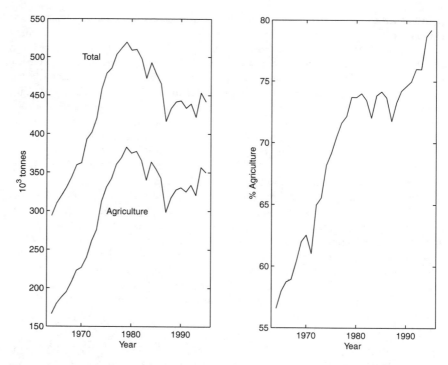

FIGURE 10.7   Pesticide use in agriculture in the United States from 1964 to 1994. *Source*: EPA (1999).

Combined with this feeling is the realization that the present heavy use of pesticides in agriculture is associated with significant undesirable side effects. It is estimated, for example, that 400,000–2 million pesticide poisonings occur worldwide each year, most of them among farmers in developing countries. These poisonings result in 10,000–40,000 deaths each year (Postel, 1987). Chlorinated organics are not particularly dangerous to humans in this context, but some organophosphorus compounds and carbamates such as parathion, methylparathion, and aldicarb are acutely toxic and must be applied with great care. According to the Centers for Disease Control, aldicarb is the most toxic agricultural chemical currently used in the United States. One drop absorbed through the skin is enough to kill an adult (PAN, 1994). Unfortunately, many Third World farmers lack the training and equipment needed to apply such pesticides safely. "A 1985 survey in one county of the Brazilian state of Rio de Janeiro found that 6 out of 10 farmers using pesticides had suffered acute poisonings, two-thirds of them from organophosphates" (Postel, 1987, p. 19). In addition to the overt poisoning of the individuals involved in applying the pesticides, there are the somewhat less obvious impacts on nontarget species, including humans. In a survey conducted in India's Punjab region, for example, samples of breast milk from 75 women were all found to contain residues of DDT and BHC, the average concentrations being over 20 times WHO guidelines (Postel, 1987). Similar studies of breast milk from Nicaraguan women have revealed DDT levels 45 times the tolerance limits (Postel, 1987). In 1985 roughly 1,000 people were poisoned by aldicarb-tainted watermelon, and more than 300 persons in British Columbia suffered aldicarb poisoning after eating tainted cucumbers (PAN, 1994). Pesticides can obviously find their way into human bodies through the consumption of contaminated crops, but in some cases pesticides leach into groundwater and later appear in drinking water. Monitoring wells and groundwater supplies for pesticides is estimated to cost about $1.2 billion per year in the United States (Pimentel et al., 1991). The consensus now is that pesticides do have a place in the control of agricultural pests, but that the heavy reliance on pesticides that developed during the Green Revolution years is undesirable, and that if a full accounting of the costs and benefits of pesticide use in agriculture were made, a considerable scaling back of pesticide use would be called for. Alternatives do exist to the present large-scale use of pesticides in agriculture (see the later discussion).

Forestry

Damage to forests from pests, mainly insects, is difficult to estimate. These losses include reduced growth rates, loss of reproductive ability, and direct mortality of trees, as well as general watershed damage[3] and loss of recreational value. In 1958 the U.S. Forest Service estimated timber losses from insects at 5 billion board-feet, and growth losses were put at 3.6 billion board-feet (Rudd, 1964). The most serious forest insect pests in the United States are the gypsy moth, the tussock moth, the spruce budworm, and the Engelmann spruce beetle. An outbreak of spruce budworms in parts of Canada during the 1930s and 1940s affected over 400,000 km^2 and resulted in over 70% timber mortality in some areas (Rudd, 1964). The gypsy moth, which first appeared in the New World in 1869, has been steadily expanding its range and is now found throughout much of the northeastern United States. In 1981 an outbreak of gypsy moths resulted in the defoliation of over 40,000 km^2 of timber (Marshall, 1981).

During the 1950s, efforts to control such outbreaks involved widescale aerial applications of pesticides such as DDT. The adverse effects of such aerial spraying soon became apparent, however, as large numbers of nontarget species were frequently killed. A major and fundamental problem with such aerial spraying is the fact that only about 25–50% of the pesticide actually reaches the target area (Akesson and Yates, 1984). As a result, significant impacts on nontarget species are difficult to avoid. The consensus is that widescale aerial application of pesticides is not a satisfactory way to deal with insect pests in forestry management and that alternative controls must be sought. Satisfactory alternatives have not, however, been readily forthcoming (Evans, 1978; Marshall, 1982); as a result, synthetic pesticides continue to be used in forestry management, though on a much less grandiose scale. In the case of gypsy moth control, for example, "the campaign has tapered off to a sporadic application of pesticides now considered safe" (Marshall, 1981, p. 992).

PESTICIDE EFFECTS ON NONTARGET SPECIES

There is no doubt that the adverse effects of pesticides on nontarget species have been a major cause of both the public's and governments' disenchantment with the widescale use of pesticides. The adverse effects or their consequences include human pesticide poisonings, poisoning of agricultural livestock, the reduction of natural pest enemies, reduced pollination due to the poisoning of bees, losses of crops and trees, and fishery and wildlife losses. Pimentel et al. (1991) have estimated the cost of these adverse effects to be roughly $650 million per year in the United States. The following examples illustrate two of the numerous well-documented adverse effects of DDT use on nontarget species.

---

[3]For example, increased erosion due to loss of vegetative cover.

Forest Spraying with DDT to Control Spruce Budworms in New Brunswick, Canada

Beginning in 1952, large areas of forest in northern New Brunswick were sprayed each spring with DDT in an effort to control an outbreak of spruce budworms. The DDT was applied at a rate of about 50 kg km^{-2}, a typical dosage for control of many forest pests. Despite these efforts, the outbreak continued to spread, and by 1957 over 20,000 km^2 were sprayed, much of the area for the second or third time. The Fisheries Research Board of Canada had been studying salmon populations in the major streams of the affected watershed for several years prior to the initiation of spraying. Therefore, a reliable estimate of salmon populations before 1952 could be made. Continued monitoring of salmon populations after 1952 led to the conclusions that application of DDT at a rate of 50 kg km^{-2} produced the following results (Rudd, 1964):

1. Loss of 90% of under-yearling salmon
2. Loss of 70% of salmon parr (juveniles over one year old)
3. Complete loss of aquatic insects

Although aquatic insects began to reappear in affected streams about three weeks after spraying, the composition of the insect population was radically changed. The population was dominated by very small insects, which were too small to be utilized satisfactorily as food by the larger young salmon (Ide, 1956). The larger insects, which would normally have provided food for the juvenile salmon, did not reappear. As a result, the salmon fingerlings were forced to turn to alternate sources of food and fed largely on snails (Ide, 1956). Insect populations did not return to normal for five to six years after spraying. Since the coho salmon is an anadromous fish that normally spends three years in freshwater, repeated spraying in successive years would affect the same year class more than once. Simply restocking the affected streams with salmon would not allow salmon populations to recover until their food supply had also returned to normal.

As a result of this experience, a spraying program for blackheaded budworms on Vancouver Island was carried out in a different manner (Rudd, 1964). A special effort was made to avoid contact with streams by using the following techniques:

1. Not using streams as boundaries for spray plots so that streams would not receive double spray doses
2. Spraying parallel to major streams but keeping one swath away
3. Shutting off the spray when streams were crossed

Despite these precautions, the mortality of coho salmon fry approached 100% in four major streams and ranged down to 10% in others (Rudd, 1964). These results point out that streams may be contaminated with pesticides via surface runoff as well as by direct application. The contamination of streams during large-scale pesticide spraying operations in forestry management is an unavoidable problem that can obviously have serious consequences for aquatic species.

DDD Treatment to Control Gnat Populations on Clear Lake, California

Clear Lake is a rather shallow (maximum depth 10 m), highly productive lake located about 160 km north of San Francisco. The large number of fish in the lake has made it an attractive resort area. Unfortunately for the resort business, Clear Lake is normally populated during the summer months by an enormous number of gnats. Although the species is not blood-sucking, its numbers are so great as to make it a considerable nuisance, particularly at night, when the gnats are attracted to lights. Between 1916 and 1941 several studies were done on the gnat population in an effort to discover a suitable control method, but no satisfactory means was found.

In 1946 laboratory experiments with DDT and DDD were begun. These experiments indicated that DDD would provide effective gnat control with less danger to fish and other organisms in the lake than would DDT. On the basis of test applications in two nearby bodies of water, it was determined that a concentration of about 14 ppb DDD would provide adequate gnat control with minimal damage to the fish population (Hunt and Bischoff, 1960). Accordingly, 14 ppb of DDT was applied to Clear Lake in September 1949. This treatment resulted in about a 99% kill of gnat larvae, but by 1951 larvae began to reappear in significant numbers, and by September 1954 a second treatment was needed. This second treatment was made at a higher concentration of 20 ppb, as was a third treatment in September 1957. The rate of gnat larvae kill in the second treatment was again about 99%, but the third treatment was not as successful as the previous two. The lower success of the third treatment was attributed by some to adverse weather, but the more rapid recovery of the gnat population following the second versus the first treatment[4] suggests that the gnats may have been developing resistance (Rudd, 1964). In December 1954, following the second treatment with DDD, 100 dead western grebes[5] were found on the lake (Hunt and Bischoff, 1960). Analysis of specimens by the Department of Fish and Game Disease Laboratory indicated no evidence of disease. Similarly, in December 1957, 75 dead grebes were found, again with no evidence of disease. However, two sections of visceral fat taken from these birds were found to contain 1,600 ppm DDD, 80,000 times higher than the DDD concentration applied to the lake (Hunt and Bischoff, 1960).

Although other explanations for the grebe deaths are possible, DDD poisoning is strongly implicated as the cause of death for the following reasons:

1. The decline in the grebe population corresponded to the period in which pesticide applications were made. Die-offs were not noted at other times.
2. There was no evidence of infectious diseases in dead grebes.
3. Clinical symptoms of poisoning (e.g., nervous tremors) were observed in some grebes on the lake.
4. Unusually high DDD concentrations were found in the fatty tissue of dead grebes.

In additional sampling of Clear Lake conducted by the California Department of Fish and Game, all fish and birds analyzed were found to contain DDD, and all parts of the lake were found to contain DDD-contaminated fish. Many of the fish taken from Clear Lake were found to contain DDD in their edible flesh at concentrations higher than the level of 5 ppm presently set by the Food and Drug Administration as the action level in marketed fish. Samples of flesh from largemouth bass and Sacramento blackfish hatched as long as seven to nine months after the last DDD application contained 22–25 and 7–9 ppm DDD, respectively. Although it is impossible to say how the grebes came to have DDD levels as high as 1,600 ppm in their fatty tissues,[6] it seems probable that the major source of input was their food, since visceral fat in fish taken from the lake contained hundreds to several thousand parts per million DDD (Rudd, 1964). Upon the recommendation of county authorities, DDD applications to Clear Lake were discontinued after the 1957 treatment, and

[4]Three years versus five years, respectively.
[5]A species of bird.
[6]They could have absorbed DDD directly from the water, for example.

methyl parathion was substituted as a control agent. Since then, the grebe population has recovered (Jukes, 1974).

EXAGGERATED AND/OR ERRONEOUS CHARGES AGAINST PESTICIDE USE

In the previous two examples, there is little doubt that pesticide use harmed nontarget species on a significant scale. An understanding and appreciation of such problems is vital to the development of a sensible pesticide use program. Unfortunately, the literature on pesticides contains some examples of exaggerated or erroneous charges regarding the impact of pesticides on nontarget species. Some of these claims have come to be accepted as established facts simply because they have been repeated and publicized so many times. In the interests of scientific credibility, it is important to distinguish effects that can be attributed beyond a reasonable doubt to a particular cause from effects whose cause is unclear. The following examples illustrate some of the controversial adverse effects that have been attributed to the use of DDT.

Destruction of Speckled Sea Trout in the Laguna Madre, Texas

In 1971, the Panel on Monitoring Persistent Pesticides in the Marine Environment submitted a report to the National Academy of Sciences (NAS), which was widely publicized in the popular press. In one section, the panel reported:

> In the speckled sea trout on the south Texas coast, DDT residues in the ripe eggs are about 8 ppm. This level may be compared with the residue of 5 ppm in freshwater trout that causes 100 percent failure in the development of sac fry or young fish. The evidence is presumptive for similar reproductive failure in the sea trout. Sea trout inventories in the Laguna Madre in Texas have shown a progressive decline from 30 fish per acre in 1964 to 0.2 fish per acre in 1969. It is significant that few juvenile fish have been observed there in recent years, although in less contaminated estuaries 100 miles away there is a normal distribution of sea trout year classes. (Goldberg et al., 1971, p. 10)

Table 10.2 is taken directly from the NAS report. Several major criticisms can be made of this report.

1. The report states that sea trout inventories declined from 30 fish per acre in 1964 to 0.2 fish per acre in 1969. In fact, these figures refer only to juvenile trout, as the footnote to the table indicates. There was no decline in adult populations. The figure of 0.2 juvenile trout per acre in 1969 is incorrect; the correct figure is 0.5.
2. Three hurricanes swept through the area during the study period, one in each of the years 1965, 1967, and 1968. The most destructive one was probably hurricane Beulah, which occurred in 1967. Beulah heavily flooded all coastal areas, dropping 76 cm of rain, and undoubtedly disrupted the shallow Laguna Madre ecosystem. The decline in juvenile sea trout is strongly correlated with the occurrence of these hurricanes. In 1970, when weather conditions were stable, juvenile trout numbers increased to 10 per acre, and in 1971 they reached 25–30 per acre. It is remarkable that the 1970 figures were not included in the NAS panel report, which was released in June 1971.

TABLE 10.2 CAPTURE OF SEA TROUT IN THE LAGUNA MADRE

Year	Number of Sea Trout Captured per Acre
1964	30
1965	25
1966	12
1967	—[a]
1968	2.7
1969	0.2

[a]No data exist for 1967, as hurricanes destroyed all the fishing gear. The decline in the fishery has been only in the juvenile trout, which eliminates the possibility of overfishing. The DDT residues in the gonads of the adult trout reach a maximum of 3 ppm prior to spawning.
Source: Reproduced from *Chlorinated Hydrocarbons in the Marine Environment* (1971), p. 10, with permission of the National Academy of Sciences, Washington, DC.

3. Since pesticides are used for agricultural purposes in the coastal region around the Laguna Madre, it would not be surprising to find that the runoff from the three hurricanes had introduced unusually high pesticide concentrations into the estuary. However, the figure of 8 ppm DDT in ripe eggs is not a mean value, but rather the highest value found in any of the sea trout eggs at any time. Mean values were typically in the range 1–2 ppm, and declined in 1970 and 1971 as the hurricane effects diminished (Butler, 1971).

In summary, the decline in juvenile sea trout in the Laguna Madre during the period 1965–1969 appears to have been caused by hurricane damage and not by DDT effects. Juvenile sea trout populations returned to normal by 1971, despite continued use of DDT in the watershed area.

DDT Reduces Photosynthesis by Marine Phytoplankton

The following excerpt from an article by Ehrlich (1969, p. 24) provides a good introduction to this issue.

The end of the ocean came late in the summer of 1979, and it came even more rapidly than the biologists had expected. There had been signs for more than a decade, commencing with the discovery of 1968 that DDT slows down photosynthesis in marine plant life. It was announced in a short paper in the technical journal, *Science*, but to ecologists it smacked of doomsday. They knew that all life in the sea depends on photosynthesis, the chemical process by which green plants bind the sun's energy and make it available to living things. And they knew that DDT and similar chlorinated hydrocarbons had polluted the entire surface of the earth, including the sea.

The "short paper" referred to in the above paragraph is an article by Wurster (1968) reporting the results of experiments on pure and mixed phytoplankton cultures to determine the effect of DDT exposure on photosynthesis. The DDT was added in concentrations ranging from 1 to 500 ppb to flasks containing the experimental cultures, the cultures were incubated for approximately one day more, and the photosynthetic rate was measured over the next four to five hours. The results of these experiments are shown graphically in Figure 10.8. The crucial point about these results is that the solubility limit of DDT in sea-

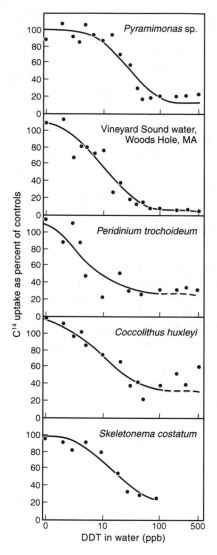

FIGURE 10.8   Photosynthetic rates of five cultures of marine phytoplankton as affected by DDT concentrations. *Source*: Reprinted with permission from C. F. Wurster. 1968. DDT reduces photosynthesis by marine phytoplankton. *Science*, **159**, 1474–1475. Copyright 1968 American Association for the Advancement of Science.

water is only 1.2 ppb. To achieve the concentrations shown in the figure, it was necessary to dissolve the DDT in alcohol before adding it to the experimental flasks. One would never see concentrations this high in the ocean. Looking at the figures, one concludes that even at its solubility limit, DDT has virtually no effect on phytoplankton photosynthesis. The findings of subsequent experiments performed by Menzel et al. (1970), Moore and Harriss (1972), Mosser et al. (1972), Luard (1973), and Fisher (1975) were consistent with the results reported by Wurster (1968). No effects of DDT on marine phytoplankton photosynthesis have been reported at concentrations below the solubility limit, and in some cases no effects have been observed at concentrations 10 times or more the solubility limit.

On the other hand, it is reasonable to assume that the DDT dissolved in the water was not directly responsible for the effects observed by Wurster and others. Rather, the DDT *absorbed by or adsorbed to the phytoplankton* was responsible for the reduction in photosynthetic rate. The fact that the percentage reduction in photosynthetic rate and the concentration of phytoplankton cells are negatively correlated at the same total DDT con-

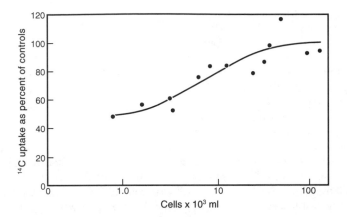

FIGURE 10.9   Effect of 10 ppb DDT on the photosynthetic rate of the marine diatom *Skeletonema costatum* as a function of cell concentration. *Source*: Reprinted with permission from C. F. Wurster. 1968. DDT reduces photosynthesis by marine phytoplankton. *Science*, **159**, 1474–1475. Copyright 1968 American Association for the Advancement of Science.

centration (Figure 10.9) is consistent with this hypothesis. In other words, the more DDT available per cell, the lower the photosynthetic rate. In the open ocean, where phytoplankton abundance in most places is low, it is therefore quite probable that DDT would depress photosynthetic rates at concentrations that were associated with no adverse effect on dense laboratory cultures. In order to circumvent this problem, Menzel et al. (1970) conducted their experiments at phytoplankton concentrations that they considered to be within the range of naturally occurring concentrations in oceanic surface waters, and Moore and Harris' (1972) experiments were conducted with natural phytoplankton communities from St. George Sound in the northeastern Gulf of Mexico. In neither case were any adverse effects of DDT on photosynthesis apparent at DDT concentrations below the solubility limit of DDT in natural waters. The conclusion would seem to be that even at the low phytoplankton concentrations typical of many natural waters, DDT at concentrations below the solubility limit does not depress photosynthetic rates. There is no convincing evidence to show that DDT concentrations in the oceans have ever had a significant effect on primary production.

DDT Residues of 5 ppm (Wet Weight) in the Eggs of Freshwater Trout Result in 100% Mortality of Fry

The basis for this statement, which we saw earlier in the Laguna Madre sea trout discussion, stems from an article published in 1964 by Dr. G. E. Burdick and colleagues concerning the failure of the Lake George fish hatchery in New York State. A complete loss of fry at this hatchery was experienced in both 1955 and 1956. "The fry floated upside down on the surface, eventually sinking and dying. Symptoms appeared after absorption of the yolk sac when the fry were about ready to feed. Pathological examination failed to show the presence of any disease" (Burdick et al., 1964, p. 127). In 1956 and 1958, eggs from the Lake George hatchery were distributed to three other hatcheries as a check on hatchery conditions. There was no survival in 1957 and negligible survival in 1958, results which indicated that the problem was associated with the Lake George eggs and was not due to water quality in the Lake George hatchery. Attempts to cross Lake George females with males from other lakes failed, although males from Lake George reproduced successfully with females from other lakes.

It was known that during the period 1951–1955 the New York State Conservation Department had distributed over 3.5 tonnes of DDT in the Lake George watershed, primarily for gypsy moth control. In 1956 and 1957 an additional 13 tonnes were distributed. Spraying and fogging with DDT for blackfly and mosquito control were also common in the watershed. Fish kills had occasionally been reported subsequent to such treatments, and a 1959 study reported that fish fry from Lake George contained measurable amounts of DDT. These observations suggested that DDT might be the cause of the fry mortality at the hatchery.

To check this hypothesis, female fish were collected from 12 different New York lakes during the period 1960–1962. It would appear from the numbering of the fish in Tables 2 and 3 of Burdick's paper that at least 95 fish were collected over this three-year period, although only 61 fish are accounted for in the later analyses. Eggs were extracted from 51 of these females. Some of the eggs were incubated following usual procedures; the remainder were analyzed for DDT and DDE. The 5 ppm figure comes entirely from the 1960 results, and we will therefore concentrate on that data set.

In 1960, DDT concentrations were measured in the eggs of 20 fish, 4 from Fourth Lake, 3 each from Seneca, Saranac, Eighth, George, and Raquette lakes, and 1 from Lake Placid. Unfortunately, the wet weights of the eggs were not determined, and it was necessary to estimate wet weights based on 1960 dry weights and the mean water loss from dehydration (70%) determined from 1961 egg analyses. An attempt was then made to correlate fry survival with the DDT residues in the eggs. The result of this exercise is shown in Figure 10.10.

In Figure 10.10, "% syndrome" simply means the percentage of fry that died, exhibiting the previously described symptoms. The authors concluded from this figure that fry mortality is severe (over 50%) when DDT concentrations in the egg exceed 5 ppm. Several comments are in order regarding this conclusion:

1. Only 16 of the original 20 fish are accounted for. The results concerning two fish each from Lake Saranac and Eighth Lake are missing.
2. The experiments were terminated when mortality reached 50%. It is impossible to say what the mortalities would have been had the experiments been allowed to con-

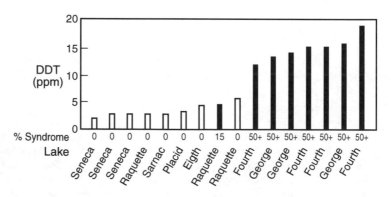

FIGURE 10.10   Total DDT of 1960 lake trout egg lots, based on wet weight, computed from dry weight and based on the average wet weight to dry weight ratio of 1961 eggs. Solid bars indicate the occurrence of the syndrome. *Source*: Redrawn from Burdick et al. (1964). Reproduced with permission of the American Fisheries Society.

tinue. There is clearly no basis for saying that 5 ppm DDT caused 100% mortality, since 100% mortality was never observed. If one accepts the results at face value, one might argue that 5 ppm DDT caused at least 50% mortality, but that would be the strongest statement one could make.

3. Establishing the breakpoint at 5 ppm DDT is completely arbitrary. The lowest DDT concentration in any of the eggs from Lake George and Fourth Lake was 10.10 ppm. Eggs from one Lake Raquette fish showed 15% mortality at a DDT level of 4.75 ppm, but a second fish from the same lake produced eggs with a DDT concentration of 5.81 ppm, yet there was no mortality. In short, one could equally well establish the breakpoint for at least 50% mortality at 10 ppm DDT as at 5 ppm DDT.

4. Only three of the fish were from Lake George, where 100% mortality of hatchery fry was observed. The lowest DDT concentration in the eggs from these fish was 13.55 ppm DDT.

5. The DDT analyses were made by the now outmoded Schechter-Haller method, which fails to distinguish between DDT and a number of other compounds, including methoxychlor, DDD, chlorobenzylate, polychlorinated biphenyls (PCBs), and most aromatic compounds. After 1966, chromatographic methods completely replaced the Schechter-Haller method in pesticide analyses. After this change in analytical methodology, the edible flesh from over half of the lake trout analyzed from Lake George was found to contain more than 5 ppm PCBs (Boyle, 1975). It is therefore possible that much of what Burdick et al. (1964) were measuring with the Schechter-Haller method was PCBs.

In summary, there is no doubt that 100% mortality of fish fry occurred at the Lake George fish hatchery in 1955 and 1956. Based on the work of Burdick et al. (1964), one can say that the concentration of something (whatever the Schechter-Haller method was measuring) in fish eggs seems to correlate positively with mortality. Considering the rather small number of samples (16) involved, the statistical basis for this statement is a bit weak. This summary is clearly a far cry from the statement that 5 ppm DDT in fish eggs causes 100% mortality of the fry. In defense of Burdick et al. (1964), one should point out that they never made this last assertion. However, in reading their paper, others have drawn this unjustified conclusion and have repeated it until it seems to have become an established fact. Indeed, the NAS report (Goldberg et al., 1971) did not even bother to reference this assertion in its summary of DDT effects on fish.

DDT Causes Cancer

DDT is suspected of being a human carcinogen. The basis for this suspicion is the fact that DDT has been shown to cause liver tumors in mice. These results are of dubious relevance to human health for reasons discussed in Chapter 8. In several other animal species, including rats, monkeys, hamsters, and dogs, no tumorigenic effects have been shown at doses of less than 50 mg DDT per kilogram of body weight per day (Ottoboni, 1972; EPA, 1980a). At doses higher than that level, evidence is equivocal for rats (EPA, 1980a). For comparison, average human DDT intake rates in the United States, primarily from the consumption of contaminated food, were about 0.0009 mg kg^{-1} in 1965 and 0.0004 mg kg^{-1} in 1970 (Duggan and Corneliussen, 1972).

There have been several studies of humans who ingested or were exposed to unusually high amounts of DDT. Ortelee (1958), for example, studied 40 men with extensive and prolonged occupational exposure to DDT in manufacturing or formulating plants. Exposure rates, which were estimated from observations on the job, ranged as high as 0.6 mg kg^{-1} d^{-1}. Physical, neurological, and laboratory findings were within normal ranges with the exception of minor skin irritations, and no correlations were found between DDT exposure and frequency and distribution of the few abnormalities seen. Laws (1967) found no evidence of adverse health effects and no cases of cancer in a group of 35 workers at the Montrose Chemical Company, which manufactured DDT exclusively. The workers had been ingesting and/or inhaling DDT for 9–19 years at rates estimated to be 0.05–0.26 mg $kg^{-1} d^{-1}$. Almeida et al. (1975) examined a group of men who had been involved for six or more years in spraying DDT as part of a malaria control program in Brazil. Although the spray workers' blood serum contained significantly higher DDT and DDE concentrations than the blood serum of a control group, there was no significant difference in the health of the two groups. Edmundson et al. (1969a, 1969b) studied 154 persons with occupational exposure to DDT over a two-year period. No clinical effects related to DDT exposure were observed. Hayes et al. (1971) administered DDT at rates of up to 0.51 mg $kg^{-1} d^{-1}$ to a group of human volunteers for periods in excess of 600 days, with absolutely no signs of pathological effects. DDT accumulated in the fatty tissues of the volunteers for a period of time until an equilibrium concentration was reached, with the equilibrium concentration being roughly proportional to the rate of DDT ingestion. When DDT feeding was terminated, tissue concentrations returned to normal.

Under certain conditions, DDT and related compounds actually appear to have anticarcinogenic properties. In fact, DDD has been used successfully as a chemotherapeutic agent in the treatment of adrenal carcinoma (Bledsoe et al., 1964; Conney et al., 1967; Sakauchi et al., 1969; Southren et al., 1966). In experiments with mice, Laws (1971) fed an experimental group 5.5 mg $kg^{-1} d^{-1}$ DDT. At two- to four-week intervals, six mice each from a control group (no DDT feeding) and the experimental group were inoculated with cancer cells under the skin of the back. These transplants continued for a period of 13 months. According to Laws, this transplant technique had, "produced a 100% take of tumor transplants over a ten-year period" (Laws, 1971, p. 182). Not surprisingly, 100% of 87 control mice developed cancer and died after a mean period of 40 days following tumor transplantation. However, 8% of 89 DDT-fed mice did not develop cancer, and those that did lived significantly longer (mean = 56.5 days) after the transplant than did the controls. The results suggest that DDT may suppress or even block the growth of at least one kind of malignant tumor. The mechanism of this effect is unknown.

Between 1989 and 1995 several scientific studies addressed the possible relationship between exposure to organochlorine pesticides and breast cancer (Austin et al., 1989; Mussalo-Rauhamaa et al., 1990; Falck et al., 1992; Wolff et al., 1993; Dewailly et al., 1994; Krieger et al., 1994; Houghton and Ritter, 1995). Because of small sample sizes, most of these studies lacked the minimum power necessary to detect possible differences between DDT levels in breast cancer patients and the corresponding levels in women without breast cancer (López-Carrillo et al., 1996). The study with the largest sample size (Krieger et al., 1994) compared women with and without breast cancer among a cohort established between 1964 and 1971. The 150 subjects and controls included 50 whites, 50 blacks, and 50 Asian-Americans. There was no statistically significant difference in the DDE levels in the women with and without breast cancer.

One of the problems with such epidemiological studies is the fact that a large number of factors can influence breast cancer. The risk factor that links most known causes of breast cancer is exposure to endogenous estrogens. Early age at menarche, late age at menopause, nulliparity, and absence of breastfeeding all increase lifetime estrogen exposure and are associated with an increased risk of breast cancer (Kelsey et al., 1993). It is difficult to set up an epidemiological study in which all of these factors are more or less constant between two groups of women who differ only with respect to their exposure to certain pesticides. Furthermore, even if such an experiment could be designed, a person who has been exposed to increased levels of DDT may also have been exposed to increased levels of other pesticides. In such cases, it would be naïve to attribute all observed health effects to DDT.

In summary, the idea that DDT causes cancer is based almost entirely on studies with mice fed DDT at rates 1,000 times or more the rates of average human exposure in the United States during the 1960s. With the exception of some equivocal results from studies with rats fed even higher doses of DDT, there is no evidence that exposure to DDT causes cancer in other animals or in humans. Recent studies exploring a possible link between DDT exposure and breast cancer have been confounded by problems associated with experimental design and small sample size. The most extensive such study showed no statistically significant relationship between DDE levels and breast cancer. In fact, evidence suggests that under certain conditions DDT and DDD may actually have anticarcinogenic effects. The EPA, however, does not consider the results of the human studies an adequate basis for reaching conclusions regarding human carcinogenicity because the sample sizes were small and the studies short relative to the average human life span (EPA, 1980a).

Implications

Rational decisions about pesticides require an informed understanding of the benefits and costs associated with their use. DDT, for example, does have harmful effects on many nontarget species, and it is true that many target species have developed resistance to DDT. It is for these reasons that DDT use has been discontinued in many parts of the world. On the other hand, if DDT were a deadly carcinogen, it is very doubtful that governments would still authorize spraying the interiors of homes with DDT to combat malaria. Problems associated with pesticide use in the past have often stemmed from inadequate knowledge of the true environmental costs. It is important not to underestimate those costs; it is also important not to overestimate or exaggerate them. Synthetic chemical pesticides will probably continue to play a role in pest control for many years. An objective assessment of both the pros and cons of pesticide use will be necessary if that role is to be properly defined.

PESTICIDE PERSISTENCE IN THE BIOSPHERE AND FOOD CHAIN MAGNIFICATION

The persistence of pesticides in the environment is one of the important factors that determines their efficacy as pesticides and their impact on nontarget species. If pesticides degraded rapidly to harmless substances through natural processes, there would be little opportunity for them to be spread throughout the food web through feeding relationships and their concentrations to increase through biological magnification. On the other hand, if pesticides degrade rapidly, repeated applications may be necessary to achieve the same level of pest control that could be obtained with one or a few applications of a more persistent pesticide. Thus, the very characteristic that makes persistent pesticides desirable from the standpoint of pest control makes them undesirable from the standpoint of their impact on nontarget species.

The subject of pesticide persistence has been a controversial one, in part because the persistence of many pesticides depends very much on the surrounding environment. We again use DDT as an example. The half-life[7] of DDT sprayed on citrus crops is estimated to be about 50 days, but only 11–15 days on peaches and 7 days on alfalfa. Similarly, parathion has a half-life of about 78 days on citrus crops, 3–6 days on apples, and 2 days on alfalfa (Gunther and Jeppson, 1960). The situation is often very different if pesticides are washed into soils. Soil plots experimentally treated with 11.2 tonnes km^{-2} of DDT in

---

[7]The half-life is the time required for half of the substance to disappear.

1947 had a residue of 3.16 tonnes km^{-2} in 1951 (Rudd, 1964). The implied half-life is about 2.2 years. Similar studies by Fleming and Maines (1953), Lichtenstein (1957), and MacPhee et al. (1960) suggest a half-life for DDT in soils of 6.6, 5.2, and 5.4 years, respectively. The persistence of pesticides in soils, especially halogenated hydrocarbons, has been of particular concern with respect to groundwater contamination. Persistent pesticides may slowly migrate down through the soil complex, eventually reaching the water table. In some cases, the contamination does not become apparent until years after the first use of the pesticide, and may continue for many years after use of the pesticide has been discontinued (Anonymous, 1981; Yoshishige, 1991). Such contamination can create very serious public health problems because groundwater supplies are not easily decontaminated (see Chapter 16).

With respect to the marine environment, Wilson et al. (1970) reported that 92% of the DDT and its metabolites disappeared from a seawater sample stored in a sealed container that had been immersed in an outdoor flowing seawater tank for 38 days. This rate of disappearance would imply a half-life for DDT of only about 10 days under the experimental conditions. On the other hand, the fact that Clear Lake fish that had hatched as long as seven to nine months after the last DDD application to the lake contained 7–25 ppm DDD in their flesh obviously implies that DDD had persisted somewhere in the Clear Lake ecosystem for at least seven to nine months. Furthermore, the fact that intact brown pelican eggs taken off the southern California coast were found to contain more than 30 ppm DDE in 1969 (Figure 8.1) suggests an impressive degree of environmental persistence, since neither the pelican eggs, nor the pelicans themselves, nor the ocean, for that matter, were ever sprayed with DDT. The DDT must have entered the ocean from land runoff or from fallout (e.g., rainfall), entered the parent bird via the marine food chain or from direct contact with the water, and finally been transferred from the parent to the egg. Suffice it to say that this process must have taken longer than 10 days. Some idea of the persistence of DDT and its metabolites in the marine environment may be gained from an examination of the historical record of total DDT residues in anchovies taken from southern California coastal waters. Values reported by Anderson et al. (1975) are given in Table 10.3. As noted in Chapter 8, the sharp drop in residues between 1969 and 1970 was probably due largely to the virtual elimination of DDT discharges from the Montrose Chemical Company outfall in 1970. The second sharp decline, between 1972 and 1973, may be attributed to the EPA's ban on DDT use in the United States, which went into effect in January 1973. Since in both cases effects were evident in the anchovies within one year of the associated discharge reductions, it seems clear that the effective half-life of DDT and its metabolites in this system could not have been more than several months. However, the half-life must have been more than a few days; otherwise, DDT residues would have been degraded long before they appeared in the anchovies.

TABLE 10.3 DDT RESIDUES IN ANCHOVIES TAKEN OFF THE SOUTHERN CALIFORNIA COAST

Year	Total DDT Residues[a] in Anchovies (ppm, Fresh Weight)
1969	4.27
1970	1.40
1971	1.34
1972	1.12
1973	0.29
1974	0.15

[a]DDT residues = DDT + DDE + DDD.
Source: Anderson et al. (1975).

Confusion over the persistence of DDT and other pesticides in aquatic systems sometimes arises because the solubility of many pesticides in water is quite low and the dissolved concentrations are therefore very small. For example, Harvey et al. (1973) found the concentration of DDT residues in the North Atlantic Ocean to be less than 1 part per trillion, over 1,000 times less than the solubility limit. Although such observations may be taken to imply a short residence time for DDT residues in the water, they do not necessarily imply that DDT residues have been degraded. It is entirely possible that most of the DDT residues in the water are present in the plankton and other living organisms rather than in the dissolved state. Indeed, we have seen that the concentration of DDT residues in some aquatic organisms may be quite high. The half-life of DDT residues in the fatty tissues of such organisms may be considerably longer than the residence time of DDT residues dissolved in the water.

The issue of pesticide persistence is closely associated with the controversy surrounding food chain magnification. The fact that some pesticides are stored primarily in fatty tissues, where they are metabolized or excreted at a rather slow rate, has suggested to ecologists that such pesticides might become greatly concentrated at higher trophic levels in a food chain. The high concentrations of pesticides sometimes found in the bodies of carnivorous fish and birds lend further support to this argument. A well-known paper by Woodwell et al. (1967) has often been cited to support the idea of pesticide magnification through food chains. This paper deals with DDT residue concentrations measured in various animals from an extensive salt marsh on the south shore of Long Island, New York. The area selected was characterized by Woodwell et al. as being "representative of relatively undisturbed marsh." Table 10.4 summarizes some of Woodwell et al.'s findings.

The increase in pesticide concentrations at higher trophic levels is clearly evident in the table. The data are consistent with the effects expected from food chain magnification, but they do not prove that this mechanism is responsible for the observations. It is possible, for example, that higher trophic level organisms simply accumulate more DDT residues from the environment than do organisms further down the food chain.

In contrast to the Woodwell et al. (1967) results, Harvey et al. (1974) found no evidence of DDT residue magnification in Atlantic Ocean pelagic food chains. In fact, the highest DDT residue concentrations were found in the plankton, whereas flying fish, which feed largely on plankton, carried about 10 times *less* DDT residues than the plankton. At

TABLE 10.4 DDT RESIDUES IN SELECTED ORGANISMS FROM A LONG ISLAND SALT MARSH

Organism	Concentration of DDT Residues (ppm, Wet Weight)
Water	0.00005
Plankton	0.04
Silverside minnow	0.23
Sheephead minnow	0.94
Pickerel (predatory fish)	1.33
Needlefish (predatory fish)	2.07
Heron (feeds on small animals)	3.57
Tern (feeds on small animals)	3.91
Herring gull (scavenger)	6.00
Fish hawk (osprey) egg	13.8
Merganser (fish-eating duck)	22.8
Cormorant (feeds on larger fish)	26.4

Source: Woodwell et al. (1967). Copyright 1967 by the American Association for the Advancement of Science.

higher trophic levels, sharks have typically been found to carry high concentrations of DDT residues; but barracuda, which are also top-level carnivores, contain about 100 times lower concentrations than sharks (Giam et al., 1972). Such observations led Harvey et al. (1974) to postulate that chlorinated hydrocarbon concentrations in aquatic organisms may largely reflect a simple physical-chemical equilibrium between chlorinated hydrocarbon concentrations in the water and the lipid tissues of the organisms. Concentration differences between aquatic organisms in the same environment might then reflect differences in the solubility of chlorinated hydrocarbons in the lipid tissues of the organisms and/or metabolic differences affecting the storage and breakdown or excretion of chlorinated hydrocarbons by the organisms.

At least one systematic study of the biological magnification question (Hamelink et al., 1971) has shown that some fish (largemouth bass, green sunfish, and pumpkinseed sunfish) may pick up DDT residues primarily from the water by adsorption on their gills rather than from their food. In one set of experiments, ponds were set up containing either algae, invertebrates, and fish or algae and fish only. In the latter case, the fish were fed laboratory-raised brine shrimp, uncontaminated with DDT. DDT was added to the sediments of both ponds and rapidly appeared in the organisms in each system. There was no significant difference, however, between the DDT residue concentrations in the fish from the first pond, which ate invertebrates containing DDT residues, and in the fish from the second pond, which ate uncontaminated brine shrimp. Since the fish in the second pond obviously picked up their DDT residues entirely from the water, the implication is that the final concentration in the fish was independent of whether the DDT came solely from the water or from a combination of water and food. Interestingly, there was almost a perfect correlation over the time of the experiment between the concentration of DDT residues in the invertebrates and the concentration of DDT residues in the water (Figure 10.11),

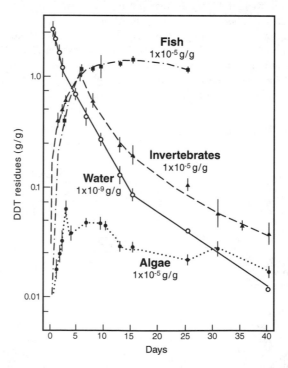

FIGURE 10.11   DDT residues in components of artificial ponds treated with 5 ppb DDT in 1966. *Source*: Redrawn from Hamelink et al. (1971). Reproduced with permission of the American Fisheries Society.

an observation that suggests that the invertebrates obtained much of their DDT directly from the water rather than from the algae. Nevertheless, in all experimental ponds that contained algae, invertebrates, and fish, there was a steady increase in the concentration of DDT residues from the algae to the invertebrates to the fish, similar to the results found by Woodwell et al. (1967) in the Long Island salt marsh. Mean concentration factors (versus water) were 7.2×10^3 for the algae, 1.8×10^4 for the invertebrates, and 1.7×10^5 for the fish. The results of the Hamelink et al. study suggest that effects similar to those attributed to food chain magnification may be caused by a very different mechanism, namely, the direct exchange of pollutants between an aquatic organism and the water, with the effective magnification determined by the metabolic characteristics of the organism. On the other hand, the results of Hamelink et al. do not rule out food chain transfers as a significant means of distributing pesticide residues within the biosphere, and indeed, both mechanisms (direct exchanges between organism and environment and food chain transfers) may play an important role in some systems. Without conducting more detailed studies, however, it is impossible to say to what extent food chain effects and direct exchanges are responsible for observations such as those made by Woodwell et al. (1967). It seems probable, however, that the transfer of pesticides through food chains has been a major factor in some cases, such as the poisoning of raptorial and fish-eating birds (e.g., western grebes on Clear Lake, California), a subject to which we now turn.

PESTICIDE EFFECTS ON BIRDS

Pesticides, particularly some chlorinated organics, have been blamed as the cause of reproductive failure among certain birds of prey (raptorial birds). These effects, perhaps more than any other single factor, aroused public concern over the indiscriminate use of pesticides during the 1960s and 1970s. According to Peakall (1970b) and Peakall and Peakall (1973), reproductive failure was associated with three general types of effects:

1. A remarkable thinning of the eggshells and much breakage of the eggs that were laid
2. Modification of parental behavior, leading to delayed breeding or failure to lay eggs altogether, improper incubation of eggs, and/or failure to produce more eggs after earlier clutches were lost
3. High mortality of the embryos and among fledglings due to elevated pesticide levels in the offspring

While acknowledging that some bird populations have been, at least temporarily, adversely affected by the unwise use of pesticides, Jukes (1974) has pointed out that some of the reports of declining bird populations are not borne out by bird census data,[8] and that in other cases population declines might have been caused by human harassment[9] or invasion of natural bird habitats. Peakall (1970b), however, reported that some brown pelican eggshells sampled off the California coast "were so thin that the eggs could not be picked up without denting the shells," a condition that can hardly be considered healthy.

The allegations made against pesticides are based both on field studies and on laboratory experiments that involved controlled feeding of pesticides to birds. The former have been concerned mainly with the first of the above-mentioned effects and have attempted to show whether or not significant correlations existed between pesticide residues in bird eggs and egg weight or shell thickness. The latter experiments have tended to be wider in scope, and have included studies of pesticide effects on the number of eggs laid, percent-

---

[8]For example, the Audubon Christmas Bird Count.
[9]For example, shooting of bald eagles.

age of eggshells cracked, mortality of fledglings, and the activity of certain enzymes within the parent birds. The literature dealing with pesticide effects on birds is so voluminous that it would be impossible to survey all the relevant articles here. In the following discussion we review some of the more important papers, pointing out areas where controversy has arisen. This discussion provides an excellent example of an issue on which supposedly knowledgeable scientists have sometimes reached very different conclusions, although their judgments have been based on essentially the same set of experimental facts and observations.

Field Observations

Changes either in eggshell weight or shell thickness of certain birds were reported both in England (Ratcliffe, 1970) and in the United States (Hickey and Anderson, 1968). The birds involved were peregrine falcons, sparrowhawks, and golden eagles in England and peregrine falcons, ospreys, and bald eagles in the United States. In England, Ratcliffe (1970) reported that the relative weight of peregrine falcon, sparrowhawk, and golden eagle eggs had declined by 19%, 17%, and 10%, respectively, since 1946, and that since 1950 egg breakage had occurred with unprecedented frequency in the nests of these birds. Ratcliffe observed no effect, however, on the frequency of egg breakage or relative egg weight in a number of other species studied. His observations indicated that the problem was peculiar to certain species or that some species were much more susceptible than others. The onset of the high frequency of egg breakage followed by a few years the introduction of DDT and BHC use in England (1946–1948). Lockie et al. (1969) attributed the thinning of golden eagle eggshells to the use of dieldrin for dipping sheep.[10] Indeed, a marked decline in the breeding success of English golden eagles did not begin until 1960, after the introduction of dieldrin (Peakall, 1970b).

Figures 10.12 and 10.13 show the results of two often-cited studies that relate DDE residues in eggs to eggshell thickness. In both cases, there appears to be a significant negative correlation between eggshell thickness and DDE residues. Actually, no one believes that thin eggshells are caused by the pesticides in the eggs. The physiological processes of the female bird control the thickness of the eggshell. The pesticide residues in the egg, however, are commonly assumed to reflect the residues in the female bird, and the latter residues may affect the bird's physiological processes.

Results such as those shown in Figures 10.12 and 10.13 have frequently been cited as implicating DDE as a causative agent of eggshell thinning. According to Spitzer et al. (1978), eggshell thinning of about 15–20% appears to be a critical level associated with hatching failure due to cracking of eggs during incubation. Others, however, have criticized these field studies as being statistically meaningless. Hazeltine (1972), for example, pointed out that the data in Figure 10.13 consist of results obtained from three separate regions involving two subspecies of pelicans. He argued that if the overall correlation were significant, each region "should show a similar trend in itself" (p. 410). If the data are broken down by region, only the results from the Florida study show a significant correlation between eggshell thickness and DDE levels.

The banning of DDT use in the United States at the end of 1972 provided scientists with an opportunity to study correlations between DDT residues (mainly DDE) and eggshell thickness in single geographic regions as residues in eggs declined. Figure 10.14 shows such results for southern California brown pelicans based on the data shown in Figure 8.1 from 1969 to 1986. The negative correlation between DDE residues and eggshell thickness in this figure is unmistakable. As noted in Chapter 8, the reproductive success of

[10]Golden eagles feed on dead sheep.

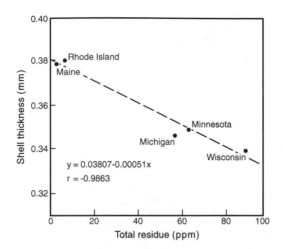

FIGURE 10.12   Variation in shell thickness and DDE concentrations in the eggs of herring gulls in 1967. The eggs were taken off Block Island, Rhode Island; Green Island in Penobscot Bay, Maine; Rogers City, Michigan, on Lake Huron; near Knife River, Minnesota; on Lake Superior; and the Sister Islands in Green Bay, Wisconsin. Some PCBs probably also occurred in these eggs, but they have not been identified. *Source*: Reprinted with permission from J. J. Hickey and D. W. Andersen. 1968. Chlorinated hydrocarbons and eggshell changes in raptorial and fish-eating birds. *Science*, **162**, 271–273. Copyright 1968 American Association for the Advancement of Science.

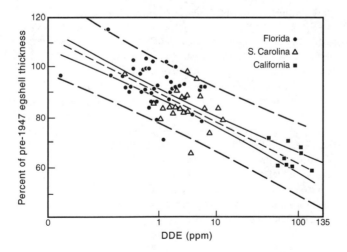

FIGURE 10.13   Association of DDE residues in 80 pelican eggs from Florida (solid circles), South Carolina (triangles), and California (solid squares) with the percentage of pre-1947 eggshell thickness. The regression line (narrow dashed line) is shown together with the 95% confidence limits for both the regression line (solid lines) and the individual ordinate values (thick dashed lines). *Source*: Redrawn from Blus et al. (1972b). Reproduced with permission.

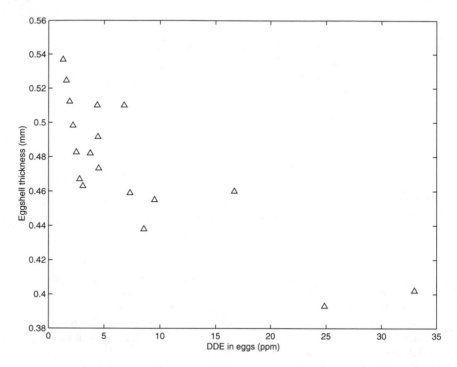

FIGURE 10.14   Relationship between DDE concentrations in southern California brown pelican eggs and eggshell thickness based on data shown in Figure 8.1 for the period 1969–1986.

these birds increased dramatically as the DDE levels in the eggs declined. Between 1970 and 1976 the DDE concentrations in the eggs of ospreys breeding in the Connecticut-Long Island area decreased fivefold, and mean eggshell thickness increased from about 0.41 to 0.45 mm (Spitzer et al., 1978). These results offer strong support to the idea that DDE is associated with eggshell thinning in at least some species of birds. The existence of simple correlations in field data, however, by no means proves a cause-and-effect relationship. In order to examine the possible role of pesticides in eggshell thinning, as well as to look for possible effects on parental behavior and juvenile mortality, scientists have performed a number of laboratory experiments on bird populations under controlled conditions in which the only significant variable has been the degree of pesticide exposure. Such experiments, it is hoped, eliminate much of the environmental "noise" caused by natural variations in environmental parameters other than pesticide exposure levels. Some of these experiments have been designed simply to determine whether exposure to pesticides influences the reproductive success of certain birds; other experiments have been aimed at elucidating the mechanism or mechanisms by which pesticides might affect reproduction.

Laboratory Studies

Why would we expect pesticides to interfere with the reproduction of birds? First of all, it is generally agreed that certain pesticides can upset the normal chemical balance in an organism by inducing the production of liver enzymes. Studies have shown that such enzymes can hydroxylate a number of substances, including the sex hormones estrogen, testosterone, and progesterone (Peakall, 1970b). The inference was then made that pesticides might be interfering with bird reproduction by upsetting the birds' hormone balance. Interference with a bird's sex hormones could affect parental behavior with respect to egg laying, incubation of eggs, and/or rearing of hatchlings. With respect to eggshell thickness, it is known that estrogen promotes the storage in the bird's bone marrow of a large percentage of the Ca needed to produce the eggshell (Taylor, 1970). Carbonic anhydrase (CA) and Ca-dependent adenosine triphosphatase (Ca-ATPase), enzymes found in the uterus of birds, are also important in eggshell formation (Peakall, 1970b; Pocker et al., 1971; Miller et al., 1976). Inhibition of CA by sulfanilamide, for example, results in poorly calcified eggs (Gutowska and Mitchell, 1945). Thus, pesticides could affect eggshell thickness either by depressing estrogen levels or by inhibiting CA or Ca-ATPase activity.

Peakall (1970a) performed a series of experiments with ringdoves to examine some of these hypotheses. He found that estrogen levels were depressed by 33% in the blood of birds given 10 ppm DDT in their feed for three weeks prior to breeding, and the DDT-fed birds were found to have stored only about 40% as much Ca in their bone marrow as had birds given uncontaminated feed. Furthermore, the time from mating to the laying of the first egg was 5 days longer (21.5 versus 16.5 days) for the DDT-fed birds. Both the reduced amount of Ca deposited in the bone marrow and the lengthened time to egg laying may have resulted from the lower estrogen levels in the DDT-fed birds.

The possible role of DDT in causing birds to produce thin eggshells through inhibition of CA is a somewhat controversial subject. CA is usually present in at least 100-fold excess in biological systems (Maren, 1967), so that inhibition of enzyme activity by at least 99% would be needed to produce a significant physiological effect. Both Peakall (1970a) and Bitman et al. (1970) conducted experiments in which birds were exposed to either DDT and DDE, and then the birds' blood or shell gland (uterus) was assayed for CA activity. They observed inhibition ranging from 20% to 59%. Inhibition by such amounts would not be likely to produce a noticeable physiological effect. Straus and Goldstein (1943), however, have noted that the degree of enzyme inhibition is often dependent on dilution effects, and since the method of sample preparation used by Peakall (1970a) and Bitman et al. (1970) involved dilution, there is a good chance that the inhibition they reported is an underestimation of the inhibition that may have occurred *in vivo*.

A number of other workers have attempted to observe CA inhibition *in vitro* (Anderson and March, 1956; Dvorchik et al., 1971; Pocker et al., 1971). These experiments included studies of CA from human red blood cells, from insects, and from cattle. The chemicals studied included DDT, DDE, and dieldrin. In no case was any chemical inhibition of CA activity observed. Dvorchik et al. (1971) and Pocker et al. (1971) both reported that in tests with very high pesticide concentrations, some inhibition of activity occurred due to physical precipitation of the enzyme with the pesticide. It seems unlikely, however, that this mechanism would underlie possible *in vivo* effects. Earlier, Torda and Wolff (1949) and Keller (1952) had reported inhibition of bovine and human red blood cell CA by DDT, but later attempts (above) failed to reproduce these results. It now appears that the earlier results were in error.

Miller et al. (1976) gave Pekin ducks and leghorn hens feed containing 40 ppm DDE. The ducks produced eggshells that were 18% thinner than those of controls after one to three months. The Ca-ATPase in whole gland homogenates from the ducks was depressed by 31% compared to controls after a similar time period, and there was a 19% depression in CA activity. Miller et al. (1976, p. 124) noted, "Since Ca in many tissues seems to be present in great excess, 10–20% reduction in the activity of duck shell gland suggests a minor role for CA in shell thinning caused by DDE." However, they noted that transport enzymes such as Ca-ATPase do not seem to be present in excess and that partial inhibition is generally associated with a reduction in ion transport. They concluded, "All the available evidence points to a major role for Ca-ATPase inhibition by DDE in the shell gland of the duck" (p. 124). In contrast, 40 ppm in the feed of the leghorn hens had no effect on eggshell thickness, ATPase, or CA. Clearly, some species of birds are more susceptible than others to eggshell thinning, and it appears that there may be several shell-thinning mechanisms, their relative importance being species dependent.

In addition to the papers of Peakall (1970a) and Bitman et al. (1970), a number of other reports (Bitman et al., 1969; Heath et al., 1969; Porter and Wiemeyer, 1969; Wiemeyer and Porter, 1970; Cecil et al., 1971) have often been cited as implicating DDT or DDE in the reproductive failure of birds. Unfortunately, some of these studies have been conducted in a manner that makes unambiguous interpretation of the results difficult, and as a result, the studies have occasionally been the object of some rather harsh criticism.[11] It is of interest to review those reports.

Bitman et al. (1969) fed Japanese quail a low-Ca diet containing 100 ppm DDT. The quail produced eggs with about 4% less Ca than controls and had a lag in egg production analogous to the lag reported by Peakall (1970a) for ringdoves. Edwards (1971) and others criticized the experiments on the grounds that the birds were artificially stressed by the low-Ca diet. In response to this criticism, Cecil et al. (1971) performed a second set of experiments in which Japanese quail were given an adequate amount of Ca and either 100 ppm DDT or DDE. The experimental birds produced over twice as many broken eggs as the controls. There was no significant difference, however, between the shell Ca content of the DDE-fed birds and those of controls, nor was there any significant difference in the eggshell thickness of the controls and either the DDT- or DDE-fed birds.

Porter and Wiemeyer (1969) fed combined doses of DDT and dieldrin to captive American sparrow hawks. They found a marked reduction in reproductive success, particularly among first-generation (yearling) hawks.[12] The eggshells produced by the yearling hawks were 15–17% thinner than those of controls. Unfortunately, it is impossible to tell from these experiments whether DDT had anything to do with the results. Peakall (1970b) found dieldrin to be an even more powerful inducer of liver enzymes than DDT, and in England, Lockie et al. (1969) found that golden eagle breeding success increased by 38% as soon as dieldrin sheep dips were banned. As Edwards (1971, p. 6) asked, "Why set up experiments feeding only mixtures of DDT and dieldrin, then blame the DDT for all ill effects?" Wiemeyer and Porter (1970) later reported the results of a study on captive American kestrels in which the experimental birds were given a diet containing 10 ppm DDE. No other pesticides were added to the feed. The eggshells produced during the first seven weeks of feeding were virtually identical in thickness to control eggshells, but after one year of feeding, the eggshells produced by the DDE-fed birds were about 10% thinner.

The laboratory study that most convincingly demonstrated that DDT residues in feed could adversely affect bird reproduction is probably that of Heath et al. (1969), who conducted a two-year study of the effects of feeding DDT, DDE, and DDD to mallard ducks. Several weeks before the onset of the first laying season, test ducks began receiving either 10 or 40 ppm DDE or DDD in their feed, or 2.5 or 10 ppm DDT. The results of the DDE feeding were discussed briefly in Chapter 8; a more complete summary of the results is shown in Table 10.5. Ducks fed 10 or 40 ppm DDE produced only 20–51% as many offspring as controls and produced eggshells about 10% thinner than those of controls. Ducks fed 10 or 40 ppm DDD produced only 44–57% as many offspring as controls, and although the experimental eggshells were slightly thinner than those of controls, the difference was not statistically significant. DDT-fed ducks produced slightly fewer offspring than controls

[11]For example, Edwards (1971) and the letter to the *Barometer* quoted at the beginning of this chapter.

[12]In other words, the offspring of hawks fed DDT and dieldrin.

TABLE 10.5 REPRODUCTION AND EGGSHELL THICKNESS DATA FROM PENNED MALLARDS MAINTAINED FOR TWO SEASONS ON FEED CONTAINING DDE, DDD, OR DDT[a]

| | | Pesticide Added to Feed (ppm) | | | | | |
| | | DDE | | DDD | | DDT | |
	Year	10	40	10	40	2.5	10
14-day ducklings per hen	1	51	35	57	51	97	89
	2	24	20	53	44	137	123
Shell thickness	1	92	87	97	97	—	—
	2	89	86	95	95	95	92

[a]aNumbers are results as a percentage of control results.
Source: Heath et al. (1969).

in the first study year but, surprisingly, 23–37% more offspring in the second study year. Eggshell thinning for the DDT-fed birds was not statistically significant.

The fact that the DDT-fed ducks produced more offspring than controls during the second study year has been offered by some as evidence that DDT has little or no adverse effect on bird reproduction (Edwards, 1971). The results of this study, however, leave little doubt that DDE and DDD at concentrations as low as 10 ppm in feed can greatly reduce mallard duck reproduction. Since DDE is a rather stable metabolite of DDT, the inference can be made that release of DDT to the environment could interfere with mallard duck reproduction through the effects of the metabolite DDE. It is therefore noteworthy that over three times as much DDE as DDT was found in southern California anchovies by Anderson et al. (1975), despite the fact that DDE itself has never been used as a pesticide. The anchovy is a principal food source of southern California brown pelicans, and according to Anderson et al. (1975), the DDE in southern California anchovies in 1969 was about 3.2 ppm on a wet weight basis. This figure is equivalent to about 9.7 ppm on a dry weight basis, a concentration virtually identical to the 10 ppm dry weight concentration that Table 10.5 indicates reduced mallard duck reproduction by at least 50%.

Summary of Pesticide Effects on Birds

Now that we have reviewed most of the better-known studies of pesticide effects on birds, what conclusions can we reach? First, the field studies of Blus et al. (1971, 1972a, 1972b), Hickey and Anderson (1968), Anderson et al. (1975), Spitzer et al. (1978), and Gustafson (1990) all show significant negative correlations between DDE residues in eggs and eggshell thickness. The last three studies are particularly relevant, since the data were collected from birds in the same geographical area and were therefore not subject to possible interregional biases. Other studies have shown that dieldrin is probably more effective than DDE in interfering with bird reproduction.

Second, Peakall (1970a) has convincingly demonstrated that DDT can depress estrogen levels in the bird's blood, and hence cause a delay in reproduction and a reduction in the amount of Ca stored in the bone marrow. Dieldrin was found to be even more potent than DDT in this respect.

Third, the evidence linking DDE to CA inhibition is weak. Several studies have reported no effect at all *in vitro*. In the two cases where some inhibition has been reported

(Peakall, 1970a; Bitman et al., 1970), the inhibition was almost certainly insufficient to produce a physiological effect. However, the issue is still open due to the difficulty of comparing *in vitro* and *in vivo* systems.

Fourth, the study of Miller et al. (1976) offered convincing evidence that DDE-induced depression of Ca-ATPase in the uterus of Pekin ducks could be a significant cause of eggshell thinning in that species.

Fifth, at least one laboratory experiment (Heath et al., 1969) involving controlled feeding of DDE to birds demonstrated that DDE has a significant effect on reproduction and eggshell thinning. The effects were evident at DDE concentrations of 10 ppm dry weight (about 3 ppm wet weight), a level that unfortunately could have been easily encountered in the natural food of some birds prior to the EPA ban on DDT use in 1972.

Finally, in the United States, there has been a dramatic reduction in the release of DDT and other pesticides to the environment following restrictions on their use (Figure 10.15). These reductions have been closely correlated with the dramatic recovery of several species of raptorial birds. In 1963, for example, there were fewer than 500 pairs of bald eagles in the contiguous 48 states. Today there are more than 5,000 pairs. In 1975, there were only 39 breeding pairs of peregrine falcons, all of them in the western half of the country. In 1996, 993 pairs were counted in the contiguous 48 states, including 153 pairs reestablished in the eastern United States. Ospreys increased from 8,000 breeding pairs in 1981 to 14,246 pairs in 1994. On the East Coast, brown pelicans were taken off the endangered species list in 1985 (EDF, 1997). They remain on the list elsewhere, but their numbers have steadily increased (Anderson and Gress, 1983) and their range has expanded. In England, golden eagles recovered dramatically following restrictions on the use of dield-

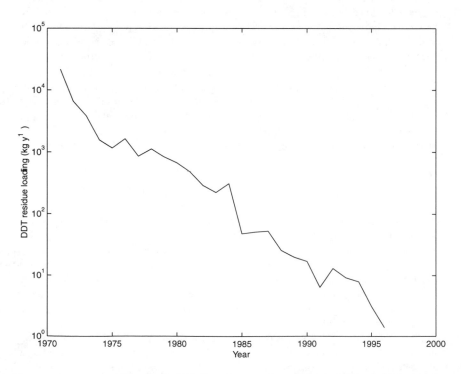

FIGURE 10.15   Annual loading of DDT residues to the southern California bight by the four municipal sewer outfalls that account for 89% of the point source discharges to the bight. *Source*: Raco-Rands (1997).

rin in sheep dips (Lockie et al., 1969). These observations strongly support the implications of the laboratory studies that some pesticides or their residues can have serious effects on bird populations when passed up through the food chain.

PEST RESISTANCE

Disenchantment with the widespread use of synthetic pesticides is often perceived as being largely the result of the adverse effects on nontarget species associated with pesticide use. Development of resistance in pest populations, however, has also been a major factor in the decline of pesticide popularity. In 1938, only seven insect and mite species were known to have acquired resistance to synthetic pesticides. By 1991, this figure had increased to 504 and included most of the world's major pests (Pimentel, 1991). Resistance to herbicides, which was virtually nonexistent before 1970, had become apparent in 273 weed species by 1991 (Pimentel, 1991).

In the case of insects, resistance, once acquired, appears to be more or less permanent. Resistance of Danish houseflies to the DDT and chlordane pesticide groups, for example, has persisted for more than 20 years, and according to Metcalf (1981, p. 472), "When these compounds were tried again they became useless within 2 months." A major problem with pest resistance is the fact that the effect of resistance genes is often not specific to a particular pesticide, but instead confers resistance to several or even many pesticides simultaneously. In northeastern Mexico, for example, the tobacco budworm, a major cotton pest, has developed resistance to every registered insecticide; and in Suffolk County, Long Island, New York, the Colorado potato beetle is now resistant to all major insecticides registered for use on potatoes (Postel, 1987).

Mechanisms of Resistance

Understanding of the mechanisms responsible for the development of pest resistance does much to explain this behavior. The principal mechanisms involved appear to be enhancement of the capacity to detoxify pesticides metabolically and alterations in target sites that prevent pesticides from binding to them (Arntzen et al., 1982; Brattsten et al., 1986). Certain herbicides that interfere with photosystem II, for example, compete with quinone for binding to thylakoid membranes. It has been proposed that herbicide resistance results when an alteration in the membrane reduces herbicide binding and hence allows quinone to bind even in the presence of the herbicide (Vermaas et al., 1983). Erickson et al. (1985) have identified three amino acid residues in the thylakoid membrane protein that can be independently altered to produce resistance to such herbicides.

In the case of insects, it is known that lipophilic insecticides are primarily detoxified by microsomal oxidases (Brattsten et al., 1986). With some exceptions, these enzymes convert the pesticides to polar metabolites that can be excreted. An important exception is the enzyme DDT dehydrochlorinase, which rapidly converts DDT to DDE (Metcalf, 1981). Considering the adverse effects of DDE on bird reproduction, it is ironic that DDE is noninsecticidal. Of the various microsomal oxidases, glutathione transferases are important in organophosphate detoxification (Montoyama and Dauterman, 1980), and carboxylesterases are often involved in pyrethroid resistance (Soderlund et al., 1983).

Target site modification is believed to be less common than metabolic detoxification as a mechanism of insect resistance, but several types of target site modification have been documented among insects. So-called knockdown resistance is associated with target site modification in the Na channel of the nervous system and was responsible for the rapid development of DDT resistance in houseflies (Winteringham et al., 1951). The gene responsible for knockdown resistance also confers resistance to pyrethroids (Sawicki and Farnham, 1967). A second type of target site resistance is a modification in the synaptic acetylcholinesterase that renders it far less sensitive to inhibition by organophosphates and carbamates (Brattsten et al., 1986).

The Cost of Pest Resistance

The cost of pest resistance appears in two forms. The first is the obvious fact that certain pesticides or classes of pesticides become virtually worthless for controlling pests. DDT, for example, was initially used with great success to control the insect vectors of the pathogens causing malaria, typhus, yellow fever, and plague. Largely because of pest resistance, DDT is either no longer used to control these vectors or, in the case of malaria, is used with much less success than initially. The second cost is quite simply the amount of money that must be spent to achieve a satisfactory level of pest control. Corn rootworm beetles, for example, were controlled from about 1954 to 1964 by soil applications of aldrin and heptachlor, but resistance to these insecticides became almost complete during the mid-1960s. A switch to organophosphates at that time was associated with almost a five-fold increase in pesticide cost per kilogram (Metcalf, 1981). Resistance of the Colorado potato beetle to insecticides has caused potato growers on Long Island to spray up to 10 times per season, and pesticide costs have risen as high as $700 per hectare (Postel, 1987). These heavy pesticide applications have caused serious contamination of groundwater, the region's only source of drinking water (Dover and Croft, 1986; Holden, 1986). The pesticides currently being used to replace DDT in malaria control programs cost 5–20 times as much per capita, and as a result, malaria control has become too costly for some developing countries. In addition to the previously discussed case of Sri Lanka, major recrudescences of the disease have occurred in India and Pakistan (Metcalf, 1981).

ALTERNATIVES TO SYNTHETIC PESTICIDE USE

The foregoing discussion leads one to ask, "Isn't there a better way?" In many cases, the answer is yes. In fact, there are several better ways. Most scientists now agree that widespread use of synthetic chemicals to control pests was a mistake and in some cases even counterproductive. There is now a consensus that reliance on such pesticides should be minimized and that alternative pest control methods should be utilized to the maximum extent possible. This realization has stimulated much basic and applied research into alternative control strategies. In the following discussion, we review the progress that has been made in this area.

Biological Control

The introduction of natural pest predators, competitors, or pathogens has in some cases proven an effective means of controlling pest populations. The idea is not new. Postel (1987) claims that since the 1860s, scientists have introduced roughly 300 organisms worldwide in biological control programs, but Holcomb (1970) maintains that almost 700 insect enemies alone have been introduced in such control efforts. Unfortunately, many such introductions do not produce the desired results. Holcomb (1970), for example, states that less than 25% of the introduced insect enemies actually became established. Even if the pest enemy does become established, there is no guarantee that it will reduce the pest population to an acceptably low level. Mongooses, for example, were introduced into Hawaii to control the rats that had taken up residence in sugar cane fields. The mongooses were quite successful at establishing themselves in the islands, but because the cane rats are nocturnal and the mongooses are active primarily during the day, the introduction had virtually no effect on the rat population. When biological control works, however, the effects can be dramatic and far less costly than the use of synthetic pesticides. The following sections summarize some of the more noteworthy successes of biological control.

Natural Predators and Parasites

In 1868 an insect called the cottony cushion scale (*Icerya purchasi*) was accidentally introduced into California on acacia plants from Australia. The spread of the scale onto citrus trees threatened to destroy the California citrus industry. In 1886 ladybug beetles (*Rodalia carbinalis*), a natural predator of the insect, were introduced in an effort to achieve biological control. The beetles reduced the pest population so effectively that no further effort to control the cottony cushion scale has proven necessary. Similar use of a small parasitic wasp, *Aphytis melinus*, has helped to control California red scale on citrus orchards in southern California (Marx, 1977). A related wasp introduced into Florida citrus orchards to control the snow scale has saved as much as $10 million per year in reduced insecticide costs.

In Australia, control of several species of prickly pear (genus *Opuntia*) has been successfully achieved by the accidental introduction of a moth (*Cactoblastis cactorum*) from Argentina. The prickly pear, which originated in the United States, had spread over 400,000 km^2 of Australian grazing land by 1925. In many areas the prickly pear had rendered the grazing land unusable, and in some areas the cactus was impenetrable (Gunn, 1976). The introduction of about 50 species of natural insect enemies of the prickly pear had proven largely unsuccessful. By 1930, however, following the accidental introduction of *C. cactorum*, the number of prickly pear had been so reduced that no further control measures were necessary. Deliberate introduction of the European beetle (*Chrysolina quadrigemina*) to control Klamath weed (*Hypericum perforatum*) on California rangeland in 1946 met with similar success. Roughly 8,000 km^2 were cleared for grazing. About $750,000 was spent to locate and introduce the beetle; accumulated savings from the biological control have been estimated at more than $100 million (Postel, 1987).

Cassava is a major food crop of some 200 million Africans, and much of it is grown on small plots by subsistence farmers (Postel, 1987). Mealybugs, which first appeared in Zaire in 1973, had infiltrated a large portion of the cassava belt by 1982 and, together with the spider mite, were causing cassava crop losses estimated at $2 billion per year. A pesticide program was ruled out because the infrastructure was lacking to deliver chemicals to the subsistence farmers. A search in Latin America, cassava's place of origin, turned up about 30 natural enemies of the mealybug, and one of these, a tiny wasp called *Epidinocarsis lopezi*, has produced remarkable results. By parasitizing the eggs of the mealybug, the wasp now effectively controls the population of the pest over 650,000 km^2 in 13 countries (Postel, 1987).

One of the principal insect pests of onions is the onion maggot, and without some form of pest control, onion losses in the United States average about 40%. Pimentel et al. (1991), however, reported that losses could be reduced to 2–3% by simply raising cattle adjacent to the onion field and mulching the onions with straw. A parasitic wasp uses the maggots in the cattle manure as an alternate host, and the straw protects a beetle that preys on the onion maggot.

In China, a parasitic wasp is used to control stem borers, a sugar cane pest, at one-third the cost of chemical control; and a fungus and parasitic wasp provide 80–90% control of a major corn pest (Postel, 1987). In Sri Lanka, pest damage to coconuts estimated at $11.3 million per year has been prevented with the introduction of a parasite found and shipped for $32,250 in the early 1970s (Postel, 1987). In Costa Rica, the use of pesticides on bananas was simply stopped. Natural enemies reinvaded to control the banana pests.

One variation of the predator/parasite control strategy is to release large numbers of a natural enemy at a critical time of the year. *Trichogramma*, a tiny wasp that parasitizes the eggs of certain butterflies and moths, is used extensively in this sort of control. The idea is to prevent the eggs from developing into crop-damaging caterpillars. *Trichogramma* is presently used to control moth pests on about 170,000 km^2 of cropland worldwide (Postel, 1987).

Pathogens and Natural Toxins

Insects are attacked by a great many pathogens, and about 450 viruses, 80 bacteria, 460 fungi, 250 protozoa, and 20 rickettsial diseases are effective natural enemies (Metcalf, 1981). A number of these are adaptable for use as insecticides, and in recent years, much interest has developed in using these pathogens or toxins derived from them to control insect populations. Such natural pesticides tend to be specific to one or a few pests and are generally harmless to other animals. The bacterium *Bacillus popilliae*, for example, has been used to control the Japanese beetle (*Popilbia japonica*) on lawns and golf courses. The strategy has been to collect *B. popilliae* spores from infected larvae of the beetle and to mix the spores with talc to produce a standardized powder, which is then applied to the soil. *B. popilliae* causes a disease (milky disease) that effectively controls the beetle population under certain conditions, but the control is not so effective in rough grass. Parasites taken from Europe to Canada to control the European spruce sawfly (*Neodiprion sertifier*) apparently carried a virus that proved highly effective in controlling the sawfly population (Gunn, 1976). The virus has since been deliberately spread to other parts of Canada and the United States. The first insect virus to be commercially developed in the United States was the nuclear polyhedrosis virus (NPV) of the cotton bollworm. Since then, NPV insecticides have been developed to control several other insect pests, including the corn earworm, the primary insect pest in sweet corn (Pimentel et al., 1991) and the gypsy moth (Marshall, 1981).

Undoubtedly the best-known example of a bacterial control agent has been the use of the bacterium *Bacillus thuringiensis*. *B. thuringiensis* was first identified as an insecticide in 1927. The bacterium is lethal to insects because it produces spores that release toxic proteins when eaten by insect larvae. It is nontoxic, however, to mammals and birds. *B. thuringiensis* can be used on a wide variety of agricultural crops to control insect pests, and spraying a mixture of spores and protein crystals on tree foliage has proven effective in controlling the gypsy moth (Moffatt, 1991). The bacterium has many subspecies, and each is effective against only a narrow range of insects belonging to the same order. This characteristic is desirable from the standpoint of impact on nontarget species, but it is undesirable from the perspective of companies contemplating commercialization. It is difficult to make money marketing a pesticide that affects only a very narrow range of pests. The specificity problem may well be overcome with genetic engineering techniques.

Most biological control methods developed to date have been aimed at insects, but in some cases, biological control methods are available for weed control as well. The development of such alternatives is particularly important in countries such as the United States, where herbicides now account for a major fraction of pesticide use. Two bioherbicides that rely on a fungus as the biological agent have entered the U.S. market. DeVine, which is produced by Abbott Laboratories, is used to control the milkweed vine in Florida citrus groves. The weed, also known as the *strangler vine*, climbs a tree's trunk, covers the crown, and eventually smothers the tree (Postel, 1987). Treatment with DeVine has been found to give at least five years' protection from this pest. Use of Collego, a bioherbicide marketed by the Upjohn Company, has produced 90% control of northern jointvetch, a weed commonly found in Arkansas rice and soybean fields (Postel, 1987). Another form of biological control of weeds is the planting of cover crops that inhibit the germination or growth of weeds. This approach takes advantage of a phenomenon called *allelopathy*, the inhibition of one plant by another through the release of natural toxins. Leaving residues of rye, sorghum, wheat, or barley on a field, for example, can provide as much as 95% weed control for up to two months (Postel, 1987).

Genetic Control

Resistant Plants

Through genetic manipulations such as cross-breeding and selection of desirable mutant strains, it has been possible to develop crop strains that are wholly or largely resistant to predation or infection by certain pests. For example, the European wine industry was saved from ruin a century ago by the practice of grafting nonresistant but high-quality European vines onto American root stocks resistant to the attacking aphids, *Phylloxera vastatix* (Gunn, 1976). Pest-resistant strains are available for a number of crops of agricultural importance in the United States, including alfalfa, beans, cole, corn, cotton, cucumbers, grapefruit, lemons, oranges, potatoes, rice, sweet potatoes, and wheat (Pimentel et al., 1991). In the case of wheat, for example, only about 7% of the hectarage is treated with insecticides, because host-plant resistance, crop rotations, and other agricultural practices have succeeded in minimizing insect damage. On the other hand, in the case of corn, which accounts for over 40% of agricultural insecticide use in the United States, farmers appear not to have taken full advantage of the availability of insect-resistant strains. Pimentel et al. (1991) estimate that the use of insecticides on corn in the United States could be reduced by 80% by rotating crops and planting corn resistant to the corn borer and chinch bug.

An important variation on the theme of resistant plants has been the use of genetic engineering to insert specific genes into the plant or into microorganisms that live in or on the plant. For example, the Rohm and Haas company has developed a genetically engineered strain of tobacco that contains a gene from the bacterium *Bacillus thuringiensis* (*Bt*). The gene triggers production of a *Bt* protein that is toxic to a broad spectrum of caterpillars that feed on plant leaves (Crawford, 1986). Similar insertions of *Bt* genes into cotton, corn, tomato, and potato plants have produced strains resistant to the cotton bollworm, pink bollworm and tobacco budworm (all cotton pests), European corn borer, tomato pinworm, and the Colorado potato beetle, respectively (Moffatt, 1991). A related strategy is to insert the *Bt* toxin gene into other bacteria such as *Pseudomonas fluorescens*, which grows in roots, and *Clavibacter xyli*, which grows in the vascular tissue of corn and other plants. Plants infected with the genetically engineered bacteria receive a continuous supply of insect toxins to protect them from attackers (Moffatt, 1991).

As with conventional chemical pesticides, development of pest resistance to the *Bt* toxin is a major concern (McGaughey and Whalon, 1992). There is fear that heavy reliance on crops genetically engineered to produce the *Bt* toxin will soon lead to insect populations dominated by *Bt*-resistant individuals. This fear was heightened by infestations of thousands of hectares of *Bt* cotton in Texas during the summer of 1996. Monsanto, the producer of the transgenic cotton, argued that the infestations were the result of an unusually large population of bollworms caused by a combination of very hot weather and the fact that many Southern farmers had planted corn—a breeding ground for bollworms—that year to take advantage of high corn prices (Kaiser, 1996a). Critics, such as the Union of Concerned Scientists, have suggested that current resistance management plans are inadequate and that large-scale planting with *Bt* corn and cotton will lead to the rapid emergence of resistant insect populations (Wadman, 1997).

Sterile Males

Releasing large numbers of sterile males into an insect population at the appropriate time may effectively reduce reproductive success, particularly if each female pest mates with only one male. One of the best-known and most successful examples of this type of pest control has been the eradication of the screwworm fly population from the United States. The female screwworm fly lays her eggs in open wounds on warm-blooded animals. She can utilize an opening as small as a tick bite. The eggs hatch in about one day and release hundreds of larval worms that burrow into the wound and feed on the animal's flesh. A full-grown steer can be killed in a week if the worms enter a dehorning wound and penetrate the brain, and infestation of the umbilical cord of a newborn calf can lead to death within a few days. The fly has even been known to kill humans by entering the brain or lungs (Palea, 1990). Prior to the advent of the control program, livestock losses to screwworm flies in the United States were estimated to be $40 million per year (Carson, 1962).

During the late 1950s, scientists at the U.S. Department of Agriculture began raising hundreds of millions of screwworm flies on artificial media and irradiating pupae with gamma rays just before they became adult flies. In this way, the scientists produced vast numbers of sterile male screwworm flies, which they then released into infested areas. The sterile males compete with wild males for female mates, and since a female screwworm fly typically mates only once, the sterile male release program can drastically reduce reproduction as long as the wild male population is greatly outnumbered by the sterile males. In the case of the screwworm fly program, the ratio of sterile males to wild males was about 10:1 (Bush et al., 1976). The control program was initiated in Florida in 1958–1959, with remarkable success. The screwworm fly was completely eradicated at a total cost of $10 million. The focus of the program then shifted to other parts of the southern United States, and by 1962 the screwworm fly population in the United States was confined almost entirely to the State of Texas.

The effort to control the screwworm fly population in Texas was initiated in 1962 and at first met with remarkable success (Figure 10.16). The number of annual infestations dropped from almost 50,000 in 1962 to fewer than 500 in 1969–1971. There was a remarkable epizootic, however, from 1972 to 1976, during which screwworm fly infestations in Texas averaged over 30,000 per year. The cause of this epizootic has never been identified beyond a reasonable doubt, but it appears to have been due to a combination of factors including local abiotic conditions, matters related to sterile fly production and release, seasonal effects, and general program execution (Krafsur, 1985). The success of the program returned in 1977, and the last screwworm fly infestations in the United States were reported in 1982. A barrier zone of sterile flies was set up along the U.S.-Mexican border to prevent reinfestation from Mexico, and an effort was initiated to eradicate the fly from Mexico. In 1976 a sterile screwworm plant was established at Tuxtla Gutierrez, Chiapas, Mexico, with a production capacity of 500 million sterile flies per week. Mexico was officially declared free of screwworms in 1991, Belize and Guatamala in 1994, and El Salvador in 1995. Screwworm flies have not been detected in Honduras since January 1995, and efforts are now underway to eradicate the pest from Nicaragua, Costa Rica, and Panama (USDA, 1999).

One of the chief problems with the sterile-male approach to insect control is getting enough sterile males into the population at the right time. In the Texas control program, for example, release rates averaged 1.0–9.2 million sterile males per day (Krafsur, 1985). Given that the density of this pest is only about 40–80 km^{-2} (Bush et al., 1976), one can imagine the problem with using sterile males to control a pest whose population density may be many orders of magnitude higher. For many insects, the job of rearing and then releasing adequate numbers of sterile males to check the pest population is simply impossible. The technique has been used, however, with some success as part of a program

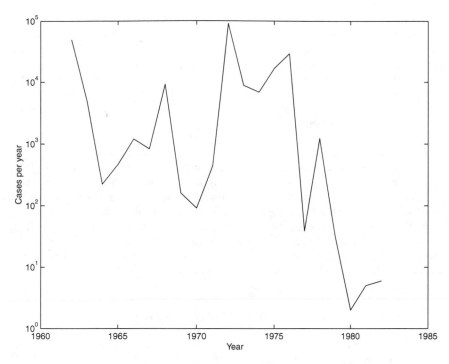

FIGURE 10.16   Incidence of screwworm fly infestations in Texas from 1962 to 1982.
Source: Krafsur (1985).

to prevent fruit flies from becoming a serious problem in southern California (Anonymous, 1990).

Chemical Control

Other than the synthetic pesticides, there are two other classes of chemical methods to
control pests. The first method is the use of hormones or natural biochemicals that affect
hormones to alter the growth and/or development processes of a pest and consequently
render it unable to grow, feed, or reproduce. For example, two chemicals isolated from the
East African medicinal plant *Ajuga remota* have been found to kill certain agricultural pests
by inhibiting molting. The chemicals cause insect caterpillars to grow as many as three
new head coverings or cuticles without shedding any old ones; as a result, their mouths
become so deeply buried inside the coverings that the caterpillars are unable to eat and
starve to death (Maugh, 1981). Unfortunately, the cost of developing such products can
be substantially higher than the cost of developing a conventional pesticide (Holcomb,
1970), and in some cases the chemical must be applied at just the right time in the life cycle of the pest to be effective. Some such products are available for mosquito and aphid
control, but it does not appear that this approach will become a major factor in the effort
to control insect pests without the use of conventional synthetic pesticides.

The second class of nonconventional chemical control involves the use of pheromones,
chemicals secreted by animals to influence other members of the same species. Pheromones
may be used to control insect pests in basically three ways. First, pheromones may be used
to attract insects into traps. This approach, for example, is used to control fire ants, root-

worms on corn, the sweet potato weevil on sweet potatoes, and the grape berry moth on grapes (Maugh, 1981; Pimentel et al., 1991). In the case of corn, this strategy is so successful that conventional insecticide use is reduced by 99% (Paul, 1989). Second, pheromones can again be used in traps, but with the purpose of providing an indication of the abundance of the insect pest. This approach is used in so-called scouting, the idea being to determine when best to apply pesticide treatments. The use of scouting can dramatically reduce requirements for conventional pesticides by confining their use to those times when pest populations are approaching nuisance levels. It is clear from empirical studies that without the use of scouting, conventional pesticides are often applied at times when their use is unnecessary (Pimentel et al., 1991). The third approach is to introduce sex-attractant pheromones into the air in amounts sufficient to confuse male insects, which are then unable to locate females for mating. This approach has been used with some success, for example, in gypsy moth control. In laboratory studies, the male gypsy moths become so confused by the smell of the pheromone that they "have attempted copulation with chips of wood, vermiculite, and other small, inanimate objects" (Carson, 1962, p. 252), so long as the objects were suitably impregnated with the pheromone. This control method breaks down, however, if the pest population density is too great because males encounter females an adequate number of times purely by chance.

Integrated Pest Management

Integrated pest management (IPM) has been defined by the United Nations Food and Agriculture Organization as "a pest management system that, in the context of the associated environment and the population dynamics of the pest species, utilizes all suitable techniques and methods in as compatible a manner as possible and maintains the pest populations at levels below those causing economic injury" (Batra, 1982, p. 135). What this statement means in practice is that one tries to minimize the use of conventional synthetic pesticides by using other means of pest control as much as possible, and one resorts to conventional pesticide applications in small doses if the pest population starts to get out of hand. The other methods used to control the pests may include one or more of the biological, genetic, or chemical control methods just discussed, and in the case of agricultural pests may also include the following:

1. Crop rotation to check the buildup of a pest specific to one crop
2. Fallowing for a year to similarly retard the development of crop pests
3. Use of ridge-till to facilitate weeding
4. Choosing the date of planting so as to be able to harvest the crop before the pest population has had a chance to peak
5. Use of rope-wick application of herbicides to increase the amount of herbicide that reaches the target weeds and hence reduce the amount of herbicide used

Recent developments in cotton growing provide a good example of IPM. Before the advent of IPM, cotton accounted for about 45% of all insecticides applied to agricultural crops in the United States (Marx, 1977). In the 1950s and 1960s, the boll weevil and cotton fleahopper were the key pests responsible for crop losses. Unfortunately, the insecticides used to control these two pests also killed the natural enemies of the bollworm and tobacco budworm, which then became serious pests (Adkisson et al., 1982). Furthermore, both the bollworm and the tobacco budworm began to develop remarkable resistance to conventional pesticides, and even though many fields were treated 15–20 times during the growing season, the crops were often severely damaged by insects. A strain of tobacco budworm began to emerge that was resistant to virtually all insecticides (Adkisson et al., 1982). It was obvious that some form of control involving minimal use of insecticides was needed.

The key element of the new control program has been the development of a new strain of cotton that matures quickly and can be harvested in late July or early August, about a month before other kinds of cotton. The population dynamics of the boll weevil is such that this short-season strain of cotton has been harvested before the boll weevil population becomes a serious problem. After harvesting, the cotton stalks are shredded and plowed under, and insecticides applied to the harvested cotton fields. This procedure kills many diapausing boll weevils and hence minimizes the number that emerge in the spring. Insecticides are again applied to the fields in the spring to reduce the number of surviving adults before they have a chance to reproduce. The springtime insecticide applications, which also control the cotton fleahopper, are limited and timed to have the least impact on the natural enemies of the bollworm and tobacco budworm (Adkisson et al., 1982). The overall strategy, then, is to let natural enemies control the bollworm and tobacco budworm, and to minimize the amount of pesticides required to control the boll weevil and fleahopper by growing short-season cotton, plowing under the stalks after harvesting, and carefully controlling the amount and timing of pesticide applications in order to minimize harm to the natural enemies of the bollworm and tobacco budworm. Cotton grown in the short-season system requires only about 25% as much insecticide as conventionally grown cotton, and in addition, only 25% as much N fertilizer and 50% as much irrigation water. The result is a substantial increase in profitability to the farmer (Table 10.6).

Experience with the use of IPM has been highly favorable. Not only is pesticide use greatly reduced, but in many cases profitability is substantially increased as well. Tables 10.6 and 10.7 summarize a few of the results. There is, however, considerable room for further improvement. In the United States, for example, there is presently minimal use of IPM on corn, which now accounts for almost twice as much agricultural insecticide use as any other single crop. According to Pimentel et al. (1991), simply rotating corn with another high-value product such as soybeans and using strains of corn resistant to the corn borer and chinch bug would not only lower insect losses but would also reduce the use of insecticides on corn by 80%.

TABLE 10.6 ESTIMATED AVERAGE ANNUAL ECONOMIC BENEFITS FROM USE OF IPM, SELECTED CASES, UNITED STATES, EARLY 1980s

State	Crop	Increase in Net Returns to IPM Users	
		Farm Level (Dollars ha^{-1})	*Statewide* (Thousand Dollars)
California	Almonds	769	96,580
Georgia	Peanuts	154	62,600
Indiana	Corn	72	134,230
Kentucky	Stored grain	<1	890
Massachusetts	Apples	222	400
Mississippi	Cotton	122	29,680
New York	Apples	528	33,000
North Carolina	Tobacco	6	780
Northwest[a]	Alfalfa seed	132	2,420
Texas	Cotton	282	215,830
Virginia	soybeans	10	2,570
Total			578,980

[a]Idaho, Nevada, Montana, Oregon, and Washington.
Source: Postel (1987).

TABLE 10.7 SUCCESSFUL APPLICATIONS OF IPM
IN AGRICULTURE

Country or Region	Crop	Effect
Brazil	Soybeans	Pesticide use decreased 80–90% over seven years
Jiangsu Province, China	Cotton	Pesticide use decreased 90%; pest control costs decreased 84%; yields increased
Orissa, India	Rice	Insecticide use cut by one-third to one-half
Nicaragua	Cotton	Efforts in early to mid-1970s cut insecticide
United States	Grain sorghum	Pesticide use reduced by 41% between 1971 and 1982
United States	Cotton	Pesticide use reduced by 75% between 1971 and 1982
Unites States	Peanuts	Pesticide use reduced by 81% between 1971 and 1982

Source: Postel (1987).

Another important area for improvement is the use of herbicides, which now account for 60% of agricultural use of conventional pesticides use in the United States (EPA, 1999). In the past, much of the development of alternative pest control techniques and implementation of IPM focused on insect pests, but there are techniques available to reduce the use of herbicides as well. In the United States, for example, corn accounts for over half of agricultural herbicide use, and there is good reason to believe that crop rotation combined with mechanical weeding could reduce the use of herbicides on corn by about 60% (Pimentel et al., 1991). In the so-called ridge-till system, corn and other row crops are planted in ridges about 20–25 cm high. The ridges are spaced about 75 cm apart and are aligned perpendicular to the slope of the land in order to minimize soil erosion. Mechanical weeding is greatly facilitated by planting the crops in ridges, and reliance on herbicides can therefore be greatly reduced. Another important development in weed control is the use of rope-wick applicators rather than spraying to apply herbicides. In the rope-wick technique, a 10- to 13-cm-diameter pipe with small holes punched into it functions as the reservoir for a liquid pesticide. Segments of rope inserted into adjacent holes in the pipe serve as wicks that deliver the pesticide to the weeds as the applicator, positioned behind a tractor, is pulled between rows of crops. Use of the applicator allows for much more direct and efficient application of pesticides to weeds than spraying. In the case of soybeans, for example, use of rope-wick applicators has reduced herbicide use by about 90% and increased soybean yields by 51% compared to conventional treatments (Pimentel et al., 1991).

COMMENTARY

It should be clear from the foregoing discussion that the indiscriminate and widespread use of pesticides can create and has created serious problems. The adverse effects of pesticides on nontarget species and the development of resistance in target species have both contributed to the decline in popularity of conventional synthetic pesticides and to the search for alternative means of pest control. Many of the alternative methods developed to date have focused on agricultural insect pests. Some remarkable success has been achieved with these alternatives, but it is clear that their potential has not been fully realized. We have a long way to go in terms of implementing alternative methods of controlling agricultural insect pests. There is also much to be done with respect to the identification and implementation of alternative methods of weed control in agriculture, and pest control in general in forestry management and public health. In the United States, for example, Dutch elm disease, chestnut blight, and the gypsy moth remain serious forestry pests, and we have yet to identify a satisfactory way to control tropical diseases such as malaria and river blindness. Despite the progress that has been made in the development of alternatives to conventional synthetic pesticides, it therefore seems likely that these synthetic pesticides will continue to be a significant component of many pest control programs in the foreseeable future. If we are to use such compounds wisely, it is important that we have a clear understanding of both the benefits and the problems associated with their use.

Two examples serve to illustrate the dilemma faced by government officials who must decide whether or not to authorize the use of conventional synthetic pesticides. Following the ban on DDT use in the United States, the EPA granted requests to the states of Washington and Idaho and to the Forest Service to use DDT on the basis of economic emergency and no effective alternative to DDT being available (EPA, 1980a). In accord with this policy, permission was granted in June 1974 to spray over 1,700 km^2 of Douglas fir trees in Oregon with DDT to control an outbreak of tussock moths. The government evidently felt that an emergency existed and that there was no effective alternative to DDT spraying. The decision was made just 18 months after the ban on DDT use in the United States went into effect, and was reached with a full awareness of the fact that large-scale killing of salmon and aquatic insects had accompanied forest spraying with DDT in parts of Canada. After evaluating the facts, the government was convinced, in this case at least, that the benefits of spraying outweighed the cost.

During the Vietnam War, the U.S. military used the herbicide Agent Orange extensively to defoliate areas of Vietnam. Unfortunately, this herbicide turned out to be contaminated with dioxin.[13] Dioxin is formed primarily as a contaminant during the production of 2,4,5-trichlorophenol, which is the major chemical feedstock in the production of several herbicides, including 2,4,5-trichlorophenoxyacetic acid (2,4,5-T), 2,4,5-T esters, and 2-(2,4,5-trichlorophenoxy)propionic acid (Silvex). According to the EPA (1984), dioxin is one of the most toxic substances known. It affects the immune system in mammals and has been shown to be acnegenic, fetotoxic, teratogenic, mutagenic, and carcinogenic. Dioxin has caused birth defects in experimental animals at concentrations as low as 10–100 parts per trillion and is so toxic to humans that only 85 grams could kill the entire population of New York City (Costa, 1983). Following the Vietnam War, a number of military veterans experienced health problems that they attributed to their exposure to Agent Orange, which contained 2,4,5-T. The health problems included cancer; nerve, liver, and immune system disorders; birth defects; and chloracne. They filed a lawsuit against seven chemical companies that had produced Agent Orange for the U.S. military and in 1984 won an out-of-court settlement of $180 million (Fox, 1984). Would the U.S. government use Agent Orange again? The answer is almost certainly no. The benefits do not outweigh the costs, and alternatives are certainly available.

At this point, it is appropriate to turn back to the beginning of this chapter and reread the excerpt from Dr. Edwards' letter. In the light of what you have learned from reading this chapter, to what extent do you agree or disagree with the statements made by Dr. Edwards? The fact is that DDT continues to be used for malaria control and at times for other purposes because the benefits of its use, under certain conditions and for certain purposes, are still perceived to outweigh the costs. The same conclusion has not been reached about some other synthetic pesticides, Agent Orange being an example. If pesticides are to be used intelligently in the future, a realistic understanding of their effects, both good and

---

[13]Dioxins are, strictly speaking, any of 75 chlorinated dibenzo-para-dioxin compounds, but the dioxin of particular concern is 2,3,7,8-tetrachlorodibenzo-para-dioxin (2,3,7,8-TCDD); we here refer to 2,3,7,8-TCDD simply as *dioxin*.

bad, will be crucial to the decision-making process. Biased and extreme statements by anti-pesticide environmentalists or pro-pesticide advocates do little to clarify a sometimes complex issue. Unfortunately, such statements have a habit of creeping into the literature, particularly when the issues involved are controversial. If this fact has not yet become clear, the following verbal exchange, which took place at EPA hearings conducted on January 13, 1972 (Woodwell, 1972), should prove thought-provoking. The two papers being discussed were published in *Science* by Dr. G. M. Woodwell (Woodwell and Martin, 1964; Woodwell et al., 1967).

EPA Hearings, 13 January, 1972: Afternoon Session

MR. O'CONNOR:

Q. Doctor, in the introductory sentence in the abstract, the abstract reads, DDT residues in the soil of an extensive salt marsh on the south shore of Long Island averaged more than 13 pounds per acre, 15 kilograms per hectare. The maximum was 32 pounds per acre, 36 kilograms per hectare. What was the lowest—what were the lowest residues that you found?

A. *Well, let's see. They're listed in the—in fact, the analyses are listed in Table 1 and where the samples were taken. We . . .*

Q. Isn't it fact, Dr. Woodwell, that after you wrote this, or after you initially studied this salt marsh, that you continued your samplings, and that you found as a result of your continued samplings that you were getting around or less than one—an average of one pound per acre of DDT?

A. *No, I wouldn't agree with that.*

Q. Did you continue your samplings, Dr. Woodwell?

A. *We have sampled the marsh subsequently and . . .*

Q. What were your averages in . . .

A. *—the conclusion we came to was that the marsh contains in the range of some pounds per acre of DDT.*

Q. Tell me how many, Doctor.

A. *I can't. It would be in the range of one to probably three pounds per acre.*

Q. One to three pounds per acre?

A. *Let me . . .*

Q. That is the average that you now hold is the DDT residue in this marsh?

A. *Let me clarify.*

Q. Is that true, Doctor: Would you answer my question? Is it one to three pounds?

A. *It is not true that the marsh now contains one to three pounds per acre.*

Q. What was true at the time, Doctor? What was true at the time of your writing this article?

A. *I would like to make a true statement. Would you like me to make a true statement or would you like me to simply say yes or no to what you say and which I consider untrue?*

EXAMINER SWEENEY:

Doctor, you're under oath so I imagine that every statement you make you realize must be true.

DOCTOR WOODWELL:

That's correct, and I intend to speak the truth here.

EXAMINER SWEENEY:

All right. Can you answer the question?

MR. O'CONNOR:

Let me give you the question again, Dr. Woodwell.

Q. Isn't it a fact that since your initial sampling which resulted in this article which is Respondents'—Environmental Defense Fund's Exhibit 41, that you took additional samples of the salt marsh and concluded that the residues of DDT in that salt marsh were one to three pounds per acre?

A. *It is true that we sampled the marsh subsequently and I believe that at the time of this sampling, the average residues in that marsh, obtainable from an extensive sampling of the marsh, would have revealed residues in the range of a few pounds per acre. It is also true that this sampling was deliberately biased in order to find the highest residues we could find, because at the time we did this study, we wondered whether we could find any residues in the marsh.*

Q. What do you mean by "average of 13 pounds per acre," if what you're talking about is a bias? Is 13 pounds per acre, Doctor, an average or is it a bias?

A. *It means the average of the samples we took in this marsh was 13 pounds per acre.*

Q. That's not what this sentence says, Doctor. It says, "The DDT residues in the soil of an extensive salt marsh was 13 pounds per acre." Is that true or is it not?

A. *That's true what that says. Based on this sampling, the average residues were 13 pounds per acre.*

Q. Didn't you also find out later, Doctor, that one of the areas where you took your samples was an area of DDT dumping?

A. *I can't say that I discovered that.*

Q. Dr. Wurster perhaps?

A. *I don't believe that he knows that either. I don't believe there's any evidence to that effect.*

Q. Are you aware of the statement that Dr. Wurster made at the Wisconsin hearings—not the Wisconsin hearings, the Washington State hearings—where he said—and I'm quoting from the verbatim transcript of the proceedings, "we have since sampled that marsh much more extensively and we found that the average, the overall figure on the marsh is closer to one pound per acre. The discrepancy was caused by the fact that our initial sampling was in a convenient place, and this turned out to be a convenient place for the Mosquito Commissions's spray truck, too." Did you learn that after the fact, Doctor?

A. *That is a true statement in my experience. I did not know that Dr. Wurster had said that, but that is a true . . .*

Q. So one of the places of your sampling was a place where a dump truck had unloaded its DDT?

A. *That is not what Dr. Wurster said.*

Q. What did he say, then?

A. *If I may read the record, I'd be glad to read it to you. Could I see . . .*

Q. What is your interpretation of what we've just read, Doctor?

A. *Could I read the paper that you've just introduced?*

Q. Is what I just read to you unclear? Do you understand what Dr. Wurster said?

A. *I would like to cite it directly.*

Q. Well, I've just read it to you. Did you not understand that?

MR. BUTLER:

Excuse me. The Doctor asked to see the paper to the question he answered. I think it would be reasonable if he could do so.

EXAMINER SWEENEY:

The only point is, Mr. Butler, the Doctor has already answered the question, and then after he answered the question, all of a sudden, he apparently figures it wasn't read correctly. He answered the question. He said, "That is a true statement," and so forth. And later on, he decided for some reason or other, that he wanted to check it to see if it was. I will ask Mr. O'Connor to reread the question and we will then get the answer from the witness. Reread the question that you asked him, please.

MR. O'CONNOR:

The statement which I read, Dr. Woodwell, was this:

Q. "We have since sampled that marsh much more extensively and we found that the average, the overall figure on the marsh is closer to one pound per acre. The discrepancy was caused by the fact that our initial sampling was in a convenient place and this turned out to be a convenient place for the Mosquito Commission's spray truck, too."

A. *That's true. Now, you suggested that the spray truck had dumped there, and that is not true as far as I know.*

Q. What does it mean by the statement that it was "a convenient place for the spray truck, too?"

A. *My interpretation of what Dr. Wurster meant is that it was close to the road and the people who control mosquitoes tend to work most intensively close to the edge of the road, and the net result of that was that over a period of years they visited if very often and put in large quantities of DDT, perfectly straightforward . . .*

Q. Doctor, have you ever published a retraction of this 13 pounds per acre or a qualification, or a further article which evidences and discloses the results of your further sampling which brings the average down to around one part per million?

A. *I never felt that this was necessary.*

Q. Isn't it a fact, Doctor, that 13 parts per acre—13 pounds per acre has been cited as a figure quite extensively in the literature on DDT, your figure here, which is used in this article, which is EDF Exhibit 41?

A. *I am not sure that I would agree that it has been cited extensively. It has certainly been . . .*

Q. You know that the figure "13 pounds per acre" has been cited in the literature, and you never took it upon yourself, Doctor, to write any kind of letter to the editor, a correction, further article, anything to clarify that error?

A. *I don't consider that an error. It is not an error in the context of this paper. If I . . .*

Q. How many samples did you take in this salt marsh, Doctor?

A. *The sampling program is outlined in this paper, and . . .*

Q. How many samples did you take, Doctor?

A. *—the samples are listed . . .*

Q. Was it six samples?

A. *It looks as though there are six samples plus one in the bay bottom.*

Q. A routine survey by a department such as the Department of Agriculture of a lake that has 2 to 300 acres to it, a marsh that has that number of acres to it, would be about 50 samplings, would it not?

A. *It could be. It could be larger. You have to recognize that the objectives in this study were to determine whether we could find DDT in the marsh. We reported what we found.*

Q. Didn't you also publish another article, Doctor, called "Persistence of DDT in Soils of Heavily Sprayed Forests?"

EXAMINER SWEENEY:

Before you go on, Mr. O'Connor, I'm kind of confused here as to what samplings we're talking about. Let me ask the Doctor—now, with reference to Intervenor EDF Number 41, where in the abstract, in the beginning, they refer to 13 pounds per acre, it is your testimony that you continued the experiment or you renewed the experiment in this same salt marsh and determined that there was something other than 13 pounds per acre?

DOCTOR WOODWELL:

We sampled more extensively over a period of a couple of years, a couple of years after that, and found that an estimate of average levels in that marsh based on a more extensive sampling, would give an average in the range of a pound or two pounds per acre, with some spots rising to much higher levels and some spots having less.

EXAMINER SWEENEY:

Is that difference of between 13 pounds and an average of one to three pounds a significant difference, in your opinion?

DOCTOR WOODWELL:

Oh, yes.

MR. O'CONNOR:

Mr. Sweeney, I move to strike this article as misleading.

EXAMINER SWEENEY:

No, I don't think so. They have offered it as an exhibit and I think the testimony will show what this Doctor has said regarding this article. I think it should stay in.

POLYCHLORINATED BIPHENYLS

Polychlorinated biphenyls (PCBs) are complex mixtures of Cl substituted biphenyls (Figure 10.17). Although PCBs are not pesticides, their similarity in many ways to chlorinated hydrocarbon pesticides makes their inclusion in this chapter appropriate. PCBs are no longer produced in the United States, but they were manufactured by the Monsanto Chemical Corporation from 1929 to 1977. As of 1975, nine different PCB formulations were being produced by the company (Boyle, 1975), but all were sold under the trade name Aroclor. Most PCB formulations are liquids or resins, although a few are powders. PCBs are exceptionally stable compounds that are nonflammable and highly resistant to strong reagents and heat. Destruction by burning requires a temperature of over 1300°C (Martin, 1977). Because of these characteristics, PCBs have been used in a variety of industrial applications. Prior to 1970, PCBs were used primarily in closed or semiclosed systems in electrical transformers, capacitors, heat transfer systems, and hydraulic fluids. They were used to a minor extent in paints, adhesives, caulking compounds, plasticizers, inks, lubricants, carbonless copy paper, sealants, coatings, and dust control agents (EPA, 1979). Foreign imports of PCBs generally accounted for only a small percentage of PCB use in the United States and were used primarily as plasticizers.

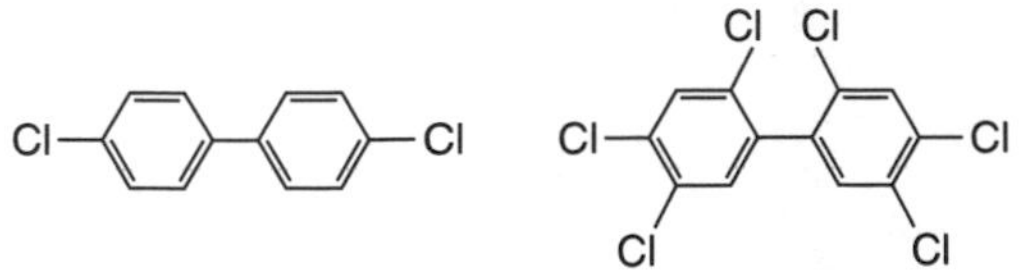

FIGURE 10.17   Two PCB compounds.

Problems with PCBs

PCBs have been found to display a degree of toxicity to certain organisms comparable to that displayed by some pesticides. Peakall (1970b), for example, found PCBs to be even more potent inducers of liver enzymes in birds than DDE, and Peakall and Peakall (1973) found that PCBs at 10 ppm in feed significantly lowered reproduction in ringdoves. In fact, there is good evidence that PCBs interfere with the reproduction of a variety of organisms, including rodents, fish, fowl, and primates (Maugh, 1975). The reproduction of mink, for example, is completely halted if the PCB concentration in their food exceeds 5 ppm (Maugh, 1972). In experiments at the University of Wisconsin, eight female rhesus monkeys were given a diet containing 2.5 ppm of Aroclor 1248 (a PCB formulation) for six months. The females were then bred to healthy males. Three of the females resorbed their fetuses shortly after conception. The five babies born to the other females were all undersized, and two died while still nursing. The remaining three showed signs of being hyperactive (Boyle, 1975). The mothers' breast milk was found to contain 7–16 ppm PCBs on a lipid weight basis (EPA, 1980b). In experiments with chickens, hens were fed a diet containing 20 ppm of certain PCB formulations for five weeks. Only 8% of the fertilized eggs hatched, and many of the embryos exhibited teratogenic abnormalities. The effect on the hens, however, appeared to be specific to only certain PCB formulations (Maugh, 1972).

As with chlorinated hydrocarbon pesticides, PCB concentrations in most parts of the ocean have always been quite low. Harvey et al. (1973), for example, reported mean PCB concentrations in the upper 200 m of the North Atlantic Ocean to be about 20 parts per trillion, although concentrations as high as 100 parts per trillion were sometimes measured (Harvey et al., 1974). The latter observation is significant, since Fisher et al. (1974) found that PCB concentrations as low as 100 parts per trillion could cause substantial disruption of phytoplankton communities grown in continuous culture.

As would be expected, PCB concentrations are much higher in the lipid tissue of aquatic organisms than in the water. Harvey et al. (1974), for example, reported PCB levels in North Atlantic plankton lipids as high as several parts per thousand in samples rich in phytoplankton. Whether PCB levels this high in phytoplankton lipids might be adversely affecting photosynthetic rates is unknown because the appropriate laboratory studies have never been done.

PCB concentrations in marine fish have never been a serious problem, at least insofar as commercial catch utilization is concerned. PCB levels in the muscle tissue of fish such as cod, haddock, and halibut sampled by Harvey et al. (1974) ranged from 2 to 190 ppb, well below the present Food and Drug Administration (FDA) guideline of 2 ppm for edible fish flesh. These results are typical for marine fish. During the 1970s, however, PCB levels in some freshwater fish, particularly fatty species, were found to exceed 5 ppm. The contaminated fish included lake trout and coho and chinook salmon, as well as anadromous species such as striped bass (Martin, 1977). Particularly contaminated freshwater systems included the Great Lakes and the Hudson and Mississippi rivers.

With respect to human health, PCBs rank third in toxicity behind dioxins and furans when the most toxic isomer of each group is considered (Sun, 1983), but how dangerous PCBs in general are to human health is a bit unclear. In Yusho, Japan, 1,291 persons became ill after eating rice bran oil contaminated with PCBs that had leaked from a heat exchanger. Twenty-nine of those persons died over a two-month period in 1968, and many developed chloracne, a disfiguring acne-like disorder (Sun, 1983). The PCB concentrations in the rice oil averaged 2,500 ppm, and persons affected by the sickness were estimated to have ingested about 2 grams of PCBs. The rice oil, however, was later found to have contained 1.6–5 ppm of polychlorinated dibenzofurans, which are estimated to be several orders of magnitude more toxic than PCBs (Martin, 1977). Although many of the symptoms of illness displayed by persons who consumed the rice oil were typical of PCB poisoning, it is possible that the polychlorinated dibenzofurans contributed significantly to the health problems of the victims. One of the problems in evaluating the human health effects of PCBs is the fact that commercial PCB mixtures often do contain small amounts of polychlorinated dibenzofurans (EPA, 1980b).

A group of Michigan sport fishermen who had been consuming about 10 kg of PCB-contaminated fish per year over a two-year period exhibited no adverse health effects, although PCB levels in their blood were higher than normal. The fishermen were estimated to have consumed about 46.5 mg of PCBs per person per year (Martin, 1977). These results are reminiscent of the experimental results of Hayes et al. (1971) in feeding DDT to human volunteers. However, in both cases, it is possible that the PCB- or DDT-contaminated food might have induced long-term effects that did not become apparent during the course of the study.

There have been several studies of persons whose health was affected by occupational exposure to PCBs. Workers who used PCBs to make capacitors at two General Electric plants near Albany, New York, for example, complained of allergic dermatitis, nausea, dizziness, eye irritation, and asthmatic bronchitis (Maugh, 1975). Workers involved in testing dielectric fluids at a Westinghouse plant similarly complained of sore throats, skin rash, gastrointestinal disturbances, eye irritations, and headaches (EPA, 1980b). The blood PCB concentrations of the Westinghouse workers were substantially above the range of values reported for the general population (EPA, 1980b). Although the levels of PCBs to which these workers were exposed is a bit unclear, some of the same symptoms have been produced in rhesus monkeys fed a diet containing 300 ppm PCBs (Allen and Norback, 1973).

The issue of PCB pollution is particularly relevant to water pollution because for most persons the principal source of exposure is the consumption of contaminated fish; and until the mid-1970s, the concentrations of PCBs in the human diet were comparable to the levels that proved toxic to rhesus monkeys. A survey reported in 1977, for example, that the average PCB concentration in the breast milk of women in the United States was 1.8 ppm on a lipid weight basis (42 *Federal Register*, 17487), a figure that is disturbingly close to the range of 7–16 ppm that proved toxic to infant rhesus monkeys. During the first half of the 1970s, the average daily intake of PCBs by persons in the United States was about 10 μg (EPA, 1980b). It is noteworthy that if one actually consumed 17.7 grams of fish and shellfish per day (Chapter 8) containing the present FDA tolerance level of 2 ppm PCBs, one would be consuming 35.4 μg of PCBs per day.

Persistence of PCBs

One of the chief concerns over PCBs has been their high degree of persistence in the environment. Because of this persistence, it was feared that serious environmental contamination with PCBs would be difficult to reverse in a short time, since the compounds would continue to cycle in the ecosystem for perhaps many years. Some indication of the persistence of PCBs is evident from the extensive sampling by Harvey et al. (1974) in the North Atlantic Ocean. Based on estimates of DDT and PCB production and loss rates to the environment, Harvey et al. (1973) estimated that the ratio of PCBs to DDT residues in the environment should be about 0.1. Harvey et al. (1974), however, found the ratio to be about 30 in the marine atmosphere, surface seawater, and plankton of the North Atlantic. The implication is that PCBs are 300 times more persistent than DDT residues in the marine environment.

Realizing the potential seriousness of the PCB contamination problem, the Monsanto Chemical Corporation voluntarily restricted its manufacture of PCBs in 1971 to include only uses in electrical capacitors and transformers. The company reasoned that PCB leakage to the environment from capacitors and transformers would be much less of a problem than PCB leakage from such previous sources as discarded lubricants and hydraulic fluids. There is good evidence that this restriction on PCB usage greatly reduced the extent of environmental loading with PCBs. Figure 10.18 shows annual loading rates of PCBs to the southern California bight from the four major sewer outfalls in the Los Angeles-San Diego area. The loading rate declined dramatically during the 1970s, and since 1987, PCB concentrations in the wastewater have been below the limit of detection.

Despite Monsanto's voluntary restriction on PCB production and despite reductions in discharge rates, the total amount of PCBs in the environment continued to increase during the first half of the 1970s (Maugh, 1975; Figure 10.19). As a result of this discovery and because of the demonstrated toxic effects of PCB formulations, Monsanto voluntarily terminated all manufacture of PCBs in 1977, and the EPA, under authorization of the Toxic Substances Control Act, banned the manufacture, processing, distribution in commerce, and use of PCBs in the United States effective July 2, 1979. The EPA ruling, however, allowed the continued use of PCBs in existing enclosed electrical equipment under controlled conditions. The lifetime of such equipment is approximately 20–30 years. Although it was initially felt that such use would not pose a significant environmental danger, several accidents involving electrical transformers heightened public concern and caused utility companies to accelerate the phaseout of PCB-containing equipment rather than simply waiting for it to wear out (Sun, 1983). The cost of the phaseout has not been cheap. Pacific Gas and Electric, for example, allocated $60 million in 1983 to replace 1,000 large PCB transformers after one of them exploded and contaminated a high-rise office

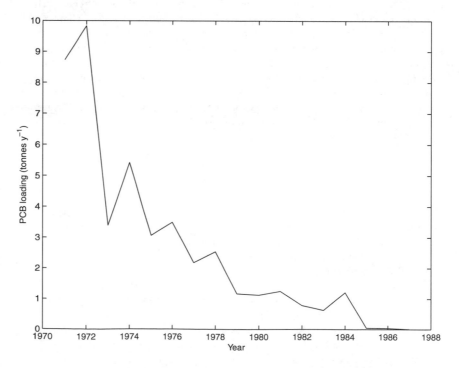

FIGURE 10.18   Annual loading of PCBs to the southern California bight by the four municipal sewer outfalls that account for 89% of the point source discharges to the bight. *Source*: Raco-Rands (1997).

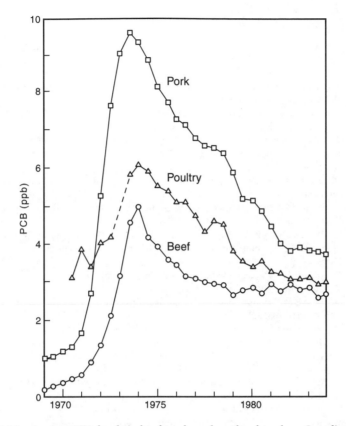

FIGURE 10.19   Average PCB levels in beef, pork, and poultry based on Canadian analyses from October 1969 to April 1985. *Source*: PCBs: A Case Study, Proceedings of a Workshop on Great Lakes Coordination Research held November 1985. Report to the Great Lakes Science Advisory 1985. Report to the Great Lakes Science Advisory Board by the Council of Great Lakes Research Managers. Windsor, Ontario, February 1988, 113 pp.

building in San Francisco. At the time, it was estimated that there were 20,000–30,000 PCB-containing transformers in the United States (Sun, 1983).

Ultimately, most PCBs discharged to aquatic systems are presumably either degraded in the water column or buried in the sediments. Elder and Fowler (1977), for example, discovered that sinking of zooplankton fecal pellets may be the major mechanism for transporting PCBs to the ocean floor. Similar mechanisms are presumably operative in freshwater systems. Gradual degradation and burial has probably accounted for much of the decline in PCB concentrations in Great Lakes fish during the 1980s (Figure 4.15). It is noteworthy, however, that in 1985 several Beluga whales died in the St. Lawrence River, and autopsies revealed PCB levels that were utterly fantastic—575 ppm in the lipid tissue and as high as 1,750 ppm in the milk (IJC, 1985). Such events dramatically underscore the persistence of PCBs and the tendency of such stable compounds to be concentrated through biological processes. It seems fair to say that the decision to terminate all use of PCBs and to phase out old PCB-containing equipment came none too soon.

REFERENCES

ADKISSON, P. L., G. A. NILES, J. K. WALKER, L. S. BIRD, and H. B. SCOTT. 1982. Controlling cotton's insect pests: A new system. *Science,* 216, 19–22.

AKESSON, N. B., and W. E. YATES. 1984. Physical parameters affecting aircraft spray application. In W. Y. Garner and J. Harvey, Eds., *Chemical and Biological Controls in Forestry.* American Chemical Society Series 238, Washington, DC. Pp. 95–115.

ALLEN, J. R., and D. H. NORBACK. 1973. Polychlorinated biphenyl- and triphenyl-induced gastric mucosal hyperplasia in primates. *Science,* 179, 498–499.

ALMEIDA, W. F., P. PIGATI, R. GAETA, and M. T. S. UNGARO. 1975. DDT residues in human blood serum in Brazil. In F. Coulston and F. Korte, Eds., *Environmental Quality and Safety, Supplement Volume III: Pesticides.* Georg Thieme Publishers, Stuttgart. Pp. 586–588.

ANDERSON, A. D., and R. MARCH. 1956. Inhibitors of carbonic anhydrase in the American cockroach, *Periplaneta americana* (L.). *Can. J. Zool.,* 34, 68–74.

ANDERSON, D. W., and F. GRESS. 1983. Status of a northern population of California brown pelicans. *Condor,* 85, 79–88.

ANDERSON, D. W., J. R. JEHL, R. W. RISEBROUGH, L. A. WOODS, L. R. DEWEESE, and W. G. EDGECOMB. 1975. Brown pelicans: Improved reproduction off the southern California coast. *Science,* 190, 806–808.

ANDERSON, D. W., R. M. JUREK, and J. O. KEITH. 1977. The status of brown pelicans at Anacapa Island in 1975. *Calif. Fish Game,* 63, 4–10.

ANONYMOUS. 1981. DBCP poison found in California wells at levels above state's safety standards. *Honolulu Advertiser,* March 31, p. B-8.

ANONYMOUS. 1990. Island fruit-fly funds OK'd. *Honolulu Star-Bulletin.* September 30, p A18.

ARNTZEN, C. J., K. PFISTER, and K. STEINBACK. 1982. The mechanisms of triazine resistance; alterations in the chloroplast site of action. In H. Lebaron and J. Gressel, Eds., *Herbicide Resistance in Plants.* Wiley-Interscience, New York. Pp. 185–214.

AUSTIN, H., J. E. KEIL, and P. COL. 1989. A prospective follow-up study of cancer mortality in relation to serum DDT. *Am. J. Public Health,* 79, 43–46.

BAGLA, P. 1997. Malaria fighters gather at site of early victory. *Science,* 277, 1437–1438.

BATRA, S. W. T. 1982. Biological control in agroecosystems. *Science,* 215, 134–139.

BITMAN, J. H., C. CECIL, and G. F. FRIES. 1970. DDT-induced inhibition of avian shell gland carbonic anhydrase: A mechanism for thin eggshells. *Science,* 168, 594–595.

BITMAN, J., C. CECIL, S. J. HARRIS, and G. F. FRIES. 1969. DDT induces a decrease in eggshell calcium. *Nature,* 224, 44–46.

BLEDSOE, T., D. P. ISLAND, R. L. NEY, and G. W. LIDDLE. 1964. An effect of o,p'-DDD on the extra-adrenal metabolism of cortisol in man. *J. Clin. Endocr.,* 24, 1303–1311.

BLUS, L. J., C. D. GISH, A. A. BELISLE, and R. M. PROUTY. 1972a. Logarithmic relationship of DDE residues to eggshell thinning. *Nature,* 235, 376–377.

BLUS, L. J., C. D. GISH, A. A. BELISLE, and R. M. PROUTY. 1972b. Further analysis of the logarithmic relationship of DDE residues to eggshell thinning. *Nature,* 240, 164–166.

BLUS, L. J., R. G. HEATH, C. D. GISH, A. A. BELISLE, and R. M. PROUTY. 1971. Eggshell thinning in the brown pelican: Implication of DDE. *Bioscience,* 21, 1213–1215.

BOYLE, R. H. 1975. The spreading menace of PCB. *Sports Illustrated,* December 1. Pp. 20–21.

BRATTSTEN, L. B., C. W. HOLYOKE, JR., J. R. LEEPER, and K. F. RAFFA. 1986. Insecticide resistance: Challenge to pest management and basic research. *Science,* 231, 1255–1260.

BURDICK, G. E., E. J. HARRIS, H. J. DEAN, T. M. WALKER, J. SKEA, and D. COLBY. 1964. The accumulation of DDT in lake trout and the effect on reproduction. *Trans. Am. Fish. Soc.,* 193, 127–137.

BUSH, G. L., R. W. NECK, and G. B. KITTO. 1976. Screwworm eradication: Inadvertent selection for noncompetitive ecotypes during mass rearing. *Science,* 193, 491–493.

BUTLER, P. A. 1971. Testimony consolidated DDT hearings. Environmental Protection Agency, Washington, DC.

CARSON, R. 1962. *Silent Spring.* Fawcett, Greenwich, CT. 304 pp.

CECIL, H. C., J. BITMAN, and S. J. HARRIS. 1971. Effects of dietary p,p'-DDT and p,p'-DDE on egg production and egg shell characteristics of Japanese quail receiving an adequate calcium diet. *Poultry Sci.,* 50, 657–659.

CONNEY, A. H., R. M. WELCH, R. KUNTZMAN, and J. J. BURNS. 1967. Effects of pesticides on drug and steroid metabolism. *Clin. Pharmacol. Ther.*, **8**, 2–10.

COSTA, P. 1983. Dioxin's deadliness breeds fear; regulation process complicated. *Honolulu Star-Bulletin*. March 20, p. F1–F2.

CRAWFORD, M. 1986. Test of tobacco containing bacterial gene approved. *Science*, **233**, 1147.

DEWAILLY, E., S. DODIN, R. VERREAULT, P. AYOTTE, L. SAUVE, J. MORIN, and J. BRISSON. 1994. High organochlorine burder in women with estrogen-receptor positive breast cancer. *J. Natl. Cancer Inst.*, **86**, 232–234.

DOVER, M. J., and B. A. CROFT. 1986. Pesticide resistance and public policy. *Bioscience*, **36**, 78–85.

DUGGAN, R. E., and P. E. CORNELIUSSEN. 1972. Dietary intake of pesticide chemicals in the United States (III), June 1968–April 1970. *Pestic. Monitor. J.*, **5**, 331–341.

DVORCHICK, B. H., M. ISTIN, and T. H. MAREN. 1971. Does DDT inhibit carbonic anhydrase? *Science*, **172**, 728–729.

EDF. 1997. 25 years after DDT ban, bald eagles, osprey numbers soar. www.edf.org/pubs/NewsReleases/1997/Jun/e_ddt.html.

EDIRISINGHE, J. S. 1988. Historical references to malaria in Sri Lanka and some notable episodes up to present times. *Ceylon Med. J.*, **33**, 110–117.

EDMUNDSON, W. F., J. E. DAVIES, M. CRANMER, and G. A. NACHMAN. 1969a. Levels of DDT and DDE in blood and DDA in urine of pesticide formulators following a single intensive exposure. *Industr. Med. Surg.*, **38**(4), 55–60.

EDMUNDSON, W. F., J. E. DAVIES, G. A. NACHMAN, and R. L. ROETH. 1969b. p,p'-DDT and p,p'-DDE in blood samples of occupationally exposed workers. *Pub. Health Rep.*, **84**, 53–58.

EDWARDS, J. G. 1971. Effects of DDT. *Chem. Eng. News*, **49**(33), 6, 59.

EHP. 1997. Mexico moves to phase out DDT and chlordane. *Environ. Health Persp.*, **105**, 790–791.

EHRLICH, P. F. 1969. Eco-catastrophe. *Ramparts*, **8**(3), 24–28.

ELDER, D. L., and S. W. FOWLER. 1977. Polychlorinated biphenyls: Penetration into the deep ocean by zooplankton fecal pellet transport. *Science*, **197**, 459–461.

EPA. 1979. EPA bans PCB manufacture; phases out uses. Environmental Protection Agency, *Environmental News*, April 19. 3 pp.

EPA. 1980a. *Ambient Water Quality Criteria for DDT*. Environmental Protection Agency, EPA 440/5-80-038. Washington, DC.

EPA. 1980b. *Ambient Water Quality Criteria for Polychlorinated Biphenyls*. Environmental Protection Agency, EPA 440/5-80-068. Washington, DC.

EPA. 1984. *Ambient Water Quality Criteria for 2,3,7,8-Tetrachlorodibenzo-p-dioxin*. Environmental Protection Agency, EPA 440/5-84-007. Washington, DC.

EPA. 1999. *Pesticides in the Hydrologic System*. http://ca.water.usgs.gov/pnsp.

ERICKSON, J. M., M. RAHIRE, J-D ROCHAIX, and L. METS. 1985. Herbicide resistance and cross-resistance: Changes at three distinct sites in the herbicide-binding protein. *Science*, **228**, 204–207.

EVANS, G. 1978. Dutch elm disease: Fighting a holding action. *TWA Ambassador Magazine*, August. Pp. 70, 72.

FALCK, R., A. RICCI, M. WOLFF, J. GODBOLD, and P. DECKERS. 1992. Pesticides and polychlorinated biphenyl residues in human breast lipids and their relation to breast cancer. *Arch. Environ. Health*, **47**, 143–146.

FISHER, N. S. 1975. Chlorinated hydrocarbon pollutants and photosynthesis of marine phytoplankton: A reassessment. *Science*, **189**, 463–464.

FISHER, N. S., E. J. CARPENTER, C. C. REMSEN, and C. F. WURSTER. 1974. Effects of PCB on interspecific competition in natural and gnotobiotic phytoplankton communities in continuous and batch cultures. *Microbial Ecol.*, **1**, 39–50.

FLEMING, W. E., and W. W. MAINES. 1953. Persistence of DDT in soils of the area infested by the Japanese beetle. *J. Econ. Entomol.*, **46**, 445–449.

FONSEKA, J., and K. N. MENDIS. 1987. A metropolitan hospital in a non-endemic area provides a sampling pool for epidemiological studies on vivax malaria in Sri Lanka. *Trans. Roy. Soc. Trop. Med. Hyg.*, **81**, 360–364.

FOX, J. L. 1984. Tentative agent orange settlement reached. *Science*, **224**, 849–850.

GARDNER, M. J., H. TETTELIN, D. J. CARUCCI, L. M. CUMMINGS, L. ARAVIND, E. V. KOONIN, S. SHALLOM, T. MASON, K. YU, C. FUJII, J. PEDERSON, K. SHEN, J. JING, C. ASTON, Z. LAI, D. C. SCHWARTZ,

M. Pertea, S. Salzberg, L. Zhou, G. G. Sutton, R. Clayton, O. White, H. O. Smith, C. M. Fraser, M. D. Adams, J. C. Ventner, and S. L. Hoffman. 1998. Chromosome 2 sequence of the human malaria parasite *Plasmodium falciparum*. *Science*, 282, 1126–1132.

Giam, C. S., A. R. Hanks, R. L. Richardson, W. M. Sackett, and M. K. King. 1972. DDT, DDE and polychlorinated biphenyls in biota from the Gulf of Mexico and the Caribbean Sea—1971. *Pestic. Monit. J.*, 6, 139–143.

Goldberg, E. D., P. Bitler, P. Meier, D. Menzel, G. Paulik, R. Risebrough, and L. F. Stickel. 1971. *Chlorinated Hydrocarbons in the Marine Environment*. Report prepared by the panel monitoring persistent pesticides in the ocean, National Academy of Sciences, Washington, DC. 42 pp.

GTI. 1999. *Pesticides in the Urban Landscape*. Guelph Turfgrass Institute. http://www.uoguelph.ca/GTI/urbanpst/urbpst.htm.

Gunn, D. L. 1976. Alternatives to chemical pesticides. In D. L. Gunn and J. G. R. Stevens, Eds., *Pesticides and Human Welfare*. Oxford University Press, Oxford. Pp. 240–255.

Gunther, F. A., and L. R. Jeppson. 1960. *Modern Insecticides and World Food Production*. Wiley, New York. 284 pp.

Gustafson, J. R. 1990. Five-Year Status Report for California Brown Pelican. California Department of Fish and Game. *Nongame Bird and Mammal Sec. Rep.* 90-4 (draft).

Gutowska, M. S., and C. A. Mitchell. 1945. Carbonic anhydrase in the calcification of the egg shell. *Poultry Sci.*, 24, 159–167.

Hamelink, J. L., R. C. Waybrant, and R. C. Ball. 1971. A proposal: Exchange equilibria control the degree chlorinated hydrocarbons are biologically magnified in lentic environments. *Trans. Am. Fish. Soc.*, 100, 207–214.

Harvey, G. R., H. P. Miclas, V. T. Bowen, and W. G. Steinhauer. 1974. Observations on the distribution of chlorinated hydrocarbons in Atlantic ocean organisms. *J. Mar. Res.*, 32, 103–118.

Harvey, G. R., W. G. Steinhauer, and J. M. Teal. 1973. Polychlorobiphenyls in North Atlantic ocean water. *Science*, 180, 643–644.

Hayes, W. J. Jr., W. E. Dale, and C. I. Prinkle. 1971. Evidence of safety of long-term high, oral doses of DDT for man. *Arch. Environ. Health*, 22, 119–135.

Hazeltine, W. 1972. Disagreements on why brown pelican eggs are thin. *Nature*, 239, 410–411.

Heath, R. G., J. W. Spann, and J. F. Kreitzer. 1969. Marked DDE impairment on mallard reproduction in controlled studies. *Nature*, 224, 47–48.

Hickey, J. J., and D. W. Anderson. 1968. Chlorinated hydrocarbons and eggshell changes in raptorial and fish-eating birds. *Science*, 162, 271–273.

Hoffman, S. L., V. Nussenzweig, J. C. Sadoff, and R. S. Nussenzweig. 1991. Progress toward malaria preerythrocytic vaccines. *Science*, 252, 520–521.

Holcomb, R. W. 1970. Insect control: Alternatives to the use of conventional pesticides. *Science*, 168, 456–458.

Holden, P. W. 1986. *Pesticides and Groundwater Quality: Issues in Four States*. National Academy Press, Washington, DC. 124 pp.

Houghton, D. L., and L. Ritter. 1995. Organochlorine residues and risk of breast cancer. *J. Am. Coll. Toxicol.*, 14, 71–89.

Hunt. E. G., and A. I. Bischoff. 1960. Inimical effects on wildlife of periodic DDD applications to Clear Lake. *Calif. Fish & Game*, 46, 91–106.

Ide, F. P. 1956. Effect of forest spraying with DDT on aquatic insects of salmon streams. *Trans. Am. Fish. Soc.*, 86, 208–219.

IJC. 1985. *PCBs: A Case Study*. Proceedings of a workshop on Great Lakes research coordination. International Joint Commission, Windsor, Ontario. 113 pp.

Jukes, T. H. 1974. Insecticides in health, agriculture, and the environment. *Naturwissenschaften*, 61, 6–16.

Kaiser, J. 1996a. Pests overwhelm *Bt* cotton crop. *Science*, 273, 423.

Kaiser, J. 1996b. Malaria genome project ready to roll. *Science*, 274, 1999.

Keller, H. 1952. Die Bestimmung Kleinster Mengen DDT auf enzymanalytischem wege. *Naturwissenschaften*, 39, 109.

Kelsey, J., M. Gammon, and E. John. 1993. Reproductive factors and breast cancer. *Epidemiol. Rev.*, 15, 36–47.

Krafsur, E. S. 1985. Screwworm flies (*Diptera: Calliphoridae*): Analysis of sterile mating frequencies and covariates. *Bull. Entomol. Soc. Am.*, 31(3), 36–40.

KRIEGER, N., M. WOLFF, R. HIATT, M. RIVERA, J. VOGELMAN, and N. ORENTREICH. 1994. Breast cancer and serum organochlorines: A prospective study among white, black, and Asian women. *J. Natl. Cancer Inst.*, **86**, 589–599.

LAWS, E. R. 1967. Men with intensive occupational exposure to DDT. *Arch. Environ. Health*, 15, 766–775.

LAWS, E. R. 1971. Evidence of antitumorigenic effects of DDT. *Arch. Environ. Health*, 23, 181–184.

LICHTENSTEIN, E. P. 1957. DDT accumulation in midwestern orchard and crop soils treated since 1945. *J. Econ. Entomol.*, 50, 545–547.

LING, L., F. W. WHITTEMORE, and E. E. TURTLE. 1972. *Persistent Insecticides in Relation to the Environment and Other Unintended Effects.* Food and Agriculture Organization of the United Nations, Misc. Pub. No. 4, Rome, May.

LOCKIE, J. D., D. A. RATCLIFFE, and R. BALHARRY. 1969. Breeding success and organochlorine residues in golden eagles in west Scotland. *J. Appl. Ecol.*, 6, 381–389.

LÓPEZ-CARRILLO, L., L. TORRES-ARREOLA, L. TORRES-SÁNCHEZ, F. ESPINOSA-TORRES, C. JIMÉNEZ, M. CE-BRIÁN, S. WALISZEWSKI, and O. SALDATE. 1996. Is DDT use a public health problem in Mexico? *Environ. Health Persp.*, 104, 584–588.

LUARD, E. J. 1973. Sensitivity of *Dunaliella* and *Scenedesmus* (Chlorophyceae) to chlorinated hydrocarbons. *Phycologia*, 12(1/2), 29–33.

MACPHEE, A. W., D. CHISHOLM, and C. R. MACEACHERN. 1960. Breeding success and organochlorine residues in golden eagles in west Scotland. *J. Appl. Ecol.*, 6, 381–389.

MAREN, T. H. 1967. Carbonic anhydrase: Chemistry, physiology, and inhibition. *Physiol. Rev.*, 47, 595–781.

MARSHALL, E. 1981. The summer of the gypsy moth. *Science*, 213, 991–993.

MARSHALL, E. 1982. USDA retreats on gypsy moth front. *Science*, 216, 716.

MARSHALL, E. 1990. Malaria research—what next? *Science*, 247, 399–403.

MARSHALL, E. 1997. African malaria studies draw attention. *Science*, 275, 299.

MARX, J. L. 1977. Applied ecology: Showing the way to better insect control. *Science*, 195, 860–862.

MARTIN, R. G. 1977. PCBs—polychlorinated biphenyls. *Sport Fishing Institute Bulletin*, No. 288, September. Pp. 1–3.

MAUGH, T. H., II. 1972. Polychlorinated biphenyls: Still prevalent, but less of a problem. *Science*, 178, 388.

MAUGH, T. H., II. 1975. Chemical pollutants: Polychlorinated biphenyls still a threat. *Science*, 190, 1189.

MAUGH, T. H., II. 1981. Starving in the midst of plenty. *Science*, 212, 430.

MCGAUGHEY, W. H., and M. E. WHALON. 1992. Managing insect resistance to *Bacillus thuringiensis* toxins. *Science*, 258, 1451–1455.

MENZEL, D. W., J. ANDERSON, and A. RANDTKE. 1970. Marine phytoplankton vary in their response to chlorinated hydrocarbons. *Science*, 167, 1724–1726.

METCALF, R. L. 1981. Insect control technology. In *Encyclopedia of Chemical Technology*, Vol. 13. Wiley-Interscience, New York. Pp. 413–485.

MILLER, D. S., W. B. KINTER, and D. B. PEAKALL. 1976. Enzymatic basis for DDE-induced eggshell thinning in a sensitive bird. *Nature*, 259, 122–124.

MILLOY, S. 1996. An olympic moment for DDT. http://www.junkscience.com/news.ddt.html.

MOFFATT, A. S. 1991. Research on biological pest control moves ahead. *Science*, 252, 211–212.

MONTOYAMA, N., and W. C. DAUTERMAN. 1980. Glutathione S-transferases: Their role in the metabolism of organophosphorus insecticides. *Rev. Biochem. Toxicol.*, 2, 49–69.

MOORE, S. A., and R. C. HARRISS. 1972. Effects of polychlorinated biphenyl on marine phytoplankton communities. *Nature*, 240, 356–357.

MOSSER, J. L., N. S. FISHER, and C. F. WURSTER. 1972. Polychlorinated biphenyls and DDT alter species composition in mixed cultures of algae. *Science*, 176, 533–535.

MUSSALO-RAUHAMAA, H. M., E. HASANEN, H. PYYSALO, K. ANTERVO, R. KAUPPILA, and P. PANTZAR. 1990. Occurrence of beta-hexachlorocyclehexane in breast cancer patients. *Cancer*, 66, 2124–2128.

ORTELEE, M. F. 1958. Study of men with prolonged intensive occupational exposure to DDT. *Am. Med. Assoc. Arch. Ind. Health*, 18, 433–440.

OTTOBONI, A. 1972. DDT: The world has been doused with it for 25 years. With what results? *California's Health*, 27(2), 1–2.

PALEA, J. 1990. Libya gets unwelcome visitor from the west. *Science*, 249, 117.

PAN. 1994. Aldicarb spill in Texas causes fire, evacuations, sends people to hospital. *Pesticide Action Network.* panna-info@igc.apc.org.

PAUL, J. 1989. Getting tricky with rootworms. *Agrichem. Age*, 33(3), 6, 25, 30.

PEAKALL, D. B. 1970a. p,p'-DDT: Effect on calcium metabolism and concentration of estradiol in the blood. *Science*, 168, 592–594.

PEAKALL, D. B. 1970b. Pesticides and the reproduction of birds. *Sci. Amer.*, 222(4), 73–78.

PEAKALL, D. B., and M. L. PEAKALL. 1973. Effect of a polychlorinated biphenyl on the reproduction of artifically and naturally incubated dove eggs. *J. Appl. Ecol.*, 10, 863–868.

PIMENTEL, D. 1991. Pesticide use. *Science*, 252, 358.

PIMENTEL, D., L. MCLAUGHLIN, A. ZEPP, B. LAKITAN, T. KRAUS, P. KLEINMEN, F. VANCINI, W. J. ROACH, E. GRAAP, W. S. KEETON, and G. SELIG. 1991. Environmental and economic impacts of reducing U.S. agricultural pesticide use. In D. Pimentel and A. A. Hanson, Eds., *CRC Handbook of Pest Management in Agriculture*, 2nd ed., Vol. I. CRC Press. Boca Raton, FL. Pp. 679–718.

PLIMMER, J. R. 1980. Herbicides. In *Encyclopedia of Chemical Technology*, Vol. 12. Wiley-Interscience, New York. Pp. 297–351.

POCKER, Y., W. M. BEUG, and V. R. AINARDI. 1971. Carbonic anhydrase interaction with DDT, DDE, and dieldrin. *Science*, 174, 1336–1338.

PORTER, R. D., and S. N. WIEMEYER. 1969. Dieldrin and DDT: Effects on sparrow hawk eggshells and reproduction. *Science*, 168, 199–200.

POSTEL, S. 1987. *Defusing the Toxics Threat: Controlling Pesticides and Industrial Waste.* Worldwatch Paper 79. Worldwatch Institute, Washington, DC. 69 pp.

RACO-RANDS, V. 1997. Characteristics of effluents from large wastewater treatment facilities in 1996. In *Southern California Coastal Water Research Project 1997–98 Annual Report.* http://www.sccwrp.org/pubs/annrpt/97/ar01.htm.

RATCLIFFE, D. A. 1970. Changes attributable to pesticides in egg breakage frequency and eggshell thickness in some British birds. *J. Appl. Ecol.*, 7, 67–107.

RUDD, R. L. 1964. *Pesticides and the Living Landscape.* University of Wisconsin Press, Madison. 320 pp.

SAKAUCHI, N., S. KUMAOKA, T. NARUKE, O. ABE, M. KUSAMA, and O. TAKATANI. 1969. A case of adrenocortical cancer treated with o,p'-DDD. *Endrocr. Jap.*, 16, 287–290.

SAWICKI, R. M., and A. W. FARNHAM. 1967. Genetics of resistance to insecticides of the SKA strain of *Musca domestica* I. Location of the main factors responsible for the maintenance of high DDT-resistance in diazinon-selected SKA flies. *Entomol. Exp. Appl.*, 10, 253–267.

SODERLUND, D. M., J. R. SANBORN, and P. W. LEE. 1983. Metabolism of pyrethrin and pyrethroids in insects. *Prog. Pesticide Biochem. Toxicol.*, 3, 401–435.

SOUTHREN, A. L., S. TOCHIMOTO, L. STROM, A. RATUSCHNI, H. ROSS, and G. GORDON. 1966. Remission in Cushing's syndrome with o,p'-DDD. *J. Clin. Endocr.*, 26, 268–278.

SPITZER, P. R., R. W. RISEBROUGH, W. WALKER, R. HERNANDEZ, A. POOLE, D. PULESTON, and I. C. NISBET. 1978. Productivity of ospreys in Connecticut-Long Island increases as DDE residues decline. *Science*, 202, 333–335.

STRAUS, O. H., and A. GOLDSTEIN. 1943. Zone behavior of enzymes. *J. Gen. Physiol.*, 26, 559–585.

SUN, M. 1983. EPA, utilities grapple with PCB problems. *Science*, 222, 32–33.

TAYLOR, T. G. 1970. How an eggshell is made. *Sci. Amer.*, 222(3), 88–95.

TORDA, C., and H. WOLFF. 1949. Effects of convulsant and anticonvulsant agents on the activity of carbonic anhydrase. *J. Pharmacol. Exp. Ther.*, 95, 444–447.

USDA. 1999. Eradicating screwworms from North America. United States Department of Agriculture. Animal and Plant Health Inspection Service. http://www.aphis.usda.gov/OA/screwworm.html.

VERMAAS, W., C. ARNTZEN, L. Q. GU, and C. A. YU. 1983. Interactions of herbicides and azidoquines at a photosystem II binding site in the thylakoid membrane. *Biochim. Biophys. Acta*, 723, 266–275.

WADMAN, M. 1997. Dispute over insect resistance to crops. *Nature*, 388, 817.

WALSH, J. 1986. River blindness: A gamble pays off. *Science*, 232, 922–925.

WALSH, J. 1991. The greening of the green revolution. *Science*, 252, 26.

WHO. 1988. Malaria control activities in the last 40 years. In *World Health Statistics Annual.* World Health Organization, Geneva. Pp. 28–29.

WHO. 1990. World malaria situation 1988. *Bull. World Health Org.*, 68(5), 667–673.

WIEMEYER, S. N., and R. D. PORTER. 1970. DDE thins eggshells of captive American kestrels. *Nature*, 227, 737–738.

WILSON, A. J., JR., J. FORESTER, and J. KNIGHT. 1070. Chemical assays. In *U.S. Department of the Interior Circular 335* (Gulf Breeze Lab., Florida, Progr. Rep. F. Y. 1969). Pp. 18–20.

WINTERINGHAM, F. P. W., P. M. LOVEDAY, and A. HARRISON. 1951. Resistance of houseflies to DDT. *Nature*, 167, 106–107.

WOLFF, M. S., P. G. TONIOLO, E. W. LEE, M. K. RIVERA, and N. DUBIN. 1993. Blood levels of organochlorine residues and risks of breast cancer. *J. Natl. Cancer Inst.*, 8, 648–652.

WOODWELL, G. M. 1972. Testimony, consolidated DDT hearings, Environmental Protection Agency, Washington, DC.

WOODWELL, G. M., and F. T. MARTIN. 1964. Persistence of DDT in soils of heavily sprayed forest stands. *Science*, 145, 481–483.

WOODWELL, G. M., C. F. WURSTER, and P. A. ISAACSON. 1967. DDT residues in an east coast estuary: A case of biological concentration of a persistent insecticide. *Science*, 156, 821–824.

WURSTER, C. F. 1968. DDT reduces photosynthesis by marine phytoplankton. *Science*, 159, 1474–1475.

YOSHISHIGE, J. 1991. DBCP in groundwater will be around for "generations." *Honolulu Advertiser*, May 8, p. A6.

YOUNG, D. R., D. J. MCDERMOTT, and T. C. HEESON. 1975. *Polychlorinated Biphenyl Inputs to the Southern California Bight.* Southern California Coastal Water Research Project. Background paper prepared for the National Conference on Polychlorinated Biphenyls, 19–21 Norvermber 1975. 50 pp.

QUESTIONS

10.1 What line of reasoning suggests that DDT has never had any significant impact on the photosynthetic rates of phytoplankton in the ocean, despite the fact that laboratory experiments have clearly demonstrated the adverse effects of DDT on marine phytoplankton photosynthetic rates?

10.2 In 1967 Dr. G. M. Woodwell and coworkers published a paper in *Science* magazine in which they reported that the concentrations of DDT residues in the soil of an extensive salt marsh on the south shore of Long Island averaged more than 13 pounds per acre. At the 1972 DDT hearings in Washington, D.C., an effort was made to strike the article as misleading. What was the logic behind the criticism of this article?

10.3 The principal reason that Sri Lanka switched from DDT to malathion in its effort to combat malaria was
   a. Publication of Rachel Carson's *Silent Spring*
   b. Pest resistance
   c. Discovery that DDT causes cancer in humans
   d. The discovery that malathion was cheaper and equally effective

10.4 In 1993 the Chinese government estimated that more than 10,000 Chinese farmers died from exposure to pesticides. The category of pesticide responsible for most of these deaths was probably
   a. Pyrethroids
   b. Chlorinated organics
   c. Carbamates
   d. Organophosphates

10.5 In the United States, pesticides are used primarily to control which one of the following types of pests?
   a. Weeds
   b. Rodents
   c. Nematodes
   d. Insects

10.6 During the second year of a two-year trial, mallard ducks given feed containing DDT actually produced more offspring per hen than control ducks. This fact has been cited by DDT advocates as evidence that "DDT probably causes wild birds . . . to be healthier if they are exposed to moderate does of it." Even if true, why is this fact irrelevant to the question of whether use of DDT has adversely impacted bird populations?

10.7 The reproductive failure of southern California brown pelicans has been linked to the concentration of DDT residues in anchovies. What are "DDT residues," and why do scientists relate adverse effects to concentrations of DDT residues as opposed to concentrations of DDT?

10.8 Describe three ways in which pheromones are used to control insect pests.

10.9 PCBs most closely resemble which one of the following categories of pesticides with respect to their chemistry and toxicity?

 a. Carbamates

 b. Chlorinated organics

 c. Pyrethroids

 d. Organophosphates

10.10 How has the use of short-season cotton allowed cotton growers in the United States to greatly reduce their use of synthetic chemical pesticides to control boll weevils?

10.11 In 1996, López-Carrillo et al. published a paper in *Environmental Health Perspectives*, **104**, 584–588, entitled "Is DDT a Public Health Problem in Mexico?" In that paper the authors state (p. 586), "Information about levels of DDT and its metabolites in human samples in Mexico is scarce . . . Mexican studies have episodically documented the presence of DDE and total DDT in breast milk." The authors report that in 1995 (p. 586), "Women living in Mexico City had DDE levels in breast milk of 0.594 mg/kg in a lipid base. In contrast, in women living in tropical Mexico these levels reached an average level of 5.02 mg/kg, which is extremely high." The authors review the epidemiological studies of DDT exposure and breast cancer and conclude (p. 587), "It is difficult to conclude whether exposure to DDT contributes to an increase in breast cancer." The paper also contains information about the relationship between DDT spraying for malaria control and the incidence of malaria in Mexico. The data reported in the paper are shown on page 333.

In summarizing their results, the authors state (p. 584), "We conclude that DDT use in Mexico is a public health problem, and suggest two solutions: identification of alternatives for the control of malaria and educational intervention to reduce DDT exposure."

Milloy (1996) has provided the following thoughts about the López-Carrillo et al. article: "In a commentary titled *Is DDT Use a Public Health Problem in Mexico?*, researchers conclude that, indeed, DDT use in Mexico is a public health problem. This is an interesting conclusion given the data presented.

In comparing the incidence of malaria with levels of DDT used in Mexico between 1959 and 1993, these researchers presented data showing that

- High rates of malaria occurred consistently when DDT use was low; and
- Low rates of malaria occurred consistently when DDT use was high.

For example, in 1959, before DDT was used in Mexico, almost 140,000 case of malaria were reported. By 1970, when 3 million houses were being sprayed annually, the number of annual cases of malaria had dropped to about 40,000.

By 1973, the number of cases of malaria had risen to more than 100,000, while the number of households sprayed decreased to about 1 million.

But by 1985, the number of cases of malaria had dropped to less than 10,000 when the number of households sprayed increased to almost 7 million.

Of course, as DDT use declined again in the late 1980s and early 1990s, the number of cases of malaria again increased.

These researchers also discussed a possible link between DDT and breast cancer but presented absolutely *no* supporting data. They even acknowledge that insufficient data exist to make such a link.

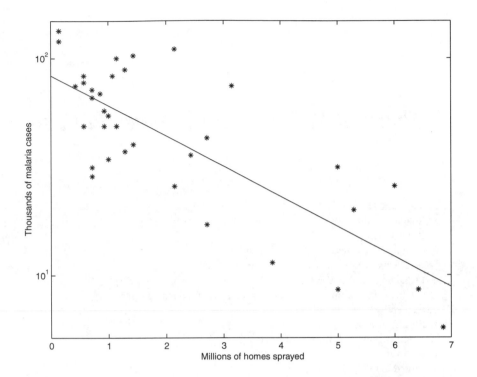

Excuse me, but exactly where's the public health problem? If DDT is so bad for people, why are there no data showing this?

I guess the more interesting junk science-related question is, why would someone conclude that DDT is a public health problem and then present data that only contradicts this conclusion?"

To what extent do you agree or disagree with the conclusions reached by López-Carrillo and the point of view expressed by Milloy? Explain your reasoning.

Note: In July 1997, the Mexican government unveiled a program designed to phase out all uses of DDT within 10 years. The malaria control program will instead rely on improvements in sanitation, increased disease surveillance, and integrated pest management schemes that focus pesticide applications on critical habitats and stages in the mosquito's life cycle. In addition, experimental research will explore the use of biological control agents such as larval parasites and adult predators, microbial products such as *BT*, and other, less persistent pesticides such as pyrethroids. (EHP, 1997).

Chapter Eleven

✧   ¤   ✧   ¤   ✧

Thermal Pollution and Power Plants

Indian Point, on the Hudson River, is the site of a nuclear power station opened by Con Ed, with great fanfare, in 1963. A lot of people don't like any nuclear power plants, and with some reason—the general idea is that utility company engineers are in a great rush to build the damned things, while nobody really knows enough yet about what their effects might be—but there wasn't any really great outcry about Indian Point. Like all reactors so far, it is a thermal polluter—it raises the temperature of the river at its side by using river water as a coolant and then dumping the warm water back into the riverbed—but the Hudson is so fouled up anyway that a lot of people have given up on it.

The only problem was that after the plant opened, somebody noticed a bunch of crows around a dump near the new power station—more crows by far than usual. When this kept up for a while, somebody got in touch with the Long Island League of Saltwater Sportsmen, a group which—in marked contrast to a lot of sportsmen's groups—happened to retain a consulting biologist. His name is Dominick Pirone, and he went to take a look at the dump.

Although Con Ed didn't want him to look, he managed to see bulldozers at work, shoving dead bass into twelve-foot-high piles, where lime was being dumped on them to hasten their decomposition, and he saw a line of trucks, bringing new loads of dead bass. Pirone decided to follow the trucks, and found himself at Indian Point.

"I saw and smelled," he said, "some 10,000 dead and dying fish under the dock."

The seven-degree rise in the temperature of the water, after the plant has sucked in Hudson River water and spat it back out again, is enough to attract spawning bass. They were trapped under the dock and ultimately suffocated. Intake pipes sucked them up into wire baskets, the baskets were dumped into the trucks—and before Con Ed, under pressure, put up a fine-mesh screen around the dock, 2,000,000 bass had been killed.

Con Ed did its best to kill the story. New York's representative Richard Ottinger (who would probably have leprosy by now if somebody on the Con Ed staff had the Evil Eye) described it this way:

> The story of the Indian Point fish kill is strangely obscure. There are reports of truckloads of fish carted away secretly; fish graveyards limed to hasten the destruction of evidence and guarded by Burns detectives to prevent witnesses' access to see the size of the kill. There are stories of pictures suppressed by state officials and state employees pressured into silence. (Marine, 1969, pp. 86–87)

M ANY INDUSTRIES and almost all nuclear and fossil fuel electric power plants discharge heated wastewater into aquatic systems. It is not hard to imagine some of the ways in which this heated effluent might adversely affect aquatic biota. In the summer and particularly in tropical climates, the ambient water temperature may already be a chronic stress to some organisms in the system. In such cases, any further increase in temperature would only increase this stress and perhaps create lethal conditions for some organisms. On the other hand, heated effluent may often attract aquatic organisms during the winter in temperate latitudes when ambient water temperatures are lower than the preferred range of many species. As noted in Chapter 8, an abrupt plant shutdown during such times may produce a large-scale kill of aquatic organisms due to cold shock. Perhaps the most serious effect associated with thermal pollution, however, is the killing or stressing of organisms that are sucked into a plant's cooling water system and are either impinged on protective screens (see above) or, if they pass through, subjected to physical buffeting and possible chemical stresses as well as thermal shock.

POWER PLANT DESIGN

Electric power plants account for roughly 75–80% of the thermal pollution in the United States. Industrial operations such as refineries, petrochemical plants, cokeries, and steel mills account for most of the remainder. We will therefore concentrate in this chapter on electric power plants, although much of what we say is relevant to industrial operations as well.

Figure 11.1 is a schematic diagram of a typical electric power plant's steam and cooling system. The plant generates electricity by first boiling liquid water to produce steam. The source of the heat may be either fossil fuel energy or nuclear energy. The expansion of the water during the transformation from the liquid to the gaseous state creates pressure, which is used to drive a turbine, much as the burning of liquid gasoline to create vapor in a car is used to drive the pistons of the car. The turning of the turbine powers the generator, which produces electricity. The steam from the turbine is then condensed back to liquid water and recycled to the boiler. Condensation is achieved by running cooling water through a long, coiled pipe exposed to the steam. Heat is transferred through the

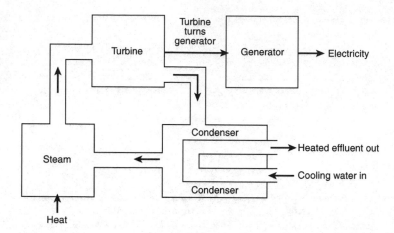

FIGURE 11.1   Schematic diagram of an electric power plant's steam and cooling system.

pipe from the hot steam to the cooling water until the steam condenses. The heated effluent from the condenser is the source of thermal pollution from such a power plant.

The increase in the temperature of the condenser cooling water as it passes through the heat exchanger depends on the steam exhaust pressure and on how fast the water is pumped. The higher the steam exhaust pressure, the higher the temperature at which steam condenses to water. In the United States, most electric power plants operate with a steam exhaust pressure of about 0.05 atmosphere, at which pressure the steam condenses at a temperature of about 33°C (Karkheck et al., 1977). At a given steam exhaust pressure, the temperature increase of the cooling water is negatively correlated with the flux of cooling water through the heat exchangers. Since organisms may be killed from physical buffeting as well as by thermal shock from the heated effluent, there is obviously some trade-off between pumping cooling water rapidly to reduce the temperature rise and pumping slowly to reduce the number of organisms sucked into the cooling system.

The characteristics of nuclear and fossil fuel power plants are such that nuclear plants usually discharge about 45% more waste heat to the cooling water than do fossil fuel plants per unit of electricity generated (GESAMP, 1984). The difference is due both to the lower efficiency of nuclear plants (35% versus 40%) and to the fact that fossil fuel plants discharge about 15% of their waste heat into the air. Nuclear plants release only about 3% of their waste heat to the atmosphere. Most power plants in the United States operate so that the effluent water is somewhere between 5°C and 15°C above ambient. The rate at which cooling water must be pumped through the heat exchangers to achieve a given temperature rise will vary from one plant to another. In round numbers, about $2.6–5.2 \times 10^3$ m^3 d^{-1} of cooling water is required per megawatt (MW) of electricity to limit the temperature increase to 10°C (GESAMP, 1984). Power plants with a capacity of 1,000 MW are by no means unusual, and power plants as large as 4,000 MW exist in the United States. A 4,000 MW power plant producing effluent 10°C above ambient requires $1.0–2.1 \times 10^7$ m^3 d^{-1} of cooling water. This rate of cooling water use equals about 75–150% of the flow of water in the Colorado River. Thus, we are talking about large fluxes of water.

Water Quality Criteria

What are the implications of a temperature increase of 5–15°C relative to federal water quality criteria? In its report to the Secretary of the Interior on April 1, 1968, the National

TABLE 11.1 MAXIMUM ELEVATIONS OF MONTHLY
MEANS OF MAXIMUM DAILY WATER TEMPERATURES OUTSIDE
DESIGNATED ZONES OF MIXING, AS RECOMMENDED BY THE
NATIONAL TECHNICAL ADVISORY COMMITTEE IN 1968

Area	Temperature Elevation
Lakes	3°F (1.7°C)
Rivers	5°F (2.8°C)
Estuaries	
Summer	1.5°F (0.8°C)
Other seasons	4°F (2.2°C)

Technical Advisory Committee made the recommendations shown in Table 11.1. The maximum recommended temperature elevations ranged from 0.8°C to 2.8°C and depended on the nature of the body of water and, in the case of estuaries, on the season of the year. The present EPA water quality criteria with respect to temperature were promulgated in 1976. In marine waters, the criteria specify that the maximum acceptable increase in the weekly average temperature resulting from artificial sources is 1.0°C during all seasons of the year. The criteria also specify that summer thermal maxima, which define the upper thermal limits for the communities in the discharge area, should be established on a site-specific basis. Recommended summer thermal maxima along the East Coast of the United States, for example, are 27.8–29.4°C for the daily mean temperature and 30.6–32.2°C for the short-term maximum (EPA, 1976). The freshwater criteria are more complicated and site-specific. They take into account the thermal tolerance of the most sensitive important species and the requirements for successful migration, spawning, egg incubation, fry rearing, and other reproductive functions. The freshwater criteria also consider impacts resulting from cold shock when plants shut down during the cooler months of the year. The conclusion that obviously emerges from an examination of the criteria is that the typical cooling water temperature increase of 5–15°C at U.S. power plants is well outside the range of acceptable temperature increases. To satisfy the criteria, the effluent water must either be cooled down before being discharged or else vigorously mixed and diluted with the receiving water in the zone of mixing.

Cooling Water System Characteristics

Two characteristics associated with cooling water systems should be mentioned at this time. First, since the cooling water is invariably drawn from a natural body of water, some effort must be made to prevent unwanted objects from being drawn through the cooling system. Figure 11.2 indicates the techniques commonly employed to keep debris out of the system. The water is drawn from under a 2–5 m deep baffle to prevent floating objects from entering the plant. The water then passes through a trash rack analogous to the trash rack in a sewage treatment plant. The trash rack consists of a grid of metal bars typically spaced about 7–8 cm apart. The water then passes through a screen with a mesh size of about 1 cm. The screen may be designed to rotate as in Figure 11.2 to facilitate periodic cleaning, or the screen may simply be removed for cleaning and replaced with another screen. Finally, the water is pumped through the condenser tube manifold. This manifold may consist of as many as several tens of thousands of tubes 2–3 cm in diameter having a length of perhaps 15 m. All of these tubes are in direct contact with the exhaust steam from the turbines.

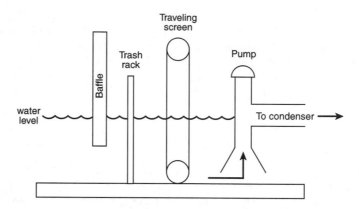

FIGURE 11.2   Diagram of the inflow system to the condenser cooling tubes in a typical electric power plant.

A second problem associated with the condenser system is the growth of fouling organisms on the interior walls of the condenser tubes. There are several approaches to solving the fouling organism problem.

1. The plant may simply introduce Cl_2 into the cooling water system, either continuously or intermittently.
2. The plant may periodically backflush the heated effluent to kill the fouling organisms by heat stress.
3. The plant may scour the tubes mechanically by periodically releasing many brushes or elastic spheres into the inflow water.

In some cases, there is no need to go to any effort to prevent fouling because natural mechanisms keep the condenser tubes clean. For example, the condenser tubes in the Hawaiian Electric power plant at Kahe Point on Oahu are kept free of fouling organisms by the scouring action of sand drawn in with the cooling water from the ocean. However, such circumstances are the exception rather than the rule in power plant operations. Most power plants use Cl_2 to control fouling. The Cl may be introduced intermittently at a concentration of typically 12–15 mg/L, generally every four to eight hours, or continuously at a concentration of 1–5 mg L^{-1}. Discharge concentrations are typically lower by a factor of 10–20 (Langford, 1983; GESAMP, 1984).

TOXIC EFFECTS OF EFFLUENT WATERS ON BIOTA

Direct kills of fish and other motile organisms due to the temperature of the effluent from power plants are fortunately rather uncommon. Data summarized by Langford (1983) indicate that about two or three voluntarily reported fish kills are associated with thermal discharges each year in the United States. Mass kills of Atlantic menhaden, for example, have been reported at three locations; on the Cape Cod Canal (Fairbanks et al., 1968), at Millstone, Connecticut (IAEA, 1972), and at Northport on Long Island Sound (Young and Gibson, 1973). In the first two cases, the maximum temperature was only about 21°C, but the temperature rose abruptly by about 15°C and the fish were unable to avoid the discharge. In the third case, the water temperature was 37–38°C, almost certainly above the upper tolerance limit of Atlantic menhaden.

In most cases, however, motile organisms are apparently able to avoid power plant discharges by simply swimming away if water temperatures become too high. Furthermore, the tendency of the heated effluent plume to rise or remain on the surface provides a refuge in deeper water for organisms that prefer cooler temperatures. This effect may be minimized, however, if the power plant uses a diffuser discharge system to mix its effluent vigorously with the receiving waters or if the receiving system is so shallow as to be effectively mixed to the bottom. In freshwater systems in the winter, if the temperature of the water is less than 4°C, the effluent plume may actually sink to the bottom, but under such conditions the temperature of the effluent water is likely to attract rather than repel most organisms.

Sessile benthic organisms are particularly susceptible to being killed by heated effluents if the effluents are mixed to the bottom. The incipient lethal temperature of most aquatic species lies within or below the range 30–35°C, so that in tropical climates where summer water temperatures may approach this limit naturally, a further temperature increase of only a few degrees centigrade may prove lethal to many organisms. Kills of shallow-water corals at Kahe Point, Oahu, Hawaii (Jokiel and Coles, 1974), and of virtually the entire benthos over a large area of Biscayne Bay, Florida (Zieman and Wood, 1975), provide two good examples of direct thermal kills of benthic organisms. Obviously, there will be no direct effect on the benthos unless the effluent plume is mixed to the bottom.

In temperate climates during the winter, organisms may be attracted to the warmer waters from a power plant's discharge. Under such conditions, direct kills of fish and other motile organisms may occur due to cold shock if the plant shuts down suddenly. The fish kill that occurred in January 1972 at the Oyster Creek power plant in Barnegat Bay, New Jersey (Chapter 8), illustrates this problem. In such cases, motility is of no advantage to an organism, since virtually all the surrounding water is lethal. The impact of such temperature changes will depend on both the magnitude and the abruptness of the temperature change. In the case of the Oyster Creek power station, the effect of the shutdown was exacerbated by the fact that the temperature in the discharge canal fell from 22°C to 15°C over a period of 48 hours prior to the shutdown. The temperature dropped to 2°C in six and a half hours when the plant shut down. As noted in Chapter 8, organisms can adapt to different temperatures, but in general, such adaptation is accomplished more rapidly at higher temperatures than at lower temperatures (Clark, 1969).

In addition to kills resulting directly from temperature stress, aquatic organisms may be killed by the discharge of chlorine used to prevent fouling in the heat exchangers. As examples of such kills, Clark and Brownell (1973) cite the killing of "a number of menhaden" at the Cape Cod Canal Plant in Massachusetts in 1968 and the killing of 40,000 blue crabs at the Chalk Point Plant in Maryland. Several other comparable incidents are reported by Brungs (1973) and Truchan (1978). In most cases, a kill results when fish swim into a discharge canal during a period of nonchlorination and are unable to escape when the chlorine concentration subsequently rises. Effluent chlorine concentrations during intermittent chlorination are typically 0.5–2.0 mg L^{-1} for periods of 20–30 minutes. For comparison, the EPA criterion maximum concentrations for chlorine are 0.013 and 0.019 mg L^{-1} in marine water and freshwater, respectively (EPA, 1986). Kills resulting from intermittent chlorination, however, appear to be of minor importance compared to some of the other problems considered in this chapter.

Organisms that are sucked through a power plant's condenser tubes during intermittent chlorination are exposed to high concentrations of chlorine as well as thermal and physical stresses. In such cases, the old adage that you can only die once applies. In the case of fish eggs and larvae, for example, thermal stress and physical buffeting appear sufficient to account for the observed mortality (GESAMP, 1984). Chlorine toxicity does, however, seem to play a role in the impact of entrainment on other types of organisms. The mortality of entrained zooplankton, for example, is positively correlated with chlorine concentration at concentrations above 0.25 mg L^{-1} (GESAMP, 1984).

There are several known instances of fish kills due to gas bubble disease associated with heated effluents from power plants (DeMont and Miller, 1971; Clark and Brownell, 1973; Schneider, 1980). The heated effluent is supersaturated with atmospheric gases, and excess gas taken into a fish's blood tends to bubble out, leading at first to disequilibrium and ultimately to death from embolism. Thousands of adult Atlantic menhaden were reportedly killed by gas bubble disease on April 9, 1973, near the discharge from the Pilgrim Power Station on Cape Cod Bay, Massachusetts. The incident was described in a Smithsonian Institution report as follows:

> There are preliminary indications that the menhaden are dying from a gas bubble disease which is caused by supersaturation of nitrogen in the water naturally. However, the water taken by the plant at a low temperature is increased roughly 26°–27°F so that when it comes out of the plant it is supersaturated.

> The fish have lesions and are hemorrhaging on their fins. They have numerous gas bubbles on their fins and in the membranes between the fin rays. (Smithsonian Institution, 1973a)

In summarizing the lethal effects of plant discharges, it is worth noting that the absence of repeated kills in the vicinity of plant outfalls does not necessarily imply the absence of a continuing problem. Motile organisms may simply avoid the area, whereas sessile benthic types may be killed once and then never replaced. Thus, the effect of the effluent would be to exclude certain organisms from the vicinity of the outfall. Actual kills occur primarily when the characteristics of the effluent suddenly change or when, for some reason, organisms are unable to escape as water quality deteriorates.

Sublethal Effects

Sublethal stresses of effluent water on biota can amount to nothing more than mild exposure to the sorts of stresses that are potentially lethal. Many sublethal stresses associated with power plant effluent are, however, the result of complex interactions between a variety of factors. For example, an increase in temperature can be expected to increase the respiration rate of an aquatic organism, at least up to the point at which the temperature approaches the lethal level for the species. Thus, the organism's demand for O_2 generally increases as the temperature is raised. Since the Q_{10} for many organisms is about 2, an increase of 10°C in the temperature of the receiving water can be expected to roughly double the respiration rate of the biological community. Increases in temperature, however, also reduce the solubility of O_2 in water. The relationship is such that a 10°C increase in temperature reduces the solubility of O_2 by about 20%. Thus, an increase in temperature generally produces an increase in respiratory demand for O_2 but a decrease in the capacity of the water to dissolve O_2. Finally, if the heated effluent plume is not thoroughly mixed with the receiving water and tends to rise to the surface, any thermal stratification of the water column will be intensified. The water column therefore becomes more stable and hence more difficult to mix. As a result, any effective exchange of O_2 between the atmosphere and subsurface water will be reduced. The heated effluent from a power plant thus exacerbates O_2 depletion problems by the following mechanisms:

1. Increasing the respiratory demand of aquatic organisms
2. Reducing the solubility of O_2 in the water

3. Stratifying or further stratifying the water column so that reoxygenation of sub-surface water is inhibited

O_2 depletion problems are most likely to develop during the summer, when the ambient water temperature is naturally high and the water column is thermally stratified.

Temperature increases invariably increase the rate of chemical reactions and therefore alter the rate of virtually all physiological processes. If the temperature becomes too high, the enzymes and hormones that catalyze and control biochemical reactions begin to break down.[1] As a result, metabolic processes begin to slow down and ultimately come to a halt if the temperature is raised sufficiently. Below this temperature range, increases in temperature speed up metabolic processes. Figure 8.8, for example, indicates the dependence of various metabolic processes on temperature for sockeye salmon. Such curves are qualitatively characteristic of virtually all aquatic organisms.

Organisms cannot carry out metabolic functions effectively if the rates of biochemical processes are sufficiently slowed. The curves in Figure 8.8 suggest why there is both an upper and a lower limit to the temperature range within which an organism can survive. As indicated in Figure 8.3, the temperature ranges for growth and reproduction are invariably narrower than the lethal temperature range. As a result, organisms may be eliminated from the thermal discharge area not by directly lethal effects, but simply because they are unable, for example, to grow efficiently or reproduce. In such cases, the area may be recolonized by species that are better suited to higher temperatures. Cyanobacteria, for example, are sometimes better able to survive in artificially warmed waters than are the natural phytoplankton communities of (usually) diatoms and/or green algae. Heated effluent from the Florida Power and Light power plant at Turkey Point, Biscayne Bay, Florida, is believed to have been largely responsible for the stimulation of cyanobacterial blooms in Biscayne Bay during the early 1970s (Clark and Brownell, 1973). Besides tending to form unsightly algal scums on the surface, some species of cyanobacteria are known to release substances that are toxic to aquatic organisms as well as humans. Toxic substances released by cyanobacteria are believed to have been the cause of a series of fish and bird kills that occurred in Biscayne Bay during the winter of 1972–1973 (Smithsonian Institution, 1973b). Thus, power plant discharges can lead both to the elimination of desirable species and possibly to their replacement by undesirable species. Cyanobacteria are a good example of the latter.

Temperature effects on reproduction of aquatic organisms are well known. Many species of fish and invertebrates initiate spawning activity at least partly in response to higher temperatures in the spring (Clark and Brownell, 1973). As a result, organisms attracted to thermal discharges during the winter may be induced to spawn earlier than usual in the spring, perhaps at a time when there is inadequate food to support their offspring. Early spawning of fish affected by thermal discharges has been reported for silver bream, tench, rudd, pike-perch, white suckers, large-mouth bass, and sauger (Langford, 1983). The reported advances in spawning have been 1–5 weeks. In one interesting case involving a geothermally heated stream, rainbow trout changed their spawning period from spring to autumn and thus avoided the hottest period of the year for hatching and fry development (Kaya, 1977).

It has been suggested that migratory patterns of fish may be disrupted by power plant discharges into rivers or estuaries that serve as migration routes for fish (Hawkes, 1969; Bush et al., 1974). The evidence in support of this hypothesis is, however, less than overwhelming. The salmon fishery in the River Severn in Great Britain was unaffected by the

---

[1] Enzymes and some hormones are composed of proteins whose complex structure is determined in part by H bonds, which are broken rather easily by thermal agitation.

Ironbridge power station, which discharged thermal effluent into the river for over 45 years, and migrations of sockeye salmon, steelhead trout, and coho salmon in the Columbia River were unaffected by thermal discharges from the Hanford nuclear facility in Washington (Langford, 1983). In the latter case, most fish avoided the thermal plume in the hottest months of the year by migrating up the opposite side of the river, but the migration rates of fish that encountered the plume were similar to those that did not. Shad migrating up the Connecticut River similarly tend to avoid the discharge from the Haddam Neck (Connecticut Yankee) power plant by migrating around or under the thermal plume (Clark and Brownell, 1973).

There is no doubt that elevated temperatures tend to increase the susceptibility of fish and shellfish to certain other stresses. As noted in Chapter 8, this interaction arises in part because increased temperatures lead to increased metabolic rates (at least below the lethal temperature range), which in turn require an increased pumping rate of water over the gills. As a result, the organism's intensity of exposure to any toxic substance in the water is increased. There is good evidence, for example, that the toxicity of chlorine is increased at elevated temperatures (Cairns et al., 1978). At 0.1 ppm residual chlorine, for example, the median survival time of chinook salmon decreases by a factor of 1,000 as the temperature is raised from 25°C to 30°C, and that of pink salmon decreases by a factor of 73 as the temperature is raised from 11°C to 22°C (Langford, 1983). Furthermore, some diseases and pathogens are known to develop faster at higher temperatures, and thermal stress may make aquatic organisms more susceptible to infection (Clark and Brownell, 1973; Langford, 1983).

Commentary

The foregoing discussion of sublethal stresses has involved few concrete examples. Instead, the discussion has revolved around interactions that we believe could affect aquatic organisms and have been postulated to contribute to the impact of power plant effluents on aquatic biota. However, the gradual elimination of a species from a system due, for example, to reduced fecundity, greater susceptibility to disease, reduced growth efficiency, or other sublethal stresses is not likely to attract the sort of attention that is drawn to a fish kill. The effects of sublethal stresses generally do not lead to spectacular events. It is possible, however, that the greatest disruption to aquatic systems from power plant effluents may be caused by the continual exposure of organisms to sublethal stresses rather than the occasional killing of large numbers of organisms due to thermal shock, chlorination, or gas bubble disease. Langford (1983), for example, cites a number of examples of changes in the species composition of benthic invertebrate communities near thermal discharges. Not surprisingly, there is a tendency for the natural fauna to be replaced by more thermotolerant species.

A final comment concerns the siting of electric power plants. For reasons of convenience and economics, power plants are often located on estuaries. Such sitings are made because large population centers are frequently found adjacent to estuaries and because the estuary obviously provides a source of cooling water for the power plant. From the standpoint of water pollution, however, estuaries are just about the last place where one would want to locate a power plant. Estuarine organisms are often exposed to a variety of natural stresses, including salinity variations; low O_2 levels due to high respiration rates; turbidity caused by the stirring up of bottom sediments and the high plankton concentrations in the water; and, in some cases, large temperature fluctuations due to the shallowness of the water and restricted circulation. Although indigenous estuarine species therefore tend to be, for example, euryhaline and eurythermal, there is a limit to the range of environmental parameters within which even these organisms can function effectively. Furthermore, as noted in Chapter 3, estuaries serve as important nursery and/or breeding grounds for a variety of aquatic organisms and may serve as part of the travel route of migratory fish. Deterioration of estuarine water quality can therefore affect a great number of aquatic species, in fact a far greater number than might be apparent from a census of organisms in the estuary at any one time. Considering this fact, and considering the magnitude of the stresses that are naturally imposed on estuarine organisms, any additional stress from thermal pollution is to be avoided if at all possible. With this introduction, we turn our attention to one of the most extreme examples of estuarine thermal pollution, the Florida Power and Light Company's Turkey Point power plant on Biscayne Bay, Florida.

CASE STUDY: THE FLORIDA POWER AND LIGHT POWER PLANT AT TURKEY POINT

The Study Area

Biscayne Bay is a subtropical estuary located just south of Miami, Florida, near the lower tip of Florida (Figure 11.3). The bay has been formed by a series of barrier islands or keys about 14 km east of the Florida peninsula. The bay is bounded on the north and south by shoals, specifically Featherbed Bank on the north and Cutter Bank on the south. The bay is shallow, with a mean depth of only 1.5–1.8 m.

Biscayne Bay is a highly productive estuary and supports a large and diverse fauna of fish, shellfish, and invertebrates. At least 27 species of fish that are important for commercial, sport, or bait fishing are found in the bay's waters, including mullet, barracuda, mackerel, jack, snapper, grouper, bonefish, snook, and pompano (FWPCA, 1970). Furthermore, at least nine invertebrate species found in the bay are of major commercial importance, including sponges, queen conch, blue crab, stone crab, shrimp, and spiny lobster. The bay also supports a wide variety of tropical birds, including some rare and endangered species. Most of the birds are wading and shore birds that are found in the shallows adjacent to the mangrove shoreline or near the mangrove islands in the southern portion of the bay. These birds include herons, pelicans, cormorants, egrets, white ibis, roseate spoonbills, frigate birds, waterfowl, gulls, and terns (FWPCA, 1970).

The food chain in Biscayne Bay is primarily a detritus food chain, the source of detritus being the extensive beds of turtle grass (*Thalassia testudinum*) that cover much of the bay's benthos. The blades of turtle grass are not grazed directly to a significant degree, but when they break off or die, their decomposition yields detritus, which forms the base of the detritus food chain. The turtle grass also serves as a refuge for the abundant benthic fauna and stabilizes the sediments against the erosive action of currents and turbulence.

Realizing the beauty and ecological significance of Biscayne Bay, the U.S. Congress authorized funds to set aside 420 km² of lower Biscayne Bay as the Biscayne National Monument on October 18, 1968. The Monument became a national park in 1980 and now includes an area of 700 km² (Figure 11.3). As stated by Congress, the original Monument was established to "preserve and protect for the education, inspiration, recreation, and enjoyment of present and future generations a rare combination of terrestrial, marine, and amphibian life in a geographical setting of great natural beauty" (U.S. Congress, 1968).

The Power Plant

In June 1964 the Florida Power and Light Company (FPL) was granted a permit to build two oil-fired electric power generators at Turkey Point (Figure 11.3). The first of these units went into operation on April 22, 1967, and the second began operating on April 25, 1968. Cooling water for the condensers was drawn from Biscayne Bay at an intake just

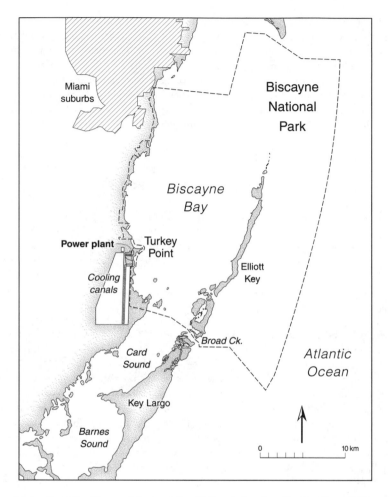

FIGURE 11.3   East Florida coast just south of Miami showing the location of Biscayne Bay, the area occupied by the Biscayne National Park, and the power plant at Turkey Point.

north of Turkey Point and was discharged back into the bay via a series of short canals (Figure 11.4) just south of Turkey Point. Each of the two units draws a total of 1.55×10^6 m^3 of cooling water per day. Under a so-called normal full load, which corresponds to 720 MW of power from the two units combined, the temperature rise of the cooling water is 6.7°C, and at a maximum capacity of 865 MW the temperature rise is 7.8°C. The actual temperature rises in 1969 exceeded 6.7°C about 5.5% and 16.4% of the time in July and August, respectively, the two hottest months of the year. During these two months in 1969, the company had been operating the Turkey Point plant under the policy that the plant would be the last plant in the South Florida network to go to full power and the first to come down from full power (FWPCA, 1970). This policy was adopted as the result of a massive fish kill that occurred in lower Biscayne Bay on June 25, 1969.

In early 1967, prior to the initial operation of the first oil-fired generator, FPL began construction of two additional generators at Turkey Point. These two units are nuclear powered, with a maximum power output of 760 MW each. Construction of the nuclear generators was completed in 1971 and 1972. When all four units are operating, the plant

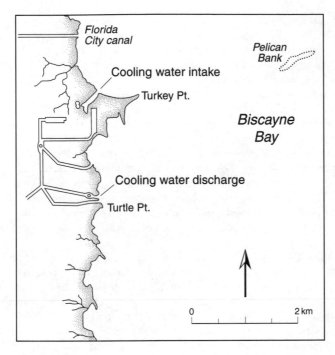

FIGURE 11.4   Detail of the original cooling canal system at Turkey Point power plant. Note the discharge into Biscayne Bay at Turtle Point. *Source*: Redrawn from FWPCA (1970).

requires 10.4×10^6 m^3 d^{-1} of cooling water. Under a normal full load of 2,080 MW the temperature of the plant's cooling water is raised 7.9°C, and at a maximum load of 2,384 MW the temperature rise is 8.8°C (FWPCA, 1970).

Effects on Biota

The average maximum daily water temperature in Biscayne Bay is normally 30–31°C during the months of July and August (FWPCA, 1970). Since very few aquatic species are able to withstand temperatures in excess of 30–35°C for more than a short time, artificially raising the temperature of Biscayne Bay by only a few degrees centigrade could be expected to have a serious impact on the bay's biota.

The first adverse effect of the heated effluent from the power plant was detected in a survey of south Biscayne Bay during late July 1968 about three months after the second oil-fired generator went on line. An area of barren sediment covering 8–12×10^3 m^2 was found off the mouth of the discharge canal. The turtle grass that flourished in similar areas of the bay was completely gone, and the denuded sediments "were covered with what appeared to be irregular mats of green or blue-green microalgae" (FWPCA, 1970, p. 28). By mid-November, the area of denuded sediments had expanded to about 4×10^5 m^2. Currents had seriously eroded unconsolidated sediments in the exposed area, and scoured depressions were filling in with mangrove peat and dead turtle grass rhizomes. Noticeable damage to the benthos, and turtle grass in particular, was apparent over an additional 8×10^5 m^2 (FWPCA, 1970).

By May 1969, regrowth of turtle grass had occurred over much of the 8×10^5 m^2 area that had been adversely affected the previous summer and fall. In this area of turtle grass regrowth, a wide variety of benthic fauna was found, including shrimp, mollusks, immature crabs, sponges, corals, and small bottom fish. However, an area of 4×10^5 m^2 near the mouth of the discharge canal still remained devoid of the usual benthic biota. The sediments in this area were covered by colonies of green algae and cyanobacteria, with occasional patches of benthic diatoms. These microalgae provided virtually no protection to the sediments from erosive currents, and additional erosion was apparent from propeller tracks cut into sediments by motorboats (FWPCA, 1970).

Further deterioration of the benthos became apparent on June 5, 1969. The coral colonies within 350–550 m of the outfall had all died, and some dead corals were found as much as 750 m from the outfall. By June 19, other benthic organisms were beginning to disappear. The turtle grass at 550 m from the outfall was losing its green color and was coated with a brown gelatinous material, and some turtle grass was beginning to lose color and vigor at distances of up to 900 m from the outfall. At 750 m from the outfall, 50–70% of the corals were dead.

A massive kill of organisms in the bay occurred approximately one week later. On June 25, 1969, the bay's water was unusually turbid and the water temperature was concomitantly high.[2] Thousands of dead fish were found south of Turkey Point on June 26. A survey on June 27 revealed numerous dead pistol shrimp, blue crabs, spider crabs, stone crabs, mollusks, and algae. The dead fish were most numerous within 900 m of the outfall. A more extensive survey up to 1.4 km from the outfall on June 29 revealed that the bottom was littered with "dead plants and animals, including sponges, spider crabs, blue crabs, corals, pistol shrimp, clams, snails, mussels, and several varieties of fish in addition to the wilted and browned algae" (FWPCA, 1970, p. 33).

Figure 11.5 indicates the area of biological damage that was apparent to scientists who surveyed the bay on June 27 and shortly thereafter. Virtually a complete kill of aquatic organisms occurred over an area of about 8.4×10^5 m^2 indicated by the densely hatched area in Figure 11.5. The shape of this area of acute effects closely follows the northeasterly flow of effluent water from the discharge canal. The total area of biological damage, somewhat arbitrarily defined in Figure 11.5, extended both north and south of the discharge canal and covered a total area of about 2.7 km^2. Subsequent temperature studies conducted during August 1969 revealed that the power plant effluent was increasing the temperature of the bay water by at least 2.2°C over an area of 2.5 km^2, and that near the mouth of the discharge canal the temperature rise was about 4.4°C (FWPCA, 1970). Although some secondary treated sewage from the Homestead sewage treatment plant (a few kilometers west of Turkey Point) is discharged into the bay via a canal and "has caused a noticeably heavier growth of bottom vegetation near the mouth of the canal" (Florida State Board of Health, 1962, p. 16), water quality studies in the bay conducted during January 1970 revealed no indication of pollution as measured by standard chemical indicators.[3] This leads to the conclusion that the June 1969 kill was caused entirely by elevated temperatures resulting from the power plant's cooling water discharge.

Modifications

Water quality standards adopted by the State of Florida require that water temperatures "shall not be increased so as to cause any damage or harm to the aquatic life or vegetation of the receiving waters or interfere with any beneficial use assigned to such waters"

---

[2]Turbid water absorbs more light energy.
[3]For example, BOD, dissolved O_2, and nutrient concentrations.

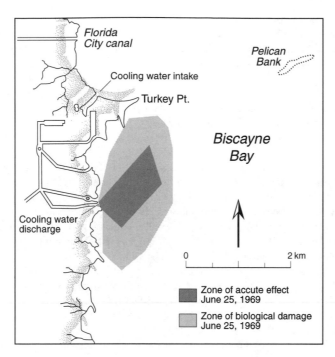

FIGURE 11.5   Biological damage zone in Biscayne Bay resulting from conditions on June 25–27, 1969. *Source*: Redrawn from FWPCA (1970).

(FWPCA, 1970, p. 15). Water uses assigned to Biscayne Bay include recreation and the propagation and management of fish and wildlife. Clearly, the deterioration of the benthos in 1968 and 1969 and the massive kill of fish and benthic organisms in late June 1969 as a result of the discharge from the Turkey Point power plant directly violated this Florida law. Dade County, which includes Biscayne Bay, further required that water temperatures not be artificially raised above 35°C, and the present EPA water quality criteria stipulate that the short-term mean temperature must not exceed 32.2°C. During June 1969, however, water temperature in Biscayne Bay exceeded 35°C and 32.2°C during part of the day at distances of up to 1.1 km and 1.4 km, respectively, from the discharge canal (FWPCA, 1970, Fig. 12). For comparison, the maximum water temperature measured at any time during August (the hottest month of the year) 1969 at the cooling water intake was 31.1°C (FWPCA, 1970).

Considering the impact of the cooling water discharge on Biscayne Bay in 1969, realizing that the nuclear generators to be installed between 1971 and 1972 would increase the amount of cooling water discharge by a factor of about 3.4 and the temperature of the effluent by about an additional degree centigrade, and knowing that lower Biscayne Bay had been declared a national monument in 1968, the State of Florida and FPL took steps to mitigate the discharge problem before installation of the nuclear generators. The solution proposed initially was to discharge the effluent by way of a 9.7 km long canal into Card Sound (Figure 11.3). The rationale behind this scheme was severalfold.

1. Card Sound is not a national monument.
2. The effluent water would have a chance to cool down by about 0.5°C during its passage through the canal (Bader and Roessler, 1971).

3. Card Sound is deeper (mean depth, 2.7 m) than Biscayne Bay and is more stable physically and chemically (FWPCA, 1970). Therefore, it was hoped that thermal discharges to Card Sound would have less of an impact on the system than the same discharges made to Biscayne Bay.
4. Studies of the mixing and flushing characteristics of Biscayne Bay had shown that effluent discharged at Turkey Point (Figure 11.4) tended to move north and be recirculated to the intake canal. This recirculation produced a cumulative effect on the temperature rise of the water returning to the plant (Lyerly and Littlejohn, 1973). Discharging the water to Card Sound would eliminate this problem.

The initial FPL proposal also called for diluting the effluent up to 150% with Biscayne Bay water to reduce the temperature of the effluent. After months of hearings this proposal was finally approved, and construction of the discharge canal begun. The project was abandoned, however, as the result of a Federal-State Conference on Biscayne Bay in February 1970. Concern at this conference was expressed over possible damage to biota in Card Sound associated with the high velocity of cooling water and the dilution water effluent that would be entering the sound (Lyerly and Littlejohn, 1973). Furthermore, studies conducted in Card Sound indicated that the water temperature in at least 35% of the Sound would still be raised over 35°C about 5% of the time in the summer if the diluted effluent were discharged to the sound as proposed (FWPCA, 1970).

As a result of these predictions, FPL adopted even more stringent measures for dealing with its effluent. The plan ultimately put into operation consists of an enormous system of 32 shallow cooling canals, each about 8.4 km long, through which cooling water from the plant flows before being recycled back in the intake system (Figure 11.3). The canal system provides about 15.6 km^2 of water surface area for heat exchange with the atmosphere, evaporation losses being made up by seepage from brackish groundwater underneath the canal system. The residence time of water in the canals is about 40 hours, during which time the water temperature drops to within 1–2°C of Biscayne Bay's ambient temperature. According to Tucker (1977, "The canal system is a closed system. There has been no surface water makeup or blowdown since the system was closed in February 1973. There are no withdrawals from or discharges to Card Sound."

Commentary

From the standpoint of thermal pollution, the physical characteristics of Biscayne Bay make it an extremely poor choice as a place to discharge large quantities of heated water. In particular:

1. Circulation in the bay is poor; therefore, heated effluent is not flushed readily from the system (FWPCA, 1970).
2. The shallowness of Biscayne Bay makes the benthos in this system particularly susceptible to thermal pollution effects. Although some stratification of the water column is apparent even during stormy weather (FWPCA, 1970), this stratification is evidently inadequate to prevent the effluent's being mixed downward sufficiently to increase bottom water temperatures. The killing of benthic biota in 1968 and 1969 attests to this fact.
3. In 1968, the mean maximum daily water temperatures in the bay during the summer months of June, July, and August were 28.3°C, 29.8°C, and 30.7°C, respectively, outside the influence of the plume (FWPCA, 1970). These temperatures are close to the lethal tolerance limit of tropical marine organisms, and an increase of water temperature of only a few degrees centigrade would be expected to have serious

consequences for the bay's biota. In short, the bay is thermally stressed naturally during the summer months.

Although the construction of a huge closed cooling canal system has eliminated the thermal pollution problem, it is not clear that this solution was the wisest choice environmentally. The construction of 15.6 km^2 of cooling canals destroyed a significant area of mangroves, a loss whose ecological impact was of some consequence. During the one year in which the cooling canal system was being dredged, effluents from the power plant were diluted with Biscayne Bay water and discharged both into Biscayne Bay by way of the original discharge canal and into Card Sound by way of a 9.7 km long canal or later only into Card Sound. According to Thorhaug et al. (1979), the discharge to Card Sound adversely affected an area of only 2–3 $\times$ 10^4 m^2. The much smaller impact on Card Sound was attributed to the greater depth of the sound, the fact that the effluent temperature never exceeded 34°C, and the fact that the canal's mouth was constructed to direct effluent water upward, so that less of the effluent impinged directly on the bottom (Thorhaug et al., 1978). In retrospect, it seems probable that a less heroic cooling canal system combined with the release of makeup water to Card Sound could have largely eliminated the thermal pollution problem while sacrificing only a small area of mangroves.

CORRECTIVES

The cooling canal system used at Turkey Point represents one of several methods for dealing with thermal pollution. If the effluent can be cooled to a temperature close to ambient, then it may be practical to recycle the water back through the cooling system. A cooling system that involves more-or-less complete recycling of the cooling water is called a *closed cooling system*, as opposed to *open* or *once-through cooling systems*, which use cooling water once and then discard it. Actually, closed cooling systems are not closed completely. In many cases, some water is lost to the atmosphere through evaporation in the process of cooling down the effluent, and in all cases, some dilution of the effluent with fresh cooling water is needed to offset the buildup of corrosion products and other dissolved and particulate substances in the cooling water. However, the so-called makeup water requirements of a closed-cycle power plant are typically only 2–4% of the cooling water demand of a once-through power plant (Clark and Brownell, 1973). As a result, closed-cycle power plants withdraw 25–50 times less cooling water from the environment than do once-through plants. There is, of course, a continuum of possibilities between strictly closed-cycle and once-through power plants. In the case of Turkey Point, for example, a partially open system involving some discharge to Card Sound and less land area devoted to cooling canals might have been preferable to either the once-through or closed-cycle alternatives.

Cooling Canals

A closed-cycle cooling system may operate with nothing more sophisticated than a long channel or lake to allow the effluent water to exchange heat with the atmosphere. Ultimately, such a system was installed at Turkey Point. The recommended design generally calls for a pond that is only 1 m or so deep at the input end and about 15 m deep at the outlet. The cooling water to be recycled is drawn from about 9 m below the surface at the outlet (Clark, 1969). Usually such canals are designed so that the residence time of water in the cooling canal is about two weeks. The major disadvantage of such systems is the large area of land required to provide adequate cooling. Typical 1,000 MW fossil fuel and nuclear power plants require about 8 and 10 km^2 of pond surface area, respectively

(Langford, 1983). The surface area required for the ponds can be reduced by as much as a factor of 20, however, if the canals are designed with spray devices that disperse the water in droplets into the air in order to speed up evaporative heat losses (Langford, 1983). About 150–200 75-horsepower pumps are required to drive the spray devices for a 1,000 MW power station (Clark and Brownell, 1973). The power required to operate the pumps represents only about 1% of the plant's output.

Cooling Towers

Effluent water may also be cooled with the use of so-called cooling towers. Cooling towers require much less space than cooling canals, but their height sometimes creates aesthetic problems. Dry cooling towers work on essentially the same principle as the radiator in an automobile. The warm water from the heat exchangers is run through a series of pipes around the interior of a tower, and heat from the water is transmitted through the pipes to a current of air rising through the tower. In a wet cooling tower, the heated water is sprayed into the tower as a mist and is allowed to fall in a thin film over a series of battles. As the water moves down, a rising current of air picks up heat from the water by way of evaporation and radiative heat loss. Figure 11.6 shows two basic types of wet cooling towers. The natural draft tower works much like a chimney. Air inside the tower rises as it is heated and is replaced by cooler air, which enters through openings in the bottom of the tower. The mechanical draft tower uses mechanical means to circulate air through the sides of the tower and out at the top.

Problems

Unfortunately, all of the closed-cycle cooling schemes have problems. Cooling canals without spray devices require enormous areas of land. Such large areas of land are frequently not available near power plants, and even if they are, it is not always obvious that devoting so much land to cooling canals is the best option from an environmental standpoint. The use of sprayers to speed up heat transfer between the water and the atmosphere greatly reduces the land area requirements but obviously increases the cost of running the cool-

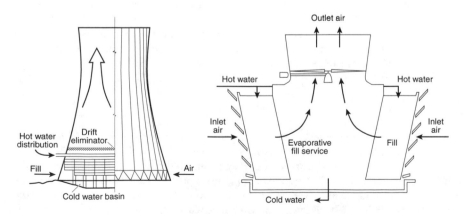

FIGURE 11.6   Natural draft (left) and mechanical draft (right) wet cooling towers. *Source*: Redrawn from Clark and Brownell, 1973. Reproduced with permission of the American Littoral Society, Highlands, NJ. 07732.

ing system. One of the chief drawbacks of cooling towers is their unsightliness. Natural draft cooling towers are typically 110–130 m high. Mechanical draft towers are generally much shorter (20 m high) but are costlier to run (Clark and Brownell, 1973). All wet cooling towers emit a plume of water vapor, which may cause problems under certain conditions. For example, fogging and/or icing may be caused or exacerbated by the plume of water vapor from a wet cooling tower. If saltwater is used for cooling, salt fallout in the vicinity of cooling towers or near cooling canals using sprayers may damage vegetation. However, salt fallout from spray ponds is generally confined to the immediate vicinity of the plant, and salt fallout from mechanical draft wet cooling towers does not extend much further (Clark and Brownell, 1973). Although salt fallout from the much taller natural draft wet cooling towers can be expected to carry for some distance, the amount of salt fallout more than a short distance from the plant is not likely to be more than a small percentage of the level that would be damaging to vegetation (Clark and Brownell, 1973). Langford (1983), for example, notes that saline water has been used in cooling towers at the Fleetwood Power Station in the United Kingdom for several years with few problems. Dry cooling towers obviously eliminate any problems associated with vapor plumes or salt fallout, but their capital cost is two to three times that of wet cooling towers, and corrosion problems are a major disadvantage.

Table 11.2 indicates the relative costs of various types of power plant cooling systems. Although the closed-cycle systems are several times more expensive than once-through cooling, use of a wet cooling tower adds no more than a small percentage to a utility customer's electric bill (Knighton, 1973). In other words, the cost of a closed cooling system is not large compared to the cost of the entire power plant. Closed cooling systems are almost mandatory at the largest power plants, since the amount of cooling water required for these installations using once-through cooling systems is prohibitive. In many cases, the impact of power plants on the environment is much less severe when closed rather than open cooling systems are used.

INTERNAL PLANT KILLS

The quotation from *America the Raped* at the beginning of this chapter provides a good introduction to the problem of internal plant kills. These kills occur when organisms are drawn with the cooling water into the power plant. The larger organisms are impinged either on the trash rack or on the screen protecting the pumps. The smaller organisms are drawn through to the inner plant. Vigorous swimmers can avoid this fate, but planktonic

TABLE 11.2 RELATIVE CAPITAL COSTS OF VARIOUS TYPES OF ELECTRIC POWER PLANT COOLING SYSTEMS

	Relative Cost	
Cooling System	Fossil Fuel Plant	Nuclear Plant
Once-through	1.0–1.5	1.5–2.5
Cooling canals or ponds	2.0–3.0	3.0–4.5
Wet cooling towers		
Mechanical draft	2.5–4.0	4.0–5.5
Natural draft	3.0–4.5	4.5–6.5
Dry cooling towers	8	12

Source: Knighton (1973) and Langford (1983).

organisms are largely at the mercy of the currents, and larval fish are often unable to resist the suction from a once-through cooling system's intake. According to Clark and Brownell (1973), the damage to aquatic biota from internal plant kills generally exceeds the damage due to external problems caused by the plant's effluent. The ecological significance of internal plant kills, however, is somewhat controversial (Langford, 1983). In the following discussion, we consider the effects of screen impingement and inner plant kills separately.

Screen Impingements

Some fish drawn in with the cooling water may be large enough to be impinged on the trash racks, but in general, fish too large to pass through the 7–8 cm openings of the trash rack can swim vigorously enough to escape from the intake suction. As a result, trash rack kills are usually insignificant (Clark and Brownell, 1973). Smaller fish that pass through the trash rack may be impinged against the screen (Figure 11.2). Most of these fish remain pinned against the screen until they die of suffocation, exhaustion, or physical damage. The screen is generally cleaned once or twice a day by rotating it and dislodging the trapped fish and debris with a jet of water. The dead fish and debris are collected in a basket or similar device and carted off for disposal. Fish still lively enough to flop off the screen when it is rotated are immediately pinned against the screen again after they fall back into the water. The number of fish killed in this way is positively correlated both with the volume of water pumped through the plant and with the intake current velocity.

Table 11.3 lists some of the estimated fish kills due to impingement at electric power plants in the United States. In most cases, the numbers for plants using once-through cooling systems are high, typically 10^5–10^7 fish per year. When these numbers are compared, however, to commercial fish catches or to the standing stock of fish in the vicinity of the plant, the percentages are generally quite low. The 3×10^7 fish impinged each year at 17 Lake Michigan power plants, for example, represent only about 0.06% of the corresponding standing stocks in Lake Michigan (IAEA, 1980). On the other hand, the fact that impingement of yellow perch at the Monroe Power Plant on Lake Erie amounts to 2.7% of the commercial catch of this species seems noteworthy, since the Monroe Power Plant is by no means the only power plant located on the shores of Lake Erie.

Measures can be and have been taken to reduce impingement losses. Bubble curtains and mechanical diversion systems have been used with partial success to divert fish away from intakes, and velocity caps, which minimize the downward flow of water toward intake structures, have been used to minimize fish entrainment. Unfortunately, no such engineering modifications have proven generally satisfactory at estuarine sites, where the impingement problem is most severe because of the high densities of juvenile and larval fish. In such cases, efforts to rescue organisms impinged on the intake screens may reduce losses significantly. At the Brunswick Steam Electric Plant, for example, a continuous rescue operation has resulted in 90% of the crabs, 80% of the shrimp, and 40% of the fish being returned to the estuary in live condition (IAEA, 1980). The most effective strategy, however, is to install a closed cooling system. This fact is illustrated dramatically in the comparison of impingement losses at the Davis-Besse, Bay Shore, and Acme power stations in Table 11.3. Although the Davis-Besse station generates more electricity than the other two stations combined, impingement losses at Davis-Besse are less than 0.1% of the losses at either of the other stations. The principal cause of the difference is the fact that Davis-Besse uses a closed cooling system that draws only 5–10% as much water as Acme or Bay Shore. The difference in impingement losses is particularly noteworthy because the adult fish populations near Davis-Besse can be twice those near Bay Shore and eight times those at Acme (Reutter and Herdendorf, 1984).

TABLE 11.3 FISH KILLS DUE TO SCREEN IMPINGEMENT AT VARIOUS POWER STATIONS

Power Plant	Impingement Event	Period	Comments
Millstone Niantic Bay, Connecticut	Massive kill of small menhaden (more than 2 million), screens clogged	1971	Occurring in late summer and early fall; plant shut down on Aug. 21; cause unknown; persistent low kill of 10 other species
P. H. Robinson Galveston Bay, Texas	7.2 million fish impinged in 1 year	1969–1970	Projected from sampling of operating plant; principal species were menhaden, anchovy, croaker; highest in March
Indian Point No. 1 Hudson River New York	Yearly kill of 1–1.5 million fish	1965–1972	Primarily white perch, with 4–10% striped bass
	Kill of 1.3 million	1969–1970	10% striped bass; Plant closed Feb. 8
Indian Point No. 2	Massive kills; maximum of 120,000 per day	Jan. 1971	Testing cooling system of new plant (no heat); white perch and other species
Port Jefferson Long Island, New York	2 truckloads (at least) of fish killed on screen in 3 days	Jan. 26–28, 1966	Mostly small menhaden; also white perch
Brayton Point Mount Hope Bay, Massachusetts	350,000 fish impinged in 1 year, mostly menhaden	1971–1972	Heaviest from November to March; flounder, silverside, and others also impinged
Oyster Creek Barnegat Bay, New Jersey	10,000 fish and 5,000 crabs destroyed per month in spring and summer	1971	Estimated from 19 days of sampling; screen kill in cold season unknown
Surrey Power Station James River, Virginia	6 million river herring destroyed in 2–3 months	Oct.–Dec. 1972	Estimated by AEC from screen sampling during partial power runs
17 Lake Michigan plants	30 million fish impinged per year	Early 1970s	Mostly alewives and rainbow smelt; biomass losses ~0.06% of standing stocks
4 Tennessee Valley Authority plants	22,000 to 3.6 million fish impinged per year.	Early 1970s	Mostly clupeoids and sciaenids Losses 0.03–0.52% of total biomass; Losses of eight major families <3%
Monroe Power Plant Lake Erie	122,000 yellow perch impinged per year	1970s	Loss equals 2.7% of Lake Erie commercial catch and 0.7% of perch population in western basin of Lake
Pilgrim New England	58,000 fish impinged per year.	1970s	Mostly Atlantic herring and alewife
Peach Bottom Pennsylvania	472,000 fish impinged per year	1973–1975	Mostly channel catfish, white crappie, and bluegill; Two units in operation
Crystal River Florida	50,000–62,000 fish impinged per year	1969–1970	Mostly batfish, catfish, and perch
Acme Power Plant Maumee River, Ohio	11.7 million fish impinged per year	1976–1977	Mostly gizzard shad, emerald shiners, and alewives; most between October and February
Bay Shore Lake Erie at mouth of Maumee River	18.3 million fish impinged per year	1976–1977	Mostly gizzard shad, emerald shiners, and alewives; most between October and February
Davis-Besse Lake Erie	4,400–6,600 fish impinged per year	1978–1979	Mostly goldfish and yellow perch; most in colder months; closed cooling system

Source: Clark and Brownell (1973), Grimes (1975), Mather et al. (1977), IAEA (1980), Reutter and Herdendorf (1984).

Inner Plant Kills

Organisms that pass through the 1 cm openings of the intake screens are drawn through the rest of the condenser cooling system. During this experience, they are subjected to a variety of stresses, including thermal shock, physical abrasion and buffeting, pressure shock, and possibly chlorination and N embolism. If an organism is not killed outright during its passage through the plant, it may be so physically damaged by the time it is discharged that it dies soon thereafter or easily falls victim to a predator.

Fish larvae and post larvae suffer particularly high mortalities due to inner plant stresses. For example, the EPA estimated that all larval menhaden were killed as a result of passage through the Brayton Point power plant on Mount Hope Bay, Massachusetts (EPA, 1972). Physical damage appeared to be the principal cause of death. Marcy (1971) found that no white perch survived passage through the Connecticut Yankee power plant at Haddam Neck if the discharge temperature exceeded 28.3°C, and no larvae or juvenile fish of any species survived if the discharge temperature exceeded 32.8°C. Most of the dead fish were mangled, an indication that physical as well as thermal stress was a factor in their demise. At the Indian Point power plant on the Hudson River, 97.5% of larval and early juvenile fish died or were severely damaged by passage through the cooling system during the summer if the temperature rise across the condensers exceeded 15°C (Clark and Brownell, 1973).

Planktonic organisms generally survive inner plant stresses better than juvenile and larval fish. Zooplankton losses have been reported to range from 0% to 100%, the average loss being about 30–35% (Clark and Brownell, 1973; Langford, 1983). The percentage kill of zooplankton is quite dependent on effluent water temperature and degree of chlorination. For example, copepods suffered 100% mortality at the Northport power plant on Long Island Sound if the discharge temperature exceeded 34°C, a condition that exists from July to early October (Suchanek and Grossman, 1971). Copepod losses at the same plant, however, were only 33% in the fall and 4% in the winter (Clark and Brownell, 1973). At four power plants along the California coast, zooplankton losses were related linearly to temperature (Icanberry and Adams, 1974). Gammarid losses in power plant cooling systems are virtually 100% if the effluent temperature exceeds 32°C, and mysid mortality is 100% at temperatures above 27°C (Clark and Brownell, 1973). Davies and Jensen (1975) reported 100% mortality of zooplankton entrained in the Marshall power plant cooling system at Cl concentrations of 1.5 mg L^{-1} but no mortality at Cl concentrations of 0.25–0.75 mg L^{-1}. These results suggest that the principal causes of inner plant zooplankton kills are thermal stress and chlorination, and that the effects of physical abrasion are of relatively minor importance. Obviously, zooplankton mortality can be expected to be highest during the summer months, when ambient water temperatures are highest.

The impact of entrainment on phytoplankton is generally negative and, as in the case of zooplankton, appears to result almost entirely from thermal stress and/or chlorination. Data summarized by Langford (1983) indicate an average reduction in photosynthetic rates of 40–50%, but in many cases it is unclear whether cells have actually been killed or just stressed temporarily. In temperate latitudes, phytoplankton production may actually be stimulated in the warmer waters of the discharge plume (Morgan and Stross, 1969). There is no doubt that chlorination adversely affects phytoplankton. Davis and Coughlan (1978), for example, show an obvious correlation at Cl concentrations above 0.1 mg L^{-1} (Figure 11.7). Fortunately, the generation time of phytoplankton is relatively short, typically 12 hours to a few days, and natural physical processes can be expected to reseed effluent waters with healthy cells within a short time. It is therefore unlikely that entrainment has a significantly negative impact on photosynthetic rates in waters surrounding electric power plants. In fact, the stimulation of undesirable cyanobacteria by the heated effluent may be more of a problem than any damage to phytoplankton caused by inner plant stresses.

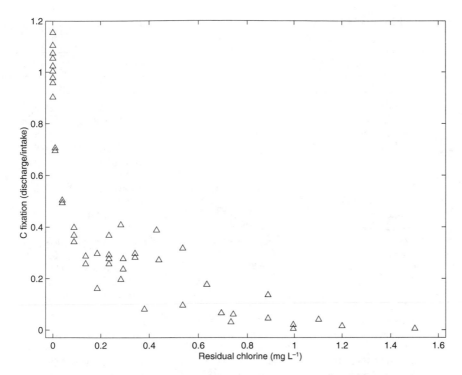

FIGURE 11.7 The effect of cooling-water chlorination on photosynthesis by marine phytoplankton at Fawley power station in the United Kingdom. Reprinted with permission from Davis and Coughlan (1976). Copyright CRC Press, Boca Raton, Florida.

Just how serious are inner plant kills to aquatic biota? Table 11.4 lists estimates of inner plant fish kills at several power plants. A comparison of Tables 11.3 and 11.4 indicates that entrainment tends to kill far more fish than impingement, and the annual losses to entrainment are roughly 5% of the at-risk population. Intelligent siting and design of intake systems help to reduce such losses, but the most effective way to minimize entrainment losses is to use a closed cooling system. The comparison of the Davis-Besse power plant with the Acme and Bay Shore facilities is again instructive. The entrainment of larval fish and fish eggs at Davis-Besse is only 3.6% and 0.01%, respectively, of the combined entrainment at the other plants, although Davis-Besse produces more power than Acme and Bay Shore combined. The difference largely reflects the use of a closed cooling system at Davis-Besse.

Commentary

Judging from Tables 11.3 and 11.4, the number of fish killed each year from internal plant stresses in a once-through cooling system can easily run into the tens or hundreds of millions. The number of fish killed from thermal stress in late June 1969 near the Turkey Point outfall was reportedly in the thousands, but not millions (FWPCA, 1970). The largest external plant fish kill on record involved 100–200 thousand menhaden that were killed by cold shock when the Oyster Creek power plant on Barnegat Bay, New Jersey, shut down

TABLE 11.4   FISH KILLS DUE TO INNER PLANT STRESSES AT VARIOUS POWER STATIONS

Power Plant	Event	Period	Comments
Brayton Point Mount Hope Bay, Massachusetts	7–165 million menhaden (some river herring) killed per day.	Summer 1971	Estimated from EPA's sampling techniques; 164.5 million killed on July 2; fish mangled
	50 million fish killed in 11 days	Aug. 10–21, 1971	Estimated from net tows at discharge; menhaden and blueback herring; tests showed that all fish died
Millstone Niantic Bay, Connecticut	36 million fish killed in 16 days (probably) menhaden and blueback herring	Nov. 2–18, 1971	Estimated by sampling of vertebrae of dead fish in discharge canal
	2.5 million flounder entrained	Apr.–June 1971	Death rate not estimated
Connecticut Yankee Connecticut River, Connecticut	179 million fish larvae killed per year	1969 and 1970	Primarily alewife and blueback herring; loss equals 4% of transported population
Cumberland Stream Cumberland River, Tennessee	57 million larvae entrained per year	1970s	Entrainment equals 3.6% of larvae transported past plant
Hanford Columbia River, Washington	142,000 chinook salmon fry entrained per year	1970s	Entrainment equals 8.8% of population
17 Lake Michigan plants	Alewife larvae	Early 1970s	Entrainment equals 0.01% of population
	Smelt larvae	Early 1970s	Entrainment equals 1% of population
Acme Maumee River	80 million larval fish entrained per year	1976–1977	Mostly gizzard shad and freshwater drum
	178 million fish eggs entrained per year	1976–1977	
Bay Shore Lake Erie	284 million larval fish entrained per year	1976–1977	Mostly gizzard shad and white bass
	426 million fish eggs entrained per year	1976–1977	
Davis-Besse Lake Erie	13 million fish larvae entrained per year	1976–1977	Mostly gizzard shad and emerald shiner
	70,000 fish eggs entrained per year	1976–1977	

Source: Clark and Brownell (1973), IAEA (1980), Reutter and Herdendorf (1984).

in late January 1972 (Chapter 8). Other external plant kills have involved thousands and in some cases tens of thousands of fish (Clark and Brownell, 1973; Langford, 1983).

The greater magnitude of the internal plant kills undoubtedly reflects the fact that fish, including larval and juvenile fish, can usually avoid discharge plumes by simply swimming away. Small fish find it more difficult to escape the suction from a power plant's water intake. One should keep in mind, however, that a large percentage of larval and juvenile fish never reach adulthood, since they are eaten by predators at an early age. Thus, the killing of 1 million juvenile fish of a particular species may remove essentially the same percentage of the species and age group as the killing of 10^4 adults. On the other hand, the killing of juvenile and larval fish not only reduces the population and the reproductive potential of the given species, but also removes a source of food for other organisms that prey on small fish. According to an Atomic Energy Commission staff analysis, entrainment of striped bass larvae by power plants along the Hudson River could result in over 60% mortality of each year's production of young striped bass (Goodyear, 1973). The impact of this loss on the adult population of striped bass has been estimated to be anywhere from 8% to 33%, depending on what assumptions are made about natural mortality and compensatory survival (Langford, 1983). Declines of 20–30% in the population of a valuable fish seem a bit much to simply ignore. What can be done to reduce the impact of internal plant kills?

Correctives

Engineering modifications such as bubble curtains, velocity caps, and mechanical diversion systems have had some success in reducing screen impingement kills but have had little or no effect on entrainment losses. There are, however, several rather obvious ways of reducing internal plant kills at power plants that use once-through cooling systems.

1. If possible, power plants should not be located in estuarine areas, since these areas are known to serve as breeding and nursery grounds for many species of aquatic organisms, and the population density of juvenile and larval fish, which are subject to entrainment, is likely to be highest in these estuarine regions.
2. According to Langford (1983, p. 192), "The greatest reductions in impingement mortalities have been at sites where water velocities have been significantly reduced." Clark and Brownell (1973) recommend that the speed of flow into the cooling system at the intake should be no more than about 8 cm s^{-1} and 15 cm s^{-1} for closed and open cooling systems, respectively. The idea is to maximize the likelihood that weakly swimming organisms will be able to resist being drawn into the cooling system. The higher limit of 15 cm s^{-1} for once-through cooling systems is established for practical rather than ecological reasons. Because of the large volumes of water needed for once-through cooling, Clark and Brownell (1973) felt that it would be impractical to expect the construction of a once-through cooling system intake structure large enough to keep flow speeds much below 15 cm s^{-1}, an opinion that is echoed by Langford (1983).
3. Impingement kills can be reduced significantly by designing a mechanism for rescuing the impinged organisms before they are injured seriously. As noted, the Brunswick Steam Electric Plant has reduced impingement losses by 40–90% with this approach.
4. The only effective way to reduce entrainment losses is to minimize the concentration of organisms in the cooling water or reduce the amount of water drawn from the environment. Judicious choice of the location of the intake structure can help greatly in the former case. The intake at the Davis-Besse power plant, for example, is located offshore and near the bottom of Lake Erie. The Acme and Bay Shore plants both utilize open shoreline intake canals, which tend to attract schooling species that follow the shoreline. In general, the concentration of organisms in the water will decrease with depth and with distance from shore.

Reducing the amount of water drawn from the environment can be accomplished with a closed cooling system, but at a price. The price includes the additional capital, operating, and maintenance costs associated with the closed cooling system and reduced efficiency in electricity production. The latter loss of efficiency amounts to about 1.5–2.0% for fossil fuel plants and 3% for nuclear plants (Monn et al., 1979). The monetary costs, of course, are passed on to the consumer. The environmental costs, however, should not be overlooked. The reduction in plant efficiency translates into more CO_2 and oxides of N and S released into the atmosphere at fossil fuel plants, as well as more radioactive wastes generated at nuclear plants. There may also be aesthetic considerations and, in some cases, significant additional environmental impacts associated with closed cooling systems, the destruction of mangroves at Turkey Point being an example of the latter. At large power stations, however, the availability of water alone may be sufficient to dictate the use of closed cooling systems. Selection of the best system will require a full accounting of the pros and cons of the various alternatives.

POSSIBLE BENEFICIAL USES OF THERMAL DISCHARGES

The idea of putting the heated water discharged by power plants to some practical use has occurred to a number of people. About 50% and 58–68% of the energy released from the fuel is rejected as waste heat with the cooling water at fossil fuel and nuclear power plants, respectively (GESAMP, 1984). What can be done with all this wasted energy?

Unfortunately, there are no uses to which a significant percentage of the heated effluent from power plants could possibly be put on a year-round basis in most parts of the world. Heated effluents have been and are being used in aquaculture and agriculture and for space and water heating, but demands for these uses drop off dramatically during certain times of the year. Furthermore, modifying present facilities to incorporate these uses will do little to help and may even aggravate water pollution problems. Nevertheless, such "beneficial" uses are worth examining to evaluate their pros and cons.

Cogeneration Power Plants

Electric power plants that produce a useful by-product in addition to electricity are called *cogeneration facilities*. In some cases, the by-products include steam and both hot and cold water. In cases where both heating and cooling are possibilities, the network served by the plant is referred to as a *district heating/cooling (DHC) system*. Power plants that serve district heating or DHC systems are designed as cogeneration facilities when they are built. They are not the result of retrofitting conventional electric power plants.

Using heated power plant effluents for space and water heating is an old idea that has been developed extensively in Europe and particularly in Russia (Table 11.5). Much of this development occurred during the rebuilding following World War II, when it was convenient to lay the necessary pipes to set up district heating networks. About 54% of all space and water heating in the former Soviet Union were provided by district heating, including 70% of all urban space and water heating (Karkheck et al., 1977).

District heating works in a very straightforward manner. The heated effluent from a power plant or a similar source of heated water is pumped to buildings, where it is used to warm rooms and heat water. In the process of heat exchange, the effluent cools down until it is no longer useful for heating, at which point it is recycled to the power plant. The principal cost of district heating networks is the cost of distributing the hot water to the customers. The hot water is typically pumped through urethane-insulated pipes constructed of polymer concrete and lined with polymer (Karkheck et al., 1977). Obviously, there is a limit to how far hot water can be economically pumped for heating, but a study conducted in the United States has indicated that steam could be transmitted economically as much as 16 km for heating (Cook and Biswas, 1974), and hot water is easier to

TABLE 11.5　PRODUCTION OF STEAM AND HOT WATER BY
FOSSIL FUEL POWER PLANTS

Location	Energy Equivalent (petajoules y^{-1})
World	8,798
Europe	8,449
Russia	5,797
Ukraine	950
Germany	385
Poland	384
Romania	216
Denmark	113
Czechoslovakia	110
Lithuania	100
Bulgaria	73
Finland	72
Hungary	48
Sweden	37
United States	329

Source: United Nations (1996).

transmit over long distances than steam (Karkheck et al., 1977). Nevertheless, district heating is not a practical scheme except in urban areas, where the population density of customers is high.

To be useful for space and water heating, hot water must have a temperature in the 50–100°C range and preferably closer to 100°C (Karkheck et al., 1977). In the United States, most electric power plants operate with a steam exhaust pressure of about 0.05 atm. At that pressure, steam condenses to water at a temperature of about 33°C. As a result, the temperature of the condenser cooling water is usually only about 5–15°C above ambient, too low to be useful for district heating. However, if the exhaust pressure to the condensers is adjusted to 1 atm, the steam condenses at 100°C and the temperature of the cooling water is raised correspondingly. Power plants to be used in a district heating network must obviously operate with a high steam exhaust pressure to produce sufficiently hot effluent water. This operational strategy sacrifices some efficiency in terms of electricity generation, but the loss of energy can be more than regained if the hot effluent water is utilized for district heating.

Although district heading is not as common in the United States as in Europe, a number of DHC systems do exist in the United States (Collins, 1991a, 1991b). The systems installed to date in the United States have been designed primarily to provide electricity and heating/cooling in high-density urban situations ideally suited for the DHC concept. The DHC system in Hempstead, New York, for example, serves a 19,000-student junior college, a 17,000-seat indoor sports arena, a 483-room hotel/conference center, a 615-bed medical center, and a large county jail complex (Collins, 1991b). A similar system in Hartford, Connecticut, serves 64 buildings, including 25% of the downtown area's high-rent office space (Collins, 1991a). A key feature of successful district heating and DHC systems is the ability to adjust the production of electricity and by-products to meet demands. In order to avoid pollution problems, the system must have some way to recycle thermal effluent when the demand for by-products drops. Because of the high temperature of the thermal effluent required to make district heating and DHC systems practical, it would be unacceptable for excess hot water to be discharged to the environment. Some method of cooling and recycling the effluent is therefore necessary when the supply of hot water exceeds the demand. The plant must therefore have either a backup closed-cooling system to recycle unneeded water or a backup conventional electricity generating capacity.

Karkheck et al. (1977) have estimated that 52% of the U.S. population could be served by district heating at a cost no greater than the present cost of heating using fossil fuel. In the process, these authors argue, the United States could cut its imports of foreign oil by about 1.1 billion barrels per year, currently about 55% of oil imports (Flaven, 1985). The cost of setting up the district heating system, about $180 billion, would be recouped in about 14 years due to the anticipated saving of $13 billion per year in the balance of payments. One obvious flaw in this argument is that most fossil fuel power plants in the United States burn coal. A more likely impact of large-scale district heating would therefore be an extension of the useful lifetime of the U.S. coal reserves. Currently, cogeneration power plants produce about 11% of the electricity generated in the United States (IEA, 1998), a percentage that has been almost invariant for the past decade. In Poland and Denmark, on the other hand, cogeneration accounts for more than 98% and 86%, respectively, of electricity production (IEA, 1998). Pressure to make more efficient use of energy resources may cause the contribution of cogeneration plants to electricity production in the United States to increase in the first half of the 21st century. However, because of the 30–40 year lifetime of existing conventional plants, the rate of increase is likely to be slow.

Agriculture

A variety of possible uses of thermal effluents in agriculture have been or are being tested. In so-called open-field thermal agriculture, heated water is used for frost protection during the spring and fall, for plant cooling during the summer, and for irrigation. In enclosed systems such as greenhouses, thermal effluents may be used in controlled-environment thermal agriculture to regulate temperature and humidity levels to provide a controlled environment for plant and/or animal growth. Unfortunately, many of the postulated uses of thermal effluents in agriculture would be highly seasonal and would fall far short of using all the effluent from even a moderately sized power plant. Nevertheless, Cook and Biswas (1974, p. 5) feel that "In a relative sense, controlled-environment agriculture is probably the most practical (technologically and economically) of all beneficial uses for low grade waste heat." The following discussion indicates some of the specific applications for which thermal effluents have been tested.

A study at Springfield, Oregon, sponsored by the Eugene Water and Electric Board investigated the feasibility of using thermal effluents for frost protection in orchards. Warm (35°C) water was sprayed onto trees when there was danger of frost, and the heat released from the water as it cooled or froze warmed the twigs and buds, thus protecting them from frost damage. If too much water was applied, however, the limbs became overloaded with ice and broke. Therefore, care had to be taken in regulating the amount of water applied (Miller, 1970).

At the same project site, heated water from the condenser of a pulp mill was piped to fields containing a variety of crops, both to irrigate the crops and to maintain optimal soil temperatures. An analogous experiment conducted in Oregon by Boersma (1970) indicated that simply regulating soil temperature by supplying heat from underground pipes could substantially increase crop growth rates and might triple the yield of some crops. In the Springfield, Oregon, study, the heated effluent used for frost protection and irrigation simply percolated through the soil and ultimately found its way to the Mackenzie River, where it entered the river at or below ambient river temperature.

The chief drawback to the foregoing type of open-field thermal agriculture is the large amount of money required to set up and maintain the system. Furthermore, water-pumping costs and heat losses with increasing distance from the power plant make this sort of scheme impractical except in the immediate vicinity of the thermal effluent. Finally, since this sort of water use would be highly seasonal, it would certainly not provide a satisfactory solution to the problem of thermal effluent disposal. On the other hand, the Springfield, Oregon, studies indicate that thermal effluents can be used to good advantage in open-field agriculture during certain times of the year and under appropriate climatological conditions.

Controlled-environment thermal agriculture has been tested in several desert areas as a possible means of growing crops in arid regions (Cook and Biswas, 1974). In studies conducted in Mexico and the Middle East, crops were grown in greenhouses made of plastic and inflated with air. Heated effluent from power plants was used to desalinate seawater and produce distilled water, which in turn was used to irrigate the crops. The use of greenhouses greatly reduced irrigation water needs by cutting down on evaporation losses and provided protection from pests, disease, and sandstorms. Capital and maintenance costs for such operations are quite high, but if cheap labor is available, Cook and Biswas (1974) indicate that crops produced by this method would be cheaper than imported crops. Because of high capital and maintenance costs, however, such controlled-environment thermal agriculture would not be cost competitive with conventional arid-land farming techniques where the latter are applicable.

In New York, the feasibility of using thermal effluent to heat greenhouses has been studied (Bell et al., 1970), and although the cost of using thermal effluent for this purpose was found to be quite competitive with conventional heating costs, the scheme was ultimately rejected by a Con Ed and Westinghouse task force because of the limited potential of greenhouses to use thermal effluent. For example, a 1,000 MW nuclear power plant could provide enough thermal effluent to heat 111.4 km^2 of greenhouses (Cook and Biswas, 1974). The idea, however, is not that farfetched. In Germany, for example, hothouses with an area of 0.2 km^2 are heated by cooling water from power plants and produce crops throughout the year (Mitzinger, 1974).

Aquaculture

The idea of using thermal discharges to stimulate aquaculture production was first explored in Japan in 1963 (Kuroda, 1979), and has been studied and proven economically practical in several countries. There are presently dozens of power stations throughout the world at which pilot-scale or commercial aquaculture facilities have been developed (Table 11.6). About two-thirds of those facilities use freshwater. The idea is to use warm discharge water either wholly or diluted with some natural water source to maintain a certain temperature for the organisms being cultured and, in particular, to extend the growing season into the colder months of the year, when growth rates would normally decline due to cold water temperatures. In this way, it may be possible to greatly reduce the time required to produce a marketable product and/or increase the number of generations produced per year.

The principal organisms cultured in freshwater systems have been catfish in the United States and carp in other countries (Table 11.6). In some cases, it has proven advantageous to culture different species at different times of the year to take advantage of natural variations in water temperature. For example, at the Mercer power station in New Jersey, a pilot farm has grown rainbow trout in the winter and freshwater prawns in the summer (Langford, 1983). Similarly, at the Flevo power station in Holland, trout are grown in the winter, and carp and eels are grown in the summer.

Japan has led the way in the culture of marine organisms in thermal effluent, the principal species of interest being shrimp, eel, yellowtail, seabream, ayu, and whitefish (Yee, 1972; Table 11.6). Use of thermal discharges in conjunction with yellowtail culture has in-

TABLE 11.6 ORGANISMS CULTURED USING THERMAL
EFFLUENT FROM POWER PLANTS IN VARIOUS COUNTRIES

Country	Power Station	Species Cultured
Japan	Ebetsu	Carp, loach, eels
	Akita	Abalone, sea bream
	Sendai	abalone
	Owase Mita	Sea bream, shrimp, crab, lobster
	Taketyo	Spawn of laver
	Tanagawa	Yellowtail, yoshiebi
	Himeji	Shrimp
	Toyama	Halibut, shrimp
	Kudamatsu	Ayu (spawn), shrimp
	Omura	Shrimp, ayu
	Karatsu	Shrimp
	Matsuyama	Shrimp
	Tanagawa	Yoshiebi
United States	Gallatin (Tennessee)	Catfish
	Colorado City (Texas)	Catfish
	Hanford (Washington)	Catfish
	Fremont (Nebraska)	Catfish and *Tilapia*
	Lake Trinidad	Catfish
	Hutchinson	Catfish
	Mercer (New Jersey)	Trout, freshwater prawns
United Kingdom	Ratcliffe-on-Soar	Carp and eels
	Trawsfynydd	Trout
	Ironbridge	Carp
Australia	Port Augusta	Red sea bream
Canada	Saint Laurent	eel
China	Zhengzhou	Red paeu
Finland	Olkiluoto	Salmon
France	Cadarache	Eels
Germany	Finkenheerd	Carp
	Rheinberg	Carp, eels, and trout
Hungary	Szazamlombatta	Carp, grass carp, silver carp, big-head carp
Korea	Seoul	Flounder, rockfish
Netherlands	Flevo	Carp, grass carp, trout
Norway	Matre	Salmon fingerlings
Poland	Konin	Carp, grass carp, silver carp, big-head carp
Russia	Burshtynsk	Bigmouth buffalo
	Klasson	Grass carp and carp
	Krivorozh	Grass carp and carp
	Primorskaya	Shrimp
	Shaktinsk	Tilapia
	Kirishi	Carp, trout

Sources: Kuroda (1979), Ingebrigtsen and Torrissen (1981), AFS (1981), Langford (1983), Andry-uschenko and Tretyak (1989), Landau (1992), Liu (1994), Svirskij et al. (1994), Park et al. (1996), and O'Sullivan and Saunders (1998).

creased growth rates by almost 50% and has nearly doubled the average feeding efficiency (Tanaka and Suzuki, 1966). In another set of experiments conducted at a power plant near Matsuyama, Japan, the wintertime weight gain of shrimp grown using thermal discharges was seven times that of control shrimp grown at ambient winter water temperatures.

In the United States, juvenile oysters were cultured year round on a commercial scale by Long Island Oyster Farms, Inc., using the thermal effluent from the Long Island Lighting Company's power plant at Northport, Long Island (Yee, 1972; Landau, 1992). Old oyster shells bearing oyster larvae were placed in racks and suspended in the power plant's discharge basin. Ambient water temperatures ranged from $-1°C$ to $24°C$, and the thermal discharge was about $14.5°C$ above ambient. During the summer, it was necessary to dilute the discharge with water of ambient temperature to avoid killing the oysters. The growth rate of the juvenile oysters in the discharge basin during the summer months was about double that of oysters growing at ambient temperatures, and growth continued during the winter in the discharge basin, whereas all growth came to a stop in surrounding waters when the temperature dropped below about $4.5°C$ (Cook and Biswas, 1974). The overall maturation time of the oysters was reduced by 1.5–2.5 years (Landau, 1992). Commercial sales amounted to $5 million in 1977 (Beall et al., 1977).[4]

Although the use of thermal effluents has proven successful in some aquaculture operations, there are significant limitations to this method of utilizing waste heat. Plant shutdowns could prove disastrous to some aquaculture operations during the winter months unless there is some way to mitigate the effect of thermal shock on the organisms being cultured. Either the aquaculture operation must have an auxiliary source of heated water or the species being cultured must be rapidly adaptable to temperature changes. Carp and eels are rather tolerant to temperature changes, but trout, for example, are not.

The extreme toxicity of Cl to many species of fish and invertebrates could weigh heavily against the use of thermal effluent for aquaculture at power plants that use chlorine to prevent fouling in the heat exchangers. Some method of dechlorinating the effluent or of bypassing the aquaculture operation when residuals become too high is essential. With respect to this issue, the most desirable power stations for aquaculture are those that use mechanical or physical means to prevent fouling. The Northport plant, for example, uses brushes and sponge balls under pressure for condenser tube cleaning.

Pollutants that might accumulate in the flesh of cultured organisms include heavy metals and, in the case of nuclear power plants, radioisotopes. The metals of particular concern are Cu, Zn, Cd, Fe, and Ni (Langford, 1983). There is no evidence that finfish cultured in power plant effluent have accumulated metals to any significant degree, perhaps because they are fed artificial diets that are essentially free of metal contaminants. Shellfish, on the other hand, filter plankton directly from the water and are well known for their tendency to bioaccumulate metals and other pollutants (see Chapter 12). In some cases, placing shellfish in clean water for a few days may reduce pollutant concentrations significantly. In general, however, careful monitoring for pollutants in the flesh of shellfish grown in power plant effluent is advisable. From the standpoint of public acceptance, it might seem doubtful whether organisms cultured in the effluent from a nuclear power plant would be marketable, even if radiation levels in the tissue were below safety limits. Nevertheless, the thermal effluent from some nuclear power plants has been put to practical use in aquaculture operations. The Japanese, for example, have grown sea bream, shrimp, abalone, sea urchins, lobsters, ayu, halibut, eels, crabs, seaweed, and several other species using the thermal effluent from nuclear power plants (Kuroda, 1979). In Finland, thermal effluent from a nuclear reactor in Olkiluoto has been used to speed up the production of Atlantic salmon smolts from fingerlings (Landau, 1992). As a result, the animals can be released as one-year-olds rather than as the typical two-year-olds. Because the salmon return in three to four years from the ocean, reducing the culture phase by one year substantially increases the annual catch because of the decrease in recruitment time (Landau, 1992).

From the standpoint of thermal pollution, it should be obvious that aquaculture operations are not likely to significantly reduce stresses to the environment resulting from

---

[4]Unfortunately, the culture facility was destroyed by a fire in 1991 (Stan Sczyk, Bluepoint Co., personal communication).

thermal discharges. Most aquaculture operations would probably be able to use only a small percentage of the effluent from a power plant anyway. For example, the catfish culturing system at the Tennessee Valley Authority's Gallatin power plant uses less than 0.5% of the effluent from the plant (Anonymous, 1974). One potential problem with thermal aquaculture is that the waste products from the cultured organisms may create pollution problems when the effluent is discharged. In such cases, it may be necessary to treat the effluent from the aquaculture operation before releasing it to the environment. Even if the effluent is recycled to the cooling system, some treatment may be necessary, since it would obviously be unwise to let the waste products accumulate continuously.

Other Uses

A variety of other possible uses for thermal discharges are discussed by Cook and Biswas (1974). These uses include desalination, melting of snow and ice, defogging airport runways, waste treatment, and warming bodies of water used for swimming. An analysis of these proposed schemes indicates that some are completely impractical,[5] and others would be practical under only rather special conditions. Power plant desalination units, for example, would be impractical unless there were a very high demand for freshwater ($\sim 5 \times 10^5$ m^3 d^{-1}) close to the power plant. On the other hand, the fact that a particular use is impractical in one part of the world need not detract from its use somewhere else. From the standpoint of thermal pollution, however, the most practical way to deal with many thermal discharges may simply be to use closed cooling systems that minimize the amount of water drawn from and discharged into the environment.

REFERENCES

AMERICAN FISHERIES SOCIETY. 1981. Intensive eel farming in a nuclear power plant effluent. In *Proceedings of the Bio-Engineering Symposium for Fish Culture*. American Fisheries Society, Traverse City, MI. Pp. 235–237.

ANDRYUSHCHENKO, A. I., and A. M. TRETYAK. 1989. On the technology of breeding and rearing bigmouth buffalo in the cooling water body of the Burshtynsk hydroelectric power plant. *Rybn. Khoz* (Kiev), **43**, 9–12.

ANONYMOUS. 1974. Power plant' fish production. *Commercial Fish Farmer*, 1(1), 10–13.

BADER, R. G., and M. A. ROESSLER. 1971. *An Ecological Study of South Biscayne Bay and Card Sound*. Progress Report to U.S. Atomic Energy Commission (AT (40-1)-3801-3) and Florida Power and Light Co.

BEALL, S. E., C. C. COUTANT, M. J. OLSZEWSKI, and J. S. SUFFERN. 1977. Energy from cooling water. *Industrial Water Eng.*, 16(6), 8–14.

BELL, R. A., W. J. CAHILL, A. S. CHIEFETZ, G. T. COWHERD, A. JOHN, J. A. NUTANT, and H. WRIGHT. 1970. Combination urban-power systems utilizing waste heat. Paper presented at the conference on beneficial uses of thermal discharges. New York Department of Environmental Conservation. Albany, September 17–18.

BOERSMA, L. 1970. Warm water utilization. Paper presented at the conference on beneficial uses of thermal discharges. New York Department of Environmental Conservation. Albany, September 17–18.

BRUNGS, W. A. 1973. Effects of residual chlorine on aquatic life. *J. Water Pollut. Contr. Fed.*, **45**, 2180–2193.

BUSH, R. M., E. B. WELCH, and B. W. MAR. 1974. Potential effects of thermal discharges on aquatic systems. *Environ. Sci. Tech.*, **8**(6), 561–568.

[5]For example, using thermal discharges to keep the St. Lawrence Seaway free of ice.

CAIRNS, J. JR., A. L. BUIKEMA, JR., A. G. HEATH, and V. C. PARKER. 1978. *Effects of Temperature on Aquatic Organism Sensitivity to Selected Chemicals.* Virginia Water Resources Research Center. Bulletin 106. Blacksburg, VA.

CLARK, J. R. 1969. Thermal pollution and aquatic life. *Sci. Amer.,* **22**(3), 19–27.

CLARK, J., and W. BROWNELL. 1973. *Electric Power Plants in the Coastal Zone: Environmental Issues.* American Littoral Society Special Publication No. 7, Highlands, NJ.

COLLINS, S. 1991a. The perfect match: Combined cycles and district heating/cooling. *Power,* **135**(4), 117–120.

COLLINS, S. 1991b. Combined-cycle cogen plant cuts costs. *Power,* **135**(5), 63–64.

COOK, B., and A. K. BISWAS. 1974. *Beneficial Uses of Thermal Discharges.* Environment Canada. Planning and Finance Service Report No. 2. Ottawa, Canada. 75 pp.

DAVIES, R. M., and L. D. JENSEN. 1975. Zooplankton entrainment at three mid-Atlantic power plants. *J. Water Pollut. Control Fed.,* **47**(8), 2130–2142.

DAVIS, M. H., and J. COUGHLAN. 1978. Response of entrained plankton to low-level chlorination at a coastal power station. In R. L. Jolley (ed.), *Water Chlorination: Environmental Impact and Health Effects,* Vol. II. Ann Arbor Science, Ann Arbor, MI. Pp. 369–376.

DeMONT, D. J., and R. W. MILLER. 1971. First reported incidence of gas-bubble disease in the heated effluent of a steam generating station. Paper delivered at the 25th Annual Southeastern Association of Game and Fish Comm. October 17–20, Charleston, South Carolina. 13 pp.

EPA. 1972. *Proceedings in the Matter of Pollution of Mount Hope Bay and Its Tributaries,* 2 vols. U.S. Environmental Protection Agency, Washington, DC. 643 pp.

EPA. 1976. *Quality Criteria for Water.* EPA-440/9-76-023. U.S. Environmental Protection Agency, Washington, DC. 501 pp.

EPA. 1986. *Quality Criteria for Water.* EPA 440/5-86-001. U.S. Environmental Protection Agency, Washington, DC.

FAIRBANKS, R. B., W. S. COLLINGS, and W. T. SIDES. 1968. *Biological Investigations in the Cape Cod Canal Prior to the Operation of a Steam Generating Plant.* Massachusetts Department of Natural Resources, Boston.

FLAVEN, C. 1985. *World Oil: Coping with the Dangers of Success.* Worldwatch Paper 66. Worldwatch Institute, Washington, DC. 66 pp.

FLORIDA STATE BOARD OF HEALTH. 1962. Survey, Biscayne Bay.

FWPCA. 1970. *Report on Thermal Pollution of Intrastate Waters of Biscayne Bay, Florida.* U.S. Department of the Interior. Southeast Water Laboratory. Federal Water Pollution Control Administration, Ft. Lauderdale, FL. 44 pp.

GESAMP. 1984. *Thermal Discharges in the Marine Environment.* Food and Agriculture Organization of the United Nations, Rome. 42 pp.

GOODYEAR, C. P. 1973. Probable reduction in striped bass in the Hudson. Testimony before the U.S. Atomic Energy Commission licensing board, Indian Point No. 2, NY, February 8.

GRIMES, C. B. 1975. Entrapment of fishes on intake water screens at a steam electric generating station. *Chesapeake Sci.,* **16**(3), 172–177.

HAWKES, H. A. 1969. Ecological changes of applied significance from waste heat. In P. A. Krenkal and F. L. Parker (Eds.), *Biological Aspects of Thermal Pollution.* Vanderbuilt University Press, Nashville, TN. Pp. 15–53.

ICANBERRY, J., and J. R. ADAMS. 1974. Zooplankton survival in cooling water systems of four thermal power plants on the California coast. Interim report, March 1971–January 1972. In J. W. Gibbons and R. R. Sharitz (Eds.), *Thermal Ecology.* Technical Information Center, U.S. Atomic Energy Commission Conference 730505, Washington, DC.

INGEBRIGTSEN, O., and O. TORRISSEN. 1981. The use of effluent water from Matre power plant for raising salmonid fingerlings at Matre aquaculture station. *Aquaculture in Heated Effluents and Recirculation Systems,* 16–17, 515–524.

INTERNATIONAL ATOMIC ENERGY AGENCY. 1972. *Thermal Discharges at Nuclear Power Stations: Their Management and Environmental Impacts.* P. J. West (Ed.). International Atomic Energy Agency, Vienna.

INTERNATIONAL ATOMIC ENERGY AGENCY. 1980. *Environmental Effects of Cooling Systems.* International Atomic Energy Agency, Vienna.

INTERNATIONAL ENERGY AGENCY. 1998. *Energy Balances of OECD Countries. 1995–1996.* Organization for Economic Co-operation and Development, Paris.

JOKIEL, P. L., and S. L. COLES. 1974. Effects of heated effluent on hermatypic corals at Kahe Point, Oahu. *Pac. Sci.*, **28**, 1–18.

KARKHECK, J., J. POWELL, and E. BEARDSWORTH. 1977. Prospects for district heating in the United States. *Science*, **195**, 948–955.

KAYA, C. M. 1977. Reproductive biology of rainbow and brown trout in a geothermally heated stream, the Firehole River of Yellowstone National Park. *Trans. Am. Fish. Soc.*, **16**(4),354–361.

KNIGHTON, G. 1973. Supporting information for staff testimony on cooling towers. U.S. Atomic Energy Commission Licensing Hearings, Indian Point No. 2, NY. 6 pp.

KURODA, T. 1979. Japanese aquaculture with thermal water from power plants—present condition and problems. In B. L. Godfriaux, A. F. Eble, A. Farmanfarmaian, C. R. Guerra, and C. A. Stephens (Eds.), *Power Plant Waste Heat Utilization in Aquaculture*. Allanheld, Osmun, & Co., Montclair, NJ. 266 pp.

LANDAU, M. 1992. *Introduction to Aquaculture*. John Wiley & Sons, New York. 440 pp.

LANGFORD, T. E. 1983. *Electricity Generation and the Ecology of Natural Waters*. Liverpool University Press, Liverpool, UK. 342 pp.

LIU, K. 1994. On technique for utilization of warm water out of the power plant to make *Colossoma brachypomum* breed in winter. *Shandong Fisheries/Qilu Yuye*, **11**(2), 12–15.

LYERLY, R. L., and C. E. LITTLEJOHN. 1973. A summary report of the Turkey Point cooling canal system. R. L. Lyerly and Associates. Report prepared for the Florida Power and Light Company.

MARCY, B. C., JR. 1971. Survival of young fish in the discharge canal of a nuclear power plant. *J. Fish. Res. Bd. Can.*, **28**, 1057–1060.

MARINE, G. 1969. *America the Raped*. Avon Books, New York. 331 pp.

MATHER, D., P. G. HEISEY, and N. C. MAGNUSSON. 1977. Impingement of fishes at Peach Bottom atomic power station, Pennsylvania. *Trans. Am. Fish. Soc.*, **106**(3), 258–267.

MILLER, H. H. 1970. The thermal water horticultural demonstration project at Springfield, Oregon. Paper presented at the conference on beneficial uses of thermal discharges. New York Department Environmental Conservation, Albany, September 17–18.

MITZINGER, W. 1974. Energetics and landscaping, exemplified by the development of one district of the German Democratic Republic. Paper 2.3–2. *Proceedings of the 9th World Energy Conference, Detroit*. World Energy Conference, London.

MONN, M. G., M. F. ROSENFIELD, and N. A. BLUM. 1979. A survey of capital costs of closed-cycle cooling systems for steam electric power plants. *Proceedings of the 41st Annual Meeting American Power Conference*. Illinois Institute of Technology, Chicago.

MORGAN, R. P., III, and R. G. STROSS. 1969. Destruction of phytoplankton in the cooling water supply of a stream electric station. *Chesapeake Sci.*, **10**, 165–171.

NATIONAL TECHNICAL ADVISORY COMMITTEE. 1968. *Water Quality Criteria*. Federal Water Pollution Control Administration, Washington, DC.

O'SULLIVAN, D., and K. SAUNDERS. 1998. Heated water from power plant used to breed Red Sea bream in Australia. *Aquaculture*, **24**(5), 63–64.

PARK, C. W., H. T. HUH, S. K. YI, and M. S. KIM. 1996. Use of heated effluents from power plant for aquaculture in Korea. In *Proceedings of the PACON Conference on Sustainable Aquaculture '95*. University of Hawaii Press, Honolulu, HI. Pp. 282–288.

REUTTER, J. M., and C. E. HERDENDORF. 1984. *Fisheries and the Design of Electric Power Plants: The Lake Erie Experience*. Ohio State University Sea Grant Technical Bulletin OHSU-TB-11. Ohio State University, Columbus, OH. 14 pp.

SCHNEIDER, M. 1980. Gas bubble disease in fish. In *Environmental Effects of Cooling Systems*. International Atomic Energy Agency, Vienna. Pp. 38–49.

SMITHSONIAN INSTITUTION. 1973a. *Rocky Point Menhaden Kill*. Report No. 1604. Smithsonian Institution Center for Short-lived Phenomena, Washington, DC.

SMITHSONIAN INSTITUTION. 1973b. *Florida Salinity Induced Fish Kill*. Report No. 1608. Smithsonian Institution Center for Short-lived Phenomena, Washington, DC.

SUCHANEK, T. H., JR., and C. GROSSMAN. 1971. Viability of zoplankton. In *Studies of the Effects of a Stream Electric Generating Plant on the Marine Environment at North Port, New York*. Technical Rept. Serial No. 9: 61–74. State University of New York, Stony Brook, Marine Science Research Center.

SVIRSKIJ, V. G., E. I. RACHEK, and I. N. ANDREEVA. 1994. Results of introduction of the freshwater shrimp Macrobrachium nipponense (DeHaan, 1849) into the cooling pond of the Primorskaya power plant. *Biotechnological Foundations of Aquaculture in the Far Eastern Region of Russia*, 113, 151–153.

TANAKA, J., and S. SUZUKI. 1966. High results of serida culture by utilization of heated effluent water from fossil fuel power plants. *Fish. Cult.*, 3(8), 13–16.

THORHAUG, A., N. BLAKE, and P. B. SCHROEDER. 1978. The effect of heated effluents from power plants on seagrass (*Thalassia*) communities quantitatively comparing estuaries in the subtropics to the tropics. *Mar. Pollut. Bull.*, 9, 181–187.

THORHAUG, A., M. A. ROESSLER, S. D. BACH, R. HIXON, I. M. BROOK, and M. N. JOSSELYN. 1979. Biological effects of power-plant thermal effluents in Card Sound, Florida. *Environ. Conserv.*, 6(2), 127–137.

TRUCHAN, J. G. 1978. Toxicity of residual chlorine to freshwater fish: Michigan's experience. In L. D. Jensen (Ed.), *Biofouling Control Procedures: Technology and Ecological Effects*. 79–89. Marcel Dekker, New York. Pp. 79–89.

TUCKER, W. S. 1977. Personal communication to the author.

UNITED NATIONS. 1996. *1994 Energy Statistics Yearbook*. New York. 490 pp.

UNITED STATES CONGRESS. 1968. Public Law 90-606. U.S. Congress, 90th Session, H. R. 551, October.

WEST, P. J., ED. 1972. *Thermal Discharges at Nuclear Power Stations: Their Management and Environmental Impacts*. International Atomic Energy Agency, Vienna.

YEE, W. C. 1972. Thermal aquaculture: Engineering and economics. *Environ. Sci. Tech.*, 6(3), 232–237.

YOUNG, J. S., and C. I. GIBSON. 1973. Effect of thermal effluent on migrating menhaden. *Mar. Pollut. Bull.*, 4(6), 94–95.

ZIEMAN, J. C., and E. J. F. WOOD. 1975. Effects of thermal pollution on tropical-type estuaries with emphasis on Biscayne Bay, Florida. In E. J. F. Wood and R. E. Johannes (Eds.), *Tropical Marine Pollution*. Elsevier, Amsterdam. Pp. 75–98.

QUESTIONS

11.1 What type of closed cooling system is most closely analogous to the radiator in an automobile?

11.2 Why are most of the conventional electric power plants in the United States ill suited for use in DHC networks?

11.3 On June 25, 1969, thousands of fish and invertebrates were killed in Biscayne Bay as a result of the cooling water from the Turkey Point power plant being discharged into the bay. In retrospect, why was Biscayne Bay a particularly bad place for the power plant to discharge its thermal effluent?

11.4 The Davis-Besse, Bay Shore, and Acme power plants are all located in the same general area on the southern shore of Lake Erie. Although Davis-Besse generates more power than the other two plants combined, and although there are more organisms in the vicinity of Davis-Besse than near either of the other plants, the adverse effects of internal plant kills are much greater at Bay Shore and Acme than at Davis-Besse. How would you account for this fact?

11.5 With respect to external plant kills, more fish have been killed when power plants are shut down than when they are operating and discharging thermal effluent. How would you account for this fact?

11.6 A power plant with a once-through cooling system has been having a serious problem with inner plant kills and has hired a consultant to recommend modifications to the plant's cooling system. The consultant has recommended the installation of bubble curtains and a velocity cap on the intake structure. Assume that you have been hired to review the consultant's recommendations. Do these recommendations seem reasonable to you? Could you recommend something more effective?

11.7 To avoid discharging hot water into Biscayne Bay, Florida Power and Light (FPL) built a system of cooling canals with an area of 15.6 km² at its Turkey Point power plant. FPL has been criticized for the large area of mangroves it destroyed to build the cooling canal system.

    a. Assuming that FPL was committed to cooling canals, what sort of modification could have been made to the canals to reduce the land area requirement?

    b. Assuming that FPL was not committed to cooling canals, what alternative closed cooling system could it have used to reduce the land area requirement for its closed cooling system at Turkey Point?

Chapter Twelve

✧ ◻ ✧ ◻ ✧

Metals

1. General Characteristics of Toxic Metals
   a. Association with Sulfur
   b. Solubility
2. Biological Magnification
3. Mercury
   a. Production and Uses
      (1) decline since 1970
      (2) chlor-alkali plants
   b. Fluxes to the Environment
      (1) anthropogenic $\geq$ natural
   c. Toxicology
      (1) inorganic and organic mercury
      (2) methyl mercury
         (a) brain damage, birth defects
         (b) conversion of other forms to methyl mercury
   d. Minamata Disease
      (1) use of inorganic Hg as a catalyst
         (a) converted to methyl Hg
      (2) efforts by Chisso to avoid liability
      (3) Niigata and Minamata court cases
   e. Pollution of U.S. and Canadian River Systems
      (1) why no Minamata disease incidents in North America
   f. Difficulty of Cleaning Up Contaminated Systems
4. Cadmium
   a. Distribution, Production, and Uses
      (1) decline in U.S. consumption
         (a) Ni-Cd batteries
   b. Emissions to the Environment
      (1) anthropogenic > natural
   c. Toxicity
      (1) kidney, liver damage
   d. Itai-Itai Disease
      (1) contaminated rice
      (2) bone softening
   e. Correctives and Prospects
      (1) recycling of Ni-Cd batteries
5. Lead
   a. Production and Use
      (1) batteries
      (2) paint and ammunition

b. Emissions
   (1) anthropogenic >> natural
   (2) impact of phaseout of leaded gasoline
c. Toxicology
   (1) general metabolic poison
   (2) effects on children
   (3) Pb shot in marshes
d. Commentary
   (1) improvements
      (a) lead paints, gasoline, shots, solder; water distribution systems

My name is Asae Fukuda, and I am a Minamata disease patient.

I am appearing in court on behalf of my daughter Itsuko, who died at the age of 11. My home is near a fishing port called Hakariishi Port, located ten odd kilometers north of the Chisso Plant.

I became pregnant in 1967. As this would be the birth of our first child, family and relatives were all very happy.

Thinking that it would be for the good of the child within me, I ate fresh fish every day. My happiness, however, was short lived.

Even seven months after birth, Itsuko could not support the weight of her own head, and could not see. Many times a day she suffered convulsions, contorted her infant features, and drooled, making it unbearable to watch.

When she became one year old, I was told at the Kumamato University Hospital that Minamata disease was suspected.

I did not think it could be so, and did not want to believe it, but the doctor at the University Hospital asked me if I was living near the sea, and if I ate a lot of fish. Everything he asked was so, and I could only accept that he was right. It was then that I began taking Itsuko to the hospital regularly.

Every time I heard of a good doctor, I took Itsuko to see him. Desperate to try anything, I also visited temples and shrines to pray.

Three or four times a month I took Itsuko to Kumamoto University Hospital for examinations, a trip that took two hours each way. On the days after examinations Itsuko was especially tired, and would cry all night. At such times I would carry Itsuko on my back to the nearby port to humor her. I cried with my child, who would not understand if I talked to her, as I thought of how great a relief it would be for us to throw ourselves in the sea and die.

When Itsuko was three, I wanted to dress her in fine clothing for the *Shichigosan* children's day, though she could not stand on her own. Even if I put the clothing on Itsuko, who was lying down, she would only be like an unmoving doll. It was so sad that I cried as I dressed her. I then put makeup on her as well. As I carried her on my back to the shrine I said, "Itsuko, when you're able to stand let's go once more to the shrine for *Shichigosan*."

At mealtime I gave her liquid food spoonful by spoonful, being very careful, as if praying, that she would not choke on it.

I was assailed by the thought that I made Itsuko this way because of the fish I ate during pregnancy. So I believed that I must care for her whatever should happen, and I nursed the child in her illness with all my heart.

However, during the winter when she became 11 years old, in February when it was still cold, Itsuko died while suffering respiratory difficulty. . . .

I have been able to persevere this long because of my 11 years of suffering.

I cannot forgive the National Government, Kumamoto Prefecture, and Chisso, who robbed Itsuko of her life, and me of my health.

Whenever I think I am about to give up, I see Itsuko's face, continuing the struggle to live even as she fights to breathe, and I can hear her voice saying, "Mommy, don't give up the fight."

I shall continue this fight as long as I live.

Statement by Ms. Asae Fukuda to the International Forum on Minamata Disease held in Kumamoto City, November 7–8, 1988 (translated from Japanese by R. Davis). (Tsuru et al., 1989, p. 338)

B ECAUSE of the devastating effects certain toxic metals can have on human health, metal pollution is potentially one of the most serious forms of aquatic pollution. The foregoing statement by Ms. Asae Fukuda is a case in point. There is evidence that lack of understanding of the toxic nature of certain metals has adversely affected human health literally since biblical times. In many cases, contaminated water has been an important conduit by which metals have been transmitted to humans, either directly or via the food chain. How do toxic metals find their way into aquatic systems, and from there into irrigated crops or fish and shellfish that humans consume?

The answer is that metals are introduced into aquatic systems by many processes, including the weathering of soils and rocks, volcanic eruptions, and a variety of human activities involving the mining, processing, or use of metals and/or substances that contain metal contaminants. Although some metals such as Mn, Fe, Cu, and Zn are essential micronutrients, others such as Hg, Cd, and Pb are not required even in small amounts by any organism. Virtually all metals, including the essential metal micronutrients, are toxic to aquatic organisms as well as to humans if exposure levels are sufficiently high. Table 12.1 gives some idea of the relative toxicity of various metals based on their maximum contaminant level goals in drinking water in the United States.

Despite the wide variation in the maximum contaminant level goals for the metals listed in Table 12.1, there are definite similarities in the toxicology of certain of these metals. For example, virtually all heavy metals show a great affinity for S, and metals such as Hg, Pb, and Cd appear to exert toxic effects largely by combining with S-containing amino acids in proteins, thus interfering with enzyme-mediated processes and/or disrupting cellular structure. Although the particular proteins that are chiefly attacked by these metals may differ from one metal to another, the underlying biochemical interaction is common to all.

Most metals are rather insoluble in water with a neutral or basic pH, and rather than dissolving, they are rapidly adsorbed to particulate matter or assimilated by living organisms. In other than acidic waters, the concentration of metals dissolved in the water may therefore give a highly misleading picture of the degree of metal pollution and in some cases may significantly underestimate the total metal concentration in the water. Hg concentrations measured by Andren (1974), for example, in the Mississippi River indicate that about 67% of the Hg in the water is associated with suspended sediments. Although it is doubtful that metals adhering to particulate materials exert a directly toxic effect on aquatic organisms, it is certainly possible that some of these metals could be desorbed in the acidic environment of a particle feeder's gut or absorbed by an organism in the process of pumping water over its gills. Lloyd (1961) found that suspended Zn was indeed toxic to rainbow trout, although apparently only dissolved Zn is toxic to Atlantic salmon (Sprague, 1964). There is unfortunately little information on the toxicity of particle-bound metals to most aquatic organisms.

Benthic organisms are likely to be most directly affected by sediment metal concentrations because the benthos is the ultimate repository of the particulate materials that wash into aquatic systems. Sediments near sewer outfalls often contain unusually high

TABLE 12.1   MAXIMUM PERMISSIBLE CONTAMINANT LEVEL
GOALS FOR VARIOUS METALS IN DRINKING WATER
IN THE UNITED STATES

| Metal | Chemical Symbol | Maximum Permissible Concentration Goal | |
		μM	$\mu g\ L^{-1}$
Lead	Pb	0	0
Mercury	Hg	0.010	2
Cadmium	Cd	0.044	5
Antimony	Sb	0.049	6
Beryllium	Be	0.444	4
Selenium	Se	0.633	50
Chromium	Cr	1.923	100
Barium	Ba	14.563	2,000
Copper	Cu	20.458	1,300

Source: *Code of Federal Regulations*, **40**, 141.51 (1999).

metal concentrations, particularly if the sewer line is the recipient of discharges from industrial operations that utilize metals in some way and do not adequately recover these metals from their wastewater. Klein and Goldberg (1970), for example, found sediment Hg concentrations near the Hyperion sewage treatment outfall off the Palos Verdes peninsula in southern California to be as much as 50 times higher than sediment Hg concentrations in control areas. The effects of pollution on sediment metal concentrations and the difference between particulate and dissolved metal concentrations is clearly evident in Table 12.2, which lists the concentrations of various metals in sediments at two locations around Oahu, Hawaii, in several unpolluted Australian bays, in river water, and in the open ocean. The enormous concentrations of metals in the sediments of the Ala Wai boat harbor undoubtedly reflect the use of certain metals such as Pb, Hg, and Cu in antifouling paints, as well as the corrosion of metals from boats docked in the harbor. The high metal concentrations in the sediments of the West Loch of Pearl Harbor reflect the discharge of about 15 million liters of sewage per day into streams that flow into the loch,[1] the weathering of antifouling paints and corrosion of metals from naval ships, and the fact that the waters of West Loch are rather stagnant.[2]

The average concentrations of the same metals in river water and seawater provide an interesting comparison to the corresponding sediment values. Sediment metal concentrations even in the unpolluted Australian bays are between 10^3 and 10^7 times higher than the dissolved concentrations in river water and seawater, and in the polluted Ala Wai boat harbor the sediment metal concentrations are between 10^5 and 10^9 times greater than the river water and seawater concentrations. Metal concentrations as high as those found in the Ala Wai boat harbor sediments could seriously contaminate and stress benthic organisms. Furthermore, under appropriate conditions, these metals could leach out of the sediments for many years after pollutant discharges have been stopped and hence continue to pollute the water column. Thus, examination of metal concentrations in suspended particles, in aquatic organisms, and in sediments will usually provide a more sen-

[1]These discharges were terminated in 1981.
[2]Hence, the metals are inefficiently flushed out.

TABLE 12.2 SEDIMENT METAL CONCENTRATIONS (ng g^{-1} DRY WEIGHT) AT TWO LOCATIONS ON OAHU, HAWAII, AND IN TWO UNPOLLUTED AUSTRALIAN BAYS, AND DISSOLVED METAL CONCENTRATIONS (ng g^{-1}) IN AVERAGE RIVER WATER AND SEAWATER

Location	Hg	As	Cu	Zn	Pb
Sediments					
Oahu					
Pearl Harbor, upper West Loch	250	11×10^3	57×10^3	8×10^4	5×10^4
Ala Wai boat harbor	58×10^3	16×10^4	7×10^6	14×10^5	5×10^6
Australia					
Halifax Bay	—	—	7×10^3	31×10^3	16×10^3
Cleveland Bay	—	—	62×10^2	3×10^4	15×10^3
Water					
Average river water	0.7	1.7	1.5	30	0.1
Average seawater	1×10^{-3}	1.5	0.1	0.1	3×10^{-3}

Source: City and County of Honolulu (1971), Knauer (1977), Martin and Whitfield (1983), Mitra (1986).

sitive and informative indicator of metal pollution than measurements of dissolved metal concentrations.

THE QUESTION OF BIOLOGICAL MAGNIFICATION

The fact that metal concentrations in aquatic organisms are typically several orders of magnitude higher than concentrations of the same metals in the water has led to some speculation that metals may become progressively concentrated at higher trophic levels in aquatic food chains due to food chain magnification. The high concentrations of Hg reported in tuna, swordfish, and marlin (Montague and Montague, 1971; DLNR, 1986), top-level carnivores, is consistent with the food chain magnification hypothesis.

Table 12.3 lists the concentrations of various metals in samples of phytoplankton and zooplankton collected at open ocean stations between California and Hawaii in 1973. Comparison with the seawater values in Table 12.2 shows that the metal concentrations in the phytoplankton are indeed about three orders of magnitude higher than the concentrations in the water, and for the first three metals at least, the concentrations in zooplankton are substantially higher than the phytoplankton concentrations. For the last four metals, however, there is little difference in concentration between the phytoplankton and zooplankton. Although the elevated metal concentrations in the zooplankton for the first three metals are suggestive of food chain magnification, we have already seen from the work of Hamelink et al. (1971) on DDT residues that such effects may be produced by entirely different mechanisms, namely, by direct uptake of a pollutant from the water and by differences in pollutant equilibria between water and organism for different classes of organisms. Therefore, it is impossible to say to what extent the increases in metal concentrations between phytoplankton and zooplankton for the first three metals in Table 12.3 reflect food chain magnification or simply differences in metal exchange equilibria between organisms and water.

In the case of Hg, Knauer and Martin (1972) have provided a thorough study of concentrations in a simple marine food chain consisting of phytoplankton, zooplankton, and

TABLE 12.3   MEDIAN CONCENTRATIONS OF METALS IN
PHYTOPLANKTON AND ZOOPLANKTON COLLECTED BETWEEN
CALIFORNIA AND HAWAII

Metal	Phytoplankton[a]	Zooplankton[b]
Copper	3.2	13.9
Zinc	19	260
Lead	<1.0	8.5
Iron	224	580
Manganese	6.1	4.4
Silver	0.2	0.25
Cadmium	1.5	2.3
Lead	0.19	0.14

[a]Phytoplankton values are μg g^{-1} organic dry weight.
[b]Zooplankton values are μg g^{-1} dry weight.
Source: Martin and Knauer (1973).

anchovies in Monterey Bay, California. Some of their results are summarized in Table 12.4. For this metal at least, there is little evidence of magnification up the food chain, and in fact, the variability in Hg concentration between some parts of the anchovies is greater than the difference in concentration between the anchovies and phytoplankton. The wide range of Hg concentrations in different parts of the anchovy reflects the importance of biochemical factors in affecting the relative tendency of different tissues to concentrate pollutants. Such biochemical or physiological differences may also play a major role in causing certain organisms to concentrate pollutants to a much higher level than other organisms, regardless of the relative position of the organisms in the aquatic food chain.

CASE STUDIES

It should be apparent from the previous discussion that some statements about metals and their effects on organisms are valid for most metals of major concern as pollutants. For example, it is true in general that the concentration of metals associated with particulate material, including living organisms, is at least several orders of magnitude higher than the concentration of metals dissolved in the water. It is also true in general that the concentration of metals may differ greatly between one organism and another and between

TABLE 12.4   Hg LEVELS IN A SIMPLE FOOD CHAIN

Trophic Level	Average Hg Concentration (ppb wet weight)
Phytoplankton	28
Zooplankton	12
Anchovy	
Skin	10
Gonads	15
Gills	30
Muscle	40
Liver	90

Source: Knauer and Martin (1972).

different parts of the same organism. In general, one ascribes these concentration differences to differences in the tendency of metals to bind to the various molecular groups found within the cells of each organism, as well as to the degree of the organism's exposure to the metal as influenced by its metabolic characteristics and its position in the food chain.

Although such generalizations are useful in understanding some of the problems associated with metal pollution, it should be clear from an examination of Tables 12.1 and 12.3 that significant differences exist in the toxicity of various metals and in the mechanisms by which these metals are transferred through the biosphere. Thus, more than a superficial understanding of metal pollution requires an in-depth examination of the characteristics and problems associated with each metal. The remainder of this chapter is therefore devoted to a detailed analysis of three of the most toxic metal pollutants: Hg, Cd, and Pb.

Mercury

Production and Uses

Hg is a highly unusual metal. It is a liquid at room temperature[3] and has a density even greater than that of Pb.[4] Hg is found in virtually all terrestrial rocks and soils, although its concentration in most soils and rocks is quite low, about 60 ppb. In some parts of the world, however, the concentration of Hg in the Earth's crust is considerably higher, particularly in regions of volcanic activity. The only ore that contains Hg in sufficient concentration to warrant extraction is cinnabar (HgS), or mercuric sulfide. World production of Hg from HgS peaked at about 10^4 tonnes y^{-1} around 1970, declined to about 2,000 tonnes per yr^{-1} during the early 1990s, and has since stabilized at about 3,000 tonnes y^{-1} (Figure 12.1). The principal producers are Spain, Kyrgyzstan, China, and Algeria. Consumption of Hg in the United States has been declining steadily in recent years, from a peak of about 2,000 tonnes y^{-1} around 1970 to roughly 350 tonnes y^{-1} at the present time.

What does the world do each year with 3,000 tonnes of Hg? Historically, Hg has been used for a great variety of purposes, including industrial and medical applications, as well as scientific research. In the United States, by far the single most important use of Hg is the production of Cl_2 and sodium hydroxide (NaOH). In the so-called chlor-alkali industry, Hg is used as a cathode in an electrolytic process by which Cl_2 and NaOH are produced from a solution of NaCl. For each tonne of Cl_2 produced, about 250–500 g of Hg are consumed (Montague and Montague, 1971). Prior to the 1970s, much of this "consumed" Hg was discharged as waste to the environment, although it is technically feasible to recover a large percentage of the Hg. Rapid expansion of the chlor-alkali industry in the United States began early in the 20th century with the increase in demand for high-quality NaOH used in the production of rayon. Since the 1940s, however, the principal users of Hg-grade NaOH have been the aluminum, glass, paper, petroleum, and detergent industries (Montague and Montague, 1971). The major user of the Cl_2 gas produced by chlor-alkali plants has been the plastics industry, for example, in the production of polyvinyl chloride (PVC). Although it is possible to produce Cl_2 and NaOH by methods other than the so-called Hg-cell technique, for economic reasons the Hg-cell technique was utilized for many years at virtually all chlor-alkali plants in the United States. However, during the last 40 years, the consumption of Hg by the chlor-alkali industry in the United States has declined by more than a factor of 4 from a peak of over 700 tonnes y^{-1}

---

[3]The melting point of Hg is $-38.9°C$.
[4]A piece of Pb will float on liquid mercury.

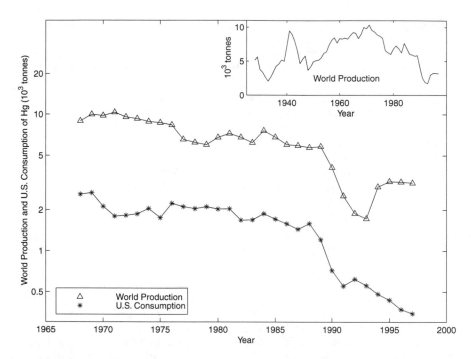

FIGURE 12.1   World production and U.S. consumption of mercury. *Source: Minerals Yearbook*, U.S. Department of the Interior, Bureau of Mines.

in the early 1960s. This dramatic decline has been due in part to recovery and recycling efforts designed to reduce Hg pollution at chlor-alkali plants and, in some cases, to conversion of plants to membrane cell technology.

Ten years ago, the second most important use of Hg in the United States was the production of Hg dry cell batteries. Historically, the batteries were used in many products, including Geiger-Muller counters, radios, digital computers and calculators, hearing aids, military hardware, and scientific and communication devices (Mitra, 1986). The cathode of the battery consists of mercuric oxide mixed with graphite; the anode is usually metallic Zn. Concern over pollution, however, has virtually eliminated the use of Hg in electrical batteries in the United States. Today mercury oxide batteries are produced only for military and medical equipment (Reese, 1997).

Because of its toxic qualities, Hg was used for many years as a component of antifouling paints and mildew-proofing paints. The use of Hg in the former category was banned in 1972 under the Federal Insecticides, Fungicides, and Rodenticides Act (FIFRA), and other components were substituted for Hg to resist fouling. The use of Hg in paints still accounted for 16% of Hg consumption in the United States in 1989, but the FDA banned the use of Hg in latex paints for interior use as of August 20, 1990 (Emond, 1990).

Hg has found use in a variety of electrical apparatus, including neon lights, arc rectifiers, power control switches, oscillators, fluorescent lights, silent light switches, and certain kinds of highway lights. In these applications, the unusual properties of Hg have often played an important role. Silent light switches, for example, consist of a hollow tube containing a puddle of Hg. Turning the switch to the "on" position inverts the tube, causing the Hg to flow to one end, where it completes a circuit between two wires embedded in the tube. Hg vapor is the form of Hg used in neon lights, fluorescent lights, arc rectifiers, and in the bright, bluish highway lights found along many streets and roadways. Largely because of environmental concerns, the use of Hg in these applications has declined by about 60% in the last decade. Hg switches, for example, are being replaced with either electronic switches or other special switches, and the Hg content of fluorescent lamps has been reduced. Light bulbs produced today contain less than half of the Hg used in those manufactured during the mid-1980s (Reese, 1997).

The unusual properties of Hg have also been important to its use in measuring and control instruments, including thermometers, manometers, barometers, diffusion pumps, Toepler pumps, and a variety of pressure gauges. Unfortunately, when such devices are broken or discarded, there is a good chance that the Hg in them will leak into the environment. According to figures cited by Aaronson (1971), in Canada alone, roughly 1.8 million Hg thermometers were broken each year in hospitals during the 1960s, and another 1.2 million were broken by the general public. If the Hg in these thermometers were simply discarded, the discharge would amount to over 6 tonnes of Hg per year. Recognition of this problem has led to almost a 75% reduction in the use of Hg in control instruments during the last decade.

Each year in the United States, up to 100 million dental fillings containing Hg combined with silver (Ag), tin (Sn), and other metals in the form of an amalgam are used to restore decayed teeth (Anonymous, 1991). This is the only use of Hg in the United States that has remained constant in recent years. The practice, however, become somewhat controversial in 1990, largely because of a documentary aired by the television program *60 Minutes* in December that year. The message conveyed by the *60 Minutes* report was that Hg vapor released from the fillings could be causing serious health problems. As an example, *60 Minutes* cited the case of a woman crippled by multiple sclerosis who went dancing the night after her amalgam fillings were removed.

The truth of the matter is that gold (Au) fillings are in fact preferable to amalgam fillings, but not because of any concern over Hg poisoning. Gold fillings last longer, are more resistant to wear, and are less likely to fracture than amalgam fillings (Anonymous, 1991). The problem with Au fillings is that they cost about six times as much as amalgam fillings. Studies have shown that when people chew, a small amount of Hg vapor is released from amalgam fillings, but current estimates indicate that even persons with moderate to large numbers of amalgam fillings receive only about 1% of the dose of Hg that might cause adverse health effects (Anonymous, 1991). Incidentally, once Hg accumulates in the human body, it is released at a rate of only about 1% per day. Thus, for example, removing one's amalgam fillings would not cause an overnight change in health due to Hg exposure. The fact that a woman crippled by multiple sclerosis went dancing the night after her fillings were removed is remarkable, but it was almost certainly not the result of any change in her body burden of Hg. What *60 Minutes* reprehensibly failed to report were the stories of numerous people who had their amalgam fillings removed to no avail. Another woman afflicted with multiple sclerosis, for example, who was unable to afford the cost of Au fillings, had all her teeth extracted and replaced with dentures. A month later, her condition worsened and she was hospitalized (Anonymous, 1991).

A more legitimate concern about the use of Hg in fillings is that only about 50% of the total amount of Hg used in dental preparations is actually applied to the teeth (Mitra, 1986). When dentists prepare to fill a cavity, they invariably make up an excess of amalgam to ensure proper execution of the restorative work. The excess amalgam is either rinsed from the mouth or aspirated, and is released to the environment unless caught in a strainer or trap in the waste drain. Although recovery of amalgam scrap is widespread in the United States, in the past many dental offices did not contain appropriate waste traps.

In scientific research, Hg is used in measuring devices such as thermometers and manometers, as a catalyst in certain chemical reactions, and as a preservative to prevent deterioration of biological specimens due to microbial breakdown. Waste Hg from such use, as in a once-used preservative, is typically discarded rather than recycled. Hg consumption in general laboratory use has declined, however, by roughly a factor of 3 in the United States during the last 30 years. This trend reflects the substitution of alternative methods and devices that do not involve the use of Hg.

Hg is used as a catalyst to speed up a variety of chemical reactions. One of its most important uses as a catalyst is in the production of several kinds of plastic, particularly urethane and polyvinyl chloride (PVC). PVC is the most widely used plastic product in the world. It is used both in its unplasticized form—for example, in bottles, pipe fittings, pipes, records, and valve parts—and in its plasticized form in rainwear, shower curtains, auto seat covers, baby pants, and so forth. Hg has also been used as a catalyst in the production of synthetic acetate fiber. Historically, discharges of spent Hg catalysts from industrial operations have been a significant cause of Hg pollution in the aquatic environment (Carter, 1977; Mitra, 1986). Consumption figures for Hg by the plastics industry are not specifically tabulated by the U.S. Bureau of Mines for proprietary reasons, but it is obvious from Table 12.5 that such use accounts for no more than 10% of U.S. Hg consumption at the present time.

Several other uses of Hg are worth mentioning, in some cases largely because of their historical significance. Beginning in 1914, Hg in various forms was used as a coating on seeds to protect them from attack by fungi during storage and during the first few days after planting. Hg-based fungicides were also used to control turf-grass disease on golf courses, parks, cemeteries, lawns, and so forth. The use of Hg to treat seed was greatly restricted in the United States under the FIFRA in 1970 and 1972, and its use as a fungicide was similarly restricted and ultimately banned by 1972 (Mitra, 1986). These actions were taken in part because of the impact of the Hg on seed-eating birds and because of concern over contamination of waterways due to runoff from areas of treated turf. Serious impacts on human health due to the practice of treating seed grain with Hg have occurred when the seed was consumed or fed to animals rather than planted. In 1969, for example, a farmer in New Mexico fed waste seed grain from a local granary to his pigs. Subsequently the pigs were slaughtered, and both the farmer and his family consumed the meat. Three of the farmer's children suffered irreparable neurological damage as a result of consuming the meat, which was later found to contain 27 ppm Hg. Several years later, in 1972, a similar but more serious incident occurred in Iraq, when thousands of persons ate homemade bread prepared from wheat that had been treated with a mercurial fungicide. Four hundred fifty persons died of Hg poisoning (Mitra, 1986).

In the past, Hg was used in pulp and paper mills to prevent the growth of slime on papermaking equipment and to prevent the growth of mold and bacteria on pulp during storage. Such use declined precipitously after 1965, when the FDA ruled that paper products used as food containers should contain absolutely no Hg, and was banned entirely in 1972 under the FIFRA (Mitra, 1986).

TABLE 12.5   U.S. Hg CONSUMPTION FIGURES FOR 1997

Use	Consumption (tonnes)	Percent of Total Consumption
Chlor-alkali	160	46
Wiring devices and switches	57	16
Dental equipment and supplies	40	12
Electric lighting	29	8
Measuring and control instruments	24	7
Other uses	36	10
Total	346	100

Source: Reese (1997).

The germ-killing properties of Hg have resulted in its use in various pharmaceutical products, dating back to at least 400 B.C. (Montague and Montague, 1971). For example, Hg came into widespread use for the treatment of syphilis as long ago as 1500. In pharmaceuticals today, Hg is used most commonly as a diuretic,[5] particularly in treating congenital heart failure patients, who frequently accumulate body fluids.

Finally, Hg will combine with almost all other common metals except Fe and platinum (Pt) to form a fusion product called an *amalgam* (e.g., dental fillings). The tendency of Hg to form amalgams with other metals may be used to purify metals by separating them through amalgamation from their nonmetal contaminants. The purified metal can be recovered from the amalgam by electrolysis or by heating to drive off the Hg. The former technique is used, for example, to produce high-purity Zn and Al. Historically and even today, the latter practice has led to serious contamination of some ecosystems. Nriagu (1993), for example, estimates that the use of Hg to extract Ag at Mexican silver mines released an average of 500 tonnes y^{-1} of Hg to the environment between 1570 and 1820. More recently, similar use of Hg to extract Au in Brazil has resulted in the release of roughly 100 tonnes y^{-1} of Hg to local ecosystems (Nriagu et al., 1992), including the Madeira River, a major source of fish for local populations.

Fluxes to the Environment

Of the Hg that is consumed each year, how much finds its way back into the environment in a manner and form that might cause pollution? In the United States at least, virtually all Hg consumption is accounted for by recycling at the present time, i.e., from recovery of Hg from secondary sources such as spent batteries, Hg vapor and fluorescent lamps, switches, dental amalgams, measuring devices, control instruments, and laboratory and electrolytic refining wastes (Reese, 1997). Nevertheless, the fact that 3,000 tonnes of Hg are extracted from ore deposits each year indicates that a comparable quantity is either accumulating in the form of supplies and manufactured products or being released back into the environment. Although there may be some net accumulation of supplies and manufactured products, it seems fair to assume that a quantity of Hg roughly equal to the amount that is mined each year is returned to the environment through emissions to the atmosphere, discharges to aquatic systems, or disposal in landfills. We conclude that about 3,000 tonnes of Hg are released to the environment each year as a direct result of human use. How does this figure compare to other fluxes of Hg through the environment?

To begin with, direct use of Hg is not the only cause of anthropogenic releases of Hg to the environment. Additional discharges result from the burning of fossil fuels and the processing of minerals and other metallic ores. Since Hg is a volatile substance, any process, such as burning or smelting that elevates the temperature of an Hg-containing substance, will tend to drive off Hg vapor into the atmosphere. Although the average concentration of Hg in crustal rocks and soils is only 60 ppb, the Hg concentration in coal is much higher, typically 0.2–20 ppm (Mitra, 1986). Coal burning is estimated to release roughly 2,000 tonnes of Hg to the atmosphere each year (Table 12.6). The other major anthropogenic source of Hg emissions to the atmosphere is solid waste incineration. Taken together, coal combustion and solid waste incineration are estimated to release about 3,000 tonnes of Hg to the atmosphere each year (Table 12.6). Other anthropogenic Hg emissions to the atmosphere are relatively small. Total estimated anthropogenic Hg emissions are about 3,500 tonnes y^{-1}.

Natural Hg emissions to the atmosphere are believed to come from three major sources (Table 12.6). Volcanic emissions are the single largest source. The estimated Hg flux of 1,000 tonnes y^{-1} is based on a global volcanic S flux of 15–20 million tonnes y^{-1}

[5]A substance used to rid the body of salt and water.

TABLE 12.6   ESTIMATED FLUXES OF Hg TO THE ATMOSPHERE

Source of Hg	Estimated Annual Flux (10^3 tonnes)
Natural	
Volcanoes	1.0
Wind-borne soil particles	0.05
Seasalt spray	0.02
Wild forest fires	0.02
Biogenic	
Marine	0.77
Continental volatiles	0.61
Continental particulates	0.02
Total natural sources	2.5
Anthropogenic	
Coal combustion	2.08
Solid waste incineration	1.12
Smelters	0.13
Wood burning	0.18
Total anthropogenic sources	3.51

Source: Nriagu and Pacyna (1988), Nriagu (1989).

and a Hg:S ratio of 6×10^{-5}. The other two major sources are biogenic in nature and are related to the release of nonmethane hydrocarbons such as isoprene and terpenes, which are known to form strong complexes with many trace metals (Nriagu, 1989). Natural emissions of Hg to the atmosphere are estimated to be about 2,500 tonnes y^{-1}, about 70% of the rate of anthropogenic emissions.

In round numbers, these estimates give some idea of the relative importance of natural and anthropogenic inputs of Hg to the atmosphere. The numbers are approximate, and in some cases, estimates are based on little more than educated guesses. As Nriagu (1989, p. 47) has candidly noted, "Current data are sufficient for only order-of-magnitude estimates." Nevertheless, the consensus of various estimates is that anthropogenic emissions of Hg to the atmosphere are at least comparable to and perhaps as much as three times greater than natural emissions (EPA, 1996a).

Inputs to Aquatic Systems. Obviously, one source of Hg input to aquatic systems is atmospheric fallout. Since rain effectively washes Hg from the atmosphere, and since the average time between rains is about 10 days (Weiss et al., 1971), the foregoing estimates imply that roughly 6,000 tonnes y^{-1} of Hg are deposited via rainfall on the surface of the Earth. A substantial percentage of that fallout undoubtedly occurs on the surface of the ocean. An additional important Hg input to aquatic systems is the direct discharge of industrial wastewaters. Nriagu and Pacyna (1988) estimated anthropogenic Hg inputs to aquatic systems to be roughly 4,600 tonnes y^{-1}, which was 74% of world Hg production at the time of their analysis. Applying a similar percentage to current world Hg production (3,160 tonnes y^{-1}) gives an estimated flux of 2,330 tonnes y^{-1} to aquatic systems. Nriagu and Pacyna (1988, p. 139) concluded, "The current rate of worldwide industrial inputs greatly exceed the baseline burdens of trace metals in the average lake and river." They further noted that most of the effluent discharges occurred in Europe, North America, and parts of Asia. In other words, global averages tend to mask the impact of anthropogenic Hg discharges to aquatic systems because many of the inputs have been localized and intense. The following examples serve to illustrate this point.

In Minamata, Japan, a factory operated by the Chisso Corporation discharged roughly

200–600 tonnes of Hg into Minamata Bay between 1932 and 1968 (Smith and Smith, 1975).[6] The company had been using mercuric oxide as a catalyst in the production of acetaldehyde, a compound used in the manufacture of plastics, drugs, perfumes, and photographic chemicals (SCEP, 1970; Smith and Smith, 1975). Concentrations of Hg in the sediments of Minamata Bay were found to be 10 to several hundred parts per million on a wet weight basis and ranged as high as 2,000 ppm in the mud near the drainage channels from the Chisso factory. Assuming the water content of the sediments in Minamata Bay to be about 30–70% by weight, these concentrations would be 1.4–3.3 times higher if reported on a dry weight basis. For comparison, Andren and Harriss (1973) found that Hg levels in sediments from the Mississippi River, Mobile Bay, and the Florida Everglades ranged from 0.08 to 0.6 ppm (dry weight); Klein and Goldberg (1970) found Hg levels in marine sediments off La Jolla and Palos Verdes, California, to be 0.02–1.0 ppm (dry weight basis); and Hg concentrations in surface sediments near mid-Atlantic sewage sludge dump sites average about 0.04 ppm (dry weight basis). Based on these comparisons, Minamata Bay was grossly polluted with Hg.

In the United States, Hg concentrations in certain species of fish taken from Lake St. Clair (between Lake Huron and Lake Erie) increased by roughly a factor of 100 between 1935 and 1970. Concentrations in 1970 were as high as 3–7 ppm (Figure 4.13; Grant, 1971). This increase in Hg levels was almost certainly due to the development of the chlor-alkali industry in the area and in particular to the establishment of the Dow Chemical chlor-alkali plant at Sarnia, Ontario, on the St. Clair River between Lake Huron and Lake St. Clair. Between 1949 and 1970 this plant discharged an estimated 91 tonnes of Hg into the St. Clair River system (Wood, 1972). Studies in Canada showed that most of the Saskatchewan River system, including Lake Winnipeg, was seriously polluted with Hg, a condition due in part to the discharge of Hg by Saskatoon Cooperatives into the South Saskatchewan River. The company discharged about 30 tonnes of Hg between 1963 and 1970 (Wood, 1972). The Dryden Cl_2 plant at Lake Wabigoon in Manitoba was similarly implicated as contributing to the Hg pollution of most lakes and rivers in central Manitoba (Wood, 1972).

Back in the United States, two rivers in Virginia were found to be heavily polluted with Hg as a result of industrial discharges (Carter, 1977). In southwestern Virginia, the North Fork of the Holston River, which feeds into the Tennessee River system, was the unfortunate recipient of Hg discharges from the Mathieson Chemical Company's[7] chlor-alkali plant at Saltville, Virginia, between 1950 and 1972. Although the plant was ultimately shut down, partly as a result of pollution control measures that were to be imposed, Hg has continued to leak into the river, both from the site of the plant itself, where the ground has been estimated by one Olin consultant to contain about 100 tonnes of Hg, and from two large "muck ponds" used for the disposal of waste from the plant (Carter, 1977). Hg is believed to be present in the soil to depths of as much as 10 m at the old plant site, and wastes in the muck ponds extend to depths of about 25 m. Discharge of Hg to the North Fork of the Holston River from the muck ponds alone is estimated to be about 35 kg y^{-1}, and there is no satisfactory method, short of massive engineering efforts that might cost hundreds of millions of dollars, to greatly reduce the discharges from the old muck ponds (Carter, 1977). In 1970 the states of Virginia and Tennessee issued a ban on the eating of fish from the North Fork, but despite the closing of the Saltville plant in 1972, 75% of the fish sampled in 1976 as much as 110 km downstream from the plant were found to contain Hg in concentrations greater than 1 ppm (wet weight basis).

The other Virginia river found to be heavily contaminated with Hg was the South Fork of the Shenandoah River. The source of the Hg in this case appears to have been the

---

[6]In other words, 33–100 pounds of Hg per day.
[7]Later the Olin Corporation.

mercuric sulfate used in the manufacture of acetate fiber at a plant operated by E. I. DuPont de Nemours and Company at Waynesboro, Virginia, between 1929 and 1950. Although the amount of Hg leaked to the stream by this plant was apparently quite small compared to the Hg discharges at Saltville, the Hg persisted in the sediments and biota of the stream long after the use of Hg at the plant ceased in 1950 and despite major floods,[8] which might have been expected to flush the Hg from the system. On the basis of fish and sediment samples taken and analyzed by the Virginia State Water Control Board (SWCB), fish in the South River (which flows into the South Fork of the Shenandoah) below Waynesboro and in the entire South Fork were banned for eating purposes in June 1977 (Carter, 1977). Sediment samples below the Waynesboro plant were found to contain as much as 240 ppm Hg. Sediments above the plant contained less than 1 ppm Hg. Bass taken as much as 124 km downstream from the plant were found to contain more than 1.0 ppm Hg. Remarkably, the SWCB failed to detect Hg pollution of the Shenandoah during sampling in 1970 because no analyses were made of sediments or biota. Only water was analyzed. This observation underscores a point noted earlier, namely, that metals tend to become concentrated in sediments and in organisms and are usually found at relatively low levels in water. As a result, the logical place to look for metal contamination is in sediments and organisms. Dissolved metal concentrations are not likely to be sensitive indicators of metal pollution.

The foregoing examples of Hg pollution, although far from inclusive, illustrate the point that Hg pollution can be a serious problem on a local scale, regardless of whether or not global fluxes of Hg have been affected significantly by human activities. It is impossible to appreciate the magnitude of the Hg pollution problem, however, solely from a knowledge of Hg levels in organisms and sediment or of the amount of Hg discharged into a particular body of water. The great concern over Hg stems not from its mere presence, but from its effects on organisms, particularly humans. To understand these effects, and in order to make intelligent decisions about how to deal with Hg pollution, we must first understand something about the toxicology of Hg.

Toxicology

The toxicity of Hg depends very much on the chemical form in which the element is found. The body absorbs some chemical forms of Hg much more effectively than others, and once inside the body, some forms of Hg are much more likely than others to produce serious damage to internal organs. Therefore, some appreciation of the chemistry of Hg is necessary for an understanding of its toxicology.

The chemical forms of Hg may be broadly classified as being either organic or inorganic. The inorganic forms of Hg include metallic Hg such as the Hg used in thermometers, and various "salts" of Hg, in which the element is considered to be in the plus one (Hg^+) or plus two (Hg^{2+}) oxidation state. Salts of Hg^+ and Hg^{2+} are referred to as *mercurous* and *mercuric salts*, respectively, and are typified by mercurous chloride (HgCl) and mercuric chloride ($HgCl_2$) or mercuric oxide (HgO). At room temperature, metallic Hg may occur either as a gas or as a liquid.

If inorganic Hg is swallowed, over 98% is rapidly excreted in the urine and feces (Grant, 1971). Unless ingested repeatedly or in massive amounts, metallic Hg is relatively harmless. According to Goldwater (1971, p. 15) a person could swallow up to half a kilogram or more of metallic Hg "with no significant adverse effects." Montague and Montague (1971, p. 18) state that "In an attempted suicide a person once actually injected two grams of metallic mercury directly into his veins without ill effect. The mercury merely formed a puddle in the right side of his heart." Ingestion of the more soluble salts of Hg,

---

[8]For example, the flooding after tropical storm Agnes in 1972.

however, can cause serious problems. Mercuric chloride is corrosive to the intestinal tract and, like other forms of inorganic Hg, can seriously damage the liver and especially the kidneys if taken in sufficient amounts (Goldwater, 1971; Hammond, 1971). In fact, the liver and kidneys are the internal organs that generally contain the highest Hg concentrations in persons who have died from Hg poisoning (Smith and Smith, 1975). Once absorbed by the body, inorganic Hg may be transported via the blood to all parts of the body. Fortunately, about 50% of the inorganic Hg transported in this way is present in the blood plasma (Grant, 1971), and in this form it is readily excreted by the kidneys.

One of the most serious consequences of Hg poisoning can be brain damage, but unless inhaled as a vapor, inorganic Hg does not readily penetrate the blood-brain barrier (Grant, 1971). When inhaled, inorganic Hg is initially deposited in the lungs, where in acute cases it may cause irritation and destruction of the lung tissue (Goldwater, 1971). From the lungs, metallic Hg may be transported by the blood to other parts of the body, including the brain. Unfortunately, any effects of Hg on the brain are likely to be permanent, since cells of the central nervous system, once damaged, do not recover. Most cases of Hg poisoning from inhalation are chronic. For many years, chronic Hg poisoning was a serious problem in the felt hat industry, for example, because mercury nitrate was used to treat fur pelts at an early stage in the felting process. The hatters were constantly exposed to the Hg vapors emanating from the pelts and often developed symptoms of chronic Hg poisoning. These symptoms include inflammation of the gums, metallic taste, diarrhea, mental instability, and tremors (Grant, 1969).[9] The practice of using Hg in felt making was banned in 1941. Inhalation of Hg remains a concern for persons who routinely work with Hg, especially those involved in the mining and extraction of Hg. From the standpoint of toxicology, one of the chief distinctions between inorganic and organic Hg is the fact that most injuries due to inorganic Hg do not involve the central nervous system, but rather organs such as the kidneys, liver, and intestines. Consequently, the effects of inorganic Hg poisoning are often reversible, except in cases of severe poisoning or when inorganic Hg penetrates the brain. Unfortunately, the most common forms of organic Hg frequently cause irreversible damage by attacking the central nervous system.

The three major kinds of organic Hg compounds are phenyl Hg (e.g., phenylmercuric acetate or PMA), methoxy Hg (e.g., methoxyethyl mercury acetate), and alkyl Hg (e.g., methylmercuric acetate). Of these compounds, by far the most common as well as the most dangerous are certain members of the alkyl Hg group, the methyl Hg compounds. The ubiquity of methyl Hg can be accounted for by the fact that methanogenic bacteria as well as certain molds in sediments are capable of converting virtually all other forms of Hg into methyl Hg (Goldwater, 1971; Grant, 1971; Wood, 1972). Relevant pathways are indicated in Figure 12.2. The final step involves methylation of Hg^{2+} with methylcobalamin, a vitamin B_{12} analogue, either by enzymatic means or by electrophilic attack of Hg^{2+} on methylcobalamin (Mitra, 1986). According to Wood (1972, p. 34), "All microorganisms capable of vitamin B_{12} synthesis are capable of methyl mercury synthesis." Methylation can take place under aerobic and anaerobic conditions, but in the latter case, formation of HgS and HgS_2^{2-} may prevent methylation (Mitra, 1986).

In addition to bacteria and molds found in sediments, there is evidence that microbes in the bodies of at least some animals may also methylate Hg (Hammond, 1971). Birds given nonmethylated Hg in their feed, for example, have been found to produce eggs containing methyl Hg (Grant, 1971), and mammalian liver contains a cobalamin-dependent transmethhylase capable of methyl Hg synthesis (Mitra, 1986). It is not known whether fish themselves can convert Hg^{2+} to methyl Hg, but studies by Knauer and Martin (1972) on a simple marine food chain in Monterey Bay, California, showed that the percentage

[9] The behavior displayed by Lewis Carroll's mad hatter in *Alice's Adventures in Wonderland* is typical of persons suffering from chronic Hg poisoning.

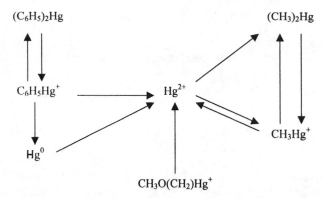

FIGURE 12.2   Transformations of Hg leading to methyl Hg. *Source*: Redrawn from Goldwater (1971). Reproduced with permission from *Scientific American* magazine.

of Hg present in organisms as organic Hg was much higher in anchovies than in their zooplankton prey (Table 12.7). The higher percentage of organic Hg in the anchovies could have resulted from the methylation of inorganic Hg within their bodies. Wood (1972), however, has noted that fish can take up at least some forms of methyl Hg from the water through diffusion across their gills, and such uptake could account in part for the high percentages of organic Hg usually found in fish. Wood (1972) indicates that about 95–100% of the Hg in fish is methyl Hg, although this generalization is not true in all cases (DLNR, 1986).

What makes methyl Hg so dangerous? First of all, if a person swallows food containing methyl Hg, about 90–95% of the methyl Hg is absorbed by the intestines (Grant, 1971). Once absorbed, methyl Hg is transported throughout the body by the circulatory system, primarily in the red blood cells. Like inorganic Hg, much of this methyl Hg is concentrated in the kidney and liver (Mitra, 1986). One factor contributing to the high rate of absorption of methyl Hg versus inorganic Hg may be the tendency of the liver to secrete absorbed methyl Hg along with bile into the small intestine, where the methyl Hg is once again absorbed through the intestinal wall back into the blood stream (Grant, 1971). Although some methyl Hg is excreted by the intestines and to a lesser degree by the kidneys, this excretory process is much less efficient than in the case of inorganic Hg excretion. In fact, some conversion of methyl Hg to inorganic Hg may occur in conjunction with methyl Hg excretion (Grant, 1971).

Although ingestion of methyl Hg may cause serious damage to organs such as the liver, kidneys, and pancreas, the most serious consequences of methyl Hg poisoning involve its effects on the central nervous system. Unlike inorganic Hg, methyl Hg effectively passes from the bloodstream into the brain, where it destroys cells, particularly in the cere-

TABLE 12.7   PERCENTAGE OF ORGANIC Hg IN A SIMPLE FOOD CHAIN

Organisms	Percent of Total Hg as Organic Hg
Phytoplankton	12–67
Zooplankton	17–78
Anchovies	76–100

Source: Knauer and Martin (1972), R. Harriss (personal communication).

bellum (leading to disturbances of equilibrium), and the frontal cortex (personality disturbances). In humans, about 10% of the total body burden of methyl Hg is found in the head, most of it presumably in the brain (Mitra, 1986). Even though the exact nature of the chemical interaction that causes Hg, and particularly methyl Hg, to destroy cells is not fully understood, the underlying interaction is believed to be the attraction between Hg and sulfhydryl groups on proteins (Goldwater, 1971). Since proteins are essential constituents of cell membranes, binding of Hg to proteins could disrupt the cell membrane sufficiently to destroy the cell or seriously interfere with its functioning. This statement applies to cells in all parts of the body, not just in the central nervous system. Although central nervous system cells are not replaced when destroyed, it is possible in the case of a limited amount of cellular destruction for other central nervous system cells to substitute functionally for the dead cells, so that overtly there is no sign of trouble. Symptoms of Hg poisoning may not become apparent until a great deal of damage has already been done to the central nervous system, including parts of the brain as well as peripheral nerves (Smith and Smith, 1975).

Methyl Hg poisoning has also been found to be the cause of serious congenital disorders. Methyl Hg apparently passes readily through the placental barrier into the fetus, where concentrations of Hg may build up to levels several times higher than those in the mother (Aaronson, 1971), reaching especially high levels in the fetal brain (Montague and Montague, 1971). Indeed, women showing no symptoms of Hg poisoning have given birth to children with serious congenital physical and/or mental defects caused by irreparable damage to the central nervous system of the fetus as a result of methyl Hg poisoning (Smith and Smith, 1975). In the case of methyl Hg poisoning in Minamata, Japan, mothers of children suffering from congenital effects generally showed less severe symptoms of Hg poisoning than other members of the same family, as did women who had miscarriages or stillbirths (Smith and Smith, 1975). These results suggest that a pregnant woman may discharge a significant portion of the methyl Hg in her body through the placenta and into the fetus.

In addition to attacking central nervous system cells of the fetus, methyl Hg may interfere with reproduction in another way. Methyl Hg (as well as phenyl Hg) is known to interfere with the process of cell division, causing daughter cells to receive an unequal number of chromosomes. This phenomenon, known as *disjunction*, has been demonstrated in plant cells (onions), fruit flies, and tissue cultures from mice and humans (Grant, 1971; Montague and Montague, 1971; Mitra, 1986). Methyl Hg has been found to be the most potent causative agent (Grant, 1971). The mechanism of interaction is again presumably the attraction of Hg for sulfhydryl groups. Although there is no definite evidence that Hg has been responsible for creating genetic abnormalities in humans, Hg has been found to concentrate in the gonads of mice and trout (Montague and Montague, 1971), and considering the demonstrated mutagenic effects of Hg on human cell cultures, the possibility certainly exists that similar effects could be produced in living persons. Montague and Montague (1971, p. 62) note that "Physicians performing autopsies on Japanese children from Minamata noted certain abnormalities that they could not attribute to direct mercury poisoning. It appeared as if the mercury eaten in fish at Minamata had entered the testes or ovaries of parents, damaged sex cells, and thus passed on genetic damage to the offspring." The other forms of organic Hg, such as phenyl Hg and methoxy Hg, can be toxic if ingested in sufficient amounts, but in general they are of less concern than methyl Hg. In particular, if phenyl Hg is ingested, a large percentage is converted to inorganic Hg by the body and is effectively excreted by way of the kidneys and intestines (Grant, 1971). Neither phenyl Hg not methoxy Hg is as effective in penetrating to the brain as methyl Hg, and hence neither class of Hg compound is as likely to cause permanent neurological damage. As far as humans are concerned, by far the most important mechanism of Hg intake is the ingestion of Hg-contaminated food (Mitra, 1986). Meat and especially fish products have in general been found to contain the highest levels of Hg (Goldwater, 1971), and at least in the case of fish, most of the Hg is usually in the form of methyl Hg (Wood, 1972). Thus, from the standpoint of human health, methyl Hg is the most dangerous form of Hg for the following reasons:

1. It has the greatest tendency to cause irreparable neurological and perhaps genetic damage.
2. As far as we know, it is more effectively retained if swallowed (more readily absorbed and more slowly excreted) than any other form of Hg.
3. It is probably the most abundant form of Hg in the food we eat.

Acceptable Levels of Exposure. What are safe levels of exposure to Hg? From the previous discussion, it should be obvious that the answer to this question depends on the chemical form of the Hg as well as the mechanism of ingestion. Furthermore, since Hg is not known to be required in any way by either plants or animals, any degree of exposure to Hg can hardly be considered beneficial. Since recent instances of large-scale Hg poisoning (Smith and Smith, 1975) have resulted from the consumption of Hg-contaminated food, much attention has been directed to establishing safe or acceptable Hg levels in food.

Assuming that virtually all the Hg we eat is methyl Hg (probably not far from the truth), it is possible to calculate the daily intake of Hg that would produce a given body burden of Hg in the steady state if we know the excretion rate of Hg by the body. The excretion rate of methyl Hg by humans is about 1% per day (Mitra, 1986), so that in the steady state a person could consume each day about 1% as much Hg as his or her body contained. In other words, the person would be excreting each day about as much Hg as he or she consumed. Based on studies of persons affected by Hg poisoning, a body burden of about 25–30 mg of Hg is "dangerous" for an adult[10] in the sense that an adult having this amount of Hg in the body is likely to show overt signs of Hg poisoning (Smith and Smith, 1975). Assuming an excretion rate of 1% per day, a Hg intake of 0.25–0.30 mg d^{-1} would lead to a dangerous concentration of Hg in the body. Based on similar reasoning, the EPA concluded that lowest observed effect levels of Hg poisoning in humans were associated with daily Hg intakes of 0.2–0.5 mg d^{-1} (EPA, 1980a). Applying the uncertainty factor of 10 appropriate to results from studies on prolonged ingestion by humans (Table 8.8), the daily intake of methyl Hg should not exceed 0.02 mg $= 20$ μg (EPA, 1980a). Recently, the EPA has revised this estimate downward by a factor of 3 for women who are pregnant or are nursing infants (EPA, 1997). The revision is based on experimental results that suggest a possible threefold increase in fetal sensitivity to methyl Hg exposure. Some persons have argued, however, that there is no safe threshold limit for Hg. Smith and Smith (1975, p. 191), for example, have stated:

> Although outward symptoms may not appear until a certain level [of mercury in the body] is reached, it can be assumed that damage is proportionate, to some extent, to the amount of mercury that is consumed and that passed through the body, whether or not the body's mercury content reaches a "dangerous" level at any given moment. If so, there is no actual "safety level." The greater the methyl mercury intake, the greater the cell damage. The lower the intake, the less damage to cells. On the cellular level there is no "threshold point."

The background level of organic Hg in food is roughly 0.02–0.05 ppm (WHO, 1973). Fruits, vegetables, grain, and milk, for example, typically contain Hg at concentrations less than 0.04 ppm (Goldwater, 1971; Mitra, 1986). Meat and freshwater fish tend to have a higher Hg content. A 1969 survey of U.S. pork and beef revealed mean Hg levels of

---

[10]Arbitrarily defined as a person weighing 70 kg.

0.1 ppm, and the background level of Hg in freshwater fish is about 0.2 ppm (Montague and Montague, 1971). Marine fish typically contain Hg at levels of 0.01–0.08 ppm. Average Hg levels in abalone, anchovies, herring, salmon, scallops, and shrimp, for example, are 0.016, 0.047, 0.013, 0.035, 0.042, and 0.047 ppm, respectively (EPA, 1996a). However, the average Hg concentrations in tuna, swordfish, and sharks are 0.21, 0.95, and 1.33 ppm, respectively (EPA, 1996a).

The high levels of Hg in tuna and swordfish have caused much concern, and on May 6, 1971, the FDA advised Americans to stop eating swordfish (Montague and Montague, 1971). As a result, the swordfish industry collapsed. Although no such recommendation was ever made with respect to tuna, it is instructive to note that consumption of a typical 6.5 ounce can of tuna containing 0.21 ppm Hg would result in the ingestion of 39 μg of Hg, twice the maximum daily intake of methyl Hg recommended by the EPA. In other words, consumption of one 6.5 ounce can of tuna every other day would lead to an average level of Hg consumption equal to the maximum level of methyl Hg intake recommended by the EPA, even if the rest of one's food contained no Hg at all. Looking at the problem another way, if one assumes an average daily rate of food consumption of 1.3 kg per person (OECD, 1975), and requires that the daily methyl Hg intake not exceed 20 μg, then the average methyl Hg concentration in food should not exceed 0.015 ppm. This figure is higher than the Hg concentration in many fruits and vegetables, but it is lower than that typically found in meat and fish.

In July 1969, the FDA established a safety level of Hg in food fish at 0.5 ppm on a wet weight basis. This figure was increased to 1.0 ppm in 1973 and was changed to 1.0 ppm methyl Hg in September 1984. From the information in the previous paragraph, it should be clear that this recommendation was based in part on the assumption that Americans do not eat food containing close to 1.0 ppm Hg on a regular basis. Thus 1.0 ppm methyl Hg in food is a safe level only if such food is eaten infrequently and if the rest of one's food averages less than 0.015 ppm Hg. According to studies conducted by the EPA, the single most important cause of methyl Hg intake for most Americans is the consumption of tuna. Methyl Hg accounts for more than 90% of the total Hg in tuna, and it is estimated that tuna accounts for about 24% of the fish consumed by Americans (EPA, 1996b). According to the FDA, only 5% of Americans consume more than 27 g of fish per day. If the assumption is that all the Hg in tuna is methyl Hg, it follows that 95% of the U.S. population consumes less than 1.4 μg d^{-1} of methyl Hg in tuna. Contributions from other fish would probably increase this figure to about 4.2 μg d^{-1} (EPA, 1980a). Perhaps the principal consequence of changing the FDA action level for fish from 0.5 ppm total Hg to 1.0 ppm methyl Hg has been to give a boost to the swordfishing industry. It is noteworthy, however, that over 50% of the swordfish tested by the EPA (1980a) contained more than 1.0 ppm total Hg.

For the protection of human health, the RfD for methyl Hg is 0.1 μg kg^{-1} d^{-1} (EPA, 1996a). The human health criterion (HHC) was calculated assuming a fish consumption (FC) rate of 18.7 g d^{-1}.[11] The practical biological concentration factor (PBCF) in the fish and shellfish was taken to be 7,337, which is a mean value calculated from EPA (1980a) data. In this case (FC)(PBCF) = (0.0187)(7,337) = 137. Since this is much greater than 2, the assumed consumption of 2 L of water per day has virtually no effect on the calculated HHC. For a 70 kg adult, the HHC for consumption of fish only is (0.1)(70)/137 = 0.051 μg L^{-1}. If water consumption is included, the HHC becomes (0.1)(70)/139 = 0.050 μg L^{-1}. Note that the acceptable daily intake in this calculation is (0.1)(70) = 7 μg of Hg, i.e., the ADI for women who are pregnant or are nursing infants. The validity of setting the HHC at 0.05 μg L^{-1} rests on the assumption that sources of exposure other than the con-

---

[11]The figure of 18.7 g d^{-1} comes from a study by Cordle et al. (1978). It differs by less than 6% from the currently accepted value of 17.7 g d^{-1} (EPA, 1996b).

sumption of fish, shellfish, and drinking water are negligible. This assumption is supported by the observation that nonfish eaters have the lowest blood concentrations of Hg (EPA, 1980a).

Most Hg in freshwater is believed to be in the form of mercuric Hg (EPA, 1980a), of which more than 98% is excreted in the feces and urine (Grant, 1971). The drinking water standard of 2 μg L^{-1} implies that it is acceptable to ingest as much as 4 μg d^{-1} of Hg in the water one drinks, but if this were all present as Hg^{2+}, the body would retain less than 80 ng. For most Americans, then, and indeed for almost all persons, the dominant source of Hg exposure is the consumption of fish and shellfish.

The discussion up to this point has been concerned with the toxicity of Hg to humans, but there is no doubt that Hg pollution can have serious consequences for aquatic organisms as well. In Minamata Bay, which received Hg discharges from an acetaldehyde plant from 1932 to 1968, Smith and Smith (1975, p. 180) report that "Unusual changes were detected in these waters [Minamata Bay] as long ago as 1950. Fish floated on the surface of the sea, shellfish frequently perished, and some of the seaweed died. In 1952 some birds such as the crow and *amedori*—a type of sea bird—began to drop into the sea while flying. The area of the sea where dead fish could be seen floating spread throughout the bay and out into the Shiranui Sea. Sometimes octopus and cuttlefish floated so weakened that children could catch them with their bare hands."

This dismal picture describes an extreme case of Hg pollution. For those charged with the responsibility of setting water quality guidelines to protect aquatic life, it is obviously necessary to know not only the levels of Hg that can be expected to produce such consequences, but also the Hg levels that could be expected to produce chronic, sublethal stresses. The problem of setting water quality standards for Hg is compounded by the fact that aquatic organisms, like humans, are much more sensitive to methyl Hg than to the other forms of Hg. Methylmercuric chloride, for example, is about 10 times more toxic to rainbow trout than mercuric chloride (EPA, 1980a).

The present EPA water quality criteria for Hg are based on data collected prior to 1985, and at that time there was enough information to determine criterion maximum concentrations only for mercuric Hg. Given that most Hg currently being discharged is Hg^{2+} (EPA, 1985a), this limitation may not be serious. The final acute values in both freshwater and marine waters were calculated as outlined in Chapter 8, and the criterion maximum concentrations listed in Table 12.8 are 50% of those final acute values. The criterion continuous concentrations are the final chronic values. One interesting point about the criterion continuous concentrations for the protection of aquatic life is the fact that they are lower than the drinking water standard of 2 μg L^{-1} (Table 12.1). Although this fact may seem surprising, one should keep in mind that, unlike aquatic organisms, humans do not spend their lives bathed in water.

TABLE 12.8 EPA WATER QUALITY CRITERIA FOR DISSOLVED Hg FOR THE PROTECTION OF AQUATIC ORGANISMS IN FRESHWATER AND SALTWATER

Criterion Maximum Concentration (μg L^{-1})		Criterion Continuous Concentration (μg L^{-1})	
Marine Water	*Freshwater*	*Marine Water*	*Freshwater*
1.8	1.4	0.94	0.77

Source: EPA (1999).

Minamata Disease: A Case Study

Minamata disease is probably the most serious and undoubtedly the most heavily publicized case of Hg pollution anywhere in the world. Although the severity and extent of the damage caused by Hg pollution at Minamata Bay make the example atypical of most cases of Hg pollution, the scientific and medical problems associated with the Hg poisoning at Minamata are nevertheless relevant to virtually all instances of Hg pollution, regardless of the extent of Hg discharges or the degree of environmental damage or human suffering. Indeed, current safety guidelines with respect to Hg ingestion by humans are based in no small part on knowledge gained from the study of Hg poisoning victims at Minamata.

Minamata Bay is a semienclosed coastal indentation on the eastern side of the Shiranui Sea,[12] an inland sea on the western part of Japan's southern island of Kyushu (Figure 12.3). About 100,000 persons depend in part on fish taken from the Shiranui Sea as a source of food, and in this respect, the people of Minamata City and its environs are typical. Historically, Minamata has been a farming and fishing area, and the population, which currently numbers about 35,000, consumes on the average between 286 g of fish in the winter and 410 g in the summer per person per day (Smith and Smith, 1975).

In 1907 the parent company of the present Chisso Corporation built a factory in Minamata. Although the Chisso Corporation was initially involved primarily in the manufacture of fertilizer and carbide products,[13] the company soon expanded into the manu-

[12]Sometimes called the Yatsushiro Sea.
[13]Chisso in Japanese means nitrogen.

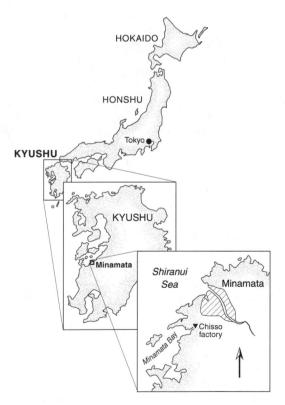

FIGURE 12.3   Location of island of Kyushu (top), Minamata (center), and the Chisso factory and Minamata Bay (bottom).

facture of petrochemicals and plastics. That discharges from the Chisso plant were creating some water pollution problems in Minamata Bay during the first few decades of the plant's operation is evident from the fact that by 1925 Chisso was compensating Minamata fishermen for damage to their fishing areas (Smith and Smith, 1975). Serious pollution of the bay presumably did not begin until 1932, however, when the plant began using mercuric oxide (HgO) as a catalyst in the production of acetaldehyde and vinyl chloride. It is not clear how much Hg the plant discharged into Minamata Bay between 1932 and 1968, when discharges were finally terminated, but one estimate has put the total figure at 200–600 tonnes (Smith and Smith, 1975). Presumably the greatest Hg discharges occurred during the 1950s, when production at the plant was apparently at a maximum and the population of Minamata had grown to 50,000 (Smith and Smith, 1975).

The first reported signs of trouble involved animals rather than humans. Cats, which were fed in part on fish taken from the bay, began to go mad and die. By 1953 some dogs and pigs were dying in a similar manner. The sickness was characterized by salivating, loss of coordination, and convulsions. Effects on aquatic organisms also began to appear during the early 1950s. Shellfish and seaweeds were killed off in certain parts of the bay, and dead fish were found floating on the surface of the bay and even out into the Shiranui Sea (Smith and Smith, 1975). Despite these overt signs of trouble, people continued to fish in Minamata Bay and to eat fish taken from the bay.

The first reported case of a human affected by the disease occurred in April 1956, when a five-year-old girl was brought to the pediatric department of the Chisso Corporation's Minamata factory. The girl was suffering from symptoms of brain damage, including delirium, disturbance of speech, and disturbance of gait. An investigation revealed that her younger sister as well as four neighbor children were suffering from the same symptoms, and "On May 1, 1956, Dr. Hajime Hosokawa, head of the Chisso factory hospital, reported to the Minamata Public Health Department: 'An unclarified disease of the central nervous system has broken out' " (Smith and Smith, 1975, pp. 180–181). This date marks the official discovery of Minamata disease.

During the summer of 1956, the occurrence of Minamata disease reached epidemic proportions, although many of the persons found to be suffering from the disease at that time had in fact been ill since about 1953. In August 1956, a research team from Kumamoto University was appointed to investigate the cause of the sickness, and on October 4, 1956, this group reported that the disease was not infectious and in fact was caused by heavy metal poisoning associated with the eating of fish and shellfish from Minamata Bay (Smith and Smith, 1975). Although it would have seemed reasonable to ban the eating of fish and shellfish from Minamata Bay at that time, the government issued only a warning that eating fish from the bay was dangerous. Two years later, when the number of victims had exceeded 50, including 21 who had died, the Kumamoto Prefecture finally ordered a ban on the selling of fish from Minamata Bay, but there was still no restriction on catching fish (Smith and Smith, 1975).

Following the October 4, 1956, report of the Kumamoto University research team, an intensive effort was begun to identify the cause of the heavy metal poisoning. Although discharges from the Chisso factory were the obvious and logical source of the pollution, no cause-and-effect relationship had been proven, and the plant was therefore allowed to continue in operation. The plant's effluent in fact contained a variety of metals, including Mn, Tl, As, Hg, Se, Cu, and Pb (Smith and Smith, 1975), and it was far from clear which metal or combination of metals was the cause of the poisoning. Initially, Mn and then Tl was suspected as being the causative agent, but experimental feeding of these metals to cats failed to reproduce the symptoms of the disease. Finally, in September 1958, Professor T. Takeuchi of Kumamoto University discovered that "Clinical and pathelogical findings in cases of Minamata Disease coincided with certain cases of methyl mercury poisoning reported in England in 1940" (Smith and Smith, 1975, p. 182). Kumamoto University researchers then found that feeding methyl Hg to cats did reproduce the symptoms of Minamata disease, and in September 1960, Professor M. Uchida of Kumamoto University extracted a methyl Hg compound (CH_3HgSCH) from Minamata Bay shellfish (Smith and Smith, 1975). Investigations of the bay following Professor Takeuchi's report revealed that Hg levels in the sediments of Minamata Bay were on the order of 10–100 ppm on a wet weight basis and that concentrations in the mud of the Chisso drainage canal ranged as high as 2,000 ppm. Hg levels in fish and shellfish sampled from the bay ranged from 5 to 40 ppm on a wet weight basis (Smith and Smith, 1975). At the time, the Chisso Corporation maintained that their Minamata plant could not be the source of the methyl Hg pollution, since the plant used only inorganic Hg in the production of acetaldehyde and vinyl chloride. Two pieces of evidence, however, indicated that the plant was indeed the cause of the poisoning.

1. Following the implication of methyl Hg as the disease-causing agent, Dr. Hajime Hosokawa, the head of the Chisso factory hospital, began experimental feeding of effluent from the acetaldehyde plant wastewater pipe to a cat. On October 7, 1959, the cat became ill, exhibiting the same symptoms—convulsions, salivating, and disoriented movements—that had been observed in cats afflicted with Minamata disease. When Dr. Hosokawa reported this result to the Chisso management, he was told to discontinue his experiments and that "There would be absolutely no more experiments connected with Minamata Disease" (Smith and Smith, 1975, p. 122). When Dr. Hosokawa's assistant went to get additional wastewater from the acetaldehyde plant, a guard stopped him. Being a good company man, Dr. Hosokawa failed to communicate his finding to the public, and it was not until he testified on his deathbed in 1969 that the knowledge of his discovery became known to those outside Chisso management.
2. Although Chisso denied scientists access to its wastewater shortly after the implication of methyl Hg as the Minamata poison in 1959, Dr. K. Irukayama of Kumamoto University discovered a bottle of sludge that had been obtained from the plant prior to that time sitting on the laboratory shelf in 1962. Analysis of the sludge revealed that it contained methyl mercury chloride, the implication being that the inorganic Hg used by the plant had been methylated prior to disposal from the plant. In fact, organic Hg was known to be a by-product of the acetaldehyde production process (Kurland, 1989).

During this time, several events occurred relevant to the Hg problem. In September 1958, Chisso began temporarily discharging its acetaldehyde wastewater into the Minamata River, which flows into the Shiranui Sea north of Minamata Bay. Within a few months, persons living in the vicinity of the river began to show symptoms of Minamata disease (Smith and Smith, 1975).

In December 1959, Chisso installed a wastewater treatment device called a *cyclator* at its Minamata plant. The cyclator was widely hailed as having made the factory wastewater safe, and at ceremonies marking the installation of the cyclator, the president of Chisso drank water taken from the outfow of the treatment system. Later it was learned that the wastewater from the cyclator on that day had not contained wastewater from the acetaldehyde process (Smith and Smith, 1975). Furthermore, testimony presented in court during the 1969–1973 trial of Chisso revealed that the cyclator had in fact not made the wastewater safe. Questioning of Eiichi Nishida, a Chisso engineer who had been head of the Minamata factory, resulted in the following exchange (Smith and Smith, 1975, p. 122):

LAWYER: in other words, did you think that [the wastewater] which went through the Cyclator was safe?
NISHIDA: Well, that is . . .
LAWYER: That is?

NISHIDA: That is . . . at that point . . . I don't know.

LAWYER: I'm asking you whether you thought it was safe or not.

NISHIDA: Well, ah, at that time, you see, I thought that putting it through the cyclator—safe, well not . . .

LAWYER: Of course you didn't think it would be, is that correct?

NISHIDA: Ah, well, yes

Thus, it is apparent that by the end of 1959, the Chisso Corporation had good reason to believe that the Hg in its wastewater was the cause of Minamata disease,[14] and it is furthermore apparent that the company was aware or at least realized later that the cyclator installed in December 1959 would not correct the problem.

On December 30, 1959, the Chisso Corporation negotiated what was, considering the information presented in the previous few paragraphs, a highly unethical contract with the Minamata disease victims known at that time. The contract provided for payments to the victims but specified that the payments were to be regarded as *mimai* (consolation) rather than as indemnity for damages. In fact, Chisso refused to accept responsibility for Minamata disease, and the *mimai* contract stipulated that Chisso would not be liable for further compensation to the victims even if the company were later proven guilty (Smith and Smith, 1975). Furthermore, Chisso continued to discharge its wastewater into Minamata Bay. A stormy protest by fishermen in October 1959, including demands for additional compensation and a cleanup of Minamata Bay, produced few tangible results. After receiving warnings from the government, the fishermen settled for meager payment. There was no cleanup (Smith and Smith, 1975).

After 1959, the furor over Minamata disease incredibly died down for several years. By 1962 there were 121 verified Minamata disease victims, including 46 dead, but although Chisso continued to discharge its wastewater into the bay, the disease was considered to be over, and people had begun to eat fish from the bay again (Smith and Smith, 1975).

Several years later, the history of Minamata disease was dramatically changed by events that occurred hundreds of kilometers away. In the city of Niigata, methyl Hg poisoning similar to Minamata disease broke out in 1965. The source of pollution was found to be an acetaldehyde plant operated by the Showa Denko Company about 65 km upstream from Niigata on the Agano River. Whereas in Minamata the Chisso plant had provided about 1,300 local jobs and hence enjoyed much sympathy, there were no such feelings in the case of Niigata. The Niigata victims, whose number eventually reached 500, began vigorous legal action against Showa Denko, resulting in a lawsuit and trial, which began on June 12, 1967 (Smith and Smith, 1975).

Aroused by the Niigata trial, some of the Minamata victims ultimately filed a lawsuit of their own against Chisso. Although the government had officially announced on September 26, 1968, that Chisso had been the cause of Minamata disease (Smith and Smith, 1975), the Minamata plaintiffs' case was weakened by the existence of the 1959 *mimai* contracts, which had stipulated that Chisso would not be held liable for further payments even if proven guilty. Furthermore, many Minamata victims declined to join in the lawsuit, but instead chose to seek compensation through a government-appointed Central Pollution Board, which was established in November 1970. In fact, only about one-third of the 1959 *mimai* signers were included among the 29 families (representing 45 victims) who sued Chisso on June 14, 1969 (Smith and Smith, 1975).

The Chisso trial lasted almost four years, with a verdict being finally reached on March 20, 1973. During the trial, several events greatly strengthened the case of the plaintiffs. First, the Niigata trial of Showa Denko ended on September 29, 1971, with a decision in favor of the plaintiffs, and on November 13, 1971, Chisso president Kenichi Shimada signed

[14]Based on Dr. Hosokawa's experiment and the report of the Kumamoto University research team.

a paper acknowledging moral (but not legal) responsibility on the part of Chisso for the Minamata poisoning. Finally, there was the July 4, 1970, testimony of Dr. Hosokawa, which revealed that Chisso had known as early as 1959 that its sludge was the probable cause of Minamata disease but had concealed the evidence.

During the course of the trial there were a number of clashes, both verbal and physical, between the Chisso Corporation and a so-called direct negotiations group, which insisted that Chisso deal directly with the Minamata victims rather than through the courts. The confrontation between Chisso and the direct negotiations group was highlighted by the beating of certain members of the direct negotiations group, as well as American photographer W. Eugene Smith at a scheduled meeting at the Chisso plant at Goi on January 7, 1972 (Smith and Smith, 1975). A year later, on January 10, 1973, the direct negotiations group was instrumental in revealing that a number of Minamata victims' signatures had been forged on Central Pollution Board documents. At the time, it was feared that the Central Pollution Board would announce its decision prior to the Minamata court decision, and it was felt that the Central Pollution Board was not disposed as favorably to the victims as was the court. In any case, the forgery scandal discredited the Central Pollution Board, forestalled its decision, and thus allowed the Minamata court to establish the precedent for compensation.

Finally, on March 20, 1973, the Minamata court announced its decision. Initial payments of $68,000 were to be made to the families of deceased victims and to severely affected individuals. Chisso made no appeal and paid out compensation checks totaling $2.2 million within a few days (Smith and Smith, 1975). However, the trial decision and payments directly affected only the small group of victims who had sued Chisso. The decision did not directly affect those who were depending on the Central Pollution Board for compensation, nor did it apply to the many newly verified patients. As a result, the direct negotiations group arranged a series of meetings with Chisso's management, beginning on March 22, 1973, to discuss compensation for patients other than the trial group, as well as to negotiate additional and regular payments for the medical needs and care of all Minamata victims. Chisso, feeling that the number of victims might ultimately number in the thousands, was reluctant to agree to all the demands of the direct negotiations group. However, on April 1, 1973, a newly verified victim by the name of Kimoto Shimada slashed his wrists with a broken ashtray in the presence of Chisso president Shimada, exclaiming that his life was worthless unless Chisso agreed to compensate him.[15] At his display of desperation, Shimada gave in. Chisso agreed to compensate all victims fully. A few weeks later, on April 27, 1973, the Central Pollution Board finally announced its decision. The initial maximum compensation was to be $68,000, the minimum $60,000, identical to the Minamata court's previous ruling. In addition, Chisso was to make monthly payments to all patients. Since that decision, the number of verified victims has steadily increased. As of 1996, 2,260 Minamata disease victims had been verified, of whom 1,198 had died (Minamata City, 1997).

The cost in terms of human suffering from Minamata disease is incalculable. The statement by Ms. Fukuda at the beginning of this chapter is a thought-provoking testimony to the frightening impact of the disease on its victims. The monetary cost of treating and caring for Minamata disease victims has been staggering, and it is instructive to examine how the Chisso Corporation has dealt with this issue. Beginning in 1959, when the Kumamoto University research team concluded that the Hg in the effluent from Chisso's Minamata plant was the probable cause of Minamata disease, the Chisso Corporation began to fragment into a large number of subsidiaries. While the formation of these subsidiaries could be rationalized to a certain extent on general principles as good business, one obvious consequence was the minimization and avoidance for the subsidiaries of the

[15]He lived.

responsibility to pay compensation (Yamaguchi, 1989). In order to assist with the payments, Kumamoto Prefecture began making loans to Chisso in 1978, the loans being financed by the sale of prefectural bonds. A condition attached to the issue of the bonds was the following (Uzawa, 1989, p. 336): "Chisso, which has received the loan of public funds, will put its management on a firmer footing, and endeavor to repay the funds, as well as make itself able in the future to pay all compensation without any assistance; in addition, the Minamata Plant shall contribute to the stability of the local economy and society." Unfortunately, there is no evidence that the Kumamoto Prefecture is monitoring Chisso to ensure that these conditions are met, and the creation of numerous Chisso subsidiaries would seem to suggest that Chisso is not taking seriously the need to repay the loan, which by 1988 amounted to 60 billion yen.[16] Furthermore, the Minamata Chisso plant has been scaled back considerably, and by 1988 the work force, which once numbered 1,300, had been reduced by more than a factor of 2 (Uzawa, 1989). The decision to reduce the work force at the plant hardly seems consistent with the goal of having the plant contribute to the stability of the local economy and society. In discussing the attitude of Chisso's management to the victims, Uzawa (1989, p. 337) made the following observation:

> Once I worked with Professor Nagai (Hosei University) in connection with the matter of compensation for Minamata disease patients to investigate Chisso's business operations. Even now I recall with a heavy heart the extremely insincere attitude of the Chisso main office during that investigation. During the investigation one of the company officers made this unforgettable statement to us: "Most of the present Chisso employees joined the company after the Minamata disease issue had already become a thing of the past, and they can't understand why they should have to take the responsibility, or be criticized, for past events." One could say that this statement reveals the true nature of Chisso, and is symbolic of the inhuman, unethical nature common to the many pollution issues in Japan, foremost among them Minamata disease.

Following the 1973 decision in the first Minamata disease lawsuit, Minamata disease victims in subsequent legal actions sued the Chisso Corporation, the Kumamoto Prefecture, and the Japanese national government. In 1979 the Kumamoto District Court found Chisso officers guilty of a criminal offense, and in 1987 it recognized the responsibility of the Kumamoto Prefecture and the Japanese national government to pay compensation (Tsuru et al., 1989). The problem of identifying funds to pay the victims was solved to a certain extent by the latter decision, which identified two culpable entities with very deep pockets. The final appeal involving Chisso's criminal liability was heard in 1988. Both the former Chisso president and the former plant manager were found guilty (Tsuru et al., 1989). The final lawsuit involving Minamata disease ended on June 6, 1996, when the 428 remaining plaintiffs accepted a court-mediated settlement of $24,000 each from Chisso (Minamata City, 1996).

In reviewing what happened at Minamata, it seems obvious that discharges from the Chisso plant should have been terminated as early as 1956, when the Kumamoto University research team reported that heavy metal poisoning was the cause of Minamata disease, since the plant was known to be discharging a variety of metals into the bay. Chisso's decision to continue operating the plant even after 1959, when Dr. Hosokawa's experiments indicated that the sludge from the plant could indeed cause Minamata disease, was highly unethical and immoral. Hg discharges from the plant were finally terminated in May 1968 because the Hg method of production had become outmoded (Smith and Smith, 1975). Why the Japanese government failed to halt the Chisso discharges earlier, at least temporarily, while a thorough investigation could be conducted, is not clear, but such gov-

---

[16]Equivalent to about $450 million.

ernment inaction is not without precedent in the history of industrialized countries, including the United States.

Efforts to clean up Minamata Bay have met with mixed success. The government initially decided to fill in or dredge all parts of the bay where sediment Hg levels exceeded 25 ppm (Smith and Smith, 1975), but 25 ppm Hg in sediments is not a very safe standard. Furthermore, dredging stirs up sediments and distributes them over a wider area. According to Wood (1972), Hg levels in fish were actually higher after the initial dredging of Minamata Bay than before. As of 1973, Hg levels in Minamata Bay shellfish averaged 0.47 ppm (Smith and Smith, 1975), a great improvement over the 5–40 ppm reported earlier but still higher than the provisional Japanese government standards of 0.4 ppm total Hg and 0.3 ppm methyl Hg (Tsuru et al., 1989, p. 415). Recognizing the extent of the problem, the government initiated a much more extensive dredging of the bay, which lasted for 13 years, from 1977 to 1990. The project cost about $160 million and removed 1.5×10^6 m^3 of sludge spread over an area of about 2,000 km^2. In 1988 some fish in the bay still violated the provisional Hg standards, yet at no time has the government enacted measures provided by law to prohibit fishing. For typical Minamata residents who eat 286–410 g of fish per day, the methyl Hg concentration in the fish would have to be less than 0.05–0.07 ppm in order to keep the intake of methyl Hg from fish consumption below the U.S. standard of 20 μg d^{-1}. It seems fair to say that there is still a Hg pollution problem in Minamata Bay.

How widespread has been the damage caused by Hg poisoning in Minamata no one can really say. The obvious victims have been identified, but mild symptoms associated with low-level Hg poisoning, particularly in congenital cases, are extremely difficult to detect. In 1970, for example, an examination of junior high school students in the most heavily contaminated areas of Minamata revealed difficulty in articulating words among 18% of the students, sensory disturbance among 21%, and clumsy movements among 9% (Harada, 1989). For comparison, the average rate of mental deficiency among Japanese junior high school students is 9.7% (Smith and Smith, 1975). To what extent the high percentage of mental and physical deficiency among these students was due to low levels of Hg poisoning is anyone's guess.

Commentary

Was Minamata an isolated incident? Not really. The number of persons who actually died of Hg poisoning at Minamata exceeds the number known to have been killed by Hg in comparable incidents elsewhere, but the 6,500 persons admitted to hospitals in Iraq as a result of methyl Hg poisoning in 1971 exceed by almost a factor of 3 the number of certified Minamata disease victims, and the number of persons who died in the Iraq incident (450) is within a factor of 3 of the number (1,200) who died at Minamata. Similar incidents, fortunately involving smaller numbers of victims, have been reported in other parts of the world. In Niigata, 690 persons have been certified to be victims of methyl Hg poisoning (Saito, 1989), and from 1958 to 1982, the Songhua River in China was seriously polluted with methyl Hg discharges, again from an acetaldehyde plant. The Chinese incident caused numerous cases of sublethal poisoning, with documented symptoms including hearing and vision impairment (Pan et al., 1989). Fortunately, there were no fatalities, but a "follow-up study showed . . . that there were many subclinical victims along the Songhua River who were easily overlooked" (Pan et al., 1989, p. 301).

High levels of Hg have been found in fish from several parts of North America, primarily in areas where chlor-alkali plants were discharging Hg-laden wastewater. Fish taken from the Canadian English-Wabigoon River system in 1970, for example, were found to contain as much as 27.8 ppm Hg, with many values between 10 and 20 ppm (Smith and Smith, 1975), and many fish taken from Lake St. Clair in the same year contained 5–7 ppm Hg (Grant, 1971). In both cases, the source of Hg was a chlor-alkali plant, the Dryden

Paper Company plant at Lake Wabigoon in the case of the English-Wabigoon River and the Dow Chemical plant at Sarnia in the case of Lake St. Clair (Wood, 1972). It is noteworthy that the Hg levels in the fish sampled from these two systems were comparable to the levels of Hg found in fish and shellfish from Minamata (5–40 ppm) in 1959 and from Niigata (1–10 ppm) in the 1960s (Grant, 1971; Smith and Smith, 1975). Nevertheless, no wide-scale Hg poisoning incident has occurred either in the United States or Canada, presumably because most people in these countries eat fish much less frequently than do Japanese. There is evidence, however, that some Canadian Indians developed symptoms of methyl Hg intoxication as a result of eating Hg-contaminated fish (EPA, 1980a).

In the United States, reduction of Hg pollution in aquatic systems has been accomplished primarily through enforcement of the 1899 Refuse Act, the FIFRA of 1947, and the Clean Water Act (CWA) of 1972. The FIFRA was used to eliminate the use of Hg compounds as antifouling agents and fungicides (Mitra, 1986). The Refuse Act was used to control discharges from chlor-alkali plants and other industrial operations prior to passage of the CWA. A chlor-alkali plant operated by Allied Chemical, for example, is estimated to have discharged about 75 tonnes of Hg into Onondaga Lake between 1946 and 1970, a discharge rate of about 8.6 kg d^{-1}. The U.S. Department of Justice took legal action against the facility in the summer of 1970, and the discharges were subsequently reduced by more than a factor of 20 through process modification (Effler and Hennigan, 1996). Likewise, a chlor-alkali plant operated by Olin Matheson on the Niagra River was able to continue operation after being taken to court in 1970 because installation of pollution control devices reduced Hg discharges to about 220 g d^{-1} (Wood, 1972). Chlor-alkali plants operated by the Oxford Paper Company in Rumsford, Maine, and by Olin Matheson at Saltville, Virginia, simply closed down in 1970 and 1972, respectively, when faced with court orders to clean up their effluent (Wood, 1972; Carter, 1977). In Canada the Law on Fisheries was amended in 1972 to limit the discharge of Hg in wastewater by chlor-alkali plants to 1.82 g per tonne of Cl_2. Discharges from U.S. plants are controlled indirectly by the criteria listed in Table 12.8, which apply to the receiving body of water. The impact of the Canadian and U.S. legislation on Hg levels in some fish populations has been quite dramatic (Figure 4.13), but further improvement is desirable.

Even when Hg pollution is detected and halted, the problem of cleaning up the damage remains. The lessons learned from Minamata Bay and from U.S. and Canadian lakes and streams have made it clear that aquatic systems contaminated with Hg are generally very difficult to clean up. Once released to the environment, Hg may continue to cycle between the sediments, water, and biota for tens, hundreds, or even thousands of years before finally being flushed from the system. For example, it has been estimated that about 5,000 years will be required for the Hg presently stored in the Lake St. Clair ecosystem to flush out effectively by natural processes (Wood, 1972), and a small amount of Hg (not much more than could be put in a Volkswagon gas tank) released to Virginia's South River and South Fork of the Shenandoah River between 1929 and 1950 has kept the system contaminated for decades, with sediment Hg levels exceeding 240 ppm in some places (Carter, 1977).

The following approaches for decontaminating Hg-polluted sediments have been suggested by Swedish workers (Mitra, 1986):

1. Dredging of sediments
2. Increasing the pH of the sediments in order to favor demethylation and increase volatilization
3. Introducing O_2-consuming materials so as to create anaerobic conditions in the sediments and hence reduce Hg methylation
4. Covering the sediments with fresh, finely divided, highly adsorptive materials such as clay or quartz
5. Covering the sediments with any inorganic, inert material

Unfortunately, all of these methods have drawbacks, either in terms of cost, permanence, effectiveness, or side effects. The Swedes, for example, dredged the sediments of polluted waterways on two occasions, and in each case found that Hg levels in fish were higher after dredging (Mitra, 1986). The Japanese experienced a similar but temporary problem in Minamata Bay. The burrowing activities of benthic infauna can frustrate efforts to seal off contaminated sediments with an overlayer of inert adsorptive material, and Swedish scientists found that uptake of Hg by fish from sediments contaminated with phenyl Hg was unaffected by ground silicates. Hg uptake was reduced by a factor of 2, however, when inorganic Hg was the contaminant (Mitra, 1986).

The problem of cleaning up Hg pollution is well illustrated by the case of the Olin Matheson chlor-alkali plant at Saltville, Virginia. So-called muck ponds at the site cover about 0.5 km^2 and extend along the Holston River for about 1 km. The ponds contain Hg to a depth of perhaps 24 m and leak Hg to the Holston River at an estimated rate of about 100 g d^{-1}. According to Carter (1977, p. 1017), "it has become increasingly apparent that this problem can never be fully overcome; anything short of gargantuan engineering remedies, undertaken at costs that might run into the hundreds of millions, may bring nothing better than a modest, perhaps trifling, amelioration."

In summarizing Hg pollution, several points seem particularly worth keeping in mind:

1. Mining of Hg has declined by about 70% since 1970, but anthropogenic emissions of Hg to the atmosphere are at least comparable to and perhaps as much as three times greater than natural emissions. On a local scale, Hg pollution has been a serious problem, particularly in some areas affected by acetaldehyde and chlor-alkali plants.
2. For humans, the most important mechanism of Hg intake is the ingestion of Hg-containing fish. Tuna, swordfish, and sharks contain naturally high concentrations of Hg, and consumption of tuna is believed to be the single most important source of Hg ingestion for most Americans. The FDA standard of 1.0 ppm methyl Hg in fish is realistic only if one consumes less than 20 g of such fish per day. The average methyl Hg content of one's food should not exceed about 0.015 ppm.
3. There is a strong tendency for all forms of Hg to be converted to methyl Hg by microbes. Methyl Hg is the most toxic form of Hg, and damage done to the central nervous system by Hg poisoing is usually permanent.
4. Decontaminating an aquatic system polluted with Hg is likely to be difficult and costly, if possible at all. Hg may continue to leach from contaminated sediments for hundreds or even thousands of years, and even in river systems, natural flushing processes may take tens or hundreds of years to decontaminate the system.
5. Considering points 3 and 4, it is evident that prevention rather than cure is by far the most effective policy for dealing with Hg pollution.

Cadmium

Distribution, Production, and Uses

Cd is a metal chemically similar to Zn. Although it is widely distributed in the lithosphere, Cd is usually found at quite low concentrations in crustal rocks and soils, typical concentrations being 100–300 ppb and 200–800 ppb, respectively (Nriagu and Sprague, 1987). Although Cd concentrations as high as 100 ppm are sometimes found in phosphatic rocks,[17] there are no ore deposits sufficiently rich in Cd to warrant extraction of the cadmium per se. Instead, Cd is invariably obtained as a by-product of mining other metals,

[17]Due to the presence of fossilized fish teeth rich in Cd.

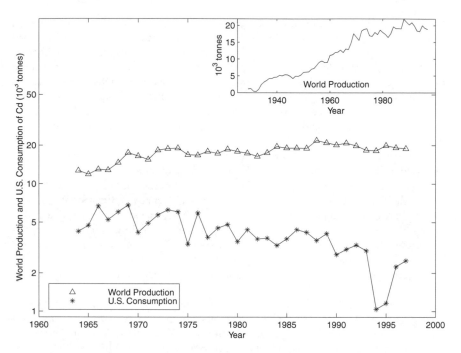

FIGURE 12.4   World production and net U.S. consumption of Cd. *Source: Minerals Yearbook*, U.S. Department of the Interior, Bureau of Mines.

primarily the sulfide ores of Zn, and to a much lesser extent of Pb and Cu. Sphalerite and wurtzite, for example, both of which are forms of ZnS, may contain up to 5% Cd by weight, although median concentrations are only about 0.3% (Nriagu, 1980). Typically, the ores are roasted to drive off gaseous S products. Cd, being more volatile than Zn, Pb, or Cu, is driven off in the roasting process. The Cd is then collected in the flue dust, which may be recycled to further concentrate the Cd. Flue dust Cd levels are generally in the range of 20–40%. From the flue dust, Cd and most other metal by-products are dissolved by acid oxidation. Final preparation of the Cd (>99.9% pure) may be effected by several different techniques,[18] the choice of method being determined by the nature of the other metals present and by whether or not other metals are to be recovered (OECD, 1975).

World production of Cd increased rather steadily from about 1,000 tonnes in the early 1930s to over 15,000 tonnes by 1970, but it has increased at an irregular rate of only about 1.3% per year since then (Figure 12.4). Since most Cd is obtained as a by-product from Zn mines, the production of Cd has little to do with the availability of or demand for Cd. Instead, production is controlled largely by activities in the Zn industry. Since 1945, world Cd production has averaged 30 ± 2% of world Zn production by weight and has accounted for about 5–10% of the income from Zn sales (Nriagu, 1980). Current production of Cd is about 19,000 tonnes y^{-1} and has averaged 20,000 ± 1,000 tonnes y^{-1} for the past 10 years. The major producing nations, in order of importance, are Japan, the United States, Belgium, China, Canada, Kazakhstan, and Germany (Plachy, 1997).

The U.S. Bureau of Mines does not keep a detailed inventory of Cd consumption in the United States, and the net consumption shown for the United States in Figure 12.4 is simply the difference between Cd imports, exports, and changes in stockpiles. Total con-

---

[18]For example, vacuum distillation or electrolysis.

TABLE 12.9 ESTIMATED END USE OF Cd IN THE WESTERN
WORLD IN 1997

Use	Percentage Use of Cd
Batteries	70%
Pigments	13%
Coatings and platings	8%
Stabilizers for plastics and similar synthetic products	7%
Alloys and others	2%

Source: Plachy (1997).

sumption equals net consumption plus the consumption of recycled Cd. Historically, recycling has accounted for less than 5% of Cd consumption (Nriagu, 1980), but a concerted effort to encourage recycling of Ni-Cd batteries has changed this picture in recent years (see the later discussion). In contrast to world production, net consumption of Cd in the United States has been declining at a rate of about 2.5% per year for the last 30 years. Although a detailed inventory of Cd consumption in the United States is unavailable, the International Cadmium Association has been making estimates of end use of Cd for the Western World. A summary of their estimates for 1997 is presented in Table 12.9. Of the various use categories, only battery use has been increasing in recent years. Use of Cd for electroplating and pigments in the United States, for example, declined by 60% between 1968 and 1988, while use of Cd in batteries increased by more than a factor of 6 during the same time period.[19] The following comments briefly describe the nature of the principal uses of Cd.

Ni-Cd Batteries. Cadmium is used as the anode in nickel-cadmium batteries. These batteries are rechargeable, have a long lifetime of approximately 3,000 cycles and a low self-discharge rate, operate over a wide temperature range, and can deliver maximal currents with a low-voltage drop. The chief disadvantages of Ni-Cd batteries are their low energy density and their high cost. They are used as sealed cells in radios, alarm systems, emergency lighting, pacemakers, calculators, motor starters, walkie-talkies, portable appliances, and tools. Larger vented units are used in buses, diesel engines, aircraft, spacecraft, military applications, and standby power and lighting systems (Nriagu, 1980).

Electroplating. Almost all Cd plating is done by electrodeposition, although some plating and coating is also done by vacuum deposition, dipping, or spraying. In electroplating, a thin layer of Cd is deposited by electrolysis on metal objects made of steel, iron, copper, brass, and other alloys to prevent corrosion. The Cd layer provides good solderability and conducts electricity well, is highly ductile,[20] and provides good corrosion resistance to tropical atmospheres, saltwater, and alkaline substances. Cd has the ability to protect steel, to which it is anodic, through sacrificial corrosion (Nriagu, 1980). Cd, however, does dissolve readily in acid solutions, including even weakly acidic solutions such as fruit juices. During the early 1940s, hundreds of persons were poisoned from using Cd-plated utensils before the practice of plating utensils with Cd was banned in the United States and Europe (McCaull, 1971). Coating and plating accounted for the largest percentage of Cd use in the United States between 1940 and 1988 (Plachy, 1997). However, because of human health and environmental concerns, consumption of Cd for coating and plating had begun to decline by as early as 1965. In 1990 coating and plating accounted

---

[19]Data for 1968 came from OECD (1975). Data for 1988 came from *Minerals Yearbook*, published by the Department of the Interior, Bureau of Mines.
[20]So that plated parts can be stamped or shaped.

for 25% of total Cd consumption in the United States, but the corresponding figure was only 9% in Europe and 1% in Japan (Plachy, 1997).

Pigments. Certain Cd compounds are used as coloring agents in a variety of products, chiefly in plastics but also including coated fabrics, textiles, rubber, glass, paints, enamels, ceramic glazes, printing inks, and artists' colors. The colors, which range from yellow to red, are produced by mixtures of cadmium sulfide and cadmium selenide (red) or of cadmium sulfide with zinc sulfide (yellow). Cd pigments have good hiding power and color intensity, do not bleed, are resistant to degradation by light, basic substances, and H_2S, and are heat stable up to about 600°C[21] (OECD, 1975).

Plastic Stabilizers. Mixtures of Cd and barium (Ba) combined with organic acid anions are used as heat stabilizers in plastics to retard degradation due to elevated temperatures. These stabilizers offer some protection against light-induced degradation as well. Compounds of Pb and organotin account for the other major forms of plastic stabilizers. Because of the toxicity of Cd, the FDA has ruled that Cd stabilizers cannot be incorporated into plastics used for food packaging (OECD, 1975).

Alloys. Alloys of Cd have found use in a variety of applications. Metals such as bismuth (Bi), Pb, and Sn have been combined with Cd to produce low-melting-point alloys that are used in fire-detection and fire door release devices, as molds for casting plastics, and as safety plugs in compressed-gas cylinders. Such alloys usually contain less than 20% Cd. Bearings made of Cd ($\geq$98%) combined with Ni, Cu, or Ag have greater heat resistance and can run at higher speeds than Sn or Pb bearings, but use of such bearings has declined since World War II. Alloys consisting of Cd (5–20%) combined with Ag, Cu, and Zn have been used as brazing compounds for soldering metals when stong, leakproof, corrosion-resistant joints are required (OECD, 1975).

Miscellaneous Products. Cd is used in certain pesticides both in agriculture[22] and in nonagricultural applications.[23] Because Cd is effective in absorbing neutrons, it has been used to make control rods for nuclear reactors. Cd is also used in smoke detection devices, in solar cells, in photocells, and as a component of the phosphors in television tubes, X-ray screens, and luminescent dials. Small concentrations of Cd are found in virtually all fossil fuels. Concentrations in coal as high as 1–2 ppm have been reported, but typical levels in oil are found in the range 0.1–0.5 ppm. Roughly 6 billion tonnes of fossil fuels are burned each year, and the concomitant release of Cd to the atmosphere amounts to about 670 tonnes (see below).

Emissions to the Environment

Table 12.10 summarizes our present estimates of natural and anthropogenic fluxes of Cd to the atmosphere and to aquatic systems. There is little doubt that human activities have substantially increased emissions of Cd to the atmosphere. The numbers in Table 12.10 indicate that the atmospheric flux has increased by a factor of 6–7 as a result of human activities. The chief cause of the increase has been the extraction and recycling of nonferrous metals, particularly Zn and Cu.

It is estimated that about half of the Cd naturally emitted to the atmosphere but only 30% of the anthropogenic atmospheric emissions are deposited in aquatic systems. The explanation is that industrial Cd emissions are associated with larger particles than those involved in natural scavenging processes (Nriagu and Sprague, 1987). A major portion of the anthropogenic emissions is therefore deposited on the land. The data, however, suggest that Cd inputs to aquatic systems have increased by a factor of about 2.6 as a result

---

[21]Hence their use in plastics, which are molded at high temperatures.

[22]For example, on broad beans, tomatoes, and wheat.

[23]For example, on golf courses.

TABLE 12.10 ESTIMATED FLUXES OF Cd TO THE
ATMOSPHERE AND OCEAN

Source of Cd	Estimated Annual Flux (10^3 tonnes)
To the atmosphere	
Natural	
Volcanoes	0.82
Wind-borne soil particles	0.21
Wild forest fires	0.11
Seasalt spray	0.06
Biogenic	
Continental particulates	0.15
Marine	0.05
Continental volatiles	<u>0.04</u>
Total natural sources	1.44
Anthropogenic	
Smelters	5.59
Solid waste incineration	0.75
Coal combustion	0.53
Cement production	0.27
Phosphate fertilizers	0.17
Oil combustion	0.14
Wood burning	<u>0.12</u>
Total anthropogenic sources	7.57
To the ocean	
Natural	
River runoff	5.11
Atmospheric deposition	<u>0.72</u>
Total natural sources	5.83
Anthropogenic	
Domestic wastewater	1.74
Electric power production	0.13
Metal mining, smelting, and refining	1.95
Manufacturing processes	2.45
Sewage sludge	0.69
Atmospheric deposition	<u>2.27</u>
Total anthropogenic sources	9.23

Source: Nriagu (1980, 1989), Nriagu and Sprague (1987), Nriagu and Pacyna (1988).

of human activities. One obvious fact about the data in Table 12.10 is that the various anthropogenic emissions of Cd account for very little of the 19,000 tonnes of Cd that are mined each year. It is apparent, then, that a major fraction of the Cd produced each year accumulates in the form of products and supplies or is discarded on land. With respect to land disposal, Nriagu (1980, p. 77) has commented, "The fate and ecological impacts of the cadmium so dissipated are essentially unknown." Cd-containing consumer products such as Ni-Cd batteries and various forms of plastic could remain stable in landfills for many years if they are not exposed to the atmosphere and do not come into contact with acidic water. Land disposal of Cd should not create environmental problems if reasonable precautions are taken, but experience has shown that many landfills are not operated in an environmentally safe manner (Chapter 16). The following discussion provides some additional information about the major fluxes listed in Table 12.10.

Natural Fluxes to the Atmosphere. The major input here is volcanic eruptions, which are also a significant source of natural Hg emissions into the atmosphere. The emissions result from the presence of the metals in volcanic rock and the volatility of the metals. At the temperature of molten lava, metals such as Cd are readily degassed into the atmosphere. Biogenic inputs are the result of plant exudates, which Nriagu (1980) assumed to contain 2.75 ppm Cd. Here again, the volatility of Cd is a factor in the emissions. Wind-blown dust contains a small amount of Cd, and forest fires introduce Cd into the atmosphere because of the presence of Cd in trees and other types of vegetation.

Anthropogenic Fluxes to the Atmosphere. Emissions resulting from the mining, extraction, and processing of Zn, Cu, and Pb account for over half of the anthropogenic inputs of Cd to the atmosphere. Zn and Cu operations are the principal culprits, despite the fact that Cd is often obtained from the flue dust. However, some of the Cd inevitably escapes from the exhaust system in the form of tiny aerosols, which may be dispersed over a wide area. Reprocessing of plated and galvanized metal also introduces significant amounts of Cd into the atmosphere because of the volatility of Cd. For example, the melting point of Fe, 1,535°C, is much higher than the boiling point of Cd, 767°C. Obvious sources of Cd emissions associated with waste incineration include the burning of sewage sludge and wood painted with Cd-containing paint. As noted in Chapter 6, sewage sludge often contains elevated metal concentrations relative to natural soils, and simply applying sewage sludge to land can lead to Cd emissions due to volatilization. The FDA recommends that sewage sludge applied to the land contain no more than 29 ppm Cd (GESAMP, 1985). The high Cd content of some phosphate deposits, particularly those formed from fossilized fish teeth, accounts for the high Cd concentrations (up to 10 ppm) measured in phosphate fertilizers (OECD, 1975). The volatilization of Cd from fertilized agricultural lands can therefore introduce significant amounts of Cd to the atmosphere, and runoff from such lands can create water pollution problems as well. Horvath et al. (1972), for example, found dissolved Cd concentrations to be seven times higher in canal water adjacent to agricultural fields in South Florida than in canal water draining underdeveloped land in the same region. A variety of metals, including Cd, had been applied to the agricultural fields in the form of pesticides as well as in fertilizer. The presence of Cd as an impurity in coal accounts for atmospheric emissions of Cd due to coal combustion.

Natural Fluxes to Aquatic Systems. The principal input here is from river runoff and reflects the erosion and weathering of soils and rocks. Rivers transport roughly 2×10^{10} tonnes of sediments to the ocean each year. The estimate of 5.11×10^3 tonnes of Cd associated with that runoff implies a Cd concentration of about 250 ppb in soils and rocks, which is consistent with analyses of crustal soils and rocks.

Anthropogenic Fluxes to Aquatic Systems. Atmospheric deposition and discharges from manufacturing processes account for roughly half of the total anthropogenic input of Cd. There is no question that the industrial use of Cd in the manufacture of finished products results in some release to the environment. The sediments below one Ni-Cd battery plant located on a tributary of the Hudson River, for example, were found to contain 16.2% Cd and 22.6% Ni (McCaull, 1971), and during the 1950s, an investigation of water supplies on Long Island revealed that a groundwater recharge basin that received effluent from an aircraft manufacturing company contained Cd at a concentration of 1.2 ppm (Lieber and Welsch, 1954). In the latter case, the company had been using Cd to anodize Al and other metals to retard corrosion. Seepage from the recharge basin was found to have contaminated local groundwater supplies to a depth of as much as 15 m for a distance of about 1 km from the recharge basin. Some test wells about 200 m from the recharge basin produced water containing 1–3 ppm Cd, 200–600 times the EPA limit for drinking water (Table 12.1). Other important anthropogenic fluxes of Cd to aquatic systems include domestic wastewater and the discharge of wastewater associated with the mining, smelting, and refining of Zn, Cu, and Pb. The story of itai-itai disease (see the later discussion) is a thought-provoking example of the latter.

Toxicity

Human Health. With respect to toxicity, there are several notable similarities and dissimilarities between Cd and Hg. Like Hg, Cd is not required even in small amounts for the maintenance of life. Unlike Hg, only a single chemical species, the Cd ion Cd^{2+}, is believed to exert a toxic effect. Transformation of Cd to more toxic compounds analogous to methyl Hg is not known to occur (OECD, 1975). The efficiencies of Cd and inorganic Hg absorption by the intestines are comparable, about 5% and 2%, respectively. Low body Fe stores or a Ca deficiency can increase intestinal absorption of Cd to 10–20% (GESAMP, 1985). Absorption of Cd vapor by the lungs is about 10–60% efficient, and about 0.1–0.2 μg of Cd is inhaled by smoking one cigarette (Friberg et al., 1986). For nonsmokers, the consumption of Cd-contaminated food normally accounts for most of the total intake of Cd (GESAMP, 1985). An analysis of Cd levels in various foodstuffs, in air, and in water, for example, has indicated that persons in the United States ingest about 30 μg d^{-1} of Cd from the food they eat, 2.6 μg d^{-1} from the water they drink, and 0.6 μg d^{-1} from the air they breathe (Table 8.12). Considerable variations exist, however, in the amount of Cd consumed with food. In Europe, for example, daily Cd intake averages only about 20 μg, but in Japan the average is 40–50 μg d^{-1} (GESAMP, 1985). Smokers and persons who are occupationally exposed to Cd may take in even more.[24]

Once ingested, Cd is transported to all parts of the body by the bloodstream. Although almost all organs probably absorb some Cd, the highest concentrations are invariably found in the liver and kidneys. Roughly one-third and one-sixth of the body burden of Cd is stored in the kidneys and liver, respectively. After long-term, low-level exposure, most of the rest is found in the muscles (Friberg et al., 1986). Unlike Hg, Cd apparently does not effectively penetrate the placental barrier, so that poisoning of the fetus is of much less concern than is the case with methyl Hg (OECD, 1975). One of the major concerns with Cd poisoning is the fact that ingested Cd has a half-life of roughly 16–33 years in the human body. As a result, Cd can be an extremely insidious poison in the sense that ingestion of only small amounts over a period of many years may lead to the accumulation of toxic levels of Cd in the body. The body burden of Cd is only about 1 μg at birth, but it steadily increases until roughly age 50, when it is typically 10–30 mg (GESAMP, 1985).

Once absorbed by the body, Cd tends to be concentrated in the kidneys and liver by a low molecular weight protein called *thionein*. This protein contains large numbers of sulfhydryl groups, which attract Cd as well as other heavy metals such as Hg, Zn, and Cu. The metal–protein complex is referred to as *metallothionein* and is synthesized in the liver. Metallothionein is transported to the kidneys by the bloodstream, where it is filtered from the plasma by glomeruli and then reabsorbed in the tubules along with other low molecular weight proteins (Friberg et al., 1986). Continuous catabolism of the Cd metallothionein occurs after reabsorption, Cd being split from the metallothionein and bound to newly formed metallothionein in the tubular cells (GESAMP, 1985). Binding heavy metals in metallothionein is believed to be a protective defense mechanism that prevents the metals from interacting with metabolically important proteins. In the case of Cd, kidney damage occurs when the amount of Cd in the kidneys overwhelms the ability of the kidneys to bind the Cd in metallothionein. The unbound Cd then damages the renal tubules, which lose their ability to reabsorb proteins. Symptoms of kidney damage include elevated urinary levels of protein (proteinuria) and Cd. In more severe cases, there may also be an increase of glucose in the urine (glucosuria) and of alkaline phosphatase in the blood (OECD, 1975). Extreme cases of Cd intoxication are associated with osteomalacia (soft-

---

[24]McCaull (1971), for example, cites one example of construction workers who inhaled Cd vapor while cutting Cd-plated bolts with oxyacetylene torches in poorly ventilated places.

ening of the bones) and osteoporosis, apparently caused by disruption of the Ca-P balance in the renal tubules (OECD, 1975). The biochemical mechanism is believed to involve, at least in part, inhibition by Cd of the formation of a metabolite of vitamin D, referred to as 1,25-DHCC. This compound, which is formed in the kidney tubular cells, stimulates Ca absorption from the intestine and is necessary for normal bone mineralization (Friberg et al., 1986).

Inhalation of Cd vapor can lead to permanent or even fatal lung damage (OECD, 1975), and in a long-term study with rats it caused an increase in lung cancer. An excess of lung cancer has also been noted among previously heavily exposed Cd workers, and it is possible that the Cd in cigarette smoke may contribute to the development of lung cancer in smokers (Friberg et al., 1986). No animal or human exposure studies suggest that Cd is carcinogenic via oral exposure, but the EPA lists Cd as a probable human carcinogen if inhaled (EPA, 1997).

Cd has been linked to hypertension, largely as the result of experimental studies with animals (Yunice and Perry, 1961). There is no evidence, however, of an unusual incidence of hypertension or cardiovascular disease among workers exposed to Cd (Friberg et al., 1986).

The derivation of the EPA water quality criteria for Cd with respect to human health has been discussed at some length in Chapter 8 and will be reviewed only briefly here. The criteria are based on kidney damage to persons who suffered long-term chronic exposure to Cd due to the consumption of Cd-contaminated food. Based on studies of such persons, the acceptable daily intake (ADI) of Cd from food and water is 20–76 μg d^{-1}. The present intake of 30 μg d^{-1} from food is due primarily to the consumption of vegetables and cereal crops, the concentrations in meat, fish, and fruits generally being quite low (GESAMP, 1985). If drinking water actually contained the EPA criterion of 5 ppb Cd, then consumption of 2 L of water per day would lead to the intake of 10 μg d^{-1} of Cd and would increase the overall intake to about 40 μg d^{-1}. Consumption of 17.7 g d^{-1} of fish and shellfish taken from such water would increase the Cd intake by roughly 3.2 μg d^{-1}. Given an ADI of 20–76 μg d^{-1}, it seems doubtful whether the intake should be allowed to increase from 30 to 43 μg d^{-1}. According to the EPA (1980c), only about 0.3% of U.S. water supplies contain more than 10 ppb Cd, and 37% contain less than 1.0 ppb.

The only extant guidelines with respect to Cd concentrations in food concern rice in Japan, the limits being 1.0 ppm in unpolished rice and 0.9 ppm in polished rice. However, daily ingestion of typically (in Japan) 335 g of rice (Yamagata and Shigematsu, 1970) containing 1.0 ppm Cd would imply a daily intake of 335 μg of Cd, roughly 5–15 times the ADI. Realizing this fact, it is difficult to understand how the Japanese government could consider 0.9–1.0 ppm Cd in rice a tolerable concentration. In Japan the mean concentration of Cd in polished rice is only 0.07 ppm (Yamagata and Shigematsu, 1970). Consumption of 335 g d^{-1} of such rice would lead to a daily Cd intake of 23.5 μg, a figure that largely accounts for the difference in Cd intake between Japanese and Americans. The average Cd concentration of food consumed in the United States is about 23 ppb.

As in the case with Hg, some aquatic species are intoxicated by Cd at concentrations substantially lower than the present public water supply standard of 5 ppb. Derivation of the EPA freshwater criteria for the protection of aquatic organisms has been discussed in Chapter 8, and the results are summarized in Table 12.11. The effect of water hardness on toxicity appears to be due to Ca, but not Mg (EPA, 1985c). The Ca ion, Ca^{2+}, bears a charge identical to that of the Cd ion, Cd^{2+}, and Ca^{2+} has virtually the same ionic radius, 0.99 angstroms, as does Cd^{2+}, 0.97 angstroms. Thus, Ca^{2+} may effectively compete with Cd^{2+} for uptake by organisms.

The marine final acute value was based on studies with 33 genera of fish and was calculated according to the procedures described in Chapter 8. The range of genus mean acute values was a factor of 2,000 for the genera studied. Several studies revealed a positive cor-

TABLE 12.11 EPA WATER QUALITY CRITERIA FOR DISSOLVED Cd FOR THE PROTECTION OF AQUATIC ORGANISMS IN FRESHWATER AND SALTWATER

Water Type	*Criterion Continuous Concentration* ($\mu g\ L^{-1}$)	*Criterion Maximum Concentration* ($\mu g\ L^{-1}$)
Marine water	9.3	42
Freshwater		
50 mg L^{-1} CaCO$_3$	1.3	2.0
100 mg L^{-1} CaCO$_3$	2.2	4.3
200 mg L^{-1} CaCO$_3$	3.7	9.0

Source: EPA (1999).

relation between 96-hour TLm values and salinity (EPA, 1985c), a trend very likely reflecting the fact that Ca^{2+} is one of the principal ions in seawater. The criteria for marine waters are based almost entirely on studies conducted in full-strength seawater. The final chronic value was calculated by dividing the final acute value by an acute/chronic ratio of 9.1, the latter being derived from studies on two species of mysids whose mean acute values were close to the final acute value. This final chronic value was lower than the final plant value and therefore became the criterion continuous concentration for marine waters.

A Case Study: Itai-Itai Disease

In 1955 two Japanese physicians reported the occurrence of a mysterious disease in the Jintsu River basin of Japan near the city of Toyama (Figure 12.5). The disease was characterized by severe pain in the back, joints, and lower abdomen, development of a wad-

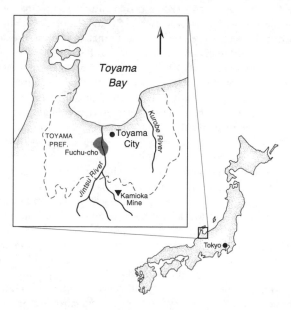

FIGURE 12.5   Toyama City and Jintsu River basin. Hatched area indicates region where disease occurred. *Source*: Redrawn from Yamagata and Shigematsu (1970). Reproduced with permission.

dling or duck-like gait, kidney lesions, proteinuria, glycosuria, and loss of Ca from the bones leading, in some cases, to multiple bone fractures (Yamagata and Shigematsu, 1970). In some bedridden patients, the skeleton became so fragile that the slightest external pressure, for example from coughing, could cause bone fractures (Nogawa, 1980). Discussions with local doctors revealed that symptoms of the disease had been noted in the region as early as 1935. The disease was largely confined to postmenopausal women who had had multiple pregnancies and who had lived in the region for all or most of their lives (Emmerson, 1970). For lack of a better name, the disease was called itai-itai (loosely in Japanese, "ouch-ouch") disease because of the severe pain experienced by the more severely affected patients.

Initial efforts to identify the cause of itai-itai disease after 1955 were unsuccessful. Some doctors suspected that the disease was caused by a local nutritional deficiency, as the victims were largely confined to a poor farming district. Others noted, however, that a mining and smelting operation had been operated since 1924 by the Mitsui Mining Company at its Kamioka mining station upstream of the affected region on the Jintsu River (Figure 12.5). The mining operation produced Zn, Pb, and Cd, and until 1955 had routinely discharged its wastewater into the Jintsu River (Anonymous, 1971). It was also known that occasional accidents or flooding resulted in sludge from the mine being washed into the river (Yamagata and Shigematsu, 1970). During World War II, pollution control measures at the mine were relaxed and the rice in downstream paddies was severely damaged by the resultant pollution (Anonymous, 1971). Cognizant of these facts, doctors began looking for a possible connection between itai-itai disease and the mine wastewater.

Initially, Zn was suspected of being the cause of the disease, but Zn-feeding experiments with animals failed to produce symptoms of itai-itai (Yamagata and Shigematsu, 1970). Analyses of metal concentrations in the bones of itai-itai victims and in rice from the affected area revealed that Cd concentrations were about 10–100 times higher than those in control samples. Zn concentrations in rice from the affected area were little different from those in controls, but as with Cd, Zn levels in the bones of dead itai-itai victims were elevated by roughly two orders of magnitude relative to those of controls. Subsequently, it was discovered that feeding rats Cd and Zn or a combination of Cd, Zn, Cu, and Pb produced bone degeneration similar to that of itai-itai disease (Anonymous, 1971). Thus, the cause of the disease was identified as being either Cd or a combination of Cd and other metals, particularly Zn. Although Cd had been implicated by 1961, the Japanese government issued no official statement regarding the cause of itai-itai disease until 1968, when an official press release stated: "The 'Itai-itai' disease is caused by chronic cadmium poisoning, on condition of the existence of such inducing factors as pregnancy, lactation, unbalanced internal secretion, aging, deficiency of calcium, etc. It sets up kidney troubles in the first stage and osteomalacia [softening of the bones] after that" (Yamagata and Shigematsu, 1970, p. 3).

The role of Cd versus other factors in causing itai-itai disease remains somewhat controversial. Cd itself had not been associated with a disruption of Ca metabolism prior to the occurrence of itai-itai disease. However, two French reports and two British studies of persons occupationally exposed to Cd have revealed bone changes similar to those seen in itai-itai victims (Chang et al., 1980). Neither Zn nor Pb, the other metals released in large amounts by the Kamioka mine, is associated with a disruption of Ca metabolism. Studies of metal concentrations in the urine of inhabitants of the Toyama Prefecture during the early 1960s revealed higher concentrations of Cd in the urine of persons from the endemic area but no geographical differences in Pb and Zn concentrations (Kjellstrom, 1986a). Studies of autopsied itai-itai victims revealed that Pb concentrations in the victims' bones were little different from those in controls, but Cd and Zn concentrations were orders of magnitude higher than those in the bones of control corpses (Yamagata and Shigematsu, 1970).

There is rather convincing evidence that nutritional deficiency played a role in itai-itai disease. Experimental studies with animals, for example, have shown that Cd in the food does not lead to bone softening under normal dietary conditions unless the dose of Cd is very high. Cd exposure, however, when coupled with a diet low in Ca and/or vitamin D, accelerates and potentiates osteoporotic and osteomalacic changes, with a disturbance in bone salt metabolism (Chang et al., 1980).

Perhaps the most convincing link with nutrition has been the results of treatment of itai-itai disease patients. Patients with mild symptoms[25] were treated as outpatients. They were given vitamin D tablets and arrangements made for more milk, eggs, meat, and vegetables to be added to their diet, as well as more sunlight exposure. Pain decreased after about two months of treatment, and the patients were able to return to everyday life after about four months, by which time the duck-like gait had disappeared. Some impairment of mobility, however, remained for a long time (Kjellstrom, 1986a). Persons with moderate symptoms[26] and severe symptoms[27] were treated as inpatients. The inpatients received regular injections of vitamin D, ate regular hospital food, and were taken out in the sunlight as much as possible. After leaving the hospital, these persons received daily vitamin D tablets and were given vitamin D injections once or twice a month when they returned for checkups. Treatment was continued for one to three years. Pain began to decrease after about three weeks of treatment. The patients with moderate symptoms were relieved of pain within about 2 months, could stand erect after 5 months, could walk inside the hospital after 8 months, and were able to return to everyday life after 13–14 months (Kjellstrom, 1986a). The patients with severe symptoms were able to sit up after 5 months, could stand after 7 months, were able to walk with a cane after 15 months, and could walk a few meters unaided after 20 months. In some cases, pseudofractures of the bones persisted for as long as one to three years (Kjellstrom, 1986a). Treatment with vitamin D and an improved diet obviously helped to alleviate the symptoms of itai-itai disease, but recovery was much slower than would be expected if the bone softening were due purely to a nutritional deficiency, and in many cases recovery was incomplete.

Studies of Cd concentrations in the Jintsu River system provide an interesting lesson in water quality monitoring. Most of the drainage water from the Kamioka mine was found to be weakly basic (Yamagata and Shigematsu, 1970), and in such water Cd is highly insoluble. As a result, virtually all the Cd discharged from the mine was transported in particulate form. Filtered water samples from the Jintsu River were found to contain less than the U.S. drinking water standard of 5 ppb Cd. Analyses of suspended materials in the river above and below the mine, however, revealed roughly an 80-fold increase in Cd concentration (Yamagata and Shigematsu, 1970). Evidently, the Cd was transported downstream in particulate form and ultimately settled out in the quiescent waters of rice paddies in the Fuchu-machi plain just south of Toyama City. Sediments from those rice fields contained about 3 ppm Cd, versus less than 1 ppm in the sediments of control fields, and rice from the contaminated paddies averaged about 1 ppm Cd versus a mean of 0.07 Cd for Japanese rice in general (Yamagata and Shigematsu, 1970).

In 1955 the Kamioka mine built a dam to retain its wastewater in a lagoon. Since that time, the incidence of itai-itai disease has dropped sharply, and evidence of damage to crops has declined similarly (Anonymous, 1971). From the standpoint of public health, it is unfortunately not clear what levels of Cd itai-itai victims were exposed to, since the extensive analyses of water, sediment, and crops in the Fuchu-machi area were made after 1955. Based on these analyses, it has been estimated that inhabitants of the area were probably ingesting at least 600 μg of Cd per day prior to 1955 (Yamagata and Shigematsu,

[25]Duck-like gait and some limitation in movement; otherwise, there were few bone symptoms.
[26]Able to crawl but barely walk.
[27]Cannot move at all.

1970). Examination of waste sludge piles near the mine, however, has indicated that pollution was probably more extreme in years past (Emerson, 1970), and there is no doubt that pollutant discharges increased during World War II (Anonymous, 1971). Thus, exposure of the Fuchu-machi population to Cd may have been substantially greater than post-1955 analyses would indicate.

Between 1939 and 1954 approximately 200 persons were afflicted with itai-itai disease, and of those nearly 100 died (Anonymous, 1971). Remarkably, the Toyama Prefecture's Department of Health did not begin mass screening of Jintsu River basin inhabitants for itai-itai disease until 1967, and since that time an additional 132 victims have been diagnosed (Kjellstrom, 1986b). Only three cases, however, have been diagnosed since 1978, and of the 217 itai-itai patients registered by the Toyama Prefecture in 1968, only 27 were less than 60 years old. Thus, almost all itai-itai victims seem to have developed the disease prior to roughly 1955, when the Kamioka mine initiated pollution control measures. It is fortunate and of considerable interest that similar outbreaks have not occurred in other parts of the world, since the number of Zn and Cd mines is large. Some examples of what appeared to be the same disease have been reported near a Zn, Pb, and Cd mine on an island between Korea and Japan (Anomymous, 1971), and in 1969 the Japanese government took steps to halt Cd pollution of Lake Suwa due to waste discharges from electroplating factories (Yamagata and Shigematsu, 1970). No itai-itai cases, however, were associated with the Lake Suwa pollution, nor have any cases been reported in other areas of Japan where per capita ingestion of Cd is reported to be 200–240 μg d^{-1} (Yamagata and Shigematsu, 1970).

Correctives and Prospects for the Future

Although the occurrence of itai-itai disease in the Jintsu River valley remains the only large-scale example of serious damage to human health from Cd pollution, the extremely long half-life of Cd in the human body and the fact that present levels of intake are comparable to ADIs has stimulated efforts to reduce Cd emissions to the environment. Naively, it might be assumed that recycling Cd or finding substitutes for it might do much to reduce the present rate of emissions. The fallacy in this logic is the fact that mining of Cd-containing ore has almost nothing to do with the demand for Cd, but instead is controlled largely by the demand for Zn.

Nevertheless, there is cause for some optimism with respect to Cd pollution. Plating wastewater must now be treated to remove Cd and other heavy metals before the water can be discharged. Electroplating sludges are no longer being landfilled, but instead are being shipped to EPA-approved reclamation facilities for metal recovery (Plachy, 1997). Although Cd continues to be used as a colorant for plastics, many producers of plastic colorants plan to phase out Cd (and Pb) to make their products more environmentally acceptable (Plachy, 1997). A number of government actions have stimulated this attitude, including a European Economic Community ban on Cd-containing pigments. Unfortunately, substitutes for Cd pigments in some applications are as much as three to four times the price of Cd pigments for equal tint strength, and in other cases they yield inferior products. Plastic stabilizers made from Sn are the principal alternatives to Cd stabilizers, and the FDA has approved certain of these Sn stabilizers for use in food packaging. In fact, organotin compounds are the most efficient stabilizers known for PVC, but they are more costly than Cd stabilizers (OECD, 1975).

Perhaps the most significant development in recent years with respect to Cd pollution has been a serious effort to recycle Ni-Cd batteries. The European Economic Community, for example, has initiated a systematic program of secondary Cd recovery from used batteries containing more than 0.025% Cd (Llewellyn, 1990). Recycling of Cd from batteries now takes place in Sweden, France, Japan, and the Republic of Korea as well

(Llewellyn, 1991). In the United States, the Rechargeable Battery Recycling Corporation (RBBC) was established in 1995 to administer the collection and recycling of rechargeable batteries. The RBBC is a nonprofit organization established "to perform a public service through the management, collection and recycling of used nickel-cadmium (Ni-Cd) batteries throughout the United States" (Fishbein, 1996). The RBRC maintains a web site (*www.rbrc.org*) to facilitate Ni-Cd battery recycling. Small Ni-Cd battery recycling rates in the United States jumped from 4% to 15% in the first year of the RBRC program, and an estimated 85% of industrial Ni-Cd batteries are now recycled (Plachy, 1997). The RBRC is targeting a small battery recycling rate of 70% by 2001.

Some improvements in the Cd pollution picture can therefore be expected from secondary recovery and substitutions, but the fact remains that Cd ores will continue to be mined as long as there is a demand for Zn and, to a lesser extent, Pb and Cu. As noted by Nriagu (1980, p. 57), "If the cadmium were not separated from the host zinc and lead ores, it would be dispersed in the environment on a much wider scale as flue particles and other industrial waste products." A significant improvement in the global Cd pollution picture therefore requires a serious effort to reduce emissions from some of the major anthropogenic sources listed in Table 12.10.

The technology for removing Cd from industrial wastewater or from flue dust is well established. In wastewater, dissolved Cd can be precipitated with sodium sulfide, cemented by the addition of Zn, or separated out by ion exchange. If the Cd is incorporated into particulates, it can be dissolved by the addition of acid and then separated by one of the above techniques, or the solids can be settled out and the Cd removed with the sludge. Cd released to the atmosphere by way of smoke stacks is primarily in the form of very fine particulate materials, which can be separated from the stack gases by wet scrubbers, fabric filters, or electrostatic precipitators (OECD, 1975).

Removal of Cd by any of the above techniques may lead to a solids disposal problem if the Cd is not to be recovered and recycled. Discharging Cd or Cd-containing compounds at sea is prohibited (OECD, 1975) and would certainly be unwise in any aquatic system. Land disposal, however, is acceptable as long as the soil is neutral or basic so that the Cd is not mobilized by percolating groundwater. Efforts to reduce industrial emissions of Cd have been underway for some time and appear to have produced at least some positive effects. Concentrations of Cd deposited in snow and ice on Greenland, for example, declined by roughly a factor of 2.5 between 1970 and 1990 (Figure 12.6; Boutron et al., 1991). The realization that living organisms have no need for Cd, that the residence time of Cd in the human body is 16–33 years, that the accumulation of Cd in the kidneys and other organs can cause serious and sometimes permanent damage, and that present daily intake rates are comparable to ADIs indicates that further reductions in anthropogenic Cd emissions are desirable.

Lead

Production and Use

Pb is widely distributed in the rocks and soils of the Earth's crust, although the mean concentration is only 12–20 ppm (Settle and Patterson, 1980). Pb is mined primarily from deposits of the mineral galena, or lead sulfide (PbS). Since metallic Pb can be separated from PbS by heating to the low temperatures easily achieved by burning wood or charcoal,[28] it was not difficult for early civilizations to extract Pb (Figure 12.7). From written and archeological evidence, it is known that Pb was used by the Egyptians as long ago as 1500 B.C.

---

[28] The overall reaction is $PbS + O_2 \rightarrow SO_2 + Pb$.

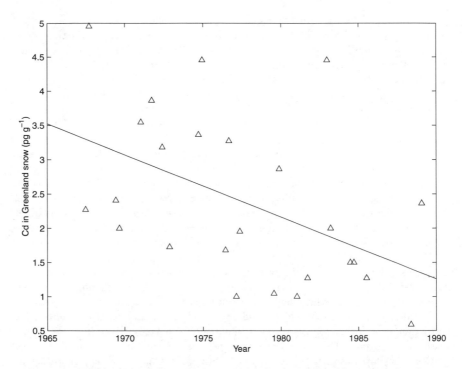

FIGURE 12.6 Decline in Cd deposition over Greenland. Reprinted by permission from *Nature* (C. F. Boutron, U. Gorbach, J. Candelone, M. A. Bolshov, and R. J. Delmas, Decrease in anthropogenic lead, cadmium and zinc in Greenland snows since the late 1960s. *Nature*, *353*, 153–156, 1991), copyright 1991 Macmillan Magazines Ltd.

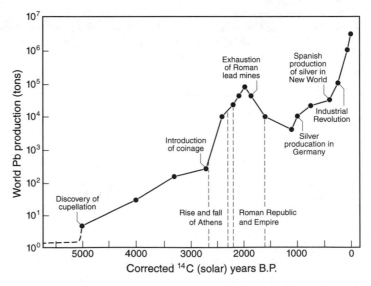

FIGURE 12.7 World Pb production during the past 5,500 years. *Source*: Redrawn from Settle and Patterson (1980). Reprinted with permission from D. M. Settle and C. C. Patterson, Lead in albacore: Guide to lead poisoning in Americans. Science, *207*, 1167–1176, 1980. Copyright 1980 American Association for the Advancement of Science.

TABLE 12.12   USE OF Pb IN THE UNITED STATES DURING 1997

Category	Tonnes
Metal products	
Batteries	1,390,000
Ammunition	55,300
Sheet lead	19,100
Casting metals	18,300
Solder	9,580
Cable covering, power, and communication	4,930
Brass and bronze billets and ingots	4,410
Bearing metals	2,490
Pipes, traps, other extruded products	1,860
Caulking lead, building construction	1,390
Other metal products	7,570
Other oxides (paints, pigments, glass and ceramic products)	67,000
Miscellaneous	8,470
Total	1,590,400

Source: Smith (1997).

The Romans used Pb extensively to line their aqueducts and water mains, and both the Greeks and Romans used Pb to line cooking vessels, since bronze pots tend to give food a bitter taste (Waldron and Stofen, 1974). At least one author (Gilfillan, 1965) has suggested that endemic Pb poisoning caused by the consumption of contaminated food and drink contributed significantly to the fall of the Roman Empire. The use of Pb or Pb salts to sweeten wine was common in Europe after the 9th century, and associated outbreaks of Pb poisoning were not infrequent. In colonial America, Pb poisoning was also a problem, mainly due to the use of Pb condensing tubes for distilling rum and to the use of earthenware with a high Pb content (Waldron and Stofen, 1974).

For obvious reasons, the principal uses of Pb today are of an industrial rather than a culinary nature. Table 12.12 summarizes the major use categories in the United States as of 1997. The major use, accounting for almost 87% of total Pb consumption in the United States, was the production of batteries. Other uses of significance from the standpoint of environmental pollution and human health were the uses of Pb in ammunition and paint pigments. The following comments provide additional information about the major use categories in Table 12.12.

1. The Pb-acid battery is used extensively as a power source for starting motors in automobiles, powerboats, lawn mowers, and a variety of other products. During 1997 about 95 million such batteries were sold in the United States. The design of the battery utilizes metallic Pb as the negative electrode and Pb dioxide (PbO_2) as the positive electrode of a voltaic cell. When the two electrodes are connected by an external electrical circuit, electrons flow from the negative to the positive electrode through the external circuit. As a result, the metallic Pb is oxidized to Pb^{2+}, while the Pb in the PbO_2 is reduced, also to Pb^{2+}. The cell can be recharged many times by applying a voltage somewhat over 2 volts across the terminals to reverse these reactions and convert Pb^{2+} back into Pb and PbO_2. The density of Pb (11.3 g cc^{-1}) and hence the weight of Pb-acid cells, are obvious drawbacks to their use, but for many purposes, no better alternative has been found. In recent years, use of Pb in batteries has been increasing at the rate of a little over 4% per year in the United States.

2. Pb has been used for many years in paints as a pigment, and in oil paints Pb naph-
    thenate is used as a drying agent. In 1918 40% of all painters were estimated to have
    symptoms of Pb poisoning because of their contact with such paint (Craig and
    Berlin, 1971). The most serious problem associated with the use of Pb in paints,
    however, has been the poisoning of small children who eat paint chips or teethe on
    windowsills or other surfaces that have been painted with Pb-based paints. Dust
    from deteriorating paint may also contaminate windowsills, carpets, furniture, and
    toys and can be ingested by children who touch the dust and subsequently put their
    fingers in their mouths. Between 1954 and 1967, a total of 2,038 children were
    treated for Pb poisoning in New York City alone, and of these, 128 died (Craig and
    Berlin, 1971). On January 1, 1973, the FDA banned the use of paint containing over
    0.5% Pb on residential surfaces accessible to children (Waldron and Stofen, 1974),
    and effective February 27, 1978, paint containing over 0.06% Pb was banned as haz-
    ardous for use in residences, schools, hospitals, parks, playgrounds, and public build-
    ings under the aegis of the Consumer Product Safety Act [*Federal Register*, 42, 44,199
    (1977)]. The only exceptions specifically written into the 1977 legislation were paints
    and coatings for motor vehicles and boats. Poisoning from Pb-based paints remains
    a serious public health problem, however, because many older buildings contain
    Pb-based paints. A 1988 Public Health Service report revealed that 52% of Ameri-
    can homes had layers of Pb-based paint on their walls and woodwork (Blackman,
    1991).
3. The use of Pb in ammunition has created some serious environmental problems
    because of the tendency of scavengers and waterfowl to inadvertently ingest the Pb.
    The decline in the population of California giant condors, for example, has been
    due in part to Pb poisoning caused by feeding on dead animals shot by hunters
    (Crawford, 1985a, 1985b). Similar poisoning of waterfowl such as ducks, geese, and
    coots is caused by the fact that the birds mistake Pb shot in the sediments of marshes
    for seeds or grit. The sediments of marshes frequented by hunters may contain as
    many as six or seven lead shots per square meter (EPA, 1972). Experiments with
    various waterfowl have shown that four to five number 4 shot are lethal to Canada
    geese and that six number 6 shot are lethal to mallard, pintail, and redhead ducks
    (EPA, 1972). During the early 1970s, the U.S. Fish and Wildlife Service (FWS) es-
    timated that about 2.4 million waterfowl out of a North American population of
    about 100 million died each year due to the ingestion of lead shot (Carter, 1977).
    As a result, the FWS initiated a program to phase out the use of Pb shot in certain
    parts of the country in 1976. The FWS action was opposed by the National Rifle
    Association (NRA) on the grounds that nontoxic steel shot was not as effective as
    Pb shot, but the NRA's protest was rejected (Carter, 1977). The result of the FWS
    action is apparent in the trend of Pb use in ammunition shown in Figure 12.8. Fed-
    eral and state action under the Migratory Bird Treaty Act of 1918 and the Endan-
    gered Species Act of 1973 ultimately resulted in restrictions on the use of Pb shot
    throughout much of the United States, and it is now estimated that about 92% of
    the total waterfowl harvest occurs in so-called nontoxic shot zones [*Federal Regis-
    ter*, 55, 33, 626–33,633 (1990)].

Although leaded gas is no longer sold in the United States, Pb was used extensively
in the United States as a gasoline additive for roughly 60 years. The use of Pb in gasoline
dates back to 1923, when tetraethyl Pb was first introduced as an antiknock additive. This
use was suspended during part of 1925 and 1926 pending the establishment of safety standards for its manufacture and handling, but after the spring of 1926, Pb additives were
commonly used in most gasoline sold in the United States and throughout the world. Beginning in 1960, organic alkyl Pb compounds were blended into gasoline along with

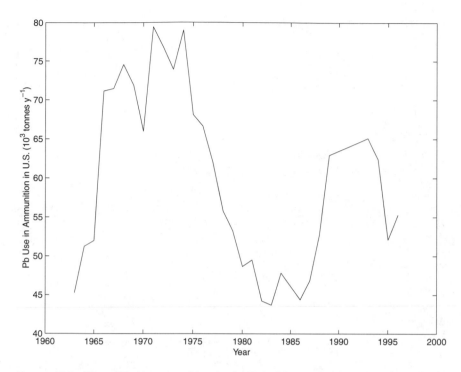

FIGURE 12.8   Use of Pb in ammunition in the United States. *Source: Minerals Yearbook,* U.S. Department of the Interior, Bureau of Mines.

tetraethyl Pb to further improve antiknock characteristics, and the additives were subsequently referred to as *Pb alkyls* rather than just tetraethyl Pb (Waldron and Stofen, 1974). As of 1973, the Pb content of gasoline sold in the United States averaged about 0.58 g L^{-1}.

Because of concern over pollution of the environment with Pb, in 1974 the EPA requested a phasedown in the use of Pb alkyls in gasoline to 0.45 g L^{-1} on January 1, 1975, and to 0.13 g L^{-1} as of January 1, 1979. In December 1974, however, the U.S. Circuit Court of Appeals in Washington ruled against the EPA, stating that evidence did not support the EPA's contention that automobile emissions contributed significantly to blood Pb levels, and that the EPA regulation was arbitrary and capricious.

Since that time, the legal status of gasoline lead additives has changed significantly. This change has resulted partly from the requirement for exhaust emission control devices on automobiles. These devices function poorly, if at all, when Pb is present in the exhaust fumes, since the Pb poisons the catalyst, which is designed to help oxidize unburned hydrocarbons. Furthermore, there is increasing awareness of the danger of environmental Pb pollution due to the use of leaded gasoline. Since July 1, 1977, all new cars sold in the United States have been required to run on unleaded gas.[29] Finally, the government required that the Pb content of leaded gas be reduced to 0.13 g L^{-1} by July 1, 1985, and to 0.10 g L^{-1} by January 1, 1986. The result of these regulations was a 10-fold decline in the use of Pb in gasoline in the United States between the early 1970s and 1986 (Figure 12.9). Leaded gasoline is now completely phased out in the United States. European countries began slowly phasing out leaded gasoline in the 1980s and mandated elimination in the 1990s. At the present time, almost 80% of gasoline sold worldwide is unleaded. This percentage is expected to increase to 84% by 2005 (Walsh, 1999).

[29]Defined as gasoline containing less than 0.013 g L^{-1} Pb.

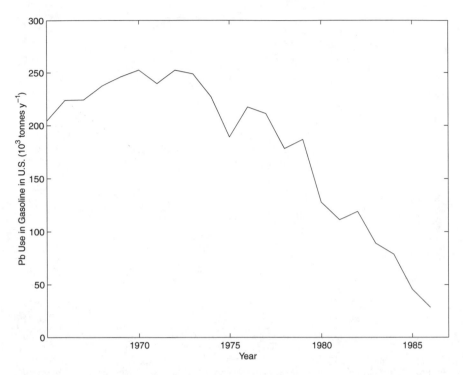

FIGURE 12.9   Use of Pb in gasoline in the United States. *Source*: *Minerals Yearbook*, U.S. Department of the Interior, Bureau of Mines.

As might be expected from this discussion, U.S. consumption of Pb declined during the 1970s (Figure 12.10). The decline was due primarily to the reduction in the use of Pb alkyls in gasoline, but it also reflected restrictions on the use of Pb shot and Pb-based paint. During the 1980s, however, U.S. consumption began to creep up as the demand for Pb-acid batteries continued to increase.

World production of Pb from mines was remarkably constant from 1970 to 1990, averaging 3.41 ± 0.06 million tonnes y^{-1} during that time (Figure 12.10). Since 1992, however, production has averaged 2.9 million tonnes y^{-1}, a decrease of about 15%. Changes in mine production reflect changes both in the demand for Pb and in the extent of Pb recycling. In the United States, for example, secondary Pb recovery as a percentage of consumption increased from $44 \pm 2\%$ between 1965 and 1977 to $56 \pm 5\%$ from 1978 to 1989 and to $69 \pm 3\%$ since 1990. This pattern in recycling has obviously tended to reduce the demand for mined Pb. The net result of all the changes in demand and recycling has been a somewhat lower rate of Pb production now than was the case 30 years ago. While there is reason to be pleased over the fact that world Pb production has declined somewhat in recent years, there is no question that the mining and use of such huge quantities of Pb has resulted in environmental pollution on a global scale. We now turn our attention to this issue.

Emissions

Several lines of evidence indicate that significant contamination of the environment with Pb has occurred on a global scale. Analyses of marine sediments, for example, indicate that

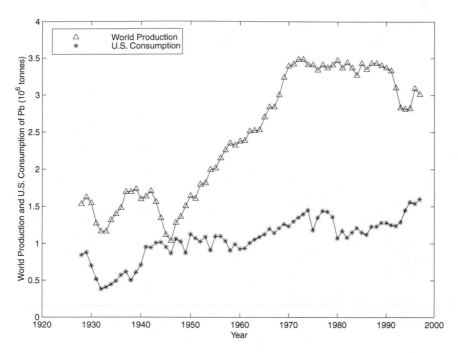

FIGURE 12.10   World production and U.S. consumption of Pb. *Source: Minerals Year-book*, U.S. Department of the Interior, Bureau of Mines.

millions of years ago, the flux of Pb to the oceans was about 13,000 tonnes y^{-1}. Lead inputs to the ocean are currently estimated to be on the order of 100,000 tonnes y^{-1} (Nriagu and Pacyna, 1988). Analyses of Pb concentrations in seawater have shown that surface waters (0–500 m) are significantly enriched with Pb relative to deep water, the cause of the enrichment presumably being atmospheric deposition (Figure 12.11; Flegal and Patterson, 1983). The lower concentrations in deep water reflect the fact that the residence time of bottom waters is approximately 500–1,000 years. Had Pb fluxes to the ocean been constant for thousands of years, surface water concentrations would be no greater than deep water concentrations. Due to industrial emissions of Pb to the atmosphere, surface water Pb concentrations in the North Pacific and North Atlantic around 1980 were conservatively estimated to be 8–20 times greater and those in the South Pacific 2 times greater than natural concentrations (Flegal and Patterson, 1983). Concentrations near Bermuda dropped by about a factor of 3 following the phaseout of leaded gasoline (Figure 12.11).

Pb concentrations in Greenland ice sheets sampled by Murozumi et al. (1969) and Boutron et al. (1991) indicate that the Pb concentration of precipitation over the area increased by several orders of magnitude between roughly 1,000 B.C. and 1950 (Figure 12.12). Much of the increase occurred after 1750 and accelerated greatly after about 1940. The increase between 1750 and 1940 presumably reflects the effect of the Industrial Revolution and associated emissions of Pb to the atmosphere, largely from lead smelters, whereas the more rapid increase after 1940 reflects the introduction of Pb alkyls from vehicle exhaust emissions. Before the phaseout of leaded gasoline began, such emissions amounted to roughly 0.28 million tonnes y^{-1}, or a little over 8% of world Pb production (Settle and Patterson, 1980). The roughly 7.5-fold decline in the Pb concentration of precipitation over Greenland from 1970 to 1990 presumably reflects the restrictions imposed on lead in gasoline (Boutron et al., 1991).

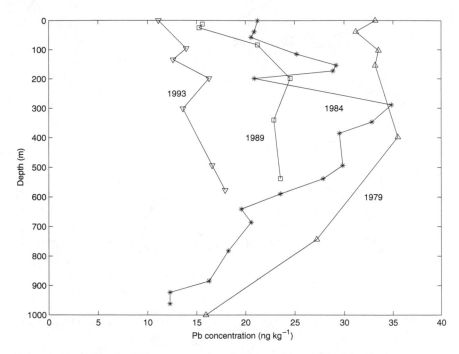

FIGURE 12.11   Dissolved Pb concentrations in the western North Atlantic near Bermuda. Reprinted from *Geochim. Cosmochim. Acta*, **61**, Wu, J., and E. A. Boyle, Lead in the western North Atlantic Ocean: Completed response to leaded gasoline phaseout, pp. 3279–3283, Copyright 1997, with permission from Elsevier Science.

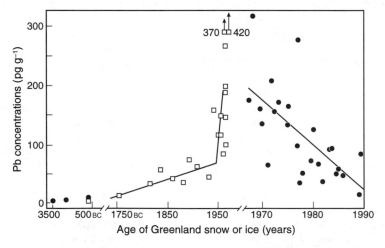

FIGURE 12.12   Pb concentrations in Greenland ice sheet. Reprinted by permission from *Nature* (C. F. Boutron, U. Gorbach, J. Candelone, M. A. Bolshov, and R. J. Delmas, Decrease in anthropogenic lead, cadmium and zinc in Greenland snows since the late 1960s. *Nature*, *353*, 153–156, 1991), copyright 1991 Macmillan Magazines Ltd.

TABLE 12.13   GLOBAL Pb EMISSIONS TO THE ATMOSPHERE
FROM NATURAL AND ANTHROPOGENIC SOURCES

Source of Hg	Estimated Annual Flux (10^3 tonnes)
Natural	
Wind-borne soil particles	3.9
Volcanoes	3.3
Wild forest fires	1.9
Seasalt spray	1.4
Biogenic	
Continental particulates	1.3
Marine	0.2
Continental volatiles	0.2
Total natural sources	12.2
Anthropogenic	
Smelters	57.5
Combustion of leaded gasoline	56.0
Coal combustion	8.2
Cement production	7.1
Oil combustion, stationary sources	2.4
Solid waste incineration	2.4
Wood burning	2.1
Phosphate fertilizers	0.2
Miscellaneous	4.5
Total anthropogenic sources	140.4

Source: Modified from Nriagu and Pacyna (1988) and Nriagu (1989).

Table 12.13 provides a revealing summary of estimated global emissions of Pb to the atmosphere from natural and anthropogenic sources. The major natural sources of emissions are windblown and volcanic dust, with Pb concentrations on the order of 10 ppm. Around 1970, by far the most important anthropogenic source was the combustion of leaded gasoline (Nriagu and Pacyna, 1988), but at the present time, only about 20% of gasoline worldwide is leaded (Walsh, 1999). As a result, Pb smelters and the combustion of leaded gasoline are of roughly equal importance, and together account for over 80% of anthropogenic Pb emissions to the atmosphere. It is thought-provoking to realize that even if the Pb content of all gasoline were reduced to the unleaded limit of 0.013 g L^{-1}, emissions of Pb to the atmosphere from the burning of gasoline would still amount to roughly 6,000 tonnes y^{-1}, about half of the natural background rate. Under these conditions, the principal anthropogenic emissions would come from ore smelting. Reducing the Pb content of gasoline has obviously done much to reduce anthropogenic emissions of Pb, but ore smelting alone emits almost five times as much Pb to the atmosphere as do natural sources. To what extent have these emissions affected human exposure to Pb?

Human Exposure. Table 12.14 summarizes estimates of human intake of Pb during prehistoric and modern times. In both cases, the most important source of exposure was the consumption of food, but present rates of absorption are believed to be at least 150 times greater than during prehistoric times. A significant cause of contamination is the fallout of Pb from the atmosphere, and in modern times the chief sources of Pb in the atmosphere have been the burning of leaded gasoline and ore smelting. Since the residence time of Pb in the atmosphere is on the order of a few days to as much as a month (Waldron and Stofen, 1974), the impact of localized emission sources can obviously be spread by winds over a wide area, including farmlands distant from urban and industrial centers.

TABLE 12.14   ESTIMATED DAILY AMOUNTS OF Pb ABSORBED
INTO THE BLOOD OF ADULT HUMANS IN PREHISTORIC
AND MODERN TIMES

Source	Intake	% Pb Absorbed	Prehistoric (ng)	Contemporary Urban (ng)
Air	20 m³	40%	0.3	6,400
Water	2 kg	10%	<3	2,250
Food	1.3 kg	7%	<182	18,200
Total			<185	26,850

Source: Modified from calculations in Settle and Patterson (1980).

Significant additional contamination of food with Pb may occur during processing, particularly if the food is dried and ground. Settle and Patterson (1980), for example, note that the Pb content of albacore tuna meat can increase by more than a factor of 50 during commercial drying and pulverizing. Other sources of Pb contamination of foodstuffs include the use of Pb arsenate as an insecticide and leaded ceramics and glazes in kitchenware. In areas such as Europe and the United States, a significant additional source of Pb contamination in the diet has been the use of Pb-soldered cans for marketing food (Shea, 1973). Canned foods account for 10–15% of the food consumed by Americans, and in 1979 the FDA estimated that about 20% of the Pb in the average daily diet of persons in the United States more than one year old came from canned food (FDA, 1979). About two-thirds of this Pb came from the solder. The implication was that Pb solder accounted for about 13% of the Pb ingested by Americans.

Pb is found in drinking water due both to its presence in raw water supplies and to the corrosion of plumbing materials in water distribution systems. The latter is usually the more important source. A survey by the EPA (1991), for example, revealed that less than 1% of public water systems in the United States had water entering the distribution system with Pb levels greater than 5 μg L^{-1}. These systems serve less than 3% of Americans who receive their drinking water from public water systems.

The principal sources of Pb in the distribution system are Pb goosenecks or pigtails, Pb service lines and interior household pipes, Pb solders and fluxes used to connect Cu pipes, and alloys containing Pb, including some faucets made of brass or bronze. Pb solder and fluxes containing as much as 50% Pb, for example, were commonly used to connect Cu pipes throughout the United States prior to the 1986 amendments to the Safe Drinking Water Act (SDWA), which limited the lead content of solder and fluxes to 0.2%. EPA (1991) estimates indicate that there are about 10 million Pb service lines/connections in the United States and that about 20% of all public water systems have some Pb service lines/connections within their distribution.[30] Even if the distribution pipes contain little or no Pb, brass and bronze in faucets and fixtures commonly contain Pb and may be a major source of Pb in drinking water that stands in faucets or fixtures (EPA, 1991). All water is corrosive to metal plumbing materials to some extent, and based on the foregoing discussion, it is not surprising to learn that Pb concentrations of tap water tend to exceed the Pb concentrations of water entering distribution systems. For example, a random survey of 782 samples of tap water in the United States, including 58 cities in 47 states, revealed average Pb levels of 13 μg L^{-1}, with 10% of the values exceeding 33 μg L^{-1} (EPA, 1991). Pb is dissolved most readily by acidic water, and in communities where tap water

---

[30]The 1986 amendments to the SDWA limited the Pb content of new pipes, fittings, and patches in municipal water systems to less than 8%.

is soft and somewhat acidic, appreciable concentrations of Pb may leach into the water, particularly if the water is left standing overnight. Tap water samples from homes served by Pb water pipes in Glasgow, Scotland, for example, where the water is extremely soft, revealed mean Pb concentrations of 350 μg L^{-1} (Waldron and Stofen, 1974), over 25 times the mean Pb concentration of tap water in the United States.

That the air people breathe is contaminated with Pb should come as no surprise by this time. Exposure to Pb through inhalation, however, depends very much on where one lives and whether or not one smokes cigarettes. The Pb concentration of rural air, for example, is one to two orders of magnitude lower than the Pb concentration of typical urban air (Settle and Patterson, 1980), and breathing rural air is therefore a very minor, if not negligible, cause of Pb exposure compared to other sources at the present time. Cigarettes contain significant amounts of Pb due to the use of Pb arsenate as an insecticide on tobacco, and smoking a single cigarette results in the absorption of about 50–250 ng of lead (Moore, 1986). Heavy smokers may therefore absorb considerable amounts of Pb due to their inhalation of cigarette smoke.

Absorption of inhaled Pb is about 40% efficient. Adults absorb about 10% of the Pb from the water they drink and about 7% from their food (Settle and Patterson, 1980). For children, gastrointestinal absorption of Pb is more efficient, typically about 50% (EPA, 1980d). Once absorbed into the bloodstream, Pb is transported to all parts of the body, primarily by red blood cells, although its incorporation into tissues apparently occurs through exchange with the blood plasma (Waldron and Stofen, 1974). Pb begins to appear in the liver and kidneys within a few hours of absorption, but ultimately 72–95% of the inorganic Pb in the body is deposited in the bones, where it replaces Ca (EPA, 1980d). Once incorporated into the bone structure, Pb is released back to the bloodstream at a steady but very slow rate of about 0.08–0.1% per day (Waldron and Stofen, 1974).[31] In contrast to inorganic Pb, organic Pb (e.g., Pb alkyls) shows no special affinity for the bones but instead tends to concentrate in lipid tissues, including those of the central nervous system. The highest concentrations of organic Pb are often found in the brain and liver (Waldron and Stofen, 1974).

Excretion of Pb occurs primarily in the feces, although some Pb is also excreted in urine and even in sweat. Absorbed Pb is transferred to the intestines, primarily in the bile fluid, although a small amount of release also seems to occur directly through the intestinal wall (Waldron and Stofen, 1974). Because of the very slow release rate of Pb from the bones, and because of the continual exposure of humans to Pb, the excretion rate of Pb is generally exceeded by the rate of Pb absorption throughout life by many persons. Hence, the total body burden of Pb tends to increase steadily with age (Figure 12.13). It is apparent from Figure 12.13 that the body burden of Pb during the first few years of life is minuscule compared to the body burden in later years, and since humans are not known to require Pb in even very small amounts, the accumulation shown in Figure 12.13 can hardly be considered beneficial.

Toxicology

Pb is a general metabolic poison. Its pathological effects are varied, but in general, they reflect the tendency of Pb to interact with proteins and hence to damage tissue and interfere with the proper functioning of enzymes (Waldron and Stofen, 1974). Pb is known to inhibit active transport mechanisms involving ATP, to depress the activity of the enzyme cholinesterase, to suppress cellular oxidation-reduction reactions, and to inhibit protein synthesis (Waldron and Stofen, 1974). Pb poisoning is also associated with the following problems.

---

[31]In round numbers, excretion rates of Hg, Pb, and Cd from the human body average 1%, 0.1%, and 0.01% per day, respectively.

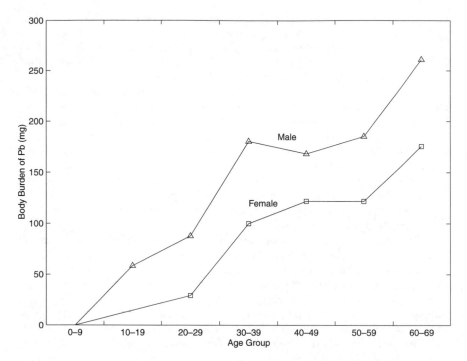

FIGURE 12.13   Total body burden of Pb in humans as a function of age. *Source*: Waldron and Stofen (1974).

Anemia. Pb is known to disrupt several enzymes involved in the production of heme, a constituent of hemoglobin and various other respiratory pigments, and has been shown to interfere with the uptake of Fe by red blood cells. Anemia associated with Pb poisoning is undoubtedly caused in part by these effects. There is also evidence, however, that Pb shortens the life of red blood cells, probably due to disruption of the red cell membrane (Waldron and Stofen, 1974).

Damage to the Central Nervous System. Damage to both the peripheral and central nervous systems, including the brain, may be caused by degeneration of nerve fibers and interference with the permeability of capillaries in the brain due to Pb intoxication. The exact mechanism responsible for these effects is not known. Several factors, including enzyme inhibition and tissue damage, may be involved (Waldron and Stofen, 1974).

Kidney Damage. Kidney damage characterized by atrophy of the renal tubules is a well-established effect of Pb poisoning. The damage is associated with elevated levels of amino acids, sugar, and phosphate in the urine (Waldron and Stofen, 1974). Gout caused by kidney damage has also been associated with Pb poisoning, the consumption of illicitly distilled whiskey (moonshine) being a common cause. Concentrations of Pb in moonshine often exceed 10 ppm due to the use of Pb in the distillation apparatus (EPA, 1980d). Waldron and Stofen (1974) cite one example of a hospital in which 37 of 43 cases of gout in 1967 were caused by the consumption of Pb-contaminated liquor.

Effects on Children and Reproduction. From the standpoint of human health, the greatest danger of Pb poisoning is undoubtedly to young children, particularly those living in urban areas. Neurological damage caused by the poisoning of such children may be permanent and can result in impaired physical as well as mental development. Even the fetus is not protected from the effects of Pb poisoning. There is evidence that the brain of the fetus is much more sensitive to Pb poisoning than the brain of the infant or young

TABLE 12.15　LOWEST BLOOD Pb LEVELS ASSOCIATED WITH OBSERVED BIOLOGICAL EFFECTS IN HUMANS

Lowest Observed Effect Level (ppb Pb in blood)	Effect	Population Group
100	ALA-d inhibition	Children and adults
	Adverse effects on intelligence, hearing, and growth	Children
100–150	Interference with vitamin D metabolism	Children
150–200	Protoporphyrin elevation in red blood cells	Women and children
250–300	Protoporphyrin elevation in red blood cells	Men
400	Increased urinary ALA excretion	Children and adults
400	Anemia	Children
400	Coproporphyrin elevation	Adults and children
500	Anemia	Adults
500–600	Cognitive deficits	Children
500–600	Peripheral neuropathies	Adults and children
800–1,000	Encephalopathic symptoms	Children
1,000–1,200	Encephalopathic symptoms	Adults

Source: EPA (1980d), CDC (1991).

child, and Pb has repeatedly been shown to cause birth defects in experimental animals (EPA, 1980d). Although there is little information to indicate that Pb has a teratogenic effect on humans, exposure of pregnant women to Pb can induce miscarriages and stillbirths (Waldron and Stofen, 1974).

Although most of the body burden of Pb is found in the bones, historically it has been much more practical to relate health effects on living persons to the level of Pb in the blood, PbB.[32] Table 12.15 summarizes the lowest PbB levels associated with various health effects, as summarized by the EPA (1980d) and CDC (1991). PbB levels as low as 100 ppb do not cause distinctive symptoms, but they are associated with decreased intelligence and impaired neurobehavorial development, decreased stature or growth, decreased hearing acuity, and decreased ability to maintain a steady posture. The activity of δ-aminolaevulic acid dehydrase (ALA-d), an enzyme involved in heme metabolism, is also adversely affected at a PbB level of 100 ppb. Interference with vitamin D metabolism occurs at PbB levels of 100–150 ppb, and increased concentrations of protoporphyrin, a substrate needed for heme synthesis, and of ALA begin to appear in red blood cells and urine, respectively, at slightly higher PbB levels. Actual decreases in hemoglobin, somewhat loosely referred to in Table 12.15 as *anemia*, occur at PbB levels of about 400–500 ppb. Based on this evidence, the Centers for Disease Control have concluded that children's PbB levels should not exceed 100 ppb (CDC, 1991).

Unfortunately, a great many children as well as adults in the United States have PbB levels greater than 100 ppb. For example, a study of urban children in seven U.S. cities between 1967 and 1970 revealed that about 29% had blood Pb levels over 400 ppb, and almost 9% had blood Pb levels over 500 ppb (Waldron and Stofen, 1974). A blood Pb level of 500–800 ppb is considered to be the threshold level of classical lead poisoning (Patterson, 1965). At the present time, one in every six American children under six years of age is estimated to have Pb poisoning (Blackman, 1991).

[32]The Pb content of bones can now be assayed using noninvasive x-ray fluorescence techniques (Blackman, 1991).

In 1975 the EPA promulgated a maximum contaminant level (MCL) of 50 ppb for Pb in drinking water as an interim drinking water standard. In 1988, however, the EPA proposed to reduce the MCL to 5 ppb and to set a maximum contaminant level goal (MCLG) of zero for Pb in drinking water (EPA, 1991). The MCLG of zero was promulgated in 1991 (Table 12.1). According to the SDWA, an MCL must be set as close to the MCLG as is feasible. The rationale for setting the MCLG at zero was in part the feeling that "Drinking water should contribute minimal lead to total lead exposures because a substantial portion of the sensitive population already exceeds acceptable blood lead levels" (EPA, 1991, p. 26,467). The EPA also considers Pb to be a probable human carcinogen, but the evidence, based largely on the development of kidney tumors in rodents exposed to levels of Pb far in excess of tolerable human doses, is of dubious relevance to humans for reasons discussed in Chapter 8. At the present time, the EPA considers that the action level of Pb in drinking water has been exceeded if more than 10% of targeted tap samples contain more than 15 ppb Pb (EPA, 1991).

Effects on Aquatic Organisms. The EPA water quality criteria for Pb for the protection of aquatic organisms are summarized in Table 12.16. As is the case with Cd, the toxicity of lead in freshwater is negatively correlated with water hardness, and for three of the four species for which data were available to the EPA (1985d), the 96-hour TLm values were almost linearly related to water hardness. Final acute values for Pb in freshwater and saltwater were calculated by the procedures outlined in Chapter 8. In the former case, the criteria are expressed in the form of an equation relating the criterion to water hardness. Final chronic values were calculated by dividing the final acute value by a mean acute:chronic ratio of 51, which was determined from studies on four different species. The final plant values were several orders of magnitude higher than the final chronic values and therefore did not enter into the calculation of the criterion continuous concentration. The criteria were updated in 1999 (EPA, 1999).

Commentary

In the case of Pb pollution, there are no wide-scale disasters comparable to Minamata disease or itai-itai disease. On the other hand, there is no doubt that many persons are today carrying blood Pb levels that are within a factor of 2 or 3 of the levels associated with classical Pb poisoning. About 15% of Americans, for example, have PbB levels above 200 ppb (Tackett, 1987). For comparison, Patterson (1965) has estimated that natural PbB levels in primitive humans were roughly 2.5 ppb. If an uncertainty factor of 10 (Chapter 8) is applied to the lowest observed effect levels reported in Table 12.15, PbB levels should not exceed 10 ppb.

TABLE 12.16 EPA WATER QUALITY CRITERIA FOR DISSOLVED Pb FOR THE PROTECTION OF AQUATIC ORGANISMS IN FRESHWATER AND SALTWATER

Water Type	Criterion Continuous Concentration ($\mu g\ L^{-1}$)	Criterion Maximum Concentration ($\mu g\ L^{-1}$)
Marine	8.1	210
Freshwater		
50 mg L^{-1} CaCO$_3$	2.0	30
100 mg L^{-1} CaCO$_3$	2.5	65
200 mg L^{-1} CaCO$_3$	2.8	136

Source: EPA (1999).

There are, to be sure, some encouraging aspects of the Pb pollution situation. The phasing out of leaded gasoline is undoubtedly the most important one because, during the peak of leaded gasoline use, emissions of Pb alkyls accounted for about 70% of the anthropogenic Pb input to the atmosphere (Figure 12.14). Fallout of Pb onto crops and onto the surface waters of the ocean has declined in recent years as leaded gas has been phased out (Figures 12.11 and 12.12). The reduction in fallout onto crops and the decreased use of Pb solder in cans have translated into lower Pb levels in human diets (CDC, 1991).

Several other developments can be expected to reduce the level of human exposure to Pb. One of the most important ones in the United States has been the sweeping restrictions passed by the Consumer Product Safety Commission in 1977 on the use of Pb in paint. This legislation has been good news for infants and young children. The bad news is that there are many old apartments and homes painted with Pb-based paint. These buildings will continue to be a hazard to youngsters for many years unless the paint is removed. The Department of Health and Human Services (HHS) has called for the implementation of a four-point plan, including the elimination of leaded paint and contaminated dust in housing, to reduce the level of Pb exposure to children. The problem is money. The HHS plan would cost about $10 billion over a period of 10 years (Blackman, 1991). The Office of Management and Budget raised allotments for Pb-screening programs from $4 million in 1990 to a proposed $41 million in 1992 but has balked at further expenditures (Blackman, 1991). However, in 1997, the Department of Housing and Urban Development announced the availability of almost $46 million for state and local governments to undertake Pb-based paint hazard control programs in eligible privately owned housing units (Smith, 1997).

Restrictions on the use of Pb solder and pipes in water distribution systems mandated by the 1986 amendments to the SDWA can be expected to reduce the contribution to Pb contamination from distribution networks. For example, in 1995 seven manufacturers of water faucets agreed to phase out the use of Pb in their products over the next

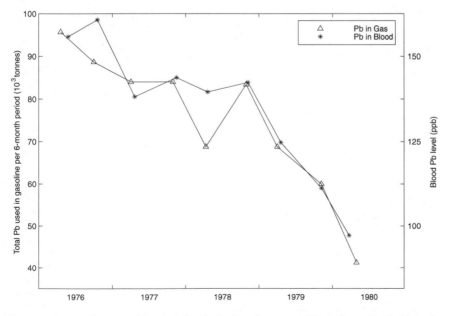

FIGURE 12.14   Change in blood Pb levels during phaseout of leaded gasoline in United States. *Source*: Annest (1983). Reproduced with permission.

four years as part of an agreement settling a lawsuit filed by the Natural Resources Defense Council and the California attorney general that accused more than a dozen principal manufacturers of water faucets of making products that leached Pb into drinking water (*Honolulu Advertiser*, 1995). Although the FDA has not officially set a Pb standard for food, it has made enough noise to cause food canners to take steps to reduce contamination problems from Pb solder. The average Pb level in evaporated milk, for example, dropped from 0.52 ppm to 0.10 ppm between 1972 and 1979, and the Pb level in canned juices similarly dropped from 0.30 ppm in 1973 to 0.05 ppm in 1979 (FDA, 1979). A major concern in this regard has been the Pb content of infant foods. The average Pb level in canned infant formula concentrate declined from 0.10 ppm to 0.06 ppm during the 1970s, and by 1979 all infant juice manufacturers had voluntarily switched from tin cans to glass jars (FDA, 1979). There are, incidentally, ways to make cans that do not involve the use of Pb solder. Sn solder can be used instead of Pb, but Sn solder is more expensive. The cans can be welded, but welding is slower than soldering. A third alternative is a two-piece can constructed of Sn plate or Al. The can is formed by forcing a "cup" of metal through a die or "redrawing" it (Moore, 1986). According to the FDA (1991), there is a definite movement away from Pb-soldered cans by the food canning industry. This is good news for people who do not like to eat Pb.

Despite the fact that anthropogenic emissions of Pb enriched the surface waters of the ocean with Pb by an order of magnitude in the Northern Hemisphere, the concentrations were always well below levels likely to exert a toxic stress. Surface water concentrations reported by Flegal and Patterson (1983), for example, were 0.010–0.035 μg L^{-1}, three orders of magnitude smaller than the EPA's criterion continuous concentration. The best-documented effect of Pb pollution in aquatic systems has been the poisoning of waterfowl

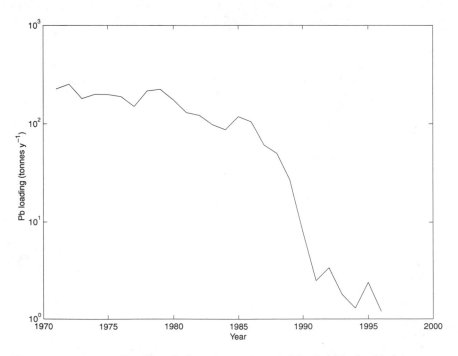

FIGURE 12.15   Annual loading of Pb to the southern California bight by the four municipal sewer outfalls that account for 89% of the point source discharges to the bight. *Source*: SCCWRP (1999).

that ingest Pb shot in marshes. Additional inputs of Pb to marshes in the United States have been largely eliminated by regulations requiring the use of nontoxic steel shot in most marsh areas, but there are still a lot of Pb shot out there from previous years' hunting.

Surface water concentrations of Pb in the open ocean have declined as leaded gasoline has been phased out, and there is evidence that changes in Pb use and emissions may be reducing Pb pollution in coastal areas in a similar way. Recently deposited sediments of the San Pedro, Santa Monica, and Santa Barbara basins off the coast of California, for example, were found to contain about 30 ppm Pb by Chow et al. (1973). Sediments deposited in the same areas prior to 1900 contained only 10 ppm Pb. The increase could be accounted for rather well by estimates of atmospheric fallout due to the burning of Pb alkyls in gasoline (Chow et al., 1973). However, Chow et al. (1973) found recently deposited sediments near White Point, California, to contain about 400 ppm Pb, a figure that could not be accounted for by atmospheric fallout. They concluded that the high Pb concentrations in the White Point sediments reflected the discharge of Pb-containing industrial and domestic wastes from the White Point sewer outfall.

Since 1971, there has been more than a 99% reduction in the reported loading of Pb to the southern California bight from sewer outfalls (Figure 12.15). As noted by Raco-Rands (1997), part of this apparent reduction is illusory. During this period, graphite furnace atomic absorption spectrophotometry (AAS) replaced flame AAS for Pb analysis. The latter method tended to overestimate Pb concentrations due to matrix interferences. However, part of the decline is undoubtedly real. Between 1987 and 1988, for example, the Hyperion wastewater treatment plant terminated discharges of sludge at its so-called 7-mile outfall. To the extent that the trends apparent in Figure 12.15 are real, they are good news for marine ecosystems in the vicinity of sewer outfalls.

REFERENCES

AARONSON, T. 1971. Mercury in the environment. *Environment*, 13(4), 16–23.

ANDREN, A. W. 1974. Ph.D. dissertation. Florida State Univeristy, Oceanography Department, Tallahassee.

ANDREN, A. W. and R. C. HARRISS. 1973. Methylmercury in estuarine sediments. *Nature*, 245, 256–257.

ANNEST, J. L. 1983. Trends in the blood lead levels of the U.S. population. In M. Rutter and R. R. Jones (Eds.), *Lead versus Health*. Chichester and New York. John Wiley and Sons. Pp. 33–58.

ANONYMOUS. 1971. Cadmium pollution and Itai-itai disease. *Lancet*, 1, 382–383.

ANONYMOUS. 1991. The mercury in your mouth. *Consumer Reports*, 56, 316–319.

BERNER, E. K., and R. A. BERNER. 1987. *The Global Water Cycle*. Simon & Schuster, Englewood Cliffs, NJ. 397 pp.

BLACKMAN, A. 1991. Controlling a childhood menace. *Time*, 137(8), 68–69.

BOUTRON, C. F., U. GORFACH, J. CANDELONE, M. A. BOLSHOV, and R. J. DELMAS. 1991. Decrease in anthropogenic lead, cadmium and zinc in Greenland snows since the late 1960s. *Nature*, 353, 153–156.

CARTER, L. J. 1977. Chemical plants leave unexpected legacy for two Virginia rivers. *Science*, 198, 1015–1020.

CHANG, L. W., K. R. REUHL, and P. R. WADE. 1980. Pathological effects of cadmium poisoning. In J. O. Nriagu (Ed.), *Cadmium in the Environment, Part II. Health Effects*. John Wiley & Sons, New York. Pp. 783–839.

CDC. 1991. *Preventing Lead Poisoning in Young Children*. Centers for Disease Control, U.S. Department of Health and Human Services, Atlanta. 105 pp.

CHOW, T. J., K. W. BRULAND, K. BERTINE, A. SOUTAR, M. KOIDE, and E. D. GOLDBERG. 1973. Lead pollution: Records in Southern California coastal sediments. *Science*, 181, 551–552.

CITY AND COUNTY OF HONOLULU. 1971. Department of Public Works. *Water Quality Program for Oahu with Special Emphasis on Waste Disposal—Final Report Work Areas 6 and 7, Analysis of Water Quality Oceanographic Studies, Parts I and II*.

CORDLE, F., P. CORNELIUSSEN, C. JELINEK, B. HACKLEY, R. LEHMAN, J. MCLAUGHLIN, R. RHODEN, and R. SHAPIRO. 1978. Human exposure to polychlorinated biphenyls and polybrominated biphenyls. *Environ. Health Perspect.*, **24**, 157–172.

CRAIG, P. P., and E. BERLIN. 1971. The air of poverty. *Environment*, **13**(5), 2–9.

CRAWFORD, M. 1985a. The last days of the wild condor? *Science*, **229**, 844–845.

CRAWFORD, M. 1985b. Condor agreement reached. *Science*, **229**, 1248.

DICKSON, D. 1985. Europe to start removing lead from gas in 1989. *Science*, **228**, 37.

DLNR. 1986. *Hawaii Fisheries Plan 1985.* Department of Land and Natural Resources, Division of Aquatic Resources, State of Hawaii, Honolulu. 163 pp.

EFFLER, S. W., and R. D. HENNIGAN. 1996. Onondaga Lake, New York: Legacy of pollution. *Lake and Reserv. Manage.*, **12**, 1–13.

EMMERSON, B. T. 1970. "Ouch-ouch" disease: The osteomalacia of cadmium nephropathy. *Ann. Intern. Med.*, **73**(5), 854–855.

EMOND, S. 1990. Still a hazard after all these years. *Harvard Health Letter*, **16**(1), 7–8.

EPA. 1972. *Water Quality Criteria.* EPA-R3-73-033. Environmental Protection Agency, Washington, DC. 594 pp.

EPA. 1980a. *Ambient Water Quality Criteria for Mercury.* EPA 440/5-80-058. Environmental Protection Agency, Washington, DC. 217 pp.

EPA. 1980b. *Seafood Consumption Data Analysis.* Final Report, Task II. Contract No. 68-01-3887 to U.S. Environmental Protection Agency. Stanford Research Institute International, Menlo Park, CA.

EPA. 1980c. *Ambient Water Quality Criteria for Cadmium.* EPA 440/5-80-025. Environmental Protection Agency, Washington, DC. 183 pp.

EPA. 1980d. *Ambient Water Quality Criteria for Lead.* EPA 440/5-80-057. Environmental Protection Agency, Washington, DC. 151 pp.

EPA. 1985a. *Ambient Water Quality Criteria for Mercury—1984.* EPA 440/5-84-026. Environmental Protection Agency, Washington, DC. 136 pp.

EPA. 1985b. *Guidelines for Deriving Numerical National Water Quality Criteria for the Protection of Aquatic Organisms and Their Uses.* PB85-227049. Environmental Protection Agency, Duluth, MN. 97 pp.

EPA. 1985c. *Ambient Water Quality Criteria for Cadmium—1984.* EPA 440/5-84-032. Environmental Protection Agency, Washington, DC. 127 pp.

EPA. 1985d. *Ambient Water Quality Criteria for Lead—1984.* EPA 440/5-84-027. Environmental Protection Agency, Washington, DC. 81 pp.

EPA. 1986. *Quality Criteria for Water.* EPA 440/5-86-001. Environmental Protection Agency, Washington, DC.

EPA. 1987. *Update #2 to Quality Criteria for Water 1986.* May 1, 1987. Environmental Protection Agency Office of Water Regulations and Standards. Criteria and Standards Division, Washington, DC.

EPA. 1991. Maximum contaminant level goals and national primary drinking water regulations for lead and copper; final rule. *Federal Register,* **56**(110), 26460–26564.

EPA. 1995. *Guidance for Assessing Chemical Contaminant Data for Use in Fish Advisories.* Vol. 1, *Fish Sampling and Analysis,* 2nd ed. EPA 823-R-95-007. Environmental Protection Agency, Washington, DC. 472 pp.

EPA. 1996a. *Mercury Study Report to Congress.* Vol. VI, *Characterization of Human Health and Wildlife Risks from Anthropogenic Mercury Emissions in the United States.* Office of Air Quality Planning and Standards and Office of Research and Development, Washington, DC. 144 pp.

EPA. 1996b. *Exposure Factors Handbook.* Vol. II, *Food Ingestion Factors.* EPA 600/P/95/002 Bb. Office of Research and Development, National Center for Environmental Assessment, Environmental Protection Agency, Washington, DC. 374 pp.

EPA. 1997. *Guidance for Assessing Chemical Contaminant Data for Use in Fish Advisories.* Vol. 2, *Risk Assessment and Fish Consumption Limits,* 2nd ed. Environmental Protection Agency, Washington, DC. 472 pp.

EPA. 1999. *National Recommended Water Quality Criteria—Correction.* EPA 822-Z-99-001. Environmental Protection Agency, Washington, DC. 25 pp.

FDA. 1979. Lead in food; advance notice of proposed rulemaking: Request for data. *Federal Register,* **44**(171), 51233–51242.

FDA. 1991. Personal communication from John Thomas.

FISHBEIN, B. 1996. Recycling nickel-cadmium batteries. http://www.informinc.org/battery.html.

FLAVIN, C. 1985. *World Oil: Coping with the Dangers of Success.* Worldwatch Paper 66. Worldwatch Institute, Washington, DC. 66 pp.

FLEGAL, A. R., and C. C. PATTERSON. 1983. Vertical concentration profiles of lead in the Central Pacific at 15°N and 20°S. *Earth and Planetary Sci. Letters,* **64,** 19–32.

FRIBERG, L., C. ELINDER, T. KJELLSTROM, and G. F. NORDBERG. 1986. *Cadmium and Health: A Toxicological and Epidemiological Appraisal,* Vol. 2. CRC Press, Boca Raton, FL. 307 pp.

FUTATSUKA, M. 1989. Epidemiological aspects of methylmercury poisoning in Minamata. In S. Tsuru, T. Suzuki, H. Shiraki, K. Miyamoto, and M. Shimizu (Eds.), *For Truth and Justice in the Minamata Disease Case,* Proceedings of the International Forum on Minamata Disease 1988, Keiso Shobo, Tokyo. Pp. 231–235.

GESAMP. 1985. *Cadmium, Lead, and Tin in the Marine Environment.* United Nations Environment Program Regional Seas Reports and Studies No. 56, Geneva. 90 pp.

GILFILLAN, S. C. 1965. Lead poisoning and the fall of Rome. *J. Occup. Med.,* 7, 53–60.

GOLDBERG, E. D. 1976. Rock volatility and aerosol composition. *Nature,* **260**(5547), 128–129.

GOLDWATER, L. J. 1971. Mercury in the environment. *Sci. Amer.,* **224**(5), 15–21.

GRANT, N. 1969. Legacy of the mad hatter. *Environment,* **11**(4), 18–23, 43–44.

GRANT, N. 1971. Mercury and man. *Environment,* **13**(4), 3–15.

HAMELINK, J. L., R. C. WAYBRANT, and R. C. BALL. 1971. A proposal: Exchange equilibria control the degree chlorinated hydrocarbons are biologically magnified in lentic environments. *Trans. Am. Fish. Soc.,* **100,** 207–214.

HAMMOND, A. L. 1971. Mercury in the environment: Natural and human factors. *Science,* **171,** 788–789.

HARADA, M. 1989. The intrauterine methylmercury poisoning known as "congenital Minamata Disease"—a 20-year serial investigation and its recent problems. In S. Tsuru, T. Suzuki, H. Shiraki, K. Miyamoto, and M. Shimizu (Eds.), *For Truth and Justice in the Minamata Disease Case,* Proceedings of the International Forum on Minamata Disease 1988, Keiso Shobo, Tokyo. Pp. 259–265.

HONOLULU ADVERTISER. 1991. Water-faucet makers ending use of lead. September 1. p. B1.

HORVATH, G., R. C. HARRISS, AND H. C. MATTRAU. 1972. Land development and trace metal distribution in the Everglades. *Mar. Pollution Bull.,* **3,** 182–183.

INTERIOR DEPARTMENT 1991. Mercury. In *Minerals Yearbook,* Vol. 1, *Metals and Minerals.* U.S. Department of the Interior, Bureau of Mines, Pp. 705–708.

KATAHIRA, K. 1989. Actual status of patients and urgency of redress. In S. Tsuru, T. Suzuki, H. Shiraki, K. Miyamoto, and M. Shimizu (Eds.), *For Truth and Justice in the Minamata Disease Case,* Proceedings of the International Forum on Minamata Disease 1988, Keiso Shobo, Tokyo. Pp. 341–343.

KJELLSTROM, T. 1986a. Itai-itai disease. In L. Friberg, C. Elinder, T. Kjellstrom, and G. F. Nordberg (Eds.), *Cadmium and Health: A Toxicological and Epidemiological Appraisal.* Vol. 2, *Effects and Response.* CRC Press, Boca Raton, FL. Pp. 257–290.

KJELLSTROM, T. 1986b. Effects on bone, on vitamin D, and calcium metabolism. In L. Friberg, C. Elinder, T. Kjellstrom, and G. F. Nordberg (Eds.), *Cadmium and Health: A Toxicological and Epidemiological Appraisal.* Vol. 2, *Effects and Response.* CRC Press, Boca Raton, FL. Pp. 111–158.

KLEIN, D. H., and E. D. GOLDBERG. 1970. Mercury in the marine environment. *Environ. Sci. Tech.,* **4**(9), 765–768.

KNAUER, G. A. 1977. Immediate industrial effects on sediment metals in a clean coastal environment. *Mar. Pollut. Bull.,* **8,** 249–254.

KNAUER, G. A., and J. H. MARTIN. 1972. Mercury in a marine pelagic food chain. *Limnol. Oceanogr.,* **17,** 868–876.

KURLAND, L. T. 1989. An epidemiological overview of Minamata Disease and a review of earlier public health recommendations. In S. Tsuru, T. Suzuki, H. Shiraki, K. Miyamoto, and M. Shimizu (Eds.), *For Truth and Justice in the Minamata Disease Case,* Proceedings of the International Forum on Minamata Disease 1988, Keiso Shobo, Tokyo. Pp. 240–249.

LIEBER, M., and W. F. WELSCH. 1954. Contamination of groundwater by cadmium. *J. Amer. Water Works Assoc.,* **46,** 541–547.

LLEWELLYN, T. O. 1990. Cadmium. In *Minerals Yearbook.* Vol. 1, *Metals and Minerals.* U.S. Department of the Interior, Bureau of Mines, Washington, D.C. Pp. 191–195.

LLEWELLYN, T. O. 1991. Cadmium. In *Minerals Yearbook.* Vol. 1, *Metals and Minerals.* U.S. Department of the Interior, Bureau of Mines, Washington, D.C. Pp. 207–210.

LLOYD, R. 1961. The toxicity of mixtures of zinc and copper sulphates to rainbow trout (*Salmo gaird-nerii* Richardson). *Ann. Appl. Biol.*, **49**, 535–538.

MARTIN, J. H., and G. A. KNAUER. 1973. The elemental composition of plankton. *Geochim. Cosmochim. Acta*, **37**, 1639–1653.

MARTIN, J. M., and M. WHITFIELD. 1983. The significance of the river input of chemical elements to the ocean. In C. S. Wong (Ed.), *Trace Metals in Sea Water*, Proceedings NATO Advanced Research Institute, March 30–April 1, 1981, Erice, Italy. Plenum Press, New York. Pp. 265–296.

MCCAULL, J. 1971. Building a shorter life. *Environment*, **13**(7), 3–14, 38–41.

MINAMATA CITY. 1996. *Chronology of Minamata Disease.* Minamata City Planning Division, Minamata, Japan. 32 pp.

MINAMATA CITY. 1997. *Learning About Minamata Disease.* Environmental Creation Development Project, Minamata, Japan. 26 pp.

MITRA, S. 1986. *Mercury in the Ecosystem.* Trans. Tech. Pub. Lancaster, PA. 327 pp.

MONTAGUE, K., and P. MONTAGUE. 1971. *Mercury.* Sierra Club, San Francisco. 158 pp.

MOORE, M. R. 1986. Sources of lead exposure. In R. Lansdown and W. Y. Yule (Eds.), *The Lead Debate.* Croom Helm, London. Pp. 131–189.

MUROZUMI, M., T. J. CHOW, and C. PATTERSON. 1969. Chemical concentrations of pollutant lead aerosols, terrestrial dusts and sea salts in Greenland and Antarctic snow strata. *Geochim. Cosmochim. Acta*, **33**, 1247–1294.

NOGAWA, K. 1980. Itai-itai disease and follow-up studies. In J. O. Nriagu (Ed.), *Cadmium in the Environment.* Part 2, *Health Effects.* John Wiley & Sons, New York. Pp. 2–37.

NRIAGU, J. O. 1980. Production, uses, and properties of cadmium. In J. O. Nriagu (Ed.), *Cadmium in the Environment.* Vol. 1, Wiley-Interscience, New York. Pp. 35–70.

NRIAGU, J. 1989. A global assessment of natural sources of atmospheric trace metals. *Nature*, **338**, 47–49.

NRIAGU, J. O. 1993. Legacy of mercury pollution. *Nature*, **363**, 589.

NRIAGU, J. O., and J. M. PACYNA. 1988. Quantitative assessment of worldwide contamination of air, water and soils by trace metals. *Nature*, **333**, 134–139.

NRIAGU, J. O., W. C. PFEIFFER, O. MALM, C. M. MAGALHAES DE SOUZA, and G. MIERLE. 1992. Mercury pollution in Brazil. *Nature*, **356**, 389.

NRIAGU, J. O., and J. B. SPRAGUE. 1987. *Cadmium in the Aquatic Environment.* Wiley-Interscience, New York. 272 pp.

OECD. 1975. *Cadmium and the Environment: Toxicity, Economy, Control.* Organization for Economic Cooperation and Development, Paris. 88 pp.

PLACHY, J. 1997. Cadmium. In *Minerals Yearbook. Vol. 1, Metals and Minerals*, U.S. Department of the Interior, Bureau of Mines. *http://minerals.usgs.gov/minerals/pubs/myb.html.*

PAN, Y., X. JIANG, and S. WANG. 1989. The proglem of methylmercury poisoning along the Songhua River in China. In S. Tsuru, T. Suzuki, H. Shiraki, K. Miyamoto, and M. Shimizu (Eds.), *For Truth and Justice in the Minamata Disease Case*, Proceedings of the International Forum on Minamata Disease 1988, Keiso Shobo, Tokyo. Pp. 299–304.

PATTERSON, C. C. 1965. Contaminated and natural lead environments and man. *Arch. Environ. Health*, **11**, 344–360.

RACO-RANDS, V. 1997. Characteristics of effluents from large wastewater treatment facilities in 1996. In Southern California Coastal Water Research Project 1997–98 Annual Report. *http://www.sccwrp.org/pubs/annrpt/97/ar01.htm.*

REESE, R. G., JR. 1997. Mercury. In *Minerals Yearbook.* Vol. 1, Metals and Minerals, U.S. Department of the Interior, Bureau of Mines. *http://minerals.usgs.gov/minerals/pubs/myb.html.*

SAITO, H. 1989. Niigata Minimata disease. In S. Tsuru, T. Suzuki, H. Shiraki, K. Miyamoto, and M. Shimizu (Eds.), *For Truth and Justice.* Proceedings of the International Forum on Minamata Disease 1988, Keiso Shobo, Tokyo. Pp. 297–298.

SCEP-Report of the Study of Critical Environmental Problems. 1970. Man's Impact on the Global Environment. Cambridge, MA. MIT Press. 319 pp.

SETTLE, D. M., and C. C. PATTERSON. 1980. Lead in albacore: Guide to lead pollution in Americans. *Science*, **207**, 1167–1176.

SHEA, K. P. 1973. Canned milk. *Environment*, **15**(2), 6–11.

SIEGEL, B. Z., S. M. SIEGEL, and F. THORARINSSON. 1973. Icelandic geothermal activity and the mercury of the Greenland icecap. *Nature*, **241**, 526.

SMITH, G. R. 1997. Lead. In *Minerals Yearbook.* Vol. 1, *Metals and Minerals.* U.S. Department of the Interior, Bureau of Mines. *http://minerals.usgs.gov/minerals/pubs/myb.html.*

SMITH, W. E., and A. M. SMITH. 1975. *Minamata.* Holt, Rinehart, & Winston, New York. 192 pp.

SPRAGUE, J. G. 1964. Lethal concentrations of copper and zinc for young Atlantic salmon. *J. Fish. Res. Bd. Can.,* **21,** 17–26.

TACKETT, S. L. 1987. Lead in the environment: Effects of human exposure. *American Laboratory,* July. Pp 32–41.

TSURU, S., T. SUZUKI, H. SHIRAKI, K. MIYAMOTO, and M. SHIMIZU (Eds.). 1989. *For Truth and Justice in the Minamata Disease Case,* Proceedings of the International Forum on Minamata Disease 1988, Keiso Shobo, Tokyo. 427 pp.

UZAWA, H. 1989. The responsibility of Chisso in the Minamata issue. In S. Tsuru, T. Suzuki, H. Shiraki, K. Miyamoto, and M. Shimizu (Eds.), *For Truth and Justice in the Minamata Disease Case,* Proceedings of the International Forum on Minamata Disease 1988, Keiso Shobo, Tokyo. Pp. 335–337.

WALDRON, H. A., and D. STÖFEN. 1974. Sub-clinical Lead Poisoning. Academic Press, New York. 224 pp.

WALSH, M. 1999. The global phaseout of leaded gasoline. *http://www.earthsummitwatch.org/gasoline.html.*

WEISS, H. V., M. KOIDE, and E. D. GOLDBERG. 1971. Mercury in a Greenland ice sheet: Evidence of recent input by man. *Science,* **174,** 692–694.

WHO. 1973. *The Hazards to Health of Persistent Substances in Water.* World Health Organization, Regional Office for Europe of the WHO, Copenhagen. 159 pp.

WOOD, J. M. 1972. A progress report on mercury. *Environment,* **14**(1), 33–39.

WOODBURY, W. D. 1991. Lead. In *Minerals Yearbook,* Vol. 1, *Metals and Minerals.* U.S. Department of the Interior, Bureau of Mines, Washington, D.C. Pp. 627–654.

WU, J., and E. A. BOYLE. 1997. Lead in the western North Atlantic Ocean: Completed response to leaded gasoline phaseout. *Geochim. Cosmochim. Acta,* **61,** 3279–3283.

YAMAGATA, N., and I. SHIGEMATSU. 1970. Cadmium pollution in perspective. *Bull. Inst. Publ. Health,* **19,** 1–27.

YAMAGUCHI, T. 1989. Chisso subsidiaries and liability. In S. Tsuru, T. Suzuki, H. Shiraki, K. Miyamoto, and M. Shimizu (Eds.), *For Truth and Justice in the Minamata Disease Case,* Proceedings of the International Forum on Minamata Disease 1988, Keiso Shobo, Tokyo. Pp. 329–334.

YUNICE, A., and H. M. PERRY, JR. 1961. The acute pressor effects of parental cadmium and mercury ions. *J. Lab. Clin. Med.,* **58,** 975.

QUESTIONS

12.1 Assume that you are hired as a consultant to review a study of the impact of a landfill suspected of being a source of toxic metals in an adjacent stream. The study concludes that the landfill is having little or no impact on the stream based on the fact that the concentrations of metals dissolved in the stream water are no different upstream and downstream of the landfill. What flaw is there in this reasoning? What sorts of follow-up studies would you suggest to address the issue of metal contamination more intelligently?

In questions 12.2–12.14, indicate which metal or metals is (are) described by the statement.

12.2 The major use of this metal in the United States is the production of batteries.

    a. Hg

    b. Cd

    c. Pb

12.3 Recycling accounts for more than 50% of the use of this metal in the United States.

    a. Hg

    b. Cd

    c. Pb

12.4 Production of this metal is closely correlated with the production of Zn.
  a. Hg
  b. Cd
  c. Pb

12.5 The best-documented effect of pollution by this metal on aquatic organisms has been associated with its consumption by waterfowl.
  a. Hg
  b. Cd
  c. Pb

12.6 One of the most serious health effects associated with consumption of this metal is softening of the bones.
  a. Hg
  b. Cd
  c. Pb

12.7 If ingested, the most toxic form of this metal is its methylated form.
  a. Hg
  b. Cd
  c. Pb

12.8 For Americans, the single most important source of exposure to this metal is the consumption of tuna.
  a. Hg
  b. Cd
  c. Pb

12.9 The EPA's maximum contaminant level goal for this metal in drinking water is zero.
  a. Hg
  b. Cd
  c. Pb

12.10 The residence time of this metal in the human body is 16–33 years.
  a. Hg
  b. Cd
  c. Pb

12.11 World production of this metal has declined by about 70% in the last 30 years.
  a. Hg
  b. Cd
  c. Pb

12.12 The single most important source of exposure to this metal in the United States was phased out during the 1970s as a result of concerns related to smog.
  a. Hg
  b. Cd
  c. Pb

12.13 Exposure to this metal can cause permanent brain damage.
  a. Hg
  b. Cd
  c. Pb

12.14 The Mad Hatter in *Alice's Adventures in Wonderland* suffered from exposure to this metal.
  a. Hg
  b. Cd
  c. Pb

12.15 The State of Hawaii's Department of Health recently conducted an exhaustive study of heavy metal concentrations in fish from the Ala Wai Canal, which receives runoff from a large area of Honolulu. Of all the metals tested, only Pb was found to be present in fish at concentrations above background levels. The average Pb concentration was 3 ppm in whole fish and 0.5 ppm in fish fillets. How would you account for the fact that the Pb concentration was so much higher in whole fish than in fish fillets?

12.16 Suppose that you are a waiter/waitress in a local seafood restaurant. A pregnant woman and her husband come into the restaurant and are trying to decide what to order. The woman has narrowed her choices to the following:

a. Tuna

b. Shrimp

c. Swordfish

d. Shark

Which one of the above would you recommend that she order and why?

Chapter Thirteen

✧   ¤   ✧   ¤   ✧

Oil Pollution

Australia's *Canberra Times* reported on a group of Boy Scouts who were asked to write about the harmful effect on marine life of oil leakages from tankers, and at least one lad seemed to have a pretty good grip on the problem. "When my mum opened a can of sardines last night," he wrote, "it was full of oil and all the sardines were dead." (Associated Press story carried in the *Honolulu Advertiser*, August 24, 1978, p. D3)

PETROLEUM IN ONE FORM or another has been used by humans since at least biblical times. Natural oil seeps provided bitumen,[1] which was used as a building material in ancient cities, and in Iraq natural gas from the Kirkuk oil field at Baba Gurger has burned continuously since the time of Nebuchadnezzar. Over 3,000 tonnes of oil per year were obtained from hand-dug pits at Baku near the Caspian Sea as early as the 19th century (Nelson-Smith, 1972).

In the United States, natural oil seeps such as those off Coal Oil Point, California, were noted as early as 1629, but the first commercial well, at Titusville, Pennsylvania, was not drilled until 1859. By 1900 world oil production from such wells had reached 20 million tonnes, and for the next 80 years it doubled almost every 10 years, reaching 3 billion tonnes by 1978 (Figure 13.1). The earliest uses of crude oil were for caulking ships, embalming, treating leather, for medicinal purposes, and, as already noted, building construction. With the advent of refineries in the mid-19th century, however, it was possible to separate kerosene from the other components of crude oil, and kerosene obtained in this way rapidly replaced whale oil and vegetable oils as a fuel in lamps. For several decades, oil was used primarily in lighting and to a minor extent as a lubricant, but the invention of the internal combustion engine and its widespread adoption in all forms of transportation during the last few decades of the 19th century rapidly changed this picture. The demand for oil as a fuel rose steadily, and today petroleum accounts for about 42% of world energy production (IEA, 1998). Over half the oil is presently used in the transportation sector of the world economy (Table 13.1).

The widespread use of petroleum and petroleum products has inevitably resulted in the discharge of oil to the environment. With respect to aquatic systems, it is the marine environment that has received the greatest attention, since the majority of the more noteworthy oil spills have involved accidents at sea. The emphasis in this chapter on marine oil pollution is not meant to imply, however, that oil pollution of freshwater systems is not a serious problem. However, some of the issues associated with marine oil pollution, such as offshore oil drilling and use of supertankers to transport oil, are peculiar to the marine environment, whereas many of the other issues are equally relevant to freshwater and marine systems. In other words, a study of marine oil pollution encompasses most if not all of the problems relevant to freshwater oil pollution as well.

---

[1] A semisolid that remains after the more volatile components of crude oil have evaporated away.

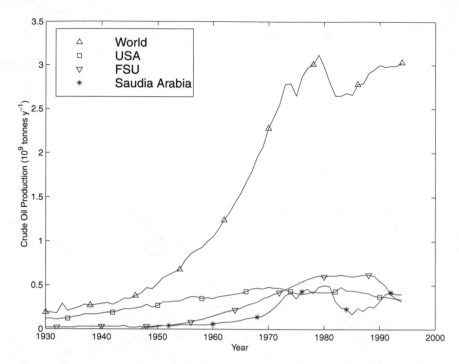

FIGURE 13.1   World crude oil production from 1930 to 1994. *Source*: Various editions of the *United Nations Energy Statistics Yearbook* and the U.S. Department of the Interior's *Minerals Yearbook*. (FSU = former Soviet Union)

OIL DISCHARGES TO THE MARINE ENVIRONMENT

Table 13.2 summarizes estimates of oil inputs to the marine environment around 1985. In most cases, the uncertainty of the estimates is rather large, and it seems reasonable to conclude that the numbers are accurate to no more than one significant figure. The following summary provides additional information on the various inputs.

TABLE 13.1   PRINCIPAL USES OF OIL IN THE WORLD ECONOMY IN 1996

Sector of Economy	% Use
Transportation	56
Industry	17
Residential/commercial	12
Electricity	6
Other	9

Source: IEA (1998).

TABLE 13.2 ESTIMATES OF PETROLEUM INPUTS TO THE MARINE ENVIRONMENT

Source	Input Rate (10^6 tonnes per year)	
	Probable Range	Best Estimate
Natural sources		
Marine seeps	0.02–2.0	0.2
Sediment erosion	0.005–0.5	0.05
Total natural sources	0.025–2.5	0.25
Anthropogenic sources		
Offshore production	0.04–0.06	0.05
Transportation		
Tanker operations	0.06–0.2	0.1
Drydocking	0.02–0.05	0.03
Marine terminals	0.01–0.03	0.02
Bilge and fuel oils	0.2–0.6	0.3
Tanker accidents	0.07–0.13	0.1
Nontanker accidents	0.02–0.04	0.02
Total transportation	0.38–1.05	0.57
Atmosphere	0.05–0.5	0.3
Municipal and industrial wastes and runoff		
Municipal wastes	0.4–1.5	0.7
Refineries	0.06–0.6	0.1
Nonrefinery industrial	0.1–0.3	0.2
Urban runoff	0.01–0.2	0.12
River runoff	0.01–0.5	0.04
Ocean dumping	0.005–0.02	0.02
Total wastes and runoff	0.585–3.12	1.18
Total	1.1–7.2	2.4

Source: Modified from NRC (1985).

Natural Sources

Marine Seeps

There are presently known to be almost 200 significant marine oil seeps, most of which are near continental coastlines. Areas of significant seepage are found offshore of California and Alaska, off the northeast coast of South America, and in the Arabian Gulf, the Red Sea, and the South China Sea. Since most petroleum is produced from sedimentary rocks at depths greater than 1 km, it is unlikely that oil from underground deposits would reach the surface through seepage other than in tectonically active areas where petroleum source or reservoir beds have been uplifted. Based on this reasoning, one can divide the ocean margins into areas of high, medium, and low seepage probability. Estimates of seepage derived in this way vary widely, from about 0.02 million tonnes (mt) per year to as much as 10 mt y^{-1} (NRC, 1985). The seepage estimate can be narrowed down by considering the age of the oil and the probable amount of oil available for seepage. The age of oil deposits is on the order of 50 million years (Blumer, 1972). At a seepage rate of 0.02 mt y^{-1}, 10^6 mt of oil would have seeped to the surface in 50 million years. This figure is roughly three to five times the amount of oil believed to have been available for exploitation before the rapid increase in oil use began in the early 20th century (Flavin, 1985). Allowing for unknowns with regard to the amount of petroleum that would have been available for seepage during geologic time and will become available in the future during the lifetimes of the seepage, an upper bound on the amount of oil available for seepage might be about 10^8 mt (NRC, 1985). A seepage rate of 2.0 mt y^{-1} would be required to deplete 10^8 mt in 50 million years. Thus, it seems reasonable to assume that natural seepage rates lie somewhere between 0.02 and 2.0 mt y^{-1}. The NRC (1985) estimate of 0.2 mt y^{-1} is simply the geometric mean of these two bounds. Given the probable range of seepage rates, 0.02–2.0 mt y^{-1}, one concludes that present underground oil deposits represent a small fraction of the quantity of oil that was formed underground over geologic time. Most of the latter has evidently seeped to the surface and been broken down by chemical and biological processes.

Sediment Erosion

Estimates of the input of petroleum due to the erosion of continental rocks are based on indirect calculations, since there are only a few places where erosional inputs can be studied in detail. The calculations of the NRC (1985) are based on estimates of the amount of organic C transported to the ocean by rivers and the percentage of that C accounted for by petroleum. Fluvial inputs of organic C are estimated to be 400 mt y^{-1}. Based on analyses of this C and of the C in ancient sedimentary rocks, petroleum is estimated to account for about 0.0125% of this figure, or 0.05 mt y^{-1}. The uncertainty of this estimate is about a factor of 10.

Offshore Production

The estimate of 0.05 mt y^{-1} is based on experience in the United States and extrapolation of this experience to other parts of the world, with allowance for differences in operating techniques. Offshore production discharges can vary greatly from one year to the next, depending on the occurrence of major offshore accidents. The Mexican Ixtoc I oil well blowout of 1979, for example, is estimated to have released about 0.48 mt of crude oil into the Gulf of Mexico, almost 10 times the estimated annual average for the entire world. Of the 0.05 mt y^{-1} average figure, about 19%, 6%, and 75% are estimated to come from offshore-produced water discharges, minor spills, and major spills, respectively.[2] The water discharges associated with oil production can vary from virtually zero to about 80% of the volume of oil produced, depending on, for example, whether water injection is used to help extract the oil (NRC, 1985). Typical hydrocarbon concentrations in the discharged water are 35–70 ppm, depending on the type of equipment used and the steps taken to separate the oil and water. Accidents include blowouts, the rupture of gathering lines, and other unpredictable occurrences.

Marine Transportation

Tanker Operations

A significant portion of the world's oil production is transported by sea. About 80 ± 5% of the oil transported by tankers is crude, the remainder being refined products (Yamaguchi, 1990). The percentage of the world's petroleum consumption accounted for by imported oil averaged about 68% during the 1970s, dropped to 41% in 1985, and since 1990 has been averaging about 54% (IEA, 1998). The percentage is expected to increase in the

---

[2] Minor and major spills are defined as spills involving less than or more than, respectively, 50 barrels of oil.

next few decades as major consuming nations such as the United States are forced to rely more and more on foreign oil. In 1988, for example, the United States imported 43% of its oil. By 1996 that figure had risen to 55% (IEA, 1998). Historically, transportation losses have been a major cause of oil discharges to the ocean.

In the past, a large percentage of the oil discharged to the ocean by tankers resulted from the practice of ballasting cargo bunkers with water after a tanker had offloaded its oil. Such ballasting was necessary to maintain the tanker in a seaworthy condition for its return voyage. Although some ports maintain facilities to accept dirty ballast water, the capacity of such facilities is limited. They are not designed to handle more than a small percentage of the water needed to ballast a large oil tanker. To avoid discharging oil-laden ballast water in port while taking on its next shipment, a tanker would therefore normally flush its ballast tanks with clean water while at sea, discharging the oily water in the process. The amount of oil discharged by this procedure was about 0.35% of the tanker's carrying capacity (NAS, 1975). In the absence of any effort to prevent such losses, oil discharges due to cargo bunker ballasting would have amounted to 4 mt y^{-1} in recent years.

Efforts to largely eliminate these losses occurred in two steps. The first was the introduction of the so-called load on top (LOT) procedure, depicted in Figure 13.2. Basically the idea is to transfer the oily wash water and ballast water to cargo holds, where the oil and water separate out.[3] The clean water at the bottom is then drawn off and discharged. Oil and water mixtures that remain are transferred to a slop tank, where further separation occurs. The clean water at the bottom of the slop tank is discharged, and at the next loading port, new cargo is loaded on top of the reclaimed oil in the slop tank. Further improvements were made with the introduction of crude oil washing systems (COWs), which involve cleaning cargo tanks with high-pressure jets of crude oil while the ship is offloading. Residues that would otherwise remain in the tanks are pumped ashore with the cargo. The combination of LOT and COW is estimated by the NRC (1985) to have reduced oil discharges from ballasting by about a factor of 50.

The problem with LOT is that it may be impractical for tankers engaged in short-haul voyages, which account for about 15% of crude oil movement at sea (NRC, 1985). In order to circumvent this problem, regulations were put into effect requiring the use of segregated ballast tanks (SBTs) or clean ballast tanks (CBTs) on the majority of the tanker fleet. The use of SBTs or CBTs eliminates the need to pump ballast water into cargo tanks under normal circumstances. The only difference between SBTs and CBTs is that the latter do not require separate pipes and pumps for taking on and discharging ballast.

The international agreement that has been primarily responsible for effecting these changes is the International Convention for the Prevention of Pollution from Ships, often referred to as MARPOL, which initially resulted from a conference on marine pollution held in 1973 under the auspices of the International Maritime Consultative Organization (IMCO). The protocol for MARPOL was modified in 1978, the same year that the U.S. Congress passed the Port and Tanker Safety Act (Public Law 95–474). The provisions of MARPOL with respect to tanker safety and the Port and Tanker Safety Act of 1978 are very similar. The MARPOL protocol, which went into force in 1983, presently requires that all new crude oil tankers over 20,000 dead-weight tonnes (dwt)[4] have SBTs and COWs, all new product tankers over 30,000 dwt have SBTs, all existing crude oil tankers over 40,000 dwt have SBTs or COWs, and all existing product tankers over 40,000 dwt have SBTs or CBTs.

The obvious loopholes in the legislation are the exclusion of small tankers and the fact that existing crude oil tankers over 40,000 dwt are not required to have SBTs. The former consideration would appear to be of minor consequence. Tankers under 40,000 dwt account for about 40% of the number of oil tankers in the world fleet but only 11% of

---

[3]Most components of oil are less dense than water.
[4]The dead weight is the weight of the ship without cargo.

1. Empty tanker takes seawater ballast into some dirty tanks.

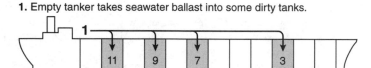

2. Other tanks cleaned by water jets.

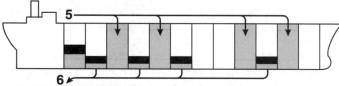

3. Washings transferred to slop tank aft.
4. Oil separates from dirty ballast.
5. Seawater pumped into clean tanks as...

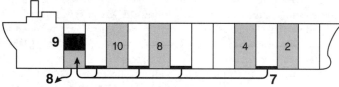

6. ...ballast beneath oil is pumped to sea.

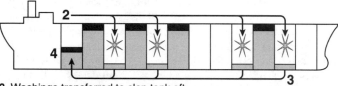

7. Oil layer transferred to slop tank.
8. After more separation, slop water pumped to sea.
9. Slop oil retained aboard when fresh cargo is loaded.

FIGURE 13.2   An outline of the LOT procedure. *Source*: Redrawn from Nelson-Smith (1972).

the dead weight tonnage (Drewry, 1991b). The U.S. Port and Tanker Safety Act, incidentally, applies to all tankers greater than 20,000 dwt, and these account for over 80% of tankers by number and 97% of the dead weight tonnage of the tanker fleet. The exclusion of existing crude oil tankers from the requirement for SBTs is of more consequence, since 70% by number and 77% by weight of the world's tanker fleet was built before 1982 (Drewry, 1991a). Thus, most crude oil tankers can, at least for the time being, operate without SBTs. However, if they lack SBTs, they must have a COW system and should be equipped with LOT if they are to comply with the requirements of the International Convention for the Prevention of Pollution of the Sea (OILPOL), which has been ratified by nations representing almost all of the world's merchant fleet (NRC, 1985). According to the provisions of OILPOL, which was last amended in 1969, tankers must not discharge more than 1/15,000 of their total cargo-carrying capacity during any one ballast voyage (NRC, 1985).

In 1985 the National Research Council (NRC) estimated the annual discharge of oil due to tanker operations to be 0.7 mt y^{-1}. That estimate was based on conditions that existed before the OILPOL protocols had been fully implemented and before MARPOL-78 had come into force. OILPOL requirements, for example, are in many cases achievable with nothing more than LOT and COW (NRC, 1985). If the 1/15,000 OILPOL criterion is applied to the present oil import figure of 1.14×10^9 tonnes y^{-1}, the implication is that the discharge from tanker operations should be no more than 0.1 mt y^{-1} (Table 13.2).

Dry Docking

At the present time, tankers go into dry dock about once every two years for maintenance and inspection (NRC, 1985). At such times, they must be clean of oil and gas-free. Rigorous cleaning is required because of the danger of explosions. Some dry docks provide facilities for receiving the washings from this cleaning, but in the absence of such facilities, tankers must discharge their washings at sea prior to entering dry dock. The use of COW systems has reduced the amount of clingage that must be removed, but some invariably remains. The NRC (1985) estimate of 0.03 mt y^{-1} for dry-docking discharges is based on the assumption that 5% of tankers discharge oil equal to 0.4% of their dead-weight tonnage when they dry dock at intervals of two years. The present tanker fleet, however, is about 30% smaller than the size assumed by the NRC (Drewry, 1991b), and the dry-docking discharges are presumably smaller by a similar percentage.

Marine Terminals

Discharges at marine terminals are in part the result of human errors, for example overfilling tanks and disconnecting hoses without adequate drainage. Other causes of spillage include line or hose failures, submarine pipeline ruptures, and storage tank ruptures. The NRC estimate of 0.02 mt y^{-1} for this category is based in part on U.S. Coast Guard records of spills at U.S. terminals and extrapolation of this information to the rest of the world. These more or less routine spills were estimated to amount to 0.01 mt y^{-1}. Major spills, for example from submarine pipeline or storage tank ruptures, are relatively infrequent but not well quantified. The oil discharged in these large but infrequent spills was simply assumed to equal, on average, the oil discharged in the better-documented smaller spills.

Bilge and Fuel Oil

Both tankers and nontankers generate fuel oil sludge and bilge oil, and nontankers may also generate oily ballast water. The oil sludge comes from the use of heavy residual bunker fuel oil to power vessels and is estimated to account for about two-thirds of the total bilge and fuel oil discharge (NRC, 1985). Bunker fuel oil contains impurities such as sludge and water, and in the case of nontankers, the lack of sufficient storage space in slop tanks may cause a large portion of the former to be discharged.

When at sea, tankers and nontankers generate 10–60 L of bilge oil per day. Most tankers retain the bilge oil in slop tanks for cargo oil or discharge it to shore reception facilities. Nontankers may discharge the bilge oil either at sea or at shore reception facilities. Use of the latter is facilitated if the ships are equipped with oily-water separators, but according to the NRC (1985), only about half of nontankers have such devices. About 30% of the estimated bilge and fuel oil discharge of 0.3 mt y^{-1} comes from this bilge oil.

Nontankers, particularly fishing vessels, may need to carry large quantities of ballast water for safety reasons, and because these vessels are not equipped with SBT, the ballast water is carried in empty fuel tanks. When the ballast water is discharged, roughly 25% of the clingage is discharged as well. This input, however, is estimated to be only about 0.003 mt y^{-1}.

Vessel Accidents

Most documented oil spills involve relatively small amounts of oil (<7 tonnes). The majority of these spills are associated with loading and discharging operations as opposed to

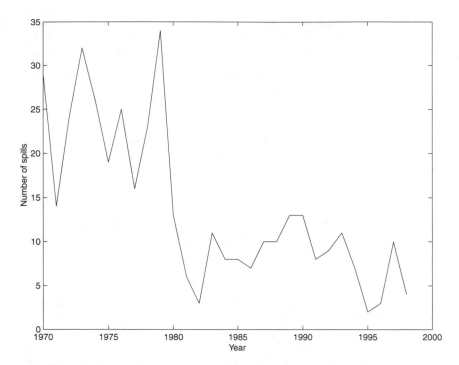

FIGURE 13.3   Frequency of oil spills that released more than 700 tonnes of oil to the marine environment. *Source*: ITOPF (1999).

vessel accidents (ITOPF, 1999). The vast majority of the spilled oil, however, is accounted for by a relatively small number of large spills (>700 tonnes), most of which involve vessel accidents (Figure 13.3). Not surprisingly, tanker accidents are by far the most important contributor in this category. During the 1970s, oil discharges from tanker accidents averaged about 0.3 mt y^{-1}, and spills involving more than 700 tonnes of oil occurred an average of 24 times per year. Since 1980, the corresponding figures have been 0.1 mt y^{-1} and eight spills per year (ITOPF, 1999). The decline in the frequency of large spills is probably accounted for by actions taken by the industry in response to some very unfortunate incidents that occurred during the 1970s (Table 13.3). For example, the largest tanker accident to date occurred in 1979, when the *Atlantic Empress* and *Aegean Captain* crashed in a thunderstorm east of Tobago. About 0.3 mt of oil was discharged by this single event. In the previous year, the *Amoco Cadiz* discharged 0.22 mt of oil when it grounded and split in half off the coast of Brittany (Ware, 1989). The spilled oil polluted about 400 km of the French coastline.

Although the frequency of large spills has declined by a factor of 3 since 1980, such accidents continue to account for the majority of spilled oil. Between 1988 and 1997, for example, 10 vessel accidents accounted for 70% of the oil discharged to the ocean from oil spills of all types (ITOPF, 1999). About two-thirds of large spills are associated with tanker collisions and groundings. The *Atlantic Empress/Aegean Captain* and *Amoco Cadiz* incidents are cases in point. Most of the remainder of the large spills have involved hull failures or fires and explosions. For example, the *Castillo de Bellver* released about 0.25 mt of oil when it caught fire and broke in half off the Cape of Good Hope in 1983. For comparison, the wreck of the *Exxon Valdez*, the largest tanker accident in U.S. coastal waters, released only 0.036 mt of oil into Prince William Sound in 1989. Tanker accidents are one of the most important causes of oil discharges to the ocean in terms of both the quantity

TABLE 13.3   MAJOR OIL TANKER ACCIDENTS

Ship	Year	Location	Oil Lost (10^3 tonnes)
Atlantic Empress/Aegean Captain	1979	Off Tobago, West Indies	287
ABT Summer	1991	1,300 km off Angola	260
Castillo de Beliver	1983	Off Saldanha Bay, South Africa	252
Amoco Cadiz	1978	Off Brittany, France	223
Haven	1991	Genoa, Italy	144
Odyssey	1988	1,300 km off Nova Scotia, Canada	132
Sea Star	1972	Gulf of Oman	123
Torrey Canyon	1967	Scilly Isles, United Kingdom	119
Texaco Denmark	1971	Belgium	107
Urquiola	1976	La Coruna, Spain	100
Irenes Serenade	1980	Greece	100
Hawaiian Patriot	1977	550 km off Honolulu	95
Independenta	1979	Bosphorus, Turkey	95
Jakob Maersk	1975	Oporto, Portugal	88
Braer	1993	Shetland Islands, United Kingdom	85
Khark 5	1989	220 km off Atlantic coast of Morocco	80
Aegean Sea	1992	La Coruna, Spain	74
Sea Empress	1996	Milford Haven, United Kingdom	72
Katina P.	1992	Off Maputo, Mozambique	72
Nova	1985	Persian Gulf, 37 km off Iran	70
Epic Colocotronis	1975	West Indies	58
Sinclair Petrolore	1960	Brazil	57
Assimi	1983	100 km off Muscat, Oman	53
Metula	1974	Magellan Straits, Chile	50
Andros Patria	1978	Spain	50
World Glory	1968	South Africa	48
British Ambassador?	1975	Japan	46
Pericles G.C.	1983	Persian Gulf	46
Yuyu Maru No. 10	1974	Japan	42
Ennerdale	1970	Seychelles	42
Mandoil II	1968	United States	41
Burmah Agate	1979	Gulf of Mexico	41
Wafra	1971	Off Cape Agulhas, South Africa	40
Juan Antonia Lavalleja	1980	Algeria	38
Trader	1971	Greece	36
Napier	1973	Chile	36
Corinthos	1975	United States	36
Exxon Valdez	1989	Prince William Sound, Alaska	36

Source: ITOPF (1999), Loughlin (1994)

of oil and the environmental impact. Because many tanker accidents occur near land, they tend to have a much greater impact on marine organisms than discharges from sources such as ballast water and bilge oil, which are normally released in the open ocean.

Atmosphere

The mechanisms here are vehicle exhaust and evaporation, followed by fallout in precipitation. Most of the fallout is believed to be the result of scavenging by rain of particulate material in the atmosphere. There is admittedly much uncertainty about the magnitude of the atmospheric input, the estimate of 0.3 mt y^{-1} being little more than an educated guess. The principal problems in making an estimate are lack of data, the complex nature of atmospheric photochemistry, the large number of chemical compounds to be considered, and the difficulty of distinguishing hydrocarbons of petroleum origin from hydrocarbons derived from other sources. It is noteworthy, however, that about 68 mt y^{-1} of petroleum are introduced into the atmosphere each year through the combustion of gasoline and other petroleum products (NAS, 1975). The NRC (1985) estimate that only 0.3 mt y^{-1} enters the ocean from the atmosphere implies that almost all of the petroleum emitted to the atmosphere falls out on the land or is broken down photochemically in the atmosphere. While this conclusion is probably correct, the magnitude of the small percentage that reaches the ocean can be estimated only crudely. The NRC (1985) calculations, for example, are based almost entirely on estimates of normal alkane inputs, since information on atmospheric concentrations of other types of hydrocarbons is meager.

Municipal and Industrial Wastes and Runoff

Municipal Wastes

At the present time, municipal wastewaters probably account for more petroleum hydrocarbons discharged to the oceans than any other source. The estimate of 0.7 mt y^{-1} made by the NRC (1985) is based on studies reported by Eganhouse and Kaplan (1981, 1982) on municipal wastewater in the Los Angeles area. Based on those studies, it was estimated that the per capita total hydrocarbon load in untreated wastewater was about 6.8 g d^{-1}. This figure was then normalized to petroleum consumption by multiplying by the number of persons estimated to be living within 80 km of the U.S. coastline and dividing by U.S. petroleum consumption in the year of the study. The conclusion was that the discharge of petroleum hydrocarbons in untreated wastewater amounts to about 0.032% of the petroleum consumed. Primary and secondary sewage treatment were estimated to eliminate about 33% and 40%, respectively, of the petroleum hydrocarbons. The total discharge of 0.7 mt y^{-1} of petroleum hydrocarbons in municipal wastewater was then calculated on the basis of regional petroleum consumption figures and similar estimates of the percentage of petroleum hydrocarbons in wastewater removed by sewage treatment.

Industrial Wastes

Both oil refineries and other industrial operations may discharge oil into the ocean. Most nonrefinery industrial wastewater is believed to be discharged into municipal sewer systems and hence would be accounted for under the municipal wastes category. Some nonrefinery industrial waste is discharged directly into the marine environment, but there is little quantitative information on the magnitude of this discharge. The estimate of 0.2 mt y^{-1} is little more than an educated guess. Some oil refineries also discharge wastewater into municipal sewer systems, but approximately 81% of them discharge directly into receiving waters (NRC, 1985). Studies have indicated that 5–75 g of petroleum hydrocarbons are discharged by refineries per tonne of operating capacity, the average value for the world being about 46 g t^{-1} (NRC, 1985). Since total refinery capacity is about 4,200 mt y^{-1}, the estimated discharge is $(46 \times 10^{-6})(4200) = 0.19$ mt y^{-1}. However, only about half of this oil is discharged directly to the ocean, since only about half of the refinery capacity is in coastal areas. The estimated discharge is therefore 0.1 mt y^{-1}.

Runoff

Considered in this category are both urban runoff and river discharges. The estimate for urban runoff is based on studies in the United States and Sweden, which indicate a per capita contribution to urban runoff of roughly 1 g of petroleum hydrocarbons per day. Multiplying this figure by the estimated 120 million persons who live within 80 km of the U.S. coastline gives a discharge of 0.04 mt y^{-1}. Given that the United States accounts for about 30% of the world's consumption of petroleum hydrocarbons, this figure was scaled up to 0.12 mt y^{-1} for urban runoff worldwide. The river discharge figure was calculated using the per capita total hydrocarbon contribution to untreated wastewater of 6.8 g d^{-1}, which was then multiplied by the estimated 110 million persons whose wastewater enters interior rivers in the United States. On the assumption that only 5% of these hydrocarbons actually reach the ocean, the river discharge for the United States is estimated to be 0.014 mt y^{-1}. Assuming this figure to be one-third of the world total, the total river discharge is estimated to be 0.04 mt y^{-1}.

Ocean Dumping

The estimate of 0.02 mt y^{-1} is based on the assumption that about 15 million tonnes of sewage sludge are discharged to the oceans each year and that this sludge contains about 0.1% petroleum hydrocarbons. The fact is, however, that ocean dumping of sewage sludge was prohibited by U.S. legislation in 1981 (Clark et al., 1984), and roughly half of the 15 million tonne figure cited by the NRC (1985) was accounted for by U.S. sludge dumping that occurred prior to 1981. It therefore seems fair to say that oil inputs associated with ocean dumping of sludge are no more than 0.01 mt y^{-1} at the present time.

Commentary

A certain amount of oil discharge is inevitably associated with oil use. It is noteworthy that oil inputs to the ocean amount to less than 0.1% of current oil production. Furthermore, discharges from tanker accidents and ballasting operations have declined significantly in the last 20 years. Not included in Table 13.2 are estimates of oil discharges associated with military hostilities. Admittedly, these discharges are very unpredictable, but they deserve some mention. In recent years, wars in the Middle East have been a very significant cause of oil released to the Persian Gulf. In 1983, for example, a winter storm ruptured pipes around an oil platform in Iran's Nowruz offshore oil field, and thousands of barrels of oil per day began to flow into the northern Persian Gulf. Iraqi missiles subsequently damaged two additional wells. Although capping the wells would probably have taken only about three weeks in the absence of hostilities (Begley et al., 1983), the Iran-Iraq war frustrated such efforts for more than six months. Estimates of the amount of oil that was discharged before the wells were finally capped range as high as 0.54 mt (Ware, 1989). Eight years later, the largest oil discharge to date occurred in almost the same location when Iraqi forces deliberately released oil from the offshore tanker-loading station at Mina al-Ahmadi in Kuwait. Roughly 1.4 mt of oil was discharged to the Gulf (Aldhous, 1991). These two incidents alone work out to an average discharge of 0.2 mt y^{-1} over the time interval 1983–1991, and they are certainly not unique. During the first 10 months of 1990, for example, there were a total of six oil spills worldwide chalked up to either suspected sabotage or bombing by guerillas (Anonymous, 1991a), the average discharge rate being about 0.01 mt y^{-1}.

One issue of particular concern about the Persian Gulf is the fact that it is shallow, with an average depth of only about 30 m (Aldhous, 1991). The ecosystem is therefore naturally productive, and there is great potential for an oil spill to damage living resources, including benthic organisms. This sensitivity is characteristic of virtually all coastal ecosystems, and it is unfortunately into coastal waters that much discharged oil goes. Geographically, most of the remaining oil is discharged along the major tanker routes from the Middle East to Europe, the American continents, and the Far East.

The inputs of petroleum hydrocarbons to the ocean are therefore not distributed evenly, and much of the oil is released in coastal areas that are more susceptible to biological damage than the open ocean and more important to humans from the standpoint of fishery resources and recreation. What impacts are these oil discharges having on marine ecosystems, and what changes can we expect with respect to oil pollution in the next few decades? To answer these questions, we must understand a little about the nature of oil, namely, how is it formed, what are its physical and chemical properties, what are its toxic characteristics, and what is its fate when released to the environment? The following three sections summarize our current understanding of these issues.

THE GENESIS OF OIL

Underground deposits of oil have been forming in one part of the world or another for hundreds of millions of years. The process of oil formation is undoubtedly very slow, probably requiring millions of years. Although no one is sure exactly how deposits of oil come to be formed, the following sequence of events describes the consensus opinion of many scientists.

Sedimentation

Organic detritus settles to the bottom of aquatic systems and is incorporated into the sediments. In most aquatic systems, the rate of accumulation of organics in sediments is quite small compared to the primary production rate in the surface waters, since most of the organic C produced is ultimately respired. In highly productive systems, however, or in systems with estuarine-type circulation patterns or stagnant bottom waters,[5] anoxic conditions may develop in bottom waters and hence prevent or retard the oxidation of detrital C. Such anoxic systems probably play an important role in the formation of oil, although the widespread nature of oil deposits has suggested to some that significant amounts of oil may have been formed from organics that accumulated in nominally oxidizing systems (Andreev et al., 1968). Such organics would necessarily have been rather refractory in nature. Since the O_2 level in sediments generally drops to zero below a depth of no more than a few millimeters,[6] any organics that are preserved sufficiently long to be buried below a few millimeters of sediment are effectively removed from the possibility of aerobic oxidation. Thus, the first step in the formation of oil presumably requires one or more of the following conditions:

1. The existence of anoxic bottom waters
2. The production of refractory C compounds
3. Rapid sedimentation

---

[5]For example, the Black Sea and the Cariaco Trench, respectively.
[6]A caveat is that burrowing organisms may allow oxygenated water to penetrate a few tens of centimeters.

Under such circumstances, organic detritus may accumulate at the bottom of the water column and ultimately become buried in the sediments.

Metamorphosis

Once buried in the sediments, organic C is ultimately incorporated into sedimentary rocks such as shales, sandstones, and carbonate rocks. Based on analyses of sediments, it seems likely that the organic C content of recently formed sedimentary rocks is on the order of no more than 0.1–1.0% (Andreev et al., 1968). Nevertheless, the organic C trapped in these rocks is apparently the source of the world's petroleum reserves. This C is slowly transformed under conditions of elevated temperature[7] and pressure found deep underground and perhaps through the mediation of catalysts such as aluminosilicate minerals. Humic acids, which are among the principal forms of organic C in sedimentary rocks, are probably the principal source of C for petroleum (Andreev et al., 1968). The presence of N and S in virtually all crude oils, however, indicates that other types of organic substances are also involved. The sequence of transformations that convert sedimentary organic detritus into petroleum is a continuous process and undoubtedly is highly complex. However, by correlating the composition of crude oil with its age, it is possible to get some idea of the sequence of transformations that organic compounds undergo while buried in sedimentary rocks. In general, there is a tendency for the higher molecular weight compounds to be broken down with time, leading to the formation of paraffins and ultimately to the production of methane and perhaps graphite as end products (Andreev et al., 1968). Thus, oil can be viewed as an intermediate stage in the breakdown of organic detritus in the absence of O_2 and under the influence of physical and chemical conditions (e.g., temperature, pressure, presence of catalysts) peculiar to deeply buried sedimentary rocks.

Migration

When contrasted to the low concentration of organic C in sedimentary rocks, the enormous deposits of oil in many parts of the world clearly imply the existence of mechanism for concentrating or trapping the oil in a localized deposit from surrounding source rocks. In other words, the oil could not have been formed *in situ*, but instead formed over a much larger region and subsequently migrated to a localized deposit. The mode of migration is not known, but it seems reasonable to assume that liquid petroleum, once formed, could move upward through porous crustal materials until it is trapped by an impervious overlying substratum. The existence of natural oil seeps in various parts of the world testifies to the fact that such migrations can extend all the way to the surface if no impervious substratum traps the oil. Oil deposits are often overlaid by a pocket of gas and invariably underlain by a reservoir of water.[8] The existence of these two fluids in association with an oil deposit simply reflects the tendency of fluid substances to migrate upward through porous rocks until an impervious substratum is encountered. Thus, the tendency of oil to migrate and form underground pools is by no means a characteristic peculiar to oil, but is a tendency common to all fluids. The migration and pooling of oil has been essential for its storage over millions of years and hence for its abundance today.

From the standpoint of modern industrial society, the most important characteristic of oil is the energy that can be derived from burning (oxidizing) it. This energy was orig-

[7]Probably 100–150°C.
[8]The gas and water are less dense and denser, respectively, than the oil.

inally fixed via photosynthesis millions of years ago and has been stored in the chemical bonds of the organic substances of which oil is composed. The anaerobic transformations that lead to the formation of oil release some of the original stored energy, but a sufficient amount remains in the chemical bonds of petroleum to provide a highly useful energy source. In summary, the genesis of petroleum deposits requires the following:

1. The formation, via photosynthesis, of energy-rich organic compounds
2. The storage of these compounds in an anaerobic environment or a similar environment sufficiently reduced to retard or prevent their oxidation
3. The anaerobic transformation of these compounds into petroleum under conditions of high temperature and pressure
4. The migration of the petroleum to underground traps or pools where it may remain for millions of years, during which time it undergoes additional transformations

WHAT IS OIL?

Natural crude oil is an exceedingly complex substance composed of literally thousands of different kinds of organic molecules. Crude oil taken from different parts of the world may vary greatly in composition, depending on the age of the oil, the conditions of its formation, and so forth. Despite the complexity and variability of crude oil, some generalizations about its composition can be made.

Crude oil consists primarily of hydrocarbons, that is, compounds containing only C and H. Some crude oils contain as much as 98% hydrocarbons (Nelson-Smith, 1972). In addition to hydrocarbons, the organic substances in crude oil include compounds containing S, N, and/or O, with S being more abundant than N and N more abundant than O. In addition, there are small concentrations of metals such as Ni, vanadium (Va), and Fe (NAS, 1975).

Crude oils can be roughly characterized according to the relative amounts of the major kinds of hydrocarbons they contain, namely, alkanes, cycloalkanes, and aromatics.

Alkanes—Paraffins or Aliphatic Compounds

Alkanes consist of one or more C atoms joined to four other atoms, either H or other C atoms. The simplest alkanes are methane, ethane, and propane, which contain one, two, and three C atoms, respectively (Figure 13.4). Larger alkanes, such as butane (C_4H_{10}) and pentane (C_5H_{12}), may exist in more than one configuration, as indicated in Figure 13.5. All such configurations, however, are classified as alkanes. Those configurations or isomers in which the C atoms are joined in a straight chain are referred to as *normal* or *straight-*

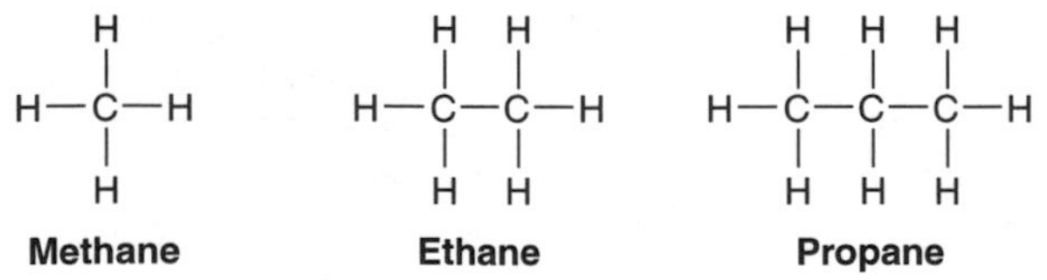

FIGURE 13.4   Structures of the simplest alkanes. Lines indicate electron-pair bonds.

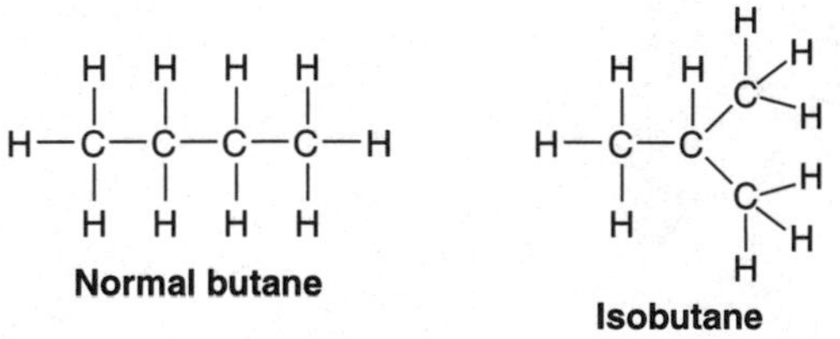

FIGURE 13.5   Structures of normal butane and isobutane.

chain alkanes, whereas molecules consisting of more complex arrangements of the C atoms are referred to as *branched-chain alkanes* (e.g., isobutane in Figure 13.5). Alkanes containing fewer than five C atoms are gases at room temperature. Alkanes containing 5 to about 17 or 18 C atoms[9] are primarily liquids at room temperature. Larger alkanes are solids at room temperature. Alkanes make up the major portion of the hydrocarbons in crude oil, accounting for 60% to well over 90% of the hydrocarbon content (Fieser and Fieser, 1961).

Cycloalkanes or Naphthenes

Cycloalkanes or naphthenes are similar to alkanes, but the C chain is joined in a ring rather than being open-ended. The simplest such cyclic compounds isolated from crude oil are cyclopentane and cyclohexane (Figure 13.6). Cyclopentane and cyclohexane are analogous to the normal alkanes pentane and hexane in the sense that the cyclic compounds have been formed by joining the two ends of the analogous normal alkanes while discarding two H atoms. Cyclic analogues of branched chain alkanes are also found in crude oil. In such cases, the C ring contains alkyl substituents as well as H atoms (e.g., methylcyclo-hexane in Figure 13.6).

Aromatics

Aromatic compounds all contain one or more benzene rings, a six-C ring structure in which the C atoms are joined by hybrid bonds intermediate in character between a single electron-pair bond and a double electron-pair bond. Some of the simpler aromatic compounds found in crude oil are indicated in Figure 13.7. These compounds may consist of a single aromatic ring (benzene) or of two or more rings joined together (e.g., naphtha-

[9]In other words, C_5 to C_{17} or C_{18} alkanes.

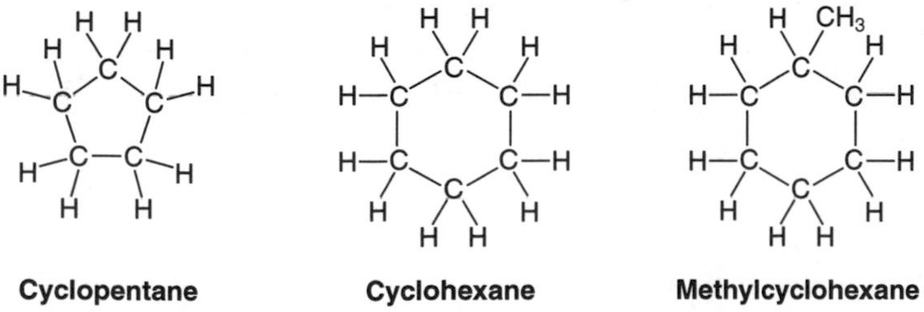

FIGURE 13.6   Structures of simple cycloalkanes.

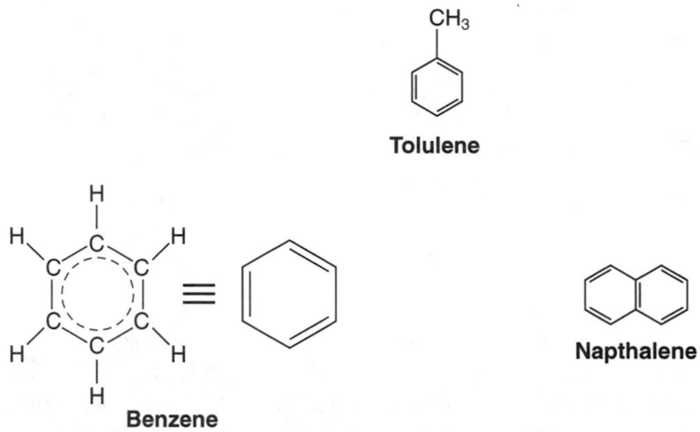

FIGURE 13.7   Simple aromatic compounds found in crude oil. The dashed circle inside the carbon hexagon indicate additional electron bonding.

lene). As is the case with the cycloalkanes, alkyl substituents may replace H atoms on the ring structure (e.g., toluene).

Since oil pollution involves both refined petroleum products and crude oil, it is of some interest to know a little about the refining process and the types of organic substances that go into various petroleum products. The basic refinery technique takes advantage of the fact that the boiling point of hydrocarbons generally increases with increasing molecular size. Therefore, one can separate hydrocarbons into various size fractions or approximate molecular weight groups by simply heating crude oil at a series of steadily increasing temperatures and collecting successively higher boiling fractions. This procedure, known as *fractional distillation,* is the basic oil refining technique. Table 13.4 indicates various fractions that may be collected in this way. The advent of the automobile created a demand for gasoline relative to the other components of crude oil that far exceeded the amount that could be obtained by simple fractional distillation. Therefore, oil companies began to look for ways to convert the higher-boiling fractions into gasoline by splitting or *cracking* the larger hydrocarbon molecules to obtain fragments in the C_4 to C_{10} range. Initially, a thermal cracking technique was used. This method called for heating the heavier hydrocarbon components under pressure to effect their breakdown. Stud-

TABLE 13.4   PETROLEUM FRACTIONS COLLECTED
BY FRACTIONAL DISTILLATION WITHIN INDICATED
TEMPERATURE RANGES

Fraction	Approximate Boiling Range	Approximate Molecular Size
Refinery gases	Up to 25°C	$C_3–C_4$
Gasoline	40–150°C	$C_4–C_{10}$
Naphtha	150–200°C	$C_{10}–C_{12}$
Kerosene	200–300°C	$C_{12}–C_{16}$
Gas oils	300–400°C	$C_{16}–C_{25}$
Residual oil	Above 400°C	Above C_{25}

Source: Reprinted from Ryan (1977).

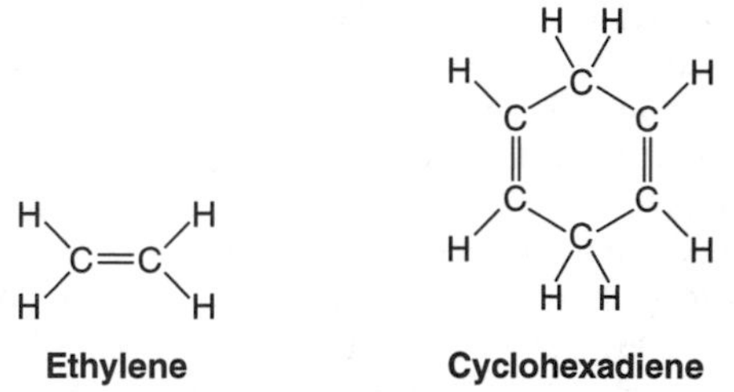

Ethylene   **Cyclohexadiene**

FIGURE 13.8   Structures of two simple olefins.

ies of the knocking characteristics of gasoline,[10] however, revealed that the hydrocarbons produced by thermally cracking crude oil were not of the sort best suited to retard knocking. Knocking is minimized in gasoline that contains primarily branched-chain alkanes, aromatics, and a class of hydrocarbons not found in crude oil, alkenes or olefins. Olefins are analogous to alkanes and cycloalkanes but contain at least some doubly bonded C atoms and correspondingly fewer H atoms. Two of the simplest olefins, ethylene and cyclohexadiene, are shown in Figure 13.8. To produce gasoline with a more desirable mixture of hydrocarbons, a catalytic cracking procedure was developed. This technique yields a product rich in branched-chain compounds, alkenes, cycloalkenes, and aromatics (Fieser and Fieser, 1961). Olefins may occur in concentrations as high as 30% in gasoline and account for about 1% of jet fuel. They are present in only minor amounts in other refined products (NRC, 1985).

With respect to aquatic pollution, it is important to note that the various hydrocarbons in crude or refined oil differ greatly in their toxicity to aquatic organisms, in their resistance to degradation, and in their tendency to be incorporated into various components of the aquatic ecosystem. Thus, the spilling of 100 tonnes of crude oil may have a very different effect on a coastal ecosystem than the spilling of 100 tonnes of gasoline or 100 tonnes of fuel oil. It is therefore appropriate at this time to examine the toxic characteristics of different petroleum fractions and products before attempting to evaluate the impact of oil pollution on aquatic systems.

TOXICOLOGY

Oiling and Ingestion

The toxic effects of oil fall into two general categories. The first category includes effects associated with coating or smothering of an organism with oil. Such effects are associated primarily with the higher molecular weight, water-insoluble hydrocarbons, the various tarry substances that coat the feathers on birds and cover intertidal organisms such as clams, oysters, and barnacles. Although some organisms such as tubeworms and barnacles are surprisingly little affected by such coatings (Nelson-Smith, 1972), the effect on organisms such as aquatic birds may be devastating. We will elaborate on this point later on. The second category of toxic effects involves disruption of an organism's metabolism due to the ingestion of oil and the incorporation of hydrocarbons into lipid or other tissues in sufficient concentrations to upset the normal functioning of the organism. With respect to this second category of effects, it is generally agreed that aromatic hydrocarbons are the

---

[10]Knocking occurs when gasoline burns explosively rather than smoothly in the cylinder of the engine.

most toxic, followed by cycloalkanes, then olefins, and lastly alkanes. There is also a definite tendency for toxicity to be positively correlated with the molecular size of the hydrocarbons (Figure 13.9). Most toxic effects caused by ingestion of oil in water, however, are believed to be due to low molecular weight (C_{12}–C_{24}) alkanes and low molecular weight aromatics, since higher molecular weight compounds are very insoluble in water (NRC, 1985). Based on studies of the concentrations of hydrocarbons in water-soluble fractions of various oils, the NRC (1985) concluded that the contribution of compounds of higher molecular weight than alkylnaphthalenes was very small and probably insignificant in terms of acutely toxic effects. Thus, the compounds of greatest concern from the standpoint of ingestion are the low molecular weight aromatics such as benzene and toluene.

Toxic hydrocarbons apparently exert their effects in part by becoming incorporated into the fatty layer that makes up the interior of cell membranes (Nelson-Smith, 1972). As a result, the membrane is disrupted and ceases to regulate the exchange of substances between the interior and exterior of the cell properly. In extreme cases the cell membrane may lyse, allowing the contents of the cell to spill out and hence destroying the cell. There

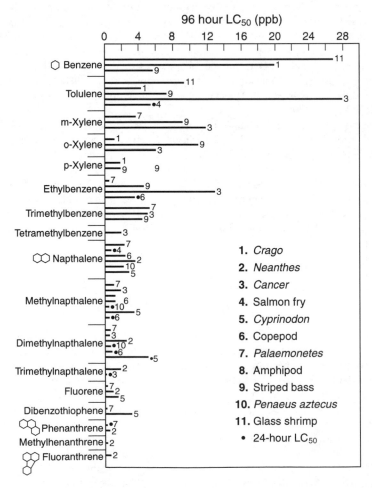

FIGURE 13.9   Median tolerance limits of some aromatic hydrocarbons for selected marine macroinvertebrates and fish. Duration of exposure was 96 hours unless otherwise noted. *Source*: Redrawn from NRC (1985).

is also evidence that hydrocarbons interact with proteins in a variety of plants and animals (Nelson-Smith, 1972). Both enzymes and structural proteins appear to be affected. Once again, aromatics seem to be more toxic than other hydrocarbon classes with respect to this effect.

Based on the foregoing information, it is possible to understand a little about the relative toxicity of various types of oil. Refined petroleum products such as gasoline or kerosene contain virtually no high molecular weight hydrocarbons (Table 13.4) and hence exert a very small smothering or coating effect. However, because refined products do contain a higher percentage of low molecular weight hydrocarbons than crude oil, these products are more dangerous from the standpoint of ingestion than a comparable amount of crude oil (NAS, 1975; Tatem et al., 1978). With respect to this point, it is noteworthy that two of the more ecologically damaging oil spills, the grounding of the tanker *Tampico Maru* off Baja, California, in 1957 and the grounding of the tanker *Florida* in Buzzards Bay, Massachusetts, in 1969, involved spillage of refined petroleum, namely, diesel oil and No. 2 fuel oil, respectively.

Crude oil, on the other hand, contains a significant fraction of high molecular weight hydrocarbons, which give it a viscous, sticky character. As a result, the greatest damage from crude oil discharges may be the coating of plants and animals with high molecular weight hydrocarbons. Undoubtedly, the organisms most affected by oil coating or *oiling* are certain kinds of aquatic birds, namely, auks, penguins, and diving sea ducks.[11] These birds are particularly susceptible to oiling for the following reasons:

1. They spend most of their lives on the surface of the sea.
2. They are poor fliers or are flightless.
3. They dive rather than fly in response to a disturbance.

Because of the third characteristic, such birds are unlikely to escape an oil slick, but instead tend to keep resurfacing in the slick after each dive. When oil is adsorbed to the feathers and down of these birds, their plumage becomes matted, and the air spaces, which normally provide buoyancy and insulation, become filled with water and oil. Because many of these birds are poor fliers at best, the additional load of oil and water in their feathers may make flight impossible, and some may literally drown due to the loss of buoyancy. Others presumably die due to their inability to maintain a proper body temperature without adequate insulation from their plumage. Invariably, oiled birds attempt to clean themselves by preening their feathers, but in the process they may ingest as much as 50% of the oil in their plumage (Nelson-Smith, 1972) and die from the toxic effects of the ingested oil.

There is no doubt that large numbers of birds are killed each year from oiling. According to figures cited by Nelson-Smith (1972), about 150,000–450,000 birds die each year from oiling in the North Sea and the North Atlantic. These estimates are based on counts of the number of dead and oiled birds that wash up on beaches, but they exclude, as far as possible, the effects of large oil spills. In other words, these losses are presumably due to small oil spills or to the deliberate discharge of small amounts of oil from unknown sources. Large spills of crude oil can result in the killing of tens or even hundreds of thousands of birds. The breakup of the tanker *Torrey Canyon* off the coast of England in 1967, for example, is estimated to have resulted in the killing of 40,000–100,000 birds (mostly auks), and the stranding of the tanker *Gerd Maersk* in the Elbe estuary in 1955 is estimated to have killed 250,000–500,000 birds (Nelson-Smith, 1972). Only about 4,500 dead and

[11]Auks include murres, guillemots, razorbills, and puffins. Diving sea ducks include scoters, scaup, eiders, golden-eye, and long-tailed ducks. Of the penguins, only the Cape penguins of South Africa and the Magellanic penguins of Argentina have been seriously involved. The distribution of other penguin species has tended to minimize the possibility of their being impacted by spilled oil.

oiled birds were recovered after the breakup of the *Amoco Cadiz* off the French coast in 1978 (Mielke, 1990), but the grounding of the *Exxon Valdez* in Prince William Sound, Alaska, in 1989 is estimated to have killed 250,000 seabirds and 150 bald eagles (Hodgson, 1990), and as many as 40,000 seabirds may have died in the Severn Estuary in 1991 as the result of an oil spill caused by a fractured oil pipe at a steel plant in Wales (Anonymous, 1991b).

Weathering

The impact of a given oil discharge is determined largely by the nature of the hydrocarbons in the oil. Other factors, however, including weather conditions and the distance of the discharge from shore, also play an important role in determining the extent of ecological damage (Kerr, 1977; Mielke, 1990). This point requires some elaboration. Because most components of oil are less dense than water, oil discharges usually tend to float on the surface. Thus, organisms that come into contact with the surface are most likely to be affected by oil pollution. However, hydrocarbons do have a finite, although small, solubility in water, and some of the oil will therefore dissolve in the water rather than simply float on the surface. As already noted, the low molecular weight hydrocarbons are also the most soluble in water. On the other hand, the low molecular weight hydrocarbons are also the most volatile and therefore tend to evaporate from the surface most readily. Thus, the characteristics of an oil slick are likely to change significantly within a day or so of the time the oil is released. Specifically, the percentage of dangerous low molecular weight hydrocarbons is likely to be greatly reduced due to evaporation and dissolution of these compounds, with evaporation accounting for most of the removal (Mielke, 1990). Most of the refinery gases, gasoline, and naphtha evaporate within a few days. Hence, a slick that washes ashore a few days after discharge is probably much less toxic to intertidal organisms than a slick that is driven ashore immediately after discharge. Kerosene and gas oils may take several weeks or more to evaporate away. Indirect evidence from compositional changes and laboratory studies indicate that ultimately 20–40% of a typical crude oil slick may be removed by evaporation (Mielke, 1990). Gundlach et al. (1983), for example, estimated that 30% of the *Amoco Cadiz* spill was removed by evaporation.

Physical mixing of the water column can disperse a surface slick throughout the mixed layer if the turbulence is sufficiently great. By thus increasing the effective contact area between oil and water, mixing helps dissolve the oil as well as dispersing small droplets of oil within the water column. Because most components of the oil are rather insoluble in water, the dispersion process tends to produce an emulsion of oil-in-water or water-in-oil. The former consist of fine particles of oil dispersed in water. The latter contain up to 70% oil and are commonly referred to as *mousses*. In either case, mixing tends to expose subsurface organisms, including even benthic organisms if the water column is shallow enough, to the impact of an oil discharge, whereas in calm weather the same organisms would feel little or no direct impact from the discharge.

Ultimately, most oil discharged into an aquatic system is degraded by a combination of biological, physical, and chemical processes collectively referred to as *weathering*. Actual degradation rates depend on a great many factors, including the characteristics of the oil, temperature, availability of nutrients, degree of physical mixing, and so forth. About 90 species of bacteria and fungi are capable of subsisting on oil and hence degrading it (Mielke, 1990), but no single species is capable of metabolizing all the different components of crude oil. Under normal circumstances, there are only about 10 bacteria capable of decomposing hydrocarbons in a liter of seawater, and growth of this bacterial population is normally slow in the period immediately following an oil spill (Mielke, 1990). As a result, bacteria are of minor importance in removing the oil during the first few days after a spill.

Bacterial degradation may, however, become significant during subsequent weeks, when the concentration of bacterial hydrocarbon decomposers may reach $5 \times 10^7 \, L^{-1}$. The principal factors limiting bacterial degradation are the availability of inorganic N and P, temperature, O_2 concentration, and the degree of dispersion of the oil. Bacterial degradation was very ineffective in the open ocean following the Ixtoc I oil well blowout, for example, because of the low concentrations of inorganic nutrients in the surface waters (Mielke, 1990). In the case of the *Amoco Cadiz* tanker accident, however, bacteria consumed about 73,000 barrels of oil before the oil could reach the shoreline. Bacterial degradation rates tend to be positively correlated with temperature and, according to Mielke (1990), are four times faster at 18°C than at 4°C. The result is that bacterial degradation takes a much longer time in polar climates than in temperate or tropical latitudes. O_2 concentrations are not likely to be limiting in the open ocean, but they may be a significant limiting factor in sediments and beach sands. The availability of O_2, for example, probably accounts in part for the fact that degradation generally occurs more rapidly in coarse than in fine-grained sediment.

In the presence of O_2 and sunlight, the organic compounds in oil may be oxidized by photochemical mechanisms, the result being the production of oxygenated compounds such as carboxylic acids, alcohols, ketones, and phenols. These compounds are generally more soluble than their precursors and in some cases are more toxic.

When the low molecular weight components of the oil have been removed through evaporation, dissolution, or degradation, there remain the tarry, viscous, residual oils. These high molecular weight compounds tend to form lumps of tar, which degrade much more slowly than the other components of the oil. Small tar balls have been noted floating on the surface of the ocean or washing up on beaches since as long ago as 1954 (Horn et al., 1970), but undoubtedly have been components of some marine ecosystems for thousands if not millions of years. The lumps range in size from a few millimeters to about 10 cm in their longest dimension and are usually composed of paraffinic hydrocarbons containing as many as 40 C atoms (Mielke, 1990). The presence of barnacles and isopods growing on such tar lumps clearly indicates that the hydrocarbons are relatively nontoxic (Horn et al., 1970). By determining the age of the attached organisms, Horn et al. (1970) inferred that at least some tar lumps float on the surface for as much as several weeks. It has been estimated, however, that 5–10% of the pelagic tar in the eastern Gulf of Mexico four years after the Ixtoc I blowout originated from that event (Mielke, 1990), a conclusion that obviously implies a residence time of several years for some pelagic tar. Ultimately, some of the tar lumps wash up on beaches (NAS, 1975); others may become negatively buoyant due to weathering or other processes and sink to the bottom. Some tar lumps have been found in the stomachs of small fish (Horn et al., 1970), and from that point on, they may either be metabolized, excreted in feces, or transferred up the food chain. There are estimated to be about 0.3 mt of pelagic tar in the oceans' surface waters at the present time (NRC, 1985).

In summary, oil discharged at sea is subject to a variety of degradative and removal processes, and if a discharge is sufficiently far from land, most of the toxic components of the oil will have been removed or reduced to insignificant concentrations by the time the oil reaches shore. The principal ecological damage in such cases probably stems from the oiling of aquatic birds with the higher molecular weight hydrocarbons and from the aesthetic damage caused by the stranding of tar on recreational beaches.

When an oil discharge occurs near shore, the ecological damage is likely to be much greater. In such cases, the oil may be washed ashore before weathering has had a chance to remove much of its toxicity. Intertidal and benthic organisms, many of which are sessile or have poor powers of locomotion, may then be destroyed in large numbers due to ingestion of the toxic hydrocarbons. The damage caused by the *Tampico Maru* and *Florida* tanker spills was intensified both by the fact that the tankers were carrying refined oil and by the fact that the spills occurred close to shore. Motile organisms such as fish can simply swim away to avoid an oil spill, but this defense mechanism is not infallible. Large numbers of dead fish washed ashore in Buzzards Bay after the grounding of the *Florida* in 1969 (NAS, 1975).

Lethal and Sublethal Effects

What concentrations of oil are toxic to marine organisms? Obviously, the answer to this question depends on what type of oil or combination of hydrocarbons one is talking about, as well as the species involved. Adult fish are rather resistant to oil pollution, since their bodies, including the mouth and gill chambers, are coated with a slimy mucus that resists wetting by oil. Crude oil becomes toxic to adult and larval finfish at concentrations of roughly 100–500 and 1–50 ppm, respectively (EPA, 1976; NRC, 1985), but individual aromatics such as benzene, toluene, and naphthalene are toxic at concentrations about 100 times lower (EPA, 1986). Whereas fish can swim away to avoid a spill, plankton cannot. Crude oil has been found to be toxic to zooplankton, including fish eggs, at concentrations ranging from 10 ppb to 10 ppm (NRC, 1985). Sublethal effects of oil on a variety of aquatic organisms have been studied. Effects on phytoplankton photosynthesis and growth have been reported at concentrations typically on the order of 0.1–10 ppm (NRC, 1985). Interference with larval development and growth occurs at crude oil concentrations ranging from 10 ppb to 100 ppm, depending on the organism and the type of oil. Adults generally tend to be more tolerant, but some sublethal effects have been reported at concentrations as low as 10 ppb (EPA, 1976).

The general conclusion is that toxic effects of oil are not observed below concentrations of 10 ppb. There are, however, a few exceptions to this generalization. A reduction in the chemotactic perception of food by the snail, *Nassarius obsoletus*, has been reported at a kerosene concentration of 1 ppb (EPA, 1976), interference with the growth of microalgae was observed at an oil (unspecified type) concentration of 0.1 ppb (EPA, 1976), and changes in the rate of development of sea urchin eggs have been observed at a benzo(a)pyrene concentration of 0.1 ppb (NRC, 1985).

Some organisms that are particularly resistant to oil pollution may flourish in oil-polluted waters or for a period of time after an oil spill due to the absence of more sensitive predators or competitors. Both the bloodworm *Tubifex* and the annelid worm *Capitella capitata* appear to thrive in oil-polluted sediments (Nelson-Smith, 1972), and Blumer et al. (1971) noted a population explosion of *C. capitata* in some oil-polluted areas of Buzzards Bay following the *Florida* spill. Thompson et al. (1977) reported that scarlet prawns, *Plesiopenaeus edwardsianus*, were three times more abundant in oiled sediments off the northwest coast of Arabia than in unpolluted control areas. Petroleum-derived compounds formed by bacterial metabolism of the oil apparently attracted the prawns. Cyanobacteria are apparently quite resistant to oil pollution, and patches of cyanobacteria have been noted on otherwise barren areas near refinery outfalls or on trickling filters designed to remove oil from effluents (Nelson-Smith, 1972).

Not a great deal is known about the effects of oil on macrophytes, most reported observations having been made following oil spills. There have been almost no laboratory studies. To a certain extent, plants are surprisingly resistant to oil toxicity, and even if extensively damaged, they tend to recover more rapidly than many other organisms (NAS, 1975). A good example of the resilience of plants is provided by events following the *Tampico Maru* spill. Giant kelps were seriously affected by the spill, but a population explosion of giant kelp occurred shortly thereafter (Nelson-Smith, 1972). The rapid recovery of the kelp was apparently due in part to the absence of the usual herbivore grazers, particularly sea urchins, which normally feed on the kelp (NAS, 1975). Evidently, these herbivores were unable to reestablish themselves in the affected area as rapidly as the kelps were. Species of *Ulva* and *Enteromorpha* have also been observed to proliferate following oil spills, again presumably due to the absence of the usual herbivore grazers (NRC, 1985).

These observations should not, however, be taken to imply that macrophytes prefer oily water. Oil does adversely affect macrophytes. For example, the oil spills resulting from the groundings of the tankers *Arrow* in 1970 and *Amoco Cadiz* in 1978 seriously impacted intertidal macrophytes. In most cases, however, macrophytes appear to recover rapidly following such spills, and because populations of their predators become reestablished slowly, a bloom of macrophytes may occur. Long-term adverse effects on macrophytes are evident, nevertheless, from some field studies. In the case of the *Arrow* spill, for example, one species of bladder wrack that disappeared during the spill had not reappeared six years later, and chronic adverse effects from oiling have been reported in certain sea grass communities (NRC, 1985). In the case of *Spartina*, a salt marsh macrophyte, adverse effects may not become apparent until a year or two after an oiling incident (NRC, 1985).

In summarizing this section, it is worthwhile to point out that the concentration of petroleum hydrocarbons in most parts of the ocean is less than 10 ppb (NRC, 1985). From the toxicity studies reviewed here, it would appear that most organisms are not likely to be affected by concentrations this low, particularly if these hydrocarbons consist predominantly of weathered oil from which many of the more toxic components have been removed. It is certainly true, however, that in some coastal areas oil pollution or various stresses or disruptions associated with oil production have significantly affected aquatic ecosystems. For example, the concentrations of petroleum hydrocarbons in polluted coastal sediments are often found to be as high as several thousand parts per million and in extreme cases as high as 19% on a dry weight basis (NRC, 1985). Natural hydrocarbon concentrations in unpolluted marine sediments are less than 250 ppm (NRC, 1985). Deep ocean sediment hydrocarbon concentrations are only 1–4 ppm (NAS, 1975).

An informative picture of the impact of oil production operations on a coastal ecosystem is provided by the case of coastal Louisiana, where more than 25,000 oil-producing wells are located in the same area as a large commercial fishery. From 1945 to 1975, it is estimated that at least 160,000 tonnes of oil were discharged into Louisiana's coastal waters, but the total production of the coastal fishery remained high (NAS, 1975). As was the case with Lake Erie, however, there were significant changes in the composition of the catch. The shrimp catch, for example, shifted from about 95% white shrimp to about a 50:50 mixture of white and brown shrimp. This shift in composition was apparently caused by destruction of the white shrimp's brackish water nursery grounds due to construction activities associated with the laying of pipelines and erection of drilling rigs. The original oyster beds were similarly destroyed, and although the total yield of oysters remained high, the yield per unit area on the new oyster grounds was only 11% of the pre-1945 figure (Howarth, 1981). Thus, a significant disruption of the coastal ecosystem occurred not so much due to the discharge of oil, but rather to construction and drilling activities that accelerated erosion, altered tidal currents, and caused the silting up of some embayments (NAS, 1975).

Human Health

With respect to human health, there seems to be little cause for concern from oil pollution of the aquatic environment. In drinking water, petroleum compounds become organoleptically objectionable at concentrations far below the levels associated with chronic toxicity (EPA, 1976). Thus, consumption of oil-polluted water is unlikely to be a significant source of exposure for humans. A more probable route of exposure is the consumption of oil-contaminated fish or shellfish, and in this regard, the chief concern is the fact that oil contains a number of carcinogenic compounds. Among these carcinogens are benzene and certain polynuclear aromatic hydrocarbons (PAHs). Benzene constitutes as much as 1.6% by volume of crude oil and up to 5% of refined products (Politzer et al., 1985). Because of benzene's volatility, however, the benzene content of other than recently discharged oil is likely to be quite small. Indeed, most human exposure to benzene comes from inhalation, the principal sources being automobile exhausts and cigarette smoke. Urban residents in the United States, for example, inhale about 0.85 mg d^{-1} of benzene (Politzer et al., 1985).

The input of PAHs to the worldwide aquatic environment is estimated to be 0.23 million tonnes per year, about 75% of which is accounted for by oil spillage (Politzer et al., 1985). It is doubtful, however, whether the consumption of PAH-contaminated fish is a significant human health concern. There is no evidence, for example, that any of the known carcinogenic PAHs cause cancer in humans by oral intake (Politzer et al., 1985). Furthermore, when the concentration of petroleum hydrocarbons in marine organisms exceeds about 200–300 ppm, the organisms acquire a distinctly tainted taste, become unpalatable, and are unlikely to be eaten anyway. Experimental studies have shown that organisms contaminated with PAHs can be depurated by simply putting them in clean water for a period of time. For example, the half-life of benzo(a)pyrene in mollusks is about 15 days (Politzer et al., 1985). Finally, seafood is by no means the only foodstuff containing PAHs. Fruits, vegetables, and cooked or smoked meats often contain PAH concentrations on the order of a few parts per billion (Politzer et al., 1985). Fish and shellfish do not contain unusually high concentrations of PAHs relative to other foodstuffs, and for most persons, the consumption of aquatic organisms is therefore a minor source of exposure to PAHs.

CASE STUDIES

Exxon Valdez

The Accident and Initial Containment Efforts

At four minutes past midnight on the morning of March 24, 1989, the supertanker *Exxon Valdez* ran aground on Bligh Reef in Price William Sound, Alaska (Figure 13.10). The tanker was carrying a full cargo (180,000 tonnes) of Alaska North Slope crude oil that it had picked up shortly before in the port of Valdez. The tanker had left the normal outbound shipping lane to avoid icebergs and subsequently failed to resume a normal course. Eight of the tanker's 11 cargo tanks and 3 of 5 ballast tanks were punctured on the reef rock. During the next five hours, approximately 20% of the *Exxon Valdez*'s cargo spilled into Prince William Sound. The Alyeska Pipeline Company was notified shortly after the accident and was instructed to activate its oil spill response plan. A barge transporting oil containment equipment left Alyeska's docks 10 hours after the spill and arrived at Bligh Reef 2 hours later. According to the oil spill response plan, the barge should have provided the containment equipment within five hours. However, the barge had been stripped of its gear prior to the spill. Reloading the barge consumed valuable time, and it was delayed when cranes loading the barge were redirected to load a tug bound for the stricken tanker (Morris and Loughlin, 1994). When the barge finally arrived at Bligh Reef, the number and size of the containment booms proved inadequate to contain the oil slick. Few skimmers were mobilized during the first 24 hours, and Alyeska did not have a tank barge into which the skimmers could offload recovered oil (Morris and Loughlin, 1994). During the first 60 hours after the spill, weather conditions were almost ideal for mechanical cleanup with booms and skimmers. The wind was no more than 5 knots, the visibility was excellent, and the seas were calm. However, any chance to contain the spill was lost because of Alyeska's delayed and inadequate response.

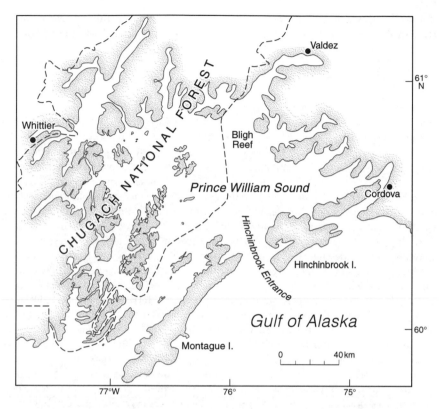

FIGURE 13.10   Map of Prince William Sound showing the location of the port of Valdez and Bligh Reef, the site of the grounding of the *Exxon Valdez*.

Within three days of the grounding, the tanker *Exxon Baton Rouge* arrived on the scene and began offloading the oil remaining in the *Exxon Valdez*. The offloading operation was completed by April 4. On April 5 the *Exxon Valdez* was refloated and towed to a sheltered harbor for temporary repairs.

In the initial days following the accident, efforts were made to burn some of the spilled oil and to break up the oil slick by adding chemical dispersants. The burning effort took place on March 26 and consumed an estimated 350 barrels of oil, about 0.14% of the quantity that had spilled from the tanker. From March 24 through the morning of March 26, several efforts were made to disperse the oil with chemicals. These efforts were largely futile because of the lack of mixing energy in the calm seas. By the afternoon of March 26 the winds had begun to pick up, and a test with chemical dispersants was successful (Morris and Loughlin, 1994). However, this window of opportunity soon closed. The winds increased to 25 knots by the late afternoon of March 26 and reached 50 knots by the end of the day. A winter windstorm was underway. The oil slick, about 6 km long by the afternoon of March 26, spread into a 64-km-long slick within hours. By the time the winds subsided two days later, the slick had spread over hundreds of square kilometers to the west and south. Three days later, the spill had contacted many kilometers of shoreline along islands in the western sector of Prince William Sound (Morris and Loughlin, 1994). The option of containment, if it ever existed, had been lost. There were in fact subsequent efforts to deploy booms and skimmers along local shorelines and embayments to prevent oil from entering strategic areas. "U.S. Coast Guard On-Scene Commander, Vice-Admiral

Robbins, later described this effort as 'trying to empty the ocean with a teaspoon' in its effect on decreasing the amounts of oil remaining in the environment" (Morris and Loughlin, 1994, p. 8).

Cleanup

When it became obvious that containment had failed, the response to the *Exxon Valdez* spill shifted to cleanup. Much of the cleanup effort focused on coastlines where the oil had washed ashore. Roughly 580 km of Prince William Sound shoreline and an additional 200 km outside the Sound were heavily oiled. The cleanup was a gargantuan effort involving thousands of workers and massive logistical support. The price tag for the *Exxon Valdez* spill was about $2 billion. More than 800 km of shoreline within Prince William Sound were treated with a variety of techniques in 1989, and there were follow-up cleanup efforts in 1990 and 1991. The cleanup efforts proved to be controversial. The standard cleanup procedures at most of the heavily oiled beaches involved cold-, warm-, and hot-water (up to 60°C) washings with a variety of devices. In some cases, high-pressure hoses were used to dislodge the oil. According to Joyce (1991), the hot water killed barnacles, mussels, clams, eelgrass, and rockweed and effectively sterilized many parts of the beaches. The washings removed only 20–25% of the beached oil, and the high-pressure hoses may actually have driven oil deeper into the beach deposits (Morris and Loughlin, 1994). In some cases, chemical dispersants and cleaning agents were used to try to remobilize the oil so that it could be more easily removed from beaches. These efforts were deemed ineffective. Finally, fertilizer was applied to many beaches in an effort to stimulate microbiological decomposition of the oil. This proved very effective for surface and near-surface oil. It was far less effective for oil buried deep in sand and gravel.

In contrast to human efforts, natural processes proved rather effective in removing oil from oiled beaches. The winter storms of 1989–1990, for example, removed about 90% of the oil from surface sediments (<25 cm) and about 40% of the subsurface oil (Morris and Loughlin, 1994). By 1992 the combination of human efforts and natural processes had removed virtually all the surface oil from oiled beaches, and today most beaches appear clean. Oil remains, however, buried in the sediments, under rocks and mussel beds, and in isolated surface patches.

Fate of the Spilled Oil

Figure 13.11 and Table 13.5 give a rough account of the fate of the oil that spilled from the *Exxon Valdez*. Given the huge effort that went into cleanup, it is thought-provoking to realize that only about 14% of the spilled oil was actually recovered and disposed of. About 20% simply evaporated away. Most of the evaporation loss occurred within the first 10 days after the spill. Fortunately, the most volatile components of crude oil include some of its most toxic constituents, including low molecular weight aromatics such as benzene, toluene, xylene, and naphthalene. As a result, the oil lost much of its toxicity within 10 days. About 50% of the oil was degraded by a combination of microbial attack and physical processes (photolysis). Only about 15% of the spilled oil remained in the ecosystem three and a half years after the spill. Most of that oil was deposited in subtidal sediments.

Effects on Organisms

Three excellent publications have documented as much as we know about the biological effects of the *Exxon Valdez* spill (Loughlin, 1994; Wells et al., 1995a; Rice et al., 1996). The short-term lethal effects of the spill on large animals have been relatively easy to document based on body counts. However, even when there is a body count, some allowance must

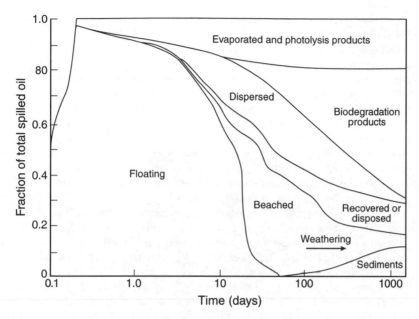

FIGURE 13.11   Fate of *Exxon Valdez* oil from the time of the spill to October 1, 1992. *Source*: Spies et al. (1996). Reproduced with permission.

be made for the fact that not all carcasses are counted. Some will sink, and others will be consumed by scavengers before they are counted. In the case of seabirds, for example, the actual carcass count was 36,000. Estimates of seabird acute mortality range from 100,000 to 645,000, with a best estimate of about 250,000 (Spies et al., 1996). Ninety species of birds were impacted by the spill, including bald eagles, several species of murres and mur-relets, pigeon guillemots, two species of puffins, four species of loons, and three species of cormorants (Spies et al., 1996). Marine mammals killed outright by the spill included sea otters and harbor seals. Sea otter mortality was estimated to be about 2,800 out of a population of 10,000 (28%). The harbor seal mortality was 300 from a population of 2,200 (14%). These are the reasonably well-documented short-term effects. What about the ef-

TABLE 13.5   FATE OF OIL SPILLED FROM THE *EXXON VALDEZ* AS OF OCTOBER 1, 1992, EXPRESSED AS A PERCENTAGE OF OIL SPILLED

Fate	Percent
Subtidal sediments	13
Beached	2
Recovered and disposed of	14
Dispersed in water column[a]	1
Aqueous biodegradation and photolysis products	50
Atmospheric photolysis products	20
Total	100

[a]Primarily as highly weathered refractory residuals.
Source: Spies et al. (1996).

fects on other species whose impacts were not so easy to measure, and what about the long-term effects? The following paragraphs summarize our current understanding.

Fish and Fisheries. Here there is both good news and bad news. The water column is at least 100 m deep in most parts of Prince William Sound (Wells et al., 1995b). As a result, the impact of the spill on the benthos was minimal in most parts of the Sound. The most serious long-term effects were expected to be caused by high molecular weight PAHs, and these were present at undetectable concentrations in the sediments of most of Prince William Sound (Wells et al., 1995b). On the other hand, the spill could not have occurred at a worse time from the standpoint of sensitive stages of fish such as eggs, larvae, and juveniles. Pacific herring spawn in the inshore waters of Prince William Sound in April, and pink salmon fry emerge from their gravel spawning redds and enter the nearshore environment at about the same time (Spies et al., 1996). Thus, there was considerable potential for the spill to impact some important fisheries.

Interestingly, the total adult returns of both hatchery and wild pink salmon to Prince William Sound were at record levels in 1990. While this might be taken to indicate the absence of an adverse impact, the return might have been even greater in the absence of an oil spill (Spies et al., 1996). Indeed, there is good evidence that juvenile salmon were directly affected by the spill in 1989, and their eggs may have been affected through 1993. Oil, for example, was visible in the stomachs of salmon fry, and chemical measurements documented the presence of PAHs in the fry (Spies et al., 1996). Reductions in growth were reported in 1989 for tagged pink salmon juveniles from hatcheries, for wild pink salmon, and for tagged cutthroat trout (Spies et al., 1996). There was no evidence of a change in food availability for the fish. The reductions in growth are therefore assumed to reflect the effects of oil exposure. Studies by the Alaska Department of Fish and Game revealed that pink salmon egg mortality was higher in oiled than in nonoiled streams from 1989 through 1993 but not after 1993 (Spies et al., 1996). Likewise, comparisons of oiled and nonoiled areas indicated that oil caused increased larval abnormalities in Pacific herring (Spies et al., 1996). Fortunately, 96% of the total 1989 herring spawn length (158 km) was free of oil, and hence only 4% of the spawn were exposed in 1989. There were no reported effects in 1990 (Wells et al., 1995b). The salmon and herring fisheries were closed in 1989 because shoreline cleanup activities resuspended oil in the nearshore areas where these fisheries are conducted, and fouling of gear was likely (Spies et al., 1996). Any adverse effect of the oil on the herring was apparently overwhelmed by the increase in stock size resulting from this relaxation of fishing pressure. The 1991 and 1992 herring harvests were large (Wells et al., 1995b). There was no evidence that sockeye salmon were directly affected by the spill, although smolts and adults returning to spawning areas had to swim through oil-contaminated waters (Spies et al., 1996).

Marine Mammals and Seabirds. Some marine mammals and almost all seabirds rely on the insulating properties of their pelage or plumage, respectively, to maintain their body temperature. When their fur or feathers become matted with oil, they are vulnerable to heat loss and can quickly die from hypothermia. Sea otters, for example, rely almost entirely on their dense fur for protection from the cold. Whales and pinnipeds (e.g., seals and walruses) are exceptions to this rule, since they rely much more on subdermal blubber for insulation. Given this information, it should not be surprising that sea otters suffered a much higher immediate mortality (28%) than harbor seals (14%). Fortunately, postspill studies of sea otter abundance, distribution, pup production, mortality, and foraging showed no relationship between oil amount and population change within Prince William Sound (Wells et al., 1995b). The impact of the spill on the sea otters therefore seems to have been short-lived, and the tentative prognosis for the otters is positive (Wells et al., 1995b). There was a reported loss of 13 killer whales from a pod between autumn 1988 and summer 1990, but it is unclear whether this loss was related to the oil spill. The pod in question had been taking fish from longlines, and there was photographic evidence of bullet holes in some of the whales (i.e., unhappy fishermen may have been shooting the whales).

Studies of bald eagle populations revealed a decrease in nesting success in oiled western Prince William Sound relative to the eastern sound, but there was no difference in the density of nests with eggs between oiled and unoiled areas in 1990 and 1991 (Wells et al., 1995b). Overall, 19 of 42 species of birds in Prince William Sound showed initial impacts from the spill. Fourteen of the 19 had recovered by the late summer of 1991 (Wells et al., 1995b); the remainder had not. On the Kenai Peninsula just west of Prince William Sound, 12 of 34 species showed negative impacts. Six of the 12 had recovered by 1991 (Wells et al., 1995b). Of all the seabirds, murres suffered the greatest population losses. Estimates of immediate mortality ranged from 74,000 to 289,000 birds (Wells et al., 1995b). Interestingly, studies of murre colonies revealed very similar historic and postspill numbers of adults as well as numbers of eggs produced (Wells et al., 1995b). Based on this evidence, it appears that the impact of the spill on murres was short-lived.

Intertidal and Subtidal Communities. The impact of the *Exxon Valdez* oil spill on intertidal and subtidal communities provided a provocative lesson for scientists and environmentalists. Remarkably, the well-intentioned efforts of many people to clean up the beaches seem to have done more harm than good. It is true that the cleaning removed large amounts of oil from intertidal areas, but it also damaged habitats. The hot-water washes were particularly destructive (Spies et al., 1996). The lower intertidal fauna on mixed soft-substrate beaches and sheltered rocky shores cleaned with hot water, for example, had recovered less by 1992 than the fauna on similar beaches that had been cleaned by other means or not at all (Spies et al., 1996).

Of the beach flora and fauna, intertidal organisms were most impacted by the spilled oil. Limpets had not recovered in the intertidal zone by 1991. The seaweed *Fucus gardneri* had recovered in the middle but not in the upper intertidal zone by 1992 (Spies et al., 1996). In contrast, small invertebrates making up the meiofauna were quite tolerant to the oil. Experimentally oiled sediments were abundantly colonized by meiofauna within a few days, and one year after the spill, epibenthic harpacticoid copepods, an important food source for juvenile salmon, were actually more abundant on heavily oiled than lightly oiled beaches (Spies et al., 1996).

No effort was made to clean mussel beds because of the potential adverse effects on predators. High concentrations of oil were found in more than 30 mussel beds three years after the spill (Spies et al., 1996). Concentrations of PAHs as high as 8 ppm were reported within tissues of some of the mussels.

Neither visible oil nor high concentrations of oil were found in subtidal sediments. Not surprisingly, then, impacts on subtidal organisms were much less apparent than was the case for intertidal organisms. The spill had no discernible effect on the abundance of subtidal fish, algae, or large epibenthic invertebrates (Spies et al., 1996).

Summary

There is no question that the *Exxon Valdez* oil spill resulted in some serious, often lethal impacts on intertidal organisms and some species of marine wildlife. At the population level, most species impacted by the spill had recovered or were well on the way to recovery within two to three years of the spill. Only about 14% of the spilled oil was cleaned up. Most of the remainder was degraded by natural processes. About 15% remains in the ecosystem, most in subtidal sediments. According to Spies et al. (1996), most of the oil remaining in the sediments has been so altered by degradation that it is no longer readily recognized as *Exxon Valdez* oil.

Arguably, the greatest long-term impact of the *Exxon Valdez* spill has been in the areas of law and politics. The State of Alaska, the federal government, and the oil industry have all implemented new prevention and response strategies to minimize the possibility of a second *Exxon Valdez* incident. For example, tug vessels now escort all outbound tankers through Prince William Sound. The U.S. Coast Guard has improved radar capabilities to track tankers in Prince William Sound and has installed a permanently lighted marker on Bligh Reef. Alyeska's response capability has been increased to include the ability to deploy high-capacity skimmers to any site within Prince William Sound within six hours and to maintain caches of boom equipment in easily accessible locations around the Sound (Morris and Loughlin, 1994). The chief response at the federal level has been passage of the Oil Pollution Act of 1990 (see the later discussion). Among other things, the act requires drug and alcohol testing for persons holding licenses to operate tankers and provides a schedule that requires virtually every vessel that carries oil in U.S. waters to have a double hull by the year 2015.

Buzzards Bay

On September 16, 1969, the oil barge *Florida* went aground within a few hundred meters of land off Fassets Point in Buzzards Bay, Massachusetts (Figure 13.12). Damage to the barge resulted in the release of its cargo of 570 tonnes of No. 2 fuel oil. A storm the following day drove the oil ashore, mixing it into water and sediments (Teal and Howarth, 1984). The spill occurred within 10 km of the Woods Hole Oceanographic Institution and therefore provided an excellent opportunity for scientists to study the impact of the spill intensively. Since the spill occurred in shallow water and consisted of a mixture of relatively low molecular weight hydrocarbons,[12] it was not surprising that the destruction of marine life in the immediate vicinity of the spill was severe. The oil penetrated into benthic sediments as much as 13 m below sea level and washed into marshes and tidal inlets, penetrating to 0.5 m or more in the inshore sediments. Sediment oil concentrations were as high as 400 ppm in sublittoral areas and over 2,000 ppm in some portions of the intertidal zone (NRC, 1985). In the area of the initial spill, which covered about 2 km^2, virtually all marine organisms were killed. The dead organisms included fish and shellfish, worms, crabs, snails, clams, lobsters, and numerous other invertebrates. Ninety-five percent of the organisms taken in trawls in the affected area were found to be dead or dying (Blumer et al., 1971), and marsh grass (*Spartina alterniflora*) was completely killed in intertidal areas where oil concentrations exceeded 2,000 ppm in the sediments.

The killing of benthic plants and animals resulted in the destabilization of the bottom sediments, a result similar to the effects of thermal pollution on the benthos of Biscayne Bay. The oil-contaminated sediments were then easily eroded by currents and wave action and were transported to other parts of the bay, carrying the oil with them. In this way the polluted area was rapidly enlarged, and by the spring of 1970 it covered about 22 km^2, roughly 10 times the area affected by the initial spill (Blumer et al., 1971).

Analyses of surviving shellfish revealed that they were contaminated with petroleum hydrocarbons for several years after the spill; as a result, shellfishing was banned in the area during that time. Placing shellfish in clean water for as much as six months failed to rid them of their hydrocarbon burden (Blumer et al., 1971). As expected, the hydrocarbons had been stored in the lipid tissue of the shellfish, where at least some of them remained, resistant to biological breakdown or excretion, much like chlorinated hydrocarbon pesticides. Damages paid by the owners of the oil for cleanup and as compensation for the

---

[12]No. 2 fuel oil is a type of diesel oil, which contains hydrocarbons that distill above 275°C. The hydrocarbon range is primarily C_{12}–C_{25}, with some higher molecular weight components (Politzer et al., 1985).

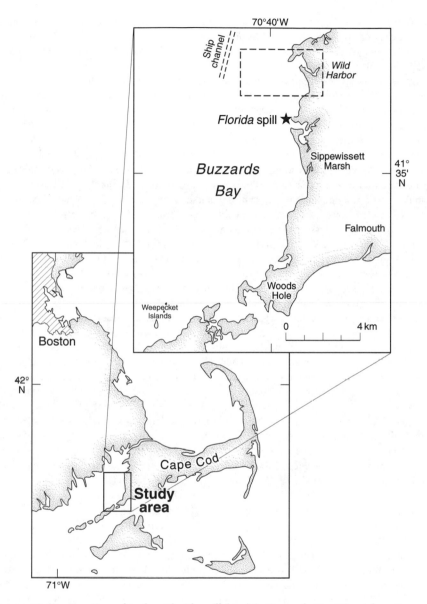

FIGURE 13.12 Location of tanker *Florida* spill (star) in Buzzards Bay, Massachusetts. The rectangle including Wild Harbor was intensively sampled by scientists. *Source*: Redrawn from Sanders et al. (1980). Reproduced with permission.

destruction of fishing and shellfishing resources totaled about $500,000 (Blumer et al., 1971).

Although some bacterial breakdown of the petroleum hydrocarbons in the sediments was evident in all parts of the bay within a year of the spill (Blumer et al., 1971), a seven-year study of fiddler crab populations in the adjacent salt marshes by Krebs and Burns (1977) revealed that some toxic components of the oil persisted in these sediments for many years. Fiddler crab populations in heavily oiled areas of the marsh were greatly reduced in number following the spill, and living crabs found in these areas displayed aberrant behavior, including impaired locomotor and burrowing behavior, lengthy escape response time, increased molting, and the display of mating colors, although the time of the spill did not correspond to the mating season. Juvenile crabs were unable to overwinter successfully in areas of the marsh where sediment petroleum hydrocarbon concentrations exceeded 200 ppm, evidently due to the abnormally shallow burrows dug by crabs in these areas. As a result, the juveniles evidently died from exposure to the cold. The abnormal burrow construction was attributed by Krebs and Burns (1977) to impairment of the crabs' locomotor abilities by ingested hydrocarbons. Some crabs were found to contain as much as 280 ppm hydrocarbons in body tissues. For at least four years after the spill, the percentage of female crabs in some areas of the marsh was abnormally low, presumably due to differential mortality or emigration from polluted areas. By the seventh year after the spill, crab populations in most areas of the marsh were returning to normal, although recovery in some parts of the marsh was still incomplete. The recovery pattern was highly correlated with the loss of the naphthalene fraction of the oil in the sediments. Oil and some of its effects persisted in part of the area impacted by the spill for at least 12 years, and although most areas of the salt marsh were normal in gross appearance by that time, complete recovery had not yet occurred (Teal and Howarth, 1984).

In summarizing the effects of the Buzzards Bay spill, several points deserve emphasis. First, the persistence of the hydrocarbons in the ecosystem and their effects on organisms far exceeded any visible evidence of the spill that a casual observer might have noted. Because the spilled oil was relatively light and contained few of the high molecular weight residual oils found in crude oil, visible evidence of the spilled oil on the beaches and salt marshes was gone within a few days. The bodies of the many dead animals initially killed by the spill had decomposed or washed away within a similar time period. Thus, to a casual observer, any evidence of the spill or its effects had disappeared within a week. Only careful chemical analyses of the sediments and organisms revealed the persistent presence of petroleum hydrocarbons in the ecosystem, and a detailed, thorough study of benthic and salt marsh organisms was required to appreciate the extent and duration of the biological damage caused by the spill.

Second, the persistence of the hydrocarbons in the ecosystem greatly exceeded what might have been expected from the characteristics of the oil and the nature of mechanisms for detoxifying or removing spilled oil from the ocean. Although some of the lighter hydrocarbons undoubtedly evaporated away, the fact that other low molecular weight hydrocarbons evidently found their way into the sediments and continued to kill organisms for several months after the spill indicates that some of the acutely toxic low molecular weight hydrocarbons persisted in these sediments for at least several months. Even though bacterial breakdown of the hydrocarbons was evident in all parts of the affected area within one year of the spill, it is generally the case that the least toxic hydrocarbons are metabolized first (Blumer et al., 1971), and it is apparent that some toxic hydrocarbons remained in the ecosystem for roughly 10 years (Krebs and Burns, 1977; Teal and Howarth, 1984). Once stored in the lipid tissues of organisms or in sediment, which is usually anoxic below the first few millimeters, petroleum hydrocarbons are subject to only slow degradation and thus mimic persistent poisons such as chlorinated pesticides in their longevity (Blumer et al., 1971). Metabolic studies revealed the induction of hydrocarbon-metabolizing enzymes in fish, and within four years of the spill, hydrocarbon burdens in the fish were down to background levels (Burns and Teal, 1979). Induction of similar enzymes was also evident in fiddler crabs, but was evidently insufficient to deal with the body burden of hydrocarbons within the lifetime of these organisms (Burns, 1976).

Finally, it is appropriate to compare the effects of the Buzzards Bay spill with the impact of the *Exxon Valdez* spill and some of the other noteworthy oil spills that have occurred in the past 25 years. In terms of the amount of oil released, the Buzzards Bay spill was certainly not one of the larger spills. The *Exxon Valdez* spill released 63 times as much oil. The Ixtoc I oil well blowout, the largest accidental oil spill of any kind, released about 840 times as much oil, and the grounding of the *Amoco Cadiz*, one of the largest tanker accidents to date, released about 400 times as much oil. The *Exxon Valdez* spill killed several hundred thousand seabirds, several thousand sea otters, and several hundred harbor seals. It impacted salmon and herring reproduction, but the impact lasted for no more than one to two years. At the population level, most species recovered from the spill within a few years. The biological impact of the Ixtoc I blowout was not very thoroughly studied, but the damage to aquatic organisms appears to have been surprisingly small, considering the amount of oil released. Shrimp landings during subsequent years in the affected area, for example, were as high as or higher than those in the years preceding the blowout (Mielke, 1990). There was some loss of aquatic birds due to oiling, but along the Texas coast, the percentage of oiled birds never exceeded 10%. The minimal impact on organisms along the U.S. coast undoubtedly reflected the extensive weathering of the oil at sea and the fact that only about 1% of the spilled oil actually reached U.S. beaches. There was a loss of revenue to the recreational and tourist industries along the Texas coast in the months following the accident, but a year later, there was almost no visual evidence of the spill. Cleanup costs along the Texas coast amounted to about $4 million. However, the lost oil revenues and estimated total cost to the Mexicans of capping the well and containing the environmental damage amounted to about $220 million (Mielke, 1990).

The *Amoco Cadiz* spill was similar to the *Florida* grounding in the sense that the oil was discharged almost on the beach and strong winds mixed the oil into the water column and drove it ashore. The quantity and type of oil discharged, however, were quite different. The *Amoco Cadiz* was carrying about 0.22 million tonnes of light Arabian crude oil, and virtually the entire cargo was lost. About 400 km of French coastline were oiled. The impact of the *Amoco Cadiz* spill on biota was thoroughly studied, but any analysis of long-term effects was seriously compromised by the fact that almost exactly two years after the *Amoco Cadiz* accident, the tanker *Tanio* broke up about 65 km off the Brittany coast and released 7,000 tonnes of oil into many of the same areas affected by the *Amoco Cadiz* spill.

Not surprisingly, there was almost a complete kill of the flora and fauna in exposed intertidal mud flats and marsh areas after the *Amoco Cadiz* spill (Teal and Howarth, 1984). In the water column, there was considerable mortality of zooplankton and an immediate kill of several tonnes of fish. About 4,500 seabirds are known to have died from oiling. There were declines in recruitment, growth rates, and catches of several commercially important bottom fish, as well as an unusual incidence of fin rot in some species during the first year after the spill. Adverse effects on plaice continued to be found for several years in heavily impacted areas, and some oysters were still contaminated with hydrocarbons, although there was little evidence of histopathological or biochemical damage (Mielke, 1990). France won a court settlement of $120 million against Amoco for the damage caused by the spill (Joyce, 1990).

Summary

The general consensus that emerges from this analysis of oil spills is that circumstances at the time of the spill play a large role in determining their environmental impact. Spills that occur close to shore are likely to cause much more environmental damage than spills in the open ocean. In the former case, there is a high probability that benthic and intertidal organisms will be impacted, and there is less time for weathering to reduce the toxicity of the oil before significant impacts occur. This fact is dramatically illustrated in the comparison of the effects of the Ixtoc I and *Amoco Cadiz* spills. Unfortunately, the concentration of marine organisms, including commercially important species, is much greater in coastal areas than in the open ocean. Thus, the potential for damage is far greater when spills occur in coastal areas. As already noted, much of the oil released to the ocean is unfortunately discharged in the nearshore environment.

Most scientists agree that, in general, the impact of a spill of refined petroleum is likely to cause more ecological damage than the release of a comparable amount of crude oil because the latter contains a smaller percentage of the more dangerous low molecular weight components. This rationale presumably accounts for the extent of biological damage caused by the comparatively small amount of refined oil released by the tanker *Florida*. While a large spill of crude oil can potentially expose organisms to the same concentrations of these toxic components as a small spill of refined product, it seems clear that most of the biological damage caused by the *Exxon Valdez* spill was associated with oiling effects. Large numbers of birds and vulnerable marine mammals died of hypothermia when their plumage or pelage became matted with oil. Given the environmental and monetary costs associated with these oil discharges, it is now appropriate to ask what can be and has been done to reduce them and to mitigate their effects when they occur.

CORRECTIVES

The wreck of the tanker *Torrey Canyon* off the coast of England in 1967 is a milestone in the history of oil pollution for several reasons. First, the spill was clearly the result of carelessness, an issue that has remained up to the present day (Hodgson, 1990). On a morning when the visibility was 13 km and his ship was within range of a lightship, three lighthouses, and a radio beacon, an experienced captain ran his tanker at top speed onto Seven Stones Reef, a well-known navigational obstacle in a channel almost 10 km wide (Ware, 1989). Second, efforts to clean up the spill dramatically illustrated the inadequacy of understanding and techniques for minimizing the environmental damage caused by such spills. It is now generally agreed, for example, that the detergents used to disperse the oil caused more damage to marine organisms than the oil itself (Nelson-Smith, 1967). Finally, the public reaction to the spill presaged an era of environmental awareness. Similar reactions to subsequent spills have been influential in pressuring governments and the oil industry to take steps to minimize the amount of oil discharged to the environment and to mitigate the adverse effects caused by the oil that is released.

Prevention

Certainly one area where significant progress has been made is in the routine operations of oil tankers. The use of LOT and COW and the requirements for clean or segregated ballast tanks mandated by MARPOL and the 1978 U.S. Port and Tanker Safety Act have done much to reduce routine discharges of oil from tankers due to ballasting operations, once a major source of oil discharges to the ocean. In addition, the Port and Tanker Safety Act mandates that tankers of 20,000 dwt or above be equipped with inert gas systems to prevent fires and explosions in the cargo tanks and that tankers of 10,000 gross tons or above be equipped with a dual radar system and two remote steering gear control systems operable separately from the navigating bridge. The requirement for backup radar and steering systems was undoubtedly influenced by the *Amoco Cadiz* grounding, which was caused in part by a faulty steering system. It is noteworthy that the Port and Tanker Safety Act applies not just to vessels of U.S. registry, but to any vessel that operates on or enters the navigable waters of the United States, or that transfers oil or hazardous materials in any port or place subject to the jurisdiction of the United States.[13] Thus, for example, any foreign oil tanker that offloads oil at U.S. ports must abide by the stipulations of the Port and Tanker Safety Act.

Although such legal developments have helped to reduce the probability of oil spillage, the efficacy of the legislation depends in part on conscientious enforcement and human motivation. The Port and Tanker Safety Act, for example, stipulates that no person may be issued a federal license to pilot any steam vessel unless he is of sound health and has no physical limitations that would hinder or prevent the performance of a pilot's duties. Nevertheless, Joseph Hazelwood, the captain of the *Exxon Valdez*, had a history of alcoholism, having been arrested several times for drunken driving and treated in 1985 for alcohol abuse. At the time of the grounding, Hazelwood was absent from the bridge, having turned over the ship to an uncertified third mate. Hazelwood had an unacceptably high blood-alcohol level nine hours after the accident (Lemonick, 1989). The fact that tanker accidents sometimes occur during inclement weather is understandable. However, accidents such as the wrecks of the *Torrey Canyon* and the *Exxon Valdez*, both of which occurred during fair weather, are inexcusable. Such events certainly raise the question of whether there is always a competent person on the bridge when tankers are underway. Tanker collisions and groundings account for 46% of all oil spills greater than 7 tonnes and 63% of all spills greater than 700 tonnes (ITOPF, 1999). How many of these spills were at least in part the result of human error or incompetence is unclear, but it seems reasonable to speculate that better training and adherence to rules might reduce the frequency of these incidents.

One area where some very significant improvement can be anticipated in the next decade is the condition and maintenance of the world's tanker fleet. As of 1991, the fleet numbered about 2,750 vessels, 55% of which were more than 15 years old (Drewry, 1991a). Over 430 of the fleet were supertankers, and these accounted for almost 50% of the fleet's dead weight tonnage. About 70% of the supertankers were more than 15 years old, most of them having been built between 1972 and 1976 (Drewry, 1991a). Although tankers do go into dry dock at approximately two-year intervals for routine maintenance and inspection, there is evidence that at least some of the tankers that put out to sea have been structurally unsound. Carter (1978, p. 514), for example, citing the number of tanker accidents chalked up to structural failure, commented that in some cases "an old rust bucket had simply broken up at sea and sunk." This problem still exists. On January 24, 1990, for example, a tanker split in two in Wakasa Bay, Japan, discharging 770 tonnes of oil. On August 4 of the same year, a tanker sank during a typhoon near Japan, discharging over 800 tonnes of oil. At least five other oil discharges during 1990 were attributable to tankers' sinking (Anonymous, 1991a).

The development that promised to change this picture was the passage of the Oil Pollution Act (OPA) of 1990 (Public Law 101–380) by the U.S. Congress. The impetus for the OPA was the *Exxon Valdez* spill in Prince William Sound the previous year, and much of the act is concerned with economic fines and penalties designed to discourage such incidents in the future. However, a portion of the OPA is intended to minimize oil spills through improved tanker design, operational changes, and greater preparedness (NRC, 1999). Section 4115 of the OPA requires that tankers that trade in U.S. waters be equipped with double hulls to prevent or minimize spillage when an accident occurs. Single-hull tankers of 5,000 gross tons or more will be excluded from U.S. waters after 2010 unless they are equipped with a double bottom or double sides, in which case they may be permitted to

---

[13]The only noteworthy exception to this restriction is vessels engaged in so-called innocent passage through the territorial sea of the United States or in transit through the navigable waters of the United States which form a part of an international strait.

trade to the United States through 2015, depending on their age.[14] Beginning in 2000, all tankers larger than 80,000 dwt without double bottoms or double sides that are older than 23 years are barred from U.S. trade. Although the OPA affects only tankers that trade to the United States, following passage of the OPA, two additions were made to MARPOL that affect the entire world tanker fleet. The changes mandate a transition to double-hull vessels or their equivalents by 2023 and specify a schedule for retrofitting or retiring single-hull vessels more than 25–30 years old. By 1994, 10% of the world's tanker fleet, which currently numbers 3,300 vessels, were equipped with double hulls (NRC, 1999).

Cleanup

Oil discharges result from human error, acts of God, sabotage, and routine operations associated with the transportation and use of oil. No matter how conscientious and careful we may be, discharges are bound to occur. Some of these discharges may be of little environmental consequence, but when a spill occurs with potentially serious environmental consequences, what can be done to mitigate the damage? Unfortunately, every cleanup method has drawbacks or limitations, but it is worthwhile to review the state of the art to see what approaches are available.

Offloading

When a tanker goes aground, collides with another ship, or for any reason is in danger of discharging its cargo, an obvious move to minimize environmental damage is to offload the oil. Much has been made, for example, of the fact that the *Exxon Valdez* discharged 0.036 mt of oil into Prince William Sound, but few people are aware of the fact that the tanker was carrying five times as much oil. Almost 80% of the cargo was successfully pumped into smaller vessels while the tanker lay atop Bligh Reef (Hodgson, 1990). Unfortunately, tankers often go aground in rough weather, and under such conditions it is difficult, if not impossible, to attempt a reasonably safe offloading operation. For example, none of the 0.025 mt of No. 6 fuel oil carried by the *Argo Merchant* was offloaded when the tanker grounded on the Nantucket Shoals in December 1976. Instead, rough seas during the following week caused the tanker to break up, releasing its entire cargo. The *Amoco Cadiz* similarly lost its entire cargo when it broke up off the coast of Brittany during heavy seas in 1978. The U.S. Coast Guard has developed special offloading systems involving hydraulically operated submersible pumps for offloading oil in high-risk situations, but the fact remains that it is simply unsafe to bring a receiving vessel alongside a stricken tanker in rough seas.

Burning

Once oil has been spilled onto the water, there are basically three short-term approaches to cleanup, namely, burning, chemical dispersal, and mechanical collection. The idea of burning an oil slick is logically appealing, at least if the slick is in a place where the fire itself is not likely to pose a threat, but in practice, it has been difficult to ignite and maintain a combustion of heavy oil, even with the aid of flame throwers. "In the case of the *Torrey Canyon* disaster, the addition of thousands of gallons of aviation fuel and Napalm combined with aerial bombing of the ship failed to produce any sustained burning" (Tully,

---

[14]A caveat is that single-hull tankers may trade to the United States if they unload their cargo offshore at deepwater ports or in designated lightering areas through 2015.

1969, p. 83). The difficulty in burning such slicks arises because many of the more volatile (and more flammable) hydrocarbons evaporate away; and because heat is rapidly transferred to the water beneath the oil, the temperature of the oil falls below the flash point. Oil can be burned if it is confined, but the tendency of spilled oil to spread out in a thin film on the surface of the water makes burning difficult. In general, maintaining a burn of a surface slick requires that the oil be relatively fresh and at least 3 mm thick (Westermeyer, 1991). It is possible, through the application of chemical additives to a slick, to maintain a controlled burn (Tully, 1969), and some small-scale experiments have produced burn percentages greater than 90% (Westermeyer, 1991).

There are some examples in which unintentional fires consumed a significant percentage of spilled oil. When the tanker *Burmah Agate* collided with a freighter in 1979, much of the 0.034 mt of oil discharged from the tanker was consumed in a fire caused by the collision that burned out of control for over two months. Similarly, when a series of explosions caused the supertanker *Mega Borg* to catch fire in the Gulf of Mexico in June 1990, most of the spilled oil either burned or evaporated (Yamaguchi, 1990). However, deliberately setting fire to a grounded tanker could prove futile, and at worst could lead to an explosion that might cause far more damage than would otherwise have occurred. Thus, for several reasons, oil burning is not currently an important oil spill response tool of any country (Westermeyer, 1991).

Chemical Dispersal

Detergents or similar surface active substances may be added to an oil slick to break up the film of oil into small droplets that become dispersed in the water column. The rationale for applying such dispersants is based in part on aesthetic considerations, since the dispersed oil is much less visible to an observer than a surface slick. It is also assumed, however, that dispersing the oil into small droplets hastens its breakdown by microorganisms, since the contact area between oil and water is increased (Canevari, 1969). However, there are a number of drawbacks to the use of chemical dispersants. Evaporation of the oil is reduced because less oil is concentrated at the surface; and because the oil is dispersed throughout the mixed layer, a greater number and variety of organisms are likely to come into contact with it. As noted by the EPA (1972, p. 262): "Because of the finer degree of dispersion, the soluble toxic fractions dissolve more rapidly and reach higher concentrations in sea water than would result from natural dispersal. The droplets themselves may be ingested by filter-feeding organisms and thus become an integral part of the marine food chain." One of the most serious problems with the early use of dispersants was the toxicity of the dispersants themselves, particularly if their solvent fractions consisted of a mixture of hydrocarbons. In the case of the *Torrey Canyon*, the dispersant applied to the oil slick contained a petroleum base solvent rich in low molecular weight aromatics (Nelson-Smith, 1972), and it is now generally agreed that this dispersant caused far more damage to marine life than the crude oil itself. As a result of such experiences, dispersants were reformulated. They are now more effective and far less toxic (Westermeyer, 1991).

At the present time, it appears that dispersants probably do serve a useful purpose in the cleanup of some oil spills. For example, where there is danger that large numbers of aquatic birds may be killed due to coating from an oil slick, it may be more desirable to break up the slick with a dispersant and risk some damage to other aquatic species than to allow the birds to become coated with oil (Straughan, 1972). In fact, the United Kingdom relies almost exclusively on dispersants to combat oil spills (Westermeyer, 1991), and the NRC has recommended that they be considered a potential first-response tool (NRC, 1989). As a general rule, however, dispersants should be applied to an oil slick while the slick is still far enough from shore that mixing processes rapidly dilute the oil-dispersant mixture to subtoxic concentrations. It is noteworthy, for example, that in the case of the

Torrey Canyon, dispersants were applied directly to oil that had come ashore, and as a result, intertidal organisms were subjected to extremely high concentrations of the oil-dispersant mixture. According to the NAS (1975), the damage to organisms was perhaps due as much to the misuse of the dispersant as to its inherent toxicity.

Mechanical Containment and Cleanup

Mechanical means for cleaning up oil spills involve the use of booms or skimmers and absorbents. Because of concern over the effectiveness and toxicity of dispersants, this approach has been used almost exclusively to combat oil spills in the United States (Westermeyer, 1991). Unfortunately, booms and skimmers are ineffective in currents greater than 1 knot, waves higher than about 2 m, or winds greater than 20 knots (Westermeyer, 1991). The devices are more effective in retrieving spills of crude oil than of refined products (NAS, 1975). If an oil spill is successfully contained by booms or similar devices, there is still the problem of removing it. For example, an oil-absorbent material in the form of a continuous belt may be used to sop up the oil. Extraction of the oil from the belt is accomplished with the use of special rollers, which scrape and squeeze the oil into a recovery vessel (Hodgson, 1990). Oil may also be sopped up by distributing straw or polyurethane foam over the oil and then collecting the oil-soaked absorbent. In general, the efficiency of absorbents in removing oil depends on the characteristics of the oil, such as viscosity and surface tension, but under appropriate conditions, mechanical containment and absorption can be a highly effective means of cleaning up an oil slick. Unfortunately, the most serious spills often occur in foul weather, when containment devices are of little use.

Sinking

A fourth cleanup method is worthy of mention not because it is recommended for use, but because it has been used in the past, and there is a need to point out why it should probably not be used in the future. The technique is to sink the oil by adding sand, talc, or chalk so as to cause the oil to agglutinate into globules more dense than water. While this method does remove the oil from sight, it should be clear that any benthic organisms beneath the slick are likely to be severely stressed, if not killed, by the blanket of oily globules. Furthermore, as Blumer et al. (1971) have noted, degradation of hydrocarbons in sediments is likely to proceed at a much slower rate than in the water column, and movements of oil-laden sediments may spread the pollution over an area many times larger than the initial spill. In this respect, it is noteworthy that "Oil from the blowout at Santa Barbara was carried to the sea bottom by clay minerals and . . . within four months after the accident the entire bottom of the Santa Barbara basin was covered with oil from the spill" (Blumer et al., 1971, p. 11). Thus, sinking oil onto the sediments does not appear to be a desirable means of cleaning up a spill, and the EPA (1972, p. 263) has stated that "Sinking of oil is not recommended." Nevertheless, one response suggested by a European scientist to a recent Persian Gulf oil spill was to spray the slick with cement powder to bind the oil and make it sink *en masse* (Aldhous, 1991). Other scientists were skeptical about this idea (Aldhous, 1991).

Bioremediation

Bioremediation of oil spills involves the stimulation of microbial breakdown of the spilled oil. In theory, bioremediation may take any or all of three forms: stimulation of indigenous microorganisms through addition of nutrients, introduction of special assemblages of natural oil-degrading microbes, and introduction of genetically engineered microorganisms with special oil-degrading properties (OTA, 1991). Bioremediation is not a first-response cleanup tactic because, even under ideal circumstances, the time required for bioremediation to work is on the order of weeks or months. Thus, for example, spraying nutrients on an offshore oil slick will not keep the oil off the beaches. Although bioremediation is still very much an experimental tactic, the successful use of this technique to help clean up almost 180 km of beaches following the *Exxon Valdez* spill has stimulated considerable interest in its potential usefulness.

There are at least 70 microbial genera known to contain organisms capable of degrading components of petroleum (OTA, 1991). Most of these organisms are either bacteria or fungi, and they are widely distributed in aquatic systems. No one organism or even a combination of organisms is capable of degrading the thousands of compounds found in crude oil, but given time, there is no doubt that a large percentage of the compounds found in most natural oils can be degraded by microbes. Normal alkanes are the hydrocarbons most easily metabolized. Microorganisms are capable of utilizing n-alkanes with up to 44 carbon atoms, with C_{10}–C_{24} compounds usually being the easiest to metabolize. Branched-chain alkanes are degraded more slowly than n-alkanes, and cycloalkanes are even more resistant, but all are subject to biodegradation. Low molecular weight aromatics may also be metabolized, but they are somewhat more resistant than the alkanes and cycloalkanes. By far the slowest degradation rates are observed for high molecular weight aromatics, asphaltenes and resins.[15]

The factor that seems most frequently to limit the rate at which microorganisms can degrade spilled oil is the availability of inorganic nutrients, in particular N, P, and perhaps trace metals. This scenario is very reminiscent of nutrient limitation in microalgae and is the rationale for the first type of bioremediation. In the case of the *Exxon Valdez* cleanup, for example, simply applying fertilizer to oiled beaches increased oil biodegradation rates by as much as a factor of 2–4 for more than 30 days. A second application after three to five weeks stimulated microbial activity 5- to 10-fold (OTA, 1991).

Once inorganic nutrients have been applied, the principal factors limiting oil biodegradation rates are the nature of the oil, the concentration and diversity of the microbial population, the availability of O_2, and the temperature. Natural weathering processes, and evaporation in particular, tend to remove some of the most easily metabolized components of the oil, and weathered crude oil is therefore much less subject to biodegradation than a fresh spill. The natural population of oil-metabolizing organisms can be expected to increase greatly during the first few days or weeks following a spill, and speeding up this process is, of course, the rationale for seeding a spill with natural oil-degrading microbes. The availability of O_2 can be a serious limiting factor on low-energy beaches and in fine-grained sediments because anaerobic biodegradation proceeds several orders of magnitude more slowly than aerobic biodegradation. There is general agreement, for example, that the success of the *Exxon Valdez* bioremediation effort was due in part to the fact that the beaches treated were all very coarse, consisting mostly of cobbles and coarse sand (OTA, 1991). The slow rates of oil degradation noted in some low-energy French beaches after the *Amoco Cadiz* spill were undoubtedly the result of O_2 limitation (OTA, 1991). Slow anaerobic degradation can also be expected if oil sinks to the sea floor and is covered by sediments, one of the principal arguments against sinking oil. Finally, over the range of temperatures likely to be encountered in natural waters, microbial rates will tend to be positively correlated with temperature, reported Q_{10} values generally being in the range 2–3. Thus, other factors being equal, bioremediation is likely to be more effective in tropical than polar climates. The cold temperatures in Prince William Sound, however, did not prevent bioremediation from greatly improving conditions on many of the beaches there.

---

[15]Asphaltenes are nonvolatile solids with molecular weights exceeding 1,000 and consisting mainly of polycyclic and heterocyclic aromatic structures with alkyl and naphthenic side chains. Resins include polar and often heterocyclic compounds containing N, S, and O as constituents.

Summary

One thought-provoking fact about oil spills is that, in most cases, very little of the spilled oil has been cleaned up. Evaporation alone may rapidly remove as much as 30–60% of the oil, and much of the remainder is slowly lost through natural weathering processes and biodegradation. Less than 10% of the oil discharged as a result of the *Amoco Cadiz* and Ixtoc I accidents and virtually none of the oil spilled by the *Florida* and the *Argo Merchant* was ever cleaned up (Teal and Howarth, 1984).

This realization does not mean that we should simply stand back and let nature take its course. It does, however, mean that nature is on our side, and this fact is important to keep in mind during cleanup operations. Some well-intentioned actions taken in an effort to clean up spilled oil have probably done more harm than good because they have adversely affected natural cleansing mechanisms. The use of hot water to clean up oiled beaches after the *Exxon Valdez* spill is a case in point. In commenting on this incident, National Oceanic and Atmospheric Administration official Sylvia Earle said, "As far as Alaska's shoreline is concerned, the environment would have been better off if there had been less aggressive hot-water treatment and we had let nature take its course. Sometimes, the best, and ironically the most difficult, thing to do in the face of an ecological disaster is to do nothing" (Joyce, 1991, p. 14).

It seems fair to say that our understanding of how to deal with spilled oil has improved considerably since the days of the *Torrey Canyon* disaster, when well-intentioned persons tried futilely to burn the slick with napalm and aerial bombing, and then spread highly toxic chemicals on the beach in order to disperse the oil that had washed ashore. We now realize that the most effective arsenal for cleaning up oil spills includes a combination of mechanical containment methods, absorbents, dispersants, and bioremediation, and that the extent to which one or another method should be used depends on the circumstances. Mechanical containment, absorption, and dispersion are first-response techniques, the last being best suited for offshore spills and rough seas. Mechanical containment and absorption work well in calm weather and nearshore environments. Bioremediation is more of a finishing or polishing tool and appears best suited for high-energy beach cleanup.

Oil Fingerprinting

When oil is present in the environment, how can we tell whether the oil is derived from petroleum or some other source, and in the former case, is there some way to identify the source of the petroleum if it is not obvious? The answer to both questions is a qualified yes.

Petroleum hydrocarbons differ in several significant ways from strictly biogenic hydrocarbons. For example:

1. In biogenic hydrocarbons there is a predominance of odd-number-C normal alkanes. In petroleum the ratio of odd- to even-number-C normal alkanes is near unity (NRC, 1985). The relative number of odd- and even-number-C normal alkanes is usually expressed as the odd-even preference ratio (OEP) or carbon preference index (CPI).
2. Biogenic hydrocarbons generally contain a much smaller percentage of cycloalkanes and aromatics than does petroleum. Aromatics, for example, account for less than 1% of the hydrocarbons in most marine organisms but up to 39% of some crude oils (Nelson-Smith, 1972; NAS, 1975).
3. Petroleum hydrocarbons contain a higher $^{13}C/^{12}C$ ratio and a much lower $^{14}C/^{12}C$ ratio than biogenic hydrocarbons.

The above list is by no means exhaustive, but it serves to make the point that there are characteristic differences between biogenic and petroleum hydrocarbons that can be used to determine the origin of the hydrocarbons, at least in a semiquantitative way. The problems with this approach are basically three. First, many of the important differences can at best be expressed in a qualitative manner. Second, there are exceptions to some of the rules. For example, the predominance of odd-number-C normal alkanes in C_{20}–C_{30} biogenic hydrocarbons has not been observed in some marine bacteria and fish (NRC, 1985). Thus, an OEP or CPI of approximately 1 could not, by itself, be taken to indicate that hydrocarbons were of petroleum origin. However, an OEP or CPI substantially greater than 1.0 would certainly imply a biogenic origin. A third issue is the fact that weathering can substantially change the chemical composition of oil, whether of petroleum or biogenic origin. However, weathering does not significantly change stable isotope ratios such as $^{13}C/^{12}C$ (NRC, 1985); hence, these may continue to serve as useful indicators of the source of the hydrocarbons.

Since hydrocarbons are produced naturally by aquatic organisms, the mere presence of hydrocarbons in the water cannot be considered a sign of petroleum pollution. Analyses based on the foregoing techniques indicate that concentrations of biogenic hydrocarbons in most aquatic systems are on the order of a few parts per billion. When there is evidence of significant petroleum inputs, a question naturally arises as to the source of the petroleum. The source could be natural (i.e., seeps), or it could be pollution. How can we determine the source of the petroleum if it is not obvious?

The answer is that the chemical composition of petroleum differs greatly from one region of the world to another, and these differences can be used in many cases to determine the source of spilled oil. Figure 13.13, for example, reveals dramatic compositional differences in oils from various sources based on a simple chromatographic analysis. The oils were first heated to drive off all the hydrocarbons having a boiling point less than 316°C to simulate the effects of weathering. They were then analyzed on a gas chromatograph, an instrument that partially separates the hydrocarbons in the oil by vaporizing the oil and passing the vapor through a fractionation column. The fractionation column acts as a series of tiny plates upon which the molecules in the oil vapor are successively adsorbed and desorbed. The time required for molecules to pass through the fractionation column is positively correlated with the molecules' molecular weight or boiling point. Thus, the method effectively separates the hydrocarbons according to approximate molecular size or C content. In the chromatogram, peak height is proportional to the concentration of organics that boil at the indicated temperature. The unresolved envelope of peaks is caused primarily by the presence of cyclic and highly branched components of the oil, which are not effectively separated by chromatography. Most of the chromatograms, however, show a series of distinct peaks that are associated with normal alkanes containing the number of C atoms indicated below each chromatogram. The differences in the chromatograms of the six oils in Figure 13.13 are obvious to even a casual observer and can be quantified for purposes of characterization. Other useful classification characteristics include the C and S isotopic composition; the total content of S, N, V, and Ni; and the OEP or CPI. Fluorescence techniques that take advantage of differences in the characteristic fluorescence spectra of different oils have also shown considerable promise as a method of oil fingerprinting (Anonymous, 1986). These methods obviously work best when there are some candidate pollution sources whose fingerprints can be examined for comparison. The problem becomes more difficult if there are no obvious suspects.

COMMENTARY

According to the 1975 NAS panel report (p. 106): "The most damaging, indisputable adverse effects of petroleum are the oiling and tarring of beaches, the endangering of seabird species, and the modification of benthic communities along polluted coastlines where petroleum is heavily incorporated in the sediments." The first effect is largely an aesthetic

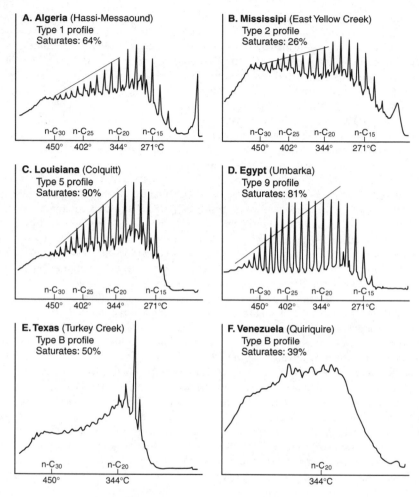

FIGURE 13.13   Typical gas chromatograms of crude oil after removal of components with a boiling point below 316°C, equivalent to the boiling point of normal octadecane ($C_{18}H_{38}$). However, the residues retain some hydrocarbons down to n-$C_{13}H_{28}$, and these appear in the gas chromatographic profile. *Source*: Redrawn from Miller (1973). Reproduced with permission.

problem, but it is a definite nuisance on recreational beaches, particularly in areas such as Bermuda that depend heavily on tourism (Butler, 1975). With respect to seabirds, Nelson-Smith (1972, p. 159) has commented that "There is no doubt that oil pollution has been the main cause of a dramatic decline in numbers of auks and sea-ducks throughout the North Atlantic and adjacent areas during the first half of this century." Auks have few natural enemies and a very low reproductive rate, and their populations therefore recover slowly from large-scale kills. It has been estimated that in the absence of oil pollution, it would take a guillemot colony about 53 years to double in size (Nelson-Smith, 1972). After the wreck of the *Torrey Canyon* in 1967, the number of puffins on the Sept Iles of Brittany declined by a factor of 6, and the razorbill population by a factor of 9 (Nelson-Smith, 1972). Despite chronic oil pollution of the North Sea and adjacent waters, 25% of the colonies of common murres and razorbills in the British Isles maintained stable numbers during the 1970s, and 50% actually increased (NRC, 1985). With few exceptions, however, other auk populations in the Atlantic have continued to decline. Although some of this decline may have been due to overhunting and drowning in fishing nets, there is reason to believe that oil pollution has been a factor, particularly in Arctic populations (NRC, 1985). Sea ducks have higher natural reproductive rates than auks, but their habit of congregating in areas adjacent to shipping lanes and other areas particularly subject to oil pollution has resulted in kill-offs. Adult populations, however, appear to recover within a few years (NRC, 1985).

In summarizing its assessment of oil pollution effects, the NRC (1985, p. 490) commented: "In contrast to offshore situations where impact may be minimal and transient . . . , there are greater immediate concerns for coastal waters (for example, biologically productive estuaries) receiving on the average a greater proportion of discharged petroleum." Certainly it is now apparent from the work of Blumer et al. (1971), Krebs and Burns (1977), Sanders et al. (1980), and others that oil pollution damage to coastal benthic communities can be of much longer duration than was once supposed due to the low degradation rate of hydrocarbons that become incorporated in sediments or the lipid tissues of organisms. Indeed, under certain conditions, it is apparent that spilled petroleum may persist in the environment for several decades (NRC, 1985). Furthermore, construction and drilling activities in coastal areas may permanently disrupt intertidal and nearshore communities by accelerating erosion and siltation and by altering tidal current patterns.

There is no doubt that some of the damage caused by oil pollution in the coastal environment could be reduced. Tanker accidents such as that of the *Exxon Valdez* are inexcusable. More strict enforcement of legislation such as the Port and Tanker Safety Act would help. Exxon now requires that tanker crews be on board at least four hours before leaving port, a move intended to allow time for sobering up (Lemonick, 1989). The economic costs of the *Exxon Valdez* spill may have been the wakeup call that the oil industry needed. From 1991 to 1995, tankers accounted for only 10% of the oil spilled in U.S. waters. Commenting on this fact, the NRC (1999, p. 20) has concluded, "The reduction in oil pollution in U.S. waters between 1991 and 1995 cannot be attributed to the requirements of Section 4115, [since] the first compulsory retirements of single-hull vessels did not occur until 1995. . . . In the view of the committee, the reduction in oil spillage . . . was the result of . . . increased awareness among vessel owners and operators of the financial consequences of oil spills and a resulting increase in attention to policies and procedures aimed at eliminating vessel accidents." Requirements that tankers be equipped with inert gas systems to prevent fires and explosions, segregated ballast tanks, and backup radar and steering systems have undoubtedly helped as well. The U.S. Oil Pollution Act of 1990 has designated funding for research aimed at reducing the impact of oil pollution, including improved cleanup methods.

Tanker accidents, however, were not perceived by the NRC (1985) as being the most serious cause of marine oil pollution. Tanker accidents, particularly involving supertankers, can have devastating effects on coastal ecosystems, and the persistence of petroleum hydrocarbons in some components of these systems is on the order of decades. Nevertheless, coastal ecosystems appear to recover remarkably fast from single incidents of oil spillage. The NRC (1985) report expressed more concern over chronic stresses associated with the long-term more or less continuous input of petroleum hydrocarbons to coastal waters. Regions associated with heavy tanker traffic and/or impacted by municipal and industrial wastewaters are the areas most likely to be affected by this sort of stress. However, impacts on biota from petroleum are in many such cases difficult to distinguish from effects caused by other mechanisms, including overfishing and other forms of pollution.

The global oil pollution picture will change dramatically during the 21st century for two reasons. In the short term, marine transportation of oil will increase substantially as the oil production capacity of the major consuming nations, in particular the United States, declines. Much of the oil will be transported from the Middle East, which currently accounts for 56% of proven reserves (Flavin, 1985).

In the longer term, however, oil production, transportation, and use will certainly decline. Proven oil reserves have changed very little since 1970, and there is no reason to believe that substantial deposits of oil remain undiscovered. At current rates of oil consumption, the world's oil resources will be largely exhausted sometime during the second half of the 21st century (Flavin, 1985). This scenario will undoubtedly translate into a reduction in oil pollution, but unless there is a major shift in the world's economy, a substitute energy source for oil will have to be found. Two of the most obvious substitutes, coal and nuclear energy, are beset with significant environmental problems of their own. The relevance of these problems to aquatic pollution is discussed in Chapters 15 and 16.

REFERENCES

ALDHOUS, P. 1991. Big test for bioremediation. *Nature,* **349,** 447.

ANDREEV, P. F., A. I. BOGOMOLOV, A. F. DOBRYANSKII, and A. A. KARTSEV. 1968. *Transformation of Petroleum in Nature.* Pergamon Press, New York. 466 pp.

ANONYMOUS. 1986. Oil spills. *New Scientist,* **111,** 27.

ANONYMOUS. 1991a. Another crude year. *Discover,* **12**(1), 70–71.

ANONYMOUS. 1991b. Oil pollution. *Honolulu Star Bulletin and Advertiser.* February 24. p. D2.

BEGLEY, S., J. CAREY, and J. CALLCOTT. 1983. Death of the Persian Gulf. *Newsweek,* **102**(4), 79.

BLUMER, M. 1972. Submarine seeps: Are they a major source of open ocean oil pollution? *Science,* **176,** 1257–1258.

BLUMER, M., H. L. SANDERS, J. F. GRASSIE, and G. R. HAMPSON. 1971. A small oil spill. *Environment,* **13**(2), 2–12.

BURNS, K. A. 1976. Microsomal mixed function oxidases in an estuarine fish, *Fundulus heteroclitus,* and their induction as a result of environmental contamination. *Comp. Biochem. Physiol.,* **53B,** 443–446.

BURNS, K. A., and J. M. TEAL. 1979. The West Falmouth oil spill; hydrocarbons in the saltmarsh ecosystem. *Estuarine Coastal Mar. Sci.,* **8,** 349–360.

BUTLER, V. N. 1975. Pelagic tar. *Sci. Amer.,* **232**(6), 90–97.

CANEVARI, G. P. 1969. The role of chemical dispersants in oil cleanup. In D. P. Hoult (Ed.), *Oil on the Sea.* Plenum Press, New York. Pp. 29–51.

CARTER, L. J. 1978. Amoco Cadiz points up the elusive goal of tanker safety. *Science,* **200,** 514–516.

CLARK, C. S., H. S. BJORNSON, C. C. LINNEMANN, JR., and P. S. GARTSIDE. 1984. *Evaluation of Health Risks Associated with Wastewater Treatment and Sludge Composting.* EPA-600/S1-84-014. Environmental Protection Agency, Research Triangle Park, NC.

DREWRY. 1991a. *Shipping Statistics and Economics,* No. 251 (Sept). Drewry Shipping Consultants, Ltd.

DREWRY. 1991b. *Shipping Statistics and Economics,* No. 252 (Oct.). Drewry Shipping Consultants, Ltd.

EGANHOUSE, R. P., and I. R. KAPLAN. 1981. Transport dynamics and mass emission rates to the ocean. *Environ. Sci. Tech.,* **15,** 310–315.

EGANHOUSE, R. P., and I. R. KAPLAN. 1982. Extractable organic matter in municipal wastewaters. 1. Petroleum hydrocarbons: Temporal variations and mass emission rates to the ocean. *Environ. Sci. Tech.,* **16,** 180–186.

EPA. 1972. *Water Quality Criteria.* EPA-R3-73-033. Environmental Protection Agency, Washington, DC. 594 pp.

EPA. 1976. *Quality Criteria for Water.* Environmental Protection Agency, Washington, DC. 256 pp.

EPA. 1986. *Quality Criteria for Water.* EPA 440/5-86-001. Environmental Protection Agency, Washington, DC.

FIESER, L., and M. FIESER. 1961. *Advanced Organic Chemistry.* Reinhold, New York. 1157 pp.

FLAVIN, C. 1985 *World Oil: Coping with the Dangers of Success.* Worldwatch Paper 66. Worldwatch Institute, Washington, DC. 66pp.

GUNDLACH, E. R., P. D. BOEHM, M. MARCHAND, R. M. ATLAS, D. M. WOOD, and D. A. WOLFE. 1983. The fate of *Amoco Cadiz* oil. *Science,* **221,** 122–129.

HODGSON, B. 1990. Alaska's big spill. Can the wilderness heal? *National Geographic,* **177**(1), 5–43.

HOLDEN, C. 1990. Spilled oil looks worse on TV. *Science*, 250, 371.

HORN, H. M., J. M. TEAL, and R. H. BACKUS. 1970. Petroleum lumps on the surface of the sea. *Science*, 168, 245–246.

HOWARTH, R. W. 1981. Fish versus fuel: A slippery quandary. *Tech. Rev.*, 83(3), 68–77.

IEA. 1998. *Energy Balances of OECD Countries. 1995–1996.* International Energy Agency, Organization for Economic Cooperation and Development, Paris.

ITOPF. 1999. Oil spill database. The International Tanker Owners Pollution Federation. *http://www.itopf.com/stats.html*

JOYCE, C. 1990. French finally clean up after *Amoco Cadiz* disaster. *New Scientist*, 127(1728), 24.

JOYCE, C. 1991. Hot water sterilises Alaska's oiled beaches. *New Scientist*, 130, No. 1765, 14.

KERR, R. A. 1977. Oil in the ocean: Circumstances control its impact. *Science*, 198, 1134–1136.

KREBS, C. T., and K. A. BURNS. 1977. Long-term effects of an oil spill on populations of the salt-marsh crab *Uca pugnax*. *Science*, 197, 484–487.

LEMONICK, M. D. 1989. The two Alaskas. *Time*, 133(16), 56–66.

LOUGHLIN, T. R. (Ed.). 1994. *Marine Mammals and the Exxon Valdez.* Academic Press, New York. 395 pp.

MIELKE, J. E. 1990. *Oil in the Ocean: The Short- and Long-term Impacts of a Spill.* Congressional Research Service Report for Congress, Washington, DC. 24 pp.

MILLER, J. W. 1973. A multiparameter oil pollution source identification system. In *Prevention and Control of Oil Spills.* American Petroleum Institute, Environmental Protection Agency, and U.S. Coast Guard, Washington, DC. Pp. 195–203.

MORRIS, B. F., and T. R. LOUGHLIN. 1994. Overview of the *Exxon Valdez* oil spill 1989–1992. In T. R. Loughlin, (Ed.), *Marine Mammals and the Exxon Valdez.* Academic Press, New York. Pp. 1–22.

NAS. 1975. *Petroleum in the Marine Environment.* National Academy of Sciences, Washington, DC. 107 pp.

NELSON-SMITH, A. 1967. Oil emulsifiers and marine life. In *The Journal of the Devon Trust for Nature Conservation, Ltd.* Supplement. *Conservation and The Torrey Canyon.* Kingsbridge, England. July. Pp. 29–33.

NELSON-SMITH, A. 1972. *Oil Pollution and Marine Ecology.* Elek Science, London. 260 pp.

NRC. 1985. *Oil in the Sea—Inputs, Fates, and Effects.* National Research Council, National Academy Press, Washington, DC. 601 pp.

NRC. 1989. *Using Oil Spill Dispersants on the Sea.* National Research Council Marine Board, National Academy Press, Washington, DC. 335 pp.

NRC. 1999. *Double-Hull Tanker Legislation. An Assessment of the Oil Pollution Act of 1990.* National Research Council Marine Board, National Academy Press, Washington, DC. 25 pp.

OTA. 1991. *Biotechnology for Marine Oil Spills.* S/N 052-003-01240-5. Congress of the United States Office of Technology Assessment, Washington, DC. 31 pp.

POLITZER, I. R., I. R. DeLEAN, and J. L. LASETER. 1985. *Impact on Human Health of Petroleum in the Marine Environment.* American Petroleum Institute, Washington, DC. 162 pp.

RICE, S. D., R. B. SPIES, D. A. WOLFE, and B. A. WRIGHT (Eds.). 1996. *Proceedings of the Exxon Valdez Oil Spill Symposium.* American Fisheries Society Symposium 18, Bethesda, MD. 931 pp.

RYAN, P. R. 1977. The composition of oil—A guide for readers. *Oceanus*, 20(4), 4.

SANDERS, H. L., J. F. GRASSLE, G. R. HAMPSON, L. S. MORSE, S. GARNER-PRICE, and C. C. JONES. 1980. Anatomy of an oil spill: Long-term effects from the grounding of the barge *Florida* off West Falmouth, Massachusetts. *J. Mar. Res.*, 38, 265–380.

SPIES, R. B., S. D. RICE, D. A. WOLFE, and B. A. WRIGHT. 1996. The effects of the *Exxon Valdez* oil spill on the Alaskan coastal environment. In S. D. Rice, R. B. Spies, D. A. Wolfe, and B. A. Wright (Eds.), *Proceedings of the Exxon Valdez Oil Spill Symposium.* American Fisheries Society, Bethesda, MD. Pp. 1–16.

STRAUGHAN, D. 1972. Factors causing environmental changes after an oil spill. *J. Pet. Tech.*, 24, 250–254.

TATEM, H. E., B. A. COX, and J. W. ANDERSON. 1978. The toxicity of oils and petroleum hydrocarbons to estuarine crustaceans. *Estuarine and Coastal Mar. Sci.*, 6, 365–373.

TEAL, J. M., and R. W. HOWARTH. 1984. Oil spill studies: A review of ecological effects. *Environ. Manage.*, 8(1), 27–44.

THOMPSON, H. C., JR., R. N. FARRAGUT, and M. H. THOMPSON. 1977. Relationship of scarlet prawns (*Plesiopenaeus edwardsianus*) to a benthic oil deposit off the northwest coast of Arabia, Dutch West Indes. *Environ. Pollut.*, 13(4), 239–253.

TIMAGENIS, G. J. 1980. *International Control of Marine Pollution.* Vols. I and II. Oceana Publications, Dobbs Ferry, NY. 877 pp.

TULLY, P. R. 1969. Removal of floating oil slicks by the controlled combustion technique. In D. P. Hoult (Ed.), *Oil on the Sea.* Plenum Press, New York. Pp. 81–91.

U.N. 1991. *United Nations Energy Statistics Yearbook.* Department of International Economic and Social Affairs, United Nations, New York.

WARE, L. 1989. Oil in the sea: The big spills and blowouts. *Audubon,* 91, 109.

WELLS, P. G., J. N. BUTLER, and J. S. HUGHES (Eds.) 1995a. *Exxon Valdez Oil Spill: Fate and Effects in Alaskan Waters.* American Society for Testing and Materials, Philadelphia. 955 pp.

WELLS, P. G., J. N. BUTLER, and J. S. HUGHES. 1995b. Introduction, overview, issues. In P. G. Wells, J. N. Butler, and J. S. Hughes (Eds.), *Exxon Valdez Oil Spill: Fate and Effects in Alaskan Waters.* American Society for Testing and Materials, Philadelphia. Pp. 3–38.

WESTERMEYER, W. E. 1991. Oil spill response capabilities in the United States. *Environ. Sci. Tech.,* 25(2), 196–200.

YAMAGUCHI, N. D. 1990. *The Changing Environment for Seaborne Oil Trade: An Overview of Fleet Developments and Forecast Tonnage Requirements Through the 1990s.* East-West Center Energy Program. Honolulu, HI. 48 pp.

QUESTIONS

13.1 Two oil supertankers run headfirst into each other in the middle of the Atlantic Ocean because their crews are below deck playing cards and are intoxicated. Several cargo compartments rupture. Large numbers of aquatic birds are in the area. Sea conditions are very rough for two weeks after the spill. If you were in charge of trying to minimize environmental damage during the first two weeks after the spill, which one of the following steps would you take?
   a. Offload the remaining oil from the tanker.
   b. Apply chemical dispersants to the spilled oil.
   c. Use booms and skimmers to contain the oil and sop up the contained oil.
   d. Burn the spilled oil.

13.2 An oil tanker breaks up in rough weather and loses its entire cargo of crude oil. The oil washes ashore onto a pebble beach. As the director of an international effort to clean up the mess and minimize the environmental damage, you are called in about a week after the spill. Which one of the following methods would you use to deal with the oil on the beaches?
   a. Apply fertilizer.
   b. Apply chemical dispersants.
   c. Remove the oil by spraying the rocks with hot water.
   d. Burn the oil.

13.3 The Buzzards Bay oil spill near Cape Cod is generally agreed to have killed more marine organisms than the Ixtoc I oil well blowout in the Gulf of Mexico, even though the latter released about 840 times as much oil. How would you account for the greater impact of the Buzzards Bay spill?

13.4 What source of oil discharges to the marine environment does the LOT procedure seek to minimize, and by what means does LOT reduce those discharges?

13.5 What is meant by the acronym COW, and how has it been used to reduce oil pollution of the ocean?

13.6 What is meant by the acronyms SBT and CBT, and how have they been used to reduce oil pollution of the ocean?

13.7 In commenting on certain aspects of the effort to clean up the oil spilled by the *Exxon Valdez,* one National Oceanic and Atmospheric administrator was quoted as saying, "Sometimes the best, and ironically the most difficult, thing to do in the face of an ecological disaster is to do nothing." What aspect of the cleanup effort prompted this statement and why?

13.8 Why was there so much criticism of the use of chemical detergents to clean up the oil that washed ashore after the *Torrey Canyon* oil spill off the coast of England? If a similar accident occurred today and you were in charge of directing the response, what actions would you recommend?

Chapter Fourteen

✧ ◻ ✧ ◻ ✧

Radioactivity

Once a bright hope, shared by all mankind, including myself, the rash proliferation of atomic power plants has become one of the ugliest clouds overhanging America. David Lilienthal, first chairman of the Atomic Energy Commission (*New York Times*, July 20, 1969).

PHYSICAL BACKGROUND

Before we become immersed in a discussion of the problems associated with the use and disposal of radioactive substances, it is appropriate to first outline briefly just what radioactivity is and how it affects living organisms. This subject is one that requires some understanding of nuclear and atomic physics, topics with which many persons concerned with aquatic pollution may not be familiar. The following discussion provides the basic physical information relevant to the subject of radioactivity.

For our purposes, we may assume that atoms are composed of nothing more than protons, neutrons, and electrons. Protons and neutrons each have a rest mass of approximately 1 atomic mass unit (amu), or 1.660×10^{-24} g. Electrons are much smaller, having a rest mass of only about 5.5×10^{-4} amu. Hence, the total mass of an atom is determined largely by the number of protons and neutrons it contains. For example, the common H atom, which contains one proton, one electron, and no neutrons, has a rest mass of about 1 amu. The common N atom, which contains seven protons, seven electrons, and seven neutrons, has a rest mass of about 14 amu. In atomic physics, the approximate mass of an atom is usually written as a superscript to the left of the chemical symbol for the element. Thus the symbols ^{1}H and ^{14}N would designate an H atom with a mass of 1 amu and an N atom with a mass of 14 amu, respectively.

The electrical charge of an atom is determined by the relative number of protons and electrons. Each proton bears a positive charge of 1 electrostatic unit (esu) or 1.6×10^{-19} coulombs; each electron bears a negative charge of 1 esu. Neutrons bear no electrical charge. Atoms containing equal numbers of protons and electrons are obviously electrostatically neutral; atoms that contain more protons than electrons bear a net positive charge and are called *cations*; atoms that contain more electrons than protons bear a net negative charge and are called *anions*. Atoms bearing either a positive or a negative charge are referred to as *ions*.

The nucleus of an atom consists of all the protons and neutrons bound in a small region at the center of the atom by a residual strong interaction associated with the so-called strong force of nuclear physics.[1] The electrons are distributed around the nucleus at distances that are large compared to the dimensions of the nucleus.[2] The binding between the electrons and the nucleus is controlled by electromagnetic interactions. The identity of the atom is determined solely by the number of protons in the nucleus. An atom containing only one proton, for example, is an H atom, regardless of the number of electrons or neutrons it contains. An atom containing seven protons is an N atom, and an atom containing eight protons is an O atom. The *atomic number* of an atom is the number of protons in its nucleus and is usually designated by a subscript to the left of the chemical symbol for the element. Thus, $^1_1$H indicates an H atom whose nucleus contains no neutrons $(1 - 1 = 0)$. The symbol $^{31}_{15}$P indicates a P atom whose nucleus contains 16 neutrons $(31 - 15 = 16)$.

Atoms of the same element whose nuclei contain different numbers of neutrons are called *isotopes* of the element. For example, ^{1}H, ^{2}H, and ^{3}H are three isotopes of H containing zero, one, and two neutrons, respectively. Collectively, isotopes are referred to as *nuclides*. There is more than one isotope of all known elements, with H having the fewest number of isotopes, three. Uranium (U), the naturally occurring element with the highest atomic number (92), has 14 isotopes ranging in atomic weight from 227 to 240 amu.

---

[1]The strong force is one of the fundamental physical forces. The other three are gravity, the electromagnetic force, and the weak force.
[2]The diameter of an atomic nucleus is on the order of 10^{-12} cm. Electrons are found with maximum probability at distances on the order of 10^{-8} cm from the nucleus.

For reasons that will shortly become clear, the common isotopes of most elements are not radioactive. However, at least one isotope of all elements is unstable or radioactive. For example, of the three H isotopes, ^{1}H and ^{2}H are stable but ^{3}H is radioactive. A radioactive isotope, or *radioisotope*, is an isotope that has a tendency to decay spontaneously or to transform itself into some other isotope. Why some nuclides are radioactive is determined by the nature of the so-called strong and weak forces of nuclear physics. An understanding of these interactions is a subject of theoretical physics and is outside the scope of this discussion. For our purposes, it is sufficient to know that there is a definite pattern to the number of protons and neutrons in stable nuclides (Figure 14.1). For the lightest elements, the number of protons and neutrons in stable nuclides is approximately the same, but in the heaviest stable nuclides there are about 50% more neutrons than protons. No nuclides with an atomic number greater than 83 are stable. Nuclides that do not fall within the zone of stability indicated by the data points in Figure 14.1 are radioactive. These nuclei tend to disintegrate spontaneously to produce "daughter" nuclei that lie within the zone of stability or at least no further away. If the daughter nuclide is itself unstable, then it will also tend to decay, producing another nuclide, and so forth. Thus, the decay of the original radionuclide can lead to a decay series involving a number of daughter nuclides. Ultimately, the decay series leads to the production of a stable nuclide. For example, the radionuclide U-238 (^{238}U) decays through a series of 13 intermediate radionuclides to produce the stable nuclide Pb-206 (^{206}Pb). On the other hand, ^{3}H decays into the stable nuclide helium-3 (^{3}He) in a single step.

Although there are quite a few mechanisms by which unstable radionuclides may decay, only three mechanisms are of direct importance insofar as the problem of radioactivity is concerned.

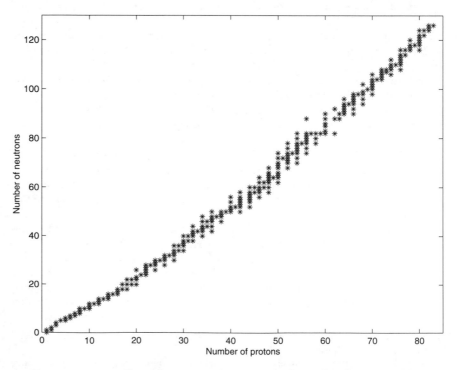

FIGURE 14.1   Numbers of protons and neutrons in the nuclei of stable nuclides.

1. The radionuclide may decay by emitting an alpha particle (α), which consists of two protons and two neutrons. This mode of decay is common for elements of atomic number greater than 83, although some radionuclides of low atomic number also decay by this mechanism. Since the α particle contains two protons, it may be regarded as the nucleus of a helium atom and designated ^{4}He. Alpha particle decay is a residual strong interaction effect. An example of α-particle decay is the decay of $^{238}_{92}$U as follows:

$$^{238}_{92}U \rightarrow {}^{234}_{90}Th + {}^4_2He \tag{14.1}$$

2. The radionuclide may decay by emitting a beta (β) particle. A beta particle has a mass identical to that of an electron and a charge of $+1$ esu (β^+) or -1 esu (β^-). If the beta particle is charged negatively, it is nothing more than an electron. If it is charged positively, it is called a *positron*. Beta particle decay is governed by the nuclear weak force. An example of beta particle decay is the decay of ^{3}H as follows:

$$^3_1H \rightarrow {}^3_2He + \beta^- \tag{14.2}$$

3. When radionuclides decay, they frequently emit gamma rays (γ). Gamma rays are nothing more than electromagnetic radiation, analogous to radio waves, visible light, and X-rays. Gamma rays, however, differ from other types of electromagnetic radiation in that they originate from the nucleus of atoms and are usually of higher energy. They are mediators of the electromagnetic force. Since gamma rays bear no charge, they are highly penetrating, whereas many alpha and beta particles can be stopped by a relatively thin layer of metal, soil, rock, or biological tissue. Returning to a previous example, the decay of $^{238}_{92}$U by alpha-particle emission is more correctly written:

$$^{238}_{92}U \rightarrow {}^{234}_{90}Th + {}^4_2He + \gamma \tag{14.3}$$

to indicate that a gamma ray is emitted along with the alpha particle.

Radiation from radionuclides therefore consists largely of the alpha and beta particles and gamma rays that are emitted when radionuclides decay. It is also true that a few radionuclides decay spontaneously by emitting neutrons, but most environmental concern over neutron radiation is related to the high-energy neutrons produced in nuclear reactors. These high-energy neutrons influence the release of radioactive wastes from the reactors.

The decay of a radionuclide is a probabilistic phenomenon. In other words, although there is a certain probability that a given radionuclide will decay within a given time, the process of decay is by no means certain. The probability of decay can be characterized by the so-called half-life of the radionuclide. The *half-life* is defined as the amount of time during which half of the atoms of a given radionuclide can be expected to decay. For example, the half-life of ^{14}C is about 5,770 years. Thus, a sample containing 10^6 atoms of ^{14}C at time zero would be expected to contain only 500,000 atoms of ^{14}C 5,770 years later, since 50% of the ^{14}C would have decayed. After a second period of 5,770 years, 50% of the 500,000 atoms would have decayed, leaving 250,000 atoms of ^{14}C. After a third period of 5,770 years, 50% of the 250,000 atoms of ^{14}C would have decayed, leaving 125,000 atoms, and so forth. Since the amount of radioactivity released by a sample of a given radioisotope is directly proportional to the amount of radioisotope present, it follows from the foregoing discussion that the amount of radioactivity released from a radioactive sample decreases steadily with time. After 10 half-lives, the number of radioactive atoms, and hence the radioactivity, has been reduced by about a factor of 1,024, and after 20 half-lives by a factor of a little over 1 million, assuming that the radionuclide has decayed to a stable nuclide. The half-lives of radionuclides vary tremendously, from less than 1 second for some very short-lived radionuclides to 5×10^{15} years for $^{144}_{60}$Nd and $^{142}_{58}$Ce. The fact that some radionuclides have half-lives on the order of a day or less is convenient from the standpoint of radioactive waste disposal because simply storing such waste for a few months results in virtually all of these radionuclides' decaying away. Unfortunately, some radionuclides produced in nuclear reactors have half-lives on the order of tens to hundreds of thousands of years or even longer. To avoid contaminating the environment with these radionuclides, some more or less permanent and isolated repository must be found for their disposal.

RADIATION TOXICOLOGY

What sort of environmental damage do radionuclides cause? On the molecular level, the effects of radiation on matter are understood in a semiquantitative way. In passing through material, radiation interacts with the atoms and molecules of which the material is composed, breaking chemical bonds to form positively and negatively charged ion pairs and free radicals.[3] These ions or radicals may then recombine in a variety of ways, leading to the production of chemical species that are not normally present in the material. In the case of living organisms, which typically contain 70% or more water, many of the radiation interactions involve water. The most important initial products resulting from the interaction of radiation with water are H_2O^- and hydroxide and hydrogen free radicals (BEIR, 1990), designated OH· and H·, respectively. Of these chemical species, the hydroxide radical is believed to be the most effective in causing damage because it is a strong oxidizing agent. Molecules such as the hydroxide radical may react with enzymes, nucleic acids, or other chemical species in the cell, perhaps leading to the destruction of the cell or causing the cell to function in an aberrant manner. What actually happens to a given cell, tissue, or organ following the initial production of ion pairs and other reactive species by the passage of radiation can be an exceedingly complex and drawn-out process, as suggested by the fact that some radiation-induced cancers may have a latency period of 30 years or more (Wilber, 1969). Recent studies have indicated that as many as 10 distinct mutations may be necessary in a cell before it becomes malignant. For example, the loss of two or more suppressor genes, which normally inhibit cell growth, and the simultaneous activation of oncogenes have been shown to be necessary to produce some of the most common forms of cancer (Marx, 1989). Regardless of the detailed long-term effects, however, radiation that causes this general sort of damage is referred to as *ionizing radiation*, due to the initial production of ion pairs associated with the interaction between radiation and matter.

Of the types of radiation we have discussed, alpha particles penetrate the shortest distance in matter because of their relatively large size and charge. In other words, alpha particles have a high probability of interacting with the atoms and molecules in a given material, and through this interaction their movement is rapidly slowed and brought to a stop. Alpha particles generally travel only a few centimeters in air and can be stopped by a piece of paper or a thin layer of skin. Thus, an alpha emitter is of relatively little concern as an external source of radiation, but it may be of great concern if ingested (internal source). In the latter case, the fact that the alpha particle is stopped over a very short

[3]Free radicals are highly reactive molecular fragments containing an unpaired electron.

distance means that ionizing damage may be quite intense in the immediate vicinity of the radionuclide, although tissue a short distance away may be virtually unaffected.

Beta particles have a charge equal to one-half the charge of an alpha particle and a rest mass about 7,200 times smaller than that of an alpha particle. Because of their smaller size and reduced charge, beta particles have a smaller probability of interacting with matter than do alpha particles; hence, they penetrate further into matter, spreading their ionizing damage over a longer distance. Beta particles may travel 1 m or more in air before coming to a stop and up to several centimeters through tissue. Obviously, beta emitters are of little concern to health unless one is standing in the immediate vicinity of the source; therefore, like alpha emitters, beta emitters are of primary concern as internal sources. The chief difference between alpha- and beta-particle damage is the fact that alpha-particle damage is relatively intense but confined to the immediate vicinity of the source. Beta-particle damage is less intense but extends over a greater distance.

Gamma rays have no electrostatic charge, but they do carry momentum and are capable of ionizing atoms and molecules, just as alpha and beta particles do. Because of their chargeless character, gamma rays penetrate matter easily, and in fact may pass through an organism without causing any damage at all. When gamma rays do interact with matter, they generally cause ionization over a long path length, and the intensity of the damage is much less than that caused by an alpha or beta particle of comparable energy. Because of their ability to penetrate matter, gamma rays are of concern from both external and internal sources of radiation.

As already noted, neutron damage is of concern primarily in the immediate vicinity of a nuclear reaction or an atomic explosion. Neutrons have a mass equal to one-fourth that of an alpha particle. Although they bear no charge, neutrons may, "like a bull in a china shop, . . . wreck local havoc by bumping atoms out of their stable states. Neutrons thus induce radioactivity in nonradioactive materials or tissues through which they pass" (Odum, 1971, p. 452).

The toxic effects of a particular radionuclide depend on a number of factors, including the intensity and energy of the emitted radiation, the location of the radionuclide, and the types of organisms affected. The intensity or activity of radiation is usually measured by simply noting the number of atoms that decay or disintegrate per unit time. The SI[4] measure of activity is the becquerel (Bq). A becquerel of radioactivity equals one disintegration per second. From the standpoint of radiation damage, however, the amount of radiation actually absorbed by an organism is more important than the number of disintegrations per second because the number of ion pairs and other excited chemical species produced by the passage of radiation through matter is directly proportional to the absorbed energy for a given type (α, β, γ) of radiation. The energy of the α, β, and γ radiation emitted by radionuclides may vary greatly from one radionuclide to another. For example, ^{14}C emits beta particles with a maximum energy of 0.156 million electron volts (meV), whereas the beta particles emitted by ^{32}P have a maximum energy of 1.71 meV. Thus, 1 Bq of ^{32}P would be expected to produce about $1.71/0.56 = 11$ times as many ion pairs in an organism as would 1 Bq of ^{14}C. The SI measure of absorbed radiation energy is the gray (Gy). One gray is defined as the absorbed dose of 1 joule of energy per kilogram of material. In radiation monitoring work, another unit of absorbed radiation, the roentgen, is often used. One roentgen equals the quantity of X-rays or gamma rays that will produce, as a consequence of ionization, 1 esu of electricity in 1 cm^3 of dry air measured at 0°C and standard atmospheric pressure. Finally, in setting radiation standards, allowance is made for the fact that the same amount of absorbed radiation energy may cause different amounts of biological damage, depending on the intensity of the ionization caused by the absorbed energy. Recall, for example, that alpha particles are stopped over a very

[4]International system of units.

TABLE 14.1   Q FACTORS FOR SELECTED TYPES OF RADIATION

Type of Radiation	Q
X rays, gamma rays, and electrons	1
High-energy protons and neutrons of unknown energy	10
Alpha particles, multiply charged particles, fission fragments, and heavy particles of unknown charge	20

Source: NRC (1991a).

short distance and therefore produce more intense ionization per unit path length than do beta particles or gamma rays, which release their energy over a longer path length. Alpha particles therefore cause greater biological damage per unit energy absorbed than do beta particles. To allow for this difference in damage intensity, a so-called quality factor (Q) is used to multiply the radiation absorbed dose in grays to estimate the relative amount of biological damage caused by the absorption of various types of radiation by humans. The product of the dose in grays times the Q factor is called the *dose equivalent* in sieverts (Sv). Table 14.1 lists the Q factors for various types of radiation. Notice that alpha particles are estimated to cause 20 times as much biological damage per unit absorbed energy as beta particles and gamma rays. Neutrons are intermediate in damage potential.

The No-Threshold and Linear Dose-Response Hypotheses

In setting radiation standards, it is now generally assumed that there is no threshold limit for radiation damage and that the amount of damage caused is a linear function of the total amount of radiation absorbed (BEIR, 1990). The assumptions of linearity and no threshold are independent, as illustrated in Figure 14.2. The curves in Figures 14.2A and 14.2C are both linear, but in the latter there is a threshold dose below which there is no effect. Neither of the curves in Figures 14.2A and 14.2B has a threshold, and only the former is linear. The combined assumptions of linearity and no threshold imply that the effect is directly proportional to the dose, an assumption that seems to hold in many cases at low levels of radiation exposure. As noted in the BEIR-V report (1990, p. 5), "The dose-dependent excess of mortality from all cancers other than leukemia, shows no departure from linearity in the range below 4 sievert." Figure 14.3A illustrates this pattern in the case of thyroid cancer. The leukemia data, however, are best described by a curvilinear function (Figure 14.3B). The assumption of no threshold for radiation damage is thought-provoking, since it implies that any artificial release of radioactivity to the environment can be expected to increase the probability of occurrence of various disorders associated with radiation damage. These disorders include cancer, leukemia, genetic defects, fetal and neonatal deaths, and a general shortening of life (Gofman and Tamplin, 1970). In setting radiation standards, one must therefore presumably make some value judgment regarding the benefits derived from the use of radionuclides versus the costs in terms of disease, death, and deterioration of the environment. For example, a figure of $100,000 per person-Sv has sometimes been used in the United States as a figure of merit in evaluating trade-offs in nuclear waste management (e.g., Adams and Rogers, 1978). Person-Sv are calculated by multiplying the total number of persons exposed by their average individual doses in sieverts. Since the recent report of the Committee on the Biological Effects of Ionizing Radiation (BEIR-V report) estimates that a dose of 1,000 person-Sv resulting from

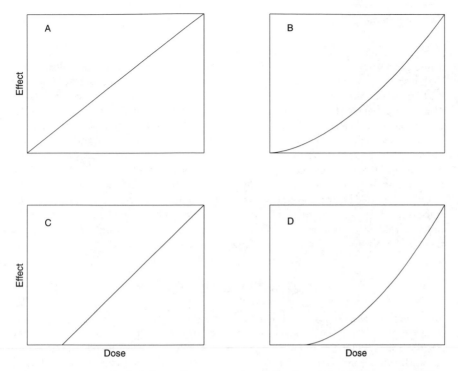

FIGURE 14.2 Hypothetical dose-effect curves describing the relationship between radiation dose and health effects.

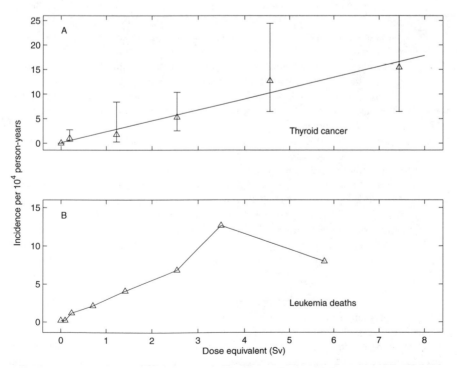

FIGURE 14.3 Actual dose-effect data describing the incidence of thyroid cancer (A) and leukemia deaths (B) caused by ionizing radiation. *Source*: ICRP (1991) and BEIR (1990).

a single exposure incident would probably produce about 80 fatal cancers (BEIR, 1990), the foregoing figure of merit would put a price tag of no more than about $1.25 million on each fatal cancer. This price tag would obviously be reduced if one also considered the likely increase in genetic defects, fetal deaths, and so on resulting from each person-Sv. Although such a balancing of human misery against social and technological advances and benefits may seem repugnant to some, it is worth pointing out that roughly 43,000 people die annually on the highways in the United States (U.S. Bureau of the Census, 1998) and that automobiles contribute significantly to the pollution of both air and water. Even though most persons would undoubtedly like to see these "costs" of automobile use reduced, it is also likely that most persons would agree that the benefits of automobile use still outweigh the price in terms of human lives and environmental deterioration. How much of this sort of price are we willing to pay for nuclear power, nuclear weapons, and the use of radiation in science and medicine?

Health Effect Estimates

To answer this question, we must first have some idea of the relationship between radiation exposure and various related diseases and disorders. Most information regarding health effects on humans has come from follow-up studies on the survivors of the bombing of Hiroshima and Nagasaki at the end of World War II and on persons who were treated with X rays for medical purposes prior to 1955. The health effects estimates in the BEIR-V report, for example, were based primarily on the following studies:

1. A mortality study of 120,321 persons who were resident in Hiroshima and Nagasaki in 1950, 91,228 of whom were exposed to radiation from the atomic bombs dropped on August 6 and 9, 1945. Most of the radiation received by the exposed persons was in the form of gamma rays, the estimated doses ranging from 10 mSv to more than 2 Sv (BEIR, 1990).
2. A study of 14,106 patients treated with radiotherapy to the spine for ankylosing spondylitis in the United Kingdom between 1935 and 1954.
3. A study of approximately 150,000 women in a number of countries treated for cervical cancer. About 70% of the women were treated with either radium implants or external radiotherapy.
4. A study of 31,710 women in Canadian tuberculosis sanatoria, a substantial number of whom were exposed to multiple fluoroscopies in conjunction with artificial pneumothorax treatments between 1930 and 1952.
5. A study of 601 women treated with radiotherapy for postpartem acute mastitis in New York State during the 1940s and 1950s, as well as 1,239 nonexposed women who also suffered from mastitis but were not treated with radiotherapy. Siblings of both groups of women were also included in the study.
6. A study of 1,742 women first treated between 1930 and 1956 in two Massachusetts sanatoria, of whom 1,044 were subjected to regular fluoroscopy in conjunction with treatment by artificial pneumothorax.

Several problems have complicated interpretation of the findings of these and similar studies. First, it has sometimes been difficult to determine to how much radiation a person was exposed. A major revision of radiation health effects in the BEIR-V report, for example, resulted from recalculation of the radiation doses received by the survivors of the bombing of Hiroshima and Nagasaki. The shielding provided by air, humidity, windows, walls, and roofs was recalculated, and the doses received by the survivors "were individually reconstructed, taking into account whether the person was facing or turned away from the blast, and, if sideways, which side of the body was exposed" (Marshall, 1990a, p. 23). Suffice it to say that such calculations are difficult, time-consuming, and involve a certain amount of educated guesswork. Another problem is the long latency period in the development of certain types of cancer, an issue that has been overlooked in some short-term studies of dose-effect relationships (Gofman and Tamplin, 1970). By focusing on persons who had been exposed to artificial radiation prior to 1955, the BEIR-V study was able to follow rather thoroughly the time course of various adverse health effects and to develop mathematical models to describe the observed results. An additional complicating factor is the tendency of rapidly dividing cells to be more sensitive to radiation damage than slowly dividing cells. As a result, young children, and especially fetuses, are much more susceptible to radiation damage than adults, and the most sensitive parts of the body are regions such as the reproductive organs and bone marrow, where cell division is relatively rapid. For example, the excess cancer mortality expected from a single dose of radiation is about twice as high if exposure occurs at age 5 than at age 30 (BEIR, 1990).

Controlled experiments conducted on plants and animals have provided much valuable additional information on radiation-effect relationships. From such studies, it has become clear that mammals are among the most sensitive organisms to acute doses of radiation and that microorganisms are the least sensitive. Figure 14.4 indicates the sensitivity range of mammals, insects, and bacteria to various acute doses of radiation. The left end of each bar indicates the acute dose that would severely affect reproduction in the most sensitive species of each group, and the right end of each bar roughly corresponds to the TLm for the more resistant species in the group. The arrows indicate doses that would kill or damage sensitive life-history stages such as embryos. The acute lethal dose for humans is about 3 Sv (EPA, 1973).

Studies of radiation damage to higher plants have shown that the acute lethal dose is directly proportional to the chromosome volume or DNA content of the cell, as indicated in Figure 14.5. These results suggest that in cases of acute lethal damage, the chromosomes in the cell represent a sort of target, the size of which is directly proportional to the probability of being hit by a particle or ray of radiation. This relationship between lethal dose and chromosome size breaks down in higher animals, presumably because effects on specific organs are of critical importance and tend to mask effects due to differences in cellular structure.

Estimates of harm to the average person from current levels of radiation exposure require a certain extrapolation of experimental data because most persons experience a chronic low-level exposure that is rather different from the acute doses received by bomb survivors and persons treated with radiation for medical reasons. It is now generally agreed that a given dose of radiation absorbed over a short time period is likely to cause more

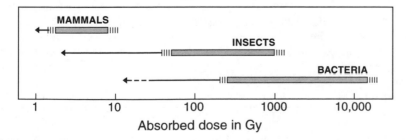

FIGURE 14.4   Relative sensitivity of three groups of organisms to acute doses of X or gamma radiation. *Source: Fundamentals of Ecology,* Third Edition, by Eugene P. Odum, copyright © 1971 by Saunders College Publishing, reproduced by permission of the publisher.

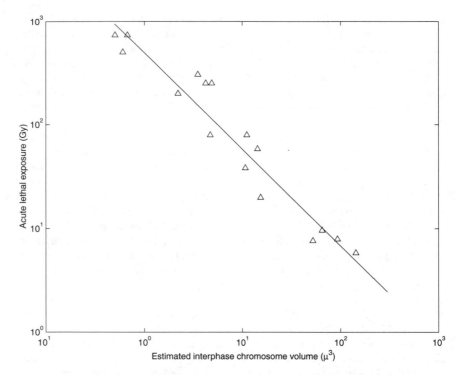

FIGURE 14.5   Relationship between chromosome volume of seed plants and acute lethal doses of radiation. *Source*: Reprinted from *Science*. Redrawn from Sparrow et al. (1963).

damage than the same dose of radiation received over a long time period. For example, Sparrow (1962) discovered that pine trees exposed to a radiation level of 1 roentgen d^{-1} for 10 years suffered about the same reduction in growth as pine trees exposed to a single dose of 60 roentgens. In other words, chronic doses of radiation administered over a long period of time had less effect per unit of radiation as a single acute dose. This effect would be expected if the single acute dose caused irreparable damage, whereas chronic doses caused only slight damage that could be repaired between exposures. The ratio of damage caused by acute and chronic doses of the same amount of radiation is referred to as the *dose-rate effectiveness factor* (*DREF*), and in the pine-tree example cited it would be $(10)(365)/60 = 61$. The DREFs for humans and animals appear to be substantially lower, with all reported values falling in the range 2–10 (BEIR, 1990). The BEIR-V report assumed a DREF of 2 for human health effects. This assumption is conservative in the sense that it overestimates the adverse health effects caused by chronic exposure if the true DREF is greater than 2.

Current Levels of Exposure

With this introduction, it is now useful to examine current levels of radiation exposure and compare these with international guidelines. Tables 14.2 and 14.3 provide the relevant information. It is apparent from Table 14.2 that, for the average person, the major sources of radiation exposure are natural, the primary radioisotope of concern being radon-222 (^{222}Rn), which is produced in the decay series of ^{238}U. The importance of ^{222}Rn in natural

TABLE 14.2 ESTIMATES OF ANNUAL AVERAGE RADIATION EXPOSURE TO A MEMBER OF THE U.S. POPULATION

Source	mSv	% of Total
Natural		
Radon	2.0	55
Cosmic	0.27	8
Terrestrial	0.28	8
Internal	0.39	11
Total natural	3	82
Artificial		
X-ray diagnosis	0.39	11
Nuclear medicine	0.14	4
Consumer products	0.10	3
Other		
Occupational	<0.01	<0.03
Nuclear fuel cycle	<0.01	<0.03
Fallout	<0.01	<0.03
Miscellaneous	<0.01	<0.03
Total natural, artificial, and other	3.6	100

Source: BEIR (1990).

radiation has been recognized only recently (BEIR, 1990). Actually, it is not so much ^{222}Rn, but rather its daughter radionuclides, that are of concern in public health. ^{222}Rn is an inert gas and therefore has no tendency to be incorporated into biological tissue. Its daughter radionuclides, which include isotopes of polonium (Po), Pb, and bismuth (Bi), are not gases and tend to become incorporated into aerosols that lodge in the lungs. Both ^{222}Rn

TABLE 14.3 RADIATION DOSE-EQUIVALENT LIMITS FOR PROTECTION OF INDIVIDUAL MEMBERS OF THE GENERAL PUBLIC DUE TO SOURCES OTHER THAN NATURAL BACKGROUND AND EXPOSURES RECEIVED AS A PATIENT FOR MEDICAL PURPOSES

Exposed Area	Dose Limit (mSv y^{-1})	Weighting Factor
Whole body		
Continuous or frequent exposure	1	
Infrequent exposure	5	
Tissues and organs[a]	50	
Gonads		0.25
Breast		0.15
Red bone marrow		0.12
Lung		0.12
Thyroid		0.03
Bone surface		0.03
Remainder		0.30

[a]The dose equivalent limit for tissues and organs is calculated from a weighted mean of the doses to individual tissues and organs using the weighting factors indicated in the right-hand column.
Source: NCRPM (1989), (1991a).

and its daughters are alpha-particle emitters. The other natural radionuclide of principal concern from the standpoint of internal exposure is ^{40}K, which emits beta particles. The primary natural sources of external exposure are cosmic rays and alpha particle-emitting radioisotopes in the decay series of ^{238}U, which includes U, thorium (Th), and Rn, among others. These latter elements are associated with soils and rocks. The principal artificial source of radiation exposure in the United States is the use of radiation in medicine. The estimate of 0.53 mSv y^{-1}, however, is about 30% smaller than the corresponding figures for 1970 and 1978 (Gillette, 1972; Marx, 1979). The difference presumably reflects efforts to reduce unnecessary exposure to radiation in medical diagnosis and therapy. The 1972 BEIR report, for example, noted that use of improved equipment, proper shielding of reproductive organs, and elimination of unnecessary X rays could reduce the "genetically significant dose" currently received by the general population by 50% (Gillette, 1972), and in 1979 Department of Health, Education, and Welfare secretary Joseph Califano directed the FDA to accelerate steps to reduce unnecessary exposure to medical and dental radiation (Marx, 1979). The principal cause of exposure from consumer products is the Rn in water supplies. Building materials, mining, agricultural products, and coal burning also contribute (BEIR, 1990). Cigarette smokers are exposed in addition to ^{210}Po in tobacco, an exposure that undoubtedly contributes to the correlation between cigarette smoking and ling cancer.

It is apparent from Table 14.2 that nuclear power in the United States presently accounts for very little of the average American's radiation exposure. Most of this exposure is concentrated in the workers producing the nuclear fuels and running the power plants (Marx, 1979). However, when evaluating exposure levels associated with the use of nuclear power, one must consider a number of potential sources, including radiation exposure to miners, risks in storage and reprocessing of spent fuel, the probability of exposure after final waste disposal, the risks of a major nuclear plant leak, and the hazards of terrorism. The nuclear power industry has existed for only about 40 years, and because of this relatively short track record, it is difficult to say how frequently certain types of events are likely to occur. In 1975, for example, the U.S. Nuclear Regulatory Commission estimated that the probability of a person's dying from a nuclear reactor accident was about 1 in 5 $\times$ 10^{11} per year per nuclear reactor (NRC, 1975). At the present time, the nuclear power industry has accumulated about 8,000 reactor-years of operation (Flavin, 1987; IAEA, 1997). Given a world population of about 6×10^9 persons, the implication is that about 96 persons should have died by now from accidents at nuclear power plants. In fact, about half of this number of people have been killed more or less outright by nuclear power plant accidents. However, estimates are that about 39,000 persons will ultimately die from radiation-induced cancers caused by the release of radionuclides from the 1986 Chernobyl accident (Marshall, 1987a). The dose of radiation to the human population from the Chernobyl accident is ultimately expected to be about 1.2 million person-Sv, or about 0.2 mSv per person. Although Chernobyl has been by far the most serious accident at a nuclear facility, it is by no means the only time a civilian population has received dangerous doses of radiation as a result of an accident. In 1990 the U.S. government revealed that large amounts of radioactive iodine leaked from fuel processing tanks at its Hanford, Washington, facility from 1944 to 1947 (Marshall, 1990b). Over 13,000 persons may have received radiation doses to the thyroid gland greater than 330 mSv, and 1,400 children may have received doses of 150–6,500 mSv from the consumption of radioactive milk. On September 30, 1999, the Japanese government advised 300,000 people within 10 km of a U processing plant in Tokaimura, Japan, to stay indoors for more than a day after workers at the plant inadvertently set off a nuclear chain reaction. How often such incidents can be expected to occur in the future is a major cause of uncertainty in the estimation of the costs associated with nuclear weapons and the nuclear power industry.

The health costs that the United States is supposedly willing to tolerate can be estimated from Table 14.3 and a knowledge of radiation dose-effect relationships. According to the BEIR-V report, for example, continuous exposure of the U.S. population to an additional 1 mSv per year of radiation would increase the frequency of cancer mortality by about 3% and reduce average life expectancy by about one month (BEIR, 1990). It is unrealistic, however, to think that all members of the U.S. population would experience a uniform increase in radiation exposure due to the operation of nuclear power plants. According to the ICRP (1984), dose limits for members of the general public are to be applied to the average effective dose equivalents to the members of the so-called critical group, which is defined as persons representative of those individuals in the population expected to receive the highest dose equivalent. The critical group is expected to include more than one but no more than a few tens of persons (ICRP, 1984). The NRC (1991a), however, stipulates that the total effective dose equivalent to the individual likely to receive the highest dose should not exceed the annual dose limit. Regardless of whether the dose limit is applied to the single member of the general public who receives the highest dose or the average dose received by a critical group of a few tens of persons, it seems fair to say that the average dose to the general public will be substantially less than the dose limit. The International Commission on Radiobiological Protection (ICRP), for example, has estimated that the average dose to individual members of the public would be less than 10% of the dose equivalent limit, "provided that the practices exposing the public are few and cause little exposure outside the critical groups" (ICRP, 1977, p. 24). The implication is that present guidelines would allow the radiation dose to individual members of the public to increase by no more than 0.1 mSv y^{-1}. The resultant increase in cancer fatalities would amount to 0.3%, too small a change to be detectable with any degree of statistical confidence. The actual body count would amount to about 1,200 additional cancer fatalities per year in the United States.

Importance of Certain Radionuclides

When guidelines on the release of radioactivity by nuclear power plants are established, a small number of radionuclides often turn out to be of special importance. These radionuclides are the ones that tend to be incorporated into living organisms, either because they are radioisotopes of essential elements or because they behave chemically very much like essential elements. In a few cases, radionuclides are of special concern because they may be inhaled and concentrated in the lungs. Table 14.4 lists some of the radionuclides that are of particular concern with respect to human health. Tritium (^{3}H), ^{14}C, ^{32}P, and radioactive iodine (^{129}I and ^{131}I) behave chemically like stable H, C, P, and I, respectively, and are utilized by the body in a similar manner. ^{3}H may be incorporated into water molecules, and ^{14}C into organic molecules, and thus may be distributed throughout the body. Both ^{3}H and ^{14}C are beta particle emitters. ^{32}P tends to concentrate in the bones, which consist of calcium phosphate, where its beta radiation may induce damage to bone marrow, possibly leading to leukemia. ^{129}I and ^{131}I concentrate in the thyroid gland, where their beta and gamma radiation may cause thyroid cancer. Both Ra and strontium (Sr) are chemically similar to Ca, and therefore both tend to concentrate in bones. Cesium-137 (^{137}Cs) is chemically similar to K, an element that is found in fluids throughout the body. U and plutonium (Pu) are not chemically similar to elements commonly found in the body, but they may be inhaled on dust particles and lodge in the lungs, where their alpha radiation may lead to lung cancer. ^{222}Rn, as already noted, is of concern because its daughter radionuclides lodge in the lungs, where their alpha-particle emissions create problems similar to those caused by ^{238}U and ^{239}Pu.

TABLE 14.4 RADIONUCLIDES OF SPECIAL CONCERN AS
HEALTH HAZARDS TO HUMANS

Radionuclide	Half-life	Cause for Concern
^{3}H	12.3 years	Behaves like stable H; assimilated by body in water
^{14}C	5.8×10^4 years	Behaves like stable C; passed up food chain
^{32}P	14.3 days	Behaves like stable P; concentrated in bones
^{90}Sr	28 years	Behaves similarly to Ca; concentrated in bones
^{129}I	1.7×10^7 years	
		Behaves like stable I; localized in thyroid
^{131}I	8.05 days	
^{137}Cs	30 years	Behaves similarly to K; found throughout the body
^{226}Ra	1.6×10^3 years	Behaves similarly to Ca; concentrated in bones
^{238}U	4.5×10^9 years	Intake likely by way of inhalation of dust; concentrated in lungs and kidneys
^{239}Pu	2.4×10^4 years	Intake likely by way of inhalation of dust; concentrated in lungs and kidneys
^{222}Rn	3.8 days	Radioactive daughters taken up by way of inhalation of dust and concentrated in lungs

Effects on Aquatic Systems

Up to this point, we have been concerned primarily with the effects of radiation on humans, but radiation may obviously affect aquatic organisms in similar ways. In the case of aquatic organisms, however, the fate of individual organisms is of less concern than the health of the species or community in general. Thus, while the deaths of 275,000 Americans out of a population of 275 million (i.e., 0.1%) might seem an alarming statistic to many persons, the loss of 0.1% of a population of crabs or tunicates would not be very likely to cause much public alarm.

Acute radiation doses that have been shown to be lethal to aquatic organisms lie in the range 2–550 Gy for invertebrates and adult fish and as low as 0.16 Gy for fish embryos (Anderson and Harrison, 1986). Adverse effects on fish reproduction have been reported at chronic dose rates in the range 6–120 mGy d^{-1} (Anderson and Harrison, 1986), but there is unfortunately not much other reliable information concerning the effects of chronic low-level radiation exposure on aquatic organisms. Sublethal effects associated with genetic defects have been reported at acute doses as low as 0.5 Gy (Anderson and Harrison, 1986). In wild populations, however, "genetic damage may be removed by natural selection, and somatically weakened individuals are probably eaten by predators. Consequently aquatic organisms adversely affected by radiation are not readily recognized in the field" (EPA, 1973, p. 273).

The aquatic organisms that have undoubtedly been subjected to the highest levels of long-term chronic stress from radiation exposure are those living in the vicinity of discharges from major nuclear power or nuclear research facilities. In White Oak Creek, Tennessee, a stream that receives radioactive wastes from the U.S. Oak Ridge National Laboratory, mosquito fish have been subjected to radiation doses of 0.6–3.6 mGy d^{-1} from contaminated sediments. In 1965, when the dose rate was about 3.6 mGy d^{-1}, the fish were found to be producing larger broods than usual, but with an increased incidence of dead and deformed embryos (Woodhead, 1984). The former response was assumed to be an adaptation to the radiation stress and the latter a manifestation of lethal mutations in the genome. Chironomoid larvae living in the same stream were estimated to be receiving a radiation dose of about 6 mGy d^{-1} and were found to have an increased frequency of chromosomal aberrations. Although the observed aberrations were apparently lethal in overall effect, the abundance of the worms was unaffected (Woodhead, 1984). Columbia River salmon spawning near the outfalls from the U.S. nuclear installation at Hanford, Washington, were apparently unaffected by radiation doses of 1–2 mGy wk^{-1} (EPA, 1973), and population and community studies of algae and invertebrates revealed no effects from radiation doses as high as 1.0 Gy d^{-1} in ponds and streams on the Hanford Reservation (Woodhead, 1984).

The examples just cited represent extreme cases of aquatic pollution, and although the list is not long, the implication is that at the population and community levels, aquatic systems are not being adversely affected by present levels of artificial radiation exposure, even in the most severe cases. This conclusion does not apply at the organismal and suborganismal levels. Aquatic species appear to be roughly comparable in sensitivity to humans with respect to radiation effects. The highest dose rate associated with the release of low-level radioactive waste in the ocean is about 2.4 mGy d^{-1} at the end of the discharge pipe from the Windscale reprocessing plant in the United Kingdom (Woodhead, 1984). Most aquatic species obviously experience far lower doses of radiation. In the ocean at roughly 100 m below the surface, the radiation dose rate is only about 0.3 mGy y^{-1}. About 93% of this radiation comes from the decay of naturally occurring ^{40}K (Odum, 1971).

At the present time, the chief concern over contamination of aquatic organisms with radionuclides involves the possible transfer of these radionuclides to human consumers. The transfer problem is compounded by the fact that certain radionuclides are greatly concentrated by some aquatic organisms. For example, the phytoplankter *Dunaliella* has been shown to concentrate ^{65}Zn by as much as a factor of 10^3–10^4 relative to the concentration in the water, and certain lamellibranchs have been found to concentrate ^{54}Mn by a factor of 10^4 (Wilber, 1969). Figure 14.6 shows the concentration of ^{90}Sr in various components of a lake food web relative to the water. From studies such as this one, it has become obvious that concentrations of radionuclides dissolved in the water may give a very misleading picture of the degree of radioactive contamination in an aquatic system. This situation is reminiscent of many cases of heavy metal or pesticide pollution, in which organisms and sediments often contain much higher concentrations of pollutants than the water. Permissible radionuclide discharge levels from nuclear installations have therefore been established only after a careful analysis of the possible pathways of the radionuclides from the discharge into humans, considering not only direct contact between humans and water but also various food chain pathways. It is then generally assumed that "controls which are applied to the discharge of radioactive wastes into the marine environment to limit the potential exposure of human populations also provide quite adequate protection for populations of marine organisms" (Woodhead, 1984, p. 1266). Stating a similar feeling, Loutit (1956, p. 3) commented, "If we take sufficient care radiobiologically to look after mankind, with few exceptions the rest of nature will take care of itself." Odum (1971, p. 457) has criticized this sort of reasoning as a "dangerous oversimplification," but such logic nevertheless remains the basis for setting radioisotope discharge limits. With this introduction to radiation toxicology, let us now examine the major applications of nuclear energy.

NUCLEAR FISSION AND FISSION REACTORS

Energy can be generated from nuclear reactions by two basic mechanisms, fission and fusion. Nuclear fission involves the splitting of a heavy radionuclide to produce two daughter nuclides of lower atomic number. Fission is ordinarily induced by bombarding a heavy

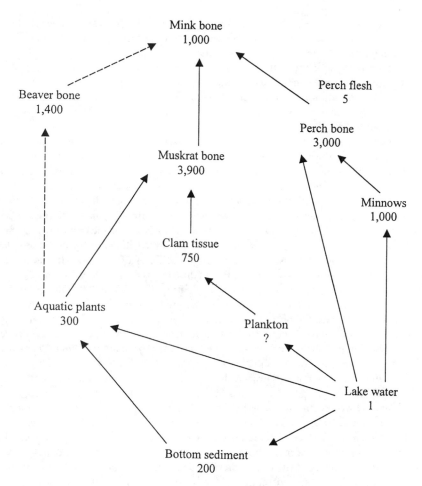

FIGURE 14.6 ^{90}Sr concentration factors in various components of a small Canadian lake receiving low-level radioactive wastes. *Source*: Ophel (1963). © Atomic Energy of Canada Limited, used with permission.

nuclide such as ^{235}U or ^{239}Pu with a neutron. For example, the fissioning of ^{235}U might occur according to the reaction

$$^{235}\text{U} + {}^{1}\text{n} \rightarrow {}^{140}\text{Ba} + {}^{93}\text{Kr} + 3{}^{1}\text{n} + \sim 200 \text{ meV of energy} \tag{14.4}$$

The fragments formed from the splitting of ^{235}U are variable, since there are obviously a great many ways to split ^{235}U. However, the distribution of daughter nuclides produced by the fissioning of ^{235}U does follow a definite pattern. As indicated in Figure 14.7, the probability distribution function is bimodal, with the most abundant daughter nuclides having masses of about 140 amu and a little over 90 amu. Two other characteristics of ^{235}U fissioning are also highly predictable:

1. Large amounts of energy are released whenever ^{235}U fissions. One gram of ^{235}U can release energy equivalent to 20 million grams of TNT.
2. On the average, between 2.4 and 2.9 neutrons are produced for each neutron used to split the ^{235}U (Metz, 1976).

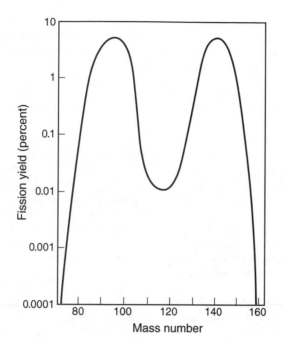

FIGURE 14.7 Probability distribution of daughter nuclides produced by the fissioning of ^{235}U. *Source*: Redrawn with permission from Hunt (1980). Copyright © 1974 by Pergamon Press.

Because of the latter characteristic, it is potentially possible to set off a chain reaction of ^{235}U fissioning, assuming that the neutrons produced by the fissioning collide with other ^{235}U atoms and induce further fissioning. Establishing a chain reaction, however, is not trivial. The neutrons produced by the fissioning of ^{235}U are of rather high energy, and as a result, they have a low probability of interacting with ^{235}U. In effect, they simply bounce off. However, if the speed of the neutrons is reduced, their probability of fissioning ^{235}U is greatly increased. To slow down the neutrons, a *moderator* is introduced into the system. The moderator is nothing more than a substance containing low atomic number atoms such as C or beryllium (Be). Water may also be used as a moderator. Atoms of low atomic number have a low probability of absorbing neutrons, but they may effectively slow down fast neutrons via collisions.

Once the neutrons are sufficiently slowed, a second problem arises. Natural U consists of about 99.3% ^{238}U and only about 0.7% ^{235}U. When ^{238}U captures a slow neutron, the dominant reaction is not fission; the neutron simply combines with the ^{238}U nucleus, forming ^{239}U. Therefore, natural U must be artificially enriched with ^{235}U to produce a fuel that can sustain a chain reaction for several years.[5] In practice, it is only necessary to produce fuel containing 2–3% ^{235}U to operate a fission reactor, since ^{235}U captures slow neutrons and fissions 200 times more efficiently than does ^{238}U (Hunt, 1980). However, the process of ^{235}U enrichment is tedious and expensive, since there is no simple chemical means for separating ^{235}U from ^{238}U. Separation techniques involve gaseous diffusion of uranium hexafluoride[6] or passage of the U ions through a magnetic field (Hunt, 1974).

[5]Natural U may be used as a fuel in reactors that use graphite or heavy water rather than ordinary water as a moderator, since graphite and heavy water absorb fewer neutrons. Heavy water is ^{2}H$_2$O; ordinary or light water is composed almost entirely of ^{1}H$_2$O.
[6]Uranium hexafluoride containing ^{235}U diffuses through a suitable membrane about 0.33% faster than ^{238}UF$_6$.

Finally, there is the problem of the critical mass. If the mass of U fuel is too small, so many neutrons will escape from the fuel before colliding with ^{235}U that it will be impossible to sustain a chain reaction. If the size of the fuel element is increased, the surface/volume ratio declines, and a smaller percentage of the neutrons escape from the fuel before interacting with ^{235}U. The mass of fuel that is just large enough to sustain a chain reaction is called the *critical mass*. Obviously, the size of the critical mass depends in part on the percentage of ^{235}U in the fuel; the larger the percentage of ^{235}U, the smaller the critical mass.

Pu may also be used as a fuel in a nuclear chain reaction, but since Pu is not a naturally occurring element, it is necessary to use somewhat extraordinary means to generate Pu fuel. If ^{238}U absorbs a neutron, it forms ^{239}U. ^{239}U is radioactive, with a half-life of 23.5 minutes, and decays via beta$^-$ emission to produce Neptunium-239 (^{239}Np). ^{239}Np has a half-life of 2.3 days and decays via beta$^-$ emission to form ^{239}Pu. Thus ^{239}Pu, a fissionable radionuclide, can be produced by bombarding ^{238}U with neutrons, and indeed, some ^{239}Pu is produced by this mechanism in conventional nuclear reactors. The efficiency of Pu production is low, however, since, as noted, ^{238}U does not effectively absorb slow neutrons.

Removing the neutron moderator can increase the efficiency of Pu production. Under these conditions, two things happen:

1. The neutrons produced by fission reactions are not slowed and therefore have a lower probability of interacting with ^{239}Pu or ^{235}U.
2. The average number of neutrons produced per fission increases from about 2.4 to 2.9 (Metz, 1976).

To sustain a chain reaction under these conditions, the percentage of fissionable material, either ^{235}U or ^{239}Pu, in the fuel must be increased from about 2–3% to 15–30% (Metz, 1976). In *breeder reactors*, a blanket of ^{238}U surrounds a compact fuel core. High-energy neutrons escaping from the fuel core combine with the ^{238}U to produce ^{239}Pu and thus generate additional fissile fuel. Because of the increased neutron yield per fission, such reactors can be designed to produce more fuel in the form of ^{239}Pu than they consume in the form of ^{235}U or ^{239}Pu, hence the name *breeders*. A breeder reactor can also be designed using ^{232}Th as a fuel. Bombarding ^{232}Th with slow neutrons produces ^{233}Th, which then decays via beta$^-$ emission with a half-life of 22.1 minutes to produce protactinium-233 (^{233}Pa), which in turn decays via beta$^-$ emission with a half-life of 27.4 days to yield ^{233}U, a radionuclide that, like ^{235}U and ^{239}Pu, can be bombarded with slow neutrons to sustain a fission chain reaction.

The first use of nuclear fission was for the production of bombs rather than for the generation of electricity. The United States began producing Pu at special reactors near Hanford, Washington, in 1944 (Seymour, 1971), and Pu from these reactors was used to build the bomb that was dropped on Nagasaki, Japan, in 1945. The bomb that was dropped on Hiroshima, Japan, was made of ^{235}U-enriched U from the government plant at Oak Ridge, Tennessee (Novick, 1969). In the case of a bomb, the goal is to release an enormous amount of energy in a short time. In a nuclear reactor, the release of energy must occur at a slow, controllable rate.

At the present time, all nuclear reactors in the United States use ^{235}U-enriched U as a fuel, and with a few exceptions, all use water as a moderator. A diagram of the core region of a typical reactor is shown in Figure 14.8. The core of such a reactor consists of literally thousands of small U oxide fuel rods, each enclosed in a cladding of stainless steel or zirconium alloy. Each fuel rod is about 1 cm in diameter and several meters long. A typical 500-MW nuclear reactor would probably contain about 20,000 such fuel rods (Adler, 1973). Although no single fuel element contains a critical mass of fissile material, when the fuel elements are assembled together in the core of the reactor, a critical mass is

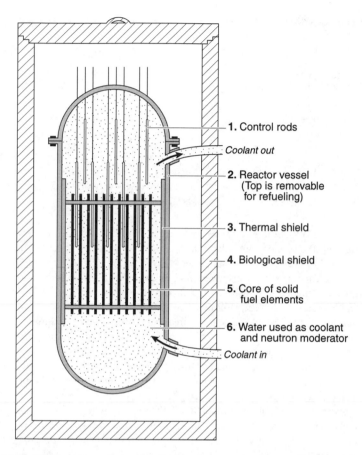

1. Control rods

Coolant out

2. Reactor vessel
   (Top is removable
   for refueling)

3. Thermal shield

4. Biological shield

5. Core of solid
   fuel elements

6. Water used as coolant
   and neutron moderator

Coolant in

FIGURE 14.8   Cross section of a pressurized nuclear reactor. *Source*: Reprinted from *Poisoned Power*. Copyright © 1979 by John W. Gofman and Arthur R. Tamplin. Permission granted by Rodale Press, Inc. Emmaus, PA.

achieved. To control the rate of the nuclear reaction, control rods made of Cd are inserted into the reactor core. Since Cd is a good absorber of neutrons, the control rods can effectively slow down or, if necessary, completely stop the chain reaction in the core. The ^{235}U-enriched fuel elements are normally used for two to three years, during which time the percentage of ^{235}U decreases from about 3% to roughly 0.8%. At that time the rods must be replaced, since the percentage of ^{235}U is becoming marginal for sustaining a chain reaction (Cohen, 1977).

The water circulating through the core serves several purposes. As already noted, it slows down, by a factor of 1,000 or more, the high-energy neutrons produced by the fissioning of ^{235}U (Novick, 1969). Second, it cools the reactor core. The reactor core is typically operated at a temperature of 540°C (Novick, 1969), but the intense heat produced by the chain reaction in the core would rapidly raise this temperature if water or a similar coolant were not constantly pumped through the core. If the cooling system were to fail and if the chain reaction were not quickly stopped, the rise in temperature could crack or melt the cladding and perhaps even melt the fuel, the result being an uncontrolled release of energy. Finally, heating the water is used to generate electricity. In a *boiling water reactor* (*BWR*), the water is converted to steam by the temperature in the core, and the steam, in turn, used to drive a turbine to generate electricity, as indicated in Figure 14.9. An al-

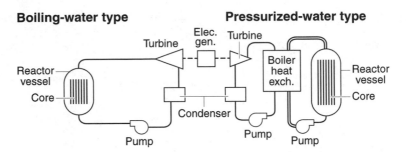

FIGURE 14.9   Schematic of two types of nuclear power reactors. *Source*: Redrawn from Joseph et al. (1973). Copyright 1971 by the National Academy of Sciences. Courtesy of the National Academy Press, Washington, D.C.

ternative to the BWR is the *pressurized water reactor* (*PWR*), in which the water that circulates through the reactor core is kept under sufficient pressure to keep it from boiling and is cooled via a heat exchanger with a secondary cooling circuit (Figure 14.9). The water in the secondary circuit is converted to steam as it passes through the heat exchanger, and this steam is then used to drive a turbine to produce electricity. Most nuclear reactors in the United States are PWRs. Although BWRs and PWRs account for almost all of the nuclear reactors in the United States, there are several other reactor designs that have been commonly used in other countries. In the United Kingdom and France, gas-cooled nuclear reactors are used (Joseph et al., 1973). In Canada, highly successful reactors have been designed using heavy water rather than ordinary water as the moderator (Robertson, 1978).

Although electricity was produced from the first breeder reactor as long ago as 1951, breeder reactor technology has never advanced much beyond the experimental stage. Small-scale prototypes were built in the United States, the former Soviet Union, Japan, the United Kingdom, West Germany, and France (Zaleski, 1980). Since the neutrons produced in such reactors must be of the fast variety, water is ruled out as a primary coolant. Instead, all breeder reactors have used liquid sodium (Na) as a primary coolant, and have hence been referred to as *liquid metal fast breeder reactors* (*LMFBRs*). Figure 14.10 shows a schematic of one possible design of a LMFBR. Based on experience to date, one of the principal problems with LMFBRs has been the design of the heat exchangers. Na reacts explosively with both air and water, so it is essential to keep the Na coolant in a completely enclosed sys-

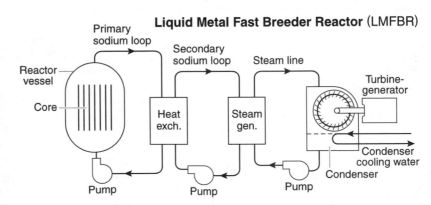

FIGURE 14.10   Diagram of a LMFBR. *Source*: Redrawn from *Poisoned Power*. Copyright © 1971 by John W. Gofman and Arthur R. Tamplin. Permission granted by Rodale Press, Inc. Emmaus, PA.

tem; a small leak in the Na circuit could lead to a serious accident. However, since power is ultimately generated by the production of steam, a heat exchanger in which liquid Na and water are in close proximity is a feature of all LMFBRs. Leaks in this system, usually associated with welding connections, led to at least three accidents in breeder reactors in the former Soviet Union (Metz, 1976) and were a problem in the French and British reactors as well. Because of this problem, a secondary Na loop has been utilized in all LMFBRs to minimize the possibility that the primary coolant will come into contact with water or air and react explosively. The magnitude of breeder reactor technological problems is dramatically illustrated by the case of the French Superphenix reactor, the largest and most advanced commercial breeder reactor ever built. From its entry into service in 1986 until the end of 1995, it produced electricity for a total of 10 months. The rest of the time it was beset by technical problems.

A second problem with breeder reactors is the large amounts of Pu involved. A large LMFBR would contain about a tonne of ^{239}Pu, only about 0.5% of which would represent a critical mass (Hammond, 1972). It is conceivable, although improbable, that terrorist groups could seize shipments of Pu or spent fuel rods and use the ^{239}Pu to manufacture a bomb. Since Pu is chemically different from U and other elements, separation of Pu is accomplished much more easily than separation of ^{235}U from ^{238}U. Furthermore, "the information necessary to construct a crude nuclear bomb is readily available—pamphlets describing the required steps have been circulated in Great Britain" (Hammond, 1972, p. 149).

Perhaps the most serious problem with breeder reactors, however, has been their cost. The French Superphenix breeder reactor produced electricity at about twice the cost of conventional PWRs (Dickson, 1987). The French eventually shut down the Superphenix on Christmas Eve, 1997, leaving Japan as the only nation actively pursuing breeder technology. While acknowledging that breeders are more expensive than conventional nuclear power plants over the roughly 40-year design lifetime of the plants, Weinberg (1986) has argued that the lifetime of a breeder reactor might be much longer than 40 years, perhaps as long as 100–150 years. If so, electricity production by breeders could become very cheap following their amortization time. However, the idea that breeder reactors might last as long 100–150 years before being decommissioned is highly speculative at this time.

The principal motivation for breeder reactors is the fact that they consume only about 2% as much U as conventional fission reactors. Hence, the lifetime of the fission nuclear power industry could be extended by about a factor of 50 with the use of breeder reactors. This fact is of great interest to a country like Japan, which does not have much U. It is of less relevance on a global scale, since at current prices world U resources could sustain present rates of nuclear power production for about 100 years (Weinberg, 1986). The United States, in particular, has been reluctant to pursue the breeder option, because U.S. deposits of both U and coal are large. Hence the United States can afford to take a cautious approach to the development of technologies that might someday replace fossil fuel and conventional nuclear power production methods.

NUCLEAR FUSION

Nuclear fusion involves combining two elements of low atomic number. To be useful from the standpoint of energy generation, the fusion reaction must release large amounts of energy. That such high-energy-yielding reactions can and do occur is well known. Nuclear fusion accounts for the energy output of the sun and stars, and so-called hydrogen bombs utilize nuclear fusion to produce an explosive release of energy. At the present time, however, nuclear fusion is not used to generate electric power because no satisfactory means has been found to control a nuclear fusion reaction. Nevertheless, it is conceivable that within the next 50 years or so, nuclear fusion reactors will become a reality, and it is therefore worthwhile to examine how such a reactor might work and later on to discuss its potential radiation problems.

Although a number of fusion reactions are potentially utilizable as a source of energy (Holdren, 1978), the most likely candidate for use in a nuclear reactor is the fusion of ^{2}H (deuterium) with ^{3}H (tritium) as follows:

$$^2H + {}^3H \rightarrow {}^4He + n \tag{14.5}$$

Although there is an abundance of ^{2}H in the world's oceans, ^{3}H, as already noted, is radioactive and virtually nonexistent in nature. The ^{3}H necessary to fuel deuterium-tritium (D-T) fusion reactors, however, could presumably be supplied by breeding from lithium (Li) according to the reaction

$$^7Li + n \text{ (fast)} \rightarrow {}^3H + {}^4He + n \text{ (slow)} \tag{14.6}$$

or

$$^6Li + n \text{ (slow)} \rightarrow {}^3H + {}^4He \tag{14.7}$$

Thus, a D-T nuclear reactor would be a breeder reactor as well as a fusion reactor. Another possible reaction for use in a fusion reactor is the fusion of two ^{2}H atoms as follows:

$$^2H + {}^2H \rightarrow {}^3He + n \tag{14.8}$$

or

$$^2H + {}^2H \rightarrow {}^3H + {}^1H \tag{14.9}$$

If a deuterium-deuterium (D-D) reaction were utilized, there would be no need to design a breeder system because of the great abundance of ^{2}H in the oceans.

Despite the fact that reactions such as the fusion of ^{2}H and ^{3}H or ^{2}H and ^{2}H can be produced in a nuclear bomb, there are great obstacles to be overcome before these reactions can be used for constructive purposes. One important limitation, and undoubtedly the most controversial one, is the fact that the temperature required to sustain these reactions is on the order of 100 million °C (Metz, 1972), about 10,000 times the temperature at the surface of the sun. This constraint was challenged, however, when in 1989 the electrochemists Stanley Pons and Martin Fleischmann reported at a press conference that electrolysis of a Li solution in heavy water, with a palladium (Pd) cathode and a platinum (Pt) anode, produced neutrons, ^{3}H, and large amounts of heat, all from the fusion of ^{2}H nuclei (Garwin, 1991). The irreproducibility of these results, however, has cast considerable doubt on their authenticity. Even if so-called cold fusion can be achieved, there is still a legitimate question as to whether cold fusion will be a practical way to generate electricity. Although it seems premature to dismiss the possibility of cold fusion altogether, for the time being, most of the emphasis in fusion research is focused on high-temperature reactions.

Unfortunately, there is no known substance that could possibly contain a reaction at 100 million °C without melting. The most promising approach to solving this problem has been to ionize the atoms involved in the reaction and to contain the ionic soup or plasma with a magnetic field (Furth, 1990). The design of such a magnetic confinement reactor, however, is a formidable engineering task, since the magnets required to produce the confinement magnetic field must be of the superconducting variety, and hence will function only at temperatures less than roughly $-150°C$. Furthermore, the system must be designed to avoid damage to the magnets from the enormous flux of neutrons[7] emerging from the plasma in the center of the magnetic field, where the fusion reaction will occur. Since the magnets would be separated from the plasma by only a few meters, it is obvious that an enormous gradient in temperature as well as neutron flux must exist.

Even if magnetic containment devices are developed that can satisfactorily produce fusion reactions, there will still be formidable engineering obstacles to the development of fusion power reactors. Some method must be found to protect critical components of the reactor from the enormous flux of high-energy neutrons that will be produced in the fusion reaction. Much of the neutron energy could be absorbed by surrounding the inner chamber with a blanket of water about 1 m thick, and indeed, this absorption of neutron energy is likely to be the principal mechanism by which energy will be extracted from the reactor. High-energy neutron damage to reactor vessel material[8] could result in the need for periodic replacement of the reactor vessel. The associated costs and shutdown time would probably make this an unacceptable operating requirement for utilities.

The facts that no heat transfer surfaces can be located near the region of the reactor where fusion occurs, and that heat must be extracted instead by absorbing high-energy neutrons at a distance from the central region of the reactor, place a minimum size restriction on the reactor design. Because of this size restriction and the high cost of the materials required to build a fusion reactor, the cost of conducting meaningful experiments has become a serious obstacle to fusion research. The United States, for example, chose to abandon plans to build a $1.9 billion fusion reactor known as the Burning Plasma Experiment (Holden, 1991). The cost of the reactor would have doubled the Energy Department's fusion budget over a five-year period. One consequence of these high costs has been an international effort to pool resources. The United States, Japan, Russia, and the European Union, for example, began the design phase of a so-called International Thermonuclear Experimental Reactor (ITER), which, it was hoped, would be capable of sustaining a fusion chain reaction (Holden, 1991). The price of the ITER was expected to be about $10 billion, with the design phase alone costing $1 billion (Hamilton, 1991). Ultimately, economic considerations caused the ITER project to be abandoned (Lawler and Glanz, 1998). The next phase of magnetic confinement experiments may instead utilize magnetic fields produced by cooled Cu coils rather than superconducting magnets (Glanz and Lawler, 1998). The Cu coils are actually capable of producing more powerful magnetic fields than superconducters, albeit for shorter periods of time. A Cu-coil reactor would be able to sustain ignition for only about 10-second bursts rather than the 1,000-second pulses theoretically possible with heavily shielded superconductors, but it would be much less expensive and would still allow critical experiments related to plasma physics.

With the emergence of serious economic obstacles to magnetic confinement fusion, other approaches are being given serious consideration. One of the most promising alternatives is to create tiny explosions in pellets of H isotopes crushed by lasers or other methods (Lawler and Glanz, 1998). This is a technology derived from weapons research, and construction is now well underway on a so-called national ignition facility (NIF) to explore the feasibility of the method at Lawrence Livermore National Laboratory in California. When completed, the NIF will be able to focus 200 laser beams on a single target. The U.S. government is currently spending in excess of $200 million per year on construction of the NIF. X rays emitted by a plasma imploding after an array of wires is vaporized by a jolt of electric current are an alternative to the use of lasers to crush the fuel pellets (Glanz, 1998). The X-ray option is being seriously studied at Sandia National Lab-

---

[7]About 10^{13} cm^{-2} s^{-1}.
[8]Loss of ductility and swelling.

oratories in New Mexico. Much work remains to be done before any of these approaches to nuclear fusion can be shown to be viable economically and technologically.

In summary, there are a number of difficult scientific and economic problems that will have to be overcome if fusion power is to become a realistic energy option. Most scenarios do not envision fusion power becoming a significant factor in electricity production until roughly the middle of the 21st century (Holdren, 1978; Furth, 1990). The appeal of fusion power is the virtually inexhaustible supply of the fuel. Because of the abundance of ^{2}H and Li in the ocean, a fusion-based electrical system would be essentially autarkic.

RADIATION RELEASES BY POWER PLANTS

There are basically two mechanisms by which radionuclides are created in power plants. In a fission reactor, some of the fission products themselves may be radioactive, and these radionuclides may decay to produce daughter radionuclides. Important examples of such radioactive fission products are ^{137}Cs, ^{131}I, and ^{90}Sr. ^{3}H is the only radionuclide directly involved in a fusion reaction. Tritium is consumed in the D-T reaction (Eq. 14.5), but it may be produced in a D-D reaction (Eq. 14.9). The second mechanism by which radionuclides are produced in a nuclear reactor is a process called *neutron activation*, in which a neutron produced by either a fission or fusion reaction combines with a stable nuclide to produce a radionuclide. An example of such a neutron activation reaction would be the formation of the radionuclide iron-55 (^{55}Fe) from the stable nuclide ^{54}Fe according to the equation

$$^{54}\text{Fe} + {}^1\text{n} \rightarrow {}^{55}\text{Fe} \tag{14.10}$$

Important neutron activation products include ^{55}Fe, ^{32}P, ^{65}Zn, ^{3}H, and the transuranic elements.[9] All transuranics are radioactive, and none exists naturally. Obviously, the nature of the neutron activation products produced by a nuclear reactor depends very much on the nature of the elements in the vicinity of the core. ^{3}H is the neutron activation product produced in greatest abundance because of the use of water as a coolant and/or moderator. Power plants attempt to minimize the production of other neutron activation products by removing impurities and corrosion products from the water and by surrounding the reactor with nonactivating neutron absorbers.

Routine Radionuclide Releases

Once fission has been induced in the fuel elements of a conventional nuclear power reactor, radioactive fission fragments as well as neutron-activated radionuclides begin to accumulate inside the fuel elements. The amount of radioactive waste produced in this manner is enormous. A single 1,000 MW nuclear reactor produces as much long-lived radioactive wastes in one year as a thousand Hiroshima bombs, and 10 such reactors operating for two years would generate as much radioactive waste as was produced by the combined U.S. and Soviet bomb tests prior to the 1963 partial test ban (Gofman and Tamplin, 1979). In fact, the heat generated by the decay of radionuclides inside a fuel element after two to three years of use is so intense that the fuel element would literally melt unless artificially cooled (Novick, 1969).

---

[9]Transuranic elements are elements with atomic numbers greater than that of U (92) and include Pu, neptunium, americium, and californium.

Except for diffusion of some radionuclides through the fuel element cladding, little radioactivity escapes from the fuel elements unless a crack develops in the cladding. Radionuclides leaked from the fuel elements to the cooling water are normally discharged to the environment. If the radionuclides are gases, they are released via stacks to the atmosphere. Nongaseous radionuclides are normally separated from the cooling water along with other impurities prior to recycling of the water. The impurities and nongaseous radionuclides are then disposed of in some convenient manner. In short, unless there has been a leak or malfunction in the reactor core, the level of radioactivity in these wastes is sufficiently low that they can be legally discharged to the atmosphere (gases) or to the nearest convenient waterway (liquids) or simply buried in the ground in some remote and government-supervised location (solids).

There is no doubt that serious leaks in fuel element cladding have developed on occasion in some reactors. In June 1965, unusually high discharges of radioactivity began to occur from the stacks of Pacific Gas and Electric's (PGE's) Humboldt Bay power plant at Eureka, California, a 68.5 MW BWR that had gone into operation in August 1963 (Novick, 1969; Gofman and Tamplin, 1979). The level of radioactivity in the stack gases continued to increase in the following months, and in August 1965, the mean discharge rate of radioactivity from the stacks was 3 billion Bq. The Atomic Energy Commission's (AEC's) limit for the mean annual discharge rate for the Humboldt plant was 1.8 billion Bq. Fortunately, most of the radioactivity was in the form of xenon (Xe) and krypton (Kr), which are noble gases and highly inert chemically, but some radioactive release occurred in the form of ^{131}I. It was apparent at that time that the fuel rods' stainless steel cladding was developing leaks, and between September and December 1965, 25% of the fuel elements were replaced with zircaloy-clad fuel elements. Although radioactive emissions initially dropped following this change, by August 1966 emissions had risen to 1.5 billion Bq, presumably due to the failure of some of the remaining stainless steel fuel claddings. At that time, a request by PGE for an increase in their radioactivity discharge rate limit to 7.8 billion Bq was denied by the AEC. Realizing that they could not continue to operate the Humboldt Bay plant with its leaky stainless steel cladding and still maintain their discharge under the 1.8 billion Bq annual average, PGE then replaced two-thirds of the remaining stainless steel-clad fuel elements with zircaloy-clad fuel elements. The plant was then run at 40% of its rated power "in order to extend the life of the remaining stainless steel-clad fuel in the core" (Novick, 1969, p. 116).[10] From an economic standpoint at least, the desire of PGE to use the stainless steel-clad fuel elements as much as possible, despite their tendency to leak, was understandable, since the initial fuel loading for the Humboldt reactor was worth about $4 million (Novick, 1969). Stainless steel rather than zircaloy was initially chosen as the cladding because stainless steel was cheaper. The design of nonleaky cladding is obviously crucial to the successful operation of a fission power plant. Although the problem at the Humboldt plant appears to have been an unusual case, containing the radioactivity inside the fuel rods is no simple matter. Some radioactivity invariably escapes through the cladding via diffusion or through small cracks. This leakage contributes to the radioactivity routinely discharged by a nuclear power plant.

The second principal source of radionuclides routinely discharged by a nuclear power plant is neutron activation of stable elements surrounding the fuel rods. Under normal circumstances, these neutron-activated radionuclides constitute the principal source of radioactive release by a power plant (Joseph et al., 1971). Atoms in the primary coolant, including any corrosion products, the reactor vessel itself, and any other nearby materials, are subject to neutron bombardment from the fuel elements. In the United States, where most reactors are water cooled, these activation products appear largely in the primary

---

[10]The plant was shut down permanently in 1976 because it lies in a seismically active zone and was not structurally engineered to withstand earthquakes.

coolant. The principal gaseous activation products are ^{13}N and ^{41}Ar. The principal non-gaseous activation products include ^{3}H, ^{32}P, ^{51}Cr, ^{55}Fe, ^{54}Mn, ^{60}Co, and ^{65}Zn (Joseph et al., 1971). Many of the nongaseous neutron activation products are metals and result from activation of stable corrosion products. In U.S. Navy nuclear-powered ships, the principal source of radioactivity in liquid wastes is ^{60}Co resulting from neutron activation of corrosion and wear products from reactor plant metal surfaces in contact with reactor cooling water (Miles et al., 1979).

Whether the nuclear reactor is a PWR or a BWR, an attempt is made to keep the concentration of impurities in the primary coolant as low as possible by filtering and demineralizing the water periodically. This procedure reduces the corrosion rate, as well as removing dissolved and particulate substances that otherwise would be subject to neutron activation. In a BWR, the radioactive gases in the cooling water are removed by simply pumping large volumes of air through the steam turbine air ejector and venting the gases to the atmosphere, usually after a 20–30 minute delay to allow for decay of short-lived radionuclides (Joseph et al., 1971). In a PWR, the radioactive gases are collected in special storage tanks and removed periodically when the coolant is withdrawn.

The quantity of radioisotopes that can be discharged by a nuclear power plant in the United States is determined by several pieces of legislation. In all cases, the intent of the legislation is to protect members of the general public who might ingest, inhale, or come into contact with the radioisotopes, but the approaches taken to address this issue and the acceptable limits of exposure differ somewhat from one piece of legislation to the next. The Nuclear Regulatory Commission (NRC), for example, has established radioisotope concentration limits for wastewater based on the assumption that a person drinks 2 L of the wastewater per day (NRC, 1991a). The maximum allowable dose is 0.5 mSv y^{-1}. The assumption of strictly additive effects is used to determine whether wastewater containing more than one radioisotope can be discharged. For example, the maximum effluent concentrations for tritium, ^{55}Fe, and ^{60}Co are 37,000, 3,700, and 111 Bq L^{-1}, respectively. If wastewater from a nuclear power plant contained only these three radionuclides and if their concentrations were 7,400, 1,110, and 11.1 Bq L^{-1}, then the concentrations expressed as fractions of the maximum effluent concentrations would be $7,400/37,000 = 0.2$, $1,110/3,700 = 0.3$, and $11.1/111 = 0.1$ for ^{3}H, ^{55}Fe, and ^{60}Co, respectively. Since the sum of the fractions is less than 1.0, it would be acceptable to discharge this water, according to the NRC (1991a).

In the foregoing example, it was assumed that a person drank 2 L of the wastewater from the power plant each day, a rather improbable scenario. The other two pieces of legislation that limit radioisotope discharges from power plants take a more sophisticated approach in the sense that they require a careful analysis of the pathways leading to human exposure and a quantitative assessment of the corresponding doses. The EPA (1991a) requires that the annual dose equivalent to any member of the public resulting from planned discharges of radioisotopes associated with the U fuel cycle[11] not exceed 0.25 mSv to the whole body, 0.75 mSv to the thyroid gland, and 0.25 mSv to any other organ. Radiation exposure resulting from Rn and its daughters, however, is excluded from this restriction. The NRC (1991b) requirement is that the amount of radioisotopes discharged with the wastewater from all nuclear power plants at a site to unrestricted areas will not result in an estimated annual dose from all pathways of exposure in excess of 0.05 mSv to the whole body of any individual in an unrestricted area.

Pathway analysis requires the ability to model the various routes by which radionuclides, once released to the environment, come into contact with humans. In applying this

---

[11]The U fuel cycle means the operations of milling U ore, chemical conversion of U, isotopic enrichment of U, fabrication of U fuel, generation of electricity by nuclear power plants, and reprocessing of spent U fuel.

approach, one must obviously be able to estimate with some accuracy the efficiency and rate at which radionuclides are transferred from one compartment in the model to another. Often only one or a few pathways are found to be much more important than any others in causing human exposure. These pathways are referred to as the *critical pathways*.

One of the best-known applications of critical pathway analysis was the determination of acceptable radioisotope discharge rates from the U.S. government's Pu production reactors at Hanford, Washington. Although Pu production and recovery at Hanford finally ceased in 1988 (Holden, 1990), for many years the Hanford reactors provided Pu needed by the Department of Defense for the production of nuclear weapons. Prior to 1971, the cooling systems for at least some of the Hanford reactors were of the once-through variety (NAS, 1978), and the quantity of radionuclides discharged to the Columbia River far exceeded discharge rates from conventional electric power plants. Although the Hanford discharge occurred about 580 km from the mouth of the Columbia River, some radionuclides did reach the sea in significant amounts. A critical pathway analysis revealed that of these radionuclides, ^{32}P and ^{65}Zn posed the greatest danger to humans. Oysters grown in a bay several kilometers from the mouth of the river were found to have elevated concentrations of both radionuclides, and a person eating 230 g of these oysters per week for a year would have received about 0.3–0.6% of the acceptable yearly radiation dose at that time (Joseph et al., 1971). The figure of 230 g wk^{-1} was estimated to be the average oyster consumption rate of the so-called critical group of persons who relied heavily on seafood for nutrition. In the Columbia River itself near the Hanford reactors, a critical pathway analysis indicated that human consumption of ^{32}P-containing fish was the most important mechanism for human exposure. Figure 14.11 depicts this critical pathway from the reactor wastewater to the river water to the fish and finally to humans. In this case, the critical group who relied heavily on fish for nutrition were assumed to consume 40 kg of fish per year. The calculated dose to the bone of the critical group was 3 mSv per year, 20% of the acceptable dose at that time.

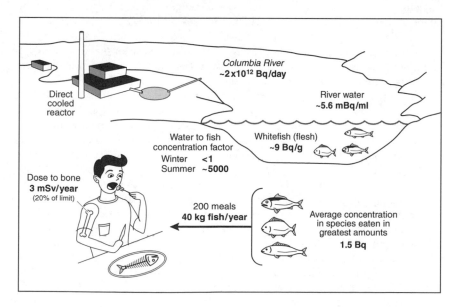

FIGURE 14.11   The critical pathway by which humans were exposed to ^{32}P from Hanford reactor wastes in the Columbia River. *Source*: Redrawn from *Radioactivity in the Marine Environment* (1973), p. 255, with the permission of the National Academy of Sciences, Washington, DC.

Accidents

Much of the concern over nuclear power has centered on the issue of power plant accidents, in particular the sorts of accidents that might lead to an uncontrolled nuclear chain reaction and perhaps release large amounts of radioactivity into populous areas. Obviously, the best method of dealing with this problem is to try, insofar as possible, to design a failproof, foolproof power plant. However, since human endeavors are subject to errors and oversights, accidents are almost inevitable. Therefore, power plants should be designed to minimize the danger to both plant personnel and the public when accidents occur. The following summaries of the most noteworthy accidents illustrate the nature of the problem.

The NRX Accident

On December 12, 1952, an operator at Canada's experimental nuclear reactor (NRX) at Chalk River, Ontario, mistakenly opened three or four (no one is sure) valves, causing 3 or 4 of the reactor's 12 shutoff rods to be withdrawn from the reactor core. At that time the reactor, which contained a natural-U-fueled, heavy-water-moderated core, was being operated on low power in conjunction with an experiment (Novick, 1969). The operator who mistakenly opened the valves was working in the basement of the reactor building, and initially at least, had no indication that he had done anything wrong. However, red lights quickly turned on at the control desk upstairs, indicating that the chain reaction in the core was out of control, and the supervisor at the desk immediately phoned the operator in the basement to tell him to stop whatever he was doing. The supervisor then went to the basement, realized the operator's mistake, and closed the valves. The red lights on the control desk then went out, indicating that the reaction in the core was again under control. Unfortunately, subsequent events indicated that closing the valves had caused the control rods to drop only partway into the core, so that although the reaction was technically under control, in fact there was only a small margin of safety. The cause of this apparent mechanical failure has never been known (Novick, 1969).

The supervisor, then thinking that he had fully corrected the problem, phoned an assistant in the control room, intending to have him return the reactor to normal operation by pushing buttons 3 and 4 on the control desk. Instead, he inadvertently told the operator to push buttons 1 and 3. Although the supervisor quickly realized his mistake and attempted to correct his message, the assistant in the control room had already laid down the phone to carry out the order. Pushing button 1 caused four more shutoff rods to be withdrawn from the core, in addition to the three or four partially withdrawn rods that had evidently failed to fall back into place. As a result, the chain reaction in the core again went out of control, the power output steadily increased, and the red lights on the control panel again flashed on (Novick, 1969).

About 20 seconds after pushing button 1, the assistant in the control room realized that something was wrong and pushed an emergency button that should have caused all the control rods to drop back into the core. However, only one of the seven or eight shutoff rods dropped back into place, and that rod moved very slowly, requiring about 90 seconds to fall the necessary 3 m. Two red lights remained on at the control desk, indicating that the nuclear reaction was still out of control. Realizing the seriousness of the situation, the assistant supervisor ordered the heavy water drained from the reactor vessel, a move that would ensure that the chain reaction came to a halt. The water required about 30 seconds to drain from the reactor, and at the end of that time, instruments in the control room indicated that the reactor was shut down (Novick, 1969). The damage, however, had already been done.

The buildup of heat in the core during the roughly 70 seconds that the shutoff rods were withdrawn had melted some of the U fuel elements, and the molten U and Al then reacted violently with the water and steam in the core. The U-water reaction produced H gas, which, at the high temperature in the core, in turn reacted violently with inflowing air, causing an explosion that lifted the 4 tonne gas-holder dome over 1 m into the air, jamming it among surrounding structures (Novick, 1969). The runaway nuclear fissions occurring in the core released energy equivalent to that produced by exploding about half a tonne of TNT, although fortunately, the release of nuclear energy occurred over a sufficiently long time span (several seconds) that no violent nuclear explosion was produced. A cloud of radioactivity was released into the air, however, setting off automatic alarm systems in the vicinity of the reactor. Although no lives were lost and the exposure of personnel to radiation was relatively mild, the reactor core was destroyed by the accident (Novick, 1969).

Windscale

In 1950 and 1951, the British government began operating two Pu-producing nuclear reactors at its facility at Windscale, Cumberland, in a sparsely populated area on the Irish Sea. The reactors were fueled with natural U fuel elements clad in steel and inserted in a 15 m cube of graphite, which served as a neutron moderator (Novick, 1969). The core of each reactor was cooled by means of air that was exhausted through two 125 m stacks to the atmosphere (Dunster et al., 1958).

In a graphite moderator, some of the energy transmitted to the graphite by the high-energy neutrons emitted from the fuel elements is not immediately released as heat, but instead is stored as chemical energy within the graphite matrix of C atoms. In effect, some of the C atoms in the graphite have been promoted to an energetically elevated or "excited" state. This excited state is thermodynamically unstable, but it may persist for a considerable time before the system reverts spontaneously to a lower-level energy state, at which time the stored energy is released as heat. This release of energy chemically stored in the graphite matrix is referred to as *Wigner release*, after a German physicist whose research contributed to the understanding of the process (Hunt, 1980).

Since unexpected releases of energy within a reactor core can lead to serious damage, some method of controlling Wigner release is necessary in a graphite-moderated reactor. Regular procedures to control Wigner release were first begun at Windscale following a spontaneous Wigner release in 1952 (Novick, 1969). By accident, the British discovered that heating the graphite slightly could trigger the release of the stored energy, so that by periodically heating the graphite, it was possible to prevent a large-scale buildup of stored energy. On October 7, 1957, such an energy release procedure was begun on Windscale Pile No. 1 while the reactor was shut down following a low-power run.

On October 8, operators noted that heating of the core was occurring at a faster rate than expected, and Cd rods were inserted to dampen the chain reaction in the U (Novick, 1969). In retrospect, it is apparent that by this time, one or more fuel elements in the core had in fact overheated, and the steel jackets had melted or cracked. As the temperature in the core continued to rise, the U in the damaged fuel elements, now exposed to the air, began to burn, causing nearby fuel elements to overheat and crack, thus spreading the fire.

The first indication of a serious problem did not occur until noon on October 10, when radioactivity levels in the stack gases from Pile No. 1 began to rise sharply, and it became apparent to operators that some of the fuel elements had failed. In attempting to locate the damaged fuel elements, operators opened a plug on the side of the reactor and observed that the four fuel elements they could see inside the core were red hot. At this time, the operator first realized that a fire had broken out in the core. Push rods subsequently inserted into the core came out dripping with molten U (Novick, 1969). Efforts to remove the damaged fuel elements failed because the fuel cartridges were badly distorted. Fuel elements surrounding the fire were removed, however, so that at least the fire was contained. However, about 150 fuel elements were on fire by the time this action was taken. Efforts to cool the system by blowing carbon dioxide into the core failed, since "by this time the region involved was too hot to be effectively cooled in this way" (Dunster et al., 1958, p. 297).

During the early morning of October 11, a decision was made to attempt to put out the fire by pumping water into the core. This decision was reached with much difficulty, since it was known that water striking molten U could cause an explosion. Pumping of water into the core began at 9 A.M. on October 11. Fortunately, no explosion occurred, and by October 12 the core was "quite cold" (Novick, 1969, p. 10). According to an official of the AEC, "They will not guarantee that they could do it a second time without an explosion" (McCullough, 1959, p. 78).

The net result of the Windscale incident was the complete loss of the nuclear reactor and the release of an amount of radioactivity equal to about 10% of the radioactivity produced by the atomic bomb dropped on Hiroshima (Novick, 1969). The plume of radioactivity emitted from the Pile No. 1 stack from October 10 to October 12 moved in roughly a south-southeast direction over the English countryside and on into northern Europe. Examination of the fallout indicated that the principal radionuclide emitted was ^{131}I, with a half-life of fortunately only 8 days. According to the report of Dunster et al. (1958), by far the most significant danger to human health from the incident was the contamination of cows' milk due to grazing in fields on which ^{131}I had been deposited. As a result, sampling of cows' milk was immediately begun over a wide area, and milk was seized over roughly a 520 km^2 area for periods ranging from 20 to 40 days until radiation levels in the milk dropped to acceptable levels (Dunster et al., 1958).

The Windscale accident occurred primarily because an unusually large amount of Wigner energy had been stored in the graphite moderator during the previous low-power run, and the release of this energy produced sufficient heat to melt the cladding on some of the fuel elements. It has since been learned that this problem occurs only after low-power runs, since at normal power the temperature reached in the graphite is high enough to give the excited C atoms sufficient "mobility" to return to their normal state during the operation of the reactor. Thus, there is much less energy stored in the graphite after a run at normal power than after a low-power run (Hunt, 1980).

The SL-1 Incident

On August 11, 1958, a small, 3 MW nuclear BWR began producing power at the U.S. National Reactor Testing Station in Idaho. The reactor, called SL-1, was fueled with ^{235}U-enriched fuel elements clad in Al. Neutron-absorbing "poison" strips consisting of thin plates of Al and boron (B) had been tack welded to one or both sides of each of the 40 fuel elements in the core (Nelson et al., 1961). These strips were incorporated into the core design to serve as a burnable poison, "the depletion of which would compensate for the burning of fuel" (Nelson et al., 1961, p. 17). The neutrons produced by the fissioning of the fuel in the core were moderated by ordinary water, and the power output was regulated by means of five Cd control rods. The reactor was designed to be a prototype of a low-power BWR for use in geographically remote locations, a type of power plant that had been requested by the Defense Department in 1955. Part of the purpose of operating the prototype reactor was to gain plant operating experience with military personnel, since it was envisioned that military personnel would one day operate such reactors in the field. Accordingly, military personnel were on the site from the time the reactor began producing power in 1958 and operated the reactor under the supervision of Combustion Engineering, the contractor responsible for overseeing the operation of the reactor (Nelson et al., 1961).

The first signs of trouble with the reactor appeared in 1959, when it was noted that the B poison strips were bowed in the 76 mm sections between tack welds. During an August 1960 inspection, large amounts of the B strips were found to be missing, and fuel elements in the center of the core were extremely difficult to remove by hand. Removal of fuel elements caused flaking of material and resulted in some plates' falling off. Because it was evident that further removal of fuel elements would cause additional loss of B, no further inspections were conducted. At that time, it was estimated that about 18% of the B originally present in the core was missing. Because of this loss of B, the reactivity of the core had increased, thus reducing the ability of the control rods to render the core subcritical. To compensate for this problem, strips of Cd were inserted in two of the control rod shrouds on November 11, 1960 (Nelson et al., 1961).

The second major problem with the reactor was the tendency for control rods to stick in their shrouds. This problem was noted early in the operation of the reactor and became more severe with time. Some personnel felt that the bowing of the B strips was producing a lateral force on the control rod shrouds sufficient to restrict free movement of the control rods, but this causative mechanism was never proven. In any case, the problem of control rod sticking became sufficiently severe during the last few months of 1960 that a program of "rod exercising" was initiated to maintain the rods in an operable status (Nelson et al., 1961).

On December 23, 1960, the reactor was shut down for maintenance on various components of the system. The only work on the core involved the insertion of 44 Co flux measuring assemblies into coolant channels between plates of the fuel elements. Removal of the control-rod drive assemblies was required to install these assemblies. The installation was accomplished by the day crew on January 3, 1961, and the night shift was assigned the job of reassembling the control-rod drives and preparing the reactor for startup.

At approximately 9 P.M. on the evening of January 3, 1961, the nuclear chain reaction in the SL-1 reactor went out of control, rapidly releasing energy roughly equivalent to that produced by detonating 10 kg of TNT. This rapid release of energy led to an explosion, probably caused by rapid vaporization of cooling water and/or to explosive reactions between O_2 and H_2 gas in the reactor. The explosion killed three men on the night shift and released a cloud of radioactivity that was detected by radiation monitoring devices around the reactor. Because of the damage to the reactor and because all three operators were killed, no one can say for sure what set off the explosion, but in light of the foregoing discussion, a probable mechanism is not difficult to deduce. In reassembling the control-rod drives, the operator on the night shift would have been required to lift the control rods. To produce an energy release of the observed magnitude, it would have been necessary only to jerk the central control rod up by about 55–60 cm in about 1 second (Nelson et al., 1961). If additional B had somehow been lost from the core during the 11-day shutdown, the uncontrolled chain reaction could have been set off more easily. That the military operators on the night shift would have deliberately jerked the central control rod in the manner described is highly unlikely, since the operators had been instructed never to raise the control rod more than 10 cm. However, the training procedure had not indicated the reason for this restriction. It is therefore conceivable that an operator, upon finding the central control rod stuck in its shroud, would have attempted to pull it out, unaware of the danger if the rod released suddenly. In testimony following the incident, it was stated that the central control rod never gave trouble, but records indicated that at least once shortly before the incident, the central control rod "failed to fall freely when called upon to scram" (Nelson et al., 1961, p. vi). Although it is possible to postulate other mechanisms as the cause of the explosion, there is no evidence to support alternative hypotheses. Considering the history of the reactor, it seems most likely that the accident was caused by a rapid withdrawal of the central control rod, with the loss of the B from the core as perhaps a contributing factor.

The SL-1 incident occurred rather early in the U.S. reactor program development, and to a certain extent it reflects the naiveté and/or inexperience of authorities in the area of reactor safety. As Nelson et al. (1961, pp. v–vi) noted, "the condition of the reactor core and the reactor control system had deteriorated to such an extent that a prudent operator would not have allowed operation of the reactor to continue without thorough analysis and review, and subsequent appropriate corrective action, with respect to the possible consequences or hazards resulting from the known deficiencies."

The Fermi Reactor Accident

The Enrico Fermi fast breeder reactor, one of only two commercial breeder reactors operated in the United States, was started up at Lagoona Beach, Michigan, 48 km from the city of Detroit, in 1963. The reactor had a maximum capacity of 200 MW, was cooled by liquid Na, and contained about half a tonne of ^{235}U in its core, enough ^{235}U to make 40 Hiroshima-sized nuclear bombs (Novick, 1969). On the evening of October 4, 1966, operators at the plant were preparing to start up the reactor after a shutdown period during which modifications had been made to correct a problem with the steam generators.

Between roughly 11 P.M. on October 4 and 8 A.M. the following morning, the reactor was run at a very low power level while the liquid Na primary coolant heated to about 290°C. At 8 A.M. on October 5, the operators began to slowly increase the power output of the reactor, with the intention of testing the newly repaired steam generator and steam system. By that afternoon at 3 P.M., the reactor was producing 20 MW of heat energy, still only 10% of its maximum capacity.

At that time, an erratic signal appeared on one of the control room instruments monitoring neutron production in the core, and although the instrument settled down again a short time later, the erratic signal reappeared when the power level had reached 34 MW. At about this time, an operator noticed that instruments monitoring the temperature in the core indicated higher than normal readings in two regions of the core, and shortly thereafter, radiation alarms sounded in the reactor building and elsewhere. Realizing that something had obviously gone wrong, the operators stopped the chain reaction at 3:20 P.M. by inserting six shutoff rods completely into the core. Subsequent sampling of the liquid Na coolant revealed high concentrations of radioactive fission products, indicating that some of the fuel elements had melted. Had a sufficient amount of molten ^{235}U dropped to the bottom of the reactor and formed a critical mass, a much larger energy release could have been touched off, but fortunately, the reactor was shut down before such a condition developed.

The cause of the partial meltdown was not determined until over a year after the accident, when a flexible periscope inserted into the core revealed and photographed a triangular piece of metal lying on the bottom of the reactor. The piece of metal was finally identified as a zirconium plate that had been welded to a cone about 30 cm high on the bottom of the reactor. The purpose of the cone was to direct the liquid Na coolant upward into the core. Six zirconium strips had been welded to the cone as a last-minute modification in the design of the reactor,[12] ironically to disperse molten fuel falling on the cone in the event of a meltdown. On October 5, the zirconium strip that had come detached from the cone had evidently blocked one or two openings that normally admitted coolant to the core, thus causing the partial meltdown.

[12]The zirconium strips did not appear on the final construction "as built" drawings (Novick, 1969).

Although the Fermi reactor accident caused no loss of life and released little radioactivity to the environment, the accident is nevertheless noteworthy if the AEC's Hazards Summary Report for the Fermi reactor is compared with what actually took place on October 5, 1966. As noted by Novick (1969, pp. 164–165):

> As is required by the AEC, the Hazards Summary Repot contains a section describing the "maximum credible accident." This is an attempt to specify an accident which is not expected to occur, but which is the worst which the designers feel could occur. Page 603.15 of the Hazards Summary states:
>
>> The maximum credible accident in the Fermi reactor is the melting of some or all of the fuel in one core subassembly, due to either complete or partial plugging of the nozzle of that subassembly or to a flow restriction within the subassembly. . . .
>
> As a result of the October 5 accident, fuel in at least two subassemblies melted. Some damage was done to at least two other subassemblies.
>
>> The reactor would probably be shut down automatically as a result of the reactivity loss due to the melting of the fuel. . . .
>
> This did not occur on October 5; the reactor was shut down manually.
>
>> . . . fission products . . . would have been released to the primary coolant system, and . . . the inert gas system. Since these systems are normally radioactive and sealed, there would be *no additional outward* effect caused by the addition of fission products from melted fuel [emphasis added].
>
> In fact, there were outward effects, including high radiation levels in the reactor containment and fission product detection buildings. It is fortunate that there was leakage of radioactivity, for it was the sounding of radiation alarms that prompted the shutdown of the reactor. (From *The Careless Atom*, copyright © 1969 by Sheldon Novick, Reprinted by permission of Houghton Mifflin Company)

The Three Mile Island Incident

At 4 A.M. on March 28, 1979, an auxiliary feedwater pump broke down on the Three Mile Island Unit 2 nuclear power plant near Harrisburg, Pennsylvania. Although the plant had officially been in operation for only three months, three of the four auxiliary feedwater pumps in the system had been taken out of commission two weeks before the accident and left out (Marshall, 1979). As a result, there was no margin of safety when the accident began at 4 A.M. Operation of the plant under these conditions was in fact a violation of federal regulations. In addition, operators had violated instructions by failing to close a pressure relief valve (PORV) in the primary coolant loop that had been leaking for weeks prior to the accident (Ahearne, 1987b) and had inadvertently left closed a pair of block valves in the discharge lines of the emergency feedwater pumps (Lewis, 1980). The block valves were supposed to be open at all times during the normal operation of the plant.

Because of the feedwater pump failure, the core of the reactor began to overheat. The increase in temperature and pressure caused the reactor's control rods to drop automatically into the core, shutting down the reactor. Unfortunately, the closed block valves caused a loss of feedwater to the steam generators, which soon boiled dry. In fact, for 13.5 hours "the reactor core was left partially exposed above the cooling water, while temperatures inside the reactor vessel climbed off the recording chart" (Marshall, 1979, p. 280). The zirconium fuel element cladding began to melt and/or crack, and some portions of the core became so hot that the uranium oxide fuel melted (Booth, 1987). The failure of the fuel elements' cladding and the melting of part of the fuel caused large amounts of radioisotopes to escape. Following the initial loss of coolant, three additional mechanical failures intensified the problem (Marshall, 1979).

1. The PORV in the primary coolant loop correctly opened so as to let out over-heated water but then failed to close. As a result, there was a dangerous drop in pressure inside the primary coolant loop.
2. A water level indicator on the pressurizing system evidently malfunctioned, causing a technician to think that the system was filled with water when in fact it was not. As a result, the technician concluded incorrectly that the situation was under control.
3. When the emergency core cooling system automatically switched on, another automatic system designed to contain radioactive leaks failed to turn on. Both systems should have been automatically activated at the same time.

Finally, technicians in the control room turned off the emergency and primary cooling pumps because they did not realize that the PORV was stuck in an open position and were concerned that the system would lose the steam bubble at the top of the pressurizer (Silver, 1987). As a result, they aggravated the overheating problem in the core. The core of the reactor was evidently in a highly unstable condition for about 16 hours.

The total amount of radioactivity released to the atmosphere as a result of the Three Mile Island accident has been estimated to be $1–5 \times 10^{17}$ Bq, but most of it was in the form of radioisotopes of Xe and Kr, which are inert gases and have half-lives of only a few days (Mynatt, 1982). Although small amounts of ^{131}I and ^{137}Cs were found in samples of milk and water taken near Three Mile Island, the plant's filters were apparently successful in preventing large-scale emissions of highly toxic radioisotopes. Cleaning up the damaged reactor, however, required more than eight years and cost about $1 billion (Booth, 1987). The reactor is now mothballed and will ultimately be disposed of around 2010 when the Unit 1 reactor at Three Mile Island is decommissioned (Booth, 1987).

Chernobyl

At 1 A.M. on April 25, 1986, electrical engineers assumed control of the former Soviet Union's (FSU's) Chernobyl No. 4 nuclear power reactor, one of four graphite-moderated, water-cooled 1,000 MW electrical power plants located in the Ukraine about 130 km from Kiev. The engineers wanted to test the generator's capacity to power emergency systems while coasting after a steam shutoff. The Chernobyl No. 4 reactor was an example of a so-called boiling-water pressure tube reactor, or RBMK. At the time, such reactors accounted for more than half of the nuclear-generated electrical capacity in the FSU (Ahearne, 1987a). One disturbing feature of RBMKs is the fact that when voids are formed in the reactor core—for example, if the coolant water boils and becomes less dense or if there is a substantial steam-air mixture—neutrons that would otherwise be captured by the water go into the graphite, where they are efficiently moderated to cause more fissions. Consequently, the absence of water increases the rate of fissioning, and the reactivity increases. In the jargon of the trade, RBMKs are said to have a *positive void coefficient*. RBMKs depend on a complex computer-run control system to handle this effect (Ahearne, 1987a). They are particularly sensitive to the positive void coefficient at low power output.

At 2 P.M. the emergency core cooling system, which would have drawn power and affected the test results, was shut off, the first of numerous safety violations. At 11:10 P.M. monitoring systems were adjusted to low power levels, but an operator failed to reprogram the computer to maintain power at the desired level, about 25% of capacity. As a result, power fell to about 1% of capacity, a dangerously low level. At that point, the majority of the control rods were withdrawn to increase power, but the accumulation in the fuel rods of Xe, a neutron absorber, frustrated the engineers' attempts to increase power.[13] Virtually

---

[13] ^{135}Xe is produced in the decay chain from tellurium, one of the direct products of ^{235}U fission.

all control rods were then withdrawn, an additional violation of safety standards. The power output then climbed to a little over 6% of capacity and stabilized. At 1:03 A.M. on April 26, all eight feedwater pumps were activated to ensure adequate cooling after the test. The combination of low power output and high water flow necessitated many manual adjustments. The operators were having difficulty getting a stable flow of water, and because of this problem, an operator turned off the emergency shutdown signals that would normally have reacted to instabilities in the water-steam separator by tripping the reactor and shutting everything down (Edwards, 1987). At 1:22 A.M. the computer indicated excess reactivity, but the operators chose to go ahead with the experiment anyway. Because of the instabilities of the reactor and the accumulation of Xe in the fuel rods, the operators realized that it would take a long time to start up the reactor if it shut down. Since they wanted to repeat the experiment immediately if it were done correctly, an operator manually disconnected the only remaining automatic trip safety system. The test began at 1:23 A.M., and power started to rise. Because of the instability of the reactor, any increase in power at this point triggered an even larger increase. Facing catastrophe, the operators tried to insert the control rods, but the rods had 5 m of graphite at their ends, and the addition of this moderator further accelerated the reaction (Edwards, 1987). Within four seconds the power output increased to more than 100 times the reactor's capacity. The U fuel disintegrated, burst through its cladding, and reacted explosively with the cooling water. The resultant steam explosion sheared 1,600 water pipes, flung the reactor's cap aside, blew through the concrete walls of the reactor hall, and threw burning blocks of graphite and fuel into the compound. Radioactive dust rose high into the atmosphere on a plume of intense heat (Edwards, 1987).

As a result of the accident, 3 persons were killed outright and 28 others died within days or weeks. Most of these casualties were firemen who fought about 30 fires caused by the explosion, most ignited by graphite. The firemen were successful in preventing the spread of the flames from reactor 4 to an adjoining reactor, and with the exception of the fire in the reactor itself, all fires were extinguished by 5 A.M. The firemen, however, received large doses of radiation. Other immediate casualties included construction workers engaged in building a fifth reactor and a physician and paramedics who aided the fire fighters.

How many casualties may ultimately result from the Chernobyl accident is a matter of speculation and debate. The Soviets estimated that about 3.7×10^{18} Bq of radioactivity were released as a result of the accident. About half of this amount was in the form of inert gases, but the remainder was in the form of biologically dangerous radionuclides, in particular ^{131}I. To a certain extent, the impact of the accident on persons living near the reactor was mitigated by the fact that the accident occurred at night, that it was not raining, and that the plume rose very high, about 5 km into the sky. Nevertheless, some Soviet citizens did receive large radiation doses, in part because the Soviets did not evacuate them from the immediate area or take other appropriate measures quickly enough. Children living in a village about 9 km from the power plant, for example, were found to have received thyroid doses as high as 2.5 Sv as a result of ingesting ^{131}I in contaminated milk (Edwards, 1987). Ultimately, much of the radioactivity released to the atmosphere fell out in the western Soviet Union, Europe, and parts of Asia. In the short term, the major concern from this fallout was the presence of ^{131}I. During the five years prior to the accident, for example, only three cases of childhood thyroid cancer were reported in Minsk, the principal Belarussian center for thyroid cancer diagnosis and treatment of children. Between 1986 and 1990 this number increased to 47, and to 286 from 1991 to 1994 (NEA, 1995). In the long term, many of the health effects from the accident will probably result from the fallout of ^{137}Cs, which has a half-life of 30 years (Goldman, 1987). The total dose to the human population from the radioisotopes released at Chernobyl will probably amount to about 1.2 million person-Sv, roughly half of which will be received over a period of many decades as ^{137}Cs levels gradually decline. The increase in fatal cancers associated with

TABLE 14.5   MONETARY LOSSES TO THE SOVIET UNION FROM THE CHERNOBYL ACCIDENT, JUNE 1986

Loss	Estimated Cost (billions of dollars)
Replacement cost of plant	1.04–1.25
Lost agricultural output	1.00–1.90
Site cleanup	0.35–0.69
Health care for victims	0.28–0.56
Lost export earnings	0.22–0.66
Relocation of residents	0.07
Total	2.96–5.13

Source: Flavin (1987).

this dose is estimated to be about 39,000. While this figure may seem large, it is trivial compared to the 630 million cancer fatalities expected over the same time frame for the affected population (Marshall, 1987a). The damaged reactor has been entombed in a sarcophagus of concrete and ideally should remain isolated from the environment for thousands of years while the radiation levels inside gradually decline. The monetary cost of the accident to the Soviet Union has been estimated at $3–5 billion (Table 14.5).

Summary

While the Chernobyl accident has correctly been characterized as the most serious accident ever to occur at a nuclear power plant, it is in some respects just another example of the sort of thing that has been happening at nuclear power plants from time to time for the past 40 years. From the foregoing discussion, one can see four general causes of accidents at nuclear power plants:

1. Human errors (e.g., an operator turning the wrong valves at the NRX reactor)
2. Violation of rules (e.g., turning off the emergency core cooling system at the Chernobyl reactor)
3. Equipment malfunctions (e.g., the zirconium plate that came detached and blocked the core cooling system of the Fermi reactor)
4. Inadequate technical knowledge (e.g., how to deal properly with Wigner energy release at the Windscale reactor)

Experience and better design of power plants can and has helped avoid accidents from the latter two causes; better training and selection of personnel would help reduce accidents from the first two. For example, in the September 30, 1999, accident at the Tokaimura, Japan, fuel reprocessing facility, workers were mixing ^{235}U with nitric acid to produce uranium dioxide. The workers inexplicably mixed about seven times as much ^{235}U as normal with the acid and then violated rules by using stainless steel buckets rather than pipes to transfer the U. During a 1998 inspection of the Oyster Creek nuclear power plant in New Jersey, Nuclear Regulatory Commission officials noted that control-room operators were playing on the Internet and doing college-course homework while on duty (AP, 1998). This sort of behavior is not acceptable at nuclear power plants. The fact that accidents and near-accidents continue to occur has suggested to many that nuclear power is simply not worth the risks. As a result, there has been growing disillusionment with the nuclear power industry (Figure 14.12) and a rapid acceleration in the cost of producing electricity from

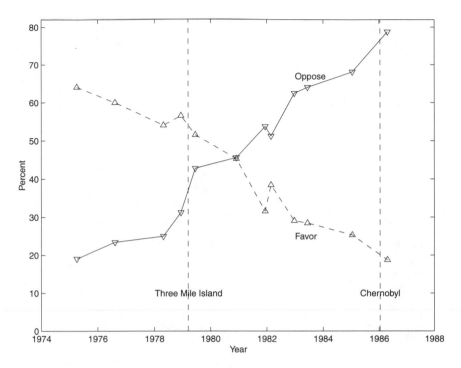

FIGURE 14.12 U.S. public opinion on building more nuclear power plants, 1975–1986.
Source: Flavin (1987). Reproduced with permission from the Worldwatch Institute,
http/www.worldwatch.org.

nuclear fission. The latter development is particularly noteworthy because it was in large
part the apparent cheapness of nuclear power that caused utilities to buy nuclear plants in
the first place. In the United States, for example, the cost of nuclear power plants, uncor-
rected for inflation, rose by a factor of 15–20 between 1970 and the late 1980s (Figure
14.13), or about a factor of 7–10 when corrected for inflation. The increased costs reflect
the economic realities of designing and operating plants in a way that minimizes the possibility of an accident and the costs associated with proper handling and disposal of the
radioactive wastes produced by the U fuel cycle. By 1987 the cost of producing electricity
in the United States from new nuclear plants was at least twice the cost of electricity produced from new fossil fuel plants (Flavin, 1987). The result was a de facto moratorium on
nuclear power plant construction in the United States.

The increase in the cost of nuclear power plants in the United States has been more
rapid than in other countries, in part because the American system is dominated by private utilities. To a certain extent, public authorities in other countries have been able to
pass on the extra financial costs to the public (Flavin, 1987). An exception to this pattern
has been in the area of power plant insurance, where the United States government has
stepped in to bail out the U.S. nuclear power industry. Obviously, nuclear power plants
must be insured, since a private utility could certainly expect to be sued if a nuclear accident resulted in loss of life or damage to health or property. In 1957 a committee appointed
by the AEC at the request of the U.S. Congress released a report called *WASH-740*, in which
monetary estimates were put on the amount of damage that might be caused by various
accidents at a nuclear power plant. The worst accident postulated by the committee involved a core meltdown that resulted in a radiation release in which about half of the fission products accumulated in the reactor fuel were emitted to the atmosphere. Since a

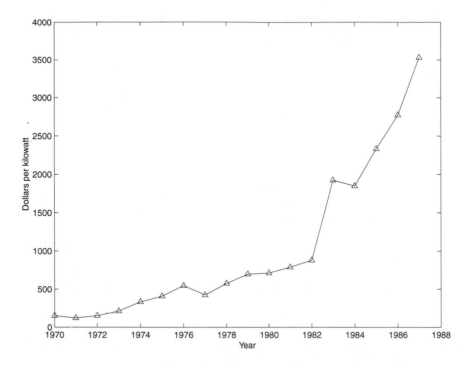

FIGURE 14.13   Average cost of new U.S. nuclear power plants entering operation, 1970–1987. *Source*: Flavin (1987). Reproduced with permission from the Worldwatch Institute, *http:/www.worldwatch.org.*

1,000 MW reactor, after operating for only one year, contains more radioactive Sr, Cs, and I than were released in all the weapon tests prior to 1969 (Novick, 1969), it is obvious that such an accident could cause enormous damage and loss of life. The committee's estimate of the monetary cost of such an accident was put at $7 billion (Gofman and Tamplin, 1979). No insurance company or consortium of insurance companies would agree to insure nuclear power plants for such a large amount of money, and the utilities made it clear to Congress that without adequate insurance there would be no commercial nuclear power plants. To break this impasse, Senator Clinton Anderson and Congressman Melvin Price introduced a bill to provide government insurance for privately operated nuclear power plants. The bill, called the *Price-Anderson Act*, was passed by Congress in 1957 and became Public Law 85-256. The Price-Anderson Act modified the Atomic Energy Act so that the liability for a single reactor accident could not exceed $500 million plus whatever amount of private insurance was available and then required that the AEC provide the $500 million (Novick, 1969). Private insurance companies were ultimately induced to put up an additional $74 million in insurance coverage and to agree to cover the liability for small accidents, but the total coverage of $574 million was still only about 8% of the maximum possible damage estimated by the 1957 AEC committee. Furthermore, as nuclear power plants became larger, the potential damage from an accident increased. In 1967 the largest commercial nuclear power plant in operation in the United States had a capacity of only 265 MW (Gofman and Tamplin, 1979). Most nuclear power plants built since 1970 have an electrical output of 800–1,000 MW. It is thought-provoking to realize that the Chernobyl accident released only 3–4% of the radioisotopes in the reactor's core (Flavin, 1987). If 50% of the fission products in the core had been released, as postulated in the *WASH-740* worst-case scenario, the economic costs in terms of health care, lost agricultural output, and relocation costs could have been roughly 15 times as great, or as much as $38 billion. The most recent estimates indicate that as many as 100,000 immediate deaths and $150 billion in economic damage could result from a serious accident at a nuclear power plant located in a densely populated area of the United States (Flavin, 1987). By 1986 the nuclear power industry's maximum liability for any single accident under the Price-Anderson Act had increased to only $665 million.

As a result of this very unsatisfactory state of affairs, Congress drastically amended the Price-Anderson Act in 1988. The amendments provide three layers of financial protection. The first layer is private insurance coverage. The amount of required private insurance coverage was set at $200 million per incident. In other words, a privately operated nuclear power plant must obtain private insurance coverage of at least $200 million per incident. The second layer of protection is provided by retrospective premiums imposed on all operators of nuclear power plants in the United States. The maximum retrospective premium is $75.5 million per licensed reactor per incident. If an accident occurs at one power plant, all licensed nuclear power plants in the United States are subject to the retrospective premium. Since there are currently 110 licensed nuclear power plants in the United States, the total layer 2 coverage is $8.3 billion. Combined with the $200 million in coverage provided by layer 1, the total coverage comes to $8.5 billion per incident. The retrospective premiums are to be paid in amounts not exceeding $10 million per year per incident. Thus, about 7.5 years would be required for a single power plant to make a $75.5 million retrospective premium payment. The third layer of coverage comes from the federal government. The federal government's indemnification can be as much as $500 million, but it applies only if the primary and secondary layers are not available to cover the damage caused by an accident.

As things stand, the economic cost of a serious accident at a nuclear power plant will be borne primarily by the nuclear power industry, at least up to a figure of $8.5 billion. If the cost were even greater, the President could, of course, declare a national disaster, in which case the additional monetary cost would be passed along to taxpayers. Not everyone agrees that this is a satisfactory state of affairs. To some at least, the potential damage from a serious nuclear power plant accident is so great as to raise a serious moral question about whether nuclear power plants should be allowed to exist at all. As noted by Yount (1957, p. 249), "It is a reasonable question of public policy as to whether a hazard of this magnitude should be permitted, if it actually exists. . . . Even if insurance could be found, there is a serious question whether the amount of damage to persons and property would be worth the possible benefits accruing from atomic development."

WASTE DISPOSAL

The problem of deciding what to do with the enormous amount of radioactive waste that has been produced since the dawn of the nuclear age in 1942 has become something of a nightmare for the 32 countries currently operating nuclear power plants. The problem is a complex one because the various problems posed by radioisotopes vary greatly from one radioisotope to another. Thus, a multifaceted approach is needed to deal with the problem.

Types of Radioactive Waste

The radioactive wastes of greatest concern are undoubtedly the *high-level wastes*. These include irradiated reactor fuel and liquid or solid wastes resulting from the reprocessing of reactor fuel to recover fissionable isotopes. In some countries, the term *intermediate-level wastes* is used to designate the more highly radioactive wastes associated with maintenance of nuclear facilities, such as the spent ion exchange resins from cleanup of reactor coolant. These wastes may include transuranic elements. The United States, however, includes transuranic elements in a separate waste category and regards as low-level waste what some other countries would regard as non-transuranic-containing intermediate-level waste (Carter, 1987). The term *low-level waste* is a catchall classification that includes all radioactive waste not assigned to some other category. The designation implies items such as contaminated glassware, lab coats, and paper trash, which require little if any shielding. The radioisotopes have relatively short half-lives (Norman, 1984). This characterization, however, can be misleading. In the United States, low-level wastes are radioactive wastes that are not high-level wastes, transuranic wastes, or U mill tailings (Robertson, 1984). As such, low-level wastes may be extremely radioactive and may contain large quantities of fission products with half-lives longer than 25 years, such as ^{90}Sr and ^{137}Cs. It is also useful, for purposes of this discussion, to keep in mind that radioactive wastes have been generated from two different sources, the private sector and military operations. The military waste has resulted primarily from the production of nuclear bombs. The private sector waste has been generated mainly by the nuclear power industry, but it also includes waste from the use of radioisotopes in medicine and scientific research.

History of Disposal

Low-Level Wastes

Low-level waste has been disposed of in several ways. From 1946 until a ban was placed on ocean dumping in 1983 by the London Dumping Convention, more than 50 sites in the North Atlantic and Pacific oceans were used for disposal of low-level radioactive wastes (Gibson, 1991). Most of the U.S. waste discharged to the ocean was packaged in concrete-lined 55 gallon drums and disposed of at two sites, one near the Farallon Islands about 50–80 km off San Francisco and the other about 210 km off Sandy Hook, New Jersey. The United States virtually ceased ocean dumping in the early 1960s, although a few barrels a year were dumped until 1970 (Norman, 1982a). The total amount of the U.S. discharge from 1946 to 1970 amounted to about 3.5×10^{15} Bq (Deese, 1977). In contrast, European countries were discharging about 3.7×10^{15} Bq of low-level waste per year at a site about 885 km off the tip of Land's End, England, just prior to the signing of the London Dumping Convention (Norman, 1982a).

Low-level wastes have also been disposed of in shallow-land-burial sites. Until 1962 the United States utilized facilities at the Oak Ridge National Laboratory in Tennessee for this purpose, but between 1962 and 1971, six commercially operated shallow-land-burial sites were opened for business (Figure 14.14). These included Beatty, Nevada; Maxey Flats, Kentucky; West Valley, New York; Richland, Washington; Sheffield, Illinois; and Barnwell, South Carolina. A seventh commercial site was opened in Utah's West Desert, 130 km west of Salt Lake City, in 1994. The federal low-level waste has similarly been disposed of at five major Department of Energy (DOE) facilities and several minor sites (Figure 14.14).

The results of the low-level waste disposal program have been less than reassuring. Several of the sites have leaked. Tritium was detected in groundwater surrounding the West Valley site, and Pu turned up about 1.5 km off-site within three years of the opening of the Maxey Flats dump (Gibson, 1991). At the present time, only three of the low-level waste sites are functional: Barnwell, South Carolina; Richland, Washington; and the West Desert, Utah site. According to Robertson (1984), the problems encountered at low-level waste disposal sites could have been avoided. The key to effective isolation of the wastes is minimizing contact of water with the waste and minimizing migration rates in groundwater. For example, in some cases at West Valley and the Oak Ridge National Laboratory, burial

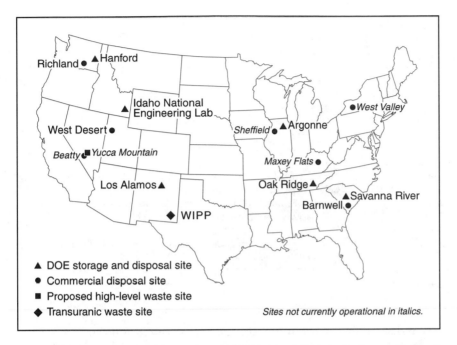

FIGURE 14.14 Location of radioactive waste sites in the United States. *Source*: Updated from Robertson (1984). Copyright 1989 by the National Academy of Sciences. Courtesy of the National Academy Press, Washington, D.C.

trenches were excavated below the water table. Proper site selection and careful attention to the design of burial trenches should allow effective isolation of low-level radioactive wastes over the time frame, no more than a few hundred years, that they remain potentially hazardous.

The 1980 Low-Level Radioactive Waste Policy Act and the 1985 amendments to that act made each state responsible for disposing of the low-level radioactive waste generated within its borders and encouraged states to enter into regional compacts in order to share waste disposal responsibilities. Although most states have formed compacts (Table 14.6), the sorts of regional disposal sites envisioned by Congress have not materialized. At the present time, all the commercial low-level waste is being sent to Barnwell, Richland, or the West Desert site. Richland is accepting waste only from the Northwest and Rocky Mountain compacts. Barnwell accepts waste from all states except North Carolina. The West Desert site accepts certain kinds of low-activity, low-level waste. Currently, about 76% by volume of the commercial low-level radioactive waste goes to the West Desert site. However, more than 99% of the low-level waste activity goes to Barnwell (Fuchs, 1999).

High-Level and Transuranic Wastes

The problem of disposing satisfactorily of high-level and transuranic wastes is much more difficult for several reasons. The intense radiation emitted by the waste poses a considerable health hazard and also produces an enormous amount of heat. Furthermore, some of the radioisotopes have very long half-lives and hence should be isolated from the environment for thousands and preferably tens of thousands of years.

In the early days of nuclear reactors, it was assumed that spent fuel elements would be recycled to recover fissionable isotopes, and in fact, this was done at several U.S. facil-

TABLE 14.6 COMPACTS FORMED BY STATES TO SHARE
RESPONSIBILITIES FOR COMMERCIAL LOW-LEVEL
RADIOACTIVE WASTE DISPOSAL

Compact	States
Appalachian	Delaware, Maryland, Pennsylvania, West Virginia
Central	Arkansas, Kansas, Louisiana, Nebraska, Oklahoma
Central Midwest	Illinois, Kentucky
Midwest	Indiana, Iowa, Minnesota, Missouri, Ohio, Wisconsin
Northeast	Connecticut, New Jersey
Northwest	Alaska, Hawaii, Idaho, Montana, Oregon, Utah, Washington, Wyoming
Rocky Mountain	Colorado, Nevada, New Mexico
Southeast	Alabama, Florida, Georgia, Mississippi, North Carolina, Tennessee, Virginia
Southwest	Arizona, California, North Dakota, South Dakota
Texas	Maine, Texas, Vermont
Unaffiliated	Massachusetts, Michigan, New Hampshire, New York, Rhode Island, South Carolina

Source: Fuchs (1999).

ities. At Hanford, for example, spent fuel elements from Pu-production reactors were initially stored underwater for several months to allow most of the ^{131}I to decay away, and the cladding was then removed with sodium hydroxide. The spent fuel inside was then dissolved in nitric acid. Several different schemes were used at different times to separate the U and Pu from the nitric acid mixture, but the so-called PUREX (Pu-U extraction) process, a solvent extraction method, was ultimately chosen and was used in the United States from 1956 until 1988 (Holden, 1990). The United States stopped recycling spent fuel elements in 1988.

After removal of most of the U and Pu, the nitric acid solution containing the other radioactive wastes was neutralized with sodium hydroxide or sodium carbonate and stored in large underground tanks constructed of C steel with a concrete outer lining. This choice of storage container proved to be unfortunate. The level of radioactivity in the waste solution stored in the tanks was sufficient to cause the contents of the tanks to boil for about 50 years (NAS, 1978). Although the boiling of the wastes had been anticipated and an appropriate cooling system had been provided on each tank to keep the boiling under control, the alkaline nature of the wastes combined with the high temperatures produced a highly corrosive solution that tended to eat through the C steel tanks. By 1978 leaks had been detected in 20 of the tanks, and a total of 5,500 m^3 of highly radioactive liquid waste was estimated to have leaked into the ground (NAS, 1978; Schumacher, 1990). About 3,800 m^3 of this radioactive water was cooling water that evidently had leaked unnoticed from a single tank during the period 1968–1978, a discovery that was not reported publicly until 1990 (Schumacher, 1990). The DOE has maintained that radioactivity from that leak remained in the soil beneath the tank and did not penetrate into the water table, 50 m below the surface. The previously reported worst spill occurred in 1973, when 435 m^3 of waste with a total activity of 10^{16} Bq escaped into the ground before the leak was detected and the tank pumped out. According to the National Academy of Sciences (1978, p. 38), the leak "was apparently the result of carelessness; a key employee was on vacation, and his substitute failed to heed indications from the monitoring devices that something had gone wrong." In this case as well, the radioisotopes remained in the soil well above the groundwater, but such incidents did little to inspire confidence in the government's waste isolation program.

Despite a 1957 recommendation by a National Academy of Sciences committee that the military wastes be solidified and a deep geologic repository sought for their storage, the AEC, mindful of the cost of such a storage program, continued to opt for on-site storage of the wastes near the generation facilities at Hanford, Oak Ridge, Savannah River, and Idaho Falls (Carter, 1977). However, as leaks from the C steel tanks began to occur, it became apparent that some other form of storage would be necessary. To date, 68 of the 277 tanks at the four facilities are known or assumed to have leaked to the surrounding soil, and some of the tank contents have reacted to form flammable gases, introducing additional safety risks.

In order to deal with the disposition of the wastes at the four facilities in a coherent manner, the DOE created a Tanks Focus Area (TFA) program in 1994. The TFA is responsible for managing, coordinating, and leveraging technology development to facilitate tank remediation, waste immobilization, and ultimately tank closure. The total cost of the program is estimated to be $49 billion over a period of roughly 35 years (Table 14.7).

Most of the waste tanks and more than 90% of the radioactivity are found at the Hanford and Savannah River facilities. Almost all the radioactivity is in the form of ^{90}Sr and ^{137}Cs or their decay products, ^{90}Y and ^{137}Ba, respectively. In 1965 a program was initiated at Savannah River to remove in the form of a sludge the principal heat-generating radionuclides, ^{90}Sr and ^{137}Cs, and then to evaporate the liquid waste down to a damp salt cake that would no longer be much of a threat to escape from the tanks. This program has been completed, but what will be done with the solidified waste? Current plans call for incorporating the highly radioactive sludge and 27% of the salt cake into glass capsules that will presumably be buried at the U.S. high-level waste disposal site, probably at Yucca Mountain, Nevada (Figure 14.14). The process of vitrification began in 1996 at the DOE's Defense Waste Processing Facility. The sludge and salt cake are mixed with sand-like borosilicate glass. The mixture is melted at a temperature of 1,150°C, and the molten mixture is poured into stainless steel canisters to cool and harden. Each canister is about 60 cm wide and 3 m long, and weighs about 2.3 tonnes. As of December 1998, 536 canisters had been produced (DOE, 1999). Until a permanent repository is available, the canisters are being stored in a bunker-like building at Savannah River, where they are cooled by natural air circulation.

The remaining salt cake at Savannah River will be stripped of most long-lived transuranic radionuclides, mixed with cement, and set in wide concrete vaults. This radioactive grout will be regarded as low-level waste. The separated transuranic isotopes, mainly ^{238}Pu, will be packaged in special drums and someday will presumably be stored at the Waste Isolation Pilot Project (WIPP) site near Carlsbad, New Mexico (Figure 14.14).

TABLE 14.7 DOE SCHEDULE FOR CLOSURE OF HIGH-LEVEL RADIOACTIVE WASTE STORED IN TANKS AT THE HANFORD, SAVANNAH RIVER, IDAHO FALLS, AND OAK RIDGE FACILITIES

Site	Cost (10^9 dollars)	Storage Units	Radioactivity (10^{15} Bq)	Completion Date
Hanford	28.5	177 tanks	7,326	2032
Savannah River	11.5	49 tanks	19,760	2028
Idaho Falls	5	11 tanks	19	2035
		7 calcine vaults	1,850	
Oak Ridge	4	40 tanks	7 .	2010

Source: DOE (1999).

The technology at Savannah River will also be used to dispose of the radioactive waste in the tanks at Hanford and Oak Ridge, as well as 2,300 m^3 of waste left by a bankrupt commercial nuclear reprocessing plant near West Valley, New York.

A somewhat different procedure for handling high-level wastes was used at the Idaho Falls facility. Instead of being neutralized with sodium hydroxide or sodium carbonate, the wastes were left in the acid state and stored in corrosion-resistant stainless steel tanks. Beginning in 1963, the wastes were run through a relatively inexpensive high-temperature (480°C) drying process to convert them to calcine, a solid with a consistency similar to that of granulated laundry detergent. By converting the waste to a granular form, the calcining process avoided the leakage problems that were experienced at Hanford. Unfortunately, the DOE does not regard calcined waste as an acceptable form for permanent disposal because of concerns that the dry waste could be easily dispersed (DOE, 1999). Plans are to retrieve the calcine from the storage vaults, dissolve it, and separate the liquid into high- and low-activity fractions. The high-activity fraction containing the ^{137}Cs, ^{90}Sr, and transuranic elements will be vitrified and disposed of in the same manner as the vitrified waste from the other facilities. The low-activity waste will be incorporated into grout and disposed of in the same was as the grouted waste from the other facilities.

The $49 billion price tag for cleaning up all the radioactive waste at DOE nuclear weapons facilities does not include the cost of disposing of all the civilian spent fuel, which by the year 2020 may be 10 times as great as all the military high-level waste (Marshall, 1987b). Although some of the spent civilian fuel was reprocessed by Nuclear Fuel Services at West Valley from 1966 to 1972, the vast majority of spent civilian fuel elements remain at the power plants that used them. The old fuel elements are stored in cooling ponds for 5–10 years and then transferred to air-cooled dry casks. Reprocessing of these fuel elements is an option, but due to the contamination of materials used in the procedure, reprocessing increases the waste volume by as much as a factor of 10 (Gibson, 1991). This fact, combined with the general decline of the nuclear power industry and hence the demand for U and Pu in the United States, probably means that the spent fuel elements will be disposed of intact (Marshall, 1987b), along with the high-level waste from the military program. One of the main problems, as with the military high-level waste, is the extreme heat produced by the ^{90}Sr and ^{137}Cs in the fuel elements. However, given the delays in identifying a suitable permanent high-level waste disposal site, some of the civilian waste may be at least 50 years old before it is placed in a permanent depository. Given the half-lives of ^{90}Sr and ^{137}Cs, the heat output of this waste will decline by about a factor of 2 every 30 years for the next few hundred years. Therefore, a workable solution to the long-term disposition of these fuel elements may be to store them on site for 100 years (Harrison, 1984) and then to bury them in appropriate caskets when their heat output has declined to an acceptable level (Krauskopf, 1990).

The Search for Long-Term Disposal Sites

The identification of a suitable disposal site for high-level and transuranic radioactive wastes has become a very difficult problem. From a scientific and public health standpoint, the principal concern is the potential contamination of groundwater. From a political standpoint, the principal problem is the fact that no one wants these wastes in his or her backyard. In 1976 the DOE authorized the construction of a facility for storage of military transuranic wastes at the WIPP site near Carlsbad, New Mexico (Figure 14.14). The facility itself was excavated from an ancient salt formation about 650 m underground at a cost of $2.2 billion. In May 1998 the EPA certified that the WIPP met all applicable federal nuclear waste disposal standards, and in March 1999 the facility began receiving shipments of transuranic waste. During the 35-year lifetime of WIPP, the DOE is expected to transport some 37,000 loads of transuranic waste to the facility from 37 locations nationwide.

The identification of a suitable high-level waste site has proven more difficult, and in 1982 Congress tried to get some movement on this issue by passing the Nuclear Waste Policy Act, which set out a detailed schedule of activities by the DOE and the NRC that would lead to the start of high-level waste disposal into a mined repository not later than 1998 (Krauskopf, 1990). The site tentatively chosen was Yucca Mountain, Nevada (Figure 14.14), but there has been considerable resistance to the designation of this site by environmentalists and the state of Nevada, and political and legal battles have pushed back the startup date to 2010 (Kerr, 1999). Because the spent commercial reactor fuel to be deposited at Yucca Mountain will contain transuranic elements, both the WIPP and Yucca Mountain sites must be capable of isolating the waste from the environment for thousands and preferably tens of thousands of years. Indeed, the EPA has mandated that the DOE design a repository to hold waste securely for at least 10,000 years (Pollock, 1986). Where could radioactive wastes be safely stored for such a long period of time?

Although aboveground storage systems could certainly be devised and would be most convenient from the standpoint of monitoring, the history and stability of sociopolitical institutions suggest that aboveground or nearground storage would not be a wise means of long-term disposal. A well-placed bomb, or even a small explosion or fire set by saboteurs, could result in significant releases of radioactivity from such a facility. Radioactive waste could be loaded aboard a spacecraft and simply fired off into outer space, but if something went wrong with the launch and the spacecraft crashed into some inconvenient place, the scattered radioactive debris might prove virtually impossible to clean up.

Current thinking therefore indicates that long-term waste disposal sites will be deep in the ground, either on continents or in the seabed. Ideally, such a location would be in a geologically stable area where the probability of crustal movements or volcanic activity's redistributing the wastes would be nil, where groundwater movements would be minimal, and where the adsorption characteristics of the sediment or rocks would tend to trap radionuclides that had leached from containment canisters. The principal geological media being considered are granite and salt, but clay, basalt, tuff, shale, and diabase are also under consideration in some countries. No single type of geological formation is clearly superior, and the choice of medium is limited by the options available in each country (Pollock, 1986).

The U.S. wastes will be packaged in corrosion-resistant canisters made of stainless steel, zirconim, or titanium. It would be optimistic and probably unrealistic to assume that the canisters would not sooner or later come into contact with groundwater. Even a dry substratum such as salt contains about 0.5% water (Cohen, 1977), and rock at a depth of a few hundred meters is nearly everywhere saturated with water (Krauskopf, 1990). Contact between groundwater and the metal canister would ultimately corrode the canister, exposing the material inside. How much time would be required for the metal canister to corrode would depend on how much groundwater came into contact with the container, as well as on the chemistry of the groundwater. If the groundwater were as corrosive as seawater, Heath (1977) indicates that confinement could probably not be guaranteed for more than a few thousand years, and de Marsily et al. (1977, p. 520) state that the canister "is not expected to survive more than a few hundred years in the geologic environment." These time periods are obviously short compared to the tens of thousands of years during which some transuranics should be isolated.

Once the metal canister has corroded away, the material inside will be directly exposed to the leaching action of groundwater. The radionuclides will probably be present in a cylinder of glass or uranium oxide in the case of reprocessed fuel and spent fuel elements, respectively. How rapidly radionuclides will leach from these cylinders is a matter of some debate, particularly since certain properties of the glass or uranium oxide matrices may be altered by the radiation from the trapped radionuclides. Obviously, if most of the radionuclides remain trapped in the cylinders for hundreds of thousands of years, the potential danger from even the long-lived transuranics will have been considerably reduced. However, if a substantial portion of the radionuclides leaches out of the cylinder within a few thousand years, the long-lived transuranics will still pose a considerable hazard. Although certain types of glass do show great resistance to leaching (Sales and Boatner, 1984), there is by no means a consensus of opinion that either glass or uranium oxide matrices will trap the radionuclides for tens of thousands of years. If the radionuclides do leach out, what then?

A radionuclide that is not adsorbed by sediment particles or temporarily trapped by ion exchange with elements in the soil or rock will presumably migrate by advection with the groundwater. Since groundwater movement in repository sites is likely to be upward (de Marsily et al., 1977), the radionuclides would ultimately be transported to the surface. How long this movement might take would depend on the extent to which adsorption and ion exchange processes retarded the migration of the radionuclides. For example, movement of the Sr ion could be slowed by as much as a factor of 100 relative to groundwater flow due to exchange with calcium ions (Cohen, 1977).

Although it is obviously impractical now to set up an experiment to study the long-term movement of various radionuclides in the ground, such an experiment was set up in what is now the southeastern part of the Gabon Republic, near the equator on the coast of West Africa about 1.9 billion years ago. At that time, natural U contained about 3% ^{235}U rather than the present 0.7%. In fact, when the Earth was formed about 4.5 billion years ago, U consisted of about 17% ^{235}U, but since the half-life of ^{235}U is about 640 times shorter than that of ^{238}U, the ratio of ^{235}U to ^{238}U has steadily decreased during the history of the Earth. It is possible to sustain a nuclear chain reaction in U using ordinary water as a moderator if ^{235}U is present at a concentration greater than about 1%, so that 1.9 billion years ago, there was no theoretical reason why a chain reaction could not occur in a nearly pure deposit of U. Evidently such a chain reaction did occur in a large underground deposit of U at a place now called Oklo in southeastern Gabon. Studies there have revealed that some of the U deposits now contain only about 0.4% ^{235}U, a reflection of the fact that much of the ^{235}U was fissioned during the probably several hundred thousand years that the chain reaction went on. According to Cowan (1976), as much as 6 tonnes of ^{235}U were consumed by the natural reactor. From the standpoint of radioactive waste disposal, it is obviously of interest to know what became of the reactor's fission products. According to a study by the U.S. Energy Research and Development Administration, Pu, the long-lived transuranic of probably greatest environmental concern, was efficiently confined within the reactor zone (Cowan, 1976). The principal radionuclides that escaped to the environment were probably ^{85}Kr, ^{90}Sr, and ^{137}Cs, but none of these has a half-life longer than 30 years.

Deep-Seabed Disposal

An alternative to disposing of high-level and long-lived radioactive wastes in an on-land disposal site is deep-seabed disposal. This option was investigated by the United States and other nations for some time (Hollister et al., 1981), but U.S. participation in the international subseabed disposal research effort terminated in 1986 (Miles et al., 1987). At the present time, Japan is the only nation actively exploring this option. The three possibilities being considered are bedrock beneath deep ocean basins and abyssal plains; ocean trenches where plate tectonics would be expected to carry the waste into the Earth's mantle; and bedrock regions in areas of high sedimentation, such as river deltas (Gibson, 1991). The appeals of deep-seabed disposal are three. First, for an earthquake-prone country such as Japan, there may be no suitable depository on land. Second, disposing of the wastes in the deep seabed obviously eliminates any concern over groundwater contamination. Third, deep-sea sediments form a very effective barrier against upward migration of isotopes of concern. Relevant to this last point is the study of Colley and Thomson (1990) of the distribution of natural radionuclides in deep-sea turbidite sediments more than 0.5 million years old. By evaluating the relationship between the distribution of ^{238}U and its daughter radionuclides in the sediments, they discovered that, with the exception of ^{226}Ra, the rate of migration of the radionuclides was negligible and that none of the radionuclides, including ^{226}Ra, had escaped from the sediments into the overlying water. One of the major problems with deep-seabed disposal is the fact that an accident or miscalculation could create an international problem, whereas pollution from an on-land disposal site would presumably be localized. Furthermore, recovering leaky canisters from the deep-seabed would be very difficult compared to recovery from an on-land site. However, because of resistance to on-land disposal and the difficulty nations such as Japan will have in identifying a suitable on-land disposal site, the deep seabed remains a possible option for long-term, high-level radioactive waste disposal.

Transmutation

The problem of high-level and transuranic waste disposal would be much easier to solve if it were necessary to isolate the wastes for only a few hundred years rather than tens of thousands of years. The radioisotopes that create the long-term problem are actinides, namely, isotopes of Pu, U, neptunium (Np), americium (Am), and curium (Cm), and these can be separated from the other waste. Indeed, the WIPP site was designed specifically to handle long-lived transuranic radioactive waste. The fact is, however, that once the actinides have been separated, they can be destroyed by a process called *transmutation*. The idea is to bombard the long-lived radioisotopes with neutrons to transform them into stable isotopes or radioisotopes with much shorter half-lives. One approach would simply be to incorporate the actinides into nuclear fuel elements and transmute them in a reactor at only a modest loss in reactor efficiency (Angino, 1977; Steinberg, 1990). Theoretically, the actinides could also be bombarded with neutrons using particle accelerators (Flam, 1991). Transmutation does not eliminate the radioactive waste disposal problem, but it greatly reduces the length of time during which the wastes must be isolated. The problem with transmutation is that it requires reprocessing of spent fuel, a procedure that costs money and generates a much greater volume of radioactive waste. Thus, there is a trade-off. The need to guarantee long-term isolation of the waste can be eliminated, but in the short term, both the volume of waste and the overall cost of waste disposal increase. At the present time, the United States is not vigorously pursuing the transmutation option, but more interest may develop if it becomes clear that no suitable long-term disposal site can be identified.

Uranium Mine Wastes

Natural U ores contain a large number of radionuclides in addition to the radioisotopes of U. These radionuclides consist primarily of the various daughter radionuclides produced in the long decay series of ^{238}U. Of the radionuclides in this decay series, particular concern is attached to ^{226}Ra due to its tendency to concentrate in bones and to its daughter ^{222}Rn, whose daughter radionuclides may be concentrated in the lungs. A study of U.S. U miners reported in 1955 that levels of radioactivity in the air of most U.S. mines far exceeded international standards, with some of the air samples having 59 times the permissible level of radiation (Novick, 1969). Remarkably, the U.S. government took no significant action until 1967, when enforceable radiation standards were finally adopted for U miners.

Although the radiation problem associated with working conditions in the mines is primarily an air pollution problem due to ^{222}Rn, the disposal of U mine wastes has become a serious water pollution problem in certain parts of the southwestern United States. Only about 1 g of uranium oxide is produced from every 400 g of raw ore processed by a

U mine, and the waste material, resembling sand and called *tailings*, has been routinely dumped in piles at a convenient location near the mine. These mine tailings contain ^{226}Ra as well as other radionuclides, and have unfortunately often been located near streams into which wind or rainwater runoff may transport them. The situation is disturbingly reminiscent of the case of Cd pollution in the Jintsu River basin (Chapter 12). The radioactive tailings have contaminated water in a number of streams draining into the Colorado River and Lake Mead, a water system that provides both drinking and irrigation water to parts of seven states: California, Nevada, Utah, Wyoming, Colorado, New Mexico, and Arizona. Water sampled from Colorado's San Miguel River in 1955 was found to contain a radiation level of 3.2 Bq L^{-1}, 17 times the Safe Drinking Water Act maximum contaminant level for radium in drinking water (EPA, 1999). Unfortunately, some of the areas where water has been contaminated by mine tailings were also exposed to considerable fallout from atmospheric bomb tests at the Nevada Test Site. According to Novick (1969), during the 1950s, people drinking water from the Animas River (Colorado) were receiving about three times the maximum permissible exposure to ^{226}Ra and ^{90}Sr.

In addition to creating water pollution problems, U mine tailings will, unless covered, continue to release Rn gas for more than 100,000 years, and according to Victor Gilinsky, a nuclear physicist and member of the NRC, these tailings could become "the dominant contribution to radiation exposure from the nuclear fuel cycle" (Carter, 1978, p. 191). "In fact, according to the American Physical Society's 1977 report on waste management and the nuclear fuel cycle, the ingestion hazard from tailings becomes greater than that from high-level wastes within the first 1,000 years" (Carter, 1978, p. 191). Incredibly, some of these tailings have been used as fill material beneath homes and other buildings. Carter (1978) reported that in Salt Lake City's Fire Station No. 1, which was built on such tailings around 1958, the exposure to Rn daughters was seven times greater than that allowed for U mine workers.

In response to this problem, Congress passed the Uranium Mill Tailings Radiation Control Act (UMTRCA) in November 1978. The purpose of the act was to require the U milling industry to clean up the mill tailings. Simply stabilizing the piles requires covering them with gravel and earth and then establishing vegetation, and indeed, such a stabilization program was begun in 1967. Proper cleanup, however, has required that the piles of tailings be graded to resist erosion and then covered with several meters of soil to reduce Rn emissions to no more than twice background levels. Fortunately, most of the roughly 200 million tonnes of tailings were in piles fairly remote from population centers, but one 1.8 million tonne pile was located only 30 blocks from Utah's state capitol in Salt Lake City, and two piles in Colorado were located in the towns of Grand Junction and Durango.

In 1978 Congress agreed to shoulder the cost of cleaning up 75 million tonnes of tailings generated prior to 1978 as a result of government contracts with the AEC. Those tailings were located at 24 sites in 10 states (Table 14.8). The cost of the cleanup has been estimated at $500 million. In-place stabilization procedures have been used at 11 of the 24 sites. However, 13 of the 24 sites, including the Salt Lake City, Grand Junction, and Durango sites, have required excavation and offsite disposal at remote disposal locations (DOE, 1996). In addition to stabilizing the tailings, the remediation program has addressed groundwater contamination because U-processing activities at most of the 24 sites resulted in contamination of groundwater beneath and, in some cases, downgradient of the sites. Unfortunately, pollution of groundwater is not easily reversed (Chapter 16), and in most cases the DOE is hoping that natural flushing processes will clean up the groundwater (DOE, 1996). Active remediation will probably be attempted at several sites, with a completion target date of 2014.

As might be expected, money has been a problem in cleaning up the remaining tailings that were not generated as a result of government contracts with the AEC. Since 1978

TABLE 14.8 LOCATION OF URANIUM MILL
TAILINGS SITES SUBJECT TO REMEDIAL
ACTION UNDER THE UMTRA PROGRAM

State	Sites
Arizona	Monument Valley
	Tuba City
Colorado	Durango
	Grand Junction
	Gunnison
	Maybell
	Naturita
	Rifle (2 sites)
	Slick Rock (2 sites)
Idaho	Lowman
New Mexico	Ambrosia Lake
	Shiprock
North Dakota	Belfield
	Bowman
Oregon	Lakeview
Pennsylvania	Canonsburg
Texas	Falls City
Utah	Green River
	Mexican Hat
	Salt Lake City
Wyoming	Riverton
	Spook

Source: DOE (1996).

the financial position of the U mining industry in the United States has deteriorated seriously, the number of persons employed by the industry having declined by more than an order of magnitude (Crawford, 1985). Because of a loophole in the 1978 legislation, U millers can delay stabilizing tailings until mills are permanently closed, and as a result, mills can defer cleanup by simply postponing the retirement dates for facilities. Overall cleanup costs have been estimated to be $2–4 billion (Crawford, 1986). A 9.5 tonne pile of tailings near Moab, Utah, illustrates the nature of the financial impasse between the federal government and the mining industry. The pile sits about 230 m from the Colorado River and is the result of years of mining by the Atlas Corporation, which declared bankruptcy under Chapter 11 of the U.S. Bankruptcy Code in 1998. In the meantime, the governor of Colorado has declared that the pile must be moved. Atlas has agreed to spend $19 million to cap the tailings, but moving the pile would cost an estimated $155 million, which is far more than Atlas can afford. In retrospect, the U mill tailings problem seems to be one that could be and should have been faced up to long ago. Certainly there can be little excuse for failing to provide some sort of stabilization for the mine tailings instead of letting wind and rain transport them into the streams of a large, important watershed.

Decommissioning Nuclear Reactors

Nuclear power plants have a useful lifetime of roughly 40 years. The constraints that radiation buildup places on routine maintenance, and the inevitable embrittlement of the reactor pressure vessel due to neutron bombardment, make it difficult to extend the operational lifetime of the plant beyond this time frame (Pollock, 1986). Since the first commercial reactor went into service in 1957, it is obvious that the problem of decommissioning nuclear power plants must now be faced by the industry, and indeed, a few reactors have already been decommissioned (Table 14.9). Until the late 1970s, it was felt that old reactors would simply be entombed in concrete until the radiation levels inside declined to safe levels. When reactors are initially shut down, most of the radioactivity in the pressure vessel and other components near the core is due to ^{60}Co, which is formed by neutron activation of stable Co, a constituent of most steel. Because ^{60}Co ha a half-life of only 5.3 years, it was felt that after a time on the order of decades, radiation levels would have dropped to relatively harmless levels.

However, the scientists had overlooked two long-lived radioisotopes formed as the result of neutron activation of trace constituents in steel. The isotopes are ^{59}Ni and niobium-94 (^{94}Nb), with half-lives of 80,000 years and 20,300 years, respectively. These radioisotopes contribute only a small percentage of the radioactivity when a plant is initially decommissioned, and for this reason, they may have been overlooked in the early studies. Because of their long half-lives, however, these radioisotopes can emit radiation well above permitted levels long after ^{60}Co has decayed to insignificance (Norman, 1982b). This fact, incidentally, was brought to public attention largely as a result of calculations done by undergraduate students.

TABLE 14.9 STATES OF DECOMMISSIONING OF NUCLEAR POWER PLANTS IN THE UNITED STATES AS OF JANUARY 1998

Power Plant	Location
Decommissioned	
Elk River Power Plant	Elk River, Minnesota
Pathfinder	Sioux Fall, South Dakota
Fort St. Vrain Generating Station	Plateville, Colorado
Shippingport Power Plant	Shippingport, Pennsylvania
Shoreham Nuclear Power Station	Suffolk County, New York
Sodium Reactor Experiment	Santa Susana, California
Decommissioning in Progress	
Big Rock Point	Charlevoix, Michigan
Haddam Neck	Haddam Neck, Connecticut
Maine Yankee	Wiscasset, Maine
Saxton Reactor	Saxton, Pennsylvania
Trojan	Rainier, Oregon
Yankee Rowe	Franklin County, Massachusetts
In Long-Term Storage	
Dresden 1	Morris, Illinois
Fermi 1	Monroe County, Michigan
Humboldt Bay 3	Eureka, California
LaCrosse	LaCrosse, Wisconsin
Indian Point 1	Buchanan, New York
Peach Bottom 1	York County, Pennsylvania
Rancho Seco	Sacramento, California
San Onofre 1	San Clemente, California
Three-Mile Island 2	Harrisburg, Pennsylvania
Vallecitos Boiling-Water Reactor	Pleasanton, California
Zion Units 1 and 2	Zion, Illinois

Source: NRC (1998).

An alternative to entombment is to remove the reactor fuel and liquids, flush out the system, and put the plant under constant guard to prevent public access for 30–100 years. By that time the ^{60}Co and other short-lived radioisotopes will have decayed away to low levels, and the plant can be dismantled with relative ease. The contaminated parts would be shipped to a low-level burial site. At the present time, 12 reactors are in such long-term storage. The NRC refers to this option as *SAFSTOR*.

A third scenario is to dismantle the reactor within a few years of the time the plant shuts down. This procedure requires the use of remotely controlled cutting and blasting equipment to dismantle the plant because of the high radiation levels in the structural components. Six reactors are in the process of being decommissioned by this process (Table 14.9). The NRC refers to this option as *DECON*.

There are pros and cons to the SAFSTOR and DECON options. The principal advantage of DECON is its lower overall cost. The principal advantage of SAFSTOR is the reduction in radioactivity as a result of radioactive decay during the storage period. Dismantling the reactor after radiation levels have declined is actually cheaper (SAFSTOR option), but the monetary savings are more than outweighed by the cost of maintaining and guarding the facility for several decades (Norman, 1982b). Initial experience has come from the dismantling of two small reactors, the Elk River plant in Minnesota and the Sodium Reactor Experiment in Santa Susana, California. Both were shut down after only a few years of operation and were decommissioned at costs of $6 million and $10 million, respectively (Norman, 1982b). The largest reactor with a significant track record[14] to be decommissioned to date has been the 72 MW reactor in Shippingport, Pennsylvania, the first commercial reactor in the United States. In that case, however, the DOE opted not to dismantle the reactor vessel, but instead encased it in concrete and barged it intact through the Panama Canal and up the Pacific coast to the Hanford reservation, where it was buried in an earthen trench (Pollock, 1986). This strategy saved money, but a valuable learning experience was lost. The Shippingport reactor had been in operation for 25 years, and the radiation levels in the reactor were far greater than in the Elk River and Sodium Reactor Experiment units, both of which were shut down after only a few years of operation and produced far less power. Transporting and burying reactor vessels intact is a questionable option for 1000 MW reactors. It would have been useful to obtain some experience dismantling the reactor vessel of a smaller unit that had been in operation for a period of time more typical of the roughly 40 year lifetime of commercial nuclear power plants.

The cost of decommissioning a typical 1000 MW nuclear reactor will probably be about $150 million and will generate roughly 17,000 m^3 of contaminated steel and concrete (Pollock, 1986). The latter figure is about 25% of the volume of low-level wastes generated annually in the United States (Norman, 1982b). These figures are quite significant because by the year 2010, 65–70 large commercial nuclear units will have ceased operation in the United States (Pollock, 1986). The price tag of $150 million per reactor is a small percentage of the roughly $5 billion cost of the most recently completed plants (Stoler, 1984), but it is noteworthy that the nuclear power industry has made almost no effort during the operation of the plants to set aside funds for decommissioning (Pollock, 1986). The result is that the decommissioning costs will be passed along to the utilities' customers at the time of decommissioning. As noted by Pollock (1986, p. 36), this approach seems "unfair to utility customers if they are charged for decommissioning a reactor from which they did not receive power."

---

[14]The Shoreham plant operated for only two effective full-power days.

COMMENTARY

It is tempting to say that the nuclear power industry is dead, and indeed, some persons have said exactly that. The year 1974 was the last year that a nuclear power plant was ordered in the United States, and the order not subsequently canceled (Flavin, 1987). In 1972 the International Atomic Energy Agency projected global nuclear power generating capacity in the year 2000 to be 3.5×10^{12} watts. In 1997 nuclear generating capacity stood at only 10% of that figure (Simpson, 1997).

Undoubtedly the most important factor in the decline of nuclear power has been economics. New nuclear power plants are no longer cost competitive with other means of generating electricity, and some would argue that nuclear power was never cost competitive. There is no doubt that the failure to take a full accounting of the costs of the U fuel cycle, including proper disposal of U mine tailings, disposal of radioactive wastes generated by power plants, and decommissioning of old plants, was a factor in making nuclear energy appear economically attractive early on. Because issues such as radioactive waste disposal and decommissioning have not yet been resolved, we still do not have a full accounting of the economic costs of nuclear power, but the calculations have gone far enough to make it clear that nuclear power, at least under present circumstances, is not viable economically.

Radioactive pollution and the economics of nuclear power are certainly not separate issues, but the public perception of the health hazards associated with nuclear power would be a serious problem for the industry at the present time even if the economics were favorable. It is thought-provoking, however, to compare the environmental and health costs of nuclear power and conventional fossil fuel power. After all, fossil-fuel-burning power plants produce wastes, and the principal waste product by weight is CO_2. A typical 1,000 MW coal-burning power plant emits about 270 kg of CO_2 per second from its exhaust stacks (Cohen, 1977). CO_2 is not considered a health hazard, but its accumulation in the atmosphere as the result of fossil fuel burning may be causing a gradual global warming as a result of the greenhouse effect (Schneider, 1989). Such a warming may have major environmental, economic, and social repercussions. From the standpoint of human health, the most directly harmful waste emitted by fossil-fuel-burning power plants is sulfur dioxide (SO_2). In the absence of efforts to control SO_2 emissions, a 1,000 MW, coal-burning power plant may emit about 4.5 kg of SO_2 per second, and such emissions could be expected to cause 25 fatalities, 60,000 cases of respiratory disease, and \$12 million in property damage per year (Cohen, 1977). Fossil fuel burning also produces N oxides, and these, together with S oxides, are the principal cancer-causing agents in cigarettes. A more complete accounting of the environmental costs of electricity derived from fossil fuel burning would include the thousands of persons who have suffered and died from black lung disease in coal mines and the various environmental costs associated with the use of oil (Chapter 13).

On the other side of the ledger, of course, are the tens of thousands of persons who will probably die from radiation-induced cancers caused by the accident at Chernobyl, and the uncertainty over how often accidents as serious as or more serious than Chernobyl can be expected to occur. Certainly the nuclear power industry as a whole has learned from its mistakes, but safety remains a major issue. On October 11, 1991, less than six years after the explosion at the Chernoby No. 4 reactor, an electrical fire broke out in the turbine hall next to the Chernobyl No. 2 reactor, blowing the roof off the building. Almost 50 firefighters worked for six hours to extinguish the blaze (Freeland, 1991).

In the near term at least, the nuclear power industry will probably remain in serious trouble until it becomes economically competitive with other forms of electricity and can come to grips with the issues of power plant accidents and the various forms of radioactive waste generated by the U fuel cycle. It is true that burning fossil fuels has environmental and public health costs, but body count comparisons will probably not convince the public that nuclear power is the way to go. The fact is that accidents such as Three Mile Island and Chernobyol should never have occurred. The nuclear power industry needs to get its act together.

If there is any appeal to nuclear power, it would seem to be in the long rather than the short term anyway. The reserves of ^{238}U, for example, could provide all of the world's electrical power for thousands of years if used in breeder reactors; and if fusion reactors someday become operable, the world's electrical energy needs could be satisfied from Li-^{2}H fusion for literally millions of years. This prospect has some appeal. Sooner or later, the wisdom of burning fossil fuels to generate electricity is going to be seriously questioned. That question has already been asked in countries such as France, which currently generates 77% of its electricity from nuclear power plants. Three other countries—Lithuania, Belgium, and Sweden—obtain more than 50% of their electricity from nuclear power (Simpson, 1997). The tendency of countries to embrace the nuclear option will, of course, depend on the alternatives available. Within 50–100 years, for example, it may be practical to generate electricity on a large scale from photovoltaics, but that time is not now. In the meantime, energy conservation and nuclear power appear to be the principal ways of reducing fossil fuel consumption. It now seems obvious that both governments and the public embraced the latter option too quickly and without asking enough questions. There is still time to examine the questions and to obtain informed answers. We should proceed with caution and continue to explore alternatives, including energy conservation.

REFERENCES

ADAMS, J. A., and V. L. ROGERS. 1978. *A Classification System for Radioactive Waste Disposal. What Waste Goes Where.* NUREG-0456. FBDU-224-10. U.S. Nuclear Regulatory Commission, Washington, D.C.

ADLER, C. 1973. *Ecological Fantasies.* Green Eagle Press, New York. 350 pp.

AHEARNE, J. F. 1987a. Nuclear power after Chernobyl. *Science,* **236,** 673–679.

AHEARNE, J. F. 1987b. TMI and Chernobyl. *Science,* **238,** 145.

ANDERSON, S. L., and F. L. HARRISON. 1986. *Effects of Radiation on Aquatic Organisms and Radiobiological Methodologies for Effects Assessments.* U.S. EPA Office of Radiation Programs, Washington, DC. 128 pp.

ANGINO, E. E. 1977. High-level and long-lived radioactive waste disposal. *Science,* **198,** 885–890.

AP. 1998. Flaw shuts Oyster Creek nuclear plant. *Associated Press,* March 22.

BEIR. 1990. *Health Effects of Exposure to Low Levels of Ionizing Radiation.* Committee on the Biological Effects of Ionizing Radiation. BEIR-V. National Research Council, National Academy Press, Washington, DC. 421 pp.

BOOTH, W. 1987. Postmorten on Three Mile Island. *Science,* **238,** 1342–1345.

CARTER, L. J. 1977. Radioactive wastes: Some urgent unfinished business. *Science,* **195,** 661–666, 704.

CARTER, L. J. 1978. Uranium mine tailings: Congress addresses a long-neglected problem. *Science,* **202,** 191–195.

CARTER, L. J. 1987. *Nuclear Imperatives and Public Trust: Dealing with Radioactive Waste.* Resources for the Future, Washington, DC. 473 pp.

COHEN, B. L. 1977. The disposal of radioactive wastes from fission reactors. *Sci. Amer.,* **236**(6), 21–31.

COLLEY, S., and J. THOMSON. 1990. Limited diffusion of U-series radionuclides at depth in deep-sea sediments. *Nature,* **346,** 260–263.

COWAN, G. A. 1976. A natural uranium reactor. *Sci. Amer.,* **235**(7), 36–47.

CRAWFORD, M. 1985. Mill tailings: A $4-billion problem. *Science,* **229,** 537–538.

CRAWFORD, M. 1986. Toxic waste, energy bills clear Congress. *Science,* **234,** 537–538.

DEESE, D. A. 1977. Seabed emplacement and political reality. *Oceanus,* **20**(1), 47–63.

DE MARSILY, G., E. LEDOUX, A. BARBREAU, and J. MARGAT. 1977. Nuclear waste disposal: Can the geologist guarantee isolation? *Science*, 197, 519–527.

DICKSON, D. 1987. Slowdown for French fast breeders? *Science*, 238, 472.

DOE. 1996. Uranium mill tailings remedial action. *http://www.em.doe.gov/bemr96/umtra.html*

DOE. 1999. Tanks Focus Area—Multiyear Program Plan FY99-FY03. http://www.pnl.gov/tfa/program/fy99mypp

DUNSTER, H. J., H. HOWELLS, and W. L. TEMPLETON. 1958. District surveys following the Windscale incident, October 1957. In *Proceedings of the Second U.N. International Conference on the Peaceful Uses of Atomic Energy*, Vol. 18. U.N. Publications, New York. Pp. 296–308.

EDWARDS, M. 1987. Chernobyl—one year after. *National Geographic*, 171, 632–653.

EPA. 1973. *Water Quality Criteria*. EPA-R3-73-033. Washington, DC. 594 pp.

EPA. 1999. Maximum contaminant levels for radium-226, radium-228, and gross alpha particle radioactivity in community water systems. *Code of Federal Regulations*, 40, 141.15 (1999).

FLAM, F. 1991. A nuclear cure for nuclear waste. *Science*, 252, 1613.

FLAVIN, C. 1987. *Reassessing Nuclear Power: The Fallout from Chernobyl*. Worldwatch Institute, Washington, DC. 91 pp.

FREELAND, K. 1991. Chernobyl fire revives Soviets' worst nightmare. *Honolulu Advertiser*, October 14, p. D1.

FUCHS, R. L. 1999. 1998 State-by-state assessment of low-level radioactive wastes received at commercial disposal sites. U. S. Department of Energy. Assistant Secretary for Environmental Restoration and Waste Management. *http://www.inel.gov/national/reports.html*.

FURTH, H. P. 1990. Magnetic confinement fusion. *Science*, 249, 1522–1527.

GARWIN, R. L. 1991. Fusion: The evidence reviewed. *Science*, 254, 1394–1395.

GIBSON, D. 1991. Confronting eternity: The long-lived problem of nuclear waste. *Calypso Log*. October. Pp. 14–16.

GILLETTE, R. 1972. Radiation standards: The last word or at least a definitive one. *Science*, 178, 966–967, 1012.

GLANZ, J. 1998. Sandia steps into the fusion race. *Science*, 280, 28.

GLANZ, J., and A. LAWLER. 1998. Planning a future without ITER. *Science*, 279, 20–21.

GOFMAN, J. W., and A. R. TAMPLIN. 1970. Radiation: The invisible casualities. *Environment*, 12(3), 12–19, 49.

GOFMAN, J. W., and A. R. TAMPLIN. 1979. *Poisoned Power*. Rodale Press, Emmaus, PA. 368 pp.

GOLDMAN, M. 1987. Chernobyl: A radiobiological perspective. *Science*, 238, 622–623.

HAMILTON, D. 1991. Geographic fission on fusion. *Science*, 253, 259.

HAMMOND, A. 1972. Fission: The pros and cons of nuclear power. *Science*, 178, 147–150.

HARRISON, J. M. 1984. Disposal of radioactive wastes. *Science*, 226, 11–14.

HEATH, G. R. 1977. Barriers to radioactive waste migration. *Oceanus*, 201(1), 26–30.

HOLDEN, C. 1990. Bailing out of the bomb business. *Science*, 250, 753.

HOLDEN C. 1991. Fusion panel lowers its sights. *Science*, 254, 193.

HOLDREN, J. P. 1978. Fusion energy in context: Its fitness for the long term. *Science*, 200, 168–180.

HOLLISTER, C. D., D. R. ANDERSON, and G. R. HEATH. 1981. Subseabed disposal of nuclear wastes. *Science*, 213, 1321–1325.

HUNT, S. E. 1974. *Fission, Fusion and the Energy Crisis*. Pergamon Press, New York. 164 pp.

IAEA. 1997. World has 443 nuclear power plants. International Atomic Energy Agency Power Reactor Information System. *http://ecolu-info.unige.ch/archives/envcee97/0254.html*.

ICRP. 1977. *Recommendations of the International Commission on Radiobiological Protection*. ICRP Publication No. 26. Pergamon Press, Oxford. 53 pp.

ICRP. 1984. A compilation of the major concepts and quantities in use by the International Commission on Radiobiological Protection. *Annals of the International Commission on Radiobiological Protection*, 14(4). Pergamon Press, Oxford.

JOSEPH, A. B., B. F. GUSTAFSON, I. R. RUSSELL, E. A. SCHUERT, H. L. VOLCHOK, and A. TAMPLIN. 1971. Sources of radioactivity and their characteristics. In *Radioactivity in the Marine Environment*. Committee on Oceanography, National Research Council, National Academy of Sciences, Washington, DC. Pp. 6–41.

KERR, R. A. 1999. Yucca Mountain panel says DOE lacks data. *Science*, 283, 1235–1236.

KRAUSKOPF, K. B. 1990. Disposal of high-level nuclear waste: Is it possible? *Science*, 249, 1231–1232.

LAWLER, A., and J. GLANZ. 1998. Competition heats up on the road to fusion. *Science*, 281, 26–29.

LEWIS, H. W. 1980. The safety of fission reactors. *Sci. Amer.*, **242**, 53–65.

LOUTIT, J. F. 1956. The experimental animal for the study of biological effects of radiation. *Proc. Int. Conf. on Peaceful Uses of Atomic Energy*, **11**, 3–6.

MARSHALL, E. 1979. A preliminary report on Three Mile Island. *Science*, **204**, 280–281.

MARSHALL, E. 1987a. Recalculating the cost of Chernobyl. *Science*, **236**, 658–659.

MARSHALL, E. 1987b. Savannah River's $1-billion glassmaker. *Science*, **235**, 1314–1317.

MARSHALL, E. 1990a. Academy panel raises radiation risk estimate. *Science*, **247**, 22–23.

MARSHALL, E. 1990b. Radiation exposure: Hot legacy of the cold war. *Science*, **249**, 474.

MARX, J. 1979. Low-level radiation: Just how bad is it? *Science*, **204**, 160–164.

MARX, J. 1989. Many gene changes found in cancer. *Science*, **246**, 1386–1388.

McCULLOUGH, C. R. 1959. The Windscale Incident. In Proceedings, 1958 AEC and Contractor Safety and Fire Protection Conference, June 24–25 TID-7591, AEC Technical Information Service, Oak Ridge, TN, May 15.

METZ, W. D. 1972. Magnetic containment fusion: What are the prospects. *Science*, **178**, 291–293.

METZ, W. D. 1976. European breeders (II): The nuclear parts are not the problem. *Science*, **191**, 368–372.

MILES, E. L., C. D. HOLLISTER, and G. R. HEATH. 1987. Subseabed waste disposal. *Science*, **238**, 144.

MILES, M. E., G. L. SJOBLOM, and J. D. EAGLES. 1979. *Environmental Monitoring and Disposal of Radioactive Wastes from U.S. Naval Nuclear-Powered Ships and Their Support Facilities*. Department of the Navy Report NT-79-1. Naval Sea Systems Command, Washington, D.C. 34 pp.

MYNATT, F. R. 1982. Nuclear reactor safety research since Three Mile Island. *Science*, **216**, 131–135.

NAS. 1978. *Radioactive Wastes at the Hanford Reservation. A Technical Review.* Panel on Hanford Wastes, Committee on Radioactive Waste Management, Commission on natural Resources, National Research Council, National Academy of Sciences, Washington, DC. 269 pp.

NATIONAL COUNCIL ON RADIATION PROTECTION AND MEASUREMENTS. 1989. *Radiation Protection for Medical and Allied Health Personnel*. National Council on Radiation Protection and Measurements, Bethesda, MD.

NELSON, C. A., K. BECK, P. A. MORRIS, D. I. WALKER, and F. WESTERN. 1961. *Report on the SL-1 Incident, January 3, 1961*. Atomic Energy Commission Investigation Board Report, Joint Committee on Atomic Energy, Congress of the United States. U.S. Government Printing Office, Washington, DC.

NORMAN, C. 1982a. U.S. considers ocean dumping of radwastes. *Science*, **215**, 1217–1219.

NORMAN, C. 1982b. A long-term problem for the nuclear industry. *Science*, **215**, 376–379.

NORMAN, C. 1984. Delay likely in high-level program. *Science*, **223**, 259.

NOVICK, S. 1969. *The Careless Atom*. Houghton Mifflin, New York. 225 pp.

NEA. 1995. *Chernobyl Ten Years On*. Radiological and health Impact. An Assessment by the NEA Committee on Radiation Protection and Public Health. OECD Nuclear Energy Agency. November. *http://www.nea.fr/html/rp/chernobyl/allchernobyl.html*.

NRC. 1975. *Reactor Safety Study. An Assessment of Accident Risks in U.S. Commercial Nuclear Power Plants*. NUREG-75/014. U.S. Nuclear Regulatory Commission, Washington, D.C.

NRC. 1991a. Standards for protection against radiation; final rule. *Federal Register*, **56**(98), 23360–23474.

NRC. 1991b. Numerical guides for design objectives and limiting conditions for operation to meet the criterion "as low as is reasonably achievable" for radioactive material in light-water-cooled nuclear power reactor effluents. *Code of Federal Regulations*, **10**(50), Appendix I.

NRC. 1998. Staff responses to frequently asked questions concerning decommissioning of nuclear power reactors (NURG-168). *http://www.nrc.gov/NRC/NUREGS/SR1628/sr1628.html*

ODUM, E. P. 1971. *Fundamentals of Ecology*. W. B. Saunders, Philadelphia. 574 pp.

OPHEL, I. L. 1963. The fate of radiostrontium in a freshwater community. In V. Schultz and W. Klement (Eds.), *Radioecology*. Reinhold, New York. Pp. 213–216.

POLLOCK, C. 1986. *Decommissioning: Nuclear Power's Missing Link*. Worldwatch Institute, Washington, DC. 54 pp.

ROBERTSON, J. A. L. 1978. The CANDU reactor system: An appropriate technology. *Science*, **199**, 657–664.

ROBERTSON, J. B. 1984. Geologic problems at low-level radioactive waste-disposal sites. In J. Bredehoeft (Ed.), *Groundwater Contamination*. National Academy Press, Washington, DC. Pp. 104–108.

SALES, B. C., and L. A. BOATNER. 1984. Lead-iron phosphate glass: A stable storage medium for high-level nuclear waste. *Science*, **226**, 45–48.

SCHNEIDER, S. H. 1989. The changing climate. *Sci. Amer.*, **261**, 70–79.

SCHUMACHER, E. 1990. Million-gallon water leakage revealed. *Honolulu Star-Bulletin*, September 30, P. A27.

SEYMOUR, A. H. 1971. *Radioactivity in the Marine Environment*. Committee on Oceanography, National Research Council, National Academy of Sciences, Washington, DC. Pp. 1–5.

SILVER, E. G. 1987. TMI and Chernobyl. *Science*, **238**, 144–145.

SIMPSON, J. 1997. World has 443 nuclear power plants. *http://ecolu-info.unige.ch/archives.envcee97/0254.html.*

SPARROW, A. H. 1962. *The Role of the Cell Nucleus in Determining Radiosensitivity*. Brookhaven Lecture Series No. 17. Brookhaven National Laboratory Publication No. 766, New York.

SPARROW, A. H., L. A. SCHAIRER, and R. C. SPARROW. 1963. Relationship between nuclear volumes, chromosome numbers, and relative radiosensitivities. *Science,* **250**, 887.

STEINBERG, M. 1990. Transmutation of high-level nuclear waste. *Science*. **250**, 887.

STOLER, P. 1984. Pulling the nuclear plug. *Time*, **123**(7), 34–42.

STROUBE, H. 1966. Letter from Hal Stroube, Coordinator of Atomic Information to Hon. Merryn M. Dymally, Chairman of the California State Assembly Committee on Industrial Relations, dated September 29, 1966, enclosing comments on the testimony of David Pesonen before the Committee.

U.S. BUREAU OF THE CENSUS. 1998. *Statistical Abstract of the United States 1998*, 118th ed. U.S. Government Printing Office, Washington, DC.

WEINBERG, A. M. 1986. Are breeder reactors still necessary? *Science*, **232**, 695–696.

WILBER, C. G. 1969. *The Biological Aspects of Water Pollution*. Charles C. Thomas, Springfield, IL. 296 pp.

WOODHEAD, D. S. 1984. Contamination due to radioactive materials. In O. Kinne (Ed.), *Marine Ecology V. Ocean management. Part 3: Pollution and Protection of the Seas—Radioactive Materials, Heavy Metals and Oil.* Wiley-Interscience, New York. Pp. 1111–1287.

YOUNT, H. W. 1957. Vice President, Liberty Mutual Insurance Co., appearing on behalf of the American Mutual Alliance, Hearings, on Governmental Indemnity before the JCAE, 84th Congress, Second Session, pp. 248–250.

ZALESKI, C. P. 1980. Breeder reactors in France. *Science*, **208**, 137–144.

QUESTIONS

14.1 The radionuclides routine released by nuclear power plants in the Unites States are primarily which one of the following?

   a. Fission products

   b. Daughters of ^{238}U

   c. Daughters of ^{235}U

   d. Neutron activation products

14.2 Which one of the following sources of radiation is believed to account for more than half of the natural radiation dose received by the average American at the present time?

   a. Daughters of ^{222}Rn

   b. ^{238}U

   c. ^{239}Pu

   d. cosmic rays

14.3 Which two of the following radionuclides are responsible for most of the heat emitted by spent nuclear power plant fuel elements?

   a. ^{60}Co

   b. ^{137}Cs

   c. ^{239}Pu

   d. ^{3}H or tritium

   e. ^{90}Sr

14.4 The site presently contemplated for disposal of the Unites States' high-level radioactive wastes is

    a. Hanford, Washington

    b. Oak Ridge, Tennessee

    c. Los Alamos, New Mexico

    d. Yucca Mountain, Nevada

14.5 The amount of radiation released to the environment as a result of the Chernobyl nuclear power plant accident was about 10 times greater than the corresponding quantity released by the Three Mile Island accident. The number of human lives lost as a result of the Chernobyl accident, however, will probably exceed the number of human lives lost as a result of the Three Mile Island accident by at least a factor of 10,000. Why are the radioisotopes released at Chernobyl expected to cause so much more harm to humans?

14.6 Why are underground salt deposits considered to be good candidates for the storage of radioactive wastes?

14.7 What are the principal concerns associated with transuranic wastes? In other words, why does the U.S. government put transuranic wastes in a separate category from other types of radioactive waste?

14.8 Pumping water through the core of conventional nuclear power plants in the United States serves several purposes. What are those purposes?

14.9 The following radionuclides are of concern from the standpoint of human health. Which ones are produced primarily by neutron activation and which ones are fission products?

    a. ^{90}Sr

    b. ^{3}H

    c. ^{55}Fe

    d. ^{137}Cs

    e. ^{131}I

    f. ^{14}C

14.10 Following accidents at nuclear power plants such as the 1957 incident at Windscale or the 1986 accident at Chernobyl, one of the major public health concerns was exposure to ^{131}I. By what mechanism does this particular radionuclide find its way into the human body, and why is it of such concern from a public health standpoint?

14.11 Why is liquid Na rather than water pumped through the core of breeder nuclear reactors?

14.12 Briefly explain the technology involved in magnetic confinement fusion, and explain why this technology is being explored as a possible means of operating nuclear power plants based on fusion.

14.13 Which one of the following nuclides is primarily responsible for maintaining the chain reaction in conventional nuclear power plants in the United States?

    a. ^{238}U

    b. ^{235}U

    c. ^{239}Pu

    d. ^{233}U

14.14 The amount of radiation emitted by a radionuclide is described by the equation $R_T = R_0(1/2)^{T/\tau}$, where R_T is the radioactivity at time T, R_0 is the radioactivity at time 0, and τ is the half-life of the radionuclide. Suppose that there has been an accident at a nuclear power plant resulting in the release of large amounts of radionuclide X. Public health authorities determine that the level of X in the area around the power plant is 8,000 times higher than acceptable levels and evacuate the area. How many half-lives will it take before the activity of X has declined to an acceptable level so that people can return to their homes?

14.15 The flux of gamma rays from a point source falls off as the square of the distance from the source of the radiation. Suppose that the flux of gamma rays at a distance of 1 m from a source is 10,000 times higher than the acceptable limit. How far would a person have to be from the source before the flux of gamma rays was equal to the acceptable limit?

Chapter 15

✧　◻　✧　◻　✧

Acid Deposition

> This most excellent canopy, the air, look you, this brave o'erhanging firmament, this majestical roof fretted with golden fire, why, it appeareth nothing to me but a foul and pestilent congregation of vapours. (William Shakespeare, *Hamlet*, II, ii, 299)

ACID DEPOSITION is a form of pollution that has drawn increasing attention during the last 40 years. The problem, however, is not new and can be traced back at least several hundred years. The term *acid deposition* refers to the transfer of acidic substances from the atmosphere to the surface of the Earth and includes a variety of processes that can be broadly classified as either dry deposition or wet deposition. The former includes both absorption and adsorption of gases and impaction and gravitational settling of fine aerosols and coarse particles (Cowling, 1982). Wet deposition or precipitation includes deposition resulting from rain, snow, hail, dew, fog, or frost. The problems caused by acid deposition are often cited as being the result of acid rain, although it is obvious from the foregoing discussion that acid rain is only one form of acid deposition. However, it has been through studies of temporal and geographical trends in the composition and acidity of rainfall that much of our appreciation of the causes and magnitude of the acid deposition phenomenon has developed, and acid rain is certainly one of the more important forms of acid deposition. We therefore begin our discussion of acid deposition with an analysis of the nature and causes of acid rain.

ACID RAIN

According to the chemical definition of acidity, water is acidic if the concentration of hydrogen ions, H^+, exceeds that of hydroxide ions, OH^-. Water is classified as basic if the hydroxide ion concentration exceeds the hydrogen ion concentration. If the concentration of H^+ and OH^- are both expressed in moles per liter, then at 25°C

$$(H^+)(OH^-) = 10^{-14} \tag{15.1}$$

where (H^+) and (OH^-) are the concentrations of H^+ and OH^-, respectively. The value of the product $(H^+)(OH^-)$ is slightly temperature dependent, but for most intents and purposes, we may assume that the product equals 10^{-14}. In neutral water, it follows that $(H^+) = (OH^-) = 10^{-7}$. The pH of water is defined as the negative of the common logarithm of the H^+ activity. For our purposes, we may assume that H^+ activity and H^+ concentration are synonymous. Therefore, in neutral water, $-\log_{10}(H^+) = pH = 7$. From the foregoing discussion, it follows that acidic water has a pH less than 7, and basic water has a pH greater than 7. For example, water with a pH of 10 would have an H^+ concentration of 10^{-10} and an OH^- concentration of 10^{-4}.

While it might therefore seem logical to define acid rain as any rainwater having a pH below 7, acid rain is not defined in this way. The reason stems from the fact that natural waters invariably contain some dissolved gases, including in particular CO_2. When CO_2 dissolves in water, it combines with H_2O to form carbonic acid H_2CO_3 by the reaction

$$H_2O + CO_2 \rightleftharpoons H_2CO_3 \tag{15.2}$$

H_2CO_3 may then dissociate to release H^+ ions by the reactions

$$H_2CO_3 \rightleftharpoons HCO^-_3 + H^+ \tag{15.3}$$

$$HCO^-_3 \rightleftharpoons CO_3^{2-} + H^+ \tag{15.4}$$

Because H_2CO_3 is a source of H^+ ions, a sample of distilled water exposed to the air will actually contain an excess of H^+ ions over OH^- ions. The pH of such a water sample would be about 5.6 to 5.7. Because of this fact, acid rain is usually defined as rainwater having a pH less than 5.6 (Cowling, 1982). In other words, a pH of 5.6 is about what one would expect if the rainwater contained no dissolved substances other than atmospheric gases. In fact, rainwater normally contains a variety of dissolved substances in addition to gases. The reason is that raindrops form on tiny atmospheric aerosols, which consist of particles of dust blown from the surface of the Earth or even salt crystals injected into the atmosphere at the surface of the ocean. Because of the presence of these other dissolved substances, the pH of rainwater may vary widely, in some cases being greater than 7 and in some cases substantially lower. Charlson and Rodhe (1982), for example, have pointed out that pH values as low as 4.5 could result from removal by rainwater of naturally occurring acids (notably H_2SO_4) from the air. However, the pH of natural precipitation normally falls in the range 5.0–5.6 (Wellford et al., 1982). For this reason, acid rain can be defined as rainwater having a pH less than this range rather than simply as rainwater with a pH less than 5.6.

Human activities have affected the pH of rainwater primarily by introducing into the atmosphere large quantities of sulfur oxides (SO_x) and various oxides of N, collectively referred to as NO_x. SO_x consists primarily of SO_2, while NO_x is a mixture of NO and NO_2. SO_x and NO_x are gases that can combine with water to form sulfuric acid (H_2SO_4) and nitric acid (HNO_3). Possible equations for such reactions are

$$3SO_2 + 3H_2O + O_3 \rightarrow 3H_2SO_4 \qquad (15.5)$$

$$SO_2 + 2OH\cdot \rightarrow H_2SO_4 \qquad (15.6)$$

$$2NO + H_2O + O_3 \rightarrow 2HNO_3 \qquad (15.7)$$

$$NO + O_3 \rightarrow NO_2 + O_2 \qquad (15.8a)$$

$$NO_2 + OH\cdot \rightarrow HNO_3 \qquad (15.8b)$$

Unfortunately, H_2SO_4 and HNO_3 are much stronger acids than H_2CO_3. Even a dilute solution of H_2SO_4 or HNO_3 will have a pH less than 1. Hence, the formation of these acids can greatly increase the acidity of rainwater. The atmospheric reactions that produce H_2SO_4 and HNO_3 from SO_x and NO_x are collectively favored by the presence of light, ozone (O_3), hydroxyl radicals ($OH\cdot$), and, of course, H_2O. In the case of S, summer rain tends to contain a higher concentration of H_2SO_4 than winter rain because summer humidities are higher and solar insolation is greatest during the summer (Galloway, 1979).

How do human activities introduce large quantities of SO_x and NO_x into the atmosphere? In the case of SO_x, we know that fossil fuels contain small but significant amounts of S. When these fuels are burned, the S is oxidized to form compounds such as SO_2. In general, coal has a higher S content than oil, which in turn has a higher S content than natural gas. Burning of fossil fuels by humans has introduced large amounts of SO_x into the atmosphere owing to the oxidation of the S in the fuels. Fossil fuel burning also explains how humans have introduced NO_x into the atmosphere, but in the case of NO_x the principal source of the N is not the N in the fuels themselves. Instead, the NO_x compounds result from the oxidation of atmospheric N_2. Because the Earth's atmosphere consists of almost 80% N_2 and because most fossil fuel burning occurs in a normal atmosphere, it is not surprising that some of the N_2 is oxidized to NO_x in the burning process. Note that while SO_x emissions can be controlled to a certain extent by switching from one fossil fuel to another, the choice of fossil fuel has very little to do with the rate of NO_x emissions.

Since what goes up usually comes down, most of the SO_x and NO_x introduced into the atmosphere ultimately returns to Earth. SO_2, for example, has a life expectancy in the atmosphere of less than five days (Wellford et al., 1982). Some of the SO_x and NO_x returns in rainfall, but a significant fraction may be returned as dry deposition. For example, Bischoff et al. (1984) estimate that in the eastern United States, dry deposition accounts for 1.4 times as much SO_x deposition and 26% as much NO_x deposition as does wet deposition. Once sulfate or nitrate particles are deposited on the ground in dry form, they may later combine with surface water or groundwater to produce H_2SO_4 or HNO_3. Thus, both wet and dry deposition can have serious impacts on the acidity of surface and ground water.

HISTORY OF THE ACID DEPOSITION PROBLEM

There is evidence to suggest that acid deposition occurred as long ago as the 17th century. S pollution became a problem in London during that time (Brimblecombe, 1977, 1978), and Koide and Goldberg (1971) have reported that sulfate deposits in the Greenland icecap began increasing during the beginning of the Industrial Revolution. Acid rain was reported in Manchester, England, in 1852 (Smith, 1852) and was described more thoroughly 20 years later (Smith, 1872). Modern awareness and concern over acid deposition developed during the 1950's, when Barrett and Brodin (1955) discovered acid rain in Scandinavia, Houghton (1955) reported finding acid fog and cloud water in New England, and Gorham (1955) discovered that rain in the English Lake District was acidic whenever the wind blew from urban/industrial areas. Recognition that the damage caused by acid rain crossed international boundaries came with the work of Oden (1968) in Sweden and Likens et al. (1972) in the United States.

The two major factors that have contributed to the acceleration of acid deposition in recent years have been the increased use of fossil fuels and the increased height of exhaust stacks at power plants and various industrial operations. The increase in stack heights is noteworthy, since the change was made to reduce local air pollution problems associated with stack emissions. For example, a suffocating smog that hovered over London for several days in 1952 resulted in 2,500–4,000 deaths in a week, three or more times the normal death rate (Hamilton, 1979). Such incidents led to serious attempts to clean up urban air in large industrialized cities. One logical remedy was to build tall smokestacks so that the emissions could be sent up into the atmosphere, where they were expected to disperse and be rendered harmless. While this goal was achieved, the increase in stack heights created acid deposition problems for many kilometers downwind of the stacks. Thus, a local fallout problem was transformed into an acid deposition problem over a much wider area.

The first reports of the seriousness of acid deposition effects came from Scandinavia. In Sweden, the pH of lakes dropped about 2 units between the 1930s and 1960. By the 1960s, about 50% of Swedish lakes were found to have a pH of less than 6, and 5,000 lakes had a pH of less than 5 (EPA, 1980). As a result of this change in pH, salmon populations were decimated in western Sweden, and other fish populations were seriously affected in central and eastern sections of the country. Norway also experienced a decline in salmon populations during the 1960s (EPA, 1980).

In North America, there has been a serious increase in the acidity of lakes in parts of both Canada and the United States. In southern Ontario, 33 of 150 lakes surveyed during the 1970s were found to have a pH of less than 4.5, and 32 lakes had a pH between 4.5 and 5.5 (EPA, 1980). Fish affected by the acidity included smallmouth bass, walleye, northern pike, lake trout, lake herring, perch, and rock bass. At that time, the largest single Canadian source of SO_x and NO_x was the enormous smelters at Sudbury, Ontario, about 50 km north of Lake Huron. Several hundred lakes within an 80 km radius of Sudbury contained few or no fish (EPA, 1980). A U.S. study during the 1960s and 1970s revealed that over 50% of remote Adirondack Mountain lakes at elevations greater than 600 m had a pH of less than 5, and 90% of these lakes contained no fish at all. For comparison, a study between 1929 and 1937 revealed that only 4% of the same lakes had a pH of less than 5 and were devoid of fish (EPA, 1980). The mean pH of Adirondack lakes during the period 1930–1938 was 6.5, but between 1969 and 1975 the mean pH had dropped to 4.8 (EPA, 1980). Later studies showed that the number of fishless lakes in the Adirondacks increased substantially during the 1970s and that the phenomenon was no longer confined to lakes at high elevations (Environment Canada, 1981). Thus, there was clear evidence of an increase in the acidity of U.S. Adirondack lakes and of a decline in their fish populations.

The cause of the increase in acidity of lakes in the eastern United States is not hard to determine. Figure 15.1 shows a plot of the pH of precipitation in the eastern United States in 1955–1956 and again in 1972–1973. During this time, the minimum pH of rainfall in this area dropped from about 4.4 to 4.0 (Bischoff et al., 1984), and the area over which the mean pH was less than 5 (acid rain) increased from about 50% of the land area to almost 100%. The problem of acid deposition has been most acute in the northeastern United States because this area lies directly downwind from major sources of SO_2 emis-

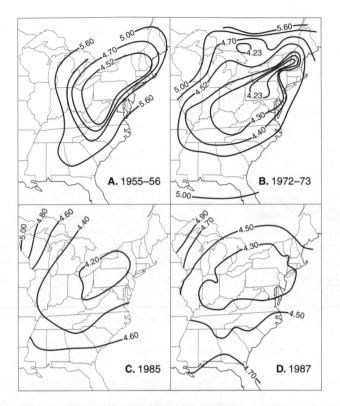

FIGURE 15.1 Annual average pH of precipitation in the eastern United States in selected years. *Source*: Likens et al. (1976), McCormick (1989), and Li (1992).

sions, the numerous coal-burning power plants and smelters in the Ohio River Valley. Figure 15.2, which shows the distribution of rainfall pH in a random month (June 1966), dramatically illustrates the difference in rainfall acidification between the western and eastern United States. Western U.S. coal contains a significantly lower S content than eastern U.S. coal, and in part because of this fact, the pH of rain tends to be higher in the West than in the East. The SO_2 emission problem is particularly acute in the summer months (Figure 15.2) because increased humidity and radiation during the summer result in more rapid oxidation of SO_2 to H_2SO_4 (Galloway, 1979). The result is apparent in Figure 15.3, which shows the concentrations of NO_3^- and SO_4^{2-} in rainwater from the Hubbard Brook watershed in New Hampshire. Note that the concentration of NO_3^- shows little seasonal dependence, but the concentration of SO_4^{2-} is two to three times higher in the summer than in the winter.

SUSCEPTIBILITY OF LAKES TO ACID DEPOSITION EFFECTS

Natural bodies of water contain a variety of dissolved and particulate substances that buffer the pH. In other words, these substances tend to neutralize acids or bases and thereby tend to maintain the pH near a certain value. Although the natural pH of different lakes may vary greatly from one region of a country to another, the organisms that inhabit these lakes

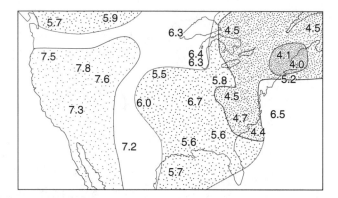

FIGURE 15.2   The acidification pattern of rainfall over the United States in June 1966. *Source*: Reprinted with permission from A. J. Vermeulen, The acidic precipitation phenomenon: A study of this phenomenon and the relationship between the acid content of precipitation and the emission of sulfur dioxide and nitrogen oxides in the Netherlands. In T. Y. Toribara, M. W. Miller, and P. E. Morrow (Eds.), *Polluted Rain*. Plenum Press, New York, 1980, p. 15.

have presumably evolved so as to be able to carry out normal activities efficiently at the lake's normal pH.

The ability of a lake to resist pH changes due to acid deposition will depend on the buffering capacity of both the lake and its watershed. In most cases, the principal buffer is the bicarbonate ion, HCO_3^- (McCormick, 1989). If the watershed consists of hard, in-

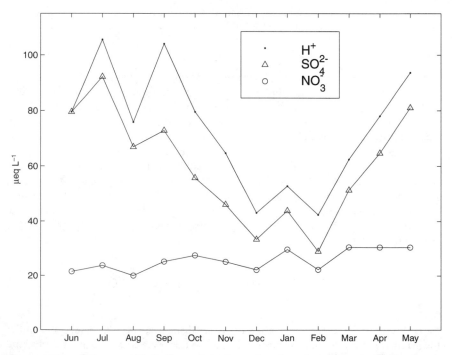

FIGURE 15.3   Mean monthly bulk precipitation values for selected ions in Hubbard Brook during 1963–1974. *Source*: Galloway (1979). Reprinted with permission of American Society of Civil Engineers.

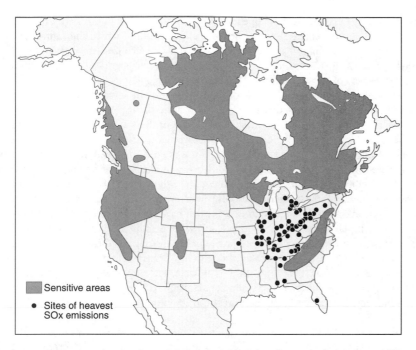

FIGURE 15.4   Regions in North America containing lakes that are sensitive to acidification by acid deposition. The shaded areas have igneous or metamorphic bedrock geology. Sources of major SO$_x$ emissions are indicated by solid dots. *Source*: Environment Canada (1981).

soluble bedrock (e.g., igneous or metamorphic rock) covered with thin, sandy, infertile soil, the buffering capacity of both the lake and the watershed is likely to be low (Johnson, 1979). On the other hand, if the soils and rocks surrounding the lake are susceptible to chemical leaching (e.g., limestone), it is likely that a high concentration of HCO$_3^-$ and other buffers will have leached into the lake, and the watershed itself will probably have considerable capacity to absorb the acidity in acid deposition. Bodies of water most susceptible to acid deposition effects are therefore those

1. that lie downwind from major sources of SO$_x$ and NO$_x$ emissions.
2. whose watersheds are underlain with chemically unreactive bedrock such as granite and basalt and contain a thin cover of soil and vegetation.
3. that receive runoff from low-order streams (i.e., near-headwater streams whose water has not been significantly buffered by reactions with soils and rocks).

Figure 15.4 illustrates the approximate distribution of land areas in the United States and Canada that are low in natural buffers and therefore particularly susceptible to acidification. As is apparent from the figure, most of the Canadian shield, an area of approximately 2.5 million km^2, is very susceptible to acid deposition. In the United States, the principal sensitive areas are the Adirondack Mountains and northern New England in the East and Oregon, most of Washington and Idaho, and the mountainous portion of California in the West.

ACID DEPOSITION TOXICOLOGY

With this introduction to the subject of acid deposition, we will now examine the effects of acidity on living organisms. With respect to fish, the most serious chronic effect of increased acidity in surface waters appears to be interference with the fish's reproductive cycle. Ca levels in the female fish may be lowered to the point where she cannot produce eggs, or (1) the eggs fail to pass from the ovaries or (2) if fertilized, the eggs and/or larvae develop abnormally (EPA, 1980). Acutely toxic effects involving kills of adult fish may be due to a variety of causes. Failure of the fish to osmoregulate properly may result in improper body salt levels and/or failure of the fish to properly metabolize elements such as Ca and Na properly. Increased acidity may also lead to the mobilization of certain metals that would otherwise remain in the soils and rocks of the watershed or in the sediments of the lake. Al, for example, is derived primarily from the weathering of rocks, and the solubility of aluminous soils increases dramatically as the pH is reduced from 7 to 5. Increases in dissolved Al concentrations due to acid deposition effects appear to be the chief cause of fish mortality in the Adirondacks (Cronan and Schofield, 1979; Schofield and Trojnar, 1980). Acutely toxic effects seem to result from severe necrosis of the fish's gill epithelium, and chronic effects such as reductions in growth have been detected at Al concentrations as low as 100 ppb (Cronan and Schofield, 1979). In Scandinavia, on the other hand, the chief cause of fish mortality has been attributed to disruption of the fishes' osmoregulatory mechanism controlling Na balance due to low pH levels and low Ca concentrations (Leivestad et al., 1976). There is no chart to indicate at exactly what pH a particular species of fish can no longer function, but when the pH drops to 5.5, most species are endangered, and when the pH reaches 4.5, almost all are dead (Environment Canada, 1981; McCormick, 1989).

The effects of acid deposition on aquatic systems are unfortunately not limited to the direct effects of increased acidity on fish populations. Acid deposition adversely affects the reproduction of frogs and salamanders. These amphibians lay their eggs in the meltwater pools of early spring. The snow that feeds these ponds may contain sulfuric and nitric acids in sufficient concentrations to prevent hatching of 80% or more of the eggs. Hatching failure under normal conditions is only about 10–15% (Environment Canada, 1981). Lake acidification generally results in a decline in algal species diversity, with a shift to more acid-tolerant forms (EPA, 1980), and generally destabilizes the plankton community (Marmorek, 1984). Acidification of lakes may result in a decline in the biomass and activity of decomposer organisms such as bacteria and fungi, with the result that organic waste products tend to accumulate in the water column and sediments (EPA, 1980; Francis et al., 1984). Thus, acid deposition may adversely affect a wide variety of organisms in aquatic systems, leading both to a serious imbalance in the cycling processes involving materials within the system and to a severe decline in the biomass of living organisms.

Under appropriate conditions, the fertilizing effects of acid deposition may actually be beneficial to the production of certain crops, but experiments conducted by the EPA have shown that acid deposition can damage the foliage of crops such as radishes, beets, carrots, mustard greens, and broccoli (Environment Canada, 1981). Acidification of soils may also render certain essential nutrients unavailable to terrestrial plants. For example, the recycling of critical elements such as N and P requires that these elements be released from decaying organic matter through the activities of soil decomposer microbes. As soil pH is reduced, the activities of these microbes are suppressed, just as the activities of aquatic microbes are suppressed by lake acidification (EPA, 1980). As a result, dead organic matter tends to accumulate on the ground, and the essential nutrients bound in these dead organics are unavailable to stimulate new growth. In addition, essential elements such as Ca and Mg, which are normally retained on soil particles, may be washed out of the soil by acid rain and hence become unavailable to plants. This process occurs because at low pH levels the chemical bond that is responsible for the binding of Ca and Mg to soil particles breaks down. Furthermore, acid deposition may actually stimulate the uptake of toxic heavy metals by vegetation. For example, lettuce exposed to acid rain has been found to retain elevated levels of Cd (EPA, 1980). Although uptake of such heavy metals may not noticeably affect the growth of potentially marketable crops, the metal concentrations in the crop may make the product unfit for human consumption.

Acid rain has been shown to adversely affect photosynthesis by terrestrial plants. The production of carbohydrates, which are among the chief components of the edible portions of the plants (seeds, fruits, roots), is often suppressed when plants are grown in an acid rain environment (EPA, 1980). Furthermore, acid is known to cause bleaching of chlorophyll *a*, the principal light-absorbing pigment in photosynthesis. Whether such bleaching has occurred to a significant extent due to the effects of acid rain is not known, but such bleaching may in part explain the reduction in carbohydrate production observed in crops exposed to acid rain.

The direct impact of acid deposition on forests is difficult to quantify because stresses caused by other pollutants and the indirect effects of SO_x and NO_x may be equally if not more damaging. For example, hydrocarbons, produced mainly by automobile engines, react with NO_x in the presence of sunlight to produce O_3. The O_3 may in turn react with SO_x and NO_x to produce sulfuric and nitric acids. Furthermore, O_3 itself can be very damaging to trees and other vegetation, and it is believed to be the principal cause of damage to pine trees in the Appalachian and Blue Ridge mountains (Postel, 1984). Acidification of forest soils can increase the concentration of soluble Al ions, which damage the fine roots of trees and prevent them from taking up adequate amounts of nutrients and water (McCormick, 1989). Deficiencies of Ca and Mg frequently result.

Concern over the effects of acid deposition on human health center on the issue of heavy metal toxicity. As we have already noted, acidification of water tends to mobilize heavy metals such as Pb and Hg. Such metals tend to remain in particulate form and sink to the bottom of aquatic systems when the water is neutral or basic, but these particulate compounds may rapidly dissolve if the water becomes acidic. In New York State, water from the Hinckley Reservoir has become so acidic that it leaches Pb from water conduit pipes, and dissolved Pb concentrations in the tap water exceed New York State public health guidelines (EPA, 1980). Other effects of acid rain on human health are suspected but not well documented. For example, how does exposure to acid rain or acid mists affect skin and hair? How does breathing in air containing mists of H_2SO_4 and HNO_3 affect our lungs? There is no doubt that SO_2 is a toxic gas (Park, 1987), and high concentrations have been shown to cause respiratory problems in humans (McCormick, 1989). Studies at the Brookhaven Laboratory in New York have suggested that increased concentrations of SO_2 in the air may be associated with an increase in mortality rates, and it has been hypothesized that increased SO_2 levels may cause chronic bronchitis and emphysema. However, not all scientists agree with these statements (Environment Canada, 1981).

One rather well-documented effect of acid deposition is its effect on building components and stone statues. SO_2, for example, transforms calcite in stone into gypsum, which is much more soluble in water than calcite. A crust of gypsum and other compounds forms on the exterior of the stone and then rapidly washes away in the next rain. The exposed fresh stone is then subject to further attack. Such deterioration causes a loss in the structural integrity of stone buildings and causes stone statues to lose their detail. SO_2 has also been implicated in the corrosion of steel (EPA, 1980). Even steel protected by Zn galvanizing is subject to corrosion because the SO_2 reacts with the Zn in the protective covering. Exposure to acid rain also has an adverse effect on paint. Oil-based and automobile paints are evidently particularly vulnerable (EPA, 1980).

MAGNITUDE OF ANTHROPOGENIC EMISSIONS

Estimates of the amount of S emitted to the atmosphere each year range from 140 to 220 million tonnes (EPA, 1980; Mackenzie et al., 1984). Some of these emissions are completely natural. The oceans, for example, are a source of dimethyl sulfide (Charleson et al., 1987), and volcanoes emit SO_2. The anthropogenic contribution to global S emissions has been increasing steadily during the last century and is roughly 75 million tonnes at the present time (Figure 15.5). Nevertheless, human activities account for only about 35–55% of total S emissions. The anthropogenic contribution is much larger in certain continental areas. In fact, most of the anthropogenic S emissions occur on only 5% of the Earth's surface, primarily in the industrial regions of Europe, eastern North America, and East Asia (Postel, 1984). Indeed, Bischoff et al. (1984, p. 18) note that "the major inputs of SO_2 and NO_x into the atmosphere of the eastern United States are anthropogenic." Of the global anthropogenic S emissions, about 70% are accounted for by the burning of coal, 16% by the combustion of petroleum, and 14% by the operations of oil refineries and nonferrous smelting (EPA, 1980). In the United States, anthropogenic SO_x emissions peaked at about 14 million tonnes of S during the 1970s and have since declined to 7–8 million tonnes (Figure 15.6). About 75% of these emissions are accounted for by stationary sources east of the Mississippi, and about 92% of the latter are located near the industrialized Ohio River Valley (EPA, 1980). Electric utilities account for about 66% of anthropogenic SO_x emissions for the United States as a whole (EPA, 1997).

Global emissions of NO_x are known only rather crudely, but it is estimated that about 50% of the NO_x in the Earth's atmosphere is anthropogenic (McCormick, 1989). Natural sources of NO_x include processes such as chemical and bacterial nitrification in soil (McCormick, 1989). Anthropogenic emissions of NO_x in the United States currently amount to about 6.5 million tonnes of N per year, roughly 25% of the world total (Figures 15.5 and 15.6). About 36% of U.S. anthropogenic emissions are contributed by transportation-related activities and 26% by the electric utility industry (EPA, 1997). There is general

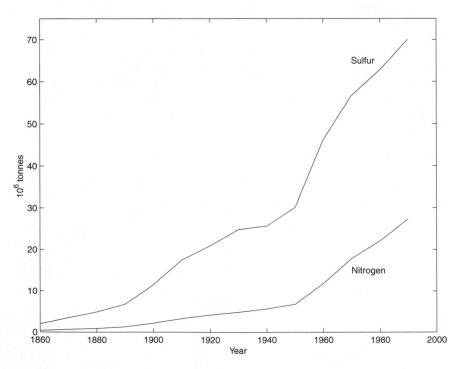

FIGURE 15.5   Historical global anthropogenic emissions SO_x (as S) and NO_x (as N). *Source*: Dignon and Hamseed (1989), Langner (1991), and Galloway (1994).

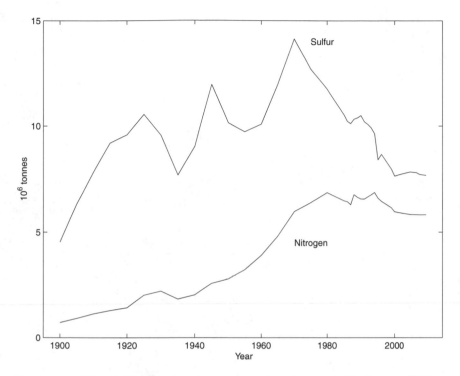

FIGURE 15.6   Historical and projected anthropogenic emissions of SO_x (as S) and NO_x (as N) for the United States. *Source*: EPA (1997).

agreement that NO_x emissions have increased in importance relative to SO_x emissions since approximately 1950, primarily because control measures were developed more rapidly for stationary SO_x sources. In areas such as New England, for example, increases in the acidity of rain from 1965 to 1975 were accounted for largely by increases in rainwater nitrate concentrations. This fact is apparent in Figure 15.7, which shows the relationship between H^+ and NO_3^- inputs to the Hubbard Brook Experimental Forest in New Hampshire over that 10-year period. Nevertheless, on a global scale, SO_x still accounts for about two-thirds of the acidity of rainfall. In the Netherlands, for example, there is a pronounced relationship between the total interior emission of SO_2 and the acidification of rainfall (Vermeulen, 1980). In the eastern United States, H_2SO_4, HNO_3, and HCl account for about 65%, 30%, and 5%, respectively, of the acidity in rainfall (Galloway, 1979). Near Denver, however, HNO_3 accounts for 80% of rainwater acidity and H_2SO_4 for only 10% (EPA, 1980). The difference reflects the relative importance of different NO_x and SO_x sources in the two areas and the effectiveness of various measures used to control SO_x and NO_x emissions.

CORRECTIVES

Methods for reducing anthropogenic emissions of NO_x and SO_x to the atmosphere may be grouped into three general categories:

1. Conservation of energy
2. Use of substitutes for fossil fuels

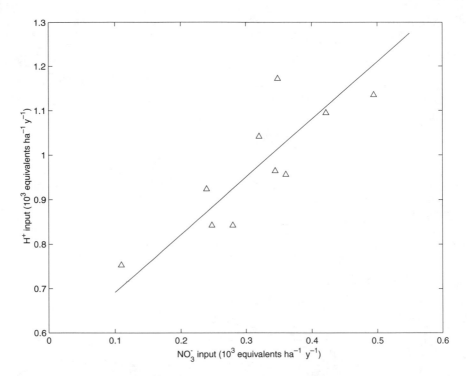

FIGURE 15.7   Relationship between the annual H^+ input and NO_3^- input to the Hubbard Brook Experimental Forest during the period 1964–1965 to 1973–1974. *Source*: Likens et al. (1976) with kind permission from Kluwer Academic Publishers.

3. Removal of the N and S from fossil fuels and/or combustion products before NO_x and SO_x escape to the atmosphere

Conservation of energy can be encouraged by either economic or political means. An obvious example was the improved fuel efficiency of new automobiles in response to the rise in oil prices during the 1970s (Flavin and Durning, 1988). Possible substitutes for fossil fuel energy include nuclear energy, geothermal energy, wind, and solar energy. At the present time, it does not appear likely that fossil fuel energy alternatives will become a major factor in the U.S. energy picture in the near future. Energy conservation has certainly helped to control the rise of fossil fuel combustion, but major reductions in emissions of SO_x and NO_x will require more than energy conservation. We conclude that some methods must be found to reduce the amount of SO_x and NO_x produced in the combustion of fossil fuels.

We will concentrate here on the problem of reducing SO_x and NO_x emissions from stationary sources such as steam-electric power plants and industrial boilers. The rationale for this approach is that only about 3% of anthropogenic SO_x emissions in the United States are accounted for by mobile sources due to the low S content of motor vehicle fuels (EPA, 1997). Historically, 45–60% of NO_x emissions associated with fossil fuel burning have been due to mobile sources (Postel, 1984). However, this percentage has decreased in recent years due to the improved fuel efficiency of new automobiles, the use of NO_x control measures on new internal combustion engines, and the phasing out of older automobiles. Automobiles sold in the United States today have catalytic converters that reduce NO_x emissions by 76% (Postel, 1984), and the fuel efficiency of these cars is 92%

higher than that of cars sold in 1973 (Flavin and Durning, 1988). As a result, stationary sources now account for more than 50% of anthropogenic NO_x emissions in countries such as the United States and Germany (Postel, 1984; EPA, 1997).

There are basically four different stages in fossil fuel use at which N and/or S may be removed. These stages are (1) fuel pretreatment, (2) fuel conversion, (3) fuel combustion, and (4) fuel postcombustion. We consider SO_x and NO_x removal separately because both the origins of the S and N and the removal processes are very different.

SO_x Removal

Pretreatment

Coal Cleaning. Coal contains anywhere from 0.5 to 5.0% S (McCormick, 1989). The S may occur as pyritic sulfur (pyrite or marcasite), as sulfatic sulfur, or as organic S. Sulfatic S is unimportant because of its low concentration. The relative amounts of pyritic and organic S vary widely among coals from different regions and even within a single vein. Mechanical coal cleaning takes advantage of the fact that pyrite has a specific gravity about twice that of coal and most of the minerals associated with coal. The approach in mechanical cleaning is to crush the coal into small particles and then to separate out the pyrite in a multistage cleaning process based on particle size and density. In effect, the coal is given a sophisticated water bath. The pulverized coal floats on the surface, and impurities including pyrite sink to the bottom. Removal of as much as 90% of the pyrite is possible if the coal contains large, readily liberated pyrite (Bomberger and Phillips, 1979), but in practice, pyritic S removal efficiencies rarely exceed 65% and average about 50% (McCandless et al., 1987). Because pyritic S accounts for about 70% of the S in coal (McCandless et al., 1987), mechanical cleaning can be expected to reduce the S content of coal by about (70%)(50%) = 35%. Since the 1990 amendments to the Clean Air Act of 1970 are intended to reduce anthropogenic SO_x emissions in the United States by about 50% (Anonymous, 1991), it is evident that something more than mechanical coal cleaning is needed to meet emission standards in the United States.

Chemical removal of S from coal can be achieved in several ways. In some processes, the coal is pulverized and the pyrite is oxidized to free S and/or sulfate. The sulfate dissolved in the leaching solution and the free S are extracted with toluene. Pyrite removal efficiencies of 90–95% can be achieved with such methods (McCandless et al., 1987). In a process developed by Battelle Laboratories, both pyritic and organic S are leached from the coal by treatment with NaOH and $Ca(OH)_2$ at elevated temperatures and pressures (Bomberger and Phillips, 1979). McCandless et al. (1987) report that up to 95% of the pyritic S and as much as 40% of the organic S can be removed from coal by such methods. At the present time, it appears that some form of coal cleaning will play a role in SO_x emission control at operations that burn other than low-S coal, in part simply to control the variability in the S content of the feed coal. Coal cleaning, however, produces its own forms of waste and generates emissions to the air, water, and land (McCandless et al., 1987).

Removal of S from Oil. The heavy fraction of crude oil known as *atmospheric residuum* has been used for years as a fuel by some power plants. This oil may contain over 4% S (Bomberger and Phillips, 1979). To meet the New Source Performance Standards (NSPS) guidelines mandated by the Clean Air Act, this S content must be reduced to less than 0.8%. A standard procedure for achieving this reduction is catalytic hydrodesulfurization. The procedure consists of treating the oil with H at elevated temperatures in the presence of a suitable catalyst. The S is captured in the form of H_2S, a gas that can be removed by standard techniques. Such procedures are capable of reducing the fuel oil S content to as low as 0.2% (Bomberger and Phillips, 1979).

Conversion

Coal Gasification. Coal gasification involves treatment of pulverized coal with steam and oxygen to produce a gas containing CO_2, CO, and H_2. The S is converted to H_2S, which is separated out by scrubbing with a solvent such as methanol. The gas produced in the process is a low- or medium-grade fuel, depending on whether air or pure O_2, respectively, is used as the O_2 source (Bomberger and Phillips, 1979). It is possible to upgrade the medium-grade gas to synthetic natural gas by removing the CO_2 and converting the CO and H_2 to methane, but the process is costly.

Coal Liquefaction. Liquefaction of coal can be achieved by hydroliquefaction, solvent extraction, or pyrolysis, or by synthesis from CO and H_2 derived from coal gasification. The net result of coal liquefaction is a product with a lower ratio of C to H (10 or less by weight) than is found in crude coal (12 to 20). In hydroliquefaction, H is added with the help of a catalyst to pulverized coal at high temperature and pressure. Cleavage of the coal releases the S as H_2S, the O as H_2O, and the N as NH_3. All three compounds are gases that can be removed in a straightforward way. The liquid product consists of fuel oil and some lighter hydrocarbons. Solvent extraction is somewhat similar to hydroliquefaction, except that no synthetic catalysts are used and the design of the reactor vessel is different. If H is added at about 1% of the weight of the coal, the product is primarily a solid known as *solvent-refined coal* (*SRC*) that melts at 148–177°C (Bomberger and Phillips, 1979). Since the SRC contains less than 1% S and is low in ash, it can be used as a solid replacement for coal. If H is added at up to 4% of the coal weight, the primary solvent extraction products are liquids. However, the cost of producing liquid fuels by solvent extraction is much higher than the cost of producing SRC. Pyrolysis of coal has been practiced for a number of years in order to produce coke. In the process, some liquid and gaseous fractions are also produced, but the standard coking process is not designed to maximize the production of liquid fuels. By suitable modifications of the pyrolysis procedure, higher yields of liquid hydrocarbons can be achieved. However, the liquids must be catalytically hydrotreated to reduce their S content. The solid residue (char) from the pyrolysis treatment also has too high an S content to be used as a fuel, but gasification of the char may produce a suitably low S fuel (Bomberger and Phillips, 1979).

Methanol Production. The CO and H_2O produced from coal gasification can be used to produce H_2 and CO_2 by the reaction $CO + H_2O \rightarrow H_2 + CO_2$. The H_2 produced by this reaction may then be used to produce methanol (CH_3OH) from either CO or CO_2 by the reactions

$$CO + 2H_2 \rightarrow CH_3OH \tag{15.9}$$

$$CO_2 + 3H_2 \rightarrow CH_3OH + H_2O \tag{15.10}$$

S is removed in the form of H_2S. Crude methanol from coal could be used as a clean utility fuel, particularly for peaking power (Bomberger and Phillips, 1979). Higher molecular weight hydrocarbons can be formed from H_2, CO and CO_2 by Fischer-Tropsch synthesis.

Combustion

Fluidized Bed Combustion. Fluidized bed combustion of coal can be used to reduce emissions of both SO_x and NO_x. The combustion chamber in such a process consists of a perforated bed of crushed limestone or dolomite, ash, and partially burned coal. A distribution system in the bottom of the bed introduces powdered coal and combustion air. The air blown into the combustion chamber suspends the particles and causes them to float freely and behave like a liquid, hence the term *fluidized bed*. The two key features of the process are as follows:

1. Combustion takes place without visible flames at a temperature of about 850–900°C, much lower than the 1,650°C needed in conventional coal-fired systems (McCormick, 1989). Because the oxidation of N_2 to NO_x requires a high activation energy, the lower combustion temperature greatly reduces NO_x formation from the air. It has less effect, however, on the formation of NO_x from N in the coal.

2. The SO_2 and SO_3 gases produced by the combustion process are in direct contact with the limestone. Absorption of these gases is about 90% complete if the bed contains adequate limestone.

From the standpoint of SO_x removal, the major drawback of the process is that limestone utilization efficiency is low. As a result, limestone must be added at up to twice the stoichiometric requirements for SO_2 and SO_3 removal. Disposal of spent limestone therefore becomes a significant solid waste disposal problem. Research currently underway may identify a means for regenerating the limestone and thereby eliminating the solids disposal problem.

Lime Injection in Multistage Burners (LIMB). This technique is similar to fluidized bed combustion in the sense that NO_x emissions are controlled by reducing the combustion temperature, and SO_x gases are absorbed with powdered limestone. LIMB, however, does not involve the use of a fluidized bed. The SO_x absorbent is injected directly into the firebox, and the combustion temperature is lowered with the use of special burners. SO_x emissions are typically reduced by 50–70% (McCormick, 1989). The chief advantage of LIMB over fluidized bed combustion is that LIMB can be cost-effectively retrofitted into existing plants. Retrofitting plants to accommodate fluidized bed combustion is impractical in some cases due to lack of space, and in any case, capital costs are far lower for LIMB than for fluidized bed combustion.

Postcombustion

Stack Gas Scrubbing. Stack gas scrubbing is a demonstrated technology for removing SO_2 and SO_3 from stack gases at up to 90–95% efficiencies. More than 1,000 plants, primarily in the United States and Japan, employ this technique (McCormick, 1989). There are basically two types of stack gas scrubbing, wet and dry. In wet scrubbing the SO_x is absorbed into water droplets, where sulfurous or sulfuric acids reacts with a base. The absorber is usually a countercurrent device in which the flue gases pass up through a spray or slurry containing the base.

Usually lime or limestone is used as the base, in which case the reaction product is a sludge of Ca sulfite and Ca sulfate with a consistency similar to that of toothpaste. The sludge must be disposed of at a landfill or in a pond. As is the case with fluidized bed systems, disposal of this sludge is a problem. In a typical 1,000 MW power plant burning coal containing 3.5% S, wet scrubbing produces about 225,000 tonnes of sludge per year (McCormick, 1989). Processes can be used to recover sulfuric acid or pure S from the sludge, but recovery is expensive and energy intensive. Air oxidation can be used to convert the Ca sulfate to gypsum, and in Japan, which has little natural gypsum, methods have been developed for purifying and marketing the converted gypsum.

Dry scrubbing methods have been developed primarily to reduce the sludge disposal problems associated with wet scrubbing. In the Saarberg-Holter process, a slurry of lime or soda ash sprayed into the flue gases absorbs the SO_x and is dried at the same time. In the Walther process, ammonia is sprayed into the gases, the end product being pelletized ammonium sulfate. The latter can be used as a fertilizer. In addition to producing less waste, dry scrubbing has several advantages over wet scrubbing. Dry scrubbers tend to have fewer operating problems, need fewer operators, have lower capital and operating and maintenance costs, and use much less water than wet scrubbers (McCormick, 1989). The principal disadvantage is that they are not as efficient in removing SO_x, the range of efficiencies being roughly 50–90% (McCormick, 1989). Consequently, they have not been widely used.

Electron Beam Method. One of the most recently developed techniques for removing both SO_x and NO_x from stack gases is the electron beam method. The strategy is to spray the flue gas with water to reduce its temperature to 60–90°C and then to spray the cooled gases with ammonia. The gas mixture is then passed through an electron beam reactor that converts the SO_x and NO_x to acids, and these in turn react with the ammonia and water vapor to produce ammonium compounds that can be used as fertilizers (McCormick, 1989). The technique is capable of removing 80–90% of the SO_x from the stack gases (McCormick, 1989).

NO_x Removal

Pretreatment and Conversion

Neither pretreatment nor conversion of fossil fuels is a cost-effective way to reduce NO_x emissions because the N in the fuels is a minor source of NO_x.

Combustion

Since most of the NO_x is produced from the oxidation of atmospheric N_2, the strategy in combustion control is to reduce the combustion temperature and/or the time the air stays in the combustion chamber. Both fluidized bed combustion and LIMB use this approach. In the former case, the heat exchanger tubes are immersed in the fluidized bed, and heat transfer from the hot solids to the tubes is much more efficient than is the case in a conventional boiler. Therefore, combustion temperatures can be lower. In the LIMB procedure, the air is split into several streams and introduced into the flames in stages (McCormick, 1989). This approach allows one to reduce both the intensity and temperature of combustion. Unfortunately, neither fluidized bed combustion nor LIMB is capable of reducing NO_x emissions by more than 35–50% (Postel, 1984; McCormick, 1989).

Postcombustion

Both wet and dry methods exist for removing NO_x from stack gases, but the former suffer from the low solubility of NO and NO_2 in water. Hence, very large absorbers and expensive catalysts are required. By far the most common dry technique for removing NO_x from stack gases is selective catalytic reduction (SCR). In this procedure, a catalyst and a reducing agent are used to convert the NO_x to N_2. The system must be designed to remove fly ash before the SCR treatment because fly ash in the flue gases tends to clog the catalyst. Fly ash removal generally requires lower flue gas temperatures than are desirable for SCR, but NO_x removal efficiencies of 90% have been achieved with this method, and at a cost roughly half that of scrubbers (Postel, 1984). One attractive feature of several SCRs is that they also remove SO_x. The other dry removal technique is the electron beam method, which is capable of removing 50–90% of the NO_x as well as 80–90% of the SO_x (McCormick, 1989).

Integrated Gasification Combined Cycle

One of the most promising new technologies for reducing both SO_x and NO_x from the stack gases of coal-burning power plants is the integrated gasification combined cycle (IGCC) process. The idea is to convert crushed coal to a useful gas mixture by reacting it under pressure and high temperature in an almost pure stream of O_2. The gaseous products consist mainly of CO and H_2. The gaseous S products, mainly H_2S, are converted to elemental S, which is sold as a by-product. The synthetic gas is then burned at a reduced temperature to minimize NO_x production. In studies reported by Spencer et al. (1986) a 100 MW IGCC plant near Barstow, California, produced SO_x and NO_x emissions comparable to or lower than those produced by a power plant burning natural gas and equal to only 10–11% of the new source performance standard limitations mandated by the Clean Air Act.

Comments

There is no question that the technology now exists to reduce SO_x and NO_x emissions from stationary sources to levels that would create virtually no adverse environmental impact. There are basically two problems with putting this technology to work. First, the most efficient methods, for example IGCC, require designing and building a power plant from scratch. The lifetime of a typical fossil fuel power plant is 30–40 years; hence, for the next few decades, one can expect that most electricity will be generated by old power plants. It is possible to retrofit old power plants to a certain extent to accommodate emission control technologies, the use of scrubbers on power plants in the United States and Japan being an obvious example. Retrofitting, however, is less effective in controlling SO_x and NO_x emissions than building a new plant with built-in control features. Effective technologies for reducing NO_x emissions have proven especially difficult to identify, both because little of the N originates from the fossil fuels themselves and because of the low solubility of NO and NO_2 in water. Hence, it is fortunate that in most parts of the world where acid rain is a problem, SO_x is responsible for about two-thirds of the acidity and NO_x for only one-third.

The second major problem in implementing SO_x and NO_x emission controls is the need for motivation in the form of either political pressure or monetary incentives. It is this second problem to which we now turn.

LEGAL ASPECTS

The legal aspects of acid deposition are a thorny issue, partly because of the long-range effects of SO_x and NO_x emissions. For example, more than 80% of the acid in rainwater falling on Sweden and Norway is generated in other European countries (Postel, 1984), and the United States contributes half of the acid rain in Canada (Smith, 1980). Canada, on the other hand, is responsible for only about 10–15% of the acid rain in the U.S. (Environment Canada, 1981).

The cost of controlling SO_x and NO_x emissions has turned out to be a pleasant surprise. Canada initially estimated cleanup costs at \$400–500 million per year through the year 2000 (Smith, 1980), and in the United States, which emits about five times as much SO_x as Canada (Postel, 1984), the cost was expected to be much higher. The Congressional Office of Technology Assessment estimated the cost of a proposed 9-million-tonne-per-year SO_2 reduction program at \$3.4 billion per year, and the National Commission on Air Quality estimated the cost of a 10-year reduction program of 6.9 million tonnes of SO_2 per year at \$2.2 billion per year (Wellford et al., 1982). The reality is that the cost of reducing SO_2 emissions in the United States has been less than \$1 billion per year (Kerr, 1998). Two factors have contributed to the low cost. First, scrubbers turned out to be about 40% cheaper than many people had thought (Kerr, 1998). Second, the cost of transporting low-S coal from the western United States to the East dropped by about 35% after passage of the Staggers Act in 1980 largely deregulated railroads. The result has been a roughly twofold drop in SO_2 emissions from many of the older power plants in the eastern United States and a decrease in the acidity of rainfall in the Ohio River Valley by 15–25% between 1983 and 1994 (Kerr, 1998). The cost of this SO_2 control has added no more than a small amount to customers' electric utility bills. In contrast, pollution control accounts for about 25% of the total cost of coal-generated power in Japan, and Japanese electricity costs are among the highest in the world (Postel, 1984).

The benefits of acid deposition reduction are not as easy to estimate in monetary terms and are a subject of some controversy. The 1980 Crocker study performed at the University of Wyoming at the direction of the EPA estimated the value of reducing acid rain in the eastern United States at $5 billion per year. This figure included "$2 billion in effects on materials, $1.75 billion in damage to forest ecosystems, $1 billion in direct effects on agriculture, $250 million in effects on aquatic systems, and $100 million in other effects, including damage to water supply systems" (Wellford et al., 1982, p. 6). If these figures were only approximately correct, the cost of control measures would be easily outweighed by the benefits from a reduction in acid deposition. Undoubtedly the most thorough study of the costs of acid deposition in the United States was carried out by the National Acid Precipitation Assessment Program (NAPAP), a 10 year, $570 million research effort funded by the U.S. government. The principal findings of the study, described in a 6,000-page report, were as follows:

1. Acid deposition has adversely affected aquatic life in about 10% of streams and lakes in the eastern United States.
2. Acid deposition has contributed to the decline of red spruce at high elevations by reducing the spruce's cold tolerance.
3. Acid deposition has contributed to erosion and corrosion of buildings and materials.
4. Acid rain and related pollutants, especially fine sulfate particles, have reduced visibility throughout the northeastern and parts of the western United States (Roberts, 1991).

How disturbing a person finds these conclusions seems to depend very much on who is making the assessment, and it is not hard to see why the conclusions of the report have been controversial. The fact that during the 1960s and 1970s roughly half of high-elevation Adirondack Mountain lakes had a pH of less than 5 and were devoid of fish suggested that a serious problem was developing. The adversely affected high-elevation lakes were nevertheless only a small percentage of eastern U.S. lakes, and the NAPAP report clearly indicated that lake and stream acidification caused by acid deposition had not become a widespread problem in the eastern United States. Does this conclusion mean that the threat of acid deposition was overestimated? At least some persons have concluded that this is indeed the case. The NAPAP report, for example, was the impetus for a December 1990 *60 Minutes* television documentary that severely criticized the Bush administration and Congress for wasting the taxpayers' money by passing legislation aimed at controlling SO_x and NO_x emissions (Roberts, 1991). Many environmentalists, however, disagreed with the tone of the *60 Minutes* program. While acknowledging that the impact of acid deposition on lakes and streams had not become as serious as was once predicted, they argued that the problems caused by acid deposition were in fact much broader than had been realized or suspected when the NAPAP study began (Roberts, 1991). As noted by NAPAP director Mahoney (1991, cited by Roberts, 1991, p. 1303), "Some of the effects are clearly less severe than the allegations, [but] there is abundant indication that we need [sulfur emission] controls."

The U.S. government has in fact taken two important steps to reduce SO_x and NO_x emissions. The first step was taken in 1979, when the government put into effect its new source performance standards (NSPS). These regulations call for removal of 70–90% of the SO_x from stack gases in all new power plants, with the percentage removal depending on whether low- or high-S fuel is burned. NO_x emissions must be no more than 38% and 50% of SO_x emissions on a weight basis for oil-fired plants and most coal-fired plants, respectively (EPA, 1991).

The NSPS were a landmark piece of legislation, because they were the first significant effort by the U.S. government to mandate a reduction in SO_x and NO_x emissions from stationary sources. However, they fell short of what environmentalists had hoped to see because projections made in 1979 indicated that even with the NSPS in effect, SO_x emissions in the United States would increase by 10% from 1975 to 1995 (Carter, 1979). Although the calculated increase would have been 28% without the NSPS, merely reducing the increase from 28% to 10% was not what environmentalists had in mind. The problem was that the NSPS applied only to power plants on which construction had begun after August 17, 1971. Since fossil fuel plants typically remain in service for about 40 years (Carter, 1979), it was obvious that NO_x and SO_x emissions from old plants would continue to be a problem for several decades unless steps were taken to control their emissions. Indeed, some of the older plants would have been in flagrant violation of NSPS requirements. Carter (1979), for example, noted that the Ohio Edison Company's Burger plant on the Ohio–West Virginia border emitted 10 times the amount of SO_x permitted by the NSPS.

This problem was finally addressed in 1990, when amendments to the Clean Air Act were passed mandating reductions in SO_x and NO_x emissions from older power plants. Under the new act, each power plant was given a reduction target, and utilities could earn credits that could be sold or traded if they exceeded the targets (Anonymous, 1991). Thus, utilities were given an incentive not only to meet but also to exceed the target reductions. It was hoped that this approach would encourage innovation and energy efficiency. The new act was expected to reduce SO_x and NO_x emissions from stationary sources by almost 50% and 15–30%, respectively (Anonymous, 1991). The actual reduction in SO_x emissions has exceeded these expectations (Kerr, 1998).

CASE STUDY: THE NETHERLANDS

This chapter has tended to focus on conditions in the United States, which emits more SO_x and NO_x than any other nation on Earth.[1] The problems caused by such emissions, however, are more severe in parts of Europe and the former Soviet Union, which collectively emit three to four times as much SO_x as the United States (Postel, 1984). It is therefore of some interest to examine how conditions in these countries have evolved and how the problem of acid deposition is being addressed. Here we examine the case of the Netherlands, which has the second highest SO_x deposition rate of any nation in Western Europe.

Following World War II, coal accounted for about 85% of Dutch fuel consumption. However, during the next 15 years, oil gradually displaced coal as the major fuel source. In 1959, the largest natural gas field in the world was discovered in the province of Groninger in the Netherlands. Exploitation of this field began in 1963 (Vermeulen, 1980). The resultant pattern in energy use is illustrated in Figure 15.8.

Between 1946 and 1956, SO_x emissions in the Netherlands increased by about 125%, and between 1956 and 1967 by another 110% (Vermeulen, 1980). By 1967, reported SO_x concentrations in Dutch air were sometimes as high as 500 to 2,000 $\mu g\ m^{-3}$. These figures may be compared to the current EPA primary standards, which require that 24-hour average SO_2 concentrations not exceed 365 $\mu g\ m^{-3}$ more than once per year and that average annual concentrations not exceed 80 $\mu g\ m^{-3}$. As noted by Vermeulen (1980, p. 27),

[1]This statement became true only after the breakup of the former Soviet Union.

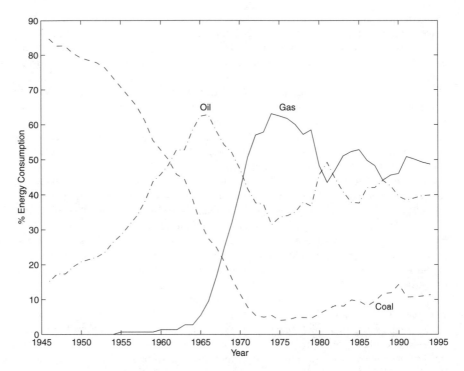

FIGURE 15.8   The share of coal, oil, and natural gas in Dutch fuel consumption from 1946 to 1978. *Source*: Vermeulen (1979) and various editions of the *United Nations Energy Statistics Yearbook*, Department of International Economic and Social Affairs, United Nations.

"In order to keep the Netherlands habitable, drastic measures were required immediately." In 1968 the Dutch Clean Air Act was passed. The act was aimed in part at reducing SO_x emissions. Although the act did not become effective until December 29, 1970, industries rapidly began cleaning up their emissions in anticipation of the act's implementation. Although some emission reductions were achieved by building desulfurization installations and by using low-S oil, the principal mechanism for reducing SO_x emissions was to switch to natural gas as a fuel. Between 1967 and 1977, the percentage of the Netherlands fuel consumption accounted for by natural gas rose from 17% to 60% (Figure 15.8). As a result, SO_x emissions dropped by almost 60%.

In response to the Arab oil boycott and the general instability in oil prices, the Dutch government decided that it would be wise to conserve the nation's natural gas as a strategic reserve. As a result of this shift in policy, Dutch power stations were required to decrease their natural gas consumption, and the percentage of Dutch energy consumption accounted for by gas rapidly dropped to 45–50% (Figure 15.8). This decline was compensated for largely by an increase in the consumption of coal and oil, both of which have a higher S content. According to Vermeulen (1980), in the absence of additional SO_x and NO_x emission controls, this switch back to oil and coal would have led to an increase of 230% in SO_x emissions by 1985 and an additional increase of 100% between 1985 and 2000. In summarizing this situation, Vermeulen (1980, p. 48) commented, "With coal and oil as the most important fuels in the years to come it is obvious that a large and costly effort will have to take place in the field of desulfurization. . . . Because most desulfurization methods have not yet been technically perfected to a degree that the efficiency is sufficiently high and the costs are acceptable for large-scale application, in my opinion hardly any positive effect can be expected before 1985. After 1985, the acidification of precipitation will strongly depend on the level of technology of desulfurization and the velocity with which these installations will be built. Except for the costs, there must be the political will to handle the problem vigorously, and that is a separate problem. . . . In my opinion, the outlook for the Netherlands with respect to the acidification of precipitation is pretty grim."

Vermeulen's bleak assessment has unfortunately proven to be qualitatively correct. A serious effort was made to reduce SO_x emissions in the Netherlands, and the result was a 41% decline in Dutch SO_x emissions between 1980 and 1985 (Wollast, 1989). Almost 70% of the decrease was due to a drop in SO_x emissions by utilities. During the same time interval, Belgium, West Germany, France, and the United Kingdom, which collectively are responsible for more than half of the acid rain that falls in the Netherlands, achieved comparable reductions in SO_x emissions of 42%, 25%, 48%, and 24%, respectively (Wollast, 1989). Unfortunately, there was very little reduction in NO_x emissions, either in the Netherlands or in neighboring countries, and as a result, the overall deposition of acid in the Netherlands declined by only about 16% between 1980 and 1986. According to Wollast (1989), about a 75% reduction in acid deposition in the Netherlands is needed to prevent damage to sensitive ecosystems, particularly coniferous forests and heathland on poor sandy soils.

It is problematic at this time whether such a reduction in acid deposition can be achieved in the Netherlands. Certainly some effort must be made to reduce NO_x emissions, about 50% of which are accounted for by motor vehicles. According to Wollast (1989), modification of combustion conditions and the fitting of three-way catalytic converters on heavier passenger automobiles would reduce NO_x emissions by 14% in the year 2000. In the case of SO_x, a limited reduction in the S content of oil, flue gas scrubbing, and a limited substitution of natural gas for heavy fuel oil would reduce SO_x emissions by 30% between 1985 and 2000. Even if neighboring countries responded with similar reductions, acid deposition in the Netherlands would still be over three times the target figure in the first decade of the 21st century. Further reductions in NO_x emissions could be achieved with the use of selective catalytic reduction for stack gases and the requirement of catalytic converters on all passenger automobiles, but these measures would still leave acid deposition at two and a half times the target level by the year 2010. The implication, according to Wollast (1989), is that technical measures alone will not solve the acid deposition problem in the Netherlands and that some effort to reduce energy consumption will probably be required to bring deposition rates within desired limits. According to Wollast (1989, p. 95), "The picture presented by these scenarios does not give direct cause for optimistic expectations for the future development of acidification and its effects."

COMMENTARY

Does the Netherlands case study presage the future of acid rain problems in other parts of the world? On a global scale, there is considerable uncertainty in the outlook for SO_x and NO_x emissions. The human population is expected to reach 8 billion by 2025, and virtually the entire increase will be in the less developed nations (PRB, 1999). Per capita emissions of SO_x and NO_x in Asia are currently about 10% of the corresponding figures for North America (Galloway, 1989). It is probably unrealistic to assume that per capita emissions in Asia will remain so low for the next 20 years, but it is very difficult to say how

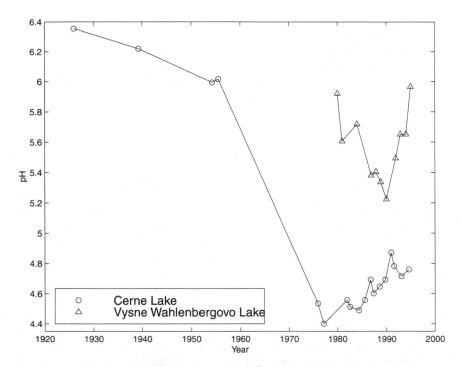

FIGURE 15.9   Changes in the pH of Cerné Lake (Šumava Mountains) and Vyšné Wahlenbergovo Lake (High Tetra Mountains) in central Europe. *Source*: Kopácek et al. (1998). Reproduced with permission.

much they will change. Given modest increases in Asian per capita SO_x and NO_x emissions, Galloway (1989) has estimated that global SO_x and NO_x emissions in 2020 could be three times the corresponding figures for 1980. That is a grim prognosis.

Even in areas where there has recently been a significant reduction in SO_x and NO_x emissions, the response of biological systems has sometimes been sluggish. Kopácek et al. (1998), for example, document the increase in pH of high-elevation lakes in central Europe associated with political and economic changes in postcommunist countries (Figure 15.9). SO_x and NO_x emissions declined by 30–40%, yet despite limnetic pH increases of 0.4–0.8, there has been no significant change in the biota of the lakes. Further increases in pH are unlikely, since NO_x emissions from mobile sources will probably increase in the near future due to economic growth.

With respect to NO_x emissions, Likens et al. (1998) note that nitric acid currently contributes about 45% of the rainfall in the Hubbard Brook watershed and is expected to be the dominant acid in precipitation during the first decade of the 21st century. They go on to point out that "There is no cap on total U.S. NO_x emissions" (p. 1991). Perhaps there should be. There is a disturbing parallel between SO_x and NO_x emissions, on the one hand, and point and nonpoint source pollution, on the other. Both SO_x emissions and point source pollution have been relatively easy to deal with technologically. We now realize, however, that control of point source discharges and SO_x emissions may not be sufficient to solve some important eutrophication and acid rain problems. Lake Erie and the Netherlands are cases in point.

REFERENCES

ANONYMOUS. 1991. New Clean Air Act will reduce acid rain, then cap emissions. *Environ. Defense Fund Letter*, **23**(1), 1, 5.

BARRETT, E., and G. BRODIN. 1955. The acidity of Scandanavian precipitation. *Tellus*, **7**, 1–257.

BISCHOFF, W. D., V. L. PATERSON, and F. T. MACKENZIE. 1984. Geochemical mass balance for sulfur- and nitrogen-bearing acid components: Eastern United States. In O. P. Bricker (Ed.), *Geological Aspects of Acid Deposition*. Acid Precipitation Series, Vol. 7. Butterworth, Pp. 1–21.

BOMBERGER, D. C., and R. C. PHILLIPS. 1979. Technological options for mitigation of acid rain. In: C. G. Gunnerson and B. E. Willard (Eds.), *Acid Rain*. American Society of Civil Engineers, New York. Pp. 132–166.

BRIMBLECOMBE, P. 1977. London air pollution, 1500–1900. *Atmos. Environ.*, **11**, 1157–1162.

BRIMBLECOMBE, P. 1978. Air pollution in industrializing England. *J. Air. Pollut. Control Assoc.* **28**, 115–118.

CANADA TODAY. 1981. How many more lakes have to die? *Canada Today*, **12**(2). Canadian Embassy, Washington, DC.

CARTER, L. J. 1979. Uncontrolled SO_2 emissions bring acid rain. *Science*, **204**, 1179–1182.

CHARLSON, R. J., J. E. LOVELOCK, M. O. ANDREAE, and S. G. WARREN. 1987. Oceanic phytoplankton, atmospheric sulphur, cloud albedo and climate. *Nature*, **326**, 655–661.

CHARLSON, R. J., and H. RODHE. 1982. Factors controlling the acidity of rainwater. *Nature*, **295**, 683–685.

COWLING, E. B. 1982. A status report on acid precipitation and its biological consequences as of April 1981. In F. M. D'Itri (Ed.), *Acid Precipitation*. Ann Arbor Science, Ann Arbor, MI, Pp. 3–20.

CRONAN, C. S., and C. L. SCHOFIELD. 1979. Aluminum leaching response to acid precipitation: Effects on high-elevation watersheds in the northeast. *Science*, **204**, 304–305.

DIGNON, J., and S. HAMEED. 1989. Global emissions of nitrogen and sulfur oxides from 1860 to 1980. *J. Air Waste Manage. Assoc.*, **39**(2), 180–186.

ENVIRONMENT CANADA. 1981. *Downwind. The Acid Rain Story*. Cat. No. En 56-561 1981E. Minister of Supply and Services Canada, Ottawa, Ontario. 20 pp.

EPA. 1980. *Acid Rain*. Environmental Protection Agency, Washington, DC. 36 pp.

EPA. 1991. Standard for sulfur dioxide. *Code of Federal Regulations*, **40**(60), 250.

EPA. 1997. *National Air Pollutant Emission Trends, 1900–1996*. EPA-454/R-97-011. United States Environmental Protection Agency Office of Air Quality Planning and Standards, Research Triangle Park, NC.

FLAVIN, C., and A. B. DURNING. 1988. *Building on Success: The Age of Energy Efficiency*. Worldwatch Paper 82. Worldwatch Institute, Washington, DC. 74 pp.

FRANCIS, A. J., H. L. QUINBY, and G. R. HENDREY. 1984. Effect of lake pH on microbial decomposition of allochthonous litter. In G. R. Hendrey (Ed.), *Early Biotic Responses to Advancing Lake Acidification*. Acid Precipitation Series, Vol. 6. Butterworth, Boston Pp. 1–21.

GALLOWAY, J. N. 1979. Acid precipitation: Spatial and temporal trends. In C. G. Gunnerson and B. E. Willard (Eds.), *Acid Rain*. American Society of Civil Engineers, New York. Pp. 1–20.

GALLOWAY, J. N. 1989. Atmospheric acidification: Projections for the future. *Ambio*, **18**(3), 161–166.

GALLOWAY, J. N., H. LEVY II, and P. S. KASIBHATLA. 1994. Year 2020: Consequences of population growth and development on deposition of oxidized nitrogen. *Ambio*, **23**, 120–123.

GORHAM, E. 1955. On the acidity and salinity of rain. *Geochim. Cosmochim. Acta*, **7**, 231–239.

GSCHWANDTNER, G., J. K. WAGNER, and R. B. HUSAR. 1988. *Comparison of Historic SO_2 and NO_x Emission Data Sets*. EPA/600/S7-88/009. Environmental Protection Agency, Research Triangle Park, NC.

HAMILTON, L. D. 1979. Health effects of acid precipitation. In *Proceedings, Action Seminar on Acid Precipitation*. Environment Canada, Toronto. Pp. 117–134.

HENDREY, G. R. 1983. Automobiles and acid rain. *Science*, **222**, 8.

HOUGHTON, H. 1955. On the chemical composition of fog and cloud water. *J. Meteorol.*, **12**, 355–357.

JOHNSON, N. M. 1979. Acid rain: Neutralization within the Hubbard Brook ecosystem and regional implications. *Science*, **204**, 497–499.

KERR, R. A. 1998. Acid rain control: Success on the cheap. *Science*, **282**, 1024–1027.

KOHOUT, E. J., D. J. MILLER, L. A. NIEVES, D. S. ROTHMAN, D. L. SARICKS, F. STODOLSKY, and D. A. HANSON. 1990. *Current Emission Trends for Nitrogen Oxides, Sulfur Dioxide, and Volatile Organic Com-*

pounds by Month and State: Methodology and Results. ANL/EAIS/TM-25. Policy and Economic Analyis Group, Environmental Assessment and Information Sciences Division, Argonne National Laboratory, Argonne, IL.

KOIDE, M., and E. D. GOLDBERG. 1971. Atmospheric sulfur and fossil fuel combustion. *J. Geophys. Res.,* **76,** 6589–6596.

KOPÁCEK, J., J. HEJZLAR, E. STUCHLÍK, J. FOTT, and J. VESELÝ. 1998. Reversibility of acidification of mountain lakes after reduction in nitrogen and sulphur emissions in Central Europe. *Limnol. Oceanogr.,* **43,** 357–361.

LANGNER, J. 1991. Sulfur in the troposphere: A global three-dimensional model study. Doctoral thesis, Department of Meteorology, Stockholm University.

LEIVESTAD, H., G. HENDRY, I. MUNIZ, and E. SNEKVIK. 1976. Effects of acid precipitation on freshwater organisms, In F. H. Braekke (Ed.), Acid Precipitation—Effects on Forest and Fish Johs. Grefslie Trykkeri A/S, Mysen, Norway SNSF Research Report 6/76. Pp. 87–111.

LI, Y. 1992. Seasalt and pollution inputs over the continental United States. *Water, Air, Soil Pollut.* **64,** 561–573.

LIKENS, G. E., H. BORMAN, J. S. EATON, R. S. PIERCE, and N .M. JOHNSON. 1976. Hydrogen ion input to the Hubbard Brook experimental forest, New Hampshire, during the last decade. *Water, Air Soil Pollut.,* **6,** 435.

LIKENS, G. E., F. H. BORMANN, and N. M. JOHNSON. 1972. Acid rain. *Environment,* **14,** 33–40.

LIKENS, G. E., K. C. WEATHERS, T. J. BUTLER, and D. C. BUSO. 1998. Solving the acid rain problem. *Science,* **282,** 1991–1992.

LIKENS, G. E., R. F. WRIGHT, J. N. GALLOWAY, and T. J. BUTLER. 1979. Acid rain. *Sci. Amer.,* **241**(4), 43–51.

MACKENZIE, F. T., W. D. BISCHOFF, and V. B. PATTERSON. 1984. *Geochemical Cycles and Trends in Estimates of Inputs of Anthropogenic Chemical Constituents to the Environment.* Ecosystems Research Center, Cornell University, Ithaca, NY.

MAHONEY, J. 1991. Quoted by Roberts (1991), p. 1303.

MARMOREK, D. R. 1984. Changes in the temporal behavior and size structure of plankton systems in acid lakes, In G. R. Hendrey (Ed.), *Early Biotic Responses to Advancing Lake Acidification.* Acid Precipitation Series, Vol. 6. Butterworth, Boston. Pp. 23–41.

MARSHALL, E. 1984. Canada goes it alone on acid rain controls. *Science,* **223,** 1275.

MCCANDLESS, L. C., A. B. ONURSAL, and J. M. MOORE. 1987. *Assessment of Coal Cleaning Technology: Final Report.* EPA/600/S7-86/037. Environmental Protection Agency, Research Triangle Park, NC.

MCCORMICK, J. 1989. *Acid Earth: The Global Threat of Acid Pollution,* 2nd ed. Earthscan Publications, Ltd., London. 225 pp.

ODEN, S. 1968. *The Acidification of Air and Precipitation, and Its Consequences in the Natural Environment.* Ecological Committee Bulletin No. 1, National Science Research Council, Stockholm (Arlington, VA: Translation Consultants, TR-1172).

PARK, C. C. 1987. *Acid Rain: Rhetoric and Reality.* Methuen, London. 272 pp.

POPULATION REFERENCE BUREAU. 1999. World Population data sheet. Washington, DC. *http://www.prb.org*

POSTEL, S. 1984. *Air Pollution, Acid Rain, and the Future of Forests.* Worldwatch Paper 58. Worldwatch Institute, Washington, DC. 54 pp.

ROBERTS, L. 1991. Learning from an acid rain program. *Science,* **251,** 1302–1305.

SCHOFIELD, C. L., and J. R. TROJNAR. 1980. Aluminum toxicity to fish in acidified waters. In T. Y. Toribara, M. W. Miller, and P. E. Morrow (Eds.), *Polluted Rain.* Plenum Press, New York. Pp. 347–366.

SMITH, R. A. 1852. On the air and rain of Manchester. *Mem. Manchester Lit. Phil. Soc.,* Ser. 2, **10,** 207–217.

SMITH, R. A. 1872. *Air and Rain.* Longmans, Green, London. 600 pp.

SMITH, R. J. 1980. Acid rain agreement. *Science,* **209,** 890.

SPENCER, D. F., S. B. ALPERT, and H. H. GILMAN. 1986. Cool water: Demonstration of a clean and efficient new coal technology. *Science,* **232,** 609–612.

VERMEULEN, A. J. 1980. The acidic precipitation phenomenon: A study of this phenomenon and the relationship between the acid content of precipitation and the meission of sulfur dioxide and nitrogen oxides in the Netherlands. In T. Y. Toribara, M. W. Miller, and P. E. Morrow (Eds.), *Polluted Rain.* Plenum Press, New York. Pp. 7–60.

WELLFORD, WEGMAN, KRULWICH, GOLD, and HOFF. 1982. *Fact Sheet on Acid Rain.* Canadian Embassy, Washington, DC. 8 pp.

WOLLAST, R. 1989. Acification. In I. F. Langeweg (Ed.), *Concern for Tomorrow*. National Institute of Public Health and Environmental Protection, Brussels. Pp. 79–96.

QUESTIONS

15.1 During the 1970s, rainfall in the northeastern United States was much more acidic during the summer than during the winter. During the same time period, the acidity of rainfall in the western United States showed little seasonal variability. How would you account for these facts?

15.2 Water is considered to be neutral if the pH is 7.0, but acid rain is often defined as rainwater with a pH of less than 5.6. Why?

15.3 Why are high-elevation lakes generally more susceptible to acidification than low-elevation lakes?

15.4 Why is wet stack gas scrubbing much more effective in removing SO_x than NO_x from stack gases?

15.5 How does fluidized bed combustion reduce the production of NO_x during the combustion of coal?

15.6 How does fluidized bed combustion reduce the production of SO_x during the combustion of coal?

15.7 Why is pretreatment and conversion of coal to a gas or liquid not a practical way to reduce emissions of NO_x from electric power plants?

15.8 What does the acronym LIMB stand for, and how does it help to reduce emissions of SO_x and NO_x from electric power plants that use coal as a fuel?

15.9 What was the principal criticism by environmental groups of the NSPS promulgated by the EPA in 1979?

Chapter 16

✦ ♦ ✦ ♦ ✦

Groundwater Pollution

GROUNDWATER POLLUTION is an environmental problem that has attracted significant national attention only in recent years. The incident that most effectively focused public interest on the problem was the contamination of the Love Canal in the city of Niagara Falls, New York. About 20,000 tonnes of hazardous wastes were buried in 55 gallon drums in the abandoned canal during the 1930s and 1940s by the Hooker Chemical Company. The accumulated waste was covered with soil, and the land was sold to the Niagara Falls Board of Education for $1 in 1953. The Board of Education proceeded to build a school and playground over the old canal site, and the surrounding area was gradually developed for residential homes. Several years of heavy precipitation leading up to 1977 saturated the old canal site with water, and toxic chemicals began to leak out of the corroding drums buried in the canal. The resultant pollution caused the New York State health commissioner to declare a state of emergency, and President Jimmy Carter shortly thereafter declared Love Canal a national disaster area. A $15 million fund was established to purchase 550 homes within a 30-square-block area surrounding the dumpsite (Epstein et al., 1982), and hundreds of residents were evacuated and resettled. The impact of this chemical waste dump on both the physical and emotional well-being of the Love Canal residents has been a subject of much controversy (Culliton, 1980) and will probably never be fully understood.

Incidents such as the pollution of Love Canal have made it clear how serious a problem the careless disposal of hazardous wastes can be. Federal regulations enacted during the 1970s have helped to restrict air and surface-water contamination, and as a result, there has been an increase in subsurface disposal. When hazardous wastes are buried in the ground, there is an excellent chance that groundwater will become contaminated unless special precautions are taken to minimize this probability. Unfortunately, it is often extremely difficult and expensive to clean up groundwater once it has become contaminated, and because dilution effects are minimal, "concentrations of contaminants are often much higher in groundwater than in surface water" (Patrick et al., 1987, p. 56). These realizations have led scientists and environmental groups to begin an examination of the degree of groundwater contamination in the United States, to study the potential for future contamination, and to carefully assess our present use of and future needs for groundwater.

RELIANCE ON GROUNDWATER

Groundwater accounts for about 96% of all freshwater in the United States, the remainder being present in lakes and streams. It is estimated that $1.2–3.8 \times 10^{17}$ L of groundwater lie within about 0.75 km of the surface in the United States (Patrick et al., 1987). Groundwater accounts for about 22% of all freshwater withdrawals in the United States, including 34% of all agricultural use (mostly for irrigation), 40% of public water withdrawals, and 53% of all drinking water (Solley et al., 1998). About 97% of the rural population depends on groundwater for drinking water, and according to Pye et al. (1983, p. 38), "of the major cities in the United States, 75% depend upon well water for most of their supplies." As Figure 16.1 indicates, in recent years, withdrawals of freshwater from the

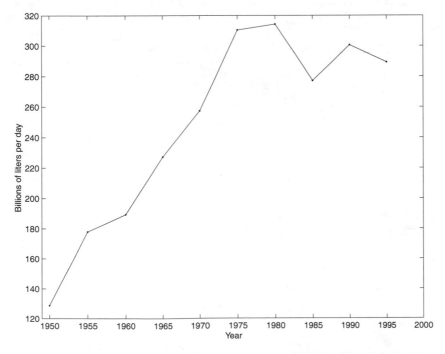

FIGURE 16.1   Groundwater withdrawals of fresh water in the United States from 1950 to 1995. *Source*: Solley (1998).

ground in the United States have been about 300 billion L d^{-1}, which is roughly 10% of the effective recharge rate. The leveling off of withdrawals since 1975 has been due largely to more frugal and efficient use of water for irrigation, which presently accounts for about 64% of all groundwater use in the United States (Solley et al., 1998). California, Texas, and Nebraska, where groundwater is used extensively for irrigation, together account for 38% of all groundwater use in the United States (GRF, 1984).

General Aquifer Information

What is groundwater? According to Pye et al. (1983, p. 2), groundwater "is water that occurs in permeable saturated strata of rock, sand, or gravel called aquifers." The two main types of aquifers, confined and unconfined, are illustrated in Figure 16.2. An *unconfined aquifer* is not overlain by impermeable material and may be recharged by percolation of rainwater through the overlying soil and rock. The water in the aquifer is at atmospheric pressure. A *perched aquifer* may occur where there is a limited layer of impermeable material above the water table. *Confined* or *artesian aquifers* are bounded at the top and bottom by geologic formations of low permeability called *aquitards*. The water in the artesian aquifer is under greater than atmospheric pressure. Such aquifers may have no recharge areas or may have "discrete and variable recharge areas where the geologic material of the aquifer forms an outcrop at the surface" (Patrick et al., 1987, pp. 2–3). Most aquifers oc-

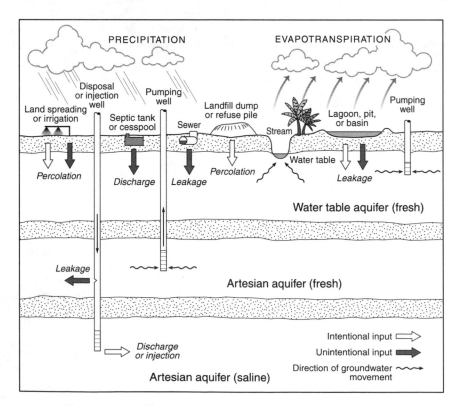

FIGURE 16.2   Diagram of confined and unconfined aquifers and disposal methods which can contaminate them. *Source*: Redrawn from Burmaster (1982). Used with permission of Heldref Publications and *Environment Magazine*.

cur within 0.75 km of the Earth's surface (Burmaster, 1982), and in areal extent they may vary greatly from very small to as large as several hundred thousand square kilometers. For example, the Ogallala aquifer in the High Plains of the United States covers an area of 4.5×10^5 km^2 and has an average thickness of 60 meters. It varies in thickness from less than 30 cm in some areas to as much as 0.30–0.45 km in parts of Nebraska.

An important point about all aquifers is that recharging of the aquifer can occur only when precipitation exceeds evapotranspiration in the aquifer's recharge area. In the arid southwestern United States, for example, aquifer recharge occurs mainly in wet or multi-year cycles (Patrick et al., 1987). If use of the aquifer is to be sustained over long periods of time, it is therefore critical that withdrawals from the aquifer not exceed average recharge rates. The Ogallala aquifer, for example, was seriously overdrawn for many years, primarily for irrigation purposes. The average thickness of the aquifer declined by about 3 m between the 1940s and 1980s, with some areas in Texas declining nearly 30 m. By 1985 it was estimated that without an imported water supply and/or change in water use, about 40% of the land then being irrigated in the High Plains of West Texas would have to be converted to dryland farming or be abandoned by the year 2000, and 60% by the year 2020. In fact, Gaines County, Texas, had completely dried up its groundwater resources by the early 1980s (Pye et al., 1983).

Realizing the severity of the problem, farmers began to implement improved irrigation practices, and during the 1980s the average thickness of the aquifer declined by only another 30 cm (HPWD, 1998). Water levels actually increased in some areas. Nevertheless, the net depletion rate of the aquifer during the 1990s averaged about 4 km^3 y^{-1}, and researchers continue to search for ways to improve water-use efficiency and increase natural recharge rates. Obviously, overdrafts can result in the eventual elimination of the groundwater resource, as in the case of Gaines County, Texas. In other cases, overdrafts result in contamination of fresh groundwater with saline groundwater, which may underlie lenses of freshwater, particularly in coastal areas.

The problem of overdrafting for irrigation purposes appears to have peaked in the United States in the early 1970s, when Burmaster (1982, p. 8) commented, "In 1975 approximately 25 percent of all groundwater withdrawals were overdrafts, mostly in agricultural regions." At that time, for example, high-pressure, hand-moved sprinklers had evaporation losses of as much as 50%. Current state-of-the-art low-pressure, full-dropline, center-pivot systems are roughly 95% efficient, and buried drip lines approach 100% efficiency (HPWD, 1998).

EXTENT OF GROUNDWATER POLLUTION

Groundwater pollution is caused by a variety of mechanisms. In some cases, entirely natural processes seriously contaminate groundwater. For example, in the southwestern and south central United States, natural leaching of chloride, sulfate, nitrate, fluoride, iron, and arsenic ions and radioactivity from naturally occurring deposits of uranium have created water quality problems in certain areas (GRF, 1984). However, most concern over groundwater contamination has centered on pollution associated with human activities. Sources of contamination associated with human activities but not directly related to waste disposal include accidents, certain agricultural activities, mining, highway deicing, acid rain, saltwater intrusion, and improper well construction and maintenance. Sources associated with waste disposal include private sewage disposal systems, land disposal of solid wastes, municipal wastewater, wastewater impoundments, land spreading of sludge, brine disposal from the petroleum industry, mine wastes, deep-well disposal of liquid wastes, animal feedlot wastes, and radioactive wastes (GRF, 1984). The following discussion gives some idea of the amount of hazardous waste generated by these activities.

Septic Tanks

Almost 25% of the U.S. population uses septic tanks for sewage treatment. These tanks handle the equivalent of about 40 billion L d^{-1} of municipal sanitary sewage. For septic tanks to work properly, it is important that a zone of unsaturated soil occur between the leach bed and the water table so that the effluent from the septic tank does not enter an unconfined aquifer. In a 1980 study, the EPA (1980a) concluded that one-third of all septic tank installations in the United States were operating improperly. Based on studies conducted by the EPA during the 1970s, improperly operating septic tanks were judged to be a major cause of groundwater contamination in the northwestern, northeastern, and southeastern United States (GRF, 1984). Elevated concentrations of pathogens and nitrate are problems typically associated with septic tank pollution, but, "more recently, widespread use of septic tank cleaners containing degreasing agents such as trichloroethylene has resulted in groundwater contamination by synthetic organic chemicals" (GRF, 1984, p. 27).

Saltwater Contamination

Contamination of fresh groundwater by saltwater can occur by several mechanisms. Overdrafting is one obvious mechanism. Most freshwater aquifers are underlain by brackish or saline groundwater, and in general, salinity can be expected to increase with depth. A U.S. Geological Survey inventory reported 114 cases of saltwater intrusion in 1977 (EPA, 1977a). Use of salt for highway deicing is another source of potential contamination. Approximately 9 million tonnes of deicing salts are applied each year to roads in the United States (Groundwater Foundation, 1999). When this salt washes off roads, it can easily move with percolating water into underground aquifers. An additional problem is created by the fact that piles of salt to be used for deicing have frequently been stored uncovered along roads. Rain or snowmelt can dissolve this salt and, through percolation, introduce it into aquifers. Groundwater contamination caused by the disposal of oil-field brines has been documented in at least 21 states (Patrick et al., 1987). At the present time, the ratio of brine to crude oil recovered is on the order of 10:1, but it may be much higher (e.g., 100:1) in older wells (Patrick et al., 1987). In the early days of oil and gas exploration, these brines were disposed of in unlined pits. Unfortunately, these so-called evaporation pits were often quite leaky, and it was easy for the brines to percolate into the ground and contaminate groundwater. Use of such pits has been wholly or at least partially eliminated in all parts of the Unitd States. Texas, for example, banned unlined pits in 1969. Today brines are usually disposed of by injection into deep underground formations considered unsuitable for other purposes, or by reinjection into oil-producing formations to enhance recovery. However, as noted by Patrick et al. (1987, p. 73), "These injection practices, if not properly controlled, pose a potential threat to groundwater." The SDWA standard for drinking water is 500 ppm of total dissolved solids (TDS), a figure which corresponds to a salinity less than 2% that of seawater.

Sewage

Improper disposal of municipal wastewater and of both human and animal waste can pose a threat to groundwater. The total volume of sewage handled by municipal sewage treatment plants in the United States is about 130 billion L d^{-1}. The principal components of domestic sewage that are regarded as a threat to groundwater are nitrate, heavy metals, and pathogens. Manure generated at feedlots is a similar threat to groundwater from the standpoint of nitrate and pathogen contamination. Between 1985 and 1996, there were 53 outbreaks of pathogen-related diseases in the United States that were attributed to consumption of untreated groundwater. More than 6,100 cases of illness were associated with these outbreaks. Mechanisms by which sewage may contaminate groundwater supplies include leaks in sewer lines due to age; disruption by tree roots, seismic activity, or poor construction; leakage from inadequately sealed sewage lagoons; and improper land disposal of treated wastewater.

Mining Activities

Mining activities can contaminate groundwater in several ways. Coal is often found in deposits that contain iron pyrites (FeS_2) and other sulfides. When these sulfides are exposed to O_2, as a result of mining activities, they become oxidized in the presence of water and form sulfuric acid (H_2SO_4). The result can be highly acidic groundwater and is a serious problem in the coal mining areas of Pennsylvania and Appalachia (Patrick et al., 1987). Drainage water and wastewater from metal mines can seriously contaminate groundwater because these waters typically contain high concentrations of toxic heavy metals. U and Cu mines are cited by Patrick et al. (1987) as producing drainage and waste water that pose the most serious threats to groundwater because the polluted water contains high concentrations of both dissolved toxic substances (e.g. As, H_2SO_4, Cu), as well as radioisotopes. Metal mine leachate has increased the concentration of Mn in wells in Washington, caused As poisoning of cattle in Idaho, and increased the radioactivity of groundwater in Wyoming (Patrick et al., 1987).

Toxic Chemicals

Contamination of groundwater by toxic organic chemicals is a problem that has only recently attracted much attention, primarily because concentrations of organic chemicals have not routinely been measured in groundwater. Recent groundwater monitoring has shown that toxic organic chemicals are an increasing problem, but according to Patrick et al. (1987, p. 254), "To date only about 10% of the organic chemicals contaminating drinking water in the United States have been identified." The combined effects of dilution, biological activity, and chemical reactions are generally much less efficient in reducing the concentrations of toxic chemicals in groundwater than in surface water. As a result, groundwater concentrations of certain toxic organic chemicals in some areas are often considerably higher than the concentrations of the same chemicals found in raw or treated drinking water drawn from the most contaminated surface supplies (Burmaster, 1982). Monitoring the concentrations of toxic chemicals in water supplies is difficult because approximately 50,000–75,000 chemicals are in use and are being distributed through the environment at the present time, and an additional 700–800 are added each year (Patrick et al., 1987). Although not all of these chemicals are toxic, merely keeping up-to-date information on the possible toxicity of so many compounds is a prodigious task. Table 16.1 gives some idea of the variety and concentration of toxic organic chemicals that have been detected in drinking-water wells in the United States. Also included are the maximum contaminant levels established by the EPA under the Safe Drinking Water Act, the potential health effects from ingestion, the principal uses of the various chemicals, and some common sources of drinking water contamination. It is obvious from the data in Table 16.1 that a number of toxic chemicals have somehow found their way into water supplies at concentrations that vastly exceed any reasonable safe level. How have certain groundwater supplies come to be so grossly contaminated with toxic chemicals?

TABLE 16.1   TOXIC ORGANIC CHEMICALS FOUND IN
DRINKING-WATER WELLS

Trichloroethylene (TCE)
Principal uses
    Mainly as a solvent for vapor degreasing in metal industries. Also used to extract caffeine from
    coffee, as a dry cleaning agent, and as a chemical intermediate in the production of pesticides,
    waxes, gums, resins, tars, paints, varnishes, and specific chemicals such as chloroacetic acid.
Potential health effects
    Liver problems; increased risk of cancer
Sources of contaminant in drinking water
    Discharge from petroleum refineries
SDWA maximum contaminant level (ppb): 5

Reported concentrations (ppb)	*State*
27,300	Pennsylvania
14,000	Pennsylvania
3,800	New York
3,200	Pennsylvania
1,530	New Jersey
900	Massachusetts

Toluene
Principal uses
    About 70% is converted to benzene. About 15% is used to produce chemicals such as toluene
    diisocyanate, phenol, benzyl and benzoyl derivatives, benzoic acid, toluene sulfonates, nitro-
    toluenes, vinyltoluene, and saccharin. Remainder used as solvent for paints and coatings and as
    a component of automobile and aviation fuel.
Potential health effects
    Nervous system, kidney, or liver problems
Sources of contaminant in drinking water
    Discharge from petroleum refineries
SDWA maximum contaminant level (ppb): 1,000

Reported concentrations (ppb)	*State*
6,400	New Jersey
260	New Jersey
55	New Jersey

1,1,1-Trichloroethane
Principal uses
    Widely used as a substitute for carbon tetrachloride. Also used in liquid form as a degreaser and
    for cold cleaning, dip-cleaning, and bucket cleaning of metals. Also used as a dry cleaning
    agent, vapor degreasing agent, and propellant.
Potential health effects
    Liver, nervous system, or circulatory problems
Sources of contaminant in drinking water
    Discharge from metal degreasing sites and other factories
SDWA maximum contaminant level (ppb): 200

Reported concentrations (ppb)	*State*
5,440	Maine
5,100	New York
1,600	Connecticut
965	New Jersey

Ethyl Benzene
Principal uses
    Used in the manufacture of cellulose acetate, styrene, and synthetic rubber. Also as a solvent or
    diluent and as a component of automotive and aviation fuel. Present in significant quantities in
    mixed xylenes, which are used as diluents in the paint industry, in insecticides, and in gasoline.

TABLE 16.1   (Continued)

Potential health effects
    Liver or kidney problems
Sources of contaminant in drinking water
    Discharge from petroleum refineries
SDWA maximum contaminant level (ppb): 700

Reported concentrations (ppb)	State
2,000	New Jersey

Tetrachloroethylene
Principal uses
    Solvent with particular use as a dry cleaning agent, degreaser, chemical intermediate, fumigant, and medically as an anthelmintic.
Potential health effects
    Liver problems; increased risk of cancer
Sources of contaminant in drinking water
    Discharge from factories and dry cleaners
SDWA maximum contaminant level (ppb): 5

Reported concentrations (ppb)	State
1,500	New Jersey
740	Connecticut
717	New York

Carbon Tetrachloride
Principal uses
    Solvent for oils, fats, lacquers, varnishes, rubber, waxes, and resins. Synthesis of fluorocarbons. Also used as an azeotropic drying agent for spark plugs, a dry cleaning agent, a fire extinguishing agent, a fumigant, and an anthelmintic.
Potential health effects
    Liver problems; increased risk of cancer
Sources of contaminant in drinking water
    Discharge from chemical plants and other industrial activities
SDWA maximum contaminant level (ppb): 5

Reported concentrations (ppb)	State
400	New Jersey
135	New York

Benzene
Principal uses
    Constituent of motor fuels, solvent for fats, inks, oils, paints, plastics, and rubber. Used in the extraction of oils from seeds and nuts and in photogravure printing. Also used as a chemical intermediate. By alkylation, chlorination, nitration, and sulfonation, chemicals such as styrene, phenols, and maleic anhydride are produced. Also used in the manufacture of detergents, explosives, pharmaceuticals, and dyestuffs.
Potential health effects
    Anemia; decrease in blood platelets; increased risk of cancer
Source of contaminant in drinking water
    Discharge from factories; leaching from gas storage tanks and landfills
SDWA maximum contaminant level (ppb): 5

Reported concentrations (ppb)	State
330	New Jersey
230	New Jersey
70	Connecticut
30	New York

TABLE 16.1   (Continued)

1,2-Dichloroethylene
Principal uses
    Solvent for waxes, resins, and acetylcellulose. Used in extraction of rubber, as a refrigerant, in the manufacture of pharmaceuticals and artificial pearls, and in the extraction of oils and fats from fish and meat.
Potential health effects
    Liver problems
Source of contaminant in drinking water
    Discharge from industrial chemical factories
SDWA maximum contaminant level (ppb): 70–100

Reported concentrations (ppb)	State
323	Massachusetts
294	Massachusetts
91	New York

Ethylene Dibromide
Principal uses
    Principally used as a fumigant for ground pest control and as a constituent of ethyl gasoline. Also used in fire extinguishers, gauge fluids, and waterproofing preparations. Also used as a solvent for celluloid, fats, oils, and waxes.
Public health effects
    Stomach problems; reproductive difficulties; increased risk of cancer
Source of contaminant in drinking water
    Treatment for ground termites; discharge from petroleum refineries
SDWA maximum contaminant level (ppb): 0.05

Reported concentrations (ppb)	State
300	Hawaii
100	Hawaii
35	California

1,1-Dichloroethylene
Principal uses
    Chemical intermediate in the synthesis of methylchloroform and in the production of polyvinylidene chloride copolymers, which are components of Saran Wrap, and of polymer coatings of ship tanks, railroad cars, fuel storage tanks, and for coating steel pipes and structures.
Potential health effects
    Liver problems
Sources of contaminant in drinking water
    Discharge from industrial chemical factories
SDWA maximum contaminant level (ppb): 7

Reported concentrations (ppb)	State
280	New Jersey
118	Massachusetts
70	Maine

1,2-Dichloroethane
Principal uses
    Used in the manufacture of ethylene glycol, diaminoethylene, PVC, nylon, viscose rayon, styrene-butadiene rubber, and various plastics. Solvent for resins, asphalt, bitumen, rubber, cellulose acetate, cellulose ester, and paint. Degreaser in engineering, textile, and petroleum industries. Extracting agent for soybean oil and caffeine. Antiknock agent in gasoline, a pickling agent, fumigant, and dry cleaning agent. Used in photography, xerography, water softening, and in the production of adhesives, pharmaceuticals, and varnishes.

TABLE 16.1   (Continued)

Potential health effects
    Increased risk of cancer
Sources of contaminant in drinking water
    Discharge from industrial chemical factories
SDWA maximum contaminant level (ppb): 5

Reported concentrations (ppb)	*State*
250	New Jersey

Vinyl Chloride
Principal uses
    Used as a vinyl monomer in the manufacture of PVC and synthetic rubber. Also used as chemical intermediate and solvent.
Potential health effects
    Increased risk of cancer
Sources of contaminant in drinking water
    Leaching from PVC pipes; discharge from plastic factories
SDWA maximum contaminant level (ppb): 2

Reported concentrations (ppb)	*State*
50	New York

Lindane (gamma isomer of benzene hexachloride [BHC])
Principal uses
    Insecticide used on seed and soil treatments, foliage application, wood protection
Potential health effects
    Liver or kidney problems
Sources of contaminant in drinking water
    Runoff/leaching from insecticide used on cattle, lumber, gardens
SDWA maximum contaminant level (ppb): 0.2

Reported concentrations (ppb)	*State*
22	New York

1,1,2-Trichloroethane
Principal uses
    Chemical intermediate and solvent, but not used as widely as the 1,1,1 isomer
Potential health effects
    Liver, kidney, or immune system problems
Sources of contaminant in drinking water
    Discharge from industrial chemical factories
SDWA maximum contaminant level (ppb): 5

Reported concentrations (ppb)	*State*
20	New York

Source: Pye et al. (1983), EPA (1999a).

Disposal of toxic chemicals as well as many other hazardous wastes in improperly constructed and/or supervised landfills and surface impoundments appears to be one major mechanism by which groundwater has become contaminated. As of 1980, the EPA estimated that there were a total of 200,000 landfills and dumps in the United States receiving 135 million tonnes y^{-1} of municipal solid wastes and 215 million tonnes y^{-1} of industrial solid wastes. In addition, about 176,000 surface impoundments located at 78,000 different sites were estimated to be receiving 38 trillion (i.e., 38×10^{12}) L y^{-1} of liquid wastes.

Certainly not all wastes are hazardous. However, the EPA (1980a) has estimated that about 55 million tonnes of hazardous waste are generated each year in the United States.

These wastes include substances that are toxic (e.g., pesticides, PCBs, toxic organic chemicals, and heavy metals), reactive (e.g., wastes that have a tendency to explode), ignitible (e.g., benzene, toluene, paint and varnish removers), corrosive (e.g., alkaline cleaners, acid liquids), infectious (e.g., improperly treated sewage sludges), or radioactive (Durso-Hughes and Lewis, 1982). According to the EPA (1980a), these hazardous wastes are generated at about 750,000 different sites in the United States. Larger industrial firms, at least in the past, tended to dispose of wastes in landfills on their own property, and according to the EPA (1980a), about 50,000 such sites have been used for the disposal of hazardous wastes. Of these 50,000 sites, about 1,200 are considered to pose a threat to the environment (Clay, 1990).

Further information on the quality of landfill and surface impoundments is revealing. A 1978 Waste Age survey identified 15,000 active municipal landfills in the United States. According to the EPA (1980b), only 31% of these landfills met state regulations. Of the industrial waste impoundments surveyed by the EPA (1980a), about one-third contained liquid wastes with potentially hazardous constituents, 70% were unlined, only 5% were known to be monitored for groundwater quality, and 30% were both unlined and located on permeable ground overlying usable aquifers. A study of over 80,000 surface impoundment sites of all types (e.g., industrial, municipal, agricultural) revealed that almost 40% were "potentially hazardous" (Pye et al., 1983, p. 60).

Roughly 10% of all industrial waste produced is hazardous (Durso-Hughes and Lewis, 1982). The industries responsible for the majority of the hazardous waste production are (in descending order by volume of hazardous waste produced) the organic chemical industry, the primary metals industry, the electroplating industry, the inorganic chemical industry, the textile industry, oil refineries, and the rubber and plastics industry (Durso-Hughes and Lewis, 1982). About 60% of industrial hazardous waste in the United States is generated by the chemical and allied products industry (EPA, 1980d). The type of hazardous waste generated varies greatly between one industry and the next. For example, cyanide wastes are associated with metallurgical operations; sulfite wastes are generated by paper and pulp manufacturing; Hg is a common waste product in the electrical industry; and the petrochemical industry produces a wide variety of toxic organic wastes ranging from pesticides and PCBs to phenol-rich tar wastes (Patrick et al., 1987).

Illegal Disposal

Unfortunately, disposal of wastes in improperly constructed and/or supervised land disposal sites is not the only mechanism by which hazardous wastes find their way into groundwater. As the costs associated with the proper disposal of hazardous wastes have increased, illegal dumping has become an increasingly attractive alternative. Some of the most widely publicized examples of illegal waste disposal operations have involved cases in New Jersey and Pennsylvania, but "illegal dumping operations have been identified in Kentucky, Texas, Ohio, Michigan, California, North Carolina, and many other states" (Epstein et al., 1982, p. 177). Patrick et al. (1987) documented 44 case histories of illegal dumping in New Jersey, although in only 5 of these cases was groundwater was clearly affected. One of the best-known examples of illegal dumping in New Jersey involved a facility operated by Chemical Control in Elizabeth. When state inspectors visited the facility in the winter of 1979, they found about 40,000 rusty and leaky 55 gallon drums of hazardous industrial waste, in some places stacked five high and packed together like sardines in a can. "In a loft in the incinerator building on the site, investigators discovered about a hundred pounds of explosive dried picric acid; several pounds of radioactive material; several large bottles of a liquid labeled 'Nitro'; over twelve hundred 'lab pacs,' packages of material wastes from research laboratories containing toxic agents; dozens of containers of explosive compressed gases; 'leaking containers of chromic acid, isopropanol, mineral spirits and petroleum naphtha, all designated hazardous substances'; and cylinders of mustard gas, 'a highly toxic nerve gas.' Also found inside the building were two other storage areas, labeled by the state as the 'pesticide room' and the 'boiler room', each crammed with rusted and leaking drums" (Epstein et al., 1982, p. 155).

The chemical wastes stored at the Chemical Control facility clearly constituted a hazard to the environment and to human health. The chemicals stored at the site included substances known to cause mucous membrane and respiratory tract irritation; cardiovascular, hepatotoxic, and neurotoxic effects; blood disorders; allergies; and skin disorders; and to be carcinogenic, mutagenic, teratogenic, and toxic to unborn children. In addition, state experts indicated that there was a serious risk of a fire or explosion at the site (Epstein et al., 1982). A 5 m high-pressure gas line was located less than 200 m from the site, and a large liquid natural gas tank, several bulk gasoline storage tanks, and 10 propane gas tanks were all located within about 600 m of the site. Since a school and a large residential area were located within 300 m of the facility, a serious fire or explosion at the site could have killed thousands of people. After several unsuccessful attempts to get the company to clean up the dump, the state finally took control of the site and initiated a cleanup. However, before the cleanup was completed, a serious fire did break out on April 11, 1980. Flames reached 100 m into the air, and smoke clouds rolled 24 km out to sea. Firemen battled the blaze for 10 hours before bringing it under control. Fortunately, 10,000 drums of the most hazardous waste had been removed from the site before the fire, and the wind was blowing offshore at the time of the blaze. The operator of the Chemical Control dump was convicted in 1978 on three counts of illegal dumping and was sentenced to a two-year prison term (Epstein et al., 1982).

This particular incident gives some idea of the magnitude of hazardous wastes that may accumulate at an illegally operated disposal site. Although the principal threats at the Chemical Control site were from fire, explosions, and air pollution, other illegal dumping operations have posed a direct threat to water supplies. For example, ABM Disposal, a Philadelphia-based waste-handling operation, was found guilty of numerous illegal dumping operations that contaminated groundwater. In 1977 ABM haulers were convicted in three separate illegal dumping incidents, and in March 1977, "ABM personnel were convicted for illegally dumping dangerous pharmaceutical wastes into a groundwater well located in a garage in Montgomery County" (Epstein et al., 1982, p. 163). In another incident, an ABM truck was discovered dumping cyanide waste containing "many, many times the lethal dose" of cyanide into Ridley Creek, a favorite swimming spot for children in the Chester, Pennsylvania, area. According to K. Welks, attorney for Pennsylvania's Department of Environmental Resources, "If anyone had been swimming at the time, it could have killed them" (cited by Epstein et al., 1982, p. 163). In another incident involving ABM, drums containing tonnes of waste and provided by ABM were stacked at a facility operated by Eastern Rubber Reclaiming in downtown Chester, Pennsylvania. The operator of the facility "made more money on the drums, 'whenever he needed it,' by opening them and pouring the contents into the ground or into lagoons on his property and selling the empty drums at $6.00 each to a local recycler. Tanker trucks filled with hazardous wastes were also allowed to discharge their contents on the grounds of Eastern Rubber Reclaiming" (Epstein et al., 1982, p. 164). The facility was raided in the fall of 1977, but before a proper inventory and cleanup could be completed, the site exploded in flames on February 2, 1978. The fire destroyed most of the drums on site, although about 4,500 remained intact, and "when the smoke cleared, inspectors also found several tanker trucks filled with waste abandoned at the site" (Epstein et al., 1982, p. 165). Cleanup costs for the Eastern Rubber Reclaiming site were about $1.5 million.

These examples give some idea of the magnitude of the illegal dumping problem. Illegal dumping is big business, and there is a strong suggestion that organized crime may have become involved in parts of the country. Furthermore, "We are now seeing a trend away from midnight dumping of drums off the backs of trucks toward more complex sewering operations, in which it is extremely difficult to locate the underground disposal line carrying the waste away from the disposal site" (Epstein et al., 1982, p. 177). Major causes of illegal dumping have undoubtedly been industry irresponsibility or ignorance, but government laxity is also a factor. The Department of Justice during the Reagan administration indicted only about 30 entities and individuals per year for improper disposal of hazardous wastes. This figure doubled during the Bush administration (E. Boling, Dept. of Justice, pers. comm.) How serious a problem is groundwater pollution? Unfortunately, no comprehensive national survey of groundwater contamination has been undertaken in the United States. Furthermore, only a small percentage of the contaminants now being found in groundwater has been tested for human health effects. Hence, it is very difficult to make a quantitative statement about the national risk associated with drinking contaminated groundwater. The EPA did commission five regional groundwater assessment studies during the 1970s (GRF, 1984) and has conducted eight national drinking water surveys since 1975 (EPA, 1987). Unfortunately, the regional studies and half of the drinking water surveys were completed before incidents such as those at Valley of the Drums[1] and Love Canal called national attention to the problems created by improper disposal of hazardous waste. There is little doubt that the conclusions of the surveys have been biased to some extent by the types of pollutants that investigators at the time of the studies considered most likely to be found in the waters they tested. If a comprehensive national study were conducted today, it seems likely that a more thorough search would be made for a variety of toxic organic chemicals that recent studies have revealed in water supplies contaminated by improper disposal of chemical wastes.

What clearly emerges from the reports and studies is that incidents of groundwater contamination have occurred in every state and that incidents are beginning to occur with increasing frequency (Patrick et al., 1987). Industrial impoundments, land disposal sites, and septic tanks and cesspools appear to be the most important sources of groundwater contamination on a national level. Municipal wastewater, petroleum exploration, and mining are of secondary importance as causes of contamination (EPA, 1980c). The four pollutants most commonly reported have been nitrates, heavy metals, microorganisms, and organic chemicals (Patrick et al., 1987). Human and animal wastes were judged to be of primary or secondary importance as causes of groundwater contamination in all five of the regional EPA surveys, and were considered to be among the top three contaminants in every state surveyed except California in a similar study conducted by the Environmental Assessment Council (GRF, 1984). Industrial wastes are also considered to be high-priority contaminants in most regions of the country and are judged to be the biggest source of groundwater contamination in the Northeast. In the south-central and southwestern portions of the country, disposal of oil-field brines appear to be a major source of groundwater pollution and accounts for a high percentage of industry-related contamination. With the exception of industrial, human, and animal wastes, the important sources of contamination vary considerably from one region of the country to another. For example, some coastal states have serious problems with saltwater intrusion; chloride contamination from road salts is a serious problem in some snowbelt states; and high concentrations of dissolved solids are a pollution problem in areas with soluble aquifers (GRF, 1984).

What percentage of the aquifers in the United States are contaminated? Several estimates have been made of the areal extent of contamination of usable surface aquifers by surface impoundments and landfills (considered the most important sources) and by sub-

---

[1]A 3 ha site in Kentucky at which 17,000 waste drums contaminated surface waters with some 200 organic chemicals and 30 metals.

surface disposal systems, petroleum exploration, and mining (secondary sources). An EPA study (EPA, 1980c) put the percentage of contaminated aquifers at 0.5–1%; a similar analysis by Lehr (1982) indicated that up to 2% of the aquifers might be contaminated. As noted by the Geophysics Research Forum (GRF, 1984, p. 4), "Although this might not seem to indicate a large problem, much of the contamination occurs in areas of heaviest reliance on groundwater." The following examples provide some indication of the extent of the problem.

1. In Nassau and Suffolk counties on Long Island, more than 36 community wells were closed in 1980 due to contamination with tetrachloroethylene (TCE), trichloroethane, trichloroethene, and other volatile synthetic organic compounds. More than 3 million people in these counties depend on groundwater as their sole source of drinking water. The Council on Environmental Quality (CEQ, 1981) estimated that over 2 million persons were affected by these well closings.
2. In January 1980, California public health officials closed 39 public wells in 13 cities in the San Gabriel Valley because of TCE contamination. Over 400,000 people were affected by the well closing (Burmaster, 1982). Nineteen wells were closed in rural California east of Sacramento (Roberts, 1981).
3. In Jackson Township, New Jersey, about 100 drinking water wells were closed because of organic chemical contamination apparently caused by illegal disposal of chemicals at a landfill (Burmaster, 1982). Throughout the state, about 400 municipal and 40 private wells have been closed because of groundwater contamination (Roberts, 1981).
4. In May 1978, four wells providing 80% of the drinking water for Bedford, Massachusetts, were closed. Toxic organic chemicals including up to 2,100 ppb of dioxane and up to 500 ppb of TCE were found in the water (Burmaster, 1982). Throughout the state, 22 public and private wells were closed because of groundwater contamination. One-third of the commonwealth's 351 communities were affected by the closings (Roberts, 1981).
5. In Washington County, Illinois, 81% of 221 dug wells and 34% of drilled wells during the 1970s had a nitrate N concentration in excess of 10 ppm, the maximum allowable concentration recommended by the EPA for public water supplies (Pye and Patrick, 1983).

These examples indicate that groundwater pollution can and has affected the water supplies of large numbers of persons in some areas, despite the fact that current estimates of the areal extent of contamination are only 1–2% for the United States as a whole. The previous discussion and Figure 16.2 indicate that there is a wide variety of mechanisms by which pollutants can be introduced into aquifers. However, both the regional studies conducted by the EPA during the 1970s and more recent information have indicated that improper disposal of hazardous waste, particularly industrial hazardous waste, is the principal mechanism responsible for groundwater contamination on a national level. For example, a report to the Committee on Environmental and Public Works of the U.S. Senate, which reviewed a number of case histories of groundwater contamination that resulted in well closings, implicated the following types of substances in the indicated number of cases: metals, 619; organics, 242; insecticides, 201; chlorides, 26; nitrates, 23 (Pye and Patrick, 1983). The nature of the most frequently cited contaminants strongly suggests that industrial waste disposal was the cause of pollution in most cases. In this respect, it is noteworthy that only about 10% of hazardous waste in the United States is disposed of in an environmentally sound method (Wood et al., 1984). Cost has been a major factor in determining the type of disposal method employed. As noted by the EPA (1980d, p. 15), "Environmentally sound technologies are available for treatment and disposal of hazardous waste. Costs vary widely, according to type and volume of waste handled, and are substantially in excess of unsound practices." The irony of this statement is that we are now beginning to realize that the true cost of improper hazardous waste disposal is usually far in excess of the cost of proper waste disposal when allowance is made for the cost of polluted water supplies, damage to the environment and human health, and cleanup (if cleanup is possible). The following case study provides a good illustration of these points.

CASE STUDY: THE ROCKY MOUNTAIN ARSENAL

The Rocky Mountain Arsenal is a 70 km^2 facility located adjacent to the northeast sector of the city of Denver, Colorado (Figure 16.3). The Arsenal was operated for 40 years by the U.S. Army for the production of chemical warfare agents. It was a principal center for the production of nerve gas and its constituents and for emptying canisters of unused mustard gas. In 1947 the Shell Oil Company began leasing part of the land for the production of herbicides and associated chemicals, a practice that continued until 1982. From 1943 to 1956, liquid wastes were discharged into one or another of several hundred unlined ponds (Figure 16.4). The water contained a variety of complex organic and inorganic compounds, including proven or suspected carcinogens such as toluene, trichlorobenzene, vinyl chloride, xylene, carbon tetrachloride, chlordane, and aldrin (Anonymous, 1983a). The waste-

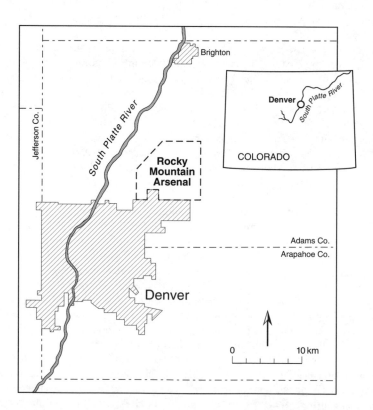

FIGURE 16.3   Location of the Rocky Mountain Arsenal and the city of Denver, Colorado. *Source*: Reprinted with permission from Konikow and Thompson (1984). Copyright 1984 by the National Academy of Sciences. Courtesy of the National Academy Press, Washington, DC.

water was rather easy to trace, since it contained a high chloride concentration, sometimes as high as 5 parts per thousand (5 $^{0}/_{00}$).

Groundwater flow in the region of the Arsenal is toward the northwest. Much of the land to the north of the Arsenal is irrigated farmland (Figure 16.4), and damage to crops from contamination of the irrigation water was apparent by 1951 (Konikow and Thompson, 1984). Similar damage was reported in 1952 and 1953. Particularly severe crop damage occurred in 1954, when precipitation was less than half the average value and use of groundwater for irrigation was therefore greater than usual. As a result of this repeated crop damage, several investigations were done to determine the extent of the groundwater contamination. A study by Petri and Smith (1956), for example, showed that contaminated groundwater extended over an area of several square kilometers to the north and northwest of the unlined disposal ponds.

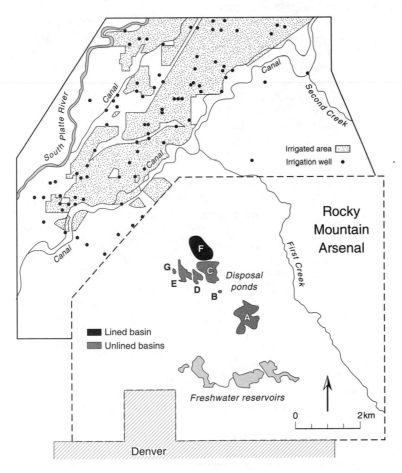

FIGURE 16.4   Location of evaporation ponds within the Rocky Mountain Arsenal and irrigated lands (stipled areas) to the north and west of the Arsenal. Dots indicate locations of irrigation wells. The evaporation pond (F) is the asphalt-lined pond. The other evaporation ponds are unlined. *Source*: Reprinted with permission from Konikow and Thompson (1984). Copyright 1984 by the National Academy of Sciences. Courtesy of the National Academy Press, Washington, DC.

As a result of these studies, several steps were taken to alleviate the groundwater pollution problem. First, a 40 ha asphalt-lined evaporation pond was constructed, and liquid wastes were discharged to that pond beginning in 1956 (Figure 16.4). Second, from 1968 or 1969 to about 1974, pond C (Figure 16.4) was maintained in a full condition most of the time by pumping in water from the freshwater reservoirs to the south. Water from pond C infiltrated into the ground at the rate of about 28 L s^{-1}, and thus helped to dilute and flush contaminated groundwater. According to Konikow and Thompson (1984, p. 95), "By 1972 the areal extent and magnitude of contamination, as indicated by chloride concentration, had significantly diminished. Chloride concentrations were then above 1,000 mg/L [1 $^0/_{00}$] in only two relatively small parts of the contaminated area and were almost at normal background levels in the middle of the affected area."

Unfortunately, the asphalt-lined pond ultimately developed leaks, and there were new claims of crop damage in 1973 and 1974. In response to these new claims, the Colorado Department of Health conducted a study that revealed the presence of a variety of contaminants in wells downgradient from the disposal ponds. For example, diisopropylmethylphosphonate (DIMP), a nerve gas by-product, was detected at a concentration of 0.57 ppb in a well located about 13 km downgradient from the disposal ponds and 1.6 km upgradient from two municipal water supply wells. A DIMP concentration of 48 ppm was measured in a groundwater sample collected near the disposal ponds. Other organic contaminants detected in wells or springs in the area included dicyclopentadiene (DCPD), endrin, aldrin, dieldrin, and several organo-sulfur compounds (Konikow and Thompson, 1984).

As a result of these discoveries, the Colorado Department of Health issued cease and desist, cleanup, and monitoring orders in April 1975 to the Rocky Mountain Arsenal and Shell Oil Company. The cease and desist order required a halt to unauthorized discharges of contaminants into both surface water and groundwater north of the Arsenal. Continued monitoring of groundwater revealed the presence of even more contaminants, including Nemagon (dibromochloro-propane) and several industrial solvents. Water samples taken from several hundred observational wells both within the Arsenal boundary and to the north and west provided a clear picture of the extent of contamination.

In response to the orders from the Colorado Department of Health, a computer model of groundwater flow was developed in order to predict the effect of possible remedial actions. The model predicted, for example, that it would probably take many decades for the contaminated aquifer to recover naturally. However, it was also predicted that certain water management policies could reduce the recovery time to a matter of years. The solution ultimately adopted consisted of several components and was implemented in several phases. First, a dike was built to intercept the flow of contaminated surface water. Second, an impervious barrier consisting of a 1 m wide, 7.6 m deep and 0.5 km long trench filled with a mixture of soil and clay was built and anchored 0.6 m deep into the bedrock along the northern boundary of the Arsenal. Groundwater was then pumped from the south side (upgradient) of the barrier using six 20 cm diameter wells spaced at equal intervals parallel to the barrier. The water was pumped at the rate of about 38 m^3 h^{-1} through two columns of granular activated C to remove organics. Each column contained about 10 tonnes of activated C. Finally, the treated water was reinjected by gravity flow into the ground on the north side (downgradient) of the barrier using twelve 0.5 m diameter injection wells. This initial system is indicated schematically in Figure 16.5. The activated C columns were replaced whenever the concentration of DIMP in the treated water reached 50 ppb. In practice, activated C use rates ranged from 100 to 150 mg of C per liter of water treated (Konikow and Thompson, 1984). A commercial vendor regenerated the exhausted carbon.

This system was operated for a total of three years, and preliminary results were sufficiently encouraging that an expanded containment system was constructed. The larger system consisted of a 2.1 km long barrier of soil and clay, 7.6 to 15.2 m deep, with 54 with-

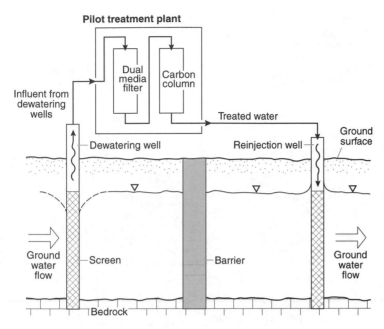

FIGURE 16.5 Schematic diagram of the barrier and treatment system installed along the northern boundary of the Rocky Mountain Arsenal. *Source*: Reprinted with permission from Konikow and Thompson (1984). Copyright 1984 by the National Academy of Sciences. Courtesy of the National Academy Press, Washington, DC.

drawal wells and 38 reinjection wells along the northern boundary of the Arsenal (Figure 16.6). Organics were removed from the water by three pulsed-bed adsorbers that contained 13.5 tonnes of activated C each. These new adsorbers were anticipated to be about four times more efficient than the cartridge filters used in the smaller pilot system. The new system became operational in 1983 and was capable of treating 136 m^3 of water per hour. The cost of the expanded system was $6 million.

Subsequently, two other containment and treatment systems were built in addition to the initial north boundary system. A system built by Shell and completed in 1983 is located at one corner of the western boundary of the Arsenal (Figure 16.6). Like the original north boundary system, this so-called Irondale system consisted of a series of dewatering wells, a liquid treatment facility, and recharge wells. However, the Irondale system did not involve the use of a soil and clay slurry trench to block groundwater flow, but instead depended entirely on the dewatering wells to intercept groundwater movement. A system built by the U.S. Army and completed in 1985 is located along a portion of the northwest boundary of the Arsenal (Figure 16.6) and consists of components similar to those used in the north boundary system. Both of these latter two systems were designed primarily to control the movement of Nemagon across the Arsenal boundary (Konikow and Thompson, 1984).

Although these containment measures worked reasonably well and despite the fact that the Rocky Mountain Arsenal was officially closed in 1982, considerable additional effort would have been required to comply with the cease and desist, cleanup, and monitoring orders issued in 1975 by the Colorado Department of Health. Federal law would have required that the site be returned to its original state (Anonymous, 1983a). Initial estimates of the cost of such a complete cleanup were $2–6 billion (Anonymous, 1983b), and in 1983 the U.S. Army sued Shell for $1.8 billion in damages, claiming that most of

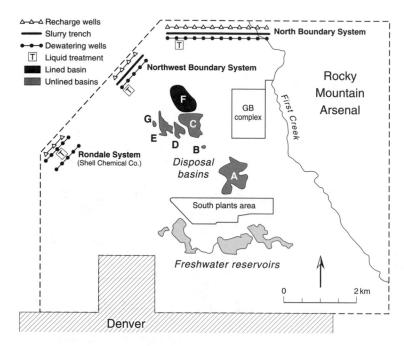

FIGURE 16.6   Location of various barrier and treatment systems at boundaries of the Rocky Mountain Arsenal. *Source*: Reprinted with permission from Konikow and Thompson (1984). Copyright 1984 by the National Academy of Sciences. Courtesy of the National Academy Press, Washington, DC.

the wastes were the result of Shell's manufacturing and packaging of pesticides. Ultimately, Shell and the Army reached an agreement whereby Shell would pay for 50% of the first $500 million in cleanup costs, 35% of the next $200 million, and 20% of any cleanup costs in excess of $700 million. The final price tag for complete restoration of the site was expected to be $1 billion (Anonymous, 1988). In 1984 the Rocky Mountain Arsenal was added to the EPA's National Priorities List (NPL), meaning that it was a cleanup priority in the eyes of the EPA. The Rocky Mountain Arsenal became (and still is) the largest military site in the United States on the NPL.

The high cost of restoring the Rocky Mountain Arsenal has led to some very creative legal and political maneuvering. In part because of Shell's influence with the Colorado congressional delegation, Congress passed legislation in 1992 designating the Arsenal a national wildlife refuge upon completion of the environmental cleanup. What was the motivation for this designation? Because the Arsenal had been turned into a wildlife refuge, cleanup standards were relaxed from a residential level of remediation to a less stringent level for wildlife (Sierra Club, 1999). That translates into reduced remediation costs. According to the Sierra Club (1999, p. 2), "The mandated refuge interferes and limits almost every cleanup decision. . . . Citizens who have worked on cleanup issues at Rocky Mountain Arsenal believe that wildlife refuge status hinders cleanup by requiring habitat to be 'protected' from cleanup activities. One of the unintended consequences of this action is that contaminated habitat will not be remediated, and will have long-lasting effects on wildlife."

As a part of the ongoing restoration, the Army transported some 76,000 drums of corrosive salts from the Arsenal to a landfill in Utah (Anonymous, 1988), and a special

2.5–5.0 km^2 landfill was constructed to store contaminated soil excavated from the Arsenal (Anonymous, 1984). The liquid wastes in Basin F proved to be a particularly difficult problem because of the tendency of the basin to leak. A temporary solution was to store the wastes in aboveground tanks, but the 32 million L of toxic liquid were so corrosive that they threatened to eat through the tanks by mid-1993 (Anonymous, 1990). An agreement was therefore reached to burn the wastes in a 1,050°C incinerator, the total cost of building the special incinerator and burning the wastes being about $35 million (Anonymous, 1990).

In considering the cost of containing the flow of contaminated groundwater from the Rocky Mountain Arsenal and ultimately restoring the site after manufacturing operations had ceased, one is reminded of a comment by Wood (1972, p. 24), who noted, "The most satisfactory cure for groundwater pollution is prevention." Furthermore, regardless of whether groundwater pollution problems are solved by prevention or by cleanup, experience has shown that in many cases legal action is required to produce a satisfactory response to the problem. The case of the Rocky Mountain Arsenal provides a good illustration of this point. The cleanup did not begin until the Colorado Department of Health issued cease and desist, cleanup, and monitoring orders. With this point in mind, it is useful to ask how well suited present laws are for dealing with groundwater pollution problems and how vigorously the government is enforcing these laws.

LEGAL CONSIDERATIONS

At the present time, there is no single agency responsible for the protection of groundwater, and as a result, effective management of contamination problems has sometimes been difficult. Furthermore, there is no federal program specifically designed to protect groundwater. Most of the concern at the federal level has been with problems associated with industrial hazardous wastes. Nevertheless, despite the lack of agencies, programs, or legislation specifically focused on the problem of groundwater pollution, there are a number of laws that, if effectively enforced, would provide all of the legal muscle needed to deal with groundwater pollution. The following is a summary of the most important pieces of legislation.

RCRA

The Resource Conservation and Recovery Act (RCRA) was passed by Congress in 1976 to provide for careful planning and management practices in the treatment, storage, and disposal of both municipal solid wastes and hazardous wastes. As noted by Pye et al., (1983, p. 243), RCRA "is foremost among federal statutes aimed at minimizing groundwater contamination." The two key features of RCRA are the Subtitle C program, which governs hazardous waste management, and the Subtitle D program, which concerns municipal solid waste disposal. Subtitle C authorizes the EPA to regulate hazardous wastes "from cradle to grave." Subtitle C directs the EPA to establish standards for all hazardous waste management facilities and to incorporate these standards into RCRA permits for individual facilities. After November 19, 1980, it was unlawful to treat, store, or dispose of any hazardous waste without an RCRA permit.

While Subtitle C sounds like a potentially very effective piece of legislation, there have been some fairly serious problems in its implementation. There were more than 10,000 hazardous waste facilities in existence at the time of the effective date of the program, and it was impossible for the EPA to thoroughly review the permit applications of these facilities overnight. Hence, interim status standards were established, and facilities that filed a timely application were eligible to continue operating under an interim status until their permit applications were processed. The interim status standards included requirements for waste analysis and inspection, groundwater monitoring, and other "housekeeping" measures, and also established certain technical standards.[2] However, with Anne Burford running the EPA as of May 1981, even these interim status standards were not effectively enforced. For example, in April 1982, the EPA suspended the requirement that companies submit an annual report describing in detail how much and what kind of hazardous waste they handled, and suspended the requirement that industries file a yearly report on groundwater conditions near a waste site (Sun, 1982). Congress reacted in 1984 by amending the RCRA so as to cancel the interim status for land disposal facilities that had failed to submit a final permit application by November 8, 1985 (Patrick et al., 1987). As a result of this requirement, and in particular because of the difficulty in obtaining commercial pollution liability insurance, about two-thirds of the interim status land disposal facilities were forced to close down in 1985.

After much delay, the EPA finally promulgated hazardous waste facility permitting standards on July 26, 1982 (the regulations were due in 1978). The criteria became effective in January 1983. The permitting standards contain two key elements: a liquids management strategy and a groundwater monitoring and response program. The liquids management strategy requires the use of liners and leachate collection systems on new surface impoundments, waste piles, and landfills. Although existing facilities are not required to employ liners and leachate collection systems, both new and existing facilities are subject to groundwater monitoring requirements, "which can lead to the imposition of treatment or removal requirements when groundwater is contaminated" (Pye et al., 1983, p. 249).

Congress intended that administration of the Subtitle C program would ultimately pass to the states, and this has now been done. However, the EPA maintains oversight over the Subtitle C program. State-administered Subtitle C programs require EPA approval and must be equivalent to the federal Subtitle C program, consistent with the federal program and other state programs, and must provide for adequate enforcement.

The Subtitle D program involving municipal solid waste is also envisioned as being administered largely by the states, although with guidance from the EPA. Under Subtitle D the states are required to develop solid waste management plans, and the EPA is required to provide technical and financial assistance in the development of those plans. The key element in the Subtitle D program is the establishment of guidelines for classifying facilities as either open dumps or sanitary landfills. A facility may be classified as a sanitary landfill "only if there is no reasonable probability of adverse effects on health or the environment from the disposal of solid waste at such facility." States with approved Subtitle D plans are to classify all municipal solid waste facilities as either open dumps or sanitary landfills. Any facilities that are classified as open dumps must be either upgraded or closed. Contamination of groundwater at a facility is considered to have occurred if the concentration of any pollutant at the "solid-waste boundary" exceeds the maximum contaminant levels (MCLs) specified by the SDWA. An obvious flaw in the law is that MCLs have been set for only 16 inorganic and 54 organic chemical contaminants and a handful of radionuclides and microbial indicator organisms. This is a small fraction of the contaminants that may turn up in groundwater.

A potentially powerful feature of RCRA is the imminent-hazard provision, which authorizes the EPA administrator to bring suit in district court to restrain handling of solid or hazardous waste at a facility if such handling poses an "imminent and substantial endangerment to health or the environment." The EPA may also issue administrative orders "as may be necessary to protect public health and the environment." Clearly, RCRA is a

---

[2]For example, containers holding hazardous waste must be in good condition and must not leak.

potentially powerful tool for dealing with groundwater pollution from hazardous and municipal solid waste.

SDWA

Congress enacted the Safe Drinking Water Act (SDWA) in 1974 to provide for unpolluted drinking water supplies. Three features of the SDWA are of particular importance.

First, the SDWA establishes a set of primary and secondary drinking water standards (i.e., MCLs). The primary drinking water standards apply to contaminants that the EPA has determined to have an adverse effect on human health. The 1986 amendments to the SDWA and subsequent modifications have resulted in standards for a total of 33 organic pesticides, 21 volatile organics, 16 inorganic chemicals, radioactivity, and microbiological indicator organisms (EPA, 1999a). The secondary drinking water standards apply to contaminants or characteristics that might adversely affect the odor or appearance of drinking water to such an extent that a substantial number of persons would not drink the water. To date, the EPA has established a total of 15 secondary MCLs.

Second, the SDWA contains an underground-injection-control (UIC) program that gives the EPA direct authority over underground waste disposal wells. The UIC program was the first federal effort directly aimed at the control of groundwater pollution. Although the provisions of the RCRA would be sufficient to regulate underground waste disposal wells, in practice the UIC program of the SDWA has been used to regulate underground injection of wastes. Like the Subtitle C and D sections of RCRA, the UIC program is designed to be administered by the states. According to the provisions of UIC, underground injection is prohibited without an authorized state permit, and in order to obtain a state permit, an applicant must satisfy the state that drinking water sources will not be endangered by the injection of waste. The state, in turn, must adopt drinking water standards in conformance with the federal standards and implement procedures for the enforcement of those standards in order to obtain EPA authorization to enforce the drinking water standards and the provisions of UIC.

The third important feature of SDWA is the "sole-source aquifer" program. This program is specifically aimed at groundwater protection and applies to areas that have only one aquifer as a principal source of drinking water. No new underground injection wells can be drilled in such areas without a permit, and if the EPA determines that a sole-source aquifer would, if contaminated, create "a significant hazard to public health," then "no commitment of federal financial assistance through grants, contracts, or loan guarantees may be given to any program which EPA determines may contaminate such an aquifer so as to create a significant hazard to human health" (Patrick et al., 1987, p. 383).

CWA

The Clean Water Act (CWA), which was passed by Congress in 1972, is the most comprehensive federal water pollution control program in existence, but it is concerned primarily with the reduction and control of pollutant discharges into navigable[3] waters. The CWA can, however, affect groundwater pollution problems indirectly to the extent that certain instances of groundwater contamination result from leaching of polluted surface waters into an aquifer. Furthermore, the CWA contains several provisions that are directly relevant to groundwater pollution. First, Section 104 of the CWA requires the EPA to "estab-

---

[3]*Navigable* was defined in 1986 in a very broad way to include waters having some association with commerce. The definition includes, for example, mud flats and intermittent streams.

lish, equip and maintain a water quality surveillance system for the purpose of monitoring navigable waters and ground waters." In practice, however, Section 104 has been implemented primarily with respect to surface waters, with little attention given to groundwater. Second, Section 208 of the CWA calls for the development and implementation of areawide plans for the management of waste treatment. These plans are to be worked out through a cooperative effort between the EPA and the states. Section 208 is potentially the most effective portion of the CWA for controlling groundwater pollution, but in practice, "the programs implemented under Section 208 have, unfortunately, received lower priority and achieved much more limited results than have other parts of the Clean Water Act" (Patrick et al., 1987, p. 387). Furthermore, the areawide planning authorized under Section 208 is no longer funded. Third, Section 402 of the CWA establishes a National Pollution Discharge Elimination System (NPDES), which allows the EPA or authorized states to issue permits for pollutant discharges if certain conditions are met. These conditions require compliance with specific water quality standards and "are the principal mechanism for enforcing measures to reduce and control the discharge of pollutants into surface waters" (Pye et al., 1983, p. 253). An important feature of Section 402 is a subsection that requires that a state have adequate authority to control the discharge of pollutants into wells as a condition for authorizing a state NPDES program. Thus, a state with an authorized NPDES program could clearly control groundwater pollution under the CWA to the extent that the state could control the disposal of pollutants in wells. However, the CWA defines the term *pollutant* in a way that specifically excludes wastes associated with oil and gas production. Hence, states could not, for example, regulate the disposal of oil field brines in injection wells under Section 402. Finally, the CWA contains a provision that gives the EPA an imminent-hazard authority to restrain the discharge of pollutants where such discharge poses an imminent and substantial danger to the health or livelihood of human beings. This provision gives the EPA broad authority to protect against both environmental and economic injury, and it "could be effective in restraining activities threatening to contaminate groundwater supplies" (Patrick et al., 1987, p. 388).

CERCLA

The Comprehensive Environment Response, Compensation and Liability Act (CERCLA) was enacted by Congress in 1980. CERCLA was designed to allow the federal government to respond immediately to the release or threatened release of hazardous substances into the environment. Specifically, CERCLA requires that remedial action be taken at inactive waste disposal sites that pose a threat to the environment or human health.[4] CERCLA authorizes the federal government to clean up the contamination at an inactive site or spill if it is impossible to obtain a satisfactory response from the private sector. Such federal cleanups are financed by a Hazardous Substance Response Trust Fund, commonly referred to as the "Superfund," seven-eighths of which is provided by taxes on industries and the remaining one-eighth from general revenues. The original Superfund contained $1.6 billion, but this figure was increased to $8.5 billion after passage of the 1986 Superfund Amendments and Reauthorization Act and by an additional $5.1 billion when CERCLA was subsequently extended. However, CERCLA specifically prohibits the government from using Superfund monies for cleanup if a responsible private party will conduct the work. Private parties such as owners or operators of dumps may be held responsible for cleanup under CERCLA without regard to fault or negligence, and they may avoid liability only with a few narrowly drawn defenses. Subsequent to the enactment of CERCLA, Congress passed the National Contingency Plan (NCP) on July 16, 1982, to ensure an effective re-

---

[4]Active waste disposal sites are covered by RCRA.

sponse to both CERCLA and CWA. The NCP is basically a program to establish priorities for cleanup operations and to determine the most effective remedial response. A key component of the NCP is the Hazard Ranking System (HRS), which is used to establish cleanup priorities. Considerations taken into account include the population at risk, the hazard potential of the pollutants, the potential for groundwater contamination, and so on. As of November 1999, there were a total of 1,219 sites throughout the country on the active Superfund National Priorities List (Table 16.2). It is clear from Table 16.2 that hazardous waste sites that pose a threat to human health or the environment are widely distributed throughout the United States, but they are most concentrated in the Northeast.

It would be an understatement to say that CERCLA has proved to be a controversial piece of legislation. Although roughly 1,200 abandoned hazardous waste sites were initially identified as priorities for cleanup, only 54 of those sites had been permanently cleaned up by the end of 1990 (Rubin and Setzer, 1990). In addition to the small number of sites that had been restored, a major criticism leveled at the Superfund program concerned its reliance on pumping and treating as a means of aquifer restoration. Travis and Doty (1990, p. 1465), for example, noted that "A recent EPA study involving 19 sites where pumping and treating had been ongoing for up to 10 years concluded that although significant mass

TABLE 16.2 NUMBER OF ACTIVE AND COMPLETED SUPERFUND SITES IN THE UNITED STATES AS OF NOVEMBER 1999

Location	Number of Sites		Location	Number of Sites	
	Active	Completed		Active	Completed
Alabama	13	3	Montana	10	3
Alaska	7	3	Nebraska	10	3
Arizona	10	3	Nevada	1	0
Arkansas	12	9	New Hampshire	18	10
California	93	37	New Jersey	113	42
Colorado	15	5	New Mexico	11	8
Connecticut	14	6	New York	83	38
Delaware	17	13	North Carolina	26	9
District of Columbia	1	0	North Dakota	0	2
Florida	49	38	Ohio	32	20
Georgia	14	9	Oklahoma	12	6
Guam	2	1	Oregon	9	4
Hawaii	4	1	Pennsylvania	95	56
Idaho	6	4	Puerto Rico	9	4
Illinois	39	14	Rhode Island	11	3
Indiana	28	21	South Carolina	25	18
Iowa	16	12	South Dakota	1	3
Kansas	10	6	Tennessee	13	10
Kentucky	16	15	Texas	35	17
Louisiana	15	9	Utah	12	7
Maine	12	5	Vermont	7	5
Maryland	18	3	Virgin Islands	2	0
Massachusetts	30	9	Virginia	29	9
Michigan	68	49	Washington	45	34
Minnesota	26	32	West Virginia	8	1
Mississippi	1	3	Wisconsin	39	30
Missouri	24	14	Wyoming	3	1

Source: EPA (1999b).

removal of contaminants had been achieved, there had been little success in reducing concentrations to the target levels. The typical experience is an initial drop in concentrations by a factor of 2–10, followed by a leveling with no further decline. To exacerbate the problem, once pumps are turned off, concentrations rise again." Such results led some persons to question whether the Superfund program was not a waste of taxpayers' money. According to Stipp (1991, p. A1), "The Superfund program has little success to show for the $7.5 billion in taxpayers money it has spent so far." In an interview in November, 1990, EPA Administrator William Reilly was more upbeat about the future of Superfund, but he acknowledged that the program was plagued with problems during its early years (Rubin and Setzer, 1990).

The Superfund's track record improved during the second decade of its existence. Although the number of active sites has not changed much, this fact by no means reflects a static situation. By January, 1993, the number of completed sites had increased from 54 to 155, and by the end of fiscal year 1997 this figure at stood at 498. As of October 1999, construction activities had been completed at a total of 670 Superfund sites (Table 16.2). Although it is fair to say that not all Superfund cleanup efforts have been an unqualified success, there have certainly been a respectable number of success stories. The Arkansas City Dump is a case in point. The dump spreads over about 80 ha in southwestern Arkansas City, ironically in the state of Kansas. From 1916 until the mid 1920s, an oil refinery operated at the site and disposed of large quantities of toxic sludge in a 1 ha pit. An explosion and a fire destroyed the refinery in 1927, and the property was used as an illegal dump for domestic and municipal solid waste. When the dump was added to the NPL in 1983, an industrial park housing several manufacturing and warehouse businesses were located on the site, and 1,100 households and a city park were located within about 3 km. The cleanup process involved removing and treating the sludge wastes, covering the site with clean soil, and planting grass for erosion control. Through careful planning and coordination of work, it was possible to keep the industrial park open for business throughout the cleanup, which was completed in 1992 (EPA, 1999b).

Other Legislation

RCRA, SDWA, CWA, and CERCLA are the principal pieces of federal legislation that can be used to control groundwater pollution. However, several other federal laws are also relevant to groundwater pollution in one way or another and are worthy of mention at this time. The Toxic Substances Control Act (TSCA) was enacted in 1976 to regulate the manufacture, use, and disposal of hazardous chemicals and chemical mixtures. Although protection of groundwater supplies is not a specific objective of TSCA, the fact that TSCA gives the EPA authority to regulate the disposal of hazardous chemicals or substances containing hazardous chemicals is obviously relevant to groundwater pollution. Furthermore, since TSCA authorizes the EPA to limit the manufacture, processing, distribution, and use of chemicals, it is obvious that TSCA can have a very significant impact on groundwater pollution by limiting the quantities and kinds of hazardous chemicals that require disposal. The Uranium Mine Tailings Radiation Control Act (UMTRCA) of 1978 and the National Low-Level Radiation Waste Policy Act (NLRWPA) of 1980 are both relevant to the contamination of groundwater by radionuclides. As previously noted (Chapter 14), it had been customary prior to UMTRCA for U mining and milling operations to leave U-tailing piles uncovered and exposed to wind and rain (Hileman, 1982). A full assessment of the impact of this practice on radionuclide concentrations in groundwater as well as surface water began shortly after passage of the UMTRCA (Gallaher and Good, 1981). Some degree of groundwater contamination was found at 75% of the active U mills licensed in the United States (Patrick et al., 1987). Under NLRWPA each state was to assume responsibility for the disposal of low-level radioactive wastes generated by commercial operations within its borders by 1986 (Pye et al., 1983). However, the 1985 amendments to the NLRWPA extended this deadline to 1993 and stipulated that it was the policy of the federal government that the responsibilities of the states for the disposal of low-level radioactive wastes could be managed most safely and effectively on a regional basis. Prior to that time, these wastes had been stored at a few commercial and defense sites around the country, and as noted in Chapter 14, there had been leakage of radioactivity at almost all of these sites. The federal government envisioned that there would be a total of about a dozen regional low-level radioactive waste sites, but this has never come to pass. No one wants radioactive waste in his or her backyard. Although 10 regional compacts have been formed (Table 14.6), the commercial low-level wastes continue to be disposed of at only three sites—Barnwell, West Desert, and Richland (Figure 14.14). "It is generally agreed that the technology exists for siting and safe packaging, handling, transport, and isolation of [low-level radioactive] wastes" (Pye et al., 1983, p. 69). However, the incident of Pu migration from shallow trenches containing low-level radioactive wastes at Maxey Flats, Kentucky (Chapter 14), indicates that when the subject is radioactive wastes disposal, we do not always know as much as we think we know.

Finally, the National Environmental Policy Act (NEPA) of 1969 is worth mentinoing. The critical provision of NEPA is the requirement that an environmental impact statement be prepared for any proposed legislation or action that might significantly affect the quality of the environment. Thus, under NEPA, any major projects sponsored or permitted by the federal government may be evaluated for their potentially adverse effects on the environment, including groundwater.

Enforcement

It should be apparent from the foregoing discussion that there exist federal statutes which, if rigorously enforced, could be used to deal effectively with groundwater pollution problems. However, it should also be clear that at least some of the provisions of these laws have not been effectively enforced, either because of lack of personnel or lack of desire on the part of the government. This inability or unwillingness to take action was apparent during the Reagan administration, particularly during the time that Anne Gorsuch Burford headed the EPA. For example, Rita Lavelle, who headed the CERCLA hazardous waste program and Superfund under Burford as of March 31, 1982, was appointed to her post after previously serving as director of communications for subsidiaries of the Aerojet-General Corporation of California. Aerojet-General's liquid fuel plant had previously been cited by EPA as one of the 40 most hazardous chemical waste sites in the United States, and in 1979 California officials accused the company of discharging almost 80 m³ d⁻¹ of hazardous waste into a swamp and pond (Sun, 1982). Lavelle was fired by EPA head Burford on February 7, 1983, and in December 1983 was convicted by a U.S. district court in Washington, D.C., of perjury and obstructing a congressional inquiry. Lavelle had denied under oath that she knew Aerojet-General was being investigated by the EPA in a hazardous waste disposal case, but the prosecution established that not only had Lavelle been aware of the investigation, she in fact had called Aerojet-General officials to warn them (Lowther, 1983). In March 1983 Burford herself resigned, and ultimately 21 top EPA officials resigned in response to allegations of perjury, conflict of interest, and political manipulation.

On the state and local levels, lack of enforcement of aquifer protection laws has also been a problem. As noted by Pye et al. (1983, p. 272), "Ineffective enforcement appears to be a much more pervasive feature of groundwater regulation than of surface water regulation. Officials in state agencies regularly complain that monitoring and enforcement programs are understaffed and underfunded. . . . Some regulations are enforced minimally, others not at all (Dawson 1979). Most violations are handled informally at the agency level, with only light sanctions imposed." The cost of establishing and maintaining an adequate groundwater monitoring program, and the difficulty of obtaining accurate measurements of the wide variety of possible contaminants when concentrations are in the parts per billion range, appear to be major factors that discourage vigorous enforcement of existing statutes. Maugh (1982), for example, cites a study conducted by the Centers for Disease Control in which 29 laboratories were sent samples containing a known concentration of PCBs. Only three of these laboratories produced results that were within 2 standard deviations of the correct value. Given this result, one is naturally led to wonder just how accurate measurements of low concentrations of toxic substances in water supplies really are, and how much time and expense would be required to establish a monitoring program that produced reliable results for the wide variety of hazardous substances that might be found in groundwater. Undoubtedly, the same question has occurred to state and local officials.

CORRECTIVES

Under the category of correctives, it is appropriate to distinguish between prevention and cleanup. It should be clear by now that prevention of groundwater pollution should be the preferred strategy for dealing with the problem, but we must also deal with aquifers that became contaminated in the past and be prepared to deal with incidents of contamination in the future. Do we now have at our disposal effective methods for cleaning up contaminated groundwater?

Cleanup

If the goal of cleanup is to restore aquifers to a pristine condition, then the answer to this question in many cases will apparently be no. We do, however, have methods for containing the contaminated plume of groundwater and reducing the mass of the contaminant. In some cases, these measures may be sufficient. For example, the water need not be potable if it is to be used exclusively for irrigation. However, even if the goal is not to produce pristine water quality, cleanup costs are usually high. Current estimates are that it will cost about $64 billion to clean up the almost 2,000 waste sites that have appeared on the NPL between 1980 and 2000 (Rubin and Setzer, 1990), an estimate that works out to about $32 million per site.

There is no question that the high cost of cleaning up contaminated aquifers and the failure of groundwater pumping and treating efforts to restore potable water quality have led some scientists to seriously question the wisdom of large scale groundwater restoration programs. Muller (1982), for example, has stated that cleanup of an aquifer that has been contaminated by organic chemicals is almost never physically or economically feasible, and Pye and Patrick (1983, p. 717) note that "Recent studies of remedial action have concluded that it is complicated, time-consuming, expensive, and often not feasible, and that the best solution to ground water contamination is prevention. . . . Often it is more cost-effective to locate a new source of water than to attempt treatment." Unfortunately, a new source of water is not always available, at least not at a reasonable cost. For example, an Army Corps of Engineers study concluded that it would cost $3.6–$22.6 billion to provide irrigation water to Great Plains farmland currently irrigated with water from the Ogallala aquifer by building a system of canals to import water from South Dakota, Missouri, and Arkansas (Patrick et al., 1987). Hence it is sometimes necessary to consider the cleanup option.

There are potentially a number of methods available for restoring groundwater quality. These methods include (1) eliminating the source of contamination and allowing restoration to proceed by natural processes, (2) accelerating the rate of removal of contaminants by the use of withdrawal wells, drains, or trenches, (3) accelerating the rate of flushing by recharging with clean water, (4) installing impermeable barriers to block the spread of contamination, (5) inducing chemical or biological reactions to neutralize or immobilize the contaminant, and (6) excavating and removing the contaminated portion of the aquifer (GRF, 1984). Note that options 1–5 have all been employed at one time or another at the Rocky Mountain Arsenal. Methods of treating contaminated water include reverse osmosis, ultrafiltration, use of ion-exchange resins, wet-air oxidation, ozonation and ultraviolet radiation, coagulation and precipitation, aerobic biological treatment, and activated carbon (Patrick et al., 1987). The nature of the aquifer and of the contaminants will obviously dictate, to a large extent, what sorts of remedial measures are appropriate and whether cleanup is feasible at all. In some cases, for example, grout curtains are used to block the flow of contaminated groundwater. The curtains are formed by injecting grout[5] under pressure through a number of closely spaced wells. However, there is always a question of whether the grout curtain will really form an impervious barrier, and the practical depth limit of the curtain is about 15–18 m (Patrick et al., 1987).

The remedial measures taken at Love Canal provide a good illustration of the application of cleanup techniques. The problems that developed at Love Canal are associated with a problem known as the *bathtub* effect (GRF, 1984). When landfills are located in low-permeability rocks and in humid areas, the trenches in which the wastes are buried tend to become filled with water from rain or snow. "This surface water seeps downward through the landfill cover, fills the trenches, and eventually overflows—the so-called 'bathtub' effect" (GRF, 1984, p. 8). In order to deal with this problem, a mounded clay cap was placed over the old canal. The cap was compacted to maximize its resistance to water seepage and graded to divert stormwater into surface drains. The clay cap also helps to block the emission of fumes from volatile chemicals. About 3.5–6 m belowground, a 1 m wide, 2.1 km long barrier drain filled with crushed stone and sand was installed to isolate the canal from the surrounding environment. A pipe at the bottom of this trench was installed to carry leachate to a treatment plant (Wood et al., 1984). The Rocky Mountain Arsenal and Love Canal cases illustrate that remedial action to eliminate or at least minimize the effects of groundwater pollution is possible, but the costs can be very high and the results may leave something to be desired. However, when public health is threatened, a cleanup may be demanded. Such has been the case in "Silicon Valley" near San Jose, California, where organic solvents from the local electronics industry were detected in groundwater (GRF, 1984), and of course remedial action has been required at the Rocky Mountain Arsenal, despite the high price tag. Clearly, prevention of groundwater pollution in the first place is much more cost effective than trying to decontaminate a polluted aquifer.

Prevention

There are many methods for preventing groundwater contamination, and one technique would be the use of improved waste disposal methods. According to the Geophysics Research Forum panel (GRF, 1984, p. 18), "Most wastes can safely be disposed of in the sub-

---

[5]Grout may be composed of a variety of substances, including cement, fly ash, epoxy resins, and so on.

surface if repositories are selected, designed, and engineered on the basis of the nature of the wastes and adequate knowledge of the hydrology, geology and hydrogeochemistry of the site. There should be a more thorough search of disposal sites that can be used safely to isolate toxic wastes from the biosphere for long periods. . . . A strategy should be developed that provides for the segregation, treatment, and disposal of wastes according to their hazards and their chemical affinities." The diversity of possible contaminants and disposal options means that each situation must be evaluated separately in its own particular context.

For example, one useful strategy that has been cited by both the U.S. military and the DOE for the storage of radioactive wastes is the use of multiple barriers for waste containment.

1. The waste should be stored in a form that is not readily soluble (e.g., incorporation of radioactive wastes into borosilicate glass).
2. The containers holding the wastes should be highly resistant to the development of leaks and cracks.
3. The waste repositories should be backfilled with material that is chemically highly sorptive and of low permeability (e.g., clay).
4. The storage facility should be located in a medium of low permeability; to the extent that groundwater is present, the natural flow should be away from the biosphere.
5. The material surrounding the storage facility should contain minerals such as zeolites that tend to sorb the contaminants, or it should consist of highly porous rocks in which diffusion into the matrix could retard flow (GRF, 1984).

Another point relevant to the siting of waste disposal facilities is the fact that a number of closed hydrologic basins having only internal drainage exist in parts of Nevada, Utah, and adjacent states. These basins are located in arid regions where leaching effects from rainfall are minimal, and even if the contaminants were moved as a result of groundwater flow, they would remain within the basin. Finally, Abelson (1984) has noted that most drinking water is drawn from wells that are less than 100 m deep, so that in some areas, waste liquids with a density greater than that of water could be safely discharged through deep wells if injected at a depth greater than 100 m.

There are a number of more or less obvious and commonsense techniques that could be used to prevent or at least reduce groundwater contamination from certain sources. For example, installation of sewer systems or the use of waterless toilets could, under appropriate conditions, be a feasible method for eliminating pollution from septic tanks. However, sewer systems are costly to build and maintain, and they may not be cost effective in areas of low population density. Waterless toilets, which operate by acid incineration, gas incineration, or composting, would eliminate the organic solids and liquids generated by septic tanks, but they would obviously not deal with domestic wastewater resulting from bathing, laundry, and so on, and would produce an ash or compost that would ultimately require disposal. Where overdrafting of an aquifer for irrigation purposes has produced groundwater pollution problems due to high salinity or has seriously depleted the groundwater resource, growing cool-season crops, crops with low water requirements, or use of subsurface irrigation systems can significantly reduce water withdrawal rates. Economic incentives may be used to encourage such practices. Land spreading of sludge need not create groundwater pollution problems if the application site is carefully chosen and if the rate of sludge application is closely matched with the ability of the system to absorb organic waste. "Sludge disposal may best be sited on land where there is a mild gradient and a good soil cover that both allows slow percolation and has a high attenuation capacity, where there is a low water table, and where the vegetation is not to be used for food"

(Patrick et al., 1987, p. 270). In fact, land farming of sludge has been used for 30 years by the petroleum industry, and the Solid Waste Management Committee of the American Petroleum Institute has concluded that such sludge application will not significantly contaminate the soil and groundwater so long as the area to be farmed is closely matched to the type and application rate of the sludge (Patrick et al., 1987). In fact, careful application of sludge can be a highly effective way to upgrade strip-mined areas and similar unproductive land (see Chapter 6).

Perhaps the most fundamental way to prevent groundwater pollution from hazardous waste disposal is simply to reduce the amount of hazardous waste. Modification in manufacturing procedures and/or recycling are the obvious mechanisms for reducing industrial hazardous wastes. Estimates of the degree of reduction that could be achieved by recovery and recycling vary widely. A study by the Arthur D. Little Company concluded that only about 3% of industrial waste is potentially salable or reusable (Durso-Hughes and Lewis, 1982). However, the EPA has estimated that the total volume of hazardous waste could be reduced by 20% if available recycling and recovery technologies were used. "Dr. Paul Palmer of Zero Waste Systems, Inc., a California-based commercial chemical recycling company, ... claims that 80 percent of the chemical waste now buried at landfills or processed through other disposal methods could be recycled" (Durso-Hughes and Lewis, 1982, p. 16). Polsgrove (1982) has gone so far as to suggest the idea of a waste exchange, in which waste materials are transferred from seller to buyer through the hands of the exchange, which earns its income from commissions charged on completed transactions. According to Polsgrove (1982), waste raw materials such as iron and plastics would be the top-priority wastes, while processing residues such as solvents from pharmaceutical and paint processing and rejected Pb plate from Pb acid batteries would receive lower priority.

One of the big success stories in the effort to reduce the generation of hazardous waste has been the EPA's 33/50 program. The program targeted 17 high-priority toxic chemicals (Table 16.3) for reduction through voluntary partnerships. The program's name was derived from its intermediate goal of a 33% reduction in the generation of these chemicals by 1992 and a 50% reduction by 1995, using 1988 as a baseline (Koshland, 1991). A total of 1,300 corporations participated in the 33/50 program, and the results exceeded the target goals (Figure 16.7). The final target goal was actually achieved a year early.

TABLE 16.3   PRIORITY CHEMICALS TARGETED
BY THE EPA's 33/50 PROGRAM

Benzene
Cadmium and compounds containing cadmium
Carbon tetrachloride
Chloroform
Chromium and compounds containing chromium
Cyanide compounds
Dichloromethane
Lead and compounds containing lead
Mercury and compounds containing mercury
Methyl ethyl ketone
Methyl isobutyl ketone
Nickel and compounds containing nickel
Tetrachloroethylene
Toluene
1,1,1-Trichloroethane
Trichloroethylene
Xylenes

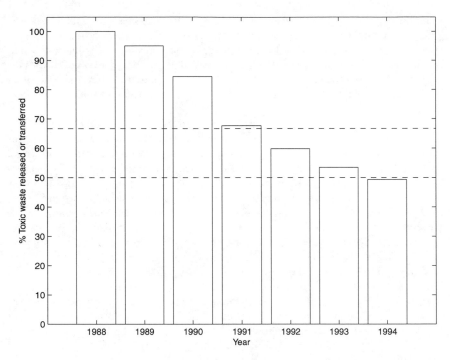

FIGURE 16.7   Reduction in the generation of toxic waste by 1,300 companies participating in the EPA 33/50 program. *Source*: EPA (1995).

Regardless of the percentage of hazardous waste that could be recycled economically on a national level, there is no question that some companies have greatly reduced waste generation and at the same time have saved considerable money by recycling. As noted in Chapter 9, for example, the "Pollution Prevention Pays" program begun by the 3M Corporation in 1975 has reduced its waste stream by a factor of 2 and in the process has saved the company over $300 million (Pollock, 1987). In another case, the Reliable Plating Company of Milwaukee, Wisconsin, was forced by the EPA in 1980 to purchase a $30,000 ion-transfer system to recover Cr from the company's nickel/chrome plating operation. Although the company at first considered the device a regulatory burden, reliable officials were pleasantly surprised to learn that in its first eight months of operation, the ion-transfer system reduced the company's purchases of Cr by a factor of 10 and cut water use by a factor of 70 (Durso-Hughes and Lewis, 1982).

The EPA Groundwater Protection Strategy

One important mechanism for preventing groundwater contamination would be a well-coordinated program involving the federal, state, and local governments to establish priorities and policies for groundwater use, to establish groundwater classification standards, to develop innovative approaches to groundwater protection, and to gather needed scientific and engineering information on groundwater contamination, assessment, enhancement, and protection. In fact, just such a program was proposed by the EPA in November 1980 and was implemented in August 1984, when the EPA's Office of Ground-Water Protection published its Groundwater Protection Strategy. The goal of the program is the assessment, protection, and enhancement of "the quality of groundwater to the levels necessary for current and projected future uses and for the protection of public health and significant ecological systems" (Pye et al., 1983, p. 191). The strategy has the following four major objectives (EPA, 1984):

1. Strengthen state groundwater programs
2. Cope with currently unaddressed groundwater problems
3. Create a policy framework for guiding EPA programs
4. Strengthen internal groundwater organization

A key feature of the program has been the development of a groundwater classification system that recognizes different levels of protection for different aquifers. Aquifers are classified primarily on the basis of current or projected uses, and it is recognized that not all aquifers have high-quality yields and that not all uses require the same level of quality (EPA 1980b). Aquifers are classified in one of three categories as follows:

1. Class I aquifers are groundwaters of unusually high value that are in need of special protective measures because they have a relatively high potential for contaminants to enter and/or to be transported within the groundwater flow system (Patrick et al., 1987). They are also regarded either as irreplaceable, in the sense that no alternative source of drinking water is available for a substantial population, or as ecologically vital, in the sense that they provide the base flow for a particularly sensitive ecological system that, if polluted, would destroy a unique habitat (Patrick et al., 1987). One might expect that groundwaters classified as sole-source aquifers under the SDWA would be considered Class I groundwaters, but such is not automaticallythe case. The criteria for Class I groundwaters are more strict than those for sole-source aquifers. The goal of EPA protection strategy for Class I aquifers is to keep pollutant concentrations below the MCLs specified in the SDWA.
2. Class II groundwaters are all other aquifers that are currently used, or potentially available, for drinking water and other beneficial use. A Class II designation allows limited siting of hazardous waste facilities and, while ensuring quality suitable for drinking, permits exemptions to requirements under certain circumstances to allow less stringent standards when the protection of human health and the environment can be demonstrated (EPA, 1986). The EPA envisioned that most fresh groundwater would fall into Class II.
3. Class III groundwaters are not potential sources of drinking water and are of limited beneficial use, either because they are saline or because they are otherwise contaminated beyond levels that allow drinking or other beneficial use. Class III aquifers include groundwaters with a TDS greater than 10,000 ppm or "those that are so contaminated, naturally or by human activity, that cleanup using methods reasonably employed in public water system treatment is not possible" (Patrick et al., 1987, p. 299). Regulations for protection of Class III aquifers are more lenient and allow siting of hazardous waste facilities, but they include technical requirements and monitoring rules to protect public health (Pye et al., 1983).

Perhaps the most controversial aspect of the EPA's Ground-Water Protection Strategy has been giving the states the primary role in groundwater protection. This policy actually went into effect before the Ground-Water Protection Strategy was formalized in 1984 because "with the confirmation of Administrator [Anne] Gorsuch at EPA in May 1981, the Reagan administration . . . quietly blocked all further development of the proposed [Ground-Water Protection] Strategy, and . . . de facto turned over all responsibilities for groundwater-quality management to the states" (Burmaster 1982, p. 36). While the decision to shift responsibility for groundwater protection from the federal government to state and local authorities was consistent with other Reagan administration policies, "The states alone are incapable of supporting the research, policy development, and program implementation necessary to protect groundwater quality; the federal government must play a role to ensure equity and efficiency among the states" (Burmaster 1982, p. 36). In summary, while it is undoubtedly reasonable to turn over a certain amount of regulatory responsibility to the states, the federal government and the EPA in particular must continue to play an active role in setting groundwater protection policies and working with the states to help implement those policies.

REFERENCES

ABELSON, P. H. 1984. Groundwater contamination. *Science,* 224, 673.

ANONYMOUS. 1983a. Who will Shell-out to clean up nerve gas? *New Scientist,* 13, 74.

ANONYMOUS. 1983b. Toxics suit asks $1.8 billion. *Engineering News-Record,* 211, 24.

ANONYMOUS. 1984. Colorado toxics fix. *Engineering News-Record,* 213, 16.

ANONYMOUS. 1988. Costs set for Arsenal plan. *Engineering News-Record,* 220, 11.

ANONYMOUS. 1990. Superfund's worst site begins to come clean. *Engineering News-Record,* 225, 39.

BURMASTER, D. E. 1982. The new pollution. *Environment,* 24(2), 6–13, 33–36.

CEQ. 1981. *Contamination of Ground Water by Toxic Organic Chemicals.* Council on Environmental Quality, Washington, DC. 84 pp.

CEQ. 1991. *Environmental Quality.* 21st Annual Report. U.S. Government Printing Office, Washington, DC. 388 pp.

CLAY, D. R. 1990. Hazardous waste sites. *Science,* 247, 1166.

CULLITON, B. J. 1980. Continuing confusion over Love Canal. *Science,* 209, 1002–1003.

DAWSON, J. W. 1979. State groundwater protection programs—inadequate. In *Proceedings of the Fourth National Ground-water Quality Symposium,* Minneapolis, EPA-600/9-29-029. US Environmental Protection Agency, Washington, DC. September 20–22, 1978. Pp. 102–108.

DURSO-HUGHES, K., and J. LEWIS. 1982. Recycling hazardous waste. *Environment,* 24, 14–20.

EPA. 1977a. *Waste Disposal Practices and Their Effects on Ground Water.* Report to Congress prepared by the Office of Solid Waste Management programs. 512 pp.

EPA. 1977b. *Multimedia Environmental Goals for Environmental Assessment.* EPA-600/7-77-136. Environmental Protection Agency of Washington, DC.

EPA. 1980a. *Groundwater Protection.* Environmental Protection Agency, Washington, DC. 36 pp.

EPA. 1980b. *Proposed Ground Water Protection Strategy.* Office of Drinking Water, Washington, DC. 61 pp.

EPA. 1980c. Planning workshops to develop recommendations for a groundwater protection strategy. Office of Drinking Water, Washington, DC. 171 pp.

EPA. 1980d. *Everybody's Problem, Hazardous Waste.* Office of Water and Waste Management, Washington, DC. 36 pp.

EPA. 1984. *Ground-Water Protection Strategy.* Office of Ground-water Protection, Washington, DC. 56 pp.

EPA. 1986. *Guidelines for Ground-water Classification Under the EPA Ground-Water Protection Strategy,* final draft. Office of Ground-Water Protection, Washington, DC. 137 pp.

EPA. 1987. Drinking water; proposed substitution of contaminants and proposed list of additional substances which may require regulation under the Safe Drinking Water Act. *Federal Register,* 52(130), 25719–25723.

EPA. 1988. *The Safe Drinking Water Act.* Environmental Protection Agency, San Francisco. 50 pp.

EPA. 1994. 33/50 hits the mark. *http://www.epa.gov/opptintr/tri/ttch4txt.htm*

EPA. 1999a. Current drinking water standards. *http://www.epa.gov/safewater/mcl.html*

EPA. 1999b. Superfund hazardous waste sites. *http://www.epa.gov/superfund/sites/query/basic.htm*

EPSTEIN, S. S., L. O. BROWN, and C. POPE. 1982. *Hazardous Waste in America.* Sierra Club Books, San Francisco. 593 pp.

GALLAHER, B. M., and M. S. GOOD. 1981. *Water Quality Aspects of Uranium Mining and Milling in New Mexico.* Special Publication No. 10. New Mexico Geological Society, Santa Fe. Pp 85–91.

GEOPHYSICS RESEARCH FORUM. 1984. *Studies in Geophysics: Ground-Water Contamination.* National Academy Press, Washington, DC. 179 pp.

GROUNDWATER FOUNDATION. 1999. Pollution prevention. *http://www.groundwater.org/guard/appendix.htm*

HIGH PLAINS WATER DISTRICT. 1998. The Ogallala aquifer. *http://www.hub.ofthe.net/hpwd/ogallala.html*

HILEMAN, B. 1982. Nuclear waste disposal: A case of benign neglect? *Environ. Sci. Tech.* **16**(5), 271A–275A.

KONIKOW, L. F., and D. W. THOMPSON. 1984. Groundwater contamination and aquifer reclamation at the Rocky Mountain Arsenal, Colorado. In *Groundwater Contamination.* National Research Council Geophysics Study Committee, National Academy Press, Washington, DC. Pp. 93–103.

KOSHLAND, D. E., JR. 1991. Toxic chemicals and toxic laws. *Science,* **253**, 949.

LEHR, J. H. 1982. How much ground water have we really polluted? *Ground Water Monitoring Rev.,* Winter, pp. 4–5.

LOWTHER, W. 1983. Turmoil at EPA. *Macleans,* **96**, 50.

MAUGH, T. 1982. Just how hazardous are dumps? *Science,* **215**, 490–493.

MULLER, D. W. 1982. Cited by Maugh (1982), p. 491.

PATRICK, R., E. FORD, and J. QUARLES. 1987. *Groundwater Contamination in the United States,* 2nd ed. University of Pennsylvania Press, Philadelphia. 513 pp.

PETRI, L. R., and R. O. SMITH. 1956. *Investigation of the Quality of Ground Water in the Vicinity of Derby, Colorado.* U.S. Geological Survey Open File Report. US Geological Survey Reston, VA. 77 pp.

POLLOCK, C. 1987. *Mining Urban Wastes: The Potential for Recycling.* Worldwatch Paper 76. Worldwatch Institute, Washington, DC. 58 pp.

POLSGROVE, C. 1982. The waste exchange option. *Environment,* **24**, 15, 37–41.

PYE, V. I., and R. PATRICK. 1983. Ground water contamination in the United States. *Science,* **221**, 713–718.

PYE, V. I., R. PATRICK, and J. QUARLES. 1983. *Groundwater Contamination in the United States.* University of Pennsylvania Press, Philadelphia. 315 pp.

ROBERTS, P. 1981. Keeping America's water drinkable. *Mainliner Magazine,* June, pp. 58, 86, 89, 112.

RUBIN, D. K., and S. W. SETZER. 1990. The Superfund decade: Triumphs and troubles. *Engineering News-Record,* **225**, 38–44.

SIERRA CLUB. 1999. The real truth. *http://www.rmc.sierraclub.org/emg/rma.html*

SITTIG, M. 1980. *Priority Toxic Pollutants.* Noyes Data Corp., Park Ridge, NJ. 370 pp.

SITTIG, M. 1985. *Handbook of Toxic and Hazardous Chemicals and Carcinogens,* 2nd ed. Noyes Publications, Park Ridge, NJ. 950 pp.

SOLLEY, W. B. 1989. Reflections on water use in the United States. In *U.S. Geological Survey Yearbook for Fiscal Year 1988.* U.S. Geological Survey, Washington, DC. Pp. 28–30.

SOLLEY, W. B., R. R. PIERCE, and H. A. PERLMAN. 1998. *Estimated Use of Water in the United States in 1995.* United States Geological Survey Circular 1200. *http://water.usgs.gov/watuse*

STIPP, D. 1991. Throwing good money at bad water yields scant improvement. *Wall Street Journal,* May 15. pp. A1, A6.

SUN, M. 1982. EPA relaxes hazardous waste rules. *Science,* **216**, 275–276.

TRAVIS, C. C., and C. B. DOTY. 1990. Can contaminated aquifers at Superfund sites be remediated? *Environ. Sci. Tech.,* **24**(10), 1464–1466.

WOOD, E. F., R. A. FERRARA, W. G. GRAY, and G. F. PINDAR. 1984. *Ground Water Contamination from Hazardous Wastes.* Prentice-Hall, Englewood Cliffs, NJ. 163 pp.

WOOD, L. A. 1972. Groundwater degradation—causes and cures. In *Proceedings, 14th Water Quality Conference,* Engineering Publications Office, Urbana, IL. Pp. 19–25.

QUESTIONS

16.1 Why has the EPA's policy of "pump and treat" been characterized as a waste of tax-payers' money by critics of the agency's approach to restoring contaminated aquifers?

16.2 The freshwater aquifer beneath the island of Oahu could become unfit for drinking purposes if rates of groundwater withdrawal exceed recharge rates. Evidence that this was happening would be which one of the following?

   a. Increase in the salinity of the water
   b. Increase in the nitrate concentrations in the water
   c. Increase in the fecal coliform concentrations in the water
   d. Increase in concentration of heavy metals in the water

16.3 Why have some environmentalists objected to the decision to designate the Rocky Mountain Arsenal a national wildlife refuge following completion of the environmental cleanup at the site?

16.4 A person who owns a vacant lot is allowing people to dispose of items such as old refrigerators, broken-down cars, discarded batteries, and so on his lot. The person has no license to operate a solid-waste disposal facility. He could be prosecuted under which one of the following federal laws?

   a. RCRA
   b. SDWA
   c. CWA
   d. CERCLA

16.5 The Grande Crevette Company plans to grow shrimp and intends to discharge wastewater from its shrimp ponds into the ocean. In order to discharge that wastewater, it must obtain a discharge permit under the authority of which one of the following federal laws?

   a. RCRA
   b. SDWA
   c. CWA
   d. CERCLA

16.6 The Tóxica Basura Company operates a landfill for several decades and then goes bankrupt. Subsequently, the aquifer under the landfill is found to be contaminated. Which one of the following federal laws could be used to appropriate funds to clean up the contaminated aquifer?

   a. RCRA
   b. SDWA
   c. CWA
   d. CERCLA

16.7 The Peligrosos Desechos Company generates hazardous waste. In the United States, what happens to that hazardous waste is regulated/monitored under the authority of which one of the following federal laws?

   a. RCRA
   b. SDWA
   c. CWA
   d. CERCLA

Chapter 17

✧ ⌂ ✧ ⌂ ✧

Plastics in the Sea

> I hopes ya swabs won't be throwing no plastics overboard. *Popeye the Sailor*, on a poster aimed at educating the public about the adverse effects caused by disposal of plastic waste in the ocean.

THE NATURE OF THE PROBLEM

Plastics synthesized from petrochemicals were first produced more than a century ago, but the most significant growth of the commercial plastics industry did not begin until World War II, when shortages of rubber and other materials created a demand for substitute products. Plastics were used to produce items such as bugles, canteens, and dinnerware for the military, and in the private sector products such as nylon stockings and Saran Wrap became overnight sensations. The growth in the variety of uses for plastic continued after the war with the development of products such as Tupperware, nylon zippers, and acrylic dentures. This rapid expansion of the plastics industry has continued more or less unabated to the present time (Figure 17.1). By 1998, plastics production in the United States had reached about 40 million tonnes y^{-1}, almost 150 kg for every person in the United States.

The magnitude of plastics production and use has, not surprisingly, created a significant solid waste disposal problem. Until recently, virtually all plastic products were nonbiodegradable. Although plastics do break down as a result of physical and chemical weathering, under most circumstances the breakdown process is slow. Plastics discarded in the aquatic environment may remain in largely unaltered form for years, if not decades. In this respect, plastics resemble certain other synthetic organics such as PCBs in terms of persistence.

Plastic debris takes basically two forms. So-called raw plastic consists of small resin pellets a few millimeters in diameter. These small pellets are convenient to ship and are ul-

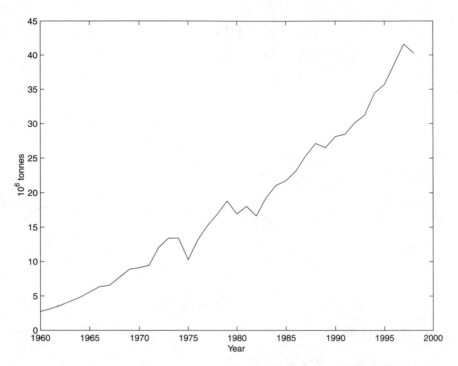

FIGURE 17.1   Plastics production in the United States since 1960. *Source*: O'Hara et al. (1988) and SPI (1999).

timately melted down and molded into manufactured products. It has been quite common, however, for raw plastic pellets to be discharged with wastewater from manufacturing plants or to be lost in the process of shipment. The resemblance of these pellets to fish eggs has undoubtedly been partly responsible for their ingestion by certain organisms that feed on plankton. Plastic pellets have been reported in concentrations as high as 3,500 and 34,000 km^{-2} in the surface waters of the Atlantic and Pacific Oceans, respectively (O'Hara et al., 1988).

The second major category of plastic debris is lost or discarded manufactured products. Only rough estimates can be made of the amount of this debris. Approximately 10 million tonnes of solid waste are discharged into the ocean each year, and of this figure, roughly 10% is plastic (O'Hara et al., 1988). Hence, about 1 million tonnes of plastic debris are introduced into the ocean each year, or about 100 tonnes h^{-1}. The nature of this debris may vary greatly from one area to another. A major problem in some parts of the ocean has been lost or discarded fishing nets and traps. Lost sections of gill nets, for example, may drift in surface waters and continue to entangle fish for years until they eventually sink due to the weight of accumulated carcasses. Lost lobster and crab pots may also attract and catch organisms, perhaps over an even longer time frame. In areas of offshore oil production, it is common for plastic debris to consist largely of items discarded or lost from oil platforms, from oil drilling operations, and from ships engaged in seismic studies. There is a general consensus that at least historically, most of the plastic debris in the ocean has come from ships, but in some areas, much of the nearshore debris has a significant land-based origin. This condition may exist, for example, near coastal municipalities that have combined sewer systems.

EFFECTS

Aesthetics

One of the most obvious adverse effects associated with plastic debris in the ocean is aesthetic. Probably more than any other factor, the appearance of plastic debris on recreational beaches has sensitized the public to the problem of plastic pollution. Annual beach cleanups sponsored by environmental groups have become common in all coastal states. The amount of debris picked up by volunteers during these cleanups is remarkable. In October 1984, for example, volunteers picked up almost 24 tonnes of debris during a three hour period along 560 km of Oregon coastline. Included in the debris were 48,900 chunks of polystyrene larger than a baseball, 5,300 food utensils, 4,900 bags or sheets of plastic, 4,800 plastic bottles, 2,000 plastic strapping bands, 1,500 six-pack rings, and 1,100 pieces of fishing gear. Along a 240 km stretch of North Carolina beach, 8,000 plastic bags were found in three hours, and more than 15,600 six-pack rings were picked up in three hours on a 480 km section of Texas coastline (O'Hara et al., 1988). Although certainly not all beach litter is plastic, national studies of the composition of beach debris indicate that 55–65% indeed consists of plastic items (O'Hara et al., 1988). Although only a small percentage of the items picked up in beach cleanups appear to have been left behind by beach users (CEE, 1987), beachgoers are not without fault. In Los Angeles County, for example, beach users leave behind about 68 tonnes of trash each week (O'Hara et al., 1988).

Perhaps the most famous incident of plastic pollution on recreational beaches was the "floatable episode" of June 1976, when large amounts of debris washed up on beaches on Long Island, New York (Swanson et al., 1978). Plastic items accounted for most of the debris and consisted mainly of tampon applicators (more than 300 per kilometer of beach), condoms, sanitary napkin liners, and disposable diapers. Other plastic items included straws, pieces of Styrofoam cups, plastic bottle caps, corks, plastic toys, and plastic cigar and cigarette tips. Within nine days after the debris first began to appear, all beaches were closed to swimming, shellfishing was banned, and the Governor of New York declared most of Long Island a disaster area. The incident was apparently caused by a combination of factors, including combined sewer overflows, ocean dumping of sewage sludge, disposal of trash by commercial and recreational vessels, unusually heavy rainfall, and onshore winds (O'Hara et al., 1988). A similar incident or, more correctly, a series of incidents, occurred during the summer of 1988, when large amounts of garbage, including needles, syringes, and other medical wastes, washed up on Long Island and New Jersey beaches. Estimates of the economic costs associated with the 1988 incident, including impacts on tourism and recreational fishing, have ranged as high as $3 billion (Wagner, 1990).

Ingestion

Marine organisms sometimes consume plastic debris, either inadvertently in the process of feeding or deliberately because they mistake the debris for food. The extent to which marine organisms have been killed or stressed as a result of plastics ingestion is not well known because an autopsy is necessary to confirm the problem. The best-documented cases of plastics ingestion have undoubtedly involved sea turtles, which presumably mistake plastic bags and sheeting for jellyfish, a common prey. At least five species of turtles are known to consume plastics (Balazs, 1985). In one case on Long Island, 11 of 15 dead leatherback turtles that washed ashore during a two week period had plastic bags blocking their stomach openings. Ten of the turtles had ingested four 7.6-L sized bags, and one had eaten 15 bags (Anonymous, 1982). Another leatherback found in New York had ingested 46 m of monofilament fishing line (CEE, 1987).

In addition to sea turtles, there is documentation of ingestion of plastics by nine species of whales (CEE, 1987). As with the turtles, most of the plastic items have been bags and sheeting, although in one case, about 1 L of tightly packed trawl net was found in the stomach of a sperm whale. Most of the information concerning plastics consumption by whales has come from dead animals that have been stranded, and in most cases it has not been clear that the ingested plastics were the cause of death. In 1985, however, "A young sperm whale was found dying on the shores of New Jersey as the result of a mylar balloon lodged in its stomach and three feet of purple ribbon wound through its intestines" (O'Hara et al., 1988, p. 22); in Hawaii a captive dolphin died from ingesting a piece of membrane plastic (CEE, 1987); and a pygmy sperm whale that was taken into captivity after being stranded along the Texas coast later died due to ingestion of paper and plastic bags (Jones et al., 1986).

Consumption of plastics by sea birds has received much attention in recent years, and a comprehensive review of the subject has been reported by Day et al. (1985). Plastic objects, most commonly raw polyethylene pellets, have been found in the digestive tracts of at least 50 of the world's 280 species of marine birds. Birds that feed primarily on crustaceans or squid tend to have the highest incidence of plastics ingestion, and the color, shape, and/or size of the ingested objects follow a pattern that suggests that the birds are selective consumers of plastics. In other words, the birds are not consuming the plastic debris by accident. The parakeet auklet, for example, which feeds mainly on planktonic crustaceans, selectively consumes light brown plastic objects that are primarily cylindrical, spherical, box-shaped, or pill-shaped (CEE, 1987). Only a small percentage of marine birds ingest dark-colored plastic debris. Of particular concern is the fact that some birds feed plastic debris to their young (Fry and Fefer, 1986). For example, 90% of Laysan albatross chicks examined on Midway and Oahu islands contained plastic objects. Ingested plastic debris, whether in adults or chicks, can fill a bird's stomach to such an extent that the bird no longer feeds properly due to a false feeling of satiation. In some cases, the debris may cause serious internal injuries.

Finfish are also known to consume plastic debris, although documentation is largely anecdotal. The problem seems to be most common with demersal fish, plastic pellets being the form of debris most frequently mentioned. Individual juvenile flounder, for example, have been reported to contain as many as thirty 2–5 cm diameter plastic pellets in their intestines (Kartar et al., 1973, 1976). The presence of such pellets may cause intestinal blockage in small fish (CEE, 1987).

Entanglement

Entanglement with plastic debris occurs when a marine organism accidentally or intentionally comes into contact with debris and becomes ensnared. Incidents of entanglement are the most easily documented biological impacts of plastic debris and undoubtedly the most publicized. Probably the most serious impacts have resulted from the accidental loss or, in some cases, intentional disposal of fishing nets or parts of fishing nets. Both gill nets and trawl nets have been involved. Much of the publicity concerning this problem came from studies in the North Pacific, where the total length of all gill nets used in the 15 major drift net fisheries during the 1980s was about 170,000 km, more than four times the circumference of the Earth (CEE, 1987). Particular attention was focused on the North Pacific high seas drift net fisheries conducted by Japan, Taiwan, and South Korea. Every night during the fishing season, each vessel involved in this fishery set a 9–27 km long drift gill net that was picked up the following day (Eisenbud, 1985). The total length of these drift nets was about 33,000 km.

The use of these nets to catch species of squid, salmon, and billfish proved to be highly controversial for a variety of reasons. Many of the fish that became entangled in the nets were not utilized in the catch, either because they were not recovered when the net was retrieved or because they were considered not worth saving and were simply thrown back into the ocean. Fisheries data furthermore indicated that the drift nets were so efficient at catching fish that some of the most important fishery resources in the North Pacific were being seriously depleted. These strictly fishery issues are not directly relevant to the problem of plastics pollution. However, when nets or pieces of nets are lost or discarded, they become marine debris.

The quantity of nets lost or discarded each year is not well known, most information on the subject being anecdotal or semiquantitative at best. Estimates of 0.05–0.06% per set have been made for the Japanese salmon pelagic gillnet fishery, and total net losses from Japanese North Pacific squid and salmon fisheries have been put at 2,500 km y^{-1} (Breen, 1990). In the New England groundfish gill net fishery, gill nets with a total length of typically 900–1,100 m are anchored on the bottom to catch demersal fish (CEE, 1987). Since these nets remain submerged when lost, observations of lost nets must be made with a submersible vessel. A 1984 survey of prime demersal fishing sites in the Gulf of Maine by the submersible *Johnson Sea-Link II* revealed 10 lost nets in an area a little over 10 ha (Carr et al., 1985).

Trawl nets account for three to four times as many fish caught by weight as do gill nets on a worldwide basis[1] and, like gill nets, are subject to loss and breakage, particularly when used in bottom trawling operations. Roughly 5,500 km of nets are used by 12 major North Pacific fisheries (CEE, 1987), and although this figure is quite small compared to the length of nets used in the gill net fisheries, studies of beach debris on Amchitka Island in the Aleutian Islands from 1972 to 1982 revealed that trawl net fragments accounted for 76–85% of the plastic litter collected (Merrell, 1985). Low et al. (1985) have estimated that during the early 1980s, 50–70 trawl nets or large portions of trawl nets were lost each year by fishing vessels off Alaska.

In addition to their use in fishing nets, plastics have become an important construction material in traps designed to catch crabs and lobsters. A standard-size wooden lobster trap, for example, typically includes about 0.4 m^2 of plastic netting (CEE, 1987). In the New England lobster fishery, about 2.5 million traps are used annually, and of these, at least 20%, or 500,000, are lost each year. King crab fishermen in Alaska lose about 10% of their crab pots each year, and in the Gulf of Mexico, annual trap losses in the stone crab fishery average 25% (CEE, 1987).

Ghost Fishing

Entanglement in lost or discarded fishing gear is usually referred to as *ghost fishing*. Plastic fishing nets, which are typically made of nylon, polypropylene, or polyethylene, are often difficult to discern in the water even during daylight hours, and gill nets may drift for literally years before washing ashore or sinking. Entanglement in most cases is undoubtedly accidental, but in some cases it appears that marine organisms are unwittingly attracted to net debris. Sea turtles, for example, may become entangled in floating net fragments because of their natural attraction to Sargassum mats and other natural floating masses that provide shelter and offer concentrated food sources (CEE, 1987). The impact of entanglement on marine organisms has undoubtedly been most thoroughly studied in the northern fur seal population of the Pribilof Islands. The population was seriously depleted during the second half of the 19th century due to overharvesting, but an inter-

[1]Purse seines, trawl nets, and gill nets account for roughly 45%, 22%, and 6% of the world fish catch, respectively. The remainder of the catch is taken by lines (9%) and by pound and other trap nets (9%).

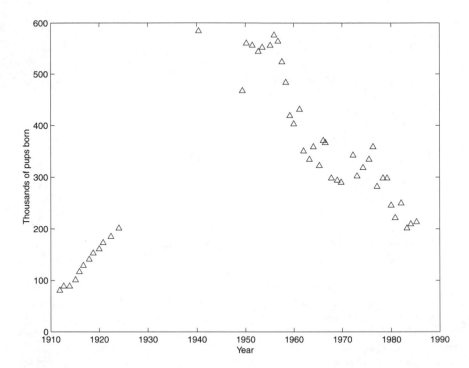

FIGURE 17.2   Reproduction of northern fur seals during the 20th century. *Source*: French and Reed (1990). Used with permission.

national ban on harvesting in 1911 allowed the population to recover during the next three decades. The population reached the apparent carrying capacity of the ecosystem of about 2 million individuals by 1940 and remained at that level throughout most of the 1950s. Some harvesting of females was permitted from 1956 to 1968. It was assumed that when the harvesting ceased, the population would return to the carrying capacity. Instead, the population declined at a rate of 4–8% per year (Figure 17.2). The decline has been attributed largely to entanglement,[2] the estimated entanglement loss being at least 50,000 seals per year (CEE, 1987). The impact appears to have been heaviest on seals younger than two years (French and Reed, 1990). Current estimates indicate that the number of northern fur seals is no more than one-third the number that existed during the 1940s and 1950s.

Although much of the concern relative to ghost fishing has been focused on the northern fur seal population, almost half of the 34 extant pinniped species (seals, sea lions, and walruses) are known to have become entangled in marine debris (Henderson, 1990). Particularly noteworthy has been the entanglement of Hawaiian monk seals, an endangered species. Almost all documented incidents have involved lost or discarded fishing gear (Henderson, 1990). As is the case with the northern fur seals, juveniles appear to be more susceptible than adults. Other pinnipeds involved in entanglement incidents include California and stellar sea lions, northern elephant seals, Cape and New Zealand fur seals, and harbor seals (CEE, 1987; Stewart and Yochem, 1990). Fortunately, entanglement does not appear to be having a significant impact on pinniped demography and population dynamics in all cases (Stewart and Yochem, 1990).

---

[2] Two other factors have been overfishing in the Bering Sea of walleye pollock, an important food item of northern fur seals, and the lack of food associated with El Niño events (SCS, 1999).

Birds and cetaceans are also known to become entangled in fishing nets, but in both cases, most losses are associated with active fishing gear rather than lost or discarded nets. During the 1980s, for example, the Japanese salmon gill net fishery was estimated to have killed more than 250,000 seabirds each year in U.S. waters during the two month fishing season (King, 1985). The birds were apparently attracted to fish caught in the net. Ghost-fishing nets attract birds in a similar manner, and 100 or more dead birds have been found entangled in large pieces of gill net debris (Jones and Ferrero, 1985). A 1.5 km section of a 3.5 km long derelict gill net retrieved in the North Pacific contained 99 dead seabirds, two salmon sharks, one ragfish, and 75 recently entangled salmon (DeGange and Newby, 1980). The net was estimated to have been adrift for about a month. Cetaceans presumably become entangled in nets or pieces of netting as a result of unintentional collisions. DeGange and Newby (1980) found two dead Dall's porpoises in a 3.5 km derelict gill net off Amchitka Island.

Lost or discarded lobster, crab, and demersal fish traps are the other form of plastic debris principally associated with ghost fishing. In the simplest form of trap ghost fishing, captured animals die in lost traps, and their bodies act as bait (Breen, 1990). However, captured live animals may also attract other organisms to a lost trap, and in some cases, organisms are apparently even attracted to empty traps, perhaps because they perceive the traps as a source of shelter (Breen, 1990). The animals presumably die as a result of either starvation or cannibalism. Some estimates of trap ghost fishing loss rates are summarized in Table 17.1. It is apparent from this table that losses to trap ghost fishing can be quite substantial.

Other Causes of Entanglement

Other than lost or discarded nets and traps, the principal types of plastic associated with entanglement are monofilament fishing line, packing straps, and six-pack rings. Monofilament fishing line, for example, is the most common type of debris reported to entangle sea turtles (Balazs, 1985), the affected species including green, hawksbill, olive ridley, and leatherback turtles, all of which are endangered or threatened species (CEE, 1987).

Brown pelicans, another protected species, become entangled in both monofilament fishing line and six-pack rings. Entanglement of these birds is a major problem in both California and Florida (Gress and Anderson, 1983). The pelicans actively collect pieces of nets and fishing line for nest material (Bourne, 1977) and apparently become entangled in six-pack rings when they try to dive or feed through the rings (CEE, 1987). Stewart and Yochem (1990) report that packing straps are the most common cause of entanglement of northern elephant seals in the Southern California Bight.

TABLE 17.1 ESTIMATED LOSS RATES DUE TO TRAP GHOST FISHING

	Annual Loss Rate	
Fishery	Weight	Value
Newfoundland snow crab	10 tonnes	
U.S. American lobster	670 tonnes	$2.5 million
British Columbia Dungeness crab	7% of landings	$80,000
British Columbia sablefish	300 tonnes[a]	

[a]Commercial catch is 1,000–4,000 tonnes y^{-1}.
Source: Breen (1990).

Damage to Vessels

Damage to vessels from marine debris results from collisions with floating objects, entanglement of debris in propeller blades, and clogging of water intakes for engine cooling systems. The amount of such damage caused by plastic debris is not well known. Takehama (1990) has estimated the annual cost of damage to Japanese fishing vessels caused by floating debris to be roughly 4 billion yen, which is about 0.2% of the total cost of operating the vessels. The contribution of plastic debris to this damage is unknown. According to the U.S. Coast Guard, about 10% of vessel accidents in U.S. waters are caused by debris (CEE, 1987). Unfortunately, plastic debris is not specifically coded for in the data analysis, and hence there is no way to judge the contribution of plastics to the problem. Anecdotal information, however, indicates that plastic garbage bags are considered to be the leading cause of engine damage of commercial and recreational vessels in New England waters, and the fact that some boating supply companies have built devices on propellers to combat entanglement indicates that the problem is not infrequent (CEE, 1987).

CORRECTIVES

Solutions to the problems caused by plastics in the sea can be broadly classified as legal, technological, and educational. The most important legal developments in recent years have been adoption of a resolution by the United Nations General Assembly requiring all nations conducting open ocean high-seas driftnet fishing to cease all such fishing by December 31, 1992 (Anonymous, 1992), and the ratification of Annex V of the International Convention for the Prevention of Pollution from Ships (MARPOL) by the United States on December 31, 1987. The end of high-seas drift net fishing was the result primarily of fisheries considerations rather than concern over pollution. The nets took many nontarget species, and their extensive use, particularly in the North Pacific, posed a serious threat of overfishing. The disappearance of this method of fishing, however, also eliminated one of the important sources of plastic debris in the ocean, namely, lost or discarded portions of drift nets. Unfortunately, the lack of quantitative information concerning the impact of these derelict gill nets on marine organisms will make it difficult to assess the effect of this development from a pollution standpoint.

MARPOL Annex V

Ratification of Annex V of MARPOL by the United States was a particularly noteworthy event because Annex V is an optional component of MARPOL and could come into force only when ratified by countries representing 50% of the world's shipping tonnage.[3] The U.S. vote put the percentage of shipping tonnage associated with ratifying nations over the 50% mark. The legislation went into effect on December 31, 1988.

As summarized in Table 17.2, Annex V concerns at-sea disposal of various forms of garbage from offshore platforms and privately owned vessels. Public vessels (e.g., military) are exempt from Annex V restrictions but are expected to comply to the extent possible. The legislation permits disposal from vessels of most forms of garbage in most parts of the ocean, the restricted regions being primarily coastal regions and special areas desig-

---

[3]Annexes I and II of MARPOL, which deal with oil discharges from ships and the transport of hazardous liquids, respectively, are automatically adopted by any country that signs onto MARPOL. Annexes III, IV, and V are optional and concern the transport of hazardous materials in packaged form, the discharge of sewage from ships, and the discharge of garbage from ships, respectively.

TABLE 17.2 SUMMARY OF MARPOL ANNEX V AT-SEA GARBAGE DISPOSAL REGULATIONS

	Private Vessels		
Garbage Type	*Outside Special Areas*	*Inside Special Areas*[a]	*Offshore Platforms*
Plastics, including synthetic ropes, fishing nets, and plastic bags	Disposal prohibited	Disposal prohibited	Disposal prohibited
Floating dunnage, lining and packing materials	>25 nautical miles offshore	Disposal prohibited	Disposal prohibited
Paper rags, metal, bottles, crockery, and similar refuse	>12 nautical miles offshore	Disposal prohibited	Disposal prohibited
All other garbage, including paper rags, glass, etc., comminuted or ground[b]	>3 nautical miles offshore	Disposal prohibited	Disposal prohibited
Food waste not comminuted or ground	>12 nautical miles offshore	>12 nautical miles offshore	Disposal prohibited
Food waste comminuted or ground	>3 nautical miles offshore	>12 nautical miles offshore	>12 nautical miles offshore
Mixed refuse types		When garbage is mixed with other harmful substances having different disposal or discharge requirements, the most stringent disposal requirements apply	

[a]Middle Eastern Gulf areas and the Mediterranean, Baltic, and Black seas have been designated special areas.
[b]Comminuted or ground garbage must be able to pass through a screen with a mesh size no larger than 1 inch.
Source: Edwards and Rymarz (1990).

nated by the Convention. Disposal of plastics, however, is prohibited by Annex V in all parts of the ocean.

While Annex V seems to be an important step in controlling plastic pollution of the ocean, it is one thing to legislate and another thing to implement. For years ships had been accustomed to disposing of their garbage, including plastics, at sea. How were vessels going to dispose of their plastic waste after December 31, 1988?

Recognizing this problem, the Marine Environment Protection Committee (MEPC) of the International Maritime Organization (IMO) developed guidelines for the implementation of Annex V at consecutive sessions in 1987 and 1988. The purpose of the guidelines was to assist governments that had ratified Annex V in developing and enacting domestic laws that would enforce and implement Annex V; assist vessel operators in complying with the requirements of Annex V and domestic laws; and assist port and terminal operators in determining the need for and providing adequate reception facilities for garbage generated by ships (Edwards and Rymarz, 1990). The guidelines fell into six general areas as follows:

1. *Training, Education, and Information.* Educating both seafarers and port authorities about the requirements of Annex V, the environmental impact of at-sea garbage disposal, and the need for the development of better vessel and port waste management facilities was perceived as essential for the successful implementation of Annex V. Knowledge of national and international regulations governing garbage disposal at sea is now, for example, expected to be a part of maritime certification examinations and requirements. Ships are expected to post a summary

declaration stating the restrictions for discharging garbage at sea and the penalties for failure to comply (Edwards and Rymarz, 1990). Curricula at maritime colleges and technical institutes are to include information on environmental impacts of garbage in the marine environment. Efforts to educate the general public about Annex V are to be made through radio and television broadcasts, articles in periodicals and trade journals, voluntary public projects such as beach cleanups, posters, brochures, and so forth. Figure 17.3, an educational sticker for boaters developed by the Center for Marine Conservation, is one example of the sort of educational material that has been developed as part of this effort. Active exchange is being encouraged between governments and the IMO of technical information on shipboard waste management, copies of current domestic laws and regulations related to Annex V, educational materials developed to raise the level of compliance with Annex V, and information on the nature and extent of marine debris.

2. *Minimizing Garbage Production.* One obvious way to reduce garbage disposal is to decrease the amount of garbage produced. In the case of plastics, this reduction may be accomplished to a certain extent by switching to reusable packaging and containers and by packaging provisions in something other than disposable plastic. Bulk packaging of consumable items may also help, at least if the product has a sufficiently long shelf life. Fishing vessel operators are encouraged to develop methods to minimize accidental encounters between ships and gear. Benthic traps, trawls, and gill nets should be designed to have degradable panels or sections made of natural fibers, wood, or wire. Efforts should be made to develop methods for recycling plastic waste returned to shore as garbage.

3. *Shipboard Garbage Handling and Storage Procedures.* Compliance with Annex V will obviously require that ships have facilities for collecting, sorting, processing, storing, and disposing of garbage. Waste management plans should be incorporated into crew and vessel operating manuals. Garbage collection devices must allow convenient separation of different forms of waste according to mandated disposal methods. There should be a clear separation, for example, between plastic waste and food

FIGURE 17.3   MARPOL Annex V educational poster. Produced by the Center for Marine Conservation. Used with permission.

waste. On-board processing facilities may include incinerators, compactors, and comminuters. Compactors are appealing because they reduce the volume of waste and hence facilitate storage. Compactors also facilitate sinking if garbage is discharged, and comminuters enhance assimilation of food wastes and other discharged garbage, but these latter considerations are irrelevant to plastics, which cannot be disposed of at sea. Incineration might seem an easy way to avoid collecting and storing plastic waste at sea, but there is a problem. According to the 1988 MEPC guidelines, "Ash from the combustion of some plastic products may contain heavy metal or other residues that can be toxic and should therefore not be discharged into the sea. Such ashes should be retained on board, where possible, and discharged at port reception facilities" (Bean, 1990, p. 993). Indeed, plastics are believed to account for 70% and 90% of the Pb and Cd, respectively, in municipal incinerator ash due to the use of these metals in stabilizers and pigments, and the fly ash from such incinerators often qualifies as a hazardous waste under the Federal Resource Conservation and Recovery Act due to its high Pb and Cd content (Bean, 1990). In general, then, ship operators should plan on retaining the ash from plastics incineration.

4. *Shipboard Equipment for Processing Garbage.* Shipboard equipment that might be used in processing plastic waste includes comminuters, compactors, and incinerators. Comminuters are relevant because it is usually impractical to compact uncomminuted plastics. Much manual labor is required to size the material for feed. Comminuted plastics, however, can be rapidly compacted to produce a medium-density material that requires minimal onboard storage space (Edwards and Rymarz, 1990). Incineration reduces the volume of plastic waste by about 95%, but it produces a potentially hazardous ash and, of course, eliminates any possibility of plastics recycling. In addition, combustion of plastics may produce dangerous gases such as hydrochloric acid vapor, and if adequate levels of O_2 are not supplied to the incinerator, high levels of soot will form in the exhaust stream (Edwards and Rymarz, 1990).

5. *Port Reception Facilities.* Upgrading port reception facilities to handle plastic waste is a key to the successful implementation of Annex V. It is critical that vessel operators know the ability of various ports to handle waste so that they can intelligently plan and direct on-board collection, processing, and storage. It would obviously be futile for a ship operator to carefully comminute, compact, and store plastic waste if the next port of call could not accept the waste. With respect to this point, it is noteworthy that of the special areas designated in MARPOL Annex V, restrictions were initially enforced only in the Baltic Sea. Dumping restrictions were not enforced in the other areas until the regions could demonstrate proper reception facilities and subsequently satisfy a one year waiting period (O'Hara et al., 1988).

6. *Ensuring Compliance.* Compliance with Annex V will to a large extent be voluntary because it is obviously impractical to place an observer on every seagoing vessel. The degree of compliance will depend on the extent to which seafarers appreciate and understand the problems caused by improper disposal of garbage at sea. This perception and understanding will come through various forms of training and education. Also important will be the need to create economic incentives to comply with Annex V, to reduce the amount of vessel-generated garbage, and to recycle. A passenger ship operator, for example, should be rewarded for switching from disposable polystyrene cups to paper cups, but the latter cost two to three times as much (Martinez, 1990). Imaginative strategies for reducing the amount of waste generated at sea, minimizing the use of plastics, and recovering and recycling plastics where possible should be rewarded economically.

Other Legislation

At about the same time as the ratification of MARPOL Annex V, several other laws relevant to the problem of plastic debris were passed in the United States. The Marine Plastic Pollution Research and Control Act (MPPRCA), Public Law 100–220, went into effect at the same time as Annex V and was intended to implement the provisions of Annex V. The MPPRCA prohibits the disposal of plastics anywhere at sea by U.S. vessels and prohibits the disposal of plastics by any vessel in U.S. waters, including bays, sounds, inland waterways, and the U.S. Exclusive Economic Zone (i.e., the coastal ocean within 200 nautical miles of the U.S. shoreline). The MPPRCA also directed all federal agencies, including the Navy and Coast Guard, to bring their vessels into full compliance with Annex V regulations by 1994.

Sometimes controversial legislation has been proposed or passed concerning the use of degradable plastics. Plastic products can be made either photodegradable or biodegradable, in the former case by changing the molecular structure of the plastic so that it decomposes in the presence of ultraviolet light and in the latter case by adding cornstarch or other vegetable additives that can be broken down by microorganisms. A number of states have passed laws requiring that six-pack rings be degradable (O'Hara et al., 1988), and similar federal legislation was passed by Congress in 1988. Other degradable plastic items, such as shopping bags, have become available in recent years.

The logic behind degradable plastics is that degradation will render the plastic harmless, or at least less of a threat to the environment. Some environmentalists, however, have questioned this wisdom. They argue that the production of degradable plastic products encourages rather than discourages the use of plastics, and that under certain conditions degradable plastics are no more degradable than ordinary plastics. The deposition of plastic debris in landfills, for example, virtually precludes photodegradation, and biodegradation is minimal in well-operated landfills due to the absence of air and water in the compacted waste (Denison, 1990). Furthermore, it is only the cornstarch or vegetable additives that actually degrade in biodegradable plastics. The plastic dust that remains following biodegradation may be less visible and pose no threat of entanglement, but it may be more of a threat from the standpoint of ingestion than the undegraded product. On the other hand, a requirement that fishing nets and traps include degradable components might do much to reduce the loss of marine life caused by ghost fishing. The consensus would seem to be that degradable plastics may prove useful in reducing the adverse effects associated with some types of plastic debris, but they are no panacea. Indeed, there are some uses of plastics for which degradability would clearly be undesirable. For example, hazardous liquids should certainly not be stored in a degradable plastic bottle.

Solutions Through Technology

An obvious way to reduce the amount of plastic debris is to recycle plastic products. This idea is not as easy to implement, however, as one might think. *Plastic* is a generic term covering hundreds of different types of products. Each type of plastic—for example, polypropylene, polyethylene, and PVC—has characteristics that make it best suited for certain applications. One cannot simply throw all sorts of plastic products into a pot, melt them down, and then mold them into useful products.

Nevertheless, the concept of plastic recycling is not new, and some forms of plastic have been recycled for more than a decade. Recycling of plastic milk bottles made from high-density polyethylene began during the mid-1970s. Recycling was facilitated by the facts that the bottles were easy to identify, they were all made from the same type of plastic, and there was a large enough supply to make recycling economical. Similar recycling of plastic soft drink bottles, all of which are made from polyethylene terephthalate (PET), began in 1979. As of 1987, plastic milk and soft drink bottles accounted for about 40% of all plastic bottles sold in the United States, and in that year almost 70,000 tonnes of PET bottles, or 20% of those sold, were recycled (O'Hara et al., 1988).

Plastic bottles are recycled in a variety of products, but because of concerns about purity and food contact, they are not recycled as beverage containers. Plastic milk bottles, for example, are recycled as plastic lumber, underground pipes, toys, pails, traffic barrier cones, garden furniture, golf bag liners, kitchen drain boards, milk bottle carriers, trash cans, and signs (O'Hara et al., 1988). PET is recycled primarily as fiberfill, which is used to stuff pillows, ski jackets, sleeping bags, and automobile seats (O'Hara et al., 1988). Obviously, recycling of certain types of plastics has proven to be feasible. What is needed now is a significant increase in the quantity and variety of plastic products that are recycled. Among the obvious problems to be solved are the development of methods to separate different types of used plastic and the identification of suitable products that can be made from recycled plastics. The importance of solving these problems was underscored in 1984 with the formation of the Plastics Recycling Foundation in the United States. The Foundation has established a Center for Plastics Recycling Research at Rutgers University to carry out research and development of methods for recycling plastic products.

A somewhat controversial issue is the significance and source of plastic resin pellets. Such pellets are a minor component of the plastic debris in the ocean, accounting for only about 0.5%, for example, of the pieces of plastic collected in the North Pacific; but they account for about 70% of the plastic found in the stomach of seabirds (Bruner, 1990). Studies conducted during the 1970s made it clear that large numbers of pellets were escaping from plastics factories, and efforts to correct this problem were undertaken by the industry. Dow Chemical, for example, estimates that pellet reclamation procedures undertaken at its Louisiana plant recover more than 80 tonnes of pellets per year that would

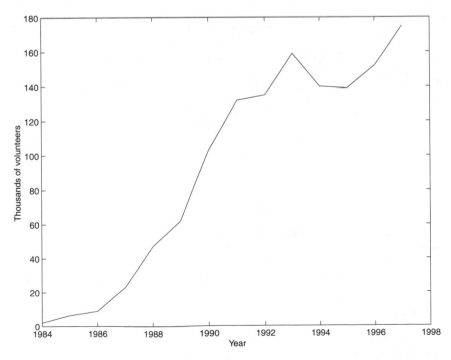

FIGURE 17.4   Volunteer participation in beach cleanups in the United States from 1984 to 1997. *Source*: Debenham (1990) and the Center for Marine Conservation (pers. comm).

otherwise escape into marine areas (O'Hara et al., 1988). Nevertheless, pellets are still detected in coastal areas of the United States, and it is unclear whether most of the escapement occurs during transportation and handling or whether losses at the manufacturing level are still a major source of pollution. Identification of the principal sources of the pellets and development of techniques for reducing pellet escapement are needed.

Education

Education of both seafarers and the general public is perceived as being crucial to a successful resolution of the plastic debris problem. One of the best examples of the impact of public education efforts has been the proliferation of annual beach cleanup days and the number of persons involved (Figure 17.4). From 2,100 persons who participated in the first beach cleanup in Oregon in 1984, beach cleanups have spread to all coastal states and involved about 175,000 persons in 1997.

A number of public and private organizations, including in particular the Center for Marine Conservation, the National Oceanic and Atmospheric Administration, and the IMO, have been actively involved in public awareness and education efforts. Since most of the derelict plastic in the offshore ocean comes from vessels, it is vital that seafarers are aware of the adverse effects caused by plastics pollution, and understand and respect laws designed to minimize this form of pollution. Although MARPOL Annex V has been an important step in reducing the input of plastics from vessels, serious problems associated with ghost fishing by lost trawl nets and traps remain. To a certain extent, these problems will probably always exist, but they can be minimized with the help of technology and by a conscientious effort by fishermen to reduce net and trap losses. Training and education will do much to create the awareness and sustain the motivation needed to further reduce pollution of the oceans by plastics.

REFERENCES

ANONYMOUS, 1982. Confused sea turtles said to be dying from diet of plastic trash. *The Baltimore Sun*, December 28, cited by CEE (1987, p. 40).

ANONYMOUS, 1992. Would anyone like a souvenir Japanese driftnet? *Mar. Cons. News*, 4(1), 1.

BALAZS, G. 1985. Impact of ocean debris on marine turtles: Entanglement and ingestion. In R. S. Shomura and H. O. Yoshida (Eds.), *Proceedings of the Workshop on the Fate and Impact of Marine Debris*. U.S. Department of Commerce, Honolulu, HI. Pp. 387–429.

BEAN, M. J. 1990. Redressing the problem of persistent marine debris through law and public policy: Opportunities and pitfalls. In R. S. Shomura and M. L. Godfrey (Eds.), *Proceedings of the Second International Conference on Marine Debris*, Vol. II. U.S. Department of Commerce, Honolulu, HI. Pp. 989–997.

BOURNE, W. R. P. 1977. Nylon netting as a hazard to birds. *Mar. Pollut. Bull.*, 8(4), 75–76.

BREEN, P. A. 1990. A review of ghost fishing by traps and gillnets. In R. S. Shomura and M. L. Godfrey (Eds.), *Proceedings of the Second International Conference on Marine Debris*, Vol. I. U.S. Department of Commerce, Honolulu, HI. Pp. 571–599.

BRUNER, R. G. 1990. The plastics industry and marine debris: Solutions through education. In R. S. Shomura and M. L. Godfrey (Eds.), *Proceedings of the Second International Conference on Marine Debris*, Vol. II. U.S. Department of Commerce, Honolulu, HI. Pp. 1077–1089.

CARR, H. A., A. W. HULBERT, and E. H. AMARAL. 1985. Underwater survey of simulated lost demersal and lost commercial gill nets off New England. In R. S. Shomura and H. O. Yoshida (Eds.), *Proceedings of the Workshop on the Fate and Impact of Marine Debris*. U.S. Department of Commerce, Honolulu, HI. Pp. 438–447.

CEE. 1987. *Plastics in the Ocean: More Than a Litter Problem*. Center for Environmental Education, Washington, DC. 127 pp.

DAY, R. H., D. H. S. WEHLE, and F. C. COLEMAN. 1985. Ingestion of plastic pollutants by marine birds. In R. S. Shomura and H. O. Yoshida (Eds.), *Proceedings of the Workshop on the Fate and Impact of Marine Debris*. U.S. Department of Commerce, Honolulu, HI. Pp. 344–386.

DEBENHAM, P. 1990. Education and awareness: Keys to solving the marine debris problem. In R. S. Shomura and M. L. Godfrey (Eds.), *Proceedings of the Second International Conference on Marine Debris*, Vol. II. U.S. Department of Commerce, Honolulu, HI. Pp. 1100–1114.

DEGANGE, A. R., and T. C. NEWBY. 1980. Mortality of seabirds and fish in a lost salmon driftnet. *Mar. Pollut. Bull.*, 11, 322–323.

DENISON, R. A. 1990. Degradable plastics: Right question, wrong answer. *Environ. Defense Fund Letter*, 21(1), 4.

EDWARDS, D. T., and E. RYMARZ. 1990. International regulations for the prevention and control of pollution by debris from ships. In R. S. Shomura and M. L. Godfrey (Eds.), *Proceedings of the Second International Conference on Marine Debris*, Vol. II. U.S. Department of Commerce, Honolulu, HI. Pp. 956–988.

EISENBUD, R. 1985. Problems and prospects for the pelagic driftnet. *Boston College Environ. Affairs Law Rev.*, 12(3), 473–490.

FRENCH, D. P., and M. REED. 1990. Potential impact of entanglement in marine debris on the population dynamics of the northern fur seal, *Callorhinus ursinus*. In R. S. Shomura and M. L. Godfrey (Eds.), *Proceedings of the Second International Conference on Marine Debris*, Vol. I. U.S. Department of Commerce, Honolulu, HI. Pp. 431–452.

FRY, D. M., and S. I. FEFER. 1986. Ingestion of floating plastic debris by seabirds in the Hawaiian Islands. Paper presented at the Sixth International Ocean Disposal Symposium, April 21–25, Pacific Grove, CA. In *Program and Abstracts*. Pp. 73–74.

GRESS, F., and D. W. ANDERSON. 1983. *California Brown Pelican Recovery Plan*. U.S. Fish and Wildlife Service, Portland, OR.

HENDERSON, J. R. 1990. Recent entanglements of Hawaiian monk seals in marine debris. In R. S. Shomura and M. L. Godfrey (Eds.), *Proceedings of the Second International Conference on Marine Debris*, Vol. I. U.S. Department of Commerce, Honolulu, HI. Pp. 540–553.

JONES, L. L., and R. C. FERRERO. 1985. Observations of net debris and associated entanglements in the North Pacific Ocean and Bering Sea, 1978–1984. In R. S. Shomura and H. O. Yoshida (Eds.), *Proceedings of the Workshop on the Fate and Impact of Marine Debris*. U.S. Department of Commerce, Honolulu, HI. Pp. 183–196.

JONES, S. C., R. J. TARPLEY, and S. FERNANDEZ. 1986. Cetacean strandings along the Texas coast, U. S. A. Paper presented at the 11th International Conference on Marine Mammals, April 2–6, Guaymas, Mexico.

KARTAR, S., R. A. MILNE, and M. SAINSBURY. 1973. Polystyrene waste in the Severn Estuary. *Mar. Pollut. Bull.*, 4, 44.

KARTAR, S., R. A. MILNE, and M. SAINSBURY. 1976. Polystyrene spherules in the Severn Estuary—a progress report. *Mar. Pollut. Bull.*, 7, 52.

KING, B. 1985. Trash and debris on the beaches of Padre Island National Seashore. Sixth Annual Minerals Management Service, Gulf of Mexico OCS Regional Office, Information Transfer Meeting. Session IV.E. Trash and Debris on Gulf of Mexico Waterfront Beaches, October 23, Metairie, LA (unpublished manuscript).

LOW, L. L., R. E. NELSON, JR., and R. E. NARITA. 1985. Net loss from trawl fisheries off Alaska. In R. S. Shomura and H. O. Yoshida (Eds.), *Proceedings of the Workshop on the Fate and Impact of Marine Debris*. U.S. Department of Commerce, Honolulu, HI. Pp. 130–153.

MARTINEZ, L. A. 1990. Shipboard waste disposal: Taking out the trash under the new rules. In R. S. Shomura and M. L. Godfrey (Eds.), *Proceedings of the Second International Conference on Marine Debris*, Vol. II. U.S. Department of Commerce, Honolulu, HI. Pp. 895–914.

MERRELL, T. R., JR. 1985. Fish nets and other plastic litter on Alaska beaches. In R. S. Shomura and H. O. Yoshida (Eds.), *Proceedings of the Workshop on the Fate and Impact of Marine Debris*. U.S. Department of Commerce, Honolulu, HI. Pp. 160–182.

O'HARA, K. J., S. IUDICELLO, and R. BIERCE. 1988. *A Citizen's Guide to Plastics in the Ocean: More Than a Litter Problem*. Center for Marine Conservation, Washington, D.C. 143 pp.

SEA CONSERVATION SOCIETY. 1999. Northern fur seal. *http://www.greenchannel.com/tec/species/norfursl.htm*

SOCIETY OF THE PLASTICS INDUSTRY. 1999. (SPI: The plastics industry trade association). *http://www.plasticsindustry.org*

STEWART, R. S., and P. K. YOCHEM. 1990. Pinniped entanglement in synthetic materials in the Southern California Bight. In R. S. Shomura and M. L. Godfrey (Eds.), *Proceedings of the Second International Conference on Marine Debris*, Vol. I. U.S. Department of Commerce, Honolulu, HI. Pp. 554–561.

SWANSON, R. L., H. M. STANFORD, J. S. O'CONNOR, S. CHANESMAN, C. A. PARKER, P. A. EISEN, and G. F. MAYER. 1978. June 1976 pollution of Long Island ocean beaches. *J. Environ. Eng. Div., ASCE,* **104**(EE6), 1067–1085.

TAKEHAMA, S. 1990. Estimation of damages to fishing vessels caused by marine debris, based on insurance statistics. In R. S. Shomura and M. L. Godfrey (Eds.), *Proceedings of the Second International Conference on Marine Debris*, Vol. II. U.S. Department of Commerce, Honolulu, HI. Pp. 792–809.

WAGNER, K. D. 1990. Medical wastes and the beach washups of 1988: Issues and impacts. In R. S. Shomura and M. L. Godfrey (Eds.), *Proceedings of the Second International Conference on Marine Debris*, Vol. II. U.S. Department of Commerce, Honolulu, HI. Pp. 811– 823.

YOUNGER, L. K. 1992. Participation soars at 1991 international coastal cleanup. *Mar. Cons. News,* **4**(1), 17.

QUESTIONS

17.1 Scientists believe that ghost fishing killed hundreds of thousands of aquatic birds every year during the 1980s. Why were so many birds killed by gear that was intended to catch fish?

17.2 Why are some environmental groups opposed to the use of biodegradable plastics?

17.3 What is the most likely explanation for the fact that sea turtles swallow plastic bags?

17.4 Raw plastic is typically shipped in the form of small pellets a few millimeters in diameter. These pellets have been found in the stomachs of certain marine organisms. Why do we think that some marine organisms deliberately ingest these small pellets?

17.5 According to Annex V of MARPOL, where is it legal to discharge plastic waste in the ocean?

✧ ⌑ ✧ ⌑ ✧

*Glossary**

acceptable daily intake (8) the amount of a food additive that can be ingested daily in the diet without appreciable risk on the basis of all facts known at the time. *Without appreciable risk* refers to the practical certainty that injury will not result, even after a lifetime of experience. The acceptable daily intake is a practical approach to determining the safety of food additives and is a means of achieving some uniformity of approach in regulatory control. It ensures that the actual human intake of a substance is well below toxic levels. In the United States, acceptable daily intake are determined by the Food and Drug Administration.

acid rain (15) rain having a pH of less than about 5.6.

actinide (14) a member of a group of chemically similar radioactive elements with atomic numbers from 89 (actinium) through 103 (lawrencium).

action level (8) the concentration of a contaminant in a product at or above which the U.S. Food and Drug Administration (FDA) will take legal action to remove the product from the market. Where no established action level exists, the FDA may take legal action against the product at the minimal detectable level of the contaminant. Action levels are established and revised according to criteria specified in Title 21, *Code of Federal Regulations*, Parts 109 and 509, and are revoked when a regulation establishing a tolerance for the same substance and use becomes effective.

activated sludge (6) a flocculent mass of microorganisms and adsorbed materials that makes up the suspended solids in a secondary wastewater treatment process referred to as the *activated sludge process*. The microorganisms in the sludge, primarily bacteria, feed off the organics in the wastewater. The activated sludge is kept in suspension by vigorous aeration.

acute/chronic ratio (8) the ratio of the concentration or level of a toxic substance or stress that produces toxic effects after a short period of exposure to the concentration or level of the same substance or stress that produces toxic effects after a long period of exposure. See *acute toxicity* and *chronic toxicity*.

acute toxicity (8) toxicity resulting from exposure to a toxic substance or stress for a relatively brief period of time, typically no more than 48 to 96 hours, but never more than 10% of the natural lifetime of an organism.

acute toxicity unit (8) for a given species and a single toxic substance, the 96-hour TLm. For a mixture of toxicants, any combination of concentrations that kills would be expected to kill half of the individuals of the same species in 96 hours.

acceptable daily intake (8) see acceptable daily intake.

adverse effect (8) any effect that results in functional impairment and/or pathological lesions that may affect the performance of a whole organism or that reduces an organism's ability to respond to an additional challenge.

alkane (13) an open-chain (straight- or branched-chain) hydrocarbon having the general formula C_nH_{2n+2}.

*Numbers in parentheses indicate the chapter in which the term is introduced.

alkene   (13) an open-chain (straight- or branched-chain) hydrocarbon containing one pair of carbon atoms joined by a double electron-pair bond and having the general formula C_nH_{2n}.

allochthonous   (2) derived from external sources. The term is most commonly used to distinguish sources of nutrients in an aquatic system. The contrasting term is *autochthonous*, which refers to something derived from internal sources.

alpha particle   (14) a particle consisting of two protons and two neutrons and hence equal to the nucleus of the common form of He. Alpha particles are commonly emitted during the radioactive decay of elements with atomic numbers greater than 83. Some radionuclides of low atomic number also decay by this mechanism.

amebic dysentery   (7) see *dysentery*.

ammonia stripping   (6) a form of tertiary wastewater treatment used to remove ammonium and ammonia. NH_4^+ is converted to NH_3 by raising the pH of the wastewater to 11, and the NH_3 gas is driven off by vigorous aeration.

anabolism   (1) the phase of metabolism in which simple organic substances are synthesized into the living biomas.

anaerobic digester   (6) a component of a conventional wastewater treatment plant designed to stabilize and reduce the volume of the waste sludge produced by the primary and secondary clarifiers. Bacteria in the anaerobic digester feed on the organic matter in the sludge and, in the absence of oxygen, convert the labile portion of the sludge to methane and carbon dioxide.

anion   (14) a negatively charged ion.

ankylosing spondylitis   (14) inflammation of the vertebrae associated with stiffening and immobility caused by disease, trauma, surgery, or abnormal bone fusion.

anoxic   (3) devoid of dissolved O_2.

anthelmintic   (16) Something that destroys or causes the expulsion of parasitic intestinal worms.

authropogenic   (4) related to or resulting from the activity of human beings.

aphotic zone   (2) the water column below the compensation depth; the light-limited portion of the water column within which the net photosynthetic rate is negative.

aquifer   (16) an underground bed or layer of earth, gravel, or porous rock that will yield water to a well or spring.

aquitard   (16) a layer of rock of low permeability that slows the movement of groundwater.

arithmetic mean   (8) the sum of n numbers divided by n.

aromatic   (13) a hydrocarbon containing one or more planar six-C ring structures analogous to benzene.

artesian aquifer   (12) see *confined aquifer*.

atomic number   (14) the number of protons in the nucleus of an element. The identity of an element is uniquely determined by its atomic number.

autotroph   (1) an organism capable of converting inorganic substances into living biomass using either light or chemical energy.

bacillary dysentery   (7) see *dysentery*.

BAF   (8) see *biological accumulation factor*.

bagasse   (9) the fibrous residue that remains after juice has been extracted from the crushed stalks of sugar cane. Bagasse may be pelletized and used as a fuel in sugar mills or incorporated into products such as particleboard or paperboard.

ballast   (13) water taken onboard a ship to prevent the ship from becoming top-heavy and unstable when its cargo is offloaded.

BCF   (8) see *biological concentration factor*.

becquerel   (14) a unit of radioactivity equal to one disintegration per second.

benthic   (2) pertaining to the bottom of an aquatic system or to organisms that live directly above, on (epifauna), or in (infauna) the sediments at the bottom.

benthos   (4) the community of organisms that live directly above, on, or in the sediments at the bottom of an aquatic ecosystem.

beta particle   (14) an electron (β^-) or positron (β^+) emitted during the radioactive decay of many radionuclides.

bilge   (13) the lowest inner portion of a ship's hull.

bilge oil   (13) oil that collects and stagnates in the bilge of a ship.

biocide   (10) a chemical that is capable of killing living organisms.

biological accumulation factor (8) the ratio of the concentration of a substance in one or more tissues of an aquatic organism to the concentration of the same substance in the water in which the organism has been living.

biological concentration factor (8) the biological accumulation factor associated with direct uptake of a substance from the water in the absence of any possible intake via the food chain.

bioremediation (13) the stimulation of biological processes to decompose, degrade, or detoxify a toxic substance.

blowdown (11) a process by which accumulated solids are removed from steam boilers. Excessive dissolved solids cause a boiler to foam and allow the formation of scale, which affects the efficiency of the unit.

BOD (5) the amount of oxygen consumed by chemical and biological processes in 300 ml of water when incubated in the dark for five days at a temperature of 20°C.

breakpoint chlorination (6) a form of tertiary wastewater treatment designed to remove ammonia by oxidizing it with Cl gas to produce N_2 and HCl.

bubble curtain (11) a wall or curtain of bubbles produced by discharging air through holes in a pipe and intended to frighten fish away from the intake structure of a power plant's cooling system.

builder (6) a component of a detergent designed to soften the water so that the surfactant will not form a gummy precipitate with Ca and Mg ions.

bunker (13) a ship's fuel storage tank.

carbamate (10) an ester or salt of carbamic acid, NH_2CO_2H. Carbamate pesticides are used to control insect and nematode populations.

carbohydrate (2) any of a group of organic compounds that contain C, M, and O in the approximate ratio 1:2:1 by atoms. Carbohydrates are important respiratory substrates and include sugars, starches, celluloses, and gums.

carnivore (1) an animal that consumes other animals as a source of food.

carrier (7) a person or animal infected with a pathogen. The immune system of a temporary carrier eventually destroys the pathogen. After the initial infection, a permanent carrier harbors the pathogen for the remainder of his or her life. The carrier of the pathogen that causes a disease may be symptomatic or asymptomatic.

catabolism (1) the phase of metabolism in which complex organic molecules are broken down into simpler ones, often resulting in a release of energy.

cation (14) a positively charged ion.

cellulose (9) a complex carbohydrate composed of glucose units. Cellulose is the main component of the cell wall in most plants and is the principal component of wood that chemical pulping methods convert to pulp for the manufacture of paper and paperboard.

cesspool (6) an underground chamber for the reception and storage of wastewater with no treatment taking place. Some cesspools are watertight and retain both the liquid and solid components of the wastewater. More typically, cesspools are constructed of porous material, in which case the liquid portion of the wastewater seeps into the surrounding leach field. For example, a cesspool may be nothing more than an underground pit lined with tile. Regardless of whether a cesspool is porous or watertight, it requires periodic pumping to remove accumulated waste. (see *septic tank*)

chemoautotroph (1) an organism that carries out chemosynthesis.

chemosynthesis (1) the synthesis of organic matter from inorganic compounds using the energy released from the oxidation of reduced inorganic chemicals as an energy source.

chironomid (14) a two-winged fly similar in appearance to a mosquito but not blood-sucking. The larvae of chironomids are aquatic and are important sources of food for some fish.

chlorophyll (2) a photosynthetic pigment found in all plants and responsible for mediating the conversion of light energy into chemical energy in the photosynthetic process. Chlorophyll occurs in several similar forms, the most common of which is designated chlorophyll *a*.

chronic toxicity (8) toxicity resulting from exposure to a toxic substance or stress for a relatively long period of time, at least 10% of the natural lifetime of an organism.

chronic toxicity unit (8) for a given species and a single toxic substance or stress, the smallest concentration of the toxic substance or level of stress that would be expected to have an adverse effect on the species after a long period of exposure. For a mixture of toxicants or stresses, any combination of concentrations or stresses that would be expected to produce a similar adverse effect after a long period of exposure.

cladding   (14) in nuclear reactors, the protective material, usually stainless steel or an alloy of zirconium, that encases the fuel elements. The cladding prevents fission products from escaping into the cooling water and prevents cooling water from coming into contact with the fuel.

clingage   (13) oil that adheres to the sides of a tanker's cargo compartments when the oil is offloaded.

cogeneration power plant   (11) a power plant that produces both electricity and hot water, the latter being hot enough to be useful in district heating or district heating/cooling networks.

coke   (15) the solid residue of C and ash obtained from bituminous coal after removal of volatile material by heating.

cold-blooded   (8) of or relating to an organism whose body temperature is determined largely by a passive exchange of heat between the organism and its surroundings.

combined sewer system   (4) a sewer system characterized by a single system of pipes and conduits for intercepting and collecting sanitary sewage and land runoff in an urban area (see *separate sewer system*).

comminute   (17) reduce to small particles; pulverize.

compensation depth   (2) the depth in the water column at which light limitation reduces the net photosynthetic rate of aquatic plants to zero.

confined aquifer   (16) an aquifer bounded above and below by materials such as clay or shale through which water moves very slowly. Groundwater in a confined aquifer is under pressure and rises in a well above the top of the aquifer.

conformer   (8) an organism that does not regulate its internal state with respect to an environmental variable such as temperature or salinity. An organism that is a conformer with respect to one environmental variable may be a regulator with respect to another. See *regulator.*

coral reef   (4) A wave-resistant structure built by marine organisms. The framework of the reef consists of Mg and Ca carbonates biochemically deposited primarily by hermatypic corals and coralline algae. The binding material includes coralline algae, hydrocorals, Bryozoa, and Foraminifera. Coral reefs are found in tropical latitudes where the winter water temperature is no less than about 18°C and at salinities between 27 and 40 psu. The reef structure provides a habitat for a wide variety of fish and invertebrates. Coral reefs are considered to be among the most productive natural ecosystems on Earth.

critical depth   (2) the depth in a water column above which net community production is zero when integrated from the surface to the critical depth.

criterion continuous concentration   (8) the four day average concentration of a toxicant not to be exceeded more than once every three years and defined by the U.S. Environmental Protection Agency to be the minimum of the final chronic and final plant values.

criterion maximum concentration   (8) the one hour average concentration of a toxicant not to be exceeded more than once every three years and defined by the U.S. Environmental Protection Agency to be half of the final acute value.

critical group   (14) in the context of radiation toxicology, the group of persons representative of those individuals in the population expected to receive the highest dose equivalent. The critical group is expected to include more than one but no more than a few tens of persons.

critical mass   (14) the minimum mass of a fissionable nuclide required to sustain a chain reaction of fissioning. For ^{235}U and ^{239}Pu, the critical mass is about 50 kg and 11 kg, respectively.

critical pathway   (14) in the context of radiation toxicology, the pathway from release to the environment to human contact responsible for the greatest amount of human exposure to the radioactivity from a particular source.

cultural eutrophication   (1) any acceleration of the eutrophication process as a result of human activities.

cyanobacterium   (2) any member of a group of photosynthetic bacteria, some species of which are capable of N fixation.

cycloalkane   (13) a hydrocarbon in which some or all of the C atoms are arranged in a ring-like structure and have the general formula C_nH_{2n}.

cyst   (7) a thick-walled reproductive body that is capable of growing into a new organism.

daughter   (14) the nuclide resulting from the radioactive decay of a radionuclide.

denitrification   (2) a process that transforms nitrate into N gas, N_2. The process is mediated by bacteria that use the O_2 from nitrate to oxidize organic mater under anoxic or nearly anoxic conditions.

depurate   (12) to clean or purify.

detritivore (1) an organism that feeds on detritus.

detritus (1) nonliving dissolved or particulate organic matter.

detritus food chain (1) a food chain whose first and second trophic levels consist of detritus and detritivores.

diarrhea (7) the frequent and excessive discharge of watery stools.

diatom (2) a phytoplanktonic organism having cell walls of silica consisting of two interlocking symmetrical halves called *valves*. Diatoms are common in most freshwater and marine systems but are most abundant in waters of the southern ocean.

dimictic (3) twice-mixing. A dimictic body of water mixes or overturns twice per year. One overturning occurs in the autumn, when the surface waters are being cooled and the temperature of the surface waters exceeds the temperature of maximum density. The other overturning occurs in the spring, when the surface waters are being warmed and the temperature of the surface waters is less than the temperature of maximum density.

dinoflagellate (2) any member of a group of microscopic, chiefly marine protozoans that possess two flagella and are therefore capable of some degree of movement. Most species of dinoflagellates have photosynthetic capabilities, but some are strictly holozoic, and others are partly holozoic and partly saprophytic. Some dinoflagellates are bioluminescent. Red tides are caused by blooms of certain species of dinoflagellates.

district heating/cooling network (11) a distribution system of insulated pipes designed to utilize the hot water discharged by a cogeneration power plant to both heat and cool buildings in the vicinity of the plant. Often referred to as a *DHC network*.

dropline (16) the piping that conveys water from the main pipeline to the discharge nozzles of a center pivot sprinkler system. In the 1960s, high-pressure center pivot sprinkler systems sprayed water into the air from nozzles on the mainline. This resulted in significant water losses due to evaporation and wind drift. The addition of droplines to center pivot sprinkler systems lowers the nozzles closer to the soil surface. This reduces losses to evaporation and wind drift.

dry dock (13) a large basin-shaped dock from which water can be pumped to facilitate building or repairing of a ship below its water line.

dry weight (8) the weight of an organism, tissue, or sediment sample after removal of water that can be evaporated off at a temperature of 105°C for 24 hours.

dual-media filter (7) a filter composed of two complementary media, such as sand and gravel or sand and activated C.

dunnage (17) loose materials used to support and protect cargo in a ship's hold.

dysentery (7) an inflammatory disorder of the lower intestines resulting in pain, fever, and severe diarrhea and often accompanied by the passage of blood and mucus. Amebic dysentery is caused by the protozoan *Entamoeba histolytica*. Bacillary dysentery is caused by bacteria of the genus *Shigella*.

EC50 (8) the concentration of a substance expected to produce a certain effect on 50% of a group of organisms after a specified period of time.

ecological efficiency (1) in general, the ratio of energy flow at different points along a food chain. As used in this book, ecological efficiency is the ratio of ingestion of food by a trophic level to the ingestion of food by the next lower trophic level. Strictly speaking, this is the trophic level energy intake efficiency.

ecological pyramid (1) schematic representations of the numbers of organisms, biomass of organisms, or energy flow at successive trophic levels in a food chain, in which the first, or producer, level forms the base and successive trophic levels the tiers that lead to the apex.

electron (14) a particle with a charge of -1 electrostatic unit and a rest mass equal to 5.5×10^{-4} atomic mass units.

electromagnetic force (14) one of the four fundamental forces of physics. The electromagnetic force is responsible for binding electrons and the nucleus of atoms together. Gamma rays are the mediators of nuclear transitions associated with electromagnetic forces.

electronegative (3) tending to attract electrons in a chemical bond.

endocrine disrupter (9) a synthetic chemical that interferes with the body's hormone (endocrine) system, which is responsible for regulating many essential functions, including growth and sexual development. A synthetic chemical can function as an endocrine disrupter by (1) mimicking naturally produced hormones such as estrogen and testosterone, (2) blocking a cell receptor and thereby preventing natural hormones from entering and performing their function, or (3) setting off reactions in the cell that would not normally be produced by a hormone. Dioxin is believed to behave in the third way.

epibenthic (13) pertaining to organisms that live on the sediments at the bottom of an aquatic ecosystem.

epilimnion (3) the mixed layer of a lake (see *mixed layer*).

epithelium (15) the membranous tissue that covers internal and external surfaces of the body, including the linings of vessels and other small cavities.

estuarine circulation (3) a circulation pattern characterized by outward flow of surface waters and inward flow of bottom waters.

estuary (3) a semienclosed coastal body of water within which seawater is measurably diluted by freshwater derived from land runoff.

euphotic zone (2) the water column above the compensation depth; the portion of the water column within which net photosynthesis is positive.

euryhaline (11) able to tolerate a wide range of salinities.

eurythermal (11) able to tolerate a wide range of temperatures.

eutrophic (3) characterized by water with a high concentration of inorganic nutrients and organic matter.

eutrophication (1) the tendency of aquatic systems to accumulate nutrients and organic matter as a result of the small excess of photosynthesis over respiration on a global scale under natural conditions.

evapotranspiration (12) the process by which water vapor is transferred from vegetated land to the atmosphere. Evapotranspiration includes both evaporation (liquid water → water vapor) and transpiration. In transpiration, liquid water is drawn from the soil into a plant's roots. The water is transported through the plant, and then evaporated from leaves and other plant surfaces into the air.

fallowing (10) the practice of allowing farmland to go unseeded during a growing season. Fallowing is a potential element of integrated pest management.

final acute value (8) the concentration of a toxic substance expected to equal or exceed the 96-hour TLm of no more than 5% of the genera in an aquatic ecosystem.

final chronic value (8) the concentration of a toxic substance expected to exert a chronic stress on no more than 5% of the genera in an aquatic ecosystem.

final plant value (8) the lowest concentration of a toxic substance that has been shown to reduce the growth of any aquatic plant.

fission (14) a process in which a relatively heavy nuclide absorbs a neutron and then splits into two smaller nuclides.

fission product (14) one of the nuclides produced by the fissioning of a heavier nuclide.

fissionable (14) a nuclide that can sustain a chain reaction of fissioning because the fission process is associated with the release of multiple neutrons with characteristics suited to inducing further fissions.

flash point (12) the lowest temperature at which the vapor of a combustible liquid can be ignited in air.

floc (7) a flocculent conglomeration of suspended particles in a fluid due to precipitation or aggregation.

flocculation (7) the formation of lumpy or fluffy masses.

fluidized bed (15) a suspension of particles supported by an upward-moving fluid.

fluoroscopy (14) examination using a fluoroscope, a device equipped with a fluorescent screen on which the internal structures of the human body may be continuously viewed as shadowy images formed by the differential transmission of X rays.

food chain (1) a simple representation of the feeding relationships in a community of organisms in which organisms are assigned to trophic levels, and the organisms in one trophic level use the organisms in the next lower trophic level as a source of food.

food chain magnification (1) the tendency for the concentration of certain substances in organisms to increase at successively higher trophic levels in a food chain due to the fact that the substances are transferred between successive trophic levels with an efficiency greater than the ecological efficiency.

food web (1) a diagrammatic representation of the feeding relationships in a community of organisms in which the position of individual organisms is determined by their approximate trophic level assignment and lines are drawn between each predator and prey pair.

free radical (14) an atom or molecule containing one or more unpaired electrons.

fresh weight (8) see *wet weight.*

fusion (14) the combining of the nuclei of two light nuclides such as ^2H and ^3H to form a heavier nuclide, with a concomitant release of energy.

gamma radiation (14) a form of electromagnetic radiation analogous to radio waves, visible light, and X rays but differing from other types of electromagnetic radiation in that they originate from the nucleus of atoms and are usually of higher energy.

gastroenteritis (7) an inflammation of the mucous membrane of the stomach and intestines.

geometric mean (8) the *n*th root of the product of *n* numbers. The geometric mean is always less than the arithmetic mean.

gray (14) a measure of radiation absorbed dose equal to 1 joule of energy per kilogram of material.

grazing food chain (1) a food chain consisting of primary producers and animals in which each successive trophic level grazes on the next lower trophic level. The grazing food chain is distinguished from the detritus food chain by the fact that primary producers occupy the first trophic level in the grazing food chain. The first trophic level in the detritus food chain consists of detritus.

Green Revolution (10) the dramatic increase in the production of agricultural crops such as corn, rice, and wheat during roughly the period 1950–1980 that resulted from the development of high-yield varieties of grain and heavy use of pesticides and fertilizers.

gross domestic product (9) the value of all final goods and services produced by an economy during a period of one year.

gross photosynthesis (2) the production of O_2 by the photosynthetic process in the absence of any respiratory losses.

gypsum (15) a colorless, white, or yellowish mineral, $CaSO_4 \cdot 2H_2O$, used as a soil amendment and to manufacture products such as drywall, plaster board, and plaster of Paris.

hard water (6) water that contains a high concentration of Ca and Mg ions. A common but arbitrary scale of water hardness is the following:

$CaCO_3$ equivalents (mg/L)	Description
0–75	Soft
75–150	Moderately hard
150–300	Hard
>300	Very hard

helminth (6) a worm, particularly a parasitic roundworm or tapeworm.

hemicellulose (9) a portion of wood fiber consisting of polysaccharides intimately associated with cellulose in the fiber wall.

herbivore (1) an animal that feeds only on plants.

heterotroph (1) an organism that cannot synthesize organic compounds from inorganic chemicals and is therefore dependent on complex organic substances for nutrition.

hill reaction (10) the production of O_2 and chemical reducing power resulting from the splitting of water molecules in photosystem II during the photosynthetic process.

hormone (10) a biochemical, usually a peptide or steroid, produced by plants and animals to regulate physiological activity, growth, or metabolism.

hydrogen bond (3) a chemical bond formed by the attraction between an H atom of one molecule and an electronegative atom, especially N or O_2, usually of another molecule.

hydrologic cycle (16) the process of evaporation, vertical and horizontal transport of vapor, condensation, precipitation, and the return of water via rivers and streams from the land to the sea.

hypolimnion (3) a region of the water column below the epilimnion within which temperature in a lake shows little depth variation. See *hypomarum.*

hypomarum (3) a region of the water column below the thermocline within which temperature in the ocean shows little depth variation. See *hypolimnion.*

hypoxic (3) having a very low dissolved O_2 concentration. Hypoxic waters are usually considered to have dissolved O_2 concentrations less than 2 mg L^{-1}.

incipient lethal level (8) the concentration of a toxic substance or level of stress expected to kill 50% of a group of organisms after a theoretically infinite period of time.

indicator organism (7) an organism whose presence can be taken as evidence of a particular condition, such as sewage pollution.

intoxication   (12) the condition of being stressed or otherwise adversely affected by a toxic chemical.

ion   (14) an atom or molecule that has acquired a positive or negative electric charge by losing or gaining, respectively, one or more electrons.

isomer   (13) one of a group of molecules all of which contain the same elements in the same proportions but whose properties differ because of differences in the arrangement of the atoms in the molecules.

isotopes   (14) different forms of the same element distinguished by different numbers of neutrons in the nucleus.

kraft process   (9) a process for producing wood pulp from wood chips that involves cooking the chips under pressure in a basic medium containing NaOH and Na$_2$S.

lamellibranch   (14) any of a group of bivalve mollusks that includes clams, scallops, and oysters.

LC50   (8) the concentration of a toxic substance or the level of stress expected to kill 50% of a group of organisms after a specified period of time.

lignin   (9) the primary noncarbohydrate component of wood. Lignin is a complex polymer that binds to cellulose fibers and hardens and strengthens the cell walls.

lipid   (2) any of a group of organic compounds that includes fats, oils, waxes, sterols, and triglycerides. Lipids are insoluble in water, soluble in common organic solvents, and oily to the touch.

LOAEL   (8) see *lowest observed adverse effect level.*

LOEL   (8) see *lowest observed effect level.*

lowest observed adverse effect level   (8) the lowest concentration of a toxic substance or level of stress observed to produce an adverse effect on humans or animals.

lowest observed effect level   (8) the lowest concentration of a toxic substance or level of stress observed to produce on humans or animals an effect that does not fit the definition of an adverse effect.

macrofauna   (9) animals large enough to be seen with the naked eye.

macronutrient   (2) an element that is required in relatively large amounts to produce living biomass.

makeup water   (11) water drawn from the environment to make up for evaporation losses and/or control the buildup of corrosion products and other impurities in the cooling water of a closed cooling system.

MARPOL   (13) the International Convention for the Prevention of Pollution from Ships. MARPOL initially resulted from a conference on marine pollution held in 1973 under the auspices of the International Maritime Consultative Organization. It was later modified by the related Protocol of 1978 and is now often referred to as *MARPOL 73/78.* MARPOL contains five annexes. Annexes I and II deal with oil discharges from ships and the transport of hazardous liquids, respectively. They are automatically adopted by any country that signs onto MARPOL. Annexes III, IV, and V are all optional and concern the transport of hazardous materials in packaged form, the discharge of sewage from ships, and the discharge of garbage from ships, respectively.

mastitis   (14) inflammation of the breast.

matrix effects   (12) changes in the apparent concentration of a compound or element due to interactions between the compound/element and the medium (matrix) within which the compound/element is located. For example, if an analytical instrument is calibrated using standards dissolved in freshwater, a correction for matrix effects may be needed if analyses are performed on seawater samples.

maximum contaminant level   (12) an enforceable standard for drinking water defined in the U.S. Safe Drinking Water Act. Maximum contaminant levels must be set as close to maximum contaminant level goals as is feasible. *Feasible* in this context is defined as "with the use of the best technology treatment techniques and other means, which the EPA Administrator finds are generally available (taking costs into consideration)."

maximum contaminant level goal   (12) a nonenforceable health goal defined in the U.S. Safe Drinking Water Act. The maximum contaminant level goal (MCLG) is defined as the concentration of a toxic substance in drinking water below which no known or anticipated adverse effects on the health of persons occur and which allows an adequate margin of safety, unless there is no safe threshold for the contaminant. In the latter case, the MCLG is set at zero.

maximum tolerated dose   (8) the highest dose of a toxic substance that does not kill experimental animals when administered in their diet for a period of time.

median tolerance limit   (8) the concentration of a toxic substance or level of stress that is tolerated by (does not kill) 50% of a group of organisms after a specified period of exposure.

median survival time   (8) the time within which 50% of a group of organisms survive exposure to a given concentration of a toxic substance or level of stress.

meiofauna   (13) animals in the size range 0.1 to 0.5 mm that live between the grains of nearshore sediments. The meiofauna include many kinds of invertebrates and protozoans.

mesotrophic   (3) having characteristics intermediate between eutrophic and oligotrophic.

metalimnion   (3) the thermocline of a lake (see *thermocline*).

methemoglobinemia   (6) an illness caused when hemoglobin in the blood is oxidized by, for example, nitrite. The oxidation product, called *methemoglobin*, contains Fe in the ferric state and cannot function as an O_2 carrier.

microbial loop   (1) a component of the detritus food chain in which the detritivores consist of heterotrophic bacteria that feed on dissolved organic matter excreted by other organisms. The bacteria are fed on by protozoans, which in turn are eaten by other protozoans or microzooplankton.

micronutrient   (2) an element that is required in relatively small amounts to produce living biomass; also known as a *trace element*.

mixed layer   (3) the upper region of the water column within which mixing caused primarily by wind-generated turbulence causes properties of the water such as temperature and salinity to be almost invariant with depth.

model II   (8) in statistics, an experimental design in which error is associated with the values of the independent variable X and X is not under the control of the investigator.

moderator   (14) a substance such as water or graphite that is used in a nuclear reactor to slow fast neutrons and thereby increase the likelihood of their fissioning ^{235}U.

monomictic   (2) once mixing. A monomictic body of water mixes or overturns once per year. Overturning occurs either (1) in the winter, when the surface waters are being cooled and the temperature of the surface waters exceeds the temperature of maximum density, or (2) in the summer, when the surface waters are being warmed and the temperature of the surface waters is less than the temperature of maximum density.

MTD   (8) see *maximum tolerated dose*.

naphthene   (13) see *cycloalkane*.

necrosis   (15) the death of cells due to injury or disease, especially in a localized area of the body.

net community production   (2) the net production of organic matter by photosynthetic or chemosynthetic organisms minus the losses associated with heterotrophic respiration.

net photosynthesis   (2) when quantified in terms of O_2, net photosynthesis equals gross photosynthesis minus the consumption of O_2 associated with the respiratory activity of plants. When quantified in terms of C, net photosynthesis is the net production of organic C by photosynthetic organisms.

neutron activation   (14) the capture of a neutron by the nucleus of a stable nuclide to produce a radionuclide.

nitrogen fixation   (2) the conversion of N_2 into forms of N such as nitrate or ammonium that can be utilized in primary production.

NOAEL   (8) the highest concentration of a toxic substance or level of stress observed to produce no adverse effect on humans or animals.

NOEL   (8) the highest concentration of a toxic substance or level of stress observed to produce no effect on humans or animals.

NTA   (6) nitrilotriacetate, an alternative to Na STP as a builder in detergents. NTA softens water by sequestering Ca and Mg ions.

NTU   (6) nephelometric turbidity unit, a standard metric of turbidity based on the intensity of light scattered by a sample of water under defined conditions with the intensity of light scattered by a standard reference suspension under the same conditions.

nuclide   (14) an atom with a specified sum of neutrons and protons in its nucleus, e.g., carbon-14.

nutricline   (2) the portion of the water column within which concentrations of inorganic nutrients such as nitrate and phosphate increase rapidly with increasing depth due to a combination of their assimilation at shallower depths within the euphotic zone and upward diffusion or advection from greater depths where they are released from decomposing organic matter.

olefin   (13) see *alkene*.

oligotrophic   (3) characterized by water with a low concentration of inorganic nutrients and organic matter.

oncogene   (14) a gene that causes the transformation of normal cells into cancerous cells.

oocyst   (7) a thick-walled egg produced by parasitic protozoans of the class *Sporozoa*. Oocysts effectively transfer zygotes that develop in the egg to new hosts.

optical depth   (2) a dimensionless number associated with depth z in a body of water and equal to $\ell n(I_0/I)$, where I_0 is the irradiance at the surface of the body of water and I is the irradiance at depth z.

organic compound   (1) a substance produced by living organisms and containing C with an oxidation number of zero. Among the most important organic compounds are carbohydrates, proteins, lipids, and nucleic acids.

organochlorine pesticide   (10) a chemical containing one or more Cl atoms attached to an organic moiety and used as a pesticide.

organoleptic   (13) relating to perception by one of the sensory organs.

organophosphate pesticide   (10) an organophosphorus pesticide containing one or more P atoms bound in phosphate groups.

organophosphorus pesticide   (10) a chemical containing one or more P atoms attached to an organic moiety and used as a pesticide.

osmoregulation   (15) the regulation of internal osmotic pressure by mechanisms such as water transport and uptake or release of ions.

osteomalacia   (12) A disease characterized by a softening of the bones, with accompanying pain and weakness. Osteomalacia most commonly results from a deficiency in vitamin D or Ca and occurs primarily in adult women. Victims of itai-itai disease suffered from osteomalacia caused by Cd intoxication.

osteoporosis   (12) A disease in which the bones become very porous, are subject to fracture, and heal slowly. Osteoporosis occurs most frequently in women following menopause and may lead to curvature of the spine from vertebral collapse.

overdraft   (16) the sustained removal of water from an aquifer at a rate that exceeds the recharge rate.

overturning   (3) vertical mixing of a water column caused by an increase in the density of surface waters.

oxygen depletion   (3) a reduction in the concentration of O_2 in the water caused by respiration.

oxygen sag   (9) the characteristic decline and subsequent rise of O_2 concentrations downstream of the point of release of BOD into a stream.

paraffin   (13) see *alkane*.

PCB   (10) any of 209 individual artificial compounds produced by chlorinating biphenyl and used primarily for industrial purposes as coolants and lubricants in transformers, capacitors, and other electrical equipment.

pelage   (13) the hair, fur, wool, or similar covering of a mammal's skin.

permanent carrier   (7) see *carrier*.

pesticide   (10) a chemical used to kill pests such as weeds, insects, rodents, nematodes, and so forth.

pheromone   (10) a chemical secreted by an animal that influences the behavior or growth of other members of the same species.

photoautotrophs   (1) organisms that are capable of synthesizing organic matter from inorganic chemicals using light as an energy source.

photosynthesis   (1) the process by which photoautotrophs produce O_2 and organic matter.

phytoplankton   (1) tiny algae, usually microscopic and often single-celled. Phytoplankton account for most of the primary production in many aquatic systems, including the ocean.

pinniped   (13) any member of a group of carnivorous aquatic mammals including the seals, walruses, and similar animals that have fin-like flippers as organs of locomotion.

pneumothorax   (14) an accumulation of air in the cavity surrounding the lungs due to disease or injury.

polymerase chain reaction   (7) a technique for replicating portions of DNA many times. The polymerase chain reaction involves heating DNA in a vial to a temperature of 90–95°C for about 30 seconds to separate the two complementary strands of DNA in the DNA double helix. The vial is then cooled to 55°C for about 20 seconds so that a short strand of nucleotides called a *primer* can bind to the end of the DNA. The vial is then warmed to 75°C so that the enzyme DNA polymerase can rapidly add nucleotides to the primer and make a complementary copy of the DNA template. This last step takes about one minute. The process can be repeated many times. After 30 cycles, more than a billion copies of a single piece of DNA can be produced.

positron   (14) a particle with a charge of $+1$ electrostatic unit and a rest mass equal to 5.5×10^{-4} atomic mass units. Positrons are the antimatter analogues of electrons.

potential temperature   (3) the temperature a parcel of water would have if it were raised to the surface without any exchange of heat. The potential temperature is always less than the *in situ* temperature, since the parcel of water expands while it is being raised and therefore does work against the surrounding water. As a result, the parcel of water loses heat and cools slightly.

POTW   (6) publicly owned treatment works, a municipal wastewater treatment plant.

primary carnivore   (1) an animal that feeds only on herbivores.

primary clarifier   (6) a sophisticated settling basin designed to remove solids from raw wastewater. Primary clarifiers typically remove 40–75% of the suspended solids from raw municipal wastewater.

primary production   (1) the transformation of inorganic chemicals into organic matter via either chemosynthesis or photosynthesis.

primary sludge   (6) the solid material skimmed from the surface or pumped from the bottom of the primary clarifier in a wastewater treatment plant. Water accounts for 95% or more of the volume of the primary sludge.

protein   (2) a complex organic macromolecule composed at least in part of one or more chains of amino acids and containing the elements C, H, O, N, and usually S. Hydrolysis of simple proteins yields only amino acids. Hydrolysis of conjugated proteins yields not only amino acids but also other organic or inorganic components. The non–amino acid portion of a conjugated protein is called its *prosthetic group*. Proteins are essential components of living organisms and have many different biological roles. They may function as enzymes (DNA polymerase), hormones (insulin), toxins (snake venom), or antibodies, or may play a role in storage (casein), transport (hemoglobin), contraction (myosin), or structure (collagen).

protozoan   (1) any member of a large group of single-celled, usually microscopic, eukaryotic heterotrophs, including amoebas, ciliates, flagellates, and sporozoans.

pyrethroid   (10) a synthetic relative of the natural pyrethrin esters obtained from *Chrysanthemum* flowers and used to control insect pests.

pyrite   (15) a brass-colored mineral, FeS_2, also known as *fool's gold*.

Q factor   (14) a quality factor used to multiply the radiation absorbed dose in grays to estimate the relative amount of biological damage caused by the absorption of various types of radiation by humans. The product of the radiation absorbed dose in grays times the Q factor equals the biologically effective absorbed radiation dose-equivalent in sieverts.

Q_{10}   (8) the rate of a reaction or metabolic process at a given temperature divided by the rate of the same reaction or process at a temperature 10°C cooler.

radioisotope   (14) a radioactive isotope of an element.

radionuclide   (14) a radioactive nuclide.

Redfield ratio   (2) the C:N:P ratio of 106:16:1 by atoms that empirical evidence has shown often closely approximates the C:N:P ratio of bulk organic matter in the ocean. Named after A. C. Redfield.

red tide   (2) a bloom of dinoflagellates that causes a reddish discoloration of coastal ocean waters, often extending for hundreds of meters or even kilometers. In some cases, red tides are accompanied by paralytic shellfish poisoning (PSP) caused by neurotoxins produced by certain species of dinoflagellates. In cases of PSP there is often no discernible effect on the shellfish, but when the shellfish are eaten by humans, the toxins may produce symptoms ranging from tingling of the extremities to paralysis and even death.

regulator   (8) an organism that regulates its internal state with respect to an environmental variable such as temperature or salinity. An organism that is a regulator with respect to one environmental variable may be a conformer with respect to another. See *conformer*.

residence time   (3) the time something remains in a given compartment of a system. See *turnover time*.

ridge tillage   (10) a system of plant cultivation whereby ridges on which the crop is planted are formed during cultivation or after harvesting and maintained from year to year in the same location. The principal advantages of ridge tillage are reduced operational costs for the farmer and reduced use of herbicides.

riparian   (3) of or related to the banks of a stream.

roentgen   (14) a unit of radiation equal to the quantity of X rays or gamma rays that will produce, as a consequence of ionization, 1 electrostatic unit of electricity in 1 cm³ of dry air measured at 0°C and standard atmospheric pressure.

rope-wick applicator   (10) a type of herbicide applicator developed during the 1970s and early 1980s that allows a liquid herbicide to drip onto weeds from rope wicks inserted into a reservoir positioned behind a tractor. The technique takes advantage of the fact that weeds often grow taller than the surrounding pasture or crop, particularly in the spring.

salinity   (3) a measure of the amount of salts dissolved in water. An early definition of salinity was the number of grams of salt dissolved in 1 kg of water when all the carbonate had been converted to oxide, the bromide and iodide to chloride, and all organic matter completely oxidized. Historically, salinity was reported in parts per thousand ($^0/_{00}$) or milligrmas per kilograms. Determinations of salinity today are based on conductivity measurements, which give results that are very close to but not identical to those based on the earlier definition of salinity. Salinity values calculated from conductivity measurements are reported in practical salinity units (psu).

salt wedge estuary   (3) an estuary in which there is a sharp transition between a surface layer of relatively freshwater and a subsurface layer of seawater.

sanitary sewage   (4) Wastewater from point sources, including homes, businesses, and public facilities such as schools, prisons, parks, and theatres. Sanitary sewage includes domestic sewage, defined as wastewater from washbasins, bathrooms, washing machines, kitchens, and lavatories. Sanitary sewage is a complex mixture of mineral and organic matter and includes living organisms, especially bacteria, viruses, and protozoa.

secondary carnivore   (1) an animal that feeds only on primary carnivores.

secondary clarifier   (6) a sophisticated settling basin designed to separate the solids from the liquid portion of the effluent from the secondary treatment process in a wastewater treatment plant.

secondary production   (1) the sum of all conversions of organic matter into living biomass.

secondary sludge   (6) the solid material pumped from the bottom and in some cases skimmed from the top of the secondary clarifier in a wastewater treatment plant. Water accounts for 95% or more of the volume of the secondary sludge.

Secchi depth   (4) the average of the depths at which a white or white-and-black disk (1) disappears from sight when lowered vertically into the water and (2) reappears again when pulled upward. Secchi disks are typically 20–30 cm in diameter (see *Secchi disk*).

Secchi disk   (4) a disk used to determine the Secchi depth in a body of water. Variations exist on the diameter (10–100 cm) and coloring (e.g., alternating black and white quadrants) of Secchi disks, but in most cases the Secchi disk is white with a diameter of typically 20 cm in limnological work and 30 cm in oceanographic work (see *Secchi depth*).

secondary sewage treatment   (4) a form of wastewater treatment designed to remove most of the suspended solids and labile organic matter in the wastewater. In the United States, secondary sewage treatment must remove at least 85% of the suspended solids and BOD from the raw sewage.

separate sewer system   (5) a sewer system characterized by a separate system of pipes and conduits for intercepting and collecting sanitary sewage and land runoff (see *combined sewer system*).

septic tank   (4) a watertight underground chamber for the partial treatment of wastewater. In a septic tank, the settleable solids accumulate in a sludge on the bottom and undergo anaerobic decomposition. The liquid supernatant seeps into a leach field via a system of underground distribution pipes (see *cesspool*).

sievert   (14) a unit of biologically effective absorbed radiation dose-equivalent defined to be the product of the dose in grays times the Q factor for the given type of radiation. The effective dose-equivalent in sieverts provides an estimate of the total risk of potential health effects from radiation exposure.

sodium channel   (10) tiny portals in the membranes of neurons that, when resting, prevent most Na ions from entering the cell. However, when opened by changes in the distribution of ions inside and outside the cell, they allow ions to flow in. This process ultimately leads to the release of neurotransmitters.

soft water   (6) water that contains a low concentration of Ca and Mg ions (see *hard water*).

STP   (6) sodium tripolyphosphate, a builder used in many laundry detergents until approximately 1985 but now almost entirely replaced by alternative builders that contain no P.

stranded filth   (4) the solid components of wastewater that settle in the pipes of a combined sewer system during low-flow conditions.

stratification   (3) a condition characterized by layering. In aquatic systems, stratification normally refers to a condition of horizontal layering associated with weak or nonexistent vertical mixing.

strong force   (14) one of the four fundamental forces in physics. The strong force manifests itself only within the nucleus of an atom. The strong force is responsible for holding together the components of particles such as neutrons and protons and is indirectly responsible for binding protons and neutrons together to form nuclei.

sulfate process   (9) see *kraft process.*

sulfhydryl group   (12) a chemical moiety consisting of an S atom bonded to an H atom (i.e., $-SH$) and found in biologically important molecules such as the amino acids cysteine and methionine.

sulfite process   (9) a process for producing wood pulp from wood chips that involves cooking the chips under pressure in an acidic medium containing H_2SO_3 and typically Ca and Mg.

surfactant   (6) generally speaking, any surface-active substance. In the case of laundry detergents, the surfactant is a bipolar compound that effectively penetrates between dirt and fabric.

symbiont   (4) an organism living together in a close relationship with another organism of a different species. If the relationship is mutually beneficial, it is referred to as *mutualism*. If the relationship benefits one symbiont but inhibits the other, it is called *parasitism*. If the relationship benefits one symbiont but has no effect on the other, it is called *commensalism*.

SI   (14) the Système International (International System) of units. Base SI units include the meter, kilogram, and second.

temporary carrier   (7) see *carrier.*

thermocline   (3) the region of a water column within which temperature changes with depth much more rapidly than is the case in the water column above or below. See metalimnion.

TL50   (8) see *median tolerance limit.*

TLm   (8) see *median tolerance limit.*

tonne   (1) a unit of mass, also called a *metric ton*, equal to 1,000 kg, 10^6 g, or approximately 2,205 lb.

toxicity unit   (8) see *acute toxicity unit* and *chronic toxicity unit.*

trace element   (2) see *micronutrient*

transuranic   (14) any element with an atomic number greater than 92 (U). All transuranics are radioactive.

trickling filter   (6) a component of a secondary wastewater treatment plant in which a complex community of decomposer organisms living within a bed of rocks feed on and decompose the organic matter in the effluent from the primary clarifier.

trophic level   (1) a group of organisms whose organic matter has undergone a similar number of predator/prey transformations since initially being converted to living biomass either from inorganic compounds (grazing food chain) or from detritus (detritus food chain).

turnover time   (3) the time required to replace the quantity of a substance equal to the amount in the compartment of the system. See *residence time.*

ultraviolet   (2) of or relating to electromagnetic radiation with a wavelength between roughly 4 nanometers (nm), on the border of the X-ray region, to about 400 nm, just beyond violet in the visible spectrum. Ultraviolet radiation is arbitrarily classified as UV-A (320–400 nm), UV-B (280–320 nm), and UV-C (4–280 nm). Because of differential attenuation by the Earth's atmosphere, virtually no UV-C radiation from the Sun reaches the surface of the Earth.

unconfined aquifer   (16) an aquifer located beneath an unsaturated layer of permeable soil and rock that extends to the land surface. Groundwater in an unconfined aquifer is not pressurized.

upwelling   (2) the upward movement of water from depths of typically 50–150 m at speeds of about 1–3 m d^{-1} resulting from the lateral advection of surface water. Coastal upwelling is associated with an offshore advection of surface water. Open-ocean upwelling may occur where Coriolis forces and/or surface winds create a divergence of surface water flow.

velocity cap   (11) a plate or similar device positioned in front of a power plant's cooling water intake in such a way as to reduce, as much as possible, the velocity of the water being drawn into the cooling system

viable but nonculturable   (7) a phrase describing the condition of a cell that is incapable of undergoing the sustained cellular division required for growth in or on a medium normally supporting growth of that cell.

vitrification   (14) the process of making glass or a glassy substance, especially by thermal fusion.

water hardness   (6) see *hard water.*

watershed   (16) the area of land that drains into a particular body of water.

water table   (16) the surface below which the ground is saturated with water.

weak force (14) one of the four fundamental forces of physics. The weak force manifests itself only within the nucleus of an atom. It is associated with the interactions that change neutrons into protons and protons into neutrons and is therefore responsible for beta particle radiation.

weathering (4) the gradual breakdown of soils and rocks by chemical and physical processes. Weathering consumes carbon dioxide and produces alkalinity in the form of anions such as bicarbonate and hydroxide. (13) The combination of biological, physical, and chemical processes responsible for the degradation of oil released to the environment

wet weight (8) the weight of a freshly collected organism, tissue, or sediment sample before any water has had a chance to evaporate.

zeolite (6) any member of a group of hydrous aluminosilicate compounds whose molecules enclose Na, K, Ca, Sr, or Ba cations. Zeolites are used mainly as molecular filters and ion-exchange agents.

✧ ◻ ✧ ◻ ✧

Units of Measurement and Abbreviations

With few exceptions, the units employed in this book correspond to those of the Système International d'Unités, i.e., Standard International units or SI units. In the SI system the standard unit of length is the meter (m), the standard unit of mass is the kilogram (kg), the standard unit of temperature is degrees kelvin (°K), the standard unit of time is the second (s), and the standard amount of matter is the mole (mol). Practical units derived from these standard units include the gram (g) as a unit of mass and Celsius degrees (°C) as a unit of temperature. In this book, temperatures are reported in degrees Celsius. The conversion between degrees Celsius and degrees Fahrenheit is $F = 9C/5 + 32$, where F is degrees Fahrenheit and C is degrees Celsius. The conversion between degrees Celsius and degrees kelvin is $K = C + 273.15$, where K is degrees kelvin and C is degrees Celsius. Standard decimal prefixes for SI units include the following:

Prefix	Abbreviation	Meaning
peta	P	10^{15}
tera	T	10^{12}
giga	G	10^{9}
mega	M	10^{6}
kilo	k	10^{3}
deci	d	10^{-1}
centi	c	10^{-2}
milli	m	10^{-3}
micro	μ	10^{-6}
nano	n	10^{-9}
pico	p	10^{-12}
femto	f	10^{-15}

Thus, for example, 10^{-6} g is abbreviated μg.

Important abbreviations to be aware of are the following:

Abbreviation	Meaning
m	meters
L	liter
M	moles per liter (molar)
ppm	parts per million, or 1 part in 10^{6}
ppb	parts per billion, or 1 part in 10^{9}

It is important not to confuse the abbreviations m and M when they stand alone with the prefixes m and M. For example, 5 millimeters is written 5 mm. The first m means 10^{-3}. The second m means meters. In aquatic ecology, concentrations may be expressed as mass per unit volume, i.e., grams per liter, amount of matter per unit volume, i.e., moles per liter, or mass per unit mass, i.e., grams per gram. A mole of matter is the mass in grams of that substance numerically equal to the atomic or molecular weight of the substance. For example, a mole of the common isotope of carbon is 12 g of carbon. Ten moles per liter would be written 10 M, and 5×10^{-6} moles per liter would be written 5 μM.

Concentrations reported in units of mass per unit mass can be confusing. A liter of water weighs almost exactly 1 kg. Thus, 5 mg of Hg per liter of water could also be reported as 5 mg per kilogram, or 5 parts per million (ppm), since 1 mg is 10^{-6} of a kilogram. One of the confusing aspects of concentrations reported in ppm or ppb is the fact that the units of the numerator and denominator are usually not the same. For example, the concentration of Hg in a swordfish might be reported as 5 ppm, and the concentration of DDT in the ocean might be reported as 5 ppb. In the case of the swordfish, 5 ppm means 5 mg of Hg per kilogram of swordfish. In the case of the ocean, 5 ppb means 5 μg of Hg per kilogram of seawater. Furthermore, when concentrations are reported in organisms or in sediments, care must be taken to indicate the condition/nature of the tissue/biomass or sediments. One of the most important distinctions is whether the organism or sediment is wet (fresh) or dry. Much of the weight of organisms and sediments is water. If concentrations are reported on a wet weight or fresh weight basis, they will be lower than if reported on a dry weight basis. For example, assume that the concentration of DDT in anchovies is 3 ppm on a fresh weight basis and that 70% of the weight of the anchovies is water. In other words, for each gram of fresh anchovies, 0.7 g is water and 0.3 g is tissue and bone. The concentration of 3 ppm on a fresh weight basis becomes $3/0.3 = 10$ ppm on a dry weight basis. Furthermore, in some cases, scientists may be interested in knowing the concentration in certain parts of an organism, such as the edible flesh, the bones, or the lipid tissue. For example, we may be told that the concentration of DDT in the eggs of brown pelicans is 5 ppm on a lipid weight basis. That means that there are 5 mg of DDT per kilogram of lipid in the egg.

The importance of keeping units straight was underscored in a recent snafu involving the U.S. National Aeronautics and Space Administration (NASA). NASA's Mars Climate Orbiter (MCO) swooped in too low as it headed for orbit around Mars, dipped too deeply into the atmosphere, and was never heard from again (Kerr, 1999). The problem was traced in part to a miscommunication between the Jet Propulsion Laboratory (JPL) in Pasadena, California, and the spacecraft team at Lockheed Martin Astronautics in Denver, Colorado. To make small adjustments to its flight path, it was necessary for the MCO to fire its thrusters once or twice per day. Lockheed Martin told the scientists at JPL how much force the thrusters had applied in English units (pounds). The JPL navigators assumed that the data were in newtons, a metric unit of force directly derivable from SI units ($1 \text{ newton} = 1 \text{ kg m s}^{-2}$). The miscommunication caused JPL navigators to think that the MCO was closer to its intended course than was the case. Apparently, the use of metric units for the MCO was spelled out in a written agreement between JPL and Lockheed Martin, but as noted by Kerr (1999), some companies in the propulsion industry have continued to use English units. Evidently, Lockheed Martin staff overlooked the memorandum.

REFERENCE

Kerr, R. A. 1999. More than missing metric doomed orbiter. *Science*, **286**, 207.

✧ ◻ ✧ ◻ ✧

Index